ADVANCES IN TURBULENCE VII

FLUID MECHANICS AND ITS APPLICATIONS
Volume 46

Series Editor: **R. MOREAU**
MADYLAM
Ecole Nationale Supérieure d'Hydraulique de Grenoble
Boîte Postale 95
38402 Saint Martin d'Hères Cedex, France

Aims and Scope of the Series

The purpose of this series is to focus on subjects in which fluid mechanics plays a fundamental role.

As well as the more traditional applications of aeronautics, hydraulics, heat and mass transfer etc., books will be published dealing with topics which are currently in a state of rapid development, such as turbulence, suspensions and multiphase fluids, super and hypersonic flows and numerical modelling techniques.

It is a widely held view that it is the interdisciplinary subjects that will receive intense scientific attention, bringing them to the forefront of technological advancement. Fluids have the ability to transport matter and its properties as well as transmit force, therefore fluid mechanics is a subject that is particulary open to cross fertilisation with other sciences and disciplines of engineering. The subject of fluid mechanics will be highly relevant in domains such as chemical, metallurgical, biological and ecological engineering. This series is particularly open to such new multidisciplinary domains.

The median level of presentation is the first year graduate student. Some texts are monographs defining the current state of a field; others are accessible to final year undergraduates; but essentially the emphasis is on readability and clarity.

For a list of related mechanics titles, see final pages.

Advances in Turbulence VII

Proceedings of the Seventh European Turbulence Conference, held in Saint-Jean Cap Ferrat, France, 30 June – 3 July 1998

Actes de la Septième Conférence Européenne de Turbulence, tenue à Saint-Jean Cap Ferrat, France, 30 Juin – 3 Juillet 1998

Edited by / Edités par

URIEL FRISCH

Laboratoire G.D. Cassini (CNRS),
Observatoire de la Côte d'Azur,
Nice, France

SPRINGER SCIENCE+BUSINESS MEDIA, B.V.

A C.I.P. Catalogue record for this book is available from the Library of Congress.

ISBN 978-94-010-6151-3   ISBN 978-94-011-5118-4 (eBook)
DOI 10.1007/978-94-011-5118-4

Printed on acid-free paper

Coverpicture:
Spiral structure of a stretched vortex (LPMMH/ESPCI, Paris)

TABLE OF CONTENTS
TABLE DES MATIÈRES

II. TRANSITION AND DYNAMICAL SYSTEMS
TRANSITION ET SYSTÈMES DYNAMIQUES

Invited Lectures/*Conférences Invitées*

Contributed Lectures/*Communications*

viii

III. NUMERICAL SIMULATION
SIMULATION NUMÉRIQUE

Contributed Lectures/*Communications*

IV. HIGH REYNOLDS NUMBERS AND INTERMITTENCY
GRANDS NOMBRES DE REYNOLDS ET INTERMITTENCE

Invited Lecture/*Conférence Invitée*

Contributed Lectures/*Communications*

V. INDUSTRIAL APPLICATIONS AND MODELLING
APPLICATIONS INDUSTRIELLES ET MODÉLISATION

Invited Lecture/*Conférence Invitée*

Contributed Lectures/*Communications*

VI. VORTEX DYNAMICS
DYNAMIQUE DU TOURBILLON

Invited Lecture/*Conférence Invitée*

Contributed Lectures/*Communications*

VII. ASTRO/GEOPHYSICAL FLOW AND CONVECTION
ÉCOULEMENTS ASTRO/GÉOPHYSIQUES ET CONVECTION

Contributed Lectures/*Communications*

VIII. TRANSPORT OF PASSIVE SCALARS
TRANSPORT DE SCALAIRES PASSIFS

Invited Lecture/*Conférence Invitée*

Contributed Lectures/*Communications*

Special Invited Lecture/*Conférence Invitée Spéciale*

PREFACE

This seventh volume in the series *Advances in Turbulence* contains an overview of the state of turbulence research with some bias towards work done in Europe. It represents an almost complete collection of the papers delivered at the *Seventh European Turbulence Conference* (ETC-7), sponsored by EUROMECH and ERCOFTAC and organized by the Observatoire de la Côte d'Azur, which will be held in Saint-Jean Cap Ferrat, June 30 - July 3, 1998. As was done for ETC-6 (Lausanne), the present volume is being produced before the conference in order to assure timely dissemination of the latest research results and replace the usual collection of abstracts which is generally accessible only to participants.

The papers presented at ETC-7 have been selected by the international EUROMECH turbulence conference committee. Besides the organizer it comprised R. Benzi, T. Bohr, I. Castro, C. Dopazo, H. Eckelmann, D. Henningson, P. Huerre and P. Monkewitz and was chaired by F.T.M. Nieuwstadt. To all of them my sincerest thanks for all the work done. This was particularly hard since the size of the conference did not permit the acceptance of more than 50% of the papers submitted.

When EUROMECH decided in late 1984 to establish the european turbulence conferences, a major goal was to bring about "turbulent mixing" of the community of physicists and that of fluid dynamicists and engineers. Over the years these communities, which are both strongly represented ETC-7, have learned much from each other and have established a fruitful dialogue.

I shall not attempt to present the reader with a catalogue of the various activities covered but I wish to convey my personal opinion that the field is truly on the move. New high-Reynolds number experiments (no more exclusively dependent on the use of large-scale facilities), combined with new techniques of imaging, non-intrusive probing, processing and simulation, provide high-quality data which put significant constraints on possible theories. In 1961 a landmark of the turbulence conference held in Marseille was the introduction by Obukhov and Kolmogorov of a phenomenological theory of intermittency, the major stumbling block of the Kolmogorov 1941 theory. Now, for the first time, we have a real theory which explains, for a class of passive scalar problems introduced by Kraichnan, why dimensional analysis sometimes gives the wrong answers and how anomalous intermittency corrections can be calculated from first principles.

After these remarks, I would like to thank all the contributors for keeping to the tight schedule for submission of manuscripts and the Kluwer Academic Publishers for producing this volume in a very short time. I would also like to thank the members of the regional (Marseille-Nice) organizing committee, F. Anselmet, P. Clavin, A. Favre, J. Pacheco, A. Pouquet, A. Pumir and P.-L. Sulem for considerable help, *Atout Organisation Science* for logistical support and all the financial sponsors (listed separately) who made the conference and this volume possible. Last, but not least, I wish to thank Valérie Chéron whose highly professional assistance has been invaluable.

April 1998,   Uriel Frisch

PRÉFACE

Ce septième volume de la série *Advances in Turbulence* présente l'état de l'art en turbulence avec un certain biais en direction des recherches européennes. Il comprend une collection presque complète des présentations à la *Septième Conférence Européenne de Turbulence* (ETC-7), qui se tient sous les auspices d'EUROMECH et d'ERCOFTAC, qui est organisée par l'Observatoire de la Côte d'Azur, et qui se tiendra à Saint-Jean Cap Ferrat du 30 juin au 3 juillet 1998. Comme à l'ETC-6 (Lausanne), ce volume est produit avant la conférence, de façon à permettre la dissémination rapide des derniers résultats. Ce volume remplace aussi l'habituelle collection de résumés accessibles aux seuls participants.

Les présentations à l'ETC-7 on été sélectionnées par le comité international EUROMECH des conférences de turbulence. Outre l'organisateur, il comprenait R. Benzi, T. Bohr, I. Castro, C. Dopazo, H. Eckelmann, D. Henningson, P. Huerre et P. Monkewitz et était présidé par F.T.M. Nieuwstadt. A tous, mes remerciements les plus sincères pour tout le travail réalisé, rendu spécialement difficile par le fait que la taille de la conférence n'a pas permis de retenir plus de 50% des soumissions.

Quand EUROMECH décida vers la fin de 1984 d'établir les conférences européennes de turbulence, un objectif majeur fut de permettre le "mélange turbulent" de la communauté des physiciens avec celle des dynamiciens des fluides et ingénieurs. Au fil des ans ces communautés, fortement représentées à l'ETC-7, ont su établir un dialogue fructueux.

Je ne tenterai pas de présenter au lecteur un catalogue des diverses activités couvertes mais je tiens à faire connaître mon sentiment personnel que le sujet est vraiment en train de bouger. De nouvelles expériences à grands nombres de Reynolds (qui ne requièrent plus l'utilisation exclusive de grandes installations), combinées avec de nouvelles techniques de traitement d'images et de données, de sondage non intrusif et de simulation, conduisent à des données de haute qualité, significativement contraignantes pour la théorie. En 1961 un fait marquant de la conférence de turbulence organisée à Marseille fut l'introduction par Obukhov et Kolmogorov d'une théorie phénoménologique de l'intermittence, la pierre d'achoppement de la théorie de Kolmogorov de 1941. Maintenant, pour la première fois nous disposons d'une véritable théorie qui explique, pour une classe de problèmes de scalaires passifs introduite par Kraichnan, pourquoi l'analyse dimensionnelle peut conduire à des réponses incorrectes et qui permet de calculer les corrections d'intermittence à partir des équations de base.

Après ces remarques, je voudrais remercier tous les auteurs et Kluwer Academic Publishers pour avoir respecté un calendrier contraignant. Je voudrais aussi remercier les membres du comité régional d'organisation (Marseille-Nice), F. Anselmet, P. Clavin, A. Favre, J. Pacheco, A. Pouquet, A. Pumir et P.-L. Sulem pour leur aide considérable, *Atout Organisation Science* pour le soutien logistique et toutes les organisations qui nous ont soutenus financièrement (voir liste séparée) et qui ont rendu possible cette conférence et le présent volume. Enfin et surtout, je voudrais remercier Valérie Chéron pour son assistance très professionnelle et inappréciable.

Avril 1998,   Uriel Frisch

SPONSORS/*PARRAINAGE*

- Conseil Régional Provence-Alpes-Côte d'Azur
- Conseil Général des Alpes-Maritimes
- Centre National de la Recherche Scientifique (Départements SDU, SPI et SPM)
- Ministère de l'Education Nationale, de la Recherche Scientifique et de la Technologie (Bureau des Colloques et Programme ACCES)
- Ministère des Affaires Etrangères
- Direction Générale de l'Armement
- Groupements de Recherche "Dynamique des Fluides Géophysiques et Astrophysiques" et "Turbulence"
- Laboratoire G.D. Cassini de l'Observatoire de la Côte d'Azur
- Association pour le Développement International de l'Observatoire de Nice
- European Research Community on Flow Turbulence and Combustion
- Russian Foundation for Basic Research
- DIGITAL France
- Municipalité de Saint-Jean Cap Ferrat

I

Experiments and Experimental Techniques

Expériences et Techniques Expérimentales

LARGE SCALE FLUCTUATIONS IN SWIRLING FLOWS

S. FAUVE AND S. AUMAITRE
Laboratoire de Physique Statistique,
Ecole Normale Supérieure,
24, rue Lhomond, 75005 Paris, France,

P. ABRY AND J.-F. PINTON
Laboratoire de Physique,
Ecole Normale Supérieure,
46, Allée d'Italie, 69364 Lyon, France,

AND

R. LABBE
Universidad de Santiago de Chile,
Casilla 307, Santiago, Chile.

We consider elementary questions about turbulence that are usually overlooked in most recent works on the subject. We think that some elementary facts, although accepted by most people working in the field, at least as empirical results, still need to be discussed and checked in careful experiments. We address simple open problems which mainly concern large scale fluctuations in turbulent flows. "Large scale" should be understood with respect to velocity fluctuations, and thus means that the integral-scale velocity i.e. the *rms* turbulent velocity is involved. Correspondingly, these fluctuations can be detected by "global" measurements. We have performed several such measurements: they concern pressure, vorticity and power consumption (or drag); we call them "global measurements" since they involve a finite volume of the flow field at a given instant, by opposition with velocity measurements which are usually made as local as possible. We discuss these global measurements on the example of swirling flows generated between two rotating co-axial disks, and show that they are efficient tools to study coherent structures in turbulent flows.

As in any other dissipative system, the amount of power needed to maintain a turbulent flow is a quantity of great interest. It is one of the first to evaluate if one has to design an experiment. Consider for instance a disk of radius R rotating about its axis with a uniform angular velocity Ω in

3

U. Frisch (ed.), Advances in Turbulence VII, 3–8.
© 1998 *Kluwer Academic Publishers.*

a fluid of density ρ and kinematic viscosity ν. From elementary dimensional analysis, we get for the injected power per unit mass of the fluid,

$$\overline{< \epsilon_I >} = R^2 \Omega^3 f(Re) = \frac{V^3}{R} f(Re),$$

where $V = R\Omega$ is the characteristic large scale velocity of the moving boundary driving the fluid, ($< \cdot >$ stands for the average in space and an overbar for the temporal average). The exact expression of $f(Re)$ is not known, but it is strongly believed, and this is an important assumption of the phenomenology of turbulence, that the transport properties of the flow behave in a well defined way in the limit $Re \to \infty$; the simplest choice $f(Re) \to constant$ is supported by empirical evidence, in the case of a rotating fan for instance, and leads to a torque proportional to the square of the rotation rate, a behavior known as the "centrifugal torque" in the engineering literature, and similar to the related law for the drag being proportional to the square of the velocity. The situation is more complex for a flat disk, or in pipe flows; in both cases $f(Re)$ decreases when Re is increased in the turbulent regime until the surface roughness becomes important (Schlichting, 1979). In these situations, one often searches for a power law of the form, $f(Re) \propto Re^{-\delta}$, and the dependence on viscosity is not removed even in the limit $Re \to \infty$; this is an example of self-similarity of the "second kind" (Barenblatt, 1996). Recent measurements of torque have been performed in the Couette-Taylor flow (Lathrop et al., 1992). A similar problem in Rayleigh-Bénard convection concerns the dependence of the convective heat flux on the Rayleigh number (for a recent review, see Siggia, 1994). In all the above studies, the problem of the temporal fluctuations of these spatially averaged quantities (power, drag, torque, heat-flux) have not been considered. This is usually overlooked since the equivalence between spatial and temporal averages is assumed. In other words, it is usually believed that, as a result of many independent random contributions from the small scales, a quantity averaged in the flow volume or on the flow boundaries, only slightly fluctuates in time. In fully developed turbulent flows, u_{rms}, the rms value of turbulent velocity fluctuations, is usually considered instead of V in the definitions of $\overline{< \epsilon_I >}$ and Re. This is an important distinction for some flow geometries, flows along a smooth plate or disk for instance, for which, as mentioned above, $f(Re)$ decreases until the surface roughness becomes important; in these regimes, we do not expect u_{rms} to be proportional to V. Anyway, it is believed that at least for homogeneous isotropic turbulence (Sreenivasan, 1984; Lohse, 1994), $\overline{< \epsilon_I >} \propto u_{rms}^3/R$ when $Re \to \infty$. This quantity plays a central rôle in the phenomenology of turbulence, which usually considers the properties of small-scale quantities such as the local viscous dissipation, $\epsilon_d(\vec{r}, t)$, or the velocity increments,

$\delta u_i \equiv u_i(\vec{r}) - u_i(\vec{r} + \vec{d})$, and studies their statistical characteristics for a fixed value of $\overline{< \epsilon_I >}$.

We address here the problem of the temporal fluctuations of $< \epsilon_I >$. The evolution equation for the kinetic energy E per unit mass is easily obtained by multiplying the Navier–Stokes equation by the velocity u_i, and taking the average over the flow volume. One obtains

$$\dot{E} = < \epsilon_I > - < \epsilon_d >,$$

where $< \epsilon_I >$ is the rate of working of the pressure and viscous forces exerted on the surface S enclosing the fluid volume $\mathcal{U}$. Both the pressure p and the viscous forces are strongly fluctuating quantities at the solid boundary S. In any experiment, $< \epsilon_I >$ thus fluctuates in time. Although it is obvious that the mean values in time are equal, $\overline{< \epsilon_I >} = \overline{< \epsilon_d >}$, large temporal fluctuations for $< \epsilon_I >$ do occur in any realistic flow. We have studied the temporal fluctuations of the total power injected in a swirling flow generated in the gap between two coaxial disks, counter-rotating at a fixed velocity (Labbé et al., 1996b). The fluctuations of the power injection are large (rms amplitude up to 10% of the mean value), and more importantly, their statistical properties do not vary with the Reynolds number. In particular, the ratio $\delta\epsilon_I / \overline{< \epsilon_I >}$ of fluctuations characteristic amplitude to the mean value, does not depend on the Reynolds number. We note that the most naïve prediction, i.e. evaluating this ratio as being proportional to the inverse square root of the number of degrees of freedom, would lead to a power law, $\delta\epsilon_I / \overline{< \epsilon_I >} \propto Re^{-9/8}$, that is in contradiction with the experimental results. In other words, there is no "smoothing effect" of the small scales on the injected power. Moreover, when the flow is confined in a cylindrical vessel, we find that the probability density function (pdf) of the power fluctuations is non-Gaussian, although we are dealing with a spatially averaged quantity. It would be interesting to check the law $\delta\epsilon_I \propto \overline{< \epsilon_I >}$ on a larger range of Re. A second aspect of this problem concerns the possible effect of the temporal fluctuations of $< \epsilon_I >$ on the energy transfers. Contrary to $< \epsilon_d >$, the dissipated power, $< \epsilon_I >$ is not constrained to be positive; its temporal mean is positive but there may exist rare events with negative $< \epsilon_I >$, i.e. events for which the flow returns energy to the driving device. We have identified such events in the case of a von Kármán flow with co-rotating disks. This flow involves a strong axial vortex and high frequency turbulent fluctuations displaying a Kolmogorov scaling. We have interpreted depletions in turbulent spectra as associated to these back transfers of energy (Labbé et al., 1996b). However, this flow is strongly anisotropic and it would be interesting to make similar observations in more isotropic turbulent flows. A last question concerns the effect of the fluctuations of $< \epsilon_I >$ on the dissipative range, for instance the measured small-scale intermittency. Although we have observed correlations between $< \epsilon_I >$ and $< \epsilon_d >$ in the case of a shell model (Poggi and Fauve, 1995), we may hope that the fluctuations of $< \epsilon_I >$ do not affect the statistical properties of ϵ_d is isotropic flows. This problem deserves more studies.

The evolution equation for the kinetic energy E per unit mass can also be written in the form

$$\mathcal{U} \dot{E} = \int_S \left(\frac{p}{\rho} + \frac{u^2}{2} \right) \vec{u}\,\hat{n}\,dS + \nu \int_S (\vec{u} \times \vec{\omega}).\hat{n}dS - \nu \int_{\mathcal{U}} \omega^2 d^3 x,$$

where $\vec{\omega}$ is the vorticity. It is therefore interesting to perform pressure and vorticity measurements and to search for correlations with injected power. Pressure measurements have been performed for a long time, mostly beneath turbulent boundary layers (Willmarth, 1975). They have been primarily used to identify and follow the dynamics of organized vortical or dissipative structures; both negative and positive pressure pulses have been observed, and the pdf of the pressure fluctuations have been found non-Gaussian with a negative skewness (Schewe, 1983). For both boundary layers (Schewe, 1983) and free shear layers (George et al., 1984), a $k^{-7/3}$ Kolmogorov scaling has been found for the power spectrum in the inertial range. It has been proposed by Brachet (1991) to use pressure to identify vorticity concentrations in fully developed turbulence and the corresponding experiment has been performed using a von Kármán flow between counter-rotating disks (Douady et al., 1991). The flow was seeded with air bubbles and the intermittent formation of filaments of bubbles was observed, thus visualizing low pressures ascribed to vorticity filaments first observed numerically by Siggia (1981). We have performed direct measurements of pressure fluctuations in von Kármán flows (Fauve et al., 1993) and showed that the whole pressure pdf scales like the square of the integral-scale velocity. With counter-rotating disks, the pressure signal displays strong pressure drops ascribed to vorticity filaments sweeping the pressure probe. The above scaling shows that the characteristic velocity increment at the core of the vortex is of the order of the integral-scale velocity of the turbulent flow. We have also studied the statistical properties of the pressure drops (Abry et al., 1994). In particular, the waiting time between two successive pressure drops displays Poisson statistics at long enough time, i.e. the waiting time pdf decays exponentially at large times, whereas an algebraic decay, i.e. a departure from Poisson statistics, is observed at short time. The important point is that the cross-over between the two behaviors occurs at an integral time scale (several rotation periods of the disks). The above results have been confirmed and pressure-velocity correlations have been studied (Cadot et al., 1995). Finally, using a wavelet technique, we have removed the localized pressure drops from the signal and compared the resynthesized and the original signals (Abry *et al.*, 1994). We have found that although the pressure drops involve all the frequencies present in the whole signal, their contribution is larger at low frequency, i.e. at the integral time scale. Pressure drops are thus large-scale events.

It should be noted that these pressure or velocity measurements (Belin *et al.*, 96), while yielding quantitative results, rely on indirect detection of the vorticity filaments. In addition, the procedure makes the assumption that the low pressure events are in exact correspondence with the presence of filamentary vortices. We have thus developed an ultrasound scattering technique that allows a non-intrusive direct measurement of the vorticity in the bulk of the flow. The sound scattering technique is based on the general principle that a wave may be used to probe a medium that interacts with it. In the case of a flow with small density variations and negligible temperature fluctuations, the only scattering mechanism is related to fluctuating velocity gradients. This approximation is realistic in the limit of small Mach numbers. Using the Born approximation, Obukhov (1941) computed the wave scattered by velocity gradients in the high-frequency and far-field limits. Later, Kraichnan (1953) obtained formulas for the angular and frequency distributions of the scattered wave in terms of the Fourier transforms in space and time of the shear velocity field. Several authors have related the scattered sound to the vorticity field of the flow; compact formulas in the far field approximation have been found by Fabrikant (1983) and Lund and Rojas (1989). It has been shown that the scattered pressure is proportional to the Fourier transform of the vorticity at the scattering wave vector. Moreover, the scattered wave is Doppler shifted, the shift giving access to the direction and magnitude of the mean velocity advecting the vorticity. The validity of vorticity measurements using ultrasound scattering has been checked in detail on simple flows (Gromov *et al.*, 1982, Baudet *et al.*, 1991). We have used this technique in order to detect vorticity concentrations in the bulk of the von Kármán flow (Dernoncourt *et al.*, 1998). In the case of counter-rotating disks, the scattered pressure reveals intense vorticity concentrations which occur intermittently in the bulk of the flow. Their statistical properties are analyzed in detail and are found identical to those of pressure fluctuations. Moreover, these large deviations in the scattered pressure signal are detected only in a range of sound wavelength; this determines the characteristic size of the scattering vorticity concentrations. We thus find that their core size scales like the Taylor microscale as the Reynolds number is varied.

Vorticity filaments have been considered for a while as small-scale objects (because of the small value of their core size) and attempts have been made to link them with intermittency. The first statement is wrong: the filaments involve all the scales of the flow and involve many features characteristic of large scale structures. No quantitative estimation of the contribution of filaments to intermittency is currently available; in the case of the pressure signal, it is known that the statistical properties of the pressure increments are not affected by the cancellation of the pressure drops (Abry et al., 1994). If the same is true for the velocity signal, vorticity filaments would have no contribution to intermittency. It has been also claimed that vorticity filaments have a negligible contribution to the energy cascade. The local cancellation of the filaments from the velocity signal cannot be used to answer this question. On the contrary, it should be noted that the filaments mean occurrence time is strongly correlated with the behavior of $f(Re)$ in the expression of $< \epsilon_I >$. It should also be kept in mind that the experimentally observed filaments depend on the driving device and are probably not universal objects, not so surprising if one keeps in mind that they are not small-scale objects. However, they do occur in many realistic flows and thus the above problems deserve more studies.

References

Abry, P., Fauve S., Flandrin, P., Laroche C.: 1994, J. Phys. II (France), 4, 725-733.
Barenblatt, G.I.: 1996, Scaling, Self-similarity, and Intermediate Asymptotics, Cambridge University Press.
Baudet, C., Ciliberto, S. and Pinton, J.-F.: 1991, Phys. Rev. Lett., 67, 193-195.
Belin, F., Maurer, J., Tabeling, P. and Willaime, H. : 1996, J. Phys. II (France), 6, 573-583.
Brachet, M.E.: 1991, Fluid Dyn. Research 8, 1-8.
Cadot, O., Douady, S. and Couder, Y.: 1995, Phys. Fluids, A 7, 630-646.
Dernoncourt, B., Pinton, J.-F. and Fauve, S,: 1998, Experimental study of vorticity filaments in in turbulent swirling flows, Physica D (to be published).
Douady, S., Couder, Y. and Brachet, M.E.: 1991, Phys. Rev. Lett. 67, 983-986.
Fabrikant, A.L.: 1982, Sov. Phys. Acoust. 28, 410-411.
Fauve, S., Laroche, C. and Castaing, B.: 1993, J. Phys. II (France), 3, 271-278.
George, W.K., Beuther, P.D. and Arndt, R.E.A.: 1984, J. Fluid Mech., 148, 155-191.
Gromov, P.R., Ezerskii, A.B. and Fabrikant, A.L.: 1982, Sov. Phys. Acoust. 28, 452-455.
Kraichnan, R.: 1953, J. Acoust. Soc. Am., 25, 1096-1104.
Labbé, R., Pinton, J.-F. and Fauve, S.: 1996a, Phys. Fluids, 8, 914-922.
Labbé, R., Pinton, J.-F. and Fauve, S.: 1996b, J. Phys. II (France), 6, 1099-1110.
Lathrop, D.P., Fineberg, J. and Swinney, H.L.: 1992, Phys. Rev.A, 46, 6390.
Lohse, D.: 1994, Phys. Rev. Lett., 73, 3223-3226.
Lund, F. and Rojas, C.: 1987, Physica D, 37, 508-514.
Obukhov, A.M.: 1941, Dokl. Akad. Nauk. SSSR, 30, 616-620.
Poggi, P. and Fauve, S.: 1995 (unpublished).
Schewe, G.J.: 1983, J. Fluid Mech., 134, 311-328.
Schlichting, H.: 1979, Boundary-Layer Theory, McGraw-Hill.
Siggia, E.D.: 1981, J. Fluid Mech., 107, 375-406.
Siggia, E.D.: 1994, Ann. Rev. Fluid Mech., 26, 137-168.
Sreenivasan, K.R.: 1984, Phys. Fluids, 27, 1048-1051.
Willmarth, W.W. : 1975, Ann. Rev. Fluid Mech. 7, 13-38.
Zandbergen, P.J. and Dijkstra, D.: 1987, Ann. Rev. Fluid Mech., 19, 465-491.

THE FLICKERING CANDLE: TRANSITION TO A GLOBAL OSCILLATION AND TURBULENCE IN A THERMAL PLUME

T. MAXWORTHY

Department of Aerospace and Mechanical Engineering,
University of Southern California,
Los Angeles, CA 90089-1191, U.S.A.

AND

PMMH-ESPCI, 10 rue Vauquelin,
Paris 05, France.

A number of experiments have been performed on the properties of propane diffusion flames at relatively low fuel flow rates and using a variety of burner types. Optical methods were used to observe the flame and plume above it. We have observed the transition from a steady flame, with the plume above it exhibiting a helical instability at low flow rates, to an axisymmetric instability at a well defined frequency, of the whole flame and the lower part of the plume, at higher flow rates. In the case of burners made of straight, constant-diameter tubing, with the fuel injected vertically, the frequency and flow rate at onset of oscillation agreed with previous measurements. For burners in which fuel is injected horizontally the flame length was very small, so that it could be considered merely as a heat source for the unstable plume above it. Under these circumstances the frequency of oscillation was robust and unaffected by perturbations caused by a moderately-strong, external sound source. Measurements of the amplitude of flame/plume oscillation with distance from the burner tip showed an exponential dependence. These results and the observation of the effects of various external modifications, e.g. the destabilising effect, on a steady flame, of an annular counterflow with the subsequent generation of, first, an axisymmetric instability followed by a low-frequency, helical oscillations; a strong pressure perturbation, etc., are consistent with the view that the transition to the axisymmetric state, at which the flame flickers, is one to a globally-excited oscillation forced by a finite region of absolutely unstable flow at or near the tip of the burners used in this study. The information feed-back inherent in this explanation could be enhanced

9

U. Frisch (ed.), Advances in Turbulence VII, 9–10.
© *1998 Kluwer Academic Publishers.*

by placing an obstacle above the flame tip, in which case the flow rate required to induce flickering was decreased. Also, increasing the flow velocity through the enclosure surrounding the burner, by manipulation of the exit conditions, stabilised the flickering flame. Such a result would be expected based on prior theoretical and experimental studies of similar types of flow exhibiting global oscillations. Finally, the upper reaches of the plume underwent a further transition to turbulence through complex sequence of vortex interactions and distortions.

The sequence of events discussed above has a strong similarity to those found in the transition to the dripping state of a stream of water issuing from a tube, see e.g. Huerre and Monkewitz, *Ann. Rev. Fluid Mech.*, vol. 22. The growth of an absolutely unstable region at the tube exit, and the propagation of the convectively unstable flow towards the exit, as the control parameters are varied, are identical in both cases and point to a common explanation for both phenomena.

The attached figure shows an example of the transition from a helical plume instability to an axisymmetric one, at an intermediate fuel flow rate. In this range of flow rate the appearance of the axisymmetric, global instability is intermittent.

The help and support of Eduardo Wesfreid and Philippe Petitjeans are gratefully acknowledged.

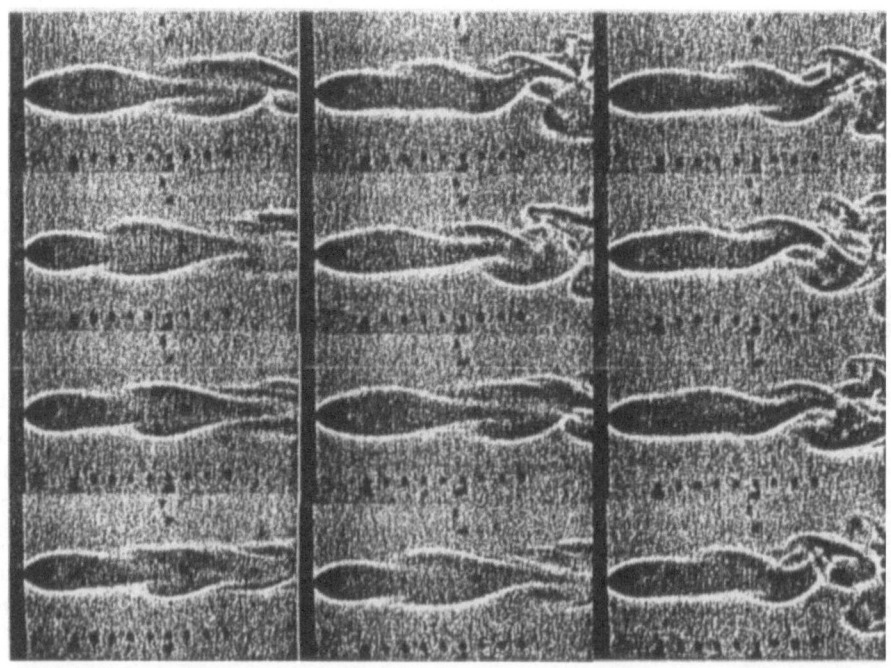

FLOW FIELD MEASUREMENTS OF
SEPARATED AND REATTACHED FLOWS

J. KOSTAS[1], A. FOURAS[1], J. SORIA[1] AND M.S. CHONG[2]
[1] *Mechanical Engineering Department*
Monash University
Clayton, VIC, Australia 3168.
[2] *Mechanical & Manufacturing Engineering Department*
Melbourne University
Parkville, VIC, Australia 3052

Abstract. A mean, 2-D separating and reattaching flow over a blunt flat plate at a Re $= \frac{Uh}{\nu} = 1000$ (plate thickness, h) has been investigated using Particle Image Velocimetry (PIV). Medium spatial resolution instantaneous velocity field measurements, as well as temporal velocity field measurements in and around the separation bubble have been obtained. The experimental results show a highly unsteady flapping of the shear layer causing considerable variation in the size of the separation bubble. This flapping occurs principally in the downstream region of the separated shear layer and results in a large uncertainty of the instantaneous reattachment point.

1. Introduction

Boundary layer separation is instigated by an adverse pressure gradient which can arise from local surface geometry and/or be imposed on the boundary layer by the outer flow. The nature of this adverse pressure gradient will determine whether the boundary layer separates, separates and reattaches or does not separate at all. The sharp edges of bluff bodies are usually responsible for flow separation because of the large pressure gradients that exist around these points. At some distance downstream of detachment the separated shear layer may reattach, if the bluff body is sufficiently long. The separated and reattached flow over a blunt flat plate has been investigated to gain insight into the dynamics of the unsteady nature of the shear layer in and around the separation bubble.

11

U. Frisch (ed.), Advances in Turbulence VII, 11–14.

2. Experimental Apparatus and Technique

Experiments on a separated flow over the blunt leading edge of a flat plate were conducted in a vertical, closed circuit, low turbulence water tunnel. The test section consists of three 250 mm square by 500 mm long modules bolted together to provide an overall test section length of 1.5 m, see Figure 1. A 4 axis (x, y, z, θ) traverse is mounted on the frame of the water tunnel, enabling positioning of a digital CCD camera anywhere in front of the working section.

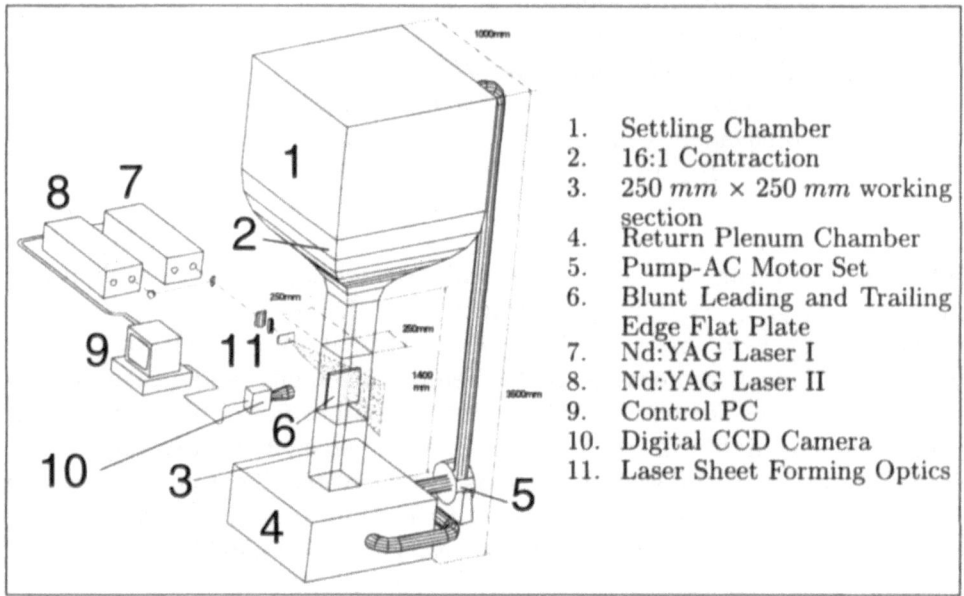

1.   Settling Chamber
2.   16:1 Contraction
3.   250 mm × 250 mm working section
4.   Return Plenum Chamber
5.   Pump-AC Motor Set
6.   Blunt Leading and Trailing Edge Flat Plate
7.   Nd:YAG Laser I
8.   Nd:YAG Laser II
9.   Control PC
10.   Digital CCD Camera
11.   Laser Sheet Forming Optics

Figure 1.   The vertical water tunnel used in the experiments. A layout of the optical components and the image acquisition equipment is also shown.

Further details and the free-stream characteristics in the test section are documented in (Nicolaides(1997)).

The model used in this investigation is a 23.22 mm thick, 230 mm long acrylic plate, machined flat with blunt leading and trailing edges spanning the width of the test section. The test plate was polished and mounted in the working section at 0° incidence to the oncoming stream.

An innovative adaptation of an image shifting technique based on that suggested by (Lourenco(1991)) and further developed and investigated by (Kostas et al. (1996)), (Fouras(1997)) and (Soria et al. (1998)) was used to acquire instantaneous 2-D velocity measurements in and around the separation bubble. Temporal data was also obtained since the gross flow time-scales were large compared to the achievable sampling time of the digital image acquisition equipment. Specific details of the technique can be found in (Fouras(1997)) and (Soria et al. (1998)).

3.   Experimental Results and Discussion

Digital particle image velocimetry was used to obtain medium spatial resolution velocity field measurements in and around the separation bubble. Vorticity field information was extracted from the velocity data using a χ^2 vorticity calculation method (Fouras & Soria(1998)).

Instantaneous velocity field measurements for a separated flow from a blunt leading edge of a flat plate are shown in Figure 2(a). The corresponding vorticity contours for the same flow are shown in Figure 2(b). Small scale structures, emanating from the separation point, are clearly visible in the shear layer and appear to be convected along it.

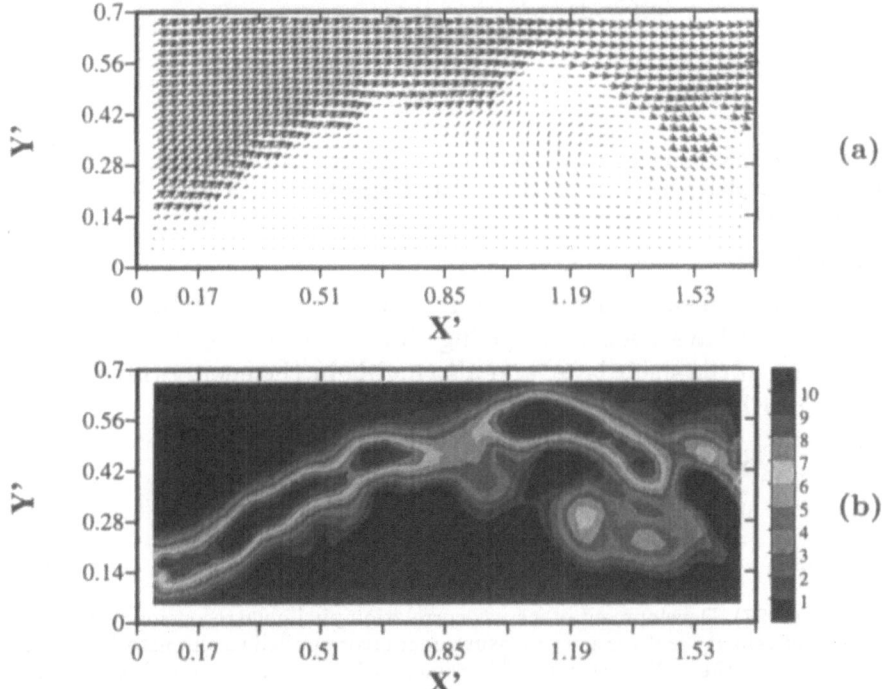

Figure 2.   Instantaneous velocity field (a) and corresponding vorticity field (b) of a separated flow over a blunt flat plate. Re = $\frac{Uh}{\nu}$ = 1000, plate thickness = h, X' = $\frac{x}{h}$, Y' = $\frac{y}{h}$; flow is from left-to-right

The spatio-temporal PIV measurements have permitted the determination of the in-plane components of the Reynolds stress tensor in and around the bubble region. Figure 3 shows that the $\overline{u'v'}$ component is found to peak in the shear layer. This is also the region of largest mean shear and hence, the region of largest energy transfer from the mean flow to the oscillating flow. The large negative value of the Reynolds stress in the initial part of the shear layer means there is a large transfer of kinetic energy into the velocity fluctuations. These fluctuations appear to amplify, become unstable and eventually roll up into the large scale structures that are visible in the instantaneous vorticity plot shown in Figure 2(b). Paring of these structures is also observed close to the reattachment region. This is consistent with the previously reported flow visualisation results of (Soria *et al.* (1991)).

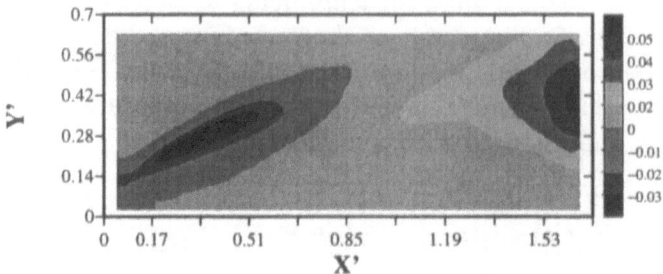

Figure 3. In-plane components of the Reynolds stress tensor, $\overline{u'v'}$, for a separated flow over a blunt flat plate. Re $= \frac{Uh}{\nu} = 1000$, plate thickness $= h$, X' $= \frac{x}{h}$, Y' $= \frac{y}{h}$; flow is from left-to-right

4. Conclusions

Spatio-temporal measurements of a separated and reattached flow over a blunt flat plate have been made using PIV. The results illustrate the dynamic nature of the shear layer and also highlight the energy transfer mechanisms occurring in the shear layer.

Acknowledgments

The financial support of the ARC for this research is greatly appreciated.

References

FOURAS, A. (1997). Development of a cross-correlation piv image recording technique and analysis of velocity and vorticity measurement error applied to an unsteady separated flow. Master's thesis Department of Mechanical Engineering, Monash University.

FOURAS, A. & SORIA, J. (1998). Accuracy of out-of-plane vorticity measurements using in-plane velocity vector field data. *Exp. Fluids.* accepted for publication.

KOSTAS, D., FOURAS, A., & SORIA, J. (1996). Application of optical image shifting in piv. In *First Australasian Conference on Laser Diagnostics in Fluid Mechanics and Combustion* pp. 113–118 Australia. The University of Sydney.

LOURENCO, L. (1991). Recent advances in lsv, piv and ptv. In *Flow Visualisation and Image Analysis* volume 279 pp. 81–99. Euromech Colloquium.

NICOLAIDES, D. (1997). Measurement of spatial quantities in grid turbulence using particle image velocimetry. Master's thesis Department of Mechanical Engineering, Monash University.

SORIA, J., KOSTAS, J., FOURAS, A., & CATER, J. (1998). A high spatial resolution and large dynamic range cross-correlation piv technique for turbulent flow measurement. In *Optical Methods and Data Processing in Heat and Fluid Flow* City University, London. Institution of Mechanical Engineers.

SORIA, J., WELSH, M., WU, J., & SHERIDAN, M. (1991). Three-dimensionalities in natural and perturbed separated unsteady flow from a square leading edge flat plate. In *44th APS/DFD* Scottsdale, Arizona, USA.

VORTEX RING FORMATION AT AN ORIFICE

J. CATER[1], J. SORIA[1] AND T.T. LIM[2]

[1] *Department of Mechanical Engineering,*
Monash University, Australia 3168.
[2] *Department of Mechanical & Production Engineering,*
National University of Singapore, Singapore 119260.

Abstract. High spatial resolution PIV measurements using film recording have yielded detailed instantaneous 2-D velocity fields of the formation of orifice generated vortex rings. The resultant high spatial resolution vorticity distributions enable the location and magnitude of regions of both primary and secondary vorticity to be determined. The experimental measurements demonstrate how secondary vorticity produced at the piston hastens the transition to turbulence.

1. Introduction

Vortex rings have been widely used to study the fundamentals of vortex motion because they are "robust, as well as simpler to generate, analyse, and isolate from end-effects than other configurations"[8]. The wide application of vortex rings to the study of vortical flow structures make it important that experimental results for vortex rings are repeatable and that the effects of generation conditions are understood. Previously there has been disagreement between various groups and between experiment and theory about the value of the circulation of rings [2]. It has been proposed that differences in formation conditions and the production of secondary vorticity is partially responsible [8].

In an attempt to standardise formation geometries and eliminate secondary vorticity, Glezer [5] proposed a simple geometry for a vortex ring generator. This geometry is initially a circular hole in a plane wall through which fluid is pushed by a piston. The piston face finishes flush with the wall so that after the ejection of the ring the generator becomes an uninterrupted plane. It is this geometry that was initially used in the present study, however it was discovered that secondary vorticity is produced at the piston face as it scrapes the boundary layer from the inside of the cylinder.

U. Frisch (ed.), Advances in Turbulence VII, 15–18.

2. Experimental Technique

In the present study vortex rings are generated by discharging water from a circular tube of 56.5 mm inner diameter through an orifice into a transparent perspex tank of internal dimensions 1100 mm × 500 mm × 500 mm using a piston. The end of the tube is mounted flush with the inner wall of the tank. The tank is filled with mains supply water initially at rest. The driving mechanism for the piston is a micro-step computer controlled stepper motor with a position encoder directly coupled to the shaft. This configuration allows the discharge mean velocity and the piston face position to be controlled.

Throughout the work presented here, the ratio of piston stroke to orifice diameter (L/D) known as the Impulse Ratio is equal to three. The distance from the piston face to the inside wall of the tank at the end of the stroke is varied from zero (the suggested standard) to a distance equal to three times the piston diameter.

Fluorescent dye visualisation using a continuous laser of wavelength 532 nm was initially used to determine the flow regions of interest. However dye is not used for quantitative measurements since a dye tracer does not capture information about the stretching, diffusion and cancellation of vorticity. Maxworthy [7] discovered that most of the ring vorticity is distributed in a volume of fluid which is much larger than the core of the ring visible with dye.

A recently developed image shifting technique that permits the use of cross–correlation PIV analysis of non-overlapping images recorded on photographic media is used to acquire images of the flow [1]. The system also allows the capture of two independant images with a small time delay (of the order of 1 ms) on one film frame. This enables the formation stages of relatively high Reynolds number rings to be analysed. The vortex ring results presented here have a Reynolds number of 7600 based on the orifice diameter. An adaptive PIV method is employed to extract the in-plane velocity field information with high velocity sampling resolution and large velocity dynamic range from the images. The resultant velocity fields are used to calculate the out-of-plane vorticity field in the flow plane using a χ^2 vorticity calculation method [4].

3. Analysis

The results of dye visualisation show that for the orifice geometry, the primary region of vorticity generation is near the lip of the orifice, where the vorticity forms a cylindrical sheet. The sheet subsequently rolls up to form a primary vortex ring. Images of dye tracer injected at the piston face also reveal secondary vorticity in the form of secondary rings that are generated at the piston face and induced at the boundary layer on the wall behind the primary ring. In low Reynolds number rings the sense of rotation of secondary vortex rings is visible.

Secondary vorticity from the roll-up of the boundary layer on the outside of the tube of a nozzle generated ring is illustrated in recent work by Fabris [3]. In this work the piston stops flush with the end of the nozzle, and a similar process can be envisioned. A region of vorticity generated at the piston face is not identified by the authors although it is clear in the figures. The identified "stopping vortex" is created by the roll-up of the boundary layer on the outside of the tube behind the primary ring. This vortex ring has vorticity of opposite sign and does not appear to translate with the primary ring. In this study core deformation is also apparent in the presented vorticity plots.

It is known that secondary vorticity of the same sign can undergo the vortex ring leapfrogging process with the primary ring [6]. The vorticity generated at the piston face is of the same sign as the primary ring. When leapfrogging occurs, the secondary ring is accelerated due to an induced velocity from the primary ring. Conversely the primary ring expands and slows down allowing the secondary ring to pass through. In this manner the core of the primary ring becomes distorted, particularly if the leapfrogging is not "clean". In some cases the rear ring is absorbed into the front one. Often misalignment of the rings can occur during the leapfrogging process, this can introduce further instabilities.

A high spatial resolution distribution of vorticity across a ring generated using the standard formation geometry can be seen in Fig. 1 . This figure reveals the motion of the secondary ring. Vortical fluid from behind the primary ring is advected rapidly to the fore of the ring. In later stages of the vortex ring evolution, this fluid "wraps around" the core. This causes core distortion of the primary ring and it is this phase that immediately precedes the transition to turbulence. It is concluded that the instability introduced by the secondary vorticity wrapping around the core leads to transition.

Thus, secondary effects are important and secondary vorticity must be recognised and accounted for. The effects of secondary vorticity from the piston face can be eliminated by having the piston face finishing position far from the inside wall. Piston face vorticity is of particular concern at high Impulse Ratios since the strength of the secondary ring is a function of the piston stroke length. It should also be noted that the effects of vorticity generated at the piston are not restricted to the impulsively started vortex ring flow.

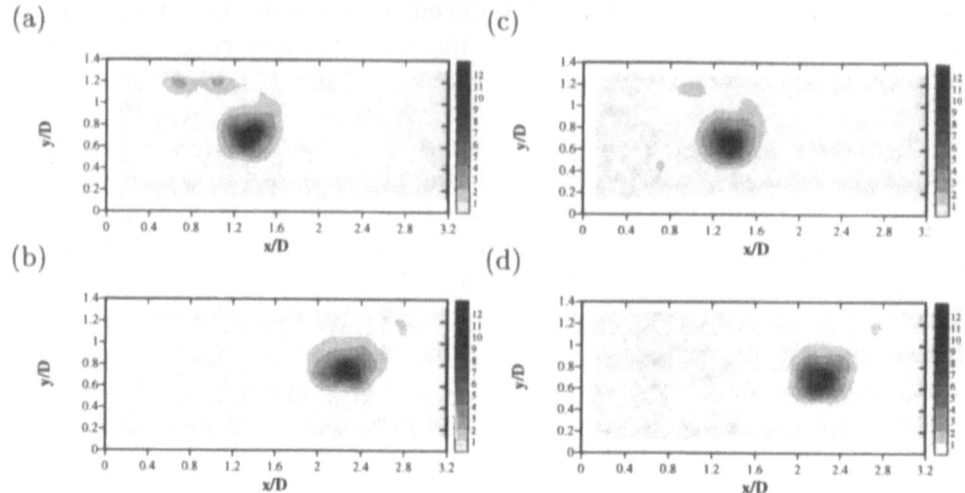

Figure 1.   Normalised vorticity distributions, Re = 7600. Flow is from left to right. (a),(b) Piston finishes flush. (c),(d) Piston ends 2D from orifice

4. Conclusions

The secondary vorticity generated at the piston face can have a significant effect on the sequence of transition to turbulence for an initially laminar ring when the piston face ends flush due to the vortex ring leapfrogging mechanism.

Acknowlegements

The technical support from Mr. John Gyura is greatly appreciated.

References

1. CATER, J., KOSTAS, D. J., FOURAS, A., & SORIA, J. (1998). High resolution cross-correlation piv on photographic film. In *International Conference on Applications of Optical Metrology* Balatonfured, Hungary.
2. DIDDEN, N. (1979). On the formation of vortex rings: rolling-up and production of circulation. *Z. Angew Math. Phys.* **30**, 101–116.
3. FABRIS, D. & LIEPMANN, D. (1997). Vortex ring structure at late stages of formation. *Phys. Fluids* **9**, 2801–2803.
4. FOURAS, A. & SORIA, J. (1998). Accuracy of out-of-plane vorticity measurements using in-plane velocity vector field data. *Exp. Fluids (accepted for publication)*.
5. GLEZER, A. (1988). The formation of vortex rings. *Phys. Fluids* **31**, 3532–3542.
6. LIM, T. (1997). On the role of kelvin-helmholtz-like instability in the formation of turbulent vortex rings. *Fluid Dyn. Res.* **21**, 47–56.
7. MAXWORTHY, T. (1972). The structure and stability of vortex rings. *J. Fluid Mech.* **51**, 15–32.
8. SHARIFF, K. & LEONARD, A. (1992). Vortex rings. *Ann. Rev. Fluid Mech.* **24**, 235–279.

USING THE HOT-FILM TECHNIQUE
FOR VELOCITY MEASUREMENTS
OVER A RIBLET STRUCTURED SURFACE

M. BRUSE[1], D.W. BECHERT[2] AND W. HAGE[1]

[1] *Hermann-Foettinger-Institute of Fluidmechanics (HFI),*
Technical University Berlin, Strasse des 17. Juni 135,
D-10623 Berlin, Germany, email: bruse@pi.tu-berlin.de
[2] *The German Aerospace Center (DLR), Institute of*
Propulsion Technology, Department of Turbulence Research,
Mueller-Breslau-Str. 8, D-10623 Berlin, Germany

The fact that riblets do reduce shear-stress has been established for more than ten years beyond any reasonable doubt. A plausible explanation is, that the longitudinal ribs rectify the turbulent flow in mean flow direction by hampering the fluctuating cross-flow velocity component w'. In a turbulent boundary layer the cross-flow w' and the flow component normal to the wall v' are connected through the low-speed-streaks and burst phenomena. If the cross-flow fluctuation w' close to the wall can be reduced, the turbulent momentum transfer close to the surface will be reduced as well and consequently, the shear stress $-\rho \cdot (\overline{u'v'})$ will be decreased.

The optimization experiments result in the optimal geometry of very thin blade-ribs (see. Fig. 1 and Refs. [2], [3] & [5]). In this case the maximum shear-stress reduction is almost 10%. However, a technologically more suitable design are riblets with a trapezoidal cross-section of the grooves (8.5%).

Assuming a viscous flow over the riblets, the theory has shown that the origins of the velocity profiles of longitudinal- and cross-flow are located underneath the rib tips (see Fig. 2 and Refs. [1] & [6]). The vertical distance to the rib tips (the protrusion height difference Δh_p) seems to affect the gradient of the shear-stress data $\Delta \tau$ for small s^+-values [$\Delta \tau = (\tau_{riblet} - \tau_{ref.})/\tau_{ref.}$ & $s^+ = s \cdot u_\tau/\nu$ with s ≡ lateral rib spacing] (see Figs. 1 & 2). A higher gradient leads to more shear-stress reduction. With increasing s^+ the ribs start to protrude out of the viscous sublayer and the riblet-surface starts to act like a rough one.

U. Frisch (ed.), Advances in Turbulence VII, 19–22.

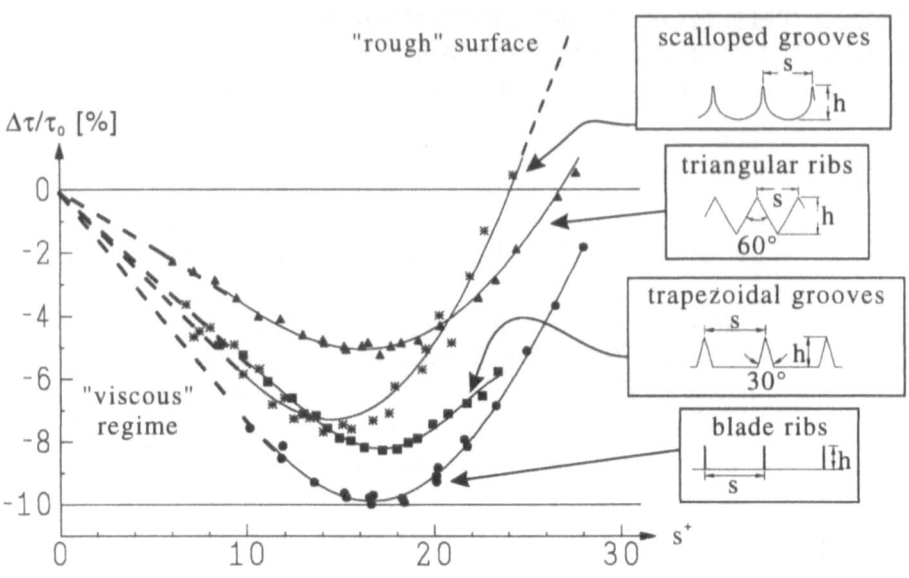

Figure 1. Shear-stress data for different riblet geometries

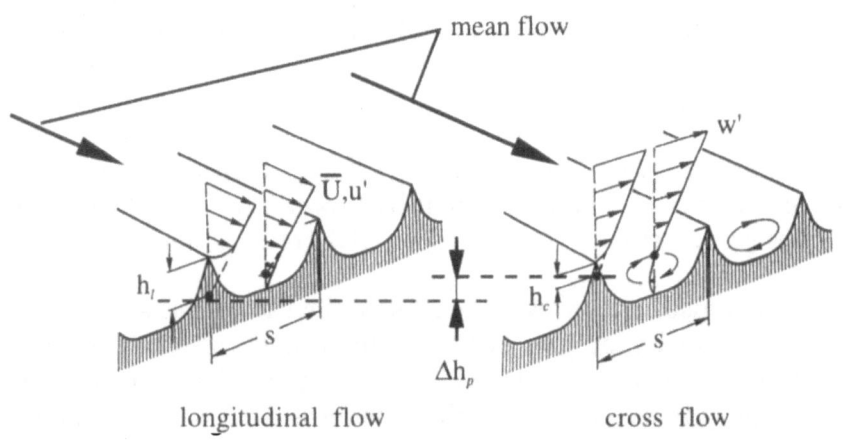

Figure 2. Velocity profiles over riblet structured surface

To check the theory we wanted to measure the velocity profiles over the rib tip and on the centerline of a groove. In the Berlin oil channel (Ref. [4]), where all the experiments were carried out, the lateral rib spacing is in the order of some millimeters. That is why we can use commercial hot-film probes for the velocity measurements very close to the surface. When we started these measurements we tried to use hot-wires, but due to the high roughness of the wires small dirt particles (diameter $< 1\mu m$) stuck to the wire and deteriorated the calibration. By contrast, using hot-film sensors, the time for this problem to occur is long enough to take data. A cleaning mechanism consisting of a small tube was installed several millimeters downstream of the sensors. It generates an oil jet which cleans the sensors during the acquisition breaks.

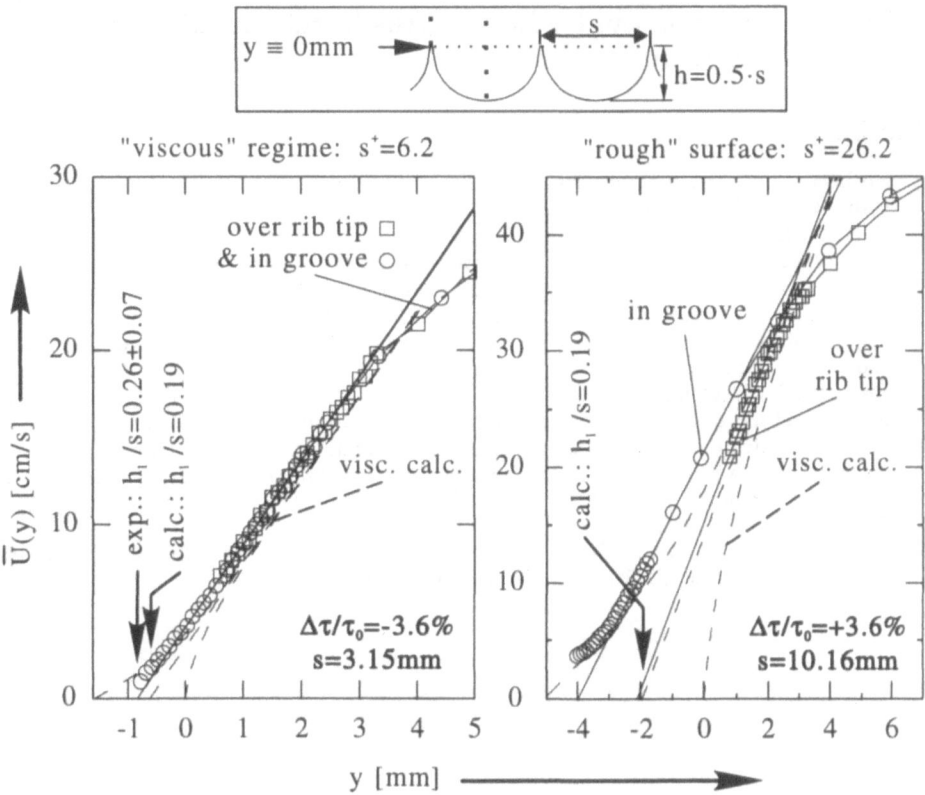

Figure 3.   Velocity measurements over riblets for small and high s^+-values

In addition to the velocity calibration of the sensors we had to carry out an angular calibration of the sensors. This is because the probes are installed slightly slanted to the wall to take data for small wall or riblet distances.

For the angular calibration a nozzle generating an oil jet (diameter 20mm) of constant speed moves in a circle around the sensors.

For low s^+-values a uniform gradient of the mean-flow velocity ($U \propto y$) can be observed (see Fig. 3, left). Here the ribs are obviously submerged in the viscous sublayer. The protrusion height calculated with this gradient agrees within the error bars of the measurement with the protrusion height of the viscous calculation ($h_{l,visc} = 0.192 \cdot s$; see Figs. 2 & 3 and Ref. [1]). To obtain more accurate protrusion height measurement a more precise determination of the wall distance would be necessary. However, the finite size of the sensitive volume of the probes produces an uncertainty of the same 0.2mm order.

For higher s^+-values (see Fig. 3, right) the rib tips protrude out of the viscous sublayer and no locally constant gradient comparable to the one of the viscous calculation can be found. Actually, this situation corresponds to the conditions of "rough" surfaces.

References

1. Bechert, D.W. & Bartenwerfer, M. (1989) The viscous flow on surfaces with longitudinal ribs, *J. Fluid Mech.* **206**, pp. 105–129.
2. Bechert, D.W., Bruse, M., Hage, W., & Meyer, R. (1997) Biological surfaces and their technological application – laboratory and flight experiments on drag reduction and separation control, *AIAA paper.* **97–1960**.
3. Bechert, D.W., Bruse, M., Hage, W., van der Hoeven, J.G.Th. & Hoppe, G. (1997) Experiments on drag-reducing surfaces and their optimization with an adjustable geometry, *J. Fluid Mech.* **338**, pp. 59–87.
4. Bechert, D.W., Hoppe, G., van der Hoeven, J.G.Th. & Makris, R. (1992) The Berlin oil channel for drag reduction research, *Exp. in Fluids* **12**, pp. 251–260.
5. Bruse, M., Bechert, D.W., van der Hoeven, J.G.Th., Hage, W. & Hoppe, G. (1993) Experiments with conventional and novel adjustable drag-reducing surfaces, in: *Near-wall turbulent flows* (eds. R.M.C. So, C.G. Speziale & B.E. Launder), pp. 719–738, ELSEVIER, Amsterdam.
6. Luchini, P., Manzo, F., & Pozzi, A. (1991) Resistance of a grooved surface to parallel flow and cross-flow, *J. Fluid Mech.* **228**, pp. 87–109.

BURSTS AND SUBGRID-SCALE ENERGY TRANSFER IN TURBULENT WALL-BOUNDED FLOW

P. S. WESTBURY[1], N. D. SANDHAM[2], J. F. MORRISON[3]

[1]*Whitby Bird & Partners*
London W1P 4DA
[2]*Department of Engineering, Queen Mary & Westfield College*
London E1 4NS
[3]*Department of Aeronautics, Imperial College*
London SW7 2BY

1. Introduction

In large-eddy simulation (LES) of wall-bounded flows, the distinction between the large, resolved scales, computed as solutions to the Navier-Stokes equations, and the small unresolved scales, represented by a subgrid-scale model, becomes unclear because all the eddies are 'small' in the near-wall region. Therefore the energy transfer between the resolved and unresolved scales is especially complicated close to the wall. Accurate modelling is required so that near-wall regions do not have to be resolved fully, so permitting simulations at high Reynolds numbers. Several workers (Horiuti 1997, Härtel *et al.* 1994, Domaradzki 1994, Piomelli *et al.* 1996) have shown that an eddy-viscosity model is incapable of representing the near-wall energy transfer accurately, it being absolutely dissipative. It is therefore unable to represent the transfer of energy from the unresolved to the resolved scales ("backscatter") properly and this is more important in the near-wall region. Mason and Thomson (1992) show that the backscatter model is intrinsically related to the surface boundary condition: simulations of the surface layer are only able to reproduce the log law by the inclusion of backscatter effects, in their case, by a stochastic model, examined using data from direct numerical simulations (DNS), by Westbury and Morrison (1998).

The aim of the present work is to investigate the possible connection between "bursts" in near-wall turbulence and the transfer of energy both out of and back in to the resolved scales, the ultimate objective being the development of a structure-based subgrid-scale model. Previous experimental work by Morrison and Bradshaw (1992) suggests that, in the logarithmic region, "ejections" ($u < 0$, $v > 0$) and "sweeps" ($u > 0$, $v < 0$) are compact, coherent, structures in both physical and wave-number space, scale on viscous variables and effect most of the universal part of the spectral transfer of energy. Therefore these structures provide a starting point for the identification of regions that contribute to both forward scatter and backscatter. Meneveau (1991a,b) has defined a

U. Frisch (ed.), Advances in Turbulence VII, 23–26.

"dual (wavelet) spectrum". Using an orthonormal wavelet decomposition of the Navier-Stokes equations, he identifies the quantity $\pi^{(m)}[i]$, that represents both the spatial flux of energy through a region of characteristic size r_m (where m is a scale index) at location $[i]$ as well as the flux of energy between different scales. It is defined so that a decrease in the energy of the large scales corresponds to positive flux.

2. Results

The VISA+LEVEL conditional sampling scheme (Morrison *et al.* 1989) has been applied to a database produced by DNS of turbulent channel flow at $Re_\tau = 180$ (Sandham and Howard 1997). The sampling is applied in the streamwise direction. Results involving event lengths and event contributions to $-\overline{uv}$ and R_{uv} are similar to the boundary-layer results of Morrison *et al.* (1992), although they are subject to the "geometry effect" described by Wei and Willmarth (1989). In order to separate the direct effects of viscosity at low wave number and the effects of production at high wave number from the genuine turbulent spectral transfer, following Härtel *et al.* (1994), quantities are split into a statistically stationary mean value (a plane average), and a fluctuating part. Thus the total dissipation rate, ε, can be broken up into the direct viscous dissipation of the resolved scales ε_v, and the net transfer out of the resolved scales. The latter can be further subdivided into $T^{ms} + T^{fs}$ so that:

$$\varepsilon \approx \varepsilon_v + T^{ms} + T^{fs} , \qquad (1)$$

where T^{ms} is the contribution to the subgrid energy by the *mean* strain rate (as the filter width tends to zero in Fourier space, this term becomes the mean production), and T^{fs} accounts for the net transfer of turbulence energy up (to higher wave number) the spectrum. In general, equation (1) is approximate owing to the effects of spatial transport which constitutes a source or sink at each wave number: in the absence of transport, it becomes exact for a Fourier sharp cut-off filter (Westbury and Morrison 1998). Thus T^{fs} is relevant to a high-Reynolds-number LES since ε_v tends to zero as the Reynolds number tends to infinity, and with the cut-off in the inertial subrange, T^{ms} will be negligibly small too at high Reynolds numbers.

Figure 1 shows conventional averages of T^{fs} calculated by using a Fourier cut-off filter at $k_1\eta \approx 0.075$, which lies in an approximate inertial subrange. T^{fs} is taken to be positive when the transfer is from the large scales to the small ones. Also shown are the contributions made by ejections and sweeps: together with the undetected quantity, these contributions sum to the conventional total. Near $y^+=10$, there does not appear to be any particular correlation between T^{fs} and either ejections or sweeps. An alternative explanation might be that these events are associated with both forward scatter and backscatter, so that the net contribution is small. Figure 2 shows the equivalent results produced by a LEVEL sampling condition only: sweeps are more strongly correlated with backscatter and ejections with forward scatter in the near-wall region. This analysis does not take account of the possibility that ejections and sweeps may be shifted in space or time relative to regions of large energy transfer.

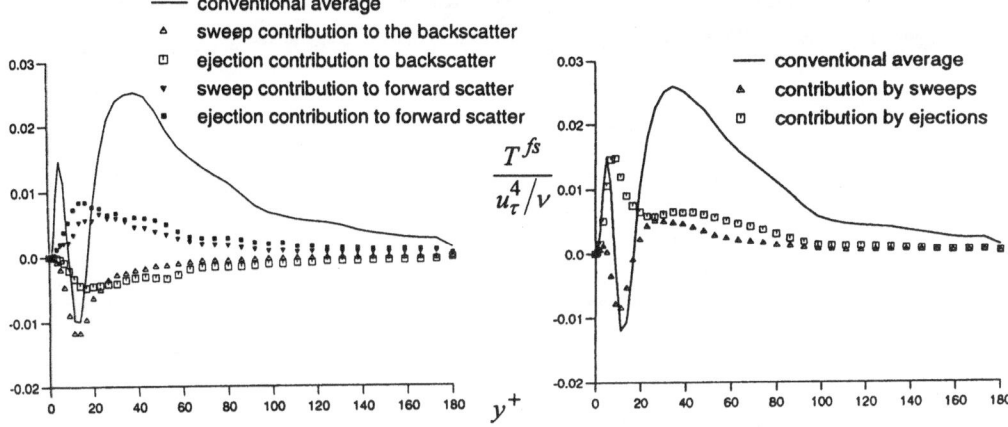

Figure 1: T^{fs} and VISA+LEVEL contributions.   *Figure 2*: T^{fs} and LEVEL contributions.

Using an orthonormal wavelet decomposition of the velocity field in two-dimensional homogeneous planes, an approximate comparison with T^{fs} may be made. The analysis uses only fluctuating quantities. Figure 3 shows contours of uv and $\pi^{(m)}[\mathbf{i}]$ at $y^+ = 11$. In the region of $x^+ = 65$, $z^+ = 25$, for example, there is large, predominantly negative, uv coinciding with large π (of both signs) at scales $r = 1, 2$ and 4. Here the uv field is 'spotty'. Contours of positive and negative π occur side by side and are elongated in x. The net transfer is negative at this location in agreement with the results of Figures 1 and 2. Figure 4 shows the equivalent data at $y^+ = 80$ where much of the flux has moved to larger scales and the net transfer is now positive.

3. Discussion and Conclusions

While T^{fs} is a filter-dependent measure of the flux in or out of the resolved scales at a fixed location, π also includes a spatial flux - the "sweeping" of small scales by the large ones. Moreover, it does not discriminate at which scales the flux is occurring. Meneveau (1991b) also defines a 'subgrid' flux, $\pi_{sg}^{(m)}[\mathbf{i}]$, caused only by interactions at scales smaller than m, and therefore excluding the effect of sweeping. It is clearly therefore this latter quantity that should be compared with T^{fs}. However, in this preliminary analysis, it is clear that there exists at least a qualitative correspondence between uv and π.

Acknowledgements: That part of the work at Imperial College was carried out under Agreement Met. 1b/2126 with the Meteorological Office, Bracknell.

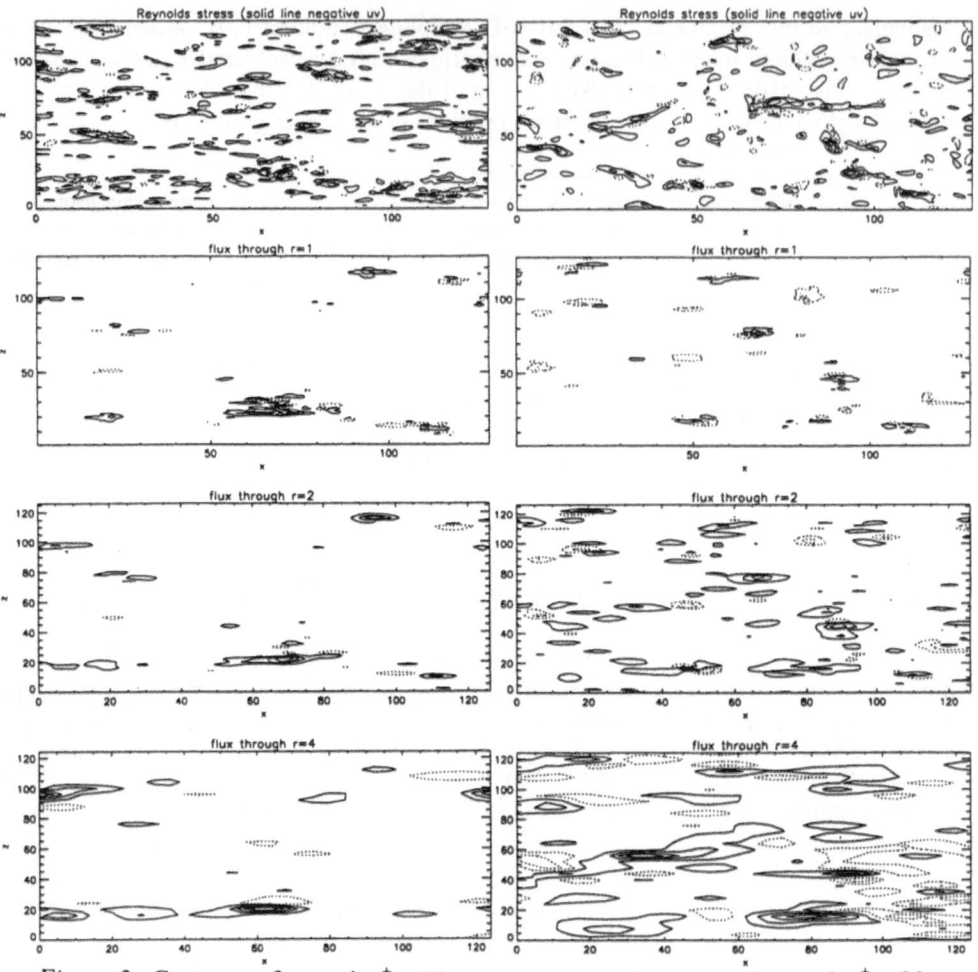

Figure 3: Contours of *uv* and $y^+ = 11$.
Solid line: negative quantities.

Figure 4: Contours of *uv* and $y^+ = 80$.
Solid line: negative quantities.

4. References

Domaradzki, J. A., Liu, W., Härtel, C. J. and Kleiser, L. 1994 *Phys. Fluids* **6**(4), 1583-1599.
Härtel, C. J., Kleiser, L., Unger, F. and Friedrich, R. 1994 *Phys. Fluids* **6**(9), 3130-3143.
Horiuti, K. 1997 *J. Physical Society of Japan* **66**(1), 91-107.
Mason, P. J. and Thomson, D. J. 1992 *J. Fluid Mech.* **242**, 51-78.
Meneveau, C. 1991a *Phys. Rev. Lett.* **66**(11), 1450-11453.
Meneveau, C. 1991b *J. Fluid Mech.* **232**, 469-520.
Morrison, J. F., Subramanian, C. S. and Bradshaw, P. 1992 *J. Fluid Mech.* **241**, 75-108.
Morrison, J. F., Tsai, H. M. and Bradshaw, P. 1989 *Expts. Fluids* **7**, 173-186.
Piomelli, U, Yu, Y. and Adrian, R. J. 1996 *Phys. Fluids* **8**(1), 215-224, 1996.
Sandham, N. D. and Howard, R. J. A. 1997 In *Parallel CFD 1997*, (ed. Ecer *et al.*) Elsevier, to appear.
Wei, T. and Willmarth, W. W. 1989 *J. Fluid Mech.* **204**, 57-95.
Westbury, P. S. and Morrison, J. F. 1998 In preparation for *Phys. Fluids*.
Westbury, P. S. and Sandham, N. D. 1996 Rep. No. QMW-EP-1111, Faculty of Engineering, Queen Mary & Westfield College.

EXPERIMENTAL STUDY ON STATISTICAL PROPERTIES OF COHERENT STRUCTURES AND INTERMITTENCY IN A CHANNEL FLOW

MIGUEL ONORATO
Dip. di Fisica Generale dell'Università di Torino
Via Pietro Giuria 1- 10125 Torino, Italy

ROBERTO CAMUSSI
Dip. di Ingegneria Meccanica e Industriale, Università Roma Tre
Via della Vasca Navale 79 - 00146 Roma, Italy

GAETANO IUSO
Dip. di Ingegneria Aerospaziale, Politecnico di Torino
Corso Duca degli Abruzzi 24 - 10129 Torino, Italy

1. Introduction

A great interest has recently been addressed to the study of the statistical properties of homogenous and isotropic turbulence and nowadays most of the experimental results agree on the deviations from the Kolmogorov 1941 laws in the inertial range. To the knowledge of the authors, extensive studies have not been performed for the case of non homogeneous and non isotropic turbulence. In this paper, the statistical properties of the stream-wise velocity fluctuation in a turbulent air channel flow are studied experimentally at various distances from the wall. Strong shear layer events, educed using the classical VITA method (Variable Interval Time Average, [1]) and the LIM method (Local Intermittency Measure, [2]), are considered and related to the deviations from the Kolmogorov theory and from the values obtained by the ESS (Extended Self Similarity, [3]) in homogeneous turbulence.

2. Experimental Details

Measurements were taken using a single hot wire (2 μm of diameter and 0.45 mm long) in a fully developed turbulent channel flow ($70\,mm$ x $300\,mm$) located at the 'M. Panetti' Laboratory of the Politecnico di Torino. The Reynolds number based on the mean velocity at the centre of the channel and on the channel height was 21000. The sampling rate was 6000 Hz. Results for statistical quantities such as

27

U. Frisch (ed.), Advances in Turbulence VII, 27–30.
© 1998 *Kluwer Academic Publishers.*

r.m.s., skewness and kurtosis are in good agreement with values found in the literature ([4], [5]) and therefore are not reported herein. In this paper, time series taken at three different distances from the wall are considered: $y^+=15$, $y^+=103$ and $y^+=310$ (quantities with the superscript + are in wall units, normalised with respect to the viscous wall length, v/u_τ, and viscous time v/u_τ^2, where u_τ is the friction velocity). The first measurement point is located in the buffer layer and corresponds to the maximum of the velocity r.m.s. value, where most of the stream-wise elongated vortices are located; the second and third measurement points are in the logarithmic region of the near wall flow.

3. Methods of Analysis and Discussion

In order to educe the events, we consider two different techniques: VITA and LIM. The first one, widely used for studying the bursting phenomena in the boundary layer, consists in looking for the instant of time in which the local variance, calculated over a window of a fixed length T^+, exceeds the quantity ku_{rms}^2, where k is a chosen threshold level and u_{rms}^2 is the variance of the signal. The LIM technique, proposed and applied to turbulent signals by Camussi and Guj [2], is based on the wavelet decomposition (we used both orthogonal and continuous wavelets) and on the computation of $[w^{(r)}(i)]^2/<[w^{(r)}(i)]^2>$, where $w^{(r)}(i)$ represents the wavelet coefficient at scale r at time (or space) i. By choosing a proper threshold level and a scale, these methods allow those events which yield the strongest energy bursts to be selected. Once VITA or LIM events are detected, they are phase aligned and then averaged. We verified that, by choosing the appropriate threshold and scales, VITA and LIM time signatures coincide. Furthermore, such events are classified as *accelerating* motions if $\partial u/\partial t > 0$ and *decelerating* motions if $\partial u/\partial t < 0$. A typical result is shown in Figure 1 for the accelerating case. At $y^+=15$ the amplitude of the events are higher than in the other two cases and also the slope at around $t^+=0$ is greater. We interpret these events using the conceptual model of wall coherent structures developed in [6] based on the analysis of direct numerical simulations of a turbulent channel flow. According to the model, internal shear layers containing positive and negative velocity derivative are seen to result from the induction of quasi-stream-wise vortices which are tilted in the (x,z) plane and inclined in the (x,y) plane (x is the stream-wise direction). Based on this model and on the considerations in [6], we suggest that the time signatures in Figure 1 are effectively due to the presence and interaction of wall structures. In order to better understand the effect of the coherent structures on the time series, we construct a "surrogate" signal, obtained by the randomisation of the Fourier phases of the measured signal. The surrogate signal has exactly the same power spectrum of the original one, but when the eduction technique is applied, the number of event is reduced by an order of magnitude and no signature is obtained. This is consistent with results that are obtained in the following analysis. In order to study the level of intermittency caused by the coherent structures at different frequencies, we consider the flatness factor defined as $<[w^{(r)}(i)]^4>/<[w^{(r)}(i)]^2>^2$. Results are shown in Figure 2. For low frequencies, the flatness factor is around 3, which indicates that at these frequencies the process is gaussian; at $y^+=15$, it starts to increase for frequencies higher than 200 Hz, indicating that the level of intermittency is higher than at $y^+=103$ and $y^+=310$. The flatness factor for the surrogate data is gaussian for all frequencies, which means that the signal is not intermittent, as the coherent structures are completely destroyed by the randomisation of the phases.

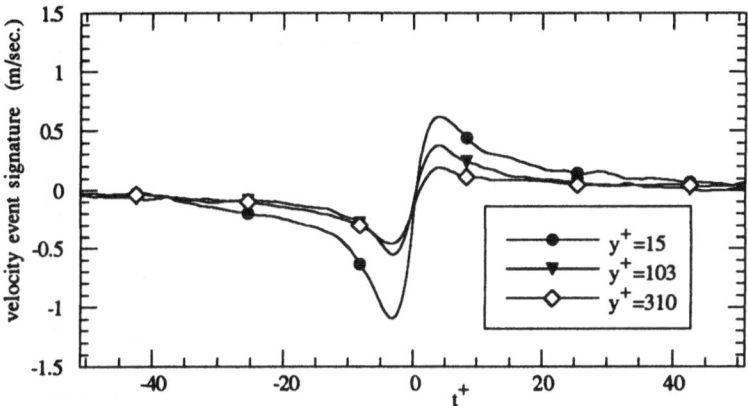

Figure 1. Turbulent signatures structures at different distances from the wall

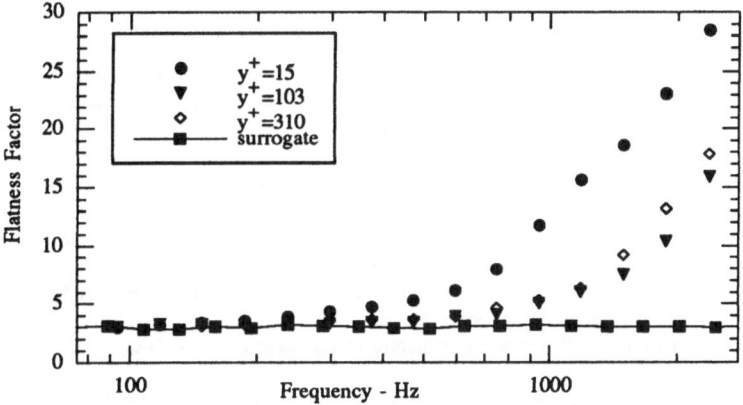

Figure 2. Flatness factor at different distances from the wall

To confirm that at $y^+=15$ the stream-wise vortices are responsible for higher intermittency, we calculate the scaling exponents of the first 6 moments of the structure function. This is done by means of the ESS, since the Reynolds number is not high enough to ensure a wide inertial range. Results are shown in Figure 3. The surrogate data follows, as expected, the Kolmogorov theory; at $y^+=103$ and $y^+=310$ the results agree with those obtained for the homogeneous case [3] and at $y^+=15$ the values of the scaling exponents are lower indicating a higher intermittency. Even though very close to the wall the Taylor hypothesis is not valid, our results are in agreement with the values of the scaling exponents calculated from a numerical simulation of a channel flow in [7] and also are well supported by a preliminary analysis conducted on PIV (Particle Image Velocimetry) measurements in which the turbulent field is spatially resolved [8].

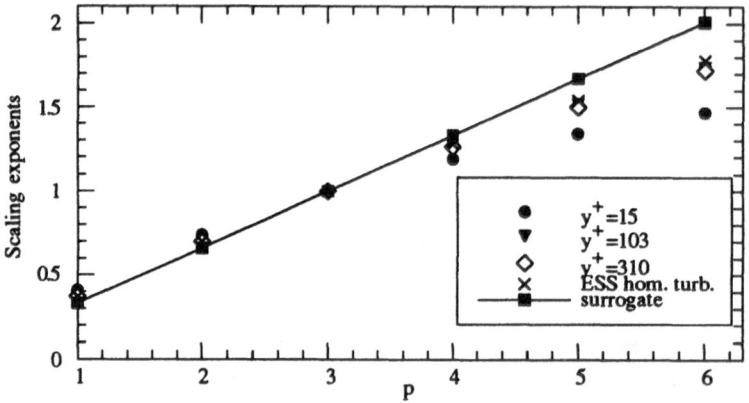

Figure 3. Scaling exponents at different distances from the wall

Acknowledgements

The experimental work was financially supported by Centro Dinamica dei Fluidi di Torino (C.N.R.). We wish to thank G. Buresti, G. Guj and M.V. Salvetti for reading and discussing the manuscript.

References

1. Alfredsson, P.H., Johansson, A.V.: On the detection of turbulence-generating events , *J. Fluid Mech.* 139, (1984) 325- 345
2. Camussi, R., Guj, G.: Orthonormal wavelet decomposition of turbulent flows: intermittency and coherent structures *J. Fluid Mech.* 348, (1997) 177 - 199
3. Benzi, R., Ciliberto, S., Tripiccione, R., Baudet, C., Massaioli, F., Succi, S.: Extended self-similarity in urbulent flows, *Phys. Rev. E* 48, n.1, (1993) 29 - 32
4. Johansson, A.V. , Alfredsson, P.H.: On the structure of turbulent channel flow, J . *Fluid Mech.* 122, (1982) 295 - 314
5. Kim, J., Moin, P. , Moser, R.D.: Turbulence statistics in fully developed channel flow, *J. Fluid Mech.* 177, (1987) 133 - 166
6. Jeong, J., Hussain, F., Schoppa, W., Kim, J., Coherent structures near the wall in a turbulent channel flow, J. *Fluid Mech.* 332, (1997) 185 - 214
7. Amati, G., Benzi, R., Succi, S.: Extended self similarity in boundary layer turbulence, *Phys. Rev. E* 55, n. 6, (1997) 6985-6988
8 Gottero M., Onorato M. in preparation, 1998

DPIV INVESTIGATION OF NEAR WALL TURBULENT FLOW STRUCTURES

MARCO GOTTERO, MICHELE ONORATO

Dip. di Ingegneria Aerospaziale, Politecnico di Torino,
Corso Duca degli Abruzzi 24 - 10129 Torino, Italy

1. Introduction

After decades of experimental and numerical research, the structure of turbulent boundary layers is only partially understood. Various organised motions have been identified as wall low-speed streaks, internal shear layers, vortical structures and ejections and sweeps [1]. Recent analysis of data bases from turbulence numerical simulation are giving important new insights about the spatial and temporal character of organised motions, their spatial relationships and their generation and evolution [2,3,4]. Numerical solutions are currently limited to simple geometries and low Reynolds number flows. Experimental investigation is still essential for flows with complex boundary conditions and high Reynolds numbers. Advanced experimental methods are becoming available to accomplish this task, most of them based on optical anemometry techniques.

A quantitative flow visualisation, suitable for turbulence investigation, has been developed at the 'Laboratorio di Anemometria Ottica' of the 'Centro di Studio per la Dinamica dei Fluidi del CNR'. The objective is the application to complex flows, such as "manipulated" boundary layers for drag and heat transfer control [5,6].

In this paper some results from the investigation of turbulent boundary layer over a flat plate will be shown. The flow in a plane parallel to the wall, in the buffer layer region, is analysed, highlighting the spatial relationship between low speed streaks and internal shear layers.

2. Experimental Set up and Procedure

Experiments were carried out in the "Hydra" water tunnel. This facility is a closed loop, open flow channel, with $350 \times 500 \times 1800$ mm^3 test section. Measurements have been performed on a flat plate positioned in the test chamber, with transition fixed at the leading edge. The observed section is characterised by $Re_x = 4 \times 10^5$, $Re_\theta = 1010$, $u_\tau/u_e = 0.046$. A double-pulsed light sheet is provided by a Nd-YAG laser source (200mj and 8ns per pulse). The flow is seeded with spherical solid particles,

U. Frisch (ed.), Advances in Turbulence VII, 31–34.
© 1998 *Kluwer Academic Publishers.*

2 μm nominal diameter. Images of the seeded flow in the illuminated plane are captured on a photographic high resolution 24x36 mm^2 film. Time statistics have been obtained from a series of digital CCD video camera images. The spatial accuracy and resolution for the analysis of turbulent flows has been verified for both techniques [7,8]. Analysis algorithms are based on two-dimensional digital particle image velocimetry technique, DPIV [9].

3. Results and Discussion

A time statistic analysis of the flow has been applied to a series of more than 500 video-camera images in a plane (x,y) normal to the wall and parallel to the mean flow and located at the centreline of the plate. Mean and fluctuating velocity profiles are in good agreement with data from the literature.

Only results from photographic frozen velocity fields in a plane parallel to the wall (x,z), at a distance from the plate of 20 wall units, $y^+=20$, are shown here.

Figure 1 shows a vector plot of the instant fluctuating velocity in the plane (x,z). The superimposed grey-level map represents the square of the longitudinal component of the velocity, u. As expected, quasi-stream wise low speed streaks characterise the flow in the buffer-layer region. In agreement with a previous research [10], visual observation indicates average span wise streak spacing of about 100 wall units, $\lambda^+=100$. Performing Fourier two-dimensional spectral analysis, averaging 15 realisations as in Figure 1, it is found that the most energetic span wise mode occurs at higher scale, $\lambda^+=260$. This confirms the results of similar analysis in [11].

The black dots in Figure 1   give the location of strong internal shear layers, detected by peaks in the local variance of u exceeding the mean variance, VISA events [12]. The local variance was averaged over about 115 viscous units in the stream wise direction.

Representations of the flow field as in Figure 1 are helpful in studying the spatial relationship between streaks and internal shear-layers. To this end, a conditional analysis has been applied to a series of instantaneous images at $y^+=20$. Detected VISA events were ensemble averaged centring individual realisations in both x and z directions. The VISA technique has been modified in order to retain the asymmetric features of the streaks. Individual events with different sign of du/dz have been ensemble averaged separately. Only accelerated events (du/dx<0) have been considered. In Figure 2 the result corresponding to du/dz<0 is shown. The ensemble averaged structure retains most of the characteristic features of single realisations. In particular the segmented character of the streaks is clearly shown by the two low-speed islands tilted in different directions seen in the iso-contour plot. The longitudinal extent of each segment composing a streak is of order of 200 viscous lengths. The position of the internal shear layer event ($x^+=0$, $z^+=0$) corresponds to the merging region of the two kinked low-speed segments, at the border between low and high speed bands. This suggests that the asymmetric features of the elongated, segmented low-speed streaks is of great importance for the generation of internal shear layer structures, these in turn are considered to be the main contributors to the production of turbulence. The structure in Figure 2 bears a close resemblance to results obtained in [13] by analysing a data base generated from direct numerical simulation of turbulent channel flow.

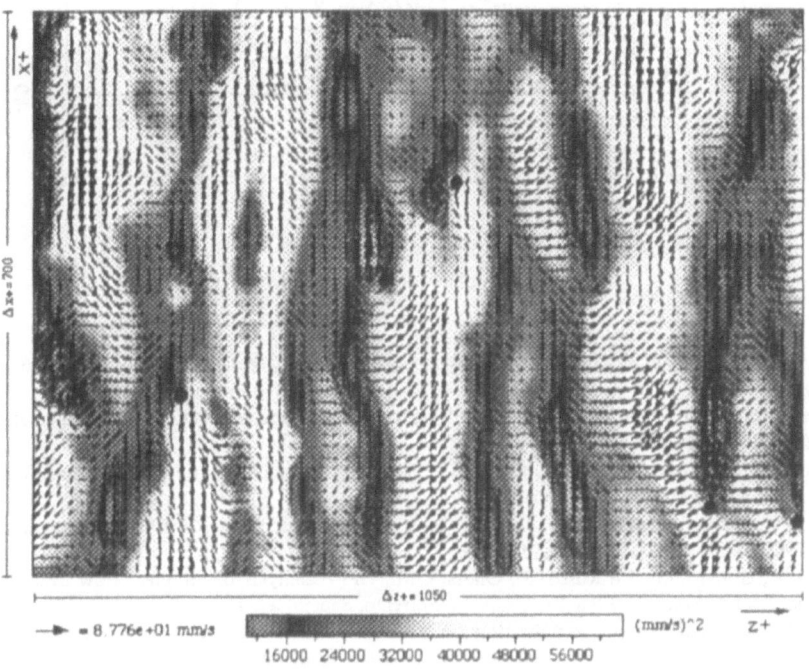

Figure 1 Instant fluctuating velocity field. $y^+ = 20$

A first attempt at interpretation of the present observed scenario suggests the conceptual model of coherent structure recently proposed by Jeong *et al.* [4]. According to this model, the inner region up to $y^+=60$ is dominated by quasi stream wise elongated vortices, slightly inclined and tilted; they appear to be organised in sequences in which vortices of alternating sign overlap in the x direction as a staggered array. The flow structure obtained from conditional analysis of the present DPIV data set may be rationally interpreted as the flow field induced by two adjacent overlapping counter-rotating vortices asymmetrically tilted due to their mutual induction. The organisation of the flow into low and high speed regions comes from the pumping of fluid away from and towards the wall respectively. Motions of opposite sign induced by the two counter-rotating vortices near the overlapping region are responsible for the creation of the very strong internal shear layer detected here by the VISA technique.

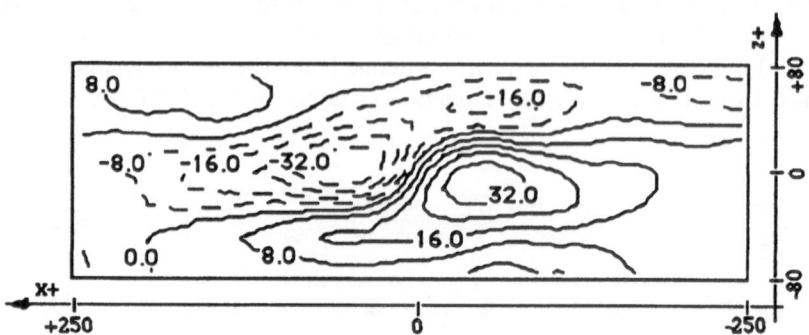

Figure 2 Conditional ensemble averaged flow structure; <u> contour levels, mm/s.

4. References

1. Robinson, S.K.: Coherent motions in the turbulent boundary layer, *Annu. Rev. Fluid Mech.* 23, (1991) 601-639
2. Bernard, P.S., Thomas, J.M., Handler, R.A.: Vortex dynamics and production of Reynolds stress, *J. Fluid Mech.* 253, (1993) 385-419
3. Blackburn, H..M., Mansour, N.N., Cantwell, B.J.: Topology of fine-scale motions in turbulent channel flow, *J. Fluid Mech.* 310, (1996) 269-292
4. Jeong, J., Hussain, F., Schoppa, W., Kim, J.: Coherent structures near the wall in a turbulent channel flow, *J. Fluid Mech.* 332, (1997) 185-214
5. Spazzini, P., Gottero, M., Lo Russo, S., Onorato, M.: DPIV analysis of turbulent flow over a back-f acing step, *8th Int. Symp. on Flow Visualisation*, Sorrento, September, 1998
6. Gottero, M., Iuso, G., Onorato, M.: The flow field structure of a turbulent boundary layer manipulated by a circular cylinder, *8th Int. Symp. on Flow Visualisation*, Sorrento, September, 1998
7. Gottero, M.: Sviluppo di un sistema di visualizzazione quantitativa per l'indagine sperimentale di flussi turbolenti, *tesi di dottorato, DIASP Politecnico di Torino*, (1997)
8. Gottero, M., Onorato, M.: DPIV analysis of wall turbulent shear flows, *21st Congress of the International Council of the Aeronautical Sciences*, Melbourne, September, 1998
9. Willert, C.E., Gharib, M.: Digital particle image velocimetry, *Exp. in Fluids*, 10, (1991)
10..Smith, C.R., Metzler, S.P.: The characteristics of low-speed streaks in the near-wall region of turbulent boundary layer, *J. Fluid Mech.* 129, (1983) 27-54
11. Liu, Z.C., Adrian, R.J., Hanratty, T.J.: A study of streaky structures in a turbulent channel flow with particle image velocimetry, *Int. Symp. on the application of laser technology to fluid mechanics*, Lisbon (1996)
12. Kim, J.: Turbulence structures associated with bursting events, *Phys. Fluids* 28, 52, (1985)
13. Johansson, A.V., Alfredsson, P.H., Kim, J.: Evolution and dynamics of shear-layer structures in near-wall turbulence, *J. Fluid Mech.* 224, (1991) 579-599

EXPERIMENTAL INVESTIGATIONS OF A JET IN COUNTERFLOW

S. BERNERO, H.E. FIEDLER
Hermann-Föttinger-Institut für Strömungsmechanik
TU Berlin, Str. des 17. Juni 135, 10623 Berlin, Germany

1. Introduction

The interaction of a jet with an external stream is of great importance in many practical applications, both in nature and in technology. Although jets in a coflow or crossflow have been widely investigated over the past years, relatively few studies are available on a jet flowing into a uniform stream of opposite direction, since this presents additional experimental and theoretical difficulties, related to flow reversal and to pronounced instability.

The enhanced mixing efficiency, also in comparison with the co- and crossflow configurations, is the most evident characteristic of a jet in counterflow and motivates further studies in view of the possible practical applications, which are to be found in environmental, chemical or process engineering (wastewater or pollutants disposal [3], mixing reactors, in-duct-burners), as well as in propulsion (as aerodynamic flameholder in afterburners of jet engines).

2. Experimental facilities and methods

The water tunnel used for this study has a vertical, 30 x 30 cm wide and 100 cm long test section and can produce a uniform counterflow with velocities up to 13 cm/s; at this velocity, the turbulence level is about 1.6 %. Three nozzles with fifth order polynomial inner contour and exit diameter $D = 10, 5$, and 2 mm respectively were used to obtain jet-to-counterflow velocity ratios up to 50.

Pointwise velocity measurements were obtained with a one-component LDA system and will be completed by measurements of instantaneous two dimensional velocity fields through a particle-image-velocimetry (PIV) system, which is currently being tested. Laser-induced-fluorescence (LIF) vi-

35

U. Frisch (ed.), Advances in Turbulence VII, 35–38.
© 1998 *Kluwer Academic Publishers.*

sualizations in an axial plane of the jet were also carried out. For both PIV and LIF experiments, the light sheet was produced from a 5 W Argon-ion laser and a scanner system. Images were acquired through a CCD camera and a Betacam VCR, then digitized through a frame grabber into a workstation for further processing.

3. Results

The present experiments confirm data from literature regarding the linearity of the relationship between the jet-to-counterflow velocity ratio $\alpha = U_j/U_0$ and the average penetration length x_p. The containment effect of the channel walls, which causes a reduction of the penetration length above a threshold value of the momentum flux ratio, was also verified and corresponds to the results in [4]. A simple model with the superposition of a jet and a uniform flow provides a linear relationship for the penetration length [6]. This procedure was further extended to model also the centerline velocity decay and the following relationship was obtained:

$$\left(\frac{U+U_0}{U_j}\right)\frac{x_p}{D} = \left(\frac{k}{2}\right)\frac{1+\sqrt{1-x/x_p}}{x/x_p}, \qquad x \le x_p \qquad (1)$$

By choosing $k = 5.83$, as in [1], the curve compares reasonably well with the experimental results: the improvement over the simple hyperbolical model suggested in [1] is negligible for low x-values (Fig. 1a), but becomes evident as x increases (Fig. 1b). The velocity decay for $x > x_p$ can on the other hand be reproduced by the potential flow model of a point source in a uniform stream [1]. For the axial profiles of velocity fluctuations (rms) no model was obtained, but it can be shown that the profiles scale with $U+U_0$ in the near field and with U_0 in the far region (Fig. 2).

Self-similarity of the radial profiles is found in the inner region of the jet for $x < x_p$. From the radial profiles the stagnation stream surface (defined as the line over which the total volume flux Q is zero) and the maximum radial extension of the mixing region were also determined. As shown in Fig. 3a, the shape of this surface seems to agree over a substantial part with the results of flow visualization, and the maximum width corresponds with that given in Fig. 3b, obtained from LIF images. Estimates of the mixing increase in comparison with a free jet in quiescent fluid can be obtained both from the axial and radial profiles: the mean centerline velocity decays by about 25% faster (same as the mean centerline concentration [6]) and the jet half-velocity width is about twice as large in the case with counterflow.

A peculiarity of the flowfield is the presence of two different flow conditions [2, 6]: the intermittent appearance of a stable case, typical for $\alpha < 1.3 \sim 1.4$, becomes progressively less frequent with increasing α, while

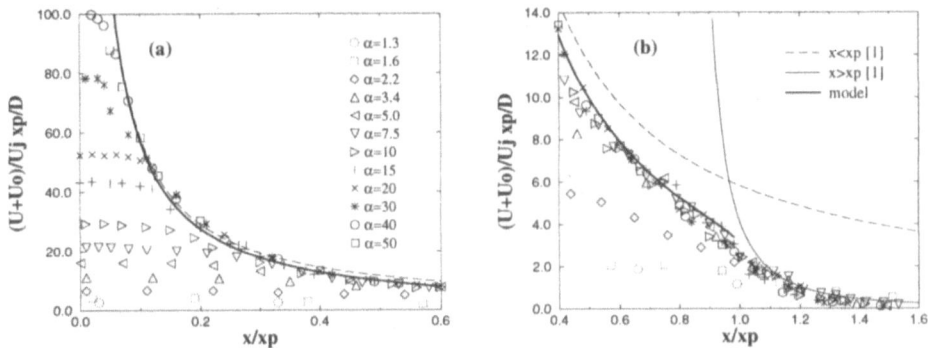

Figure 1. Axial profiles of average velocities measured with LDA for different α values, with the normalization suggested by [1] for low (a) and high (b) values of x.

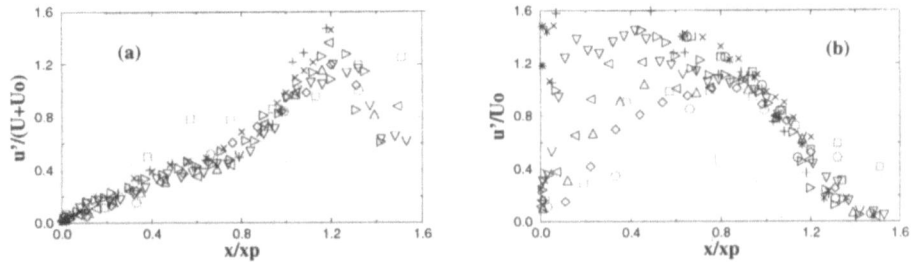

Figure 2. Axial profiles of velocity fluctuations (rms) measured with LDA for different α values, scaled with $U + U_0$ (a) and with U_0 (b) (symbols same as in Fig. 1).

an unstable condition becomes dominant, which is symmetric only in its time average (Fig. 4a). Fluctuations are recognized in the rms image (Fig. 4b) in correspondence with the initial shear layer and with the wandering of the jet tip. Even if the dynamics of this flow phenomenon is still far from being understood, the existence of these two distinct regions as well as the different scalings shown by the axial profiles suggest division of the flowfield into two parts. For $x < x_p$ and in proximity of the axis the flow is similar to a normal jet, while in the far field and at the external radial boundary the flow behavior is determined by the interaction with the counterflow and slow fluctuations of the main stream of jet fluid, characterized by asymmetry and strong mixing, can be observed.

Given the strong irregularities in the far field, the possibilities of flow control seem to be restricted only to the near field, where the flow conditions can be compared to those of a jet with annular suction around the nozzle, in which self-excited global oscillations were observed [5]. Power spectra measured in the present case also showed a similar behavior. Other possibilities of passive control, e.g. non-axisymmetrical nozzle shapes, are

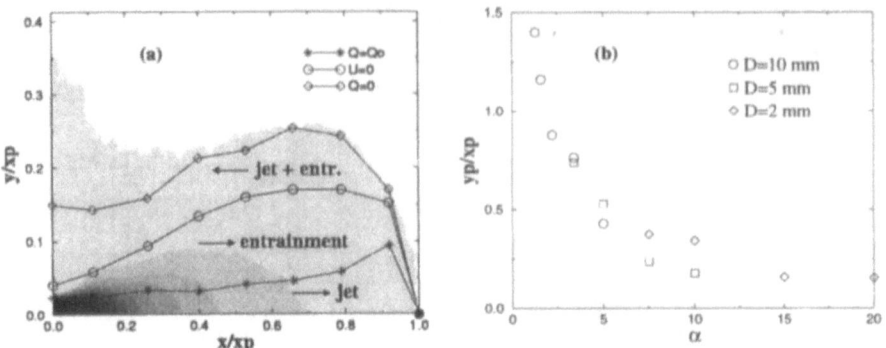

Figure 3. (a): Flow boundaries from LDA profiles superimposed to average of LIF images, $D=5$ mm, $\alpha=7.5$. (b): Maximum radial extent of mixing area vs. α from LIF.

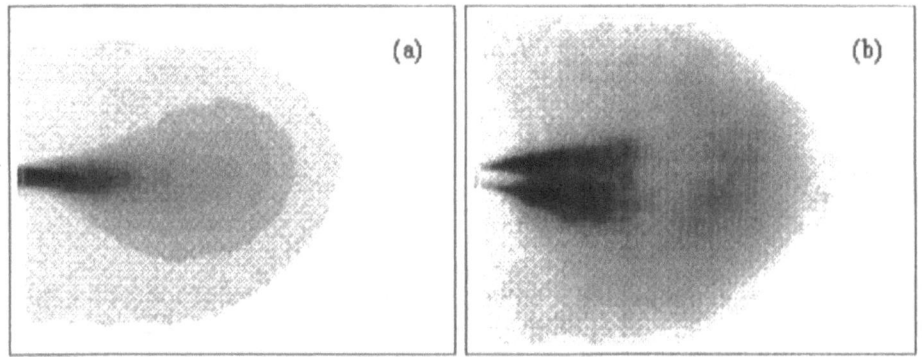

Figure 4. Time average (a) and rms values (b) of a series of instantaneous LIF images with $\alpha=4.6$; in (a) the number of gray levels was reduced to 10 for simplicity.

presently investigated, since the possibility of further mixing enhancement would be of interest for practical application.

References

1. Beltaos S., Rajaratnam N. (1973) Circular turbulent jet in an opposing infinite stream, *Proc. 1st Can. Hydr. Conf.*, Edmonton, pp.220–237.
2. König O., Fiedler H.E. (1991) The structure of round turbulent jets in counterflow: a flow visualization study, *Advances in Turbulence 3*, Johansson and Alfredsson (eds.), Springer-Verlag Berlin, pp. 61–66.
3. Lam K.M., Chan H.C. (1997) Round jet in ambient counterflowing stream, *ASCE Journ. of Hydr. Eng.*, **123. No. 10**, pp.1–8.
4. Morgan W.D., Brinkworth B.J., Evans G.V. (1976) Upstream penetration of an enclosed counterflowing jet, *Ind. Eng. Chem., Fundam.*, **15. No. 2**, pp.125–127.
5. Strykowski P.J., Wilcoxon R.K. (1993) Mixing enhancement due to global oscillations in jets with annular counterflow, *AIAA Journal* **31, No. 2**, pp. 564–570.
6. Yoda M., Fiedler H.E. (1996) The round jet in a uniform counterflow: flow visualization and mean concentration measurements, *Expts. in Fluids* **21**, pp. 427–436.

EXPERIMENTS ON THE DEVELOPMENT OF K-H BILLOWS IN STRATIFIED SHEAR LAYERS

E. KIT, *Department of Fluid Mechanics and Heat Transfer, Tel-Aviv University, Ramat-Aviv 69978, ISRAEL*
E. STRANG, H.J.S. FERNANDO AND H. NORTON, *Department of Mechanical and Aerospace Engineering, ASU, Tempe, AZ 85287 USA*

1. Introduction

The mechanism of K-H billow formation in stratified flows attracts scientist for many years and many attempts to investigate the phenomenon theoretically, experimentally and numerically have been conducted. The motive for research of this kind is to probe into microscale quantities pertinent to billows that may provide useful insights into the evolution of K-H billows in ocean, when combined with oceanic microstructure observations. The temporal evolution of K-H billows is reported in many previous experiments such as Thorpe (1968), De Silva *et al.* (1996) and Atsavapranee and Gharib (1997). A tilting tube entraining two layers has been used to produce K-H billows. Although such experiments are interesting and valuable, they do not allow detailed measurement of such important flow parameters as Reynolds stresses, density fluxes, kinetic energy dissipation and the local Richardson number. The present experiments are designed so that these parameters can be evaluated The measurements in a facility enabling a co- or/and a counter- flow of two layers of different densities and streamwise velocities are carried out to investigate the evolution of K-H billows in a stratified shear layer, from its "quaint" birth to "explosive" demise. The emphasis is placed on the initial evolution of the K-H billows. A series of homogeneous experiments was also performed as base-line cases to facilitate identification of the effects of stratification.

2. Experimental Set-Up And Instrumentation.

The experiment consisted of two layers of different densities, separated by a splitter plate, that are co-flowing in a channel of 30 cm depth and 15 cm width. The measurements include mean and fluctuating velocity and density fields in the stratified mixing layer that develops spatially in the streamwise direction. The direction of the y-coordinate is chosen from the upper layer to the lower layer to provide a positive velocity gradient of streamwise component. The velocity of the lower dense layer was varied in the range 6-10 cm/s while the velocity in the upper co-flowing layer was maintained very low and did not exceed 1 cm/s. To control the K-H billow formation and

39

U. Frisch (ed.), Advances in Turbulence VII, 39–42.
© 1998 *Kluwer Academic Publishers.*

their space-time evolution, an oscillating flapper of 3 cm width spanning across the channel was installed at the edge of the splitter plate. The stroke and frequency of the flapper were variable and their range of variations (0.5 -1.5 cm for stroke and 0.5-2 Hz for frequency) enabled the selection of appropriate flow conditions for K-H observations via simple calculations and detailed flow visualization. The flow is visualized by releasing a thin sheet of fluorescent dye in the central part of the channel over the splitter by a small aerodynamically-shaped pipe so that the released dye is advected to the leading edge of the flapper. The dye sheet was visualized by using a scanning beam of laser light. The measurement program includes the measurement of two velocity components by using a X-hot film probe, the density distribution using a gradient miniature conductivity probe (Strang and Fernando, 1997) and the velocity gradients using parallel sensor probes. It should be emphasized that the experiments in stratified flows are inherently non-stationary due to the mixing and, therefore, more detailed measurements in the lateral direction, in particular full velocity profiles across the mixing layer, are time consuming. The very possibility of such measurements stems from the advantages of the present flow configuration which allows to collect the data from several repetitive experiments using the phase locking of the controlled disturbances. The phase averaging of time traces obtained in different cross-sections enables the separation of fine-scale turbulence from coherent structures.

3. Results

Typical appearance of a train of K-H billows, obtained using the described above visualization technique, is presented in Figs. 1a and 1b. These observations show that in both homogeneous and stratified cases the billows initially emerge as laminar roll-up vortices (with very little mixing within), and then break up rather abruptly to a patch of turbulence after about two wave lengths. The further exploration of this fact was accomplished by detailed measurements of velocity and density fluctuations at 20 measuring stations along the tank.

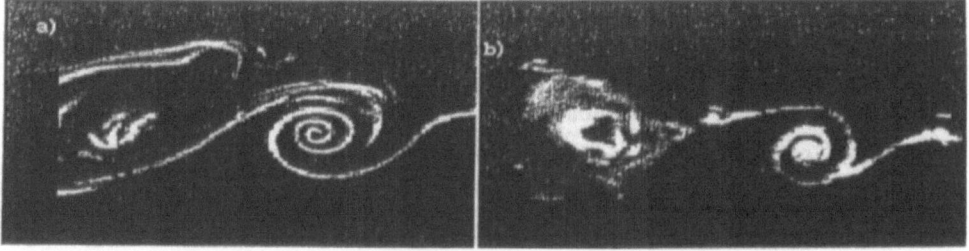

Fig. 1 The visual observation of K-H billows development in a)homogeneous flow, b)stratified flow.

The measurements in the case of a stratified layer were conducted at 20 stations in streamwise direction at two different lateral locations: i) along the 'central' line presenting continuation of the splitter plate, ii) along the line located 3mm below the central line. The first measuring station was selected at a distance 3 cm from the splitter plate and the separation between consecutive stations was 1 cm. The measurements in the homogeneous case were performed at 4 stations (3, 8, 13 and 18 cm from the splitter plate). The data acquisition rate (100 Hz) and overall time of measurements at each measuring station enables to capture about 80 periods of the flapper (~1 s), thus allowing a consistent phase averaging of turbulent fluctuations and spectral analysis. Only the data obtained along the line located 3mm below the "central" line are presented in the paper. It is well known that in mixing layer experiments, the shear layer protrudes into the low-speed region (e.g. Oster & Wygnanski, 1982, Gaster *et al.*, 1985). Therefore, it is not surprising that the mean streamwise velocities are close to the high speed measured in the bottom layer using a conical probe. This velocity is constant along the tank as expected. It can be also expected that the mixing rate should be higher in the homogeneous case since stratification causes suppression of mixing. The lateral component v in the homogeneous case is generally very small. In stratified flow case in some part of the tank it is directed upward, hence indicating mean buoyancy flux from the lower to upper layer. The streamwise velocity components measured by a probe with two parallel sensors separated by 1 mm are somewhat lower than those measured by x-probe. One of the reasons can be related to variations in spanwise direction (the separation between the probes is 25 mm), although the inaccuracy in the positioning of the probe cannot be ruled out. The experiments show that the mean velocity gradient is very low at both lateral locations, while the density gradient is significantly lower at the second location compared to the first one ("center line"). This can serve as an indication of the fact that the momentum shear layer is wider than the density interface. In Fig. 2 the variance of different velocity components is given. The left plot shows the variance of the full velocity fluctuation which includes the coherent and turbulent components, while in the right plot the coherent component was filtered out. Forward and backward FFT is used to filter out the coherent component related to the flapper frequency.

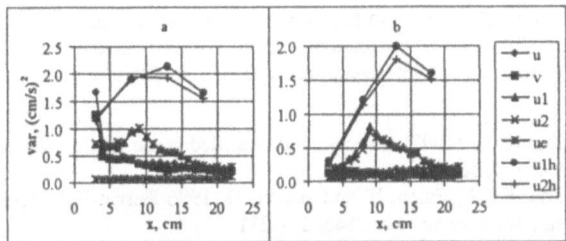

Fig. 2 Variance of velocity fluctuations in stratified and homogeneous flows:
a) raw signals, b) filtered signals. h denotes homogeneous flow

Two observation should be noticed: i) there is a clear indication that in both cases: stratified and homogeneous, the streamwise turbulent velocity component (b) starts from very low value and is growing fast in streamwise direction. The amplification of homogeneous turbulence is significantly greater; ii) the raw velocity fluctuations (a) are growing significantly slower and have high values already at the first streamwise position. This means that the coherent structure develops fast. These results prove the possibility to decompose the coherent and turbulent components and to investigate the generation of the turbulence in the homogeneous and stratified mixing layers.

The following conclusions can be drawn from Fig. 3: i) the spanwise correlation of the raw streamwise fluctuations measured by x and parallel sensors probes is high (>0.8) close to the flapper (Fig. 3a). This correlation is low for a turbulent signal only (Fig. 3b). It means that at the positions near the splitter plate the fluctuations are mostly due to two-dimensional coherent structures. The correlation decays in the streamwise direction; ii) the correlation between u and v components measured by x-probe should be negative if the fluctuations are generated by the shear. When the correlation is positive it indicates that the coherent structure transfers its energy to the mean flow. In present experiments such a behavior is revealed in both cases, stratified and homogeneous (Fig. 3a). It was also observed in mixing layers (e.g. Oster & Wygnanski, 1982). The situation is different for the turbulent fluctuations only (Fig. 3b). In homogeneous flow the correlation is negative indicating generation by a shear. In the stratified flow, at the positions near the splitter plate, the correlation is positive indicating a different mechanism for turbulence generation. This mechanism is most probably related to the turbulence generation by gravitational instability inside K-H billows. This type of instability has been recently investigated intensively theoretically and numerically (e.g. Klaassen & Peltier, 1991).

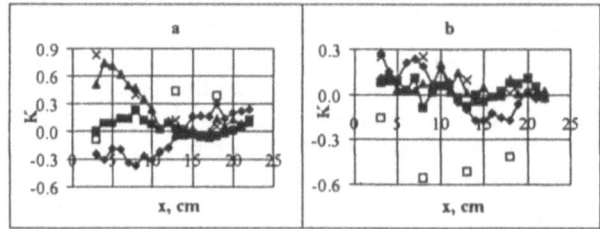

Fig. 3 Various correlation coefficients: ■, □ between u and v velocity components at the same point, ◆-between v-component and density fluctuations, ▲, x- between streamwise velocity components measured by two different hot-film probes (separation of 25 mm in spanwise direction).

4. References

Atsavapranee, P. and Gharib, M. (1997) Structures in stratified plane mixing layers and the effects of cross-shear, *J. Fluid Mech.* **342**, 53-86.

De Silva, I.P.D., Fernando. H.J.S., Eaton, F. and Hebert, D. 1996 Kelvin-Helmholtz billows in nature and laboratory, *Earth and Planetary Sci. Lett.* **143**, 217-231.

Gaster, M., Kit, E. and Wygnanski, I. (1985) Large scale structures in a forced turbulemt mixing layer, *J. Fluid Mech.* **150**, 23-39.

Oster, D. and Wygnanski, I. (1982) The forced mixing layer between parallel streams, *J. Fluid Mech.* **123**, 91-130

Strang, E.J. and Fernando, H.J.S. 1997 Entrainment and mixing in stratified shear flows. Submitted to *J. Fluid Mech.*

Thorpe, S.A. 1967 A method of producing a shear flow in a stratified shear flow. *J. Fluid Mech.* **32**, 693-704.

Klaassen, G.P. and Peltier, W.R. (1991) The influence of stratification on secondary instability in free shear layers, *J. Fluid Mech.* **227**, 71-106.

TIME-SCALE RESOLVED ACOUSTIC INTERFEROMETRY FOR THE DETECTION OF COHERENT VORTICITY STRUCTURES IN FULLY TURBULENT FLOWS.

C. BAUDET, O. MICHEL

Ecole Normale Supérieure de Lyon, 46 Allée d'Italie 69364 Lyon Cedex 07, France.

Turbulent flows are usually experimentally investigated using local pro bes: the velocity at a fixed point of the flow is recorded along time (hot wire anemometry or Laser Doppler velocimetry). In such measurements, one tries generally to use probes which are as small as possible in order to resolve the smallest length scales of interest in the flow (namely the Kolmogorov length η which is of order of a fraction of millimeters). In order to recast this information from the time domain to the space domain one has to resort to the well known Taylor hypothesis [1]. To gain information directly in the space domain, a few techniques have been proposed, relying mostly on the writing of a known pattern by an optical method [3, 2].

We address here an alternative method for probing a turbulent flow directly in the space domain, based upon the analysis of the scattering of coherent acoustic beams. Moreover, the acoustic technique which we promote is non intrusive and directly sensitive to the vorticity field, known to play a central role in the development of turbulent flows. Indeed, since the years 50 [4, 5, 6], it is known that the non-linear interaction between the longitudinal (acoustic mode) and transversal modes (vorticity mode) of the velocity field results in a scattering process, analogous to the more familiar (and linear) light scattering phenomenon. The latter has been widely used in condensed matter physics (X-ray scattering, Rayleigh scattering, ...) to investigate the statistical properties of either ordered or desordered media. As in any experimental technique based upon wave scattering, the amplitude of the scattered acoustic wave is linearly related to the spatial Fourier transform of the scatterers distribution (vortices) [7]

$$\frac{p_{scat}(\nu)}{p_{inc}} = \pi^2 i \frac{-cos(\theta_s)}{1 - cos(\theta_s)} \frac{\nu e^{i\nu D/c}}{c^2 D} (\vec{n} \wedge \vec{r}).\vec{\Omega}(\vec{q}_{scat}, \nu - \nu_o) \tag{1}$$

U. Frisch (ed.), Advances in Turbulence VII, 43–46.
© 1998 *Kluwer Academic Publishers.*

where

$$\vec{q}_{scat} = \frac{2\pi}{c}(\nu\vec{r} - \nu_o\vec{n}) \simeq \frac{4\pi\nu_o}{c}\sin(\frac{\theta_{scatt}}{2}) \text{ for } \nu \simeq \nu_o \qquad (2)$$

stands for the scattering wave vector. According to the above equations, one immediately sees that acoustic scattering enables the continuous probing (in time) of a well defined length scale (modulus and direction) of the vorticity field. Moreover, the length scale is easily varied by choosing both the frequency ν_o of the incoming sound wave and the scattering angle θ_{scatt}. Notice also that, according to equation (1), only one component of the vorticity field is probed for a given transducers configuration, this latter being perpendicular to the plane defined by the direction of the incoming and scattered wave vectors (scattering plane). By using the regular vortex array (spatial and temporal periodicity) resulting from the Bénard-von Kármán instability (around $R_e = 50$) as a test flow, we have experimentally confirmed the main features of equation (1). Indeed, according to its well defined periodicity (in time and space) and alignment (for cylinders of large aspect ratios), the Bénard-von Kármán vortex street is analogous to a 2D polarized vorticity crystal, the temperature of which being the flow Reynolds number [8, 9]. We have also applied the acoustic scattering technique to the determination of the enstrophy spectrum (second order moment) of turbulent a jet flow [10]. In the perspective of coherent structures detection, acoustic scattering is expected to be particularly efficient, as it allows the continuous probing (along time) of a finite volume of the flow, defined as the cross volume of the incoming and detected (antenna) acoustic beams. In our experimental set up the probed volume is a parallelepiped of linear size $0.15m$, close to the integral length scale of the turbulent jet flow. Recent experimental [11], numerical [12] as well as theoretical [13] results report the formation in the turbulent flow of strong elongated vortices (with a tube-like shape) having a diameter of the order of η and a length somewhere between the Taylor scale λ, and the integral scale l_o [1]. With the objective of an efficient detection (counting) of these high intensity and strongly localized events, local probes such as hot wires seem less adequate than global techniques like acoustic scattering. Rigorously, the probability of intersection of a line (vortex filament) in the flow and a point (hot wire probe) tends to zero; thus one can only expect to detect indirectly the vortex filaments by the signature of the velocity perturbation they induced in their vicinity. However, as in any spectral method based on wave vector selection, the drawback of the acoustic technique lies in its poor spatial resolution. This is a direct consequence of the well known Gabor-Heisenberg incertitude principle. Similarly, as indicated by equation (1), the time evolution of the vorticity field under investigation results in frequency shifts of the scattered wave around the frequency ν_o of the incoming sound wave (Doppler effect). Thanks to the linearity of the tranducers and to an appropriate demodulation scheme, the phase information of the scattered wave is preserved (the phase reference being that of the incoming wave). Again we are confronted to the spectral (vorticity dynamics) versus time (time occurrence of the events) resolution trade-off.

The purpose of this communication is to show how such limitations can be overcome by using an interferometric setup together with time-frequency distributions for signal analysis purposes. The interferometric setup which consists in two independent analysing channels (each defined by a pair of transducers), allows the simultaneous probing of two independent volumes of the flow. Furthermore, this double configuration, in which the two measurement channels have different scattering angles, enables the simultaneous probing of two spatial wave vectors arbitrarily close to each other (scale resolution). In the time domain, we propose to take advantage of the features of the recently introduced time-frequency analyzing methods [14]. For finite duration events of vorticity, the signal will exhibit a time-varying spectrum, which can be approached e.g. by the common spectrogram analysis. However, this latter involves a sliding time window that intends to capture a portion of the signal which is sufficiently restricted in time so that stationarity assumptions are approximately met. Furthermore, the presence of the window results in a time-frequency distribution (TFD) exhibiting both temporal and spectral leakage. The Wigner-Ville distribution (WVD) avoids the problems of windowing the signal, but the presence of interference terms between signal components (due to the quadratic nature of the WVD) often precludes its applicability. Recent studies have shown that a class of smoothed WVD (namely the Cohen's Class of TFDs) allows to reduce the amplitude of the interference terms, while preserving covariance properties of the WVD (in both the time and frequency domains) [14]. Among those, the Reduced-Interference Distribution (RID) first introduced by Choi and Williams [15], has also the desirable feature of preserving the time and frequency support of signal component (time resolution: $RID(t, f)$ is non-zero if the signal is non-zero at t and has a spectral component at f). These are highly desirable properties if one wishes to infer statistical features of the vorticity distributions from this TF analysis. The examples shown below were computed with a binomial RID, which is a good discrete approximation of the continuous distribution. On these two figures, one observes a high degree of similarity between the two representations, demonstrating the efficiency of the scale selection process. The figures also evidence the existence of strongly time localized events, corresponding to phase stationnarity in the time-frequency plane (coherence). The latter observation yields an operational definition (criterion) of coherent structures.

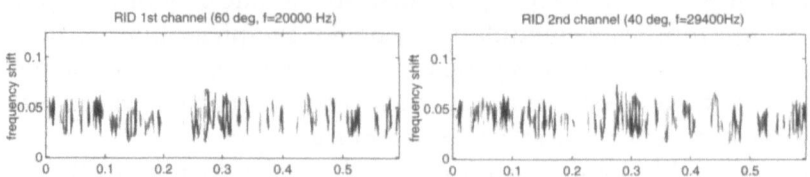

Figure 1. **Turbulent Jet Flow** $R_\lambda \simeq 500$: **2 simultaneous channels Time-Frequency distribution (RID), for nearly identical scattered wave vector. The selected length scale** $(3.7cm^{-1})$ **belongs to the inertial range.**

This work is supported by the Région Rhône-Alpes (Emergence project No. 97027229), and by Pr J.P. Hansen, member of the Institut Universitaire de France.

References

1. Frisch, U. (1995), *Turbulence: the legacy of A.N. Kolmogorov* Cambridge University Press, pp. 58–59, cambridge, New-York.
2. Noullez, A., Wallace, G., Lempert, W., Miles, R. B. & Frisch, U. (1997) Transverse velocity increments in turbulent flow using the RELIEF technique, *J. Fluid Mech.* A**339**, pp. 287–307.
3. Fermigier, M., Jenffer, P., Charmet, J.C. & Guyon, E. (1980) A non-perturbative anemometric method and flow visualization technique, *J. Phys. Lett.* A**41**, pp. 519–521.
4. Obukhov, A. M. (1941), On sound scattering in a turbulent flow, *Sov. Phys.-Dokl.* A**30**, 6, pp. 611–614.
5. Kraichnan, R.H. (1953), The scattering of sound in a turbulent medium, *J. Acoust. Soc. Am.* A**18**, 6, pp. 1096–1104.
6. Chu, B.T. & Kovásznay, L.S.G. (1953), Non-linear interactions in a viscous heat-conducting compressible gas, *J. Fluid Mech.* A**3**, pp. 494–514.
7. Lund, F. & Rojas, C. (1989), Ultrasound as a probe of turbulence, *Physica D* A**37**, pp. 508–514.
8. Baudet, C., Ciliberto, S., & Pinton, J.F. (1991), Spectral analysis of the von Kármán flow using ultrasound scattering, *Phys. Rev. Lett.* A**67**, 2, pp. 193–195.
9. Pinton, J.F. & Baudet, C. (1993), Measurement of vorticity using ultrasound scattering, *Turbulence in Spatially Extended Systems, Les Houches 1992,*eds. R. Benzi, C. Basdevant & S. Ciliberto. Nova Science, Commack, New York.
10. Baudet, C. & Hernandez, R. H. (1996), Spatial enstrophy spectrum in a fully turbulent jet, *Advances in Turbulence VI, Sixth European Turbulence Conference, Lausanne 1996,*eds. S. Gavrilakis, L. Machiels & P.A. Monkewitz. Kluwer Academic Publishers, Dordrecht.
11. Douady, S., Couder, Y. & Brachet, M.E. (1991), Direct observation of the intermittency of intense vorticity filaments in turbulence, *Phys. Rev. Lett.*, A**67**, pp. 983–986.
12. Vincent, A. & Meneguzzi, M. (1991), The spatial structure and statistical properties of homogeneous turbulence, *J. Fluid Mech.*, A**225**, pp. 1–25.
13. Moffat, H.K., Kida, S. & Ohkitani, K. (1994), Streched vortices – the sinews of turbulence; high reynolds number asymptotics, *J. Fluid Mech.*, A**259**, pp. 241–264.
14. Flandrin, P. (1993),*Temps-fréquence*, Ed. Hermès, Paris.
15. Choi, H.I. & Williams, W.J., (1989), *IEEE Trans. Acoust., Speech, Signal Proc.*, A**37**, No. 6, pp. 862–871.

EFFECTS OF WALL PROXIMITY ON ROUND JETS

A. BENAÏSSA, D. EWING, J.F. MORRISON* AND A. POLLARD
Dept. Mechanical Engineering, Queen's University at Kingston, CANADA
*Dept. Aeronautics, Imperial College, London, ENGLAND

Abstract. The influence of the proximity of a wall on the development of an axi-symmetric air jet is studied with a particular emphasis on the consequences of its height-above-the-wall on the evolution and dynamics of coherent structures. Space-time correlations are calculated between pressure at the wall using both the vertical and normal component of velocity. These correlations confirm the existence of and localize the horseshoe vortices including their inclination towards the wall.

Key Words. Pressure-velocity correlations, wall jet

1. Introduction.

Wall jets that issue from different shaped orifices have been widely investigated, Davis and Winarto (1980), Launder and Rodi (1983), Pollard and Schwab (1989), Matsuda and al. (1990) and Sullivan and Pollard (1996). Davis and Winarto (1980) examine the jet spread rates in the wall-normal and lateral directions for a variety of wall-nozzle distances. Flow structure was not considered. Matsuda et al. (1990) and Sullivan and Pollard (1996) examine the development of the structure of a jet emerging directly onto a plane wall; Matsuda et al. suggest that streamwise vortices are responsible for the anisotropic growth rate of the jet.

The pressure fluctuations generated by the jet turbulence are altered by the wall, which provides the first change to the entrainment boundary conditions for the jet. Subsequently, the large eddies, which experience the effect of the wall first, are constrained directly in the wall-normal direction, and amplify the wall parallel components of velocity and generate pressure fluctuations in the process. The changes in structural behaviour are studied primarily by the measurement of wall-pressure fluctuations and cross-hot wires. Wood and Bradshaw (1982) used the high velocity side of a mixing layer that grows towards an adjacent wall to show that the effect of wall proximity on the pressure field leads to a significant change in the velocity field before reattachment.

In the present study, the structure of a high-Reynolds-number, round air jet is investigated as it grows along an adjacent flat surface. Benaïssa et al. (1997) considered the evolution of the velocity field for nozzle-to-wall distance $1 < h/D < 2$. At larger values of h/D the Kelvin-Helmholtz primary jet instability (the production of coherent vortex rings) and the secondary three-dimensional instability that produces braids of

47

U. Frisch (ed.), Advances in Turbulence VII, 47–50.

streamwise vorticity are more or less free of wall effects. It is hypothesized that the ring-like vortex that emerges from the nozzle (for the case h/D>0) is subjected to viscous effects that cause the vortex to break and form into a hairpin-like structure, Benaissa et al. (1997). The nozzle-to-wall distance is also hypothesized to be a parameter that controls the structures: the larger the distance, the greater will be the opportunity for the jet ring-like vortices to undergo pairing and breakdown, all the while being subject to entrainment restrictions due to the presence of the plate. Here, the wall pressure for h/D=1 & 2 and more importantly the pressure-velocity correlations for h/D = 1 are considered for it is believed that these will assist to confirm the existence and the morphogenesis of the structures.

2. Experimental Conditions and Data Analysis.

A low speed air jet feeds a settling chamber through a 25:1 area ratio to a D = 7.1 cm diameter nozzle. The nozzle is attached to a moveable back plane, which can be moved vertically relative to the fixed wall. A schematic of the facility is given in Fig. 1.

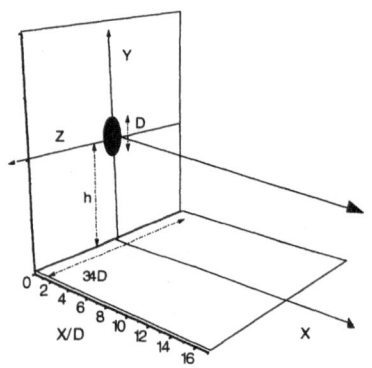

| **Figure 1:** Schematic of the wall jet facility. X, Y, Z are axial, wall-normal and wall-span directions | **Figure 2:** Streamwise variation of rms wall pressure. Vertical axis has arbitrary units |

The jet Reynolds number was $Re_D = 9.3 \times 10^4$. Velocity data were acquired using X-hot wires. Pressure fluctuations were obtained using piezometric pressure transducers (2.3 mm); the minimum pressure tap-to-wire distance is 4.5 mm (~0.06 D).

The u and v spectra were calculated for different nozzle-to-wall heights h/D, all at X/D=3. The shedding frequency of vortices is about 145 Hz. Velocity and pressure signals were acquired at a sampling rate of 5kHz.

3. Presentation and Discussion of Results.

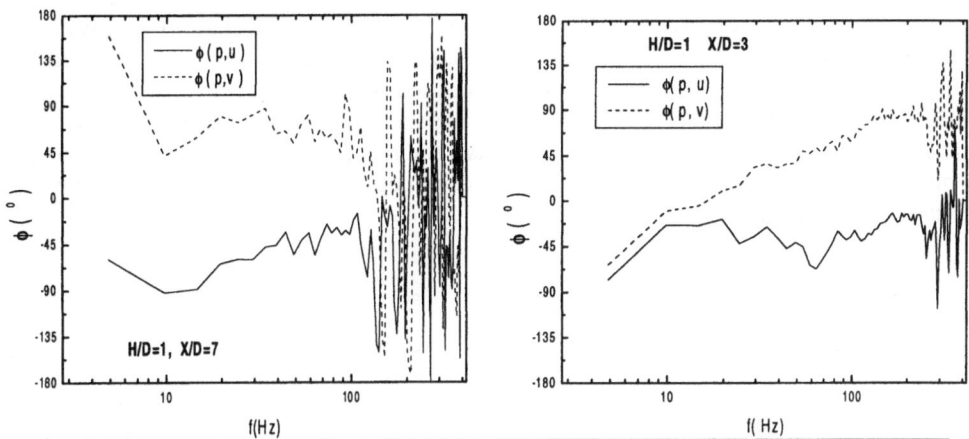

Figure 3: Phase (ϕ) between pressure at the wall and the two components of the velocity components u and v. Left: x/D=3, Right: x/D=7.

The rms wall pressure reaches a maximum at x/d = 4 before the potential core disappears by x/D= 5, see Figure 2.

The co-spectra and quad-spectra were calculated between pressure and velocity for the first position of the probe above the wall (y_o=4.5 mm); Space precludes their illustration.. The maximum correlation of the cospectrum (p-u) is obtained at the shedding frequency; the maximum of the cospectrum between p and v is obtained at lower frequency; which suggest a presence of lower frequency events near the wall (at x/D=3 position). At x/D=7, correlation between pressure and the two component of velocity is over a broad range of frequencies centred around a frequency equal to about half the shedding frequency. The phase between p and u (Fig.3) is almost the same for the two cases studied, while the phase between p and v is evolving at the position x/D=3 from negative for low frequencies to positive for higher.

The space-time correlation coefficients for the pressure-velocity fluctuations for x/D = 3 and 7 are presented in Figure 4. These correlations suggest that:

- there is no upstream propagation of disturbances in the reattachment region - unlike, for example, reattachment over a backward-facing step.
- there are sequences of alternating high and low pressure, the former caused by mixing jets or splats, the latter caused by the passage of coherent horseshoe vortices with spanwise vorticity.
- from rms wall pressure distribution, the spanwise scales of the horseshoe-type structure increase with streamwise distance as observed from lateral velocity measurements in Benaissa et al (1997), (evidence is not provided here)

4. Conclusion.

This preliminary investigation of the pressure-velocity correlations in a wall jet with axis removed from the wall suggests that at x/D=3, the wall pressure is dominated by a ring-type vortex. At x/D=7, the vortex has been sufficiently influenced by the wall to break the axisymmetry of the vortex. The data support the hypothesis for the existence of a horseshoe shaped vortex. Further analysis is required to resolve the issue of vortex morphogenisis in the near field using spatial velocity detection methods coupled with wall pressure measurements.

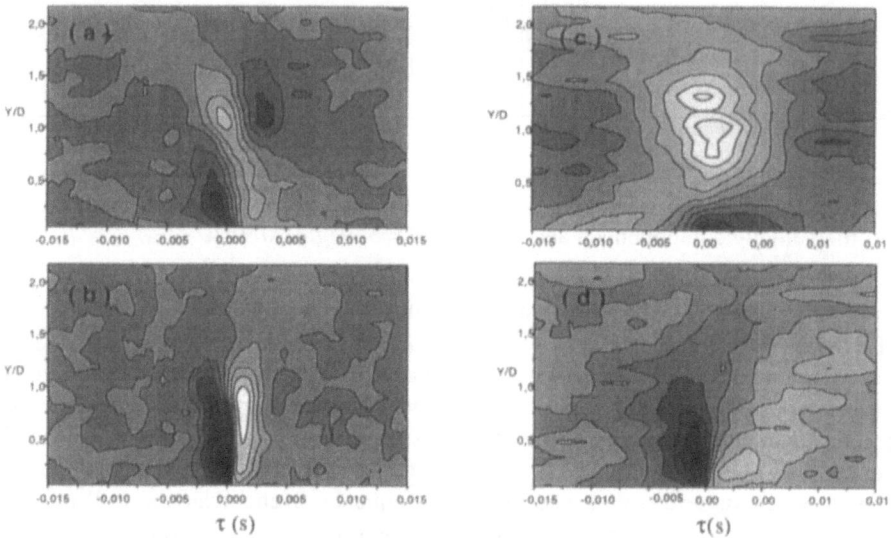

Figure 4: Space-time correlation coefficients pressure-velocity, (a) and (c) ρ_{pu}, (b) and (d) ρ_{pv}. White contour value=0.2; black contour value=-0.2. Y/D is position of hot wires. Left: x/D=3, Right: x/D=7.Jet centreline origin at Y/D=1.

Acknowledgments

JFM is indebted to the Royal Society for financial support. Research funding supplied by NSERC of Canada through grants and fellowships is gratefully acknowledged.

References
Benaissa, A., Ewing, D., Pollard, A. (1997): Proc. 11th Symp. Turb Shear Fl, Grenoble.
Davis, M.R., and Winarto, H (1980) J Fluid Mech., vol 101, part 1, pp. 201-221.
Launder, B.E. and Rodi, W, (1983): Ann. Rev. Fluid Mechanics. no. 15, pp. 429-459.
Matsuda, H., Iida S., and Hayakawa, M, (1990): ASME JFE, Vol. 112, pp. 462-465.
Pollard, A. , Schwab, R, R, (1989): Proc. 7th Symp. Turb Shear Fl, Stanford.
Sullivan, P. and Pollard, A. (1996):Meas. Sci. Technology, vol 7, pp 1498-1516.
Wood, D. H., and Bradshaw, P., (1982) : J. Fluid Mech., vol 122, pp 57-89.

TRANSVERSE STRUCTURE IN TURBULENT FLOW USING THE RELIEF TECHNIQUE

A. NOULLEZ, U. FRISCH
C.N.R.S., Observatoire de la Côte d'Azur,
B.P. 4229, F-06304 Nice Cedex 4, France

AND

G. WALLACE, W. LEMPERT, R.B. MILES
Mechanical and Aerospace Engineering Dept.,
Princeton University, Princeton, NJ 08544, USA

The investigation of the transverse structure of the velocity field in turbulent flow is of great interest because, while the incompressibility relation severely constrains the possible longitudinal structures, it gives no restriction on the transverse variations of the velocity. Also, there is no theoretical equivalent of the Kolmogorov equation that applies to the transverse structure functions, so that their scaling is *a priori* unknown. Related to this lack of theory is the dearth of experimental transverse data, due to the pointwise nature of standard techniques that can only give access to longitudinal variations using Taylor's hypothesis. It is only recently that lateral measurements could be performed, using arrays of hot-wire probes, but their spatial resolution is very limited.

We shall here present results obtained by the RELIEF technique, a new non-intrusive optical technique (Miles *et al.*, 1991) that can measure directly the spatial variations of the transverse component of the velocity along a particular line with an unprecedented resolution of typically 25 μm. Our results are also described in more details in (Noullez *et al.*, 1997).

RELIEF needs no seeding of the flow as it is based on the vibrational excitation of oxygen molecules by stimulated Raman scattering using a two-colour pulsed laser beam. The line of excited molecules is then allowed to be advected by the flow for a precise time, typically ≈ 5 μs. The flow region is then illuminated by a thick sheet of UV light, driving the excited molecules to the Schuman-Runge band, from which they decay immediately by electronic fluorescence in the near-UV. This fluorescence is imaged by a high-speed UV camera coupled to a video recorder and a frame grabber.

U. Frisch (ed.), Advances in Turbulence VII, 51–54.
© 1998 *Kluwer Academic Publishers.*

The images are then analysed to measure the center of the displaced lines with very high accuracy using the specific form of the intensity profile in the adjustment process. The distorsions of the profile with respect to a straight line then directly give the local fluctuation of the component of the velocity *transverse* to the line. Moreover, because the spatial resolution of the camera and the delay between tagging and interrogation are known, we have a direct absolute measurement of the velocity over many spatial points simultaneously.

We have chosen to apply the RELIEF technique to a standard turbulent flow, namely an axisymmetric free air jet of 10 mm diameter. This small size was dictated by the necessity to set up the experiment inside of the optics lab and does not cause any measurement problem thanks to the high resolution of RELIEF. On the other hand, to avoid compressibility effects, we could not reach very high Reynolds number and our inertial range is thus not very extended. A large number of experiments were performed around 40 diameters downstream so that the turbulence was fully developed. By changing the exit velocity, we could explore Taylor scale Reynolds numbers ranging from ≈ 350 to ≈ 600, the Kolmogorov scale being around $15\,\mu$m, so that our resolution was better than 2η. This allows us to study the transitional scales between smooth behaviour controlled by the viscosity and the inertial range. Statistical quantities were then obtained by a combination of space and time-averaging, the latter being helped by the long (in term of eddy-turnover time) time between successive RELIEF frames.

Computing the even-order structure functions of the transverse increments, we found them to exhibit reasonable scaling behaviour, switching continuously from regular to inertial-range scaling. In particular, we checked the value of the so-called 'Kolmogorov constant' C_2 giving the amplitude of the second-order transverse structure function as

$$S_2^{\perp}(r) = \frac{4}{3} C_2 \, \varepsilon^{2/3} \, r^{2/3} \, , \tag{1}$$

the 4/3 coming from the well-known isotropy constraint relating the longitudinal and transverse second-order structure functions. By compensating $S_2^{\perp}(r)$ by the factor $r^{-2/3}$, we could measure a value $C_2 = 2.1 \pm 0.5$, which is compatible with values obtained from longitudinal measurements (Sreenivasan, 1995). The odd-order structure functions were also found to vanish in the statistical noise, as they should, this being also reflected in the symmetry of the p.d.f.s of the transverse increments. Redefining the structure functions to use absolute values, we could measure scaling exponents for all orders, either direcly on the structure functions or by ESS, using one structure function (normally $S_3^{\perp}(r)$) as a reference. Plots of the local scaling exponents and the amount of available data show that we can measure

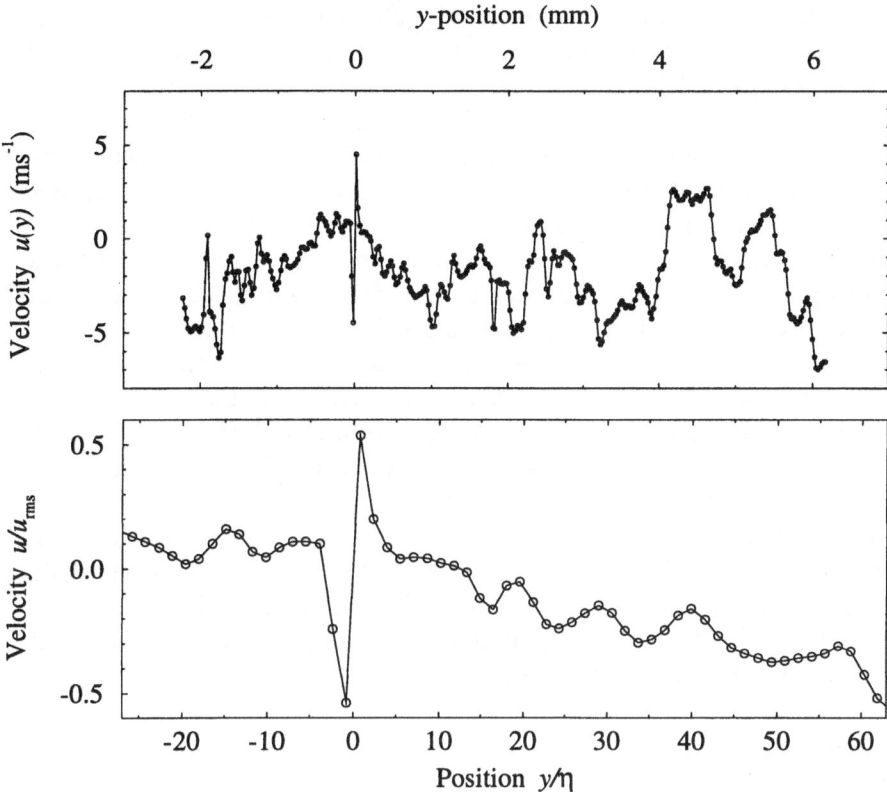

Figure 1. Example of a complete RELIEF line (about 8 mm long) showing a violent velocity fluctuation (upper) and a zoom in rescaled units centered around the event (lower).

exponents reliably up to order 8, the values found being statistically undistinguishable from those obtained for the longitudinal structure functions by other authors (Kahalerras *et al.*, 1996). The measured exponents however show clear deviations from the Kolmogorov scaling $\zeta_p = p/3$, indicating that intermittency effects are also present in transverse structure functions.

The p.d.f.s of the transverse velocity increments were also found to be very similar (except for their symmetry) to their longitudinal counterpart, showing an evolution from quasi-Gaussian at large separations to apparently nearly exponential or stretched exponential at small scales. This behaviour can only be transient however because it can be shown using homogeneity that the (cumulated) probability of having an increment Δu larger than some value v is bounded by twice the single-point probability distribution of $v/2$. The single-point p.d.f. was measured very carefully up to large velocity values (because we are not bothered by Taylor's hypothesis) and we found that its decay is actually slightly *faster* than Gaussian, precluding stretched exponential p.d.f.s for the increments. We stress that no satisfactory explanation has ever been given for either the near-Gaussian shape of the single-point p.d.f., nor for its discrepancies therefrom.

It is thus of even more interest to identify the type of structures responsible for the largest velocity increments. A procedure was then defined to extract the lines containing these extreme events. A typical example is found in figure 1, showing the full RELIEF line and a zoom around the velocity jump. It is seen that the velocity difference can reach values of the order of u_{rms} over separations as small as the Kolmogorov scale η. We observed that the structure profiles were most often nearly symmetric, which makes them compatible with *vortex filaments*, with a diameter of about η and a maximum velocity around u_{rms}, leading to a circulation $\approx \eta u_{rms}$. We note that transverse probing is much more favourable for detecting vortex filaments because it reveals symmetric profiles that also decay slower than the longitudinal ones, and also because the velocity excursion is not reduced by the factor $u_{rms}/\overline{U}$. The possible relation of these structures to intermittency effects is as yet unknown, but the high resolution and non-intrusive qualities of the RELIEF technique make it a formidable new tool to study the objects appearing at the smallest scales of turbulence.

References

H. Kahalerras, Y. Malecot and Y. Gagne, Transverse structure functions in three-dimensional turbulence, in *Advances in Turbulence* VI (Kluwer 1996), eds. S. Gavrilakis, L. Machiels and P. Monkewitz, pp. 235–238.

R.B. Miles, W. Lempert and B. Zhang, Turbulent structure measurements by RELIEF flow tagging, *Fluid Dynam. Res.* **8** (1991), pp. 9–17.

A. Noullez, G. Wallace, W. Lempert, R.B. Miles and U. Frisch, Transverse velocity increments in turbulent flow using the RELIEF technique, *J. Fluid Mech.* **339** (1997), pp. 287–307.

K.R. Sreenivasan, On the universality of the Kolmogorov constant, *Phys. Fluids* **7** (1995), pp. 2778–2784.

RECONNECTION OF A COUNTERROTATING VORTEX PAIR

T. LEWEKE[1] AND C. H. K. WILLIAMSON[2]

[1] *IRPHÉ, CNRS/Universités Aix-Marseille I & II,*
12 av. Général Leclerc, F-13003 Marseille, France.
[2] *Mechanical & Aerospace Engineering, Cornell University, Ithaca*
NY 14853-7501, USA.

1. Introduction

The dynamics of a pair of parallel counter-rotating vortices has been the object of a large number of studies in the last three decades. The continued interest in this flow is, to a great extent, due to its relevance to the problem of aircraft trailing wakes, whose far field is primarily composed of such a pair. In addition, the counter-rotating vortex pair represents one of the simplest flow configurations for the study of elementary vortex interactions, which can yield useful information for our understanding of the dynamics of more complex transitional and turbulent flows.

A prominent feature of this flow is a long-wavelength wavy instability, first treated theoretically by Crow (1970). When the amplitude of the perturbation grows sufficiently large, the vortex cores touch at the location of maximum inward displacement, leading to a break-up of the pair into a series of vortex rings. The transition between the two wavy vortices and the vortex rings happens via a cross-linking, or reconnection, of vorticity involving a change in the topology of the vortex lines. So far, experimental results concerning this process are mostly qualitative (e.g. Oshima and Asaka 1977, Lim and Nickels 1995). More detailed knowledge about vortex reconnection comes from numerical simulations. Melander and Hussain (1989) and Shelley, Meiron and Orszag (1993) chose as initial conditions configurations which are very similar to a vortex pair perturbed by the symmetric mode of the Crow instability. The interacting vortices of the present experimental study are therefore an ideal candidate for the investigation of the reconnection phenomenon.

U. Frisch (ed.), Advances in Turbulence VII, 55–58.
© *1998 Kluwer Academic Publishers.*

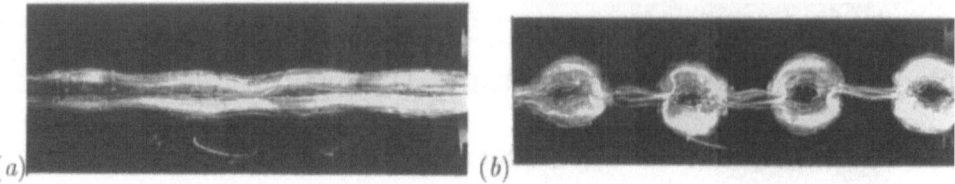

Figure 1. Visualization of the long-wavelength instability of a vortex pair for $Re = 1450$ and $a/b = 0.23$. The pair is moving towards the observer. (*a*) $t^* = 3.2$, (*b*) $t^* = 6.5$.

2. Experimental details

The vortex pair is generated in a water tank at the parallel edges of two flat plates, moved by a computer-controlled step motor. The vortices are typically separated by a distance of 2.5 cm, and their length is approximately 170 cm. Visualization is achieved using Laser-Induced Fluorescence. A vortex pair is characterized by the circulation Γ of each vortex, the separation b between the vortex centres, and a characteristic core size a, which is the radius of the tube around the vortex centre containing most of the vorticity. These characteristics were determined using Digital Particle Image Velocimetry (DPIV). These measurements also showed that the initial velocity profiles of the vortices are very well represented by the one of a Lamb-Oseen vortex with a Gaussian vorticity distribution. The Reynolds number based on the initial circulation ($Re = \Gamma_o/\nu$) was between 1500 and 2500 in this study. Time t is non-dimensionalized by the time it takes the initial vortex pair to travel one vortex spacing b, i.e. $t^* = t\,(\Gamma/2\pi b^2)$. More details are given in Leweke and Williamson (1998).

3. Long-wavelength instability

Figure 1 shows the overall features of the long-wavelength vortex pair instability. The initially straight and parallel vortices develop a waviness (Fig. 1*a*), which is symmetric with respect to the plane separating the vortices, and whose axial wavelength is about 6 times the initial vortex separation. Photographs taken simultaneously from two perpendicular directions show that the plane of the waves is inclined by about 45° with respect to the plane containing the pair (see also Fig. 2*b*), in agreement with Crow's (1970) theoretical prediction and previous observations. This waviness is amplified in time until the vortex cores touch, break up, and reconnect to form periodic vortex rings, which initially look almost circular in the front view of Fig. 1(*b*). Side-view visualizations show, however, that these rings are not contained in a plane; they are bent upwards in the transverse direction. Fig. 1(*b*) also shows that the large-scale rings are still linked by thin strands of dye. Subsequently, the rings stretch out into oval vortices exhibiting a well-known oscillatory behaviour (Lim and Nickels 1996). Measurements have shown that the mean propagation speed of the vortices is approximately constant

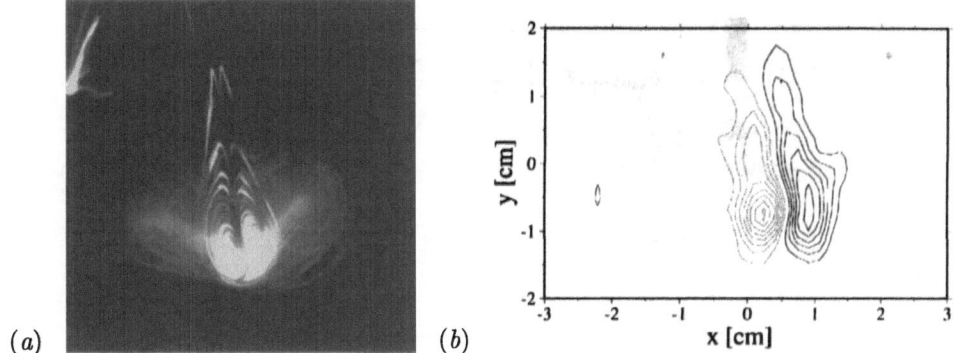

(a)   (b)

Figure 2.   Flow in the reconnection plane at $t^* \approx 5$. (a) Dye visualization using a laser sheet, (b) contours of measured axial vorticity.

in time, even after the reconnection process, which may seem surprising, considering that the geometry and topology of the late-time vortex rings are very different from the ones of the initial vortex pair. It illustrates that, even after the cross-linking, the vortices still possess most of the initial circulation and that the energy of the flow remains in the large-scale structures.

4.   Vortex reconnection

Figure 2 shows the flow structure in the reconnection plane, i.e. the cross-section of the pair at the location of minimum separation, during the cross-linking process. Initially the pair is symmetric, and the cores are well separated. In Fig. 2(a), the amplitude of the perturbation has grown quite large. The plane of the waves can be seen in the background, illuminated by scattered light from the reconnection plane. The vortex cores have come much closer and have started to elongate vertically. At the same time, a "tail" of dye is developing behind the descending pair.

This visualization is complemented by measurements of the time-dependent vorticity distribution in the same plane with the help of DPIV (Fig. 2b). These measurements confirmed the qualitative observations in Fig. 2(a), in particular the formation of a tail of vorticity during the cross-linking. At later times, a small, slowly decaying pair of counterrotating vortices remains, corresponding to the dye threads linking the oval vortex rings in Fig. 1(b). These results are in good qualitative agreement with the numerical results of Melander and Hussain (1989).

From the vorticity measurements, the circulation in each half of the reconnection plane can be calculated (Fig. 3). After the initial roll-up, the circulation remains constant for some time. Once the cores touch, it decreases rapidly due to the cancellation and reorientation of vorticity. At later times, it reaches a new, almost constant value of about 10% of the initial circulation Γ_o, which corresponds to the circulation of the threads between the large-scale vortex rings.

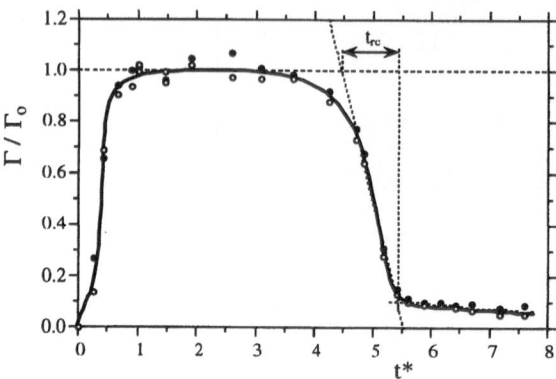

Figure 3. Evolution of the circulation in the reconnection plane for $Re = 2340$, and definition of the reconnection time t_c.

Following Melander and Hussain (1989), one can deduce from Fig. 3 a characteristic reconnection time t_c. From our experimental measurements we find $t_c \approx 0.9$. If the result found in the DNS study of Melander and Hussain (1989) is transformed to the non-dimensional units used in this study, one obtains a reconnection time $t_c \approx 0.7$ for their case, which again is in reasonably good agreement with the present result.

5. Conclusions

In our experimental study of vortex pairs we have observed and characterized the initial development and the late stages of the long-wavelength Crow instability. The initially straight vortices develop a symmetric waviness before they cut and reconnect into a system of oscillating oval vortex rings. We have obtained qualitative and quantitative information about the vortex reconnection process, including the characteristic reconnection time, which shows good agreement with previous numerical simulations.

References

Crow, S.C. (1970) Stability theory for a pair of trailing vortices, *AIAA J.* **8**, 2172–2179.

Leweke, T. and Williamson, C.H.K. (1998) Cooperative elliptic instability of a vortex pair, *J. Fluid Mech.* **360**, 85–119.

Lim, T.T. and Nickels, T.B. (1996) Vortex rings, in S.I. Green (ed.), *Fluid Vortices*, Kluwer Academic Publishers, Dordrecht, pp. 95–154.

Melander, M.V. and Hussain, F. (1989) Cross-linking of two antiparallel vortex tubes. *Phys. Fluids* A **1**, 633–636.

Oshima, Y. and Asaka, S. (1977) Interaction of two vortex rings along parallel axes in air. *J. Phys. Soc. Japan* **42**, 708–713.

Shelley, M.J., Meiron, D.I. and Orszag, S.A. (1993) Dynamical aspects of vortex reconnection of perturbed anti-parallel vortex tubes. *J. Fluid Mech.* **246**, 613–652.

LARGE SCALE STRUCTURES IN TURBULENT PLANE COUETTE FLOW

NILS. TILLMARK AND P. HENRIK ALFREDSSON

Department of Mechanics, KTH, S-100 44 Stockholm, Sweden

1. Introduction

Plane Couette flow may be viewed as a paradigm of wall bounded flows, conceptually one of the simplest non-trivial fluid dynamics systems, where the flow is solely driven by the shear at the moving walls. Previously we have used hot-film anemometry and Laser Doppler Velocimetry to establish mean flow properties of fully developed turbulent Couette flow. In the long time average sense the flow is quite similar to other wall bounded flows, however a main difference is the appearance of large scale streamwise oriented structures. These are observed both in flow visualisations and through spanwise correlation measurements but the origin of these structures is not clear (see [1],[2])

2. Experimental apparatus, measurements and data processing

In the present work Particle Image Velocimetry (PIV) was used to investigate turbulent plane Couette flow and especially the large scale structures. The flow apparatus has an "infinite" transparent plastic band moving along vertical glass plates in water [3]. The channel dimensions are 1.5 m×0.36 m, whereas the distance between the moving walls $(2h)$ was 20 mm. The PIV measurements were carried out with a DANTEC PIV 2000 system including a CCD camera and the light source was a Nd:YAG double pulse laser. Spherical plastic particles were used as tracers. The light sheet penetrated the channel from the top, through a glass cuvette to minimise optical distortions. The light sheet was parallel to the moving walls (xz-plane) and in the symmetry plane of the channel (see figure 1).

As we were primarily interested in the large scales of the flow, a fairly large object area (75 mm× 75 mm =7.5h×7.5h) was imaged by the camera.

U. Frisch (ed.), Advances in Turbulence VII, 59–62.

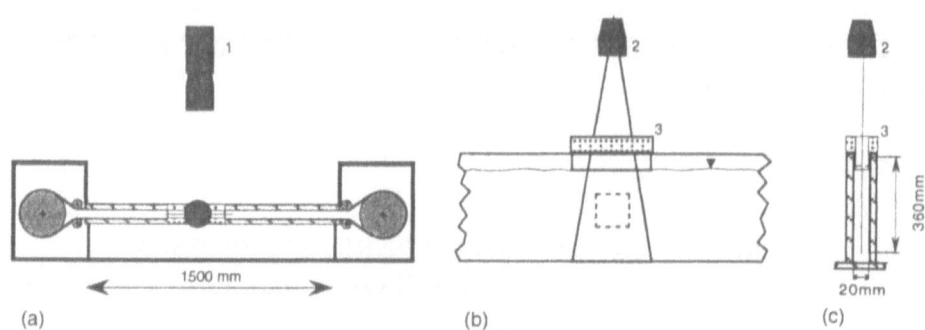

Figure 1. a) Topview of the Couette channel with CCD camera (1). b) Frontview of the channel with lightsheet generator (2), glass cuvette (3) and object area. c) Sideview.

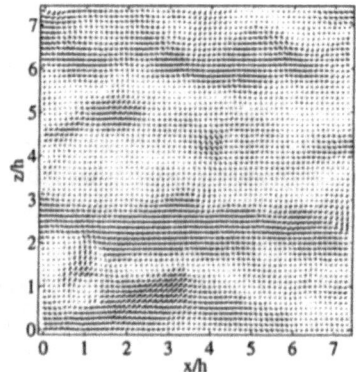

Figure 2. Instant vector picture. Re=3300, interrogation area 0.5h × 0.5h.

At this distance the interrogation area (IA) was set to 64 × 64 pixels which corresponds to an area of approximately 5 mm× 5 mm (0.5h × 0.5h) in the object plane. A 75% overlap of neighbouring IA was used to get a smooth and dense vector representation (see figure 2).

Measurements were taken at two different *Re*: 1300 and 3300. In each series 600 images were collected, each image having 3600 vectors. The separation in time between consecutive images was chosen to be at least two integral time scales to assure statistically independent flow fields. 60 spanwise and streamwise autocorrelations (unbiased) were performed in each image. This gave us a grand total of 36000 correlations in each mean correlation function. To be able to estimate the influence of spatial averaging on autocorrelation and gross turbulence statistics by the large IA, the measurements were repeated with a smaller IA and with the camera closer to the light sheet. The evaluation of data was performed in the same way as

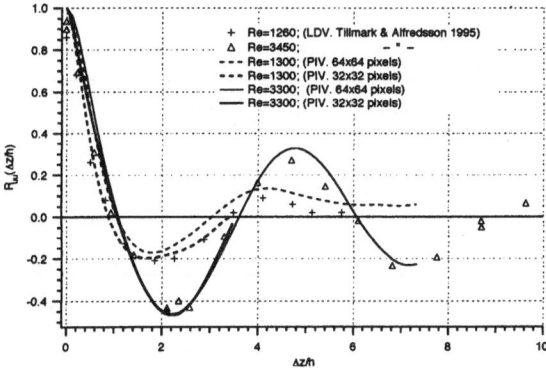

Figure 3. Spanwise correlation of the streamwise velocity at $Re=1300$ and $Re=3300$. Comparison between LDV and PIV measurements and PIV with different resolution.

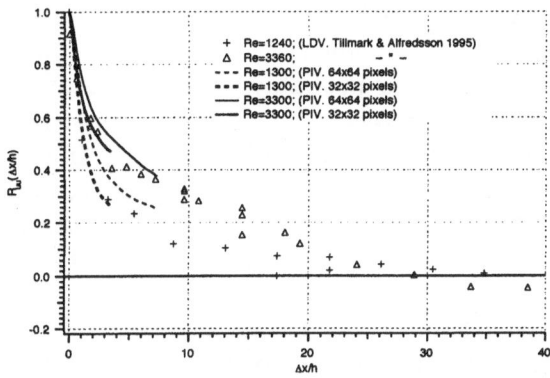

Figure 4. Spanwise correlation of the streamwise velocity at $Re=1300$ and $Re=3300$. Comparison between LDV and PIV measurements with different resolution.

in the low resolution case. The measurements were taken at the same Re but the distance between the object plane and camera was only about half of that previously used and the IA was decreased to 32×32 pixels. To get about the same numbers of vectors in each image an overlap of 50% was used. These changes increased the resolution to approximately 1 mm $\times$ 1 mm $(0.1h \times 0.1h)$.

3. Experimental results and conclusions

Figure 2 shows an instantaneous picture of the flow field at the centerline (the mean velocity is zero at this position) for $Re = 3300$. The maximum magnitude of the velocity vectors in this picture is about 30% of the wall velocity. As can be seen the velocity seems to be correlated over quite long

TABLE 1.

Re	Pixels	$\frac{U_{mean}}{U_{wall}}$	$\frac{u_{rms}}{U_{wall}}$	$\frac{w_{rms}}{U_{wall}}$	$S(u)$	$S(w)$	$F(u)$	$F(w)$
1300	32×32	-0.0078	0.112	0.079	0.022	-0.002	3.46	4.90
1300	64×64	-0.0109	0.112	0.076	0.076	0.016	2.78	3.55
3300	32×32	-0.0102	0.115	0.066	0.020	-0.032	2.60	4.66
3300	64×64	-0.0102	0.108	0.051	0.007	0.003	2.48	4.54

regions in the streamwise direction. In the spanwise direction the structures have a scale of 1-2h.

Some mean flow data are shown in Table 1. The mean velocity gives an indication on the exactness of the positioning of the laser sheet. The u_{rms} and w_{rms} values as well as the flatness F are close to those earlier obtained for LDV measurements. The skewness S should be equal to zero due to symmetry and the deviation is probably because of statistical uncertainty.

Figures 3 and 4 show correlation measurements of the streamwise velocity component obtained both with PIV and LDV. In the spanwise direction clear oscillations indicate the existence of spanwise periodic structures. Both PIV and LDV measurements show that the spanwise size of these structures is of the order of the channel width ($2h$). There is a slight difference depending on the IA, but the overall agreement between the various measurements is remarkably good. It can also be noted that the spanwise scale increases slightly with Re but also that the minimum and the second maximum become more pronounced for the higher Re. The large negative value of -0.4 is a clear indication on the existence of deterministic structures in the flow.

The present measurements confirm the earlier results that the streamwise correlations in plane Couette flow are strong over long streamwise distances as compared with other wall bounded flows. The correlation length is found to increase with Re which indicates that the structures become more dominating the higher the Re. This is also in accordance with the results of the spanwise correlations.

References

1. Bech, K.H., Tillmark, N., Alfredsson, P.H. & Andersson, H.I. (1995) An investigation of turbulent plane Couette flow at low Reynolds numbers. *J. Fluid Mech.* **286**, pp. 291-325.
2. Komminaho, J., Lundbladh, A. & Johansson, A. 1996 Very large structures in plane turbulent Couette flow. *J. Fluid Mech.* **320**, pp. 259-285.
3. Tillmark, N. & Alfredsson, P.H. (1992) Experiments on transition in plane Couette flow. *J. Fluid Mech.* **235**, pp. 89-102.

MEASUREMENTS ON THE MIXING OF A PASSIVE SCALAR IN A TURBULENT PIPE FLOW USING DPIV AND LIF.

L. AANEN[1], J. WESTERWEEL[1], F.T.M. NIEUWSTADT[1]
[1] *Laboratory for Aero and Hydrodynamics, Rotterdamseweg 145, 2628 AL Delft, The Netherlands.*

Turbulent mixing is a point of research with many applications in science and engineering. Points of interest are maximum concentration levels in turbulent mixing flows and chemical reactions. The mixing process can be described as the interaction between a flow field and a concentration field. The flow field follows from the Navier-Stokes equation, and is in many cases independent of the concentration field, whereas mass transport can be described with help of Ficks law. In the case of a time averaged steady and axisymmetric flow field, the Reynolds decomposed equation for the concentration field, only contains two cross correlation terms, that only include radial and axial velocity components and spatial derivatives. These two velocity components can be measured with Particle Image Velocimetry [1], whereas the concentration field can be measured with Planar Laser Induced Fluorescence [2]. In our experiment these two techniques are combined and applied simultaneously to a steady and axisymmetric turbulent pipe flow. Such simultaneous measurements of both the velocity field (using Particle Image Velocimetry) and the concentration field (using planar Laser Induced Fluorescence) in a plane through the pipe axis enables us to measure the full velocity-concentration correlation tensor. The results will be compared with the results of a DNS and will be used for validation of PDF-models.

Measurements are carried out on mixing of a point source at the centreline of a turbulent pipe flow. The experiments are done in water, where fluorescein is the scalar so that the Schmidt number of the dye is 2075.

After the axisymmetric measurements off-axis injection experiments will be done. Also we aim to measure the mixing of two point sources, in which one source emits acidic fluid and the other alkaline fluid. The pH dependency of the fluorescent dye is then used to determine the influence of turbulent mixing on the chemical reaction. [3] [4].

To be able to do these measurements, an experimental setup has been designed and built. The flow facility consists of a 6 meter long perspex

63

U. Frisch (ed.), Advances in Turbulence VII, 63–66.

pipe with an inner diameter of 50mm. In the pipe an injection device for fluorescence is placed. Just behind the injection mechanism a measurement section for doing PIV and LIF is mounted. In this measurement section the pipe wall is replaced by a thin glass cylinder, placed in a rectangular box filled with water and with glass windows. This reduces optical deformation by the curved pipe wall. For the remaining deformation due to the differences in index of refraction of the fluid and the thin cylinder corrections can be made.

For doing PIV measurements two YAG lasers are available. The lasers have a fixed pulse rate of 30Hz and pulse with an adjustable separation time. The light source for the LIF measurements is the 488nm line of the Argon-ion laser. Both planar and line measurements can be done by using a 1000*1000 pixel camera and a linescan camera respectively.

An optical setup has been designed to be able to combine the light sheet of the Argon-ion laser for doing LIF measurements and the light sheet of the two YAG-lasers for the DPIV measurements.

A two component LDV-measurement system is available for characterising the flow properties and controlling whether the flow is fully developed. LDV measurements have been done to determine the influence of the injection mechanism on the flow. The LDV measurements proved the pipe flow is fully developed indeed. The influence of the injection mechanism on the flow is not neglectable. Fortunately the disturbance is almost axisymmetric and decays quite fast, so within a few pipe diameters behind the injection the flow can be seen as in steady state.

The mean concentration and the spread of the point source have been determined at several distances from the injection point by measuring the concentration along a line. In figure 1 the results are compared with the results from a direct numerical simulation done by G. Brethouwer. Concentration structures are measured to determine the spatial resolution of the scalar field. The measurements show that we can resolve the concentration field with a spatial resolution of about 40 μm (see fig. 2)

The linescan camera will be used for doing more concentration line measurements with a high spatial and temporal resolution. This will give us detailed information on all spatial concentration scales at discrete positions behind the injection point. In the near future complete simultaneous LIF/PIV measurements will be done. The work is extended by measurements on other source locations and multiple source geometries.

References

1. J. Westerweel, A.A. Draad, J.G.Th. van der Hoeven, and J. van Oord. Measurement of fully-developed turbulent pipe flow with digital particle image velocimetry. *Experiments in Fluids*, 20:165–177, 1996.

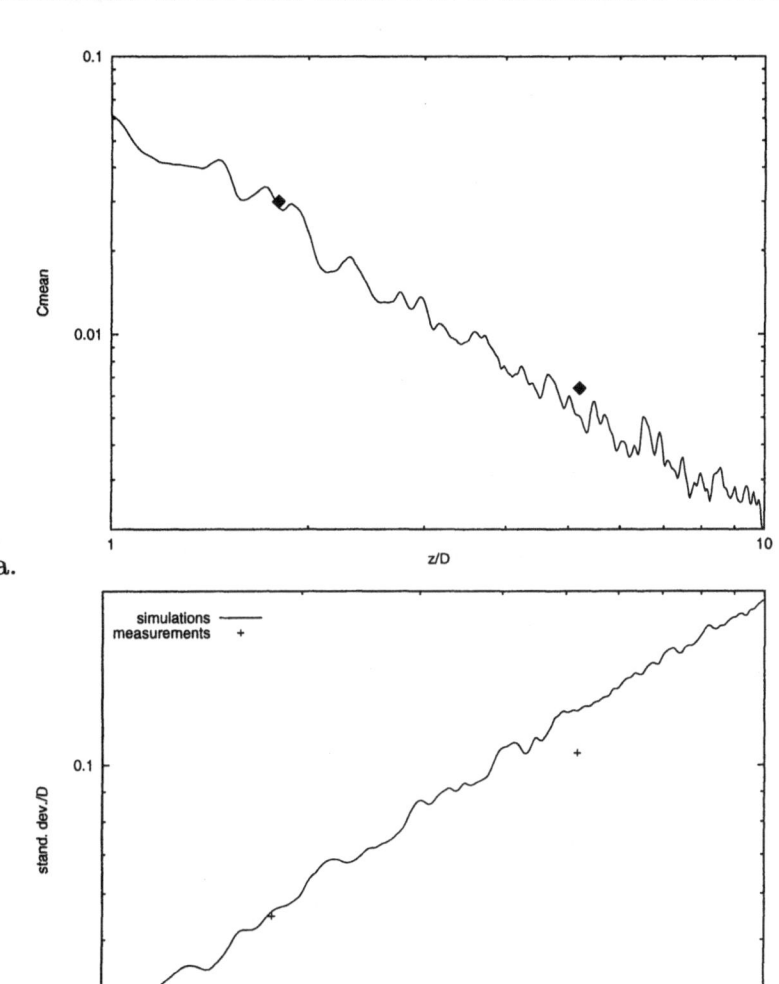

a.

b.

Figure 1. Figure a shows the normalised mean concentration as function of the distance from the point source. Figure b shows the spread of the source as function of the distance to the source. In both figures lines are results from a DNS, points are results from preliminary linescan measurements.

2. D.A. Walker. A fluorescent technique for measurement of concentration in mixing liquids. *J. Phys. E. Sci. Instrum.*, 20:217–224, 1987.
3. M.M. Koochesfahani. *Experiments on turbulent mixing and chemical reactions in a liquid mixing layer.* PhD thesis, California Institute of Technology, Pasadena, 1984.
4. H Stapountzis, J. Westerweel, J.M. Bessem, and F.T.M. Nieuwstadt. Measurement of product concentration of two parallel reactive jets using digital image processing. *Appl. Scient. Research*, 49:245–259, 1992.

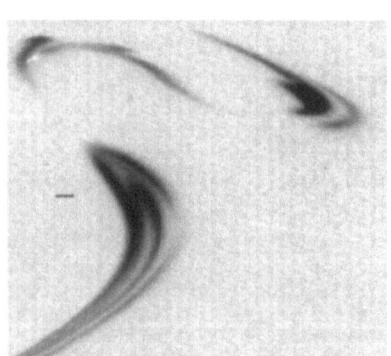
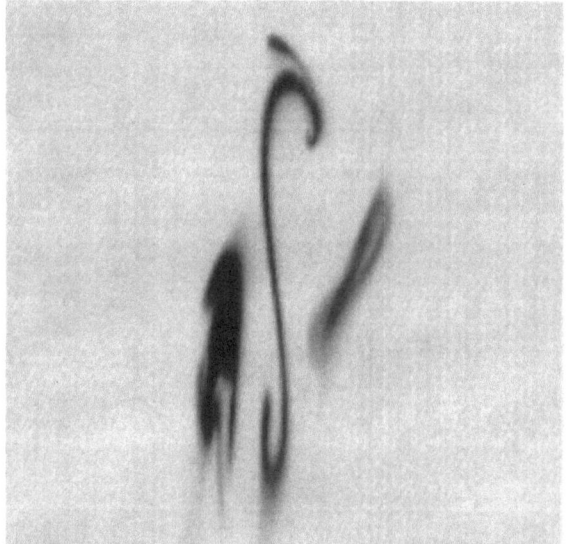

Figure 2.   These figures show some details of the concentration field, measured with the linescan camera. In these pictures the position is at the horizontal axis and the time at the vertical axis. The horizontal bar in the first picture equals the estimated Kolmogorov length scale.

II

Transition and Dynamical Systems

Transition et Systèmes Dynamiques

II

LIQUID JET INSTABILITY AND ATOMIZATION IN A COAXIAL GAS STREAM

E.J.HOPFINGER
LEGI-CNRS/UJF/INPG, B.P. 53, 38041 Grenoble Cedex

1. Introduction

The breakup and atomization of a liquid jet by a high velocity turbulent gas stream is known as airblast atomization (Lefebvre, 1989). In liquid propellant rocket engines for instance liquid oxygen is atomized by a high speed annular hydrogene gas jet. Because of these important practical applications a relatively large number of investigations have considered this problem but have to a large extent been limited to measuring the drop size at a certain distance downstream of the nozzle, as a function of the flow parameters at the nozzle exit. The correlation of the drop size, expressed for instance by the Sauter mean diameter D_{32}, with gas velocity is of the form $D_{32} \sim U^{-b}$ with b ranging from 0.8 to 1.4 and even 2. Various attempts have been made to explain such a power law (see Lasheras et al, 1998). The other important quantities are the spreading rate of the spray and the liquid intact or potential cone length. The latter has generally been correlated with the aerodynamic Weber number and the liquid Reynolds number (Arai et al., 1985). When the Weber number is large, a more relevant parameter in liquid jet breakup is the gas to liquid momentum flux ratio $M = \rho_g U_g^2 / \rho_l U_l^2$ (Hopfinger and Lasheras, 1994; Rehab et al., 1997). It has been demonstrated (Raynal, 1997) that in a certain range of M the liquid cone length is independent of Weber number and varies as $M^{-1/2}$. One of the open questions is the drop size resulting from the primary breakup.

Recent studies have focused on the details of the instability mechanisms at the liquid-gas interface with the aim to understand the instability and to model the primary breakup (Raynal et. al. 1998). High speed video images, combined with optical probe measurements, allowed to demonstrate that the instability is controlled by the vorticity layer of the gas stream and that the characteristic wavelength and frequency are primarily determined by this vorticity layer thickness and the liquid-gas density ratio. Linear stability theory supports these experimental results.

In this paper the different breakup regimes will first be illustrated on hand of a regime diagram. Then the near field mechanisms will be discussed pointing out the global flow structure, the instability and possible mechanismes of drop formation. The secondary liquid drop breakup downstream of the potential cone is shown to be primarily of the Kolmogorov type which is related with the turbulent velocity difference on the scale of the drop size. In modeling this breakup the turbulent energy dissipation has to be known. This brings up the question as to how the dissipation in the turbulent gas flow is modified by the liquid mass loading.

U. Frisch (ed.), Advances in Turbulence VII, 69–78.

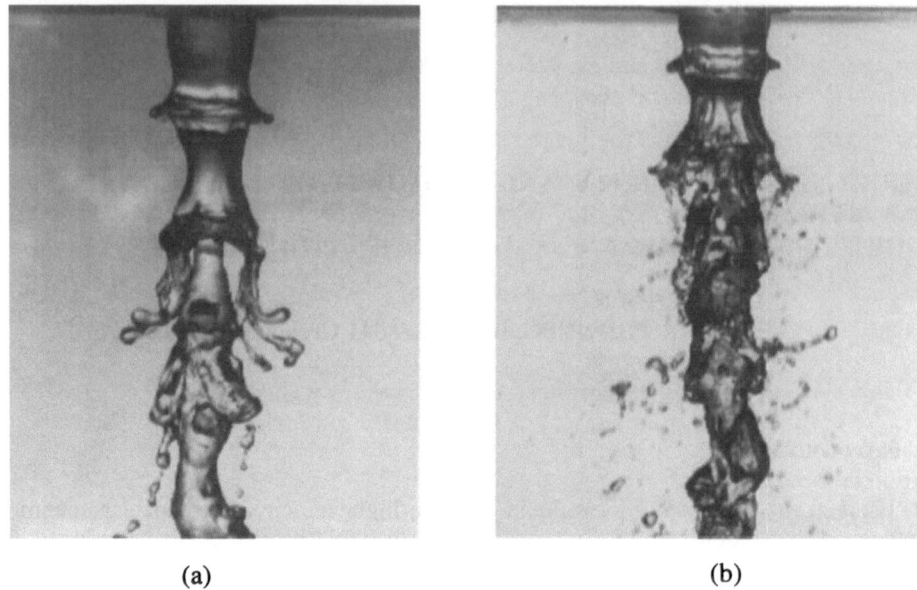

(a)   (b)

Figure 1. Visualization of liquide jet breakup and primary drop formation by a coaxial gas stream. (a), U_l=0.55ms-1, U_g=21ms^{-1}, We = 58, M =1.8; (b), U_l = 0.86 m s^{-1}, U_g=30ms^{-1}, We = 118, M =1.5. The scale is given by the nozzle diameter D_l=7.6mm.

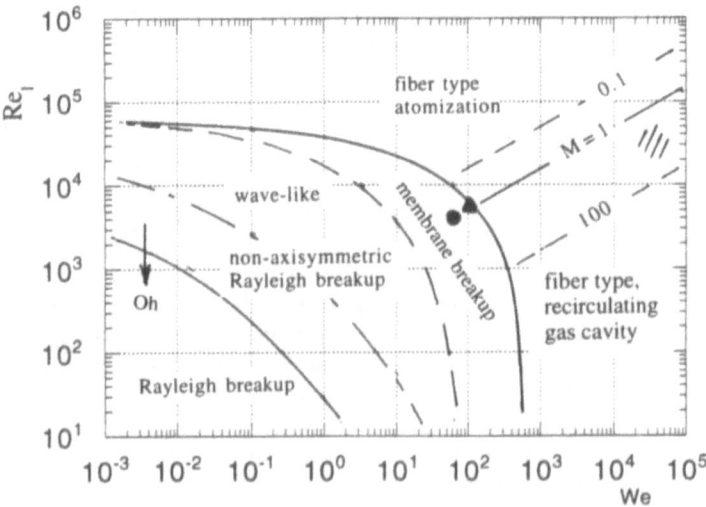

Figure 2. Breakup regimes in the parameter space Re_l-We. Lines of constant M are calculated for water-air and D_l = 5 mm. ●, conditions of Fig. 1a; ▲, Fig 1b; |||, rocket engins. We = $\rho_g U_g^2 D_l/\sigma$, Re_l = $U_l D_l/\nu_l$, M = $\rho_g U_g^2/\rho_l U_l^2$.

2. Regimes of liquid jet breakup.

The most well known breakup mechanism of a liquid jet is the Rayleigh breakup. This is due to surface tension and occurs at relatively low jet Reynolds number ($Re_l = U_lD_l/\nu_l$ of order 10^2), the value of which depends on the ratio of surface tension forces to viscous forces expressed by the viscosity or Ohnesorge number $Oh = \mu_l/(\rho_l \sigma D_l)^{1/2}$. At larger Reynolds number the jet becomes wavy due to arerodynamic effects and at large Reynolds number (order 10^5) atomization due to shear instability takes place. This regime is referred to as pressure atomization. For more detailed discussions see Lefebvre (1989).

2.1 BREAKUP MECHANISMS IN THE PRESENCE OF A CO-AXIAL GAS JET.

In this paper we are mainly interested in the instability and atomization of a liquid jet by a high velocity, annular air stream, referred to as airblast atomization. In Fig. 1 examples of the instability of the jet caused by a coaxial air jet are shown. The important point to notice is the long vavelength of the instability. Very different types of nozzles have been used in experiments. Here we focus on two generic geometries: tubular nozzles for both the liquid (water) and the gas (air), with $D_l = 3.5$ mm and $D_g = 5.6$ mm, and convergent nozzles with contraction ratios 7:1 and $D_l = 7.6$ mm, $D_g = 11.3$ mm. The images shown in Fig.1 have been taken with the convergent nozzles. In Fig 1a, $U_l = 0.55$ m s^{-1} and the annular gas velocity $U_g = 21$ m s^{-1} and in Fig. 1b, $U_l = 0.86$ ms^{-1} and $U_g = 30$ms^{-1}. Because of the contraction, the velocity profiles at the exit are flat with relatively thin, laminar boundary layers (vorticity layers). The wavelength, manifest in these images, is considerably larger than this vorticity layer thickness (order 10^2). In Fig.1a the drop size is also large because the breakup of the annular sheet is strongly affected by surface tension, the aerodynamic Weber number, $We = (\rho_g U_g^2 D_l)/\sigma$, being 58 and M =1.8. It is seen that at the tip of the sheets a bulge or rim is formed due to suface tension forces and behind it the liquide is blown up into a a bag forming a membrane by the aerodynamic pressure. For comparison, Fig. 1b shows a case of similar M but somewhat larger Weber number, $We = 118$, where surface tension forces at the tip are of lesser importance. The drop size is noticeably smaller. At large Weber numbers the breakup is in the form of fibers (Hopfinger and Lasheras, 1994; Lasheras et. al., 1998) and the drop size is much smaller. In all these experiments the gas Reynolds number is large.

In order to give an overview of the breakup and atomization, it is of interest to establish a regime diagram similar to what has been done for pressure atomizers (see Lefebvre, 1989). Airblast atomization is, however, considerably more complicated because we have in addition to Re_l and Oh to deal with the parameters We and M and also the gas Reynolds number. An attempt to establish such a diagram has been made by Farago and Chigier (1992) but without including M. In Fig.2 the liquid jet breakup regimes are presented in the liquid Reynolds number and aerodynamic Weber number space. Lines of contant M appear as straight lines and are calculated for water and $D_l = 5$mm from the relation $Re_l = (We/M)^{1/2}(D_l\sigma/\rho_l\nu_l^2)$. There are not enough experimental data available to give the different bounds to any degree of accuracy, except when $We \rightarrow 0$ (in this limit the bounds are shown for a viscosity number Oh of order 10^{-3}). Also, recent experiments performed in the range $We = 20$ to 10^3 and $Re_g = (D_g-D_l)U_g/2\nu_g = 800$ to 8.10^3 give a good indication about the upper and lower bounds of membrane breakup. Good atomization in the sense of a fine spray is achieved beyond the upper bound of membrane breakup and the farther away from the bound, the finer the spray. The mass loading, expressed by $m = \rho_l U_l A_l / \rho_g U_g A_g$, where A_l and A_g are respectively the liquid and gas nozzle cross sections, affects, for given M and We, the downstream spray formation by draining energy from the gas.

2.2 LIQUID CONE LENGTH

The momentum flux ratio M plays practically no role in the atomization as such but determines the liquid cone or intact legth. When $M \ll 1$, this length is determined by the liquid jet (Reitz and Bracco, 1982), whereas at large M it is the gas jet which is responsable for the breakup; the liquide cone length varies in this case roughly like $M^{-1/2}$ (Lasheras et al, 1998). Above a certain value M_c (of order 100) a recirculating gas cavity exists downstream of the liquid cone which is actually truncated by the recirculating motion.

The liquid cone length is predicted by a simple entrainment model (Lasheras et al, 1998, Rehab et al, 1997) which has as essential ingredient dynamic pressure continuity at the interface in the form

$$\rho_l u_e^2 = C_e \rho_g u'^2_g \ , \qquad (1)$$

where u_e is the velocity at which liquid is entrained and C_e is a proportionality factor to be determined from simple jet experiments ($C_e \cong 0.25$). From mass flux continuity $u_e S = U_l \pi D_l^2/4$, where S is the surface area of the liquid cone, and substituting from relation (1) for u_e, we get

$$L/D_l \approx 6 / M^{1/2} \qquad (2)$$

where $u_g' = 0.17 U_g$ has been used. Liquid rocket engins operate typically at $M \approx 10$, giving a liquid cone length of about $2D_l$. If relation (2) is reformulated and extrapolated to $M \to 0$, the limit would be $L/D_l \approx 6 (\rho_l/\rho_g)^{1/2}$. However, at very large Re_l the jet breakup also depends on the shear created in the nozzle (Reitz and Bracco, 1982) and the liquid cone length might be considerably shorter, the value depending on the rapidity of relaxation of the shear in the liquid.

In the above discussion of the liquid cone legth and entrainment model we have assumed that the Taylor number $T = \mu_g U_g/\sigma$ is small and that We/T is large, meaning that the gas Reynolds number is large.

2.3 CHARACTERISTIC FREQUENCIES

The frequency of the most amplified interfacial perturbations was measured with a photosensitive detector at the edge of the liquid cone close to the nozzle exit (Lasheras et al, 1998, Raynal, 1997). The Strouhal number constructed with U_g and D_l is of order 10^{-2}, with the precise value depending on the type of nozzle (convergent or tubular) and on the liquid velocity at the nozzle exit. A more relevant velocity for expressing the Strouhal number is the liquid interfacial velocity which is obtained from shear stress continuity at the interface $\tau_l = \tau_g$. When turbulent stress conditions are assumed for both streams, stress continuity can be expressed by $C_l \rho_l (U_i - U_l)^2 = C_g \rho_g (U_g - U_i)^2$ which gives, when taking $C_l = C_g$,

$$U_i = (\rho_g^{1/2} U_g + \rho_l^{1/2} U_l)/(\rho_g^{1/2} + \rho_l^{1/2}) \qquad (3)$$

When M is large and $\rho_l >> \rho_g$, as is the case for water-air, the interfacial velocity is, $U_i \cong (\rho_g/\rho_l)^{1/2} U_g$. Expressed with this velocity the Strouhal number has a value St = f $D_l/U_i \approx 0.3$ which is not unusual. This value indicates that the wavelength λ is in the experiments discussed of order D_l and a nearly constant value of St would mean that λ scales on D_l. As will be shown in the following section, this is not the case. Furthermore, the assumption of turbulent shear stress conditions leading to U_i in the form of equation (3) may not be valid when the exit profiles are laminar as expected for convergent nozzles.

3. Interfacial instability.

3.1 EXPERIMENTAL RESULTS.

In order to clarify the interfacial instability under the conditions of interest here (large gas momentum flux and large gas Reynolds number), a liquid-gas shear layer experiment has been constructed for this purpose (Raynal, 1997).

In this experimental set-up an air stream of velocity $10 < U_g < 100$ ms^{-1} merges with a water stream $0 < U_l < 1.2$ ms-1, downstream of a splitter plate. The water and air nozzle dimensions at the exits are 1cm high and 10cm wide and the contration ratios of both nozzles are 10:1. The horizontal water stream is supported by a lower plate extending downstream of the nozzle and the air stream is free on its upper end. Both streams are confined by transparent side boundaries.

In this set-up it was possible to measure with good accuracy the exit profiles, including the boundary layers, by means of hot wires and hot films. The frequency of the liquide interfacial waves was measured with a laser beam and photo diode and the convection velocity was determined by correlating the photo diode signal with the signal of a hot film placed a given distance downstream of the laser beam. The convection velocity and the frequency give the wavelength which was also determined from flow visualisations using a backlighting technique.

The main results are as follows: Although exit conditions are laminar (nearly laminar boundary layers), the measured convection velocity U_c is found to be closely approximated by U_i, expression (3). This means that the interfacial liquid velocity is generated by the dynamique pressure imposed by the gas stream. It is of interest to note that this convection velocity is the same as the one proposed by Bernal and Roshko (1986) and Dimotakis (1986) for a large density difference (helium-nitrogen) shear layer. The frequency is $f = U_c/\lambda$ with λ proportional to the vorticity layer in either the gas or liquid stream. Because of the observed long wavelength a scaling with the vorticity layer in the liquid is unlikely. The vorticity layer thickness in the gas stream at nozzle exit δ_ω =$U_g/dU/dy|_{max}$ is approximated by

$$\delta_\omega/2H_g \cong 4.25 \, Re_g^{-1/2}, \qquad (4)$$

where H_g is the height of the gas nozzle at exit and Re_g is the gas Reynolds number $Re_g = U_g 2H_g/v_g$. If λ scales on $\delta_\omega \propto U_g^{-1/2}$ and $U_c = (\rho_g/\rho_l)^{1/2} U_g (1 + M^{-1/2})$ when $\rho_g << \rho_l$, the frequency should scale as $f/(1 + M^{-1/2}) \propto U_g^{3/2}$. This is indeed what is observed (Fig.3). The Strouhal number based on δ_ω, expression (4), is

$$f \, \delta_\omega/U_c \cong 8.7 \times 10^{-3} \qquad (5)$$

This value is much less than the Strouhal number f δ_ω /U_c = 0.13 measured in a homogeneous shear layer. The difference is due to the much larger wavelength manifest in the liquid-gas shear layer. More details are given in Raynal (1997) and Raynal et al, (1998).

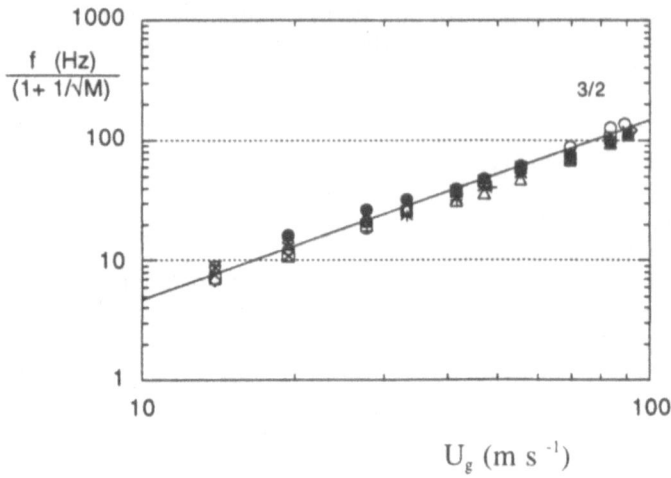

Figure 3. Measured frequency of the interfacial perturbations, weighted by $(1+M^{-1/2})^{-1}$, as a function of the gas velocity. The different symbols correspond to liquid velocities ranging from 0.069 to 1.11 ms^{-1}. The data collaps on a straight line with slope 3/2.

3.2 LINEAR STABILITY ANALYSIS.

The experiments suggest that the liquid flow is dynamically passive and acts only as a filter and damping system to the excitation imposed by the gas stream. The flow configuration, representative of the experiments, consists simply of two superposed uniform streams of velocity U_j, (j=1 liquid, j=2 gas) and densities ρ_j which are joined by a boundary layer in the gas of constant vorticity thickness δ_2. The dispersion relation for this configuration can be derived following Chandrasekhar (1961) neglecting viscosity, gravity and surface tension effects. The amplification rate as a function of wave number has been determined for values of S= ρ_2/ρ_1 ranging from 1 to 10^{-3} and for large values of the velocity ratio U_2/U_1 (Raynal et al, 1998). It is found that the most amplified wavelength increases and the amplification rate decreases as S decreases. When S < 1/10 the amplification rate varies as the density ratio ρ_2/ρ_1, and the group velocity and the selected wave number as $(\rho_2/\rho_1)^{1/2}$. The group velocity normalized by the convection velocity, equation (2), is nearly equal to 1 and the theoretical Strouhal number is fδ_2/U_c = 7.6x10^{-3} which is close to the measured value.

4. Secondary instability and primary drop formation.

An important problem concerns the drop formation in the near field. The interfacial waves are amplified and the liquid is drawn out into sheets of thickness e by the strain imposed by the gas stream. The drop size is proportional to and of the order of e (Villermaux, 1998). The problem is to predict the thickness e of this sheet. Fig.4 shows a visualization (spanwise view) of the amplification of the interfacial perturbation and breakup of the sheet in the liquid-gas shear layer experiments (Raynal, 1997). As the perturbations grow and a sheet begins to form, spanwise perturbations at the rim of the sheet also grow (compare for instance the left hand side of Fig. 4b with Fig 4c and 4d). In time, the rim radius decreases until surface tension forces σ/e become of the ordre of the aerodynamic forces. The rim is finally disrupted by aerodynamic forces and Rayleigh instability (see Fig.4e and f). Behind the rim the liquid sheet is blown out into a bag forming a membrane which breaks and forms small drops. The mechanism shown in Fig 4 is charateristic of relatively small Weber number (the Weber number in Fig.4 is about 80 (based on the liquid layer thickness), as in Fig.1, where membrane breakup occurs (see Fig 2). This breakup mechanism is similar to the scenario proposed by Li (1996). From a simple force balance between the aerodynamic forces (drag) acting on the rim and opposing surface tension force we obtain that the drop size, $d_i \propto e$, scales on the Weber number in the form $d_i/D_1 = We^{-1}$.

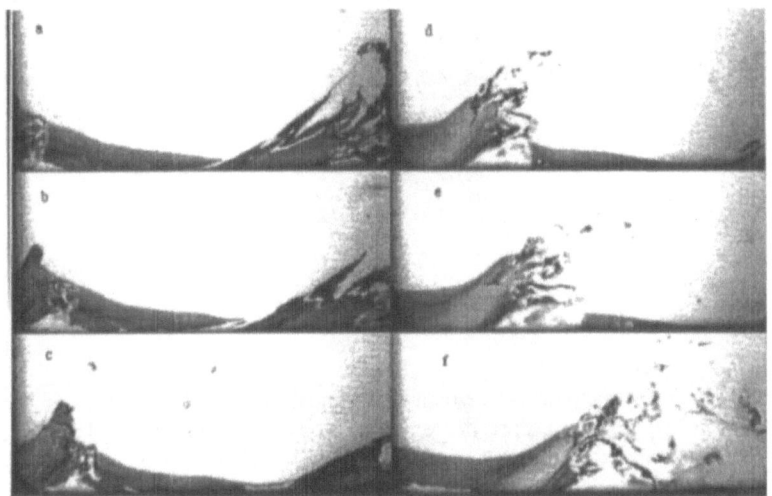

Figure 4. Visualization of the temporal growth of the interfacial waves leading to sheet formation and breakup into drops. $U_1 = 0.42ms^{-1}$, $U_g = 22ms^{-1}$.
(from Raynal, 1997).

When the Weber number is large and fiber-type breakup occurs it is likely that surface tension is unimportant in determinining the final sheet thickness and hence the primary drop size. The size of the fibers d_f may scale on the rib vortices of the interfacial gas shear layer given by the balance between viscous diffusion rate v_g/d_f^2 and strain rate $\partial u_g/\partial x \propto U_g/\lambda$. Since $\lambda \propto \delta_\omega(\rho_l/\rho_g)^{1/2}$ the size of the fibers would scale as d_f $/D_1 \propto (v_g/U_gD_1)^{1/2}(\delta_\omega/D_1)^{1/2}(\rho_l/\rho_g)^{1/4}$. With $\delta_\omega \propto (v_g/(D_g-D_l)U_g)^{1/2}$, for laminar conditions at the nozzle exit, the drop size varies as $d_f/D_1 \propto U_g^{-3/4}$. The cross-over from $d_i \propto e$ to $d_i \propto d_f$ is at a certain gas velocity or Weber number which corresponds to the upper bound of membrane breakup. In Fig.5 the drop size d_i/D_1 is presented as a function of aerodynamic Weber number based on D_1 keeping the fluid properties constant. The experimental points on this figure are, at low Weber number, evaluated from images shown in Fig. 1 and at large Weber number the experimental point corresponds to mean drop sizes, D_{32} and also D_{90} measured at the edge of the spray, typically at $x/D_1 = 20$ and $r/D_1 = 6.5$ (Préaux et al, 1998 and Préaux, private communications). Villermaux (1998) proposed an expression for the initial drop size in the form $d_i/\delta_\omega \propto We_\omega^{-1/5}(\rho_l/\rho_g)^{2/5}$. In obtaining this expression it is supposed that the rate straining of the liquid sheet is U_i/λ and continues until surface tension forces oppose the straining. This expression gives a dependency of d_i on U_g in the form $d_i \propto U_g^{-4/5}$. Villermaux further suggests that the vorticity layer grows with x and replaces δ_ω by the vorticity layer evaluated at the end of the potential cone $\Delta(L)$. In this case he finds $d_i \propto U_g^{-1}$ for laminar outlet conditions. It is clear from Fig 5 that in the range We > 200, $d_i \propto U_g^{-1/2}$ is the best fit. This power law would be obtained if we take for the strain rate $\partial u_g/\partial x \propto U_g/D_1$ Further measurements of the drop size in the near field are required to guide models.

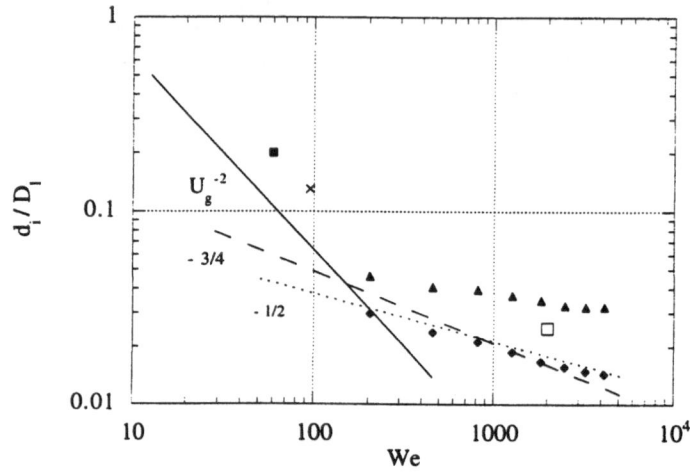

Figure 5. Primary drop size as a function of We. -----------, $d_i/D_1 \propto We^{-1}$; - - - - -, $U_g^{-3/4}$; ...., $U_g^{-1/2}$. ■, x , evaluated from images Fig.1; ❑, D_{32} from Préaux et al (1998) ; ▲, D_{90},; ◆, D_{32} (G. Préaux)

5. Secondary drop breakup.

Most of the drop size measurements in sprays are made downstream of the potential cone at $x/D_1 > 10$. The measured mean drop size is then empirically correlated with the initial conditions for given fluid properties (Lefebvre, 1998). The dependency of D_{32} on $U_R = U_g - U_1 \approx U_g$ is correlated with a power law of the form $D_{32} \sim U_g^{-b}$ and, depending on the experiments the exponent, b ranges from about 0.8 to 1.4 and occasionnally even 2. Recently, Lasheras et al (1998) showed that secondary atomization is completed at a certain distance downstream which depends on the mass loading and then the drop size increases by coallescence.

Secondary breakup may occur either by the mean aerodynamic forces related with the difference between the local gas velocity and the velocity of the drop or liquid lump, or by the turbulent velocity difference on the sclae of the drop. The drops, resulting from the primary atomization, with initial velocity U_c are rapidly accelerated by the gas drag. At the same time the local gas velocity u_g decreases due to entrainment of ambient fluid and also due to the momentum transfer to the liquid. Therefore, the drop Weber number $We_d = \rho_g(u_g - u_d)^2 d/\sigma$ falls rapidly below the critical value of about 10 below which drops remain stable. While this breakup due to the local mean velocity difference is certainly responsible for the breakup of the larger drops and lumps close to the end of the liquid cone near the jet axis, it cannot explain the minimum in drop size observed further downstream. Lasheras et al (1998) showed that the turbulence breakup model can explain the experimental results. The essential ingredient is that breakup by the turbulent stresses is active as long as the Weber number

$$We_t = \rho_g \, \overline{u(d)}^2 \, d/\sigma > 1, \tag{6}$$

where $\overline{u(d)}^2$ is the mean square of the relative velocity turbulent fluctuations on a scale equal to the drop size d. Assuming local isotropy $\overline{u(d)}^2 = C(\varepsilon \, d)^{2/3}$, where C is a constant of order 1 and ε the dissipation rate. From (6) and the expression for $u(d)^2$ we get the maximum drop size which can exist in the turbulent jet in the form

$$d_{max} = (\sigma We_{tc} / \rho_g)^{3/5} \, \varepsilon^{-2/5} \tag{7}$$

where We_{tc} is the critical turbulent Weber number equal to about 1. The problem is the expression for ε taking into account the mass loading. From energy conservation considerations, Lasheras et al proposed:

$$\varepsilon = U_g^3 / \left(1 + \frac{\rho_1 \, U_1 \, A_1}{\rho_g \, U_g \, A_g} \right) D_g \tag{8}$$

When this expression is substituted into equation (7), the dependence of d_{max} on U_g is obtained. Assuming similarity distributions so that $d_{max} \propto D_{32}$, the exponent b in $D_{32} \sim U_g^{-b}$ varies between 6/5 to 8/5 depending on liquid mass loading. Measured values of D_{90} for a given gas velocity and different liquid mass loadings are found to compare well with d_{max} (Lasheras et al). This good agreement gives strong support to the model.

It has been mentioned above that there is a minimum in drop diameter where atomization is completed. The location of this minimum can be obtained by equating the droplet breakup time $d_{max}/(\overline{u(d)}^2)^{1/2}$ to the residence time. This distance is again a function of U_g and the mass loading.

6 Conclusions.

In this paper I have tried to give first an overall picture of liquid jet breakup rangin from pressure atomization to airblast atomization. The regime diagram presented in Fig.2 shows for the first time the relation between liquid Reynolds number, aerodynamic Weber number and momentum flux ratio. The dependency of the liquid cone length on the momentum flux ratio is discussed in this context.

One of the characteristics in airblast atomization with dominant gas momentum flux is the long wavelength of the interfacial instability. Recent experiments together with linear stability analysis demonstrate that this wavelength scales with the vorticity layer in the gas stream and the square route of the liquid to gas density ratio. The dependency of the convection velocity on the liquid and gas densities is found to be identical to the relation proposed by Bernal and Roshko (1986) for a helium-nitrogen shear layer.

The mechanism of primmary drop formation discussed in section 4 is to a large extent speculative, especially in the large Weber number range of interest for practical applications. It is suggested that two regimes of drop formation exist, one, at low Weber number, being governed by surface tension-aerodynamic presuure force balance and the other, at large Weber number, by viscous diffusion-strain rate balance.

The discussion of secondary atomization depending on drop breakup by the turbulence in the jet is essentially based on material presented in a recent paper by Lasheras et al (1998). It is shown that the dependency of the drop size on gas velocity is affected by the mass loading.

ACKNOWLEDGEMENTS

The results and many ideas presented in this paper must be shared with my collegues J.C. Lasheras, E. Villermaux and L. Raynal. Particular thanks go to JCL who provided his laboratory facilities for the jet experiments and G. Préaux for making data available. The work was financially supported by the SEP under contract n° 910023.

7. References

Arai, M. and Hashimoto, H. (1985) Disintegration of a thin liquid sheet in a cocurrent gas stream. Proceedings of 3rd Int. Conf. on Liquid Atomization and Spray Systems, London. VI.B.**14**, 1-7.

Bernal, L.P. and Roshko, A.(1986) Streamwise vortex structure in a plane mixing layer. J. Fluid Mech. **170**, 499-525.

Chandrasekhar, S. (1961) Hydrodynamic Stability. Dover Publ. Inc, New York.

Farago, Z and Chigier, N. (1992) Morphological classification of disintegration of round liquid jets in a coaxial air stream. Atom.and Sprays **2**, 137-153.

Dimotakis, P.E. (1986) Two-dimensional shear layer entrainment. AIAA, **24**, 1791 1796.

Hopfinger, E.J., and Lasheras, J.C. (1994) Break-up of a water jet in high velocity coflowing air. Proceedings of 6th ICLASS, **1-15**, 110-117.

Lasheras, J.C., Villermaux, E. and Hopfinger, E.J. (1998) Break-up and atomization of a round water jet by a high speed annular air jet. J. Fluid Mech. **357**, 351-

Lefebvre, A.H. (1989) Atomization and sprays. Hemisphere Publ. Corp, New York

Li, J. (1996) Résolution numérique de l'équation de Navier-Stokes avec reconnection d'interfaces: Méthode de suivi de volume et application à l'atomization. Thèse de Doctorat, Univeristé Pierre et Marie Curie, Paris.

Préaux, G., Lasheras, J.C. and Hopfinger, E.J. (1998) Atomization of a liquid jet by a high momentum coaxial swirling gas jet. 3rd Int. Conf. Multiphase Flow, Lyon.

Raynal, L., Villermaux, E., and Hopfinger, E.J. (1998) Primary instability of a plane liquid-gas shear layer (submitted to J. Fluid Mech.).

Raynal, L. (1997) Instabilite et Entrainement à l'interface d'une couche de melange Liquide-Gaz. Thèse de Doctorat, Université Joseph Fourier, Grenoble.

Rehab, H., Villermaux, E. and Hopfinger, E.J. (1997) Flow regimes of large velocity ratio coaxial jets. J. Fluid Mech. **345**, 357-381.

Reitz, R.D. and Bracco, F.V. (1982). Mechanism of atomization of a liquid jet. Phys. Fluids, **25**, 1730-1742.

Villermaux,E. (1998) Mixing and spray formation in coaxial jets. To appear in J. Prop. and Power

CAN THE STUDY OF HYDRODYNAMIC INSTABILITIES BE USEFUL BEYOND TRANSITION ?

P.A. MONKEWITZ
Department of Mechanical Engineering,
Fluid Mechanics Laboratory,
Swiss Federal Institute of Technology Lausanne,
CH-1015 Lausanne, Switzerland

In the context of transition to turbulence, hydrodynamic stability analyses have been generally successful in describing the early stages of transition, with some notable exceptions such as pipe Poiseuille flow. Over time, the tools of stability analysis have been expanded to deal with spatially inhomogeneous (non parallel) base flows, with secondary instabilities and with finite amplitude perturbations. All these tools have permitted to carry the description of flow perturbations in terms of deterministic instability modes tantalizingly close to transition. However, before a flow becomes really "turbulent", i.e. more than just unsteady and "complicated", the stability theoretical approach usually becomes unmanageable. On the other hand, there is ample evidence from laboratory and numerical experiments, that organized structures persist in flows that are considered turbulent by the majority of researchers.

The aim of this contribution is to review some past and present efforts to transpose ideas and methods of hydrodynamic stability theory to fully turbulent flows, i.e. to leap over the transition "barrier" instead of trying to break through with brute force. At the same time, this contribution will attempt to better define some of the challenging stability problems waiting to be solved.

One of the more innovative recent ideas has been to investigate the mechanism that sustains wall-bounded turbulence. One candidate mechanism in particular, the self-sustained *"streamwise vortex - streak - streak instability cycle"* has recently gathered support from Waleffe's stability theoretical model for Couette flow, from direct numerical simulations (see e.g. Hamilton *et al.*, 1995) of channel flow and from innovative numerical manipulation experiments of Jimenez and co-workers (see e.g. Jimenez and Pinelli, 1997). In these DNS experiments it is shown convincingly that dis-

U. Frisch (ed.), Advances in Turbulence VII, 79–80.
© *1998 Kluwer Academic Publishers.*

rupting the cycle by filtering the streaks leads to the "death" of turbulence, while selectively relaxing the no-slip condition in the spanwise direction enhances it. The experiment also suggests that the creation of secondary streamwise vortices at the wall (see e.g. Haidari and Smith, 1994) is probably not an essential ingredient of the cycle. In a different DNS study of channel flow with closed loop control of the wall flux of spanwise vorticity, Koumoutsakos (1997) shows that the destruction of streamwise structures dramatically reduces the skin friction. Laboratory and numerical experiments will be presented (see the preliminary report by Cui *et al.*, 1996) that the dynamics of a highly organized flow consisting of a stable Blasius boundary layer distorted by steady streamwise vortices, produced in the manner of Acarlar and Smith (1987), does indeed show strong similarities with the viscous sublayer of a flat plate turbulent boundary layer when scaled with wall units based on the unperturbed Blasius layer. This finding strengthens the case for the relevance of Waleffe's model to the turbulent wall layer.

Since Waleffe's model applies to an equilibrium situation, the obvious next question is how the transition from a non self-sustained situation to the self-sustained turbulent flow should be modelled. Drawing on the concept of convective and absolute/global instability (see e.g. Huerre and Monkewitz, 1990), an attempt will be made to formulate the requirements for models that are capable of describing a spatial transition from a laminar to a self-sustained turbulent state and the possible structure of such models will be discussed.

References

Acarlar, M.S., and Smith, C.R.: 1987, A study of hairpin vortices in a laminar boundary layer. Part 1. Hairpin vortices generated by a hemisphere protuberance, and part 2. Hairpin vortices generated by fluid injection, *J. Fluid Mech.*, **175**, 1-83.

Cui, Y., Tanguay, B., Bottaro, A., and Monkewitz, P.A.: 1996, Spatial numerical simulations and experiments on the breakdown of streamwise vortices in a Blasius boundary layer, *Advances in Turbulence VI*, 325-328, Kluwer Academic Publishers.

Haidari, A.H., and Smith, C.R.: 1994, The generation and regeneration of single hairpin vortices, *J. Fluid Mech.*, **277**, 135-162.

Hamilton, J.M., Kim, J., and Waleffe, F.: 1995, Regeneration mechanisms of near-wall turbulence structures, *J. Fluid Mech.*, **287**, 185-212.

Huerre, P., and Monkewitz, P.A.: 1990, Local and global instabilities in spatially developing flows, *Ann. Rev. Fluid Mech.*, **22**, 473-537.

Jimenez, J., and Pinelli, A.: 1997, Wall turbulence: How it works and how to damp it, *AIAA paper* **97-2112**.

Koumoutsakos, P.: 1997, Active control of turbulent channel flow, *CTR Annual Research Briefs*, 289-297.

TRANSIENT AND FORCED REGIMES IN THE WAKE OF A SPHERE

D. ORMIERES, M. PROVANSAL, A. BARRANTES
I.R.P.H.E (Institut de Recherche sur les phénomènes hors équilibre)
UMR 6594 du CNRS, Centre Universitaire de St. Jérôme,
service 252
13397 MARSEILLE Cedex 20 France
e-mail:ormi @lrc.univ-mrs.fr

Despite its practical importance, many fundamental aspects of the wake behind a sphere remain unknown, here we have inquired the transition from steady to periodic flow.

Thanks to visualizations carried out in a water channel (see [4]) we know that in the Reynolds number range [150, 180], $(Re = Ud/\nu)$ the bubble formed downstream of the sphere is no longer axisymmetrical and one steady trail is visible. Between $Re = 180$ and the critical value $Re_c = 280$, two steady trails appear, which begin to oscillate above 280. Above $Re = 310$, vortex loops are periodically shed and their location can be changed by a very slight inclination of the upstream pipe which held the sphere (good agreement with the report of Sakamoto [5]).

Among the difficulties are the full three-dimensional character of this flow and the way of holding the sphere. In our wind-tunnel, the sphere (d = 1 cm) is hold by four wires (d = 0.08 mm), in the same way as for torus in study [3]. This configuration is used to freeze the spatial mode, allowing reproductible experiments. Quantitative laser-Doppler and hot-wire anemometry measurements are performed for Reynolds number range [100, 400]. Control of vibrations is monitored through laser-photo diode set-up.

Spectral analysis of the streamwise velocity fluctuations shows the appearance of a peak and its harmonics when the Reynolds number is above the critical value $Re_c = 280$. We were unable to detect any kind of hysteresis when increasing or decreasing the upstream velocity (at least at 1%), and near the threshold, the energy of oscillations E increases linearly. Both

81

U. Frisch (ed.), Advances in Turbulence VII, 81–84.
© *1998 Kluwer Academic Publishers.*

behaviours are in agreement with the Landau model which describes this transition as a supercritical bifurcation [2]:

M is the real amplitude of the oscillation

$$dM/dt = \sigma.M - l.M^3 \qquad (1)$$

This equation gives the energy of the oscillation:

$$E = M^2 = (\sigma/l) \rightarrow \alpha(Re - Re_c) \qquad (2)$$

The extrapolation of zero energy provides a quantitative criterion for an accurate determination of the critical Reynolds number (280).

We have searched to characterize the temporal growth rate σ from transient experiments. A periodic modulation of the upstream flow has been created by a small transverse pipe connected to a loudspeaker. The regime is subcritical and the upstream acoustic frequency is equal to 4.4 Hz, closed to the natural oscillation in the wake. The loudspeaker is suddenly switched off (with a very short response time, 0.2 s) and the wake releases to the subcritical regime with a τ-time (this transient time of hydrodynamic flow is longer, 2 s) (Fig.1).

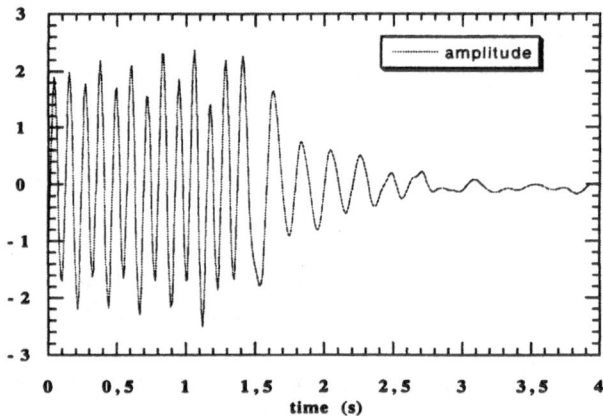

Figure 1.   Transient Phenomen: $Re = 262$ and $f_{exc} = 4.4$ Hz

The temporal growth rate $\sigma = 1/\tau$ of the instability varies linearly with the Reynolds number (Fig.2) as predicted by the Landau's theory and follows the relation:

$\sigma = (\nu/d^2).(0.9).(Recr - Re)$

The value of the critical Reynolds number based upon transient phenomena (278) is pretty closed to the value deduced from energy measurements.

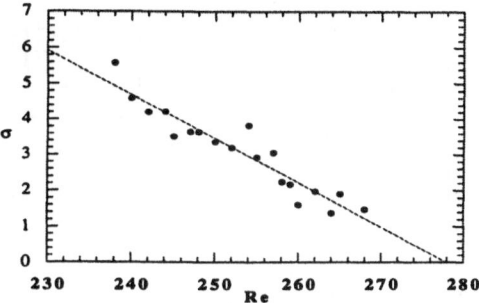

Figure 2. Variation of the temporal growth rate σ (1/sec) with the Reynolds number

In the subcritical regime, $Re = [250, 280]$, the response of the wake exhibits a resonance curve when varying the frequency of excitation with a constant amplitude. The selected frequency is independent of the probe position and closed to the natural frequency given by the Roshko-Reynolds curve of the natural oscillation (Fig.3).

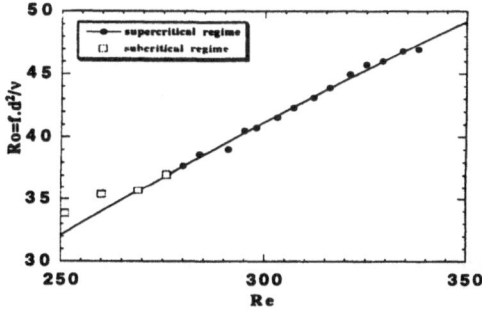

Figure 3. Roshko-Reynolds curve, 4 measurements in subcritical regime

The amplitude of the response of the wake, when forced by the upstream loudspeaker, varies along the streamwise direction x. The shape of the curve of energy as a function of x/d looks like that one obtained for a supercritical Reynolds number: the energy of oscillation increases in the near wake, from the zero value on the wall of the sphere, and decreases in the far wake. The reduced energy E/E_{max} has been plotted as function of the spatial coordinate for differents Reynolds number (Fig.4). These curves are quite similar to cylinder wake configuration [1].

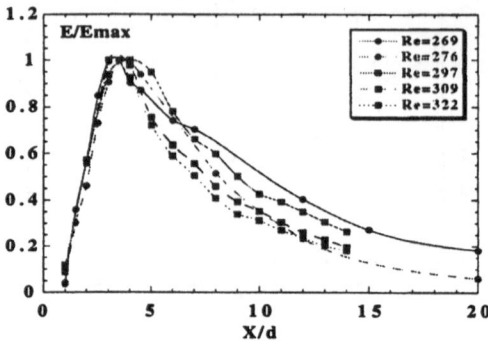

Figure 4. Dimensionless energy E/E_{max} function of x/d

We have also measured the location of the maximum amplitude along the streamwise direction and its variation with the Reynolds number. Subcritical results obtained with forced regimes are in good agreement with results obtained with natural oscillation (Fig.5). Near the threshold, this value is constant equal to 4d, very different of the cylinder configuration, where it diverges as $(Re - Re_c)^{-1/2}$.

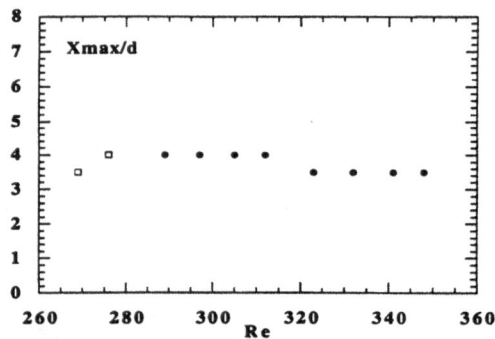

Figure 5. Location of the maximum amplitude along the x-axis

References

1. Goujon-Durand S., Jenffer P., Weisfreid J.E., (1994) Downstream evolution of the Bénard-von Karman instability *Phys. Rev. E.* **50**, 308-313
2. Landau L., Lifchitz E., (1971) *Physique theorique*, ed. Mir.
3. Leweke T.,Provansal M., (1995) The flow behind rings: bluff body wakes without end effects *JFM* **288**, 265-310
4. Provansal M., Ormières D., (1998) Bifurcation from steady to periodic flow in the wake of a sphere, *Comptes rendus de l'Académie des Sciences, Mécanique des fluides*
5. Sakamoto H. and Haniu H.,(1995) The formation mechanisme and shedding frequency of vortices from a sphere in uniform shear flow, *JFM* **287**, 151

FROM NONLINEAR WAVES TOWARDS TURBULENCE IN A LAMINAR BOUNDARY LAYER

S. BAKE[1], H.H. FERNHOLZ[1] AND Y.S. KACHANOV[2]

[1] *Hermann-Föttinger-Institut für Strömungsmechanik,*
Technische Universität Berlin, Strasse des 17. Juni 135,
10623 Berlin, Germany
[2] *Institut for Theoretical and Applied Mechanics,*
Russian Academy of Science,
630090 Novosibirsk, Russia

1. Introduction

The nonlinear stage of laminar-turbulent transition of a 2-D boundary-layer (normal profile close to the Blasius flow) has been investigated experimentally under controlled disturbance conditions. The time-harmonic instability waves were excited in the boundary-layer by means of a slit source developed in previous experiments [2]. The nonlinear development of this wave leads to a formation of the Klebanoff-regime (K-regime) of boundary layer transition.

The main object of the present investigation is, first, to improve the understanding of the mechanism responsible for the production of turbulence and, second, to delay the process of formation of the Λ-structures, hairpin vortices and ring-like vortices. Such a control of the transitional flow was achieved by using of a second disturbance source (of a similar design) positioned 100 mm downstream from the main one. A first approach consisted in influencing the transition development with a 2-D wave with the fundamental frequency.

2. Experimental Procedure

The experiment was conducted in the Laminar Wind Tunnel of the Hermann-Föttinger-Institute of Berlin Technical University at a free-stream velocity $U_\infty = 7.5$ m/s. The local Reynolds number at the position of the main source was $Re_{\delta 1} = 800$. In the present experiments the main source gen-

U. Frisch (ed.), Advances in Turbulence VII, 85-88.

erated a quasi 2-D harmonic instability wave ($f_1 = 62.5$ Hz, frequency parameter $F_1 = 115 \cdot 10^{-6}$) that had a spanwise modulation of its amplitude and phase (fig. 2, 3). The time-mean and fluctuation velocities were measured in the flow by means of hot-wire anemometry. The hot-wire signal was analysed in a way that the phase-locked ensemble-averaged time series, triggered by the reference signal, were measured and stored (fig. 1). This information was then used to reconstruct the transitional flow field at various stages of its development.

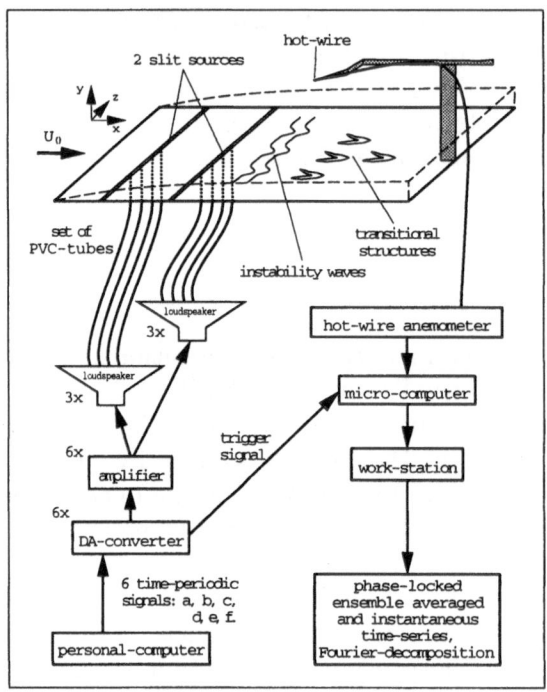

Figure 1.   Flow chart of experimental set-up

Close to the disturbance source the instability-wave front developed rather quickly into the Λ-structures, which evolved downstream in a way typical for the K-regime of boundary-layer breakdown (see e.g. [3]), including formation of the 3-D (also Λ-shaped) high-shear layer, hair-pin vortices, ring-like vortices, spikes, and subsequent growth of random (turbulent) perturbations with a continuous frequency spectrum. These phenomena were carefully documented in wall normal and parallel planes.

One of the results of the flow control is illustrated in figures 4, 5 and 6 where the normal-to-wall profiles of the total disturbance intensity and the amplitudes of the frequency Fourier-harmonics are shown at the spanwise

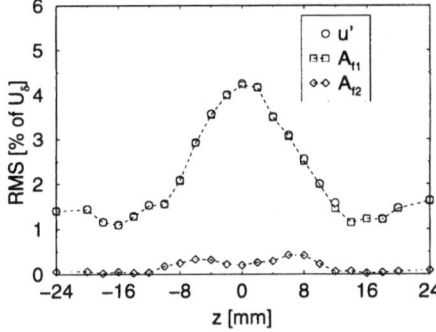

Figure 2. Initial spanwise disturbance profiles of the K-regime at $\Delta x = 90$ mm and $y = 1.36$ mm (u'_{max})

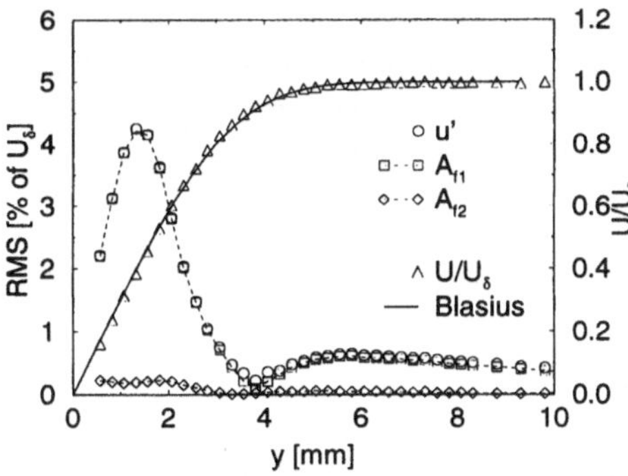

Figure 3. Initial normal disturbance and mean profiles of the K-regime at $\Delta x = 90$ mm and $z = 0$ mm

"peak" position ($z = 0$ mm) and at a distance from the main disturbance source $\Delta x = 190$ mm. It can be seen that the high disturbance amplitude of the K-regime (fig. 4) is reduced strongly (fig. 6) due to the interaction with the comparatively low amplitude 2-D control disturbance wave (fig. 5).

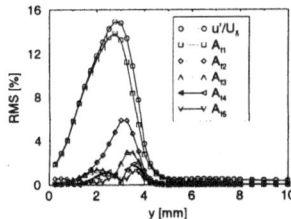 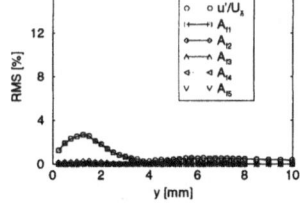 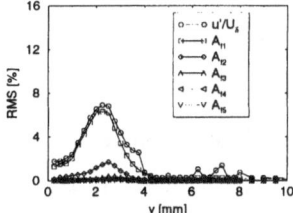

Figure 4. K-regime on the line of streamwise symmetry

Figure 5. 2-D control disturbance of the second source

Figure 6. K-regime plus 2-D control disturbance

Another illustration of the effectiveness of this method of flow control is shown in figures 7 to 10 in planes perpendicular to the mean flow direction at a relatively late stage of the transition process ($\Delta x = 270$ mm). The fundamental wave amplitude of the whole disturbance structure is reduced here by ≈ 45 % although the size of the structure is not significantly smaller. The undisturbed K-regime at this streamwise position (fig. 7) is at a well developed "3-spike-stage" indicating the formation of a row of ring-like vortices. After the 2-D control disturbance was introduced into the flow no spike occurred in the whole field at this position any longer.

The random perturbations $\hat{u}'^2$ shown in fig. 9 and fig. 10 represent the r.m.s deviation of the instantaneous disturbance from the periodic ensemble-averaged one. Looking at the two fields observed without and with control it is seen that the method of control used in the present study provides a very strong suppression of the process of turbulence generation in the boundary layer.

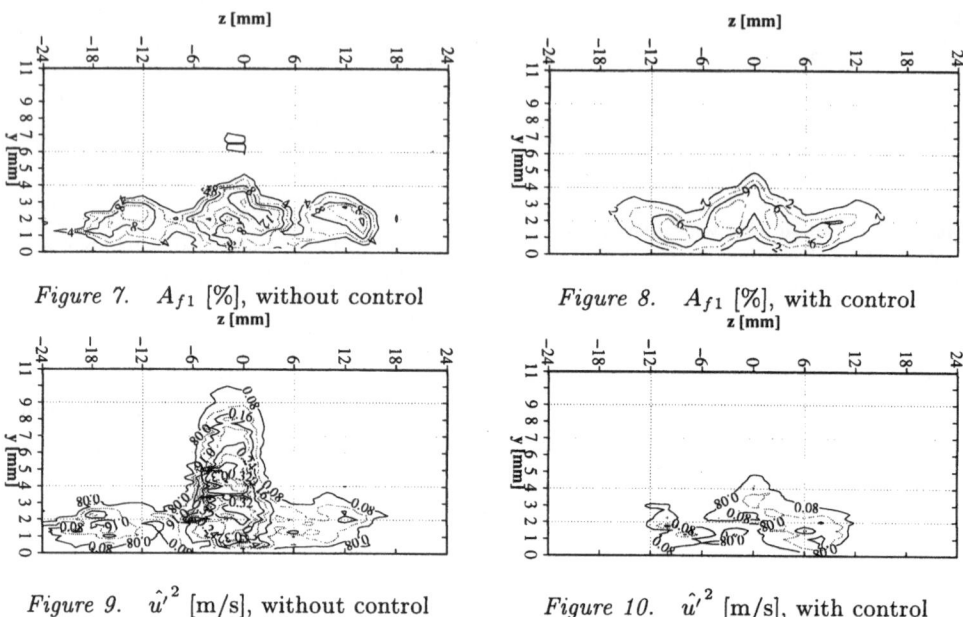

Figure 7. A_{f1} [%], without control   Figure 8. A_{f1} [%], with control

Figure 9. $\hat{u'}^2$ [m/s], without control   Figure 10. $\hat{u'}^2$ [m/s], with control

3. Conclusions

It was found that, as a result of the nonlinear interaction between the Λ-structure developing in the flow and the control disturbance generated by the second source, the transition process can be delayed significantly at an essentially nonlinear stage of the laminar flow breakdown. Further studies are planned to be directed to an improvement of the control signal, including multi-frequency excitation and 3D control disturbances.

Animations of the hot-wire signals in the transition process can be seen on our Internet web-page [1].

References

1. Bake, S., (1998) MPEG-movies of K-regime, *ftp://obiwan.pi.TU-Berlin.DE/pub/html/GSL-Crew.html*
2. Bake, S., Kachanov, Y. S., Fernholz, H. H. (1996) Subharmonic K-regime of boundary-layer breakdown, *In: Transitional Boundary Layers in Aeronautics*, Amsterdam: North-Holland, pp. 81–88.
3. Kachanov, Y.S. (1994) Physical mechanisms of laminar boundary-layer transition, *Ann. Rev. Fluid Mech.* **26**, pp. 411–482.

TURBULENCE IN PIPE FLOW OF HELIUM II

D.J. MELOTTE AND C.F. BARENGHI
Mathematics Department, University of Newcastle upon Tyne,
Newcastle upon Tyne, NE1 7RU, England

Our concern is the turbulent flow of Helium II - the quantum phase of liquid Helium. Turbulence in Helium II manifests itself as a disordered system of superfluid vortex lines, called the *vortex tangle*. The vortex tangle is an interesting state of disorder of matter near absolute zero. It is a form of turbulence which, in principle, is simpler than classical turbulence, because the eddies (the vortex lines) have all the same strength - the circulation around each vortex filament is quantised. But the study of the vortex tangle is also motivated by practical applications of cryogenics engineering: for example liquid Helium is used to cool superconducting magnets in high energy physics and infrared detectors in astrophysics, and the vortex tangle limits liquid Helium's heat transfer properties in a significant way.

Direct flow visualisation in Helium II is virtually impossible, but one can detect directly the presence of the vortex tangle by measuring the attenuation of a wave motion called second sound; another technique involves monitoring temperature differences. These measurements yield the vortex line density L_0, that is to say the length of vortex lines per cubic centimeter. From the measured L_0 and the value of the quantum of circulation, $\Gamma = 9.97 \ 10^{-4} \ cm^2/sec$, one can find the average turbulent superfluid vorticity $\omega_s = \Gamma L_0$. For a full account of the importance of vortices in Helium II see Donnelly, 1991 [1].

A great number of experiments have been performed in pipes and channels, on a flow configuration called *counterflow*, which clearly indicate that there exist two states of turbulence, denoted T1 and T2 in the literature and it is the nature of these two states that we are concerned with here. A summary of the experimental work done on counterflow can be found in Tough, 1982 [2].

The experimental setup which is used to study counterflow is shown schematically in Figure 1. One end of the pipe is closed and is provided with a resistor which dissipates a known heat flux W; the other end of the pipe is open to the helium bath. What happens at small W (Figure 1:a) can

U. Frisch (ed.), Advances in Turbulence VII, 89–92.
© 1998 *Kluwer Academic Publishers.*

be easily understood using Landau's two-fluid model. The model describes helium as the intimate mixture of two fluid components: the *superfluid* - which is related to the quantum ground state and flows without any friction and the *normal fluid* - which arises from thermally excited states (eg phonons) and carries the entropy and viscosity of the liquid. Using subscripts s and n to indicate the super and normal components respectively, we call ρ_s and ρ_n the densities, $\mathbf{v}_s$ and $\mathbf{v}_n$ the velocities, $\mathbf{j} = \rho_s \mathbf{v}_s + \rho_n \mathbf{v}_n$ the mass flux, $\rho = \rho_s + \rho_n$ helium's total density, T the temperature and S the entropy. According to Landau's model, the heat flux $W = \rho T S v_n$ is carried by the normal fluid away from the resistor, towards which some superfluid must flow in order to conserve mass. In this way a relative velocity (thermal counterflow) $v_{ns} = |\mathbf{v}_n - \mathbf{v}_s|$ is set up between the two fluids, which is proportional to the driving heat flux, $v_{ns} = W/\rho_s ST$.

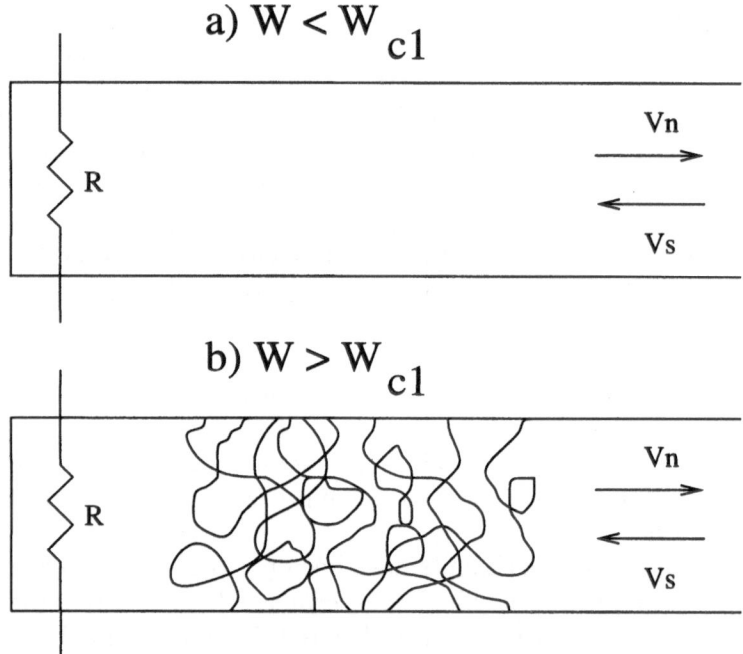

Figure 1.   Counterflow Turbulence

If the heat flux (and hence the flow speed) exceeds a critical value, W_{c1} then the superfluid forms a vortex tangle (see Figure 1:b), this is known as the T1 state. The superfluid vortex tangle affects the normal fluid via a mechanism known as *mutual friction*. Experiments show that if the heat flux is increased further to exceed a second critical value W_{c2}, the state T1 is replaced by the state T2 which is characterized by a much higher vortex lines density L_0. Until now it was not known what T1 and T2 really are and what W_{c2} represents, and progress was frustrated by the lack of direct flow visualisation in the low temperature environment.

In our model we assume that the normal fluid is subject to a frictional forcing caused by an homogeneous, isotropic superfluid vortex tangle.

$$\mathbf{F}_{ns} = \left(\frac{B\rho_n\rho_s}{2\rho}\right)\frac{2}{3}\omega_s(\mathbf{v}_n - \mathbf{v}_s), \tag{1}$$

where B is the mutual friction coefficient [3] and $\omega_s = \Gamma L_0$ is the superfluid vorticity. We study the linear stability of the normal flow in a cylindrical pipe at a given intensity of vortex tangle. The normal fluid is described by

$$\rho_n\frac{\partial \mathbf{v}_n}{\partial t} + \rho_n\mathbf{v}_n \cdot \nabla\mathbf{v}_n = -\frac{\rho_n}{\rho}\nabla p - \rho_s S\nabla T + \mu\nabla^2\mathbf{v}_n - \mathbf{F}_{ns} \tag{2}$$

together with the incompressibility condition

$$\nabla \cdot \mathbf{v}_n = 0, \tag{3}$$

where μ is helium's viscosity. The velocity and vortex line density at which the basic velocity profile $\mathbf{v}_n$ becomes unstable is determined by the analysis of both axisymmetric and nonaxisymmetric perturbations using a numerical method based on Chebychev spectral collocation. The boundary conditions and the regularity conditions on the axis are built into the spectral functions.

We find that the normal fluid becomes linearly unstable above a certain vortex line density. This compares very well to existing experimental data at various values of temperature T and channel's size d. In Figure 2 we present our results in terms of the geometrical quantity $L_0^{1/2}d$, which is the inverse of the average distance between the vortices in the tangle, in units of the channel size. The theoretical prediction (the solid line) is compared to experimental data at different temperatures. Our relatively simple model agrees quantitatively with the experiments.

In conclusion, for the first time we have theoretical evidence about the nature of the T1 and T2 states. In the state T1 the normal fluid is still in laminar motion, without being much affected by the vortex tangle; when the critical heat flux W_{c2} is exceeded the normal fluid becomes unstable. In the state T2, therefore, *both* fluid components are turbulent, which explains the extra dissipation and thus the higher vortex line density observed.

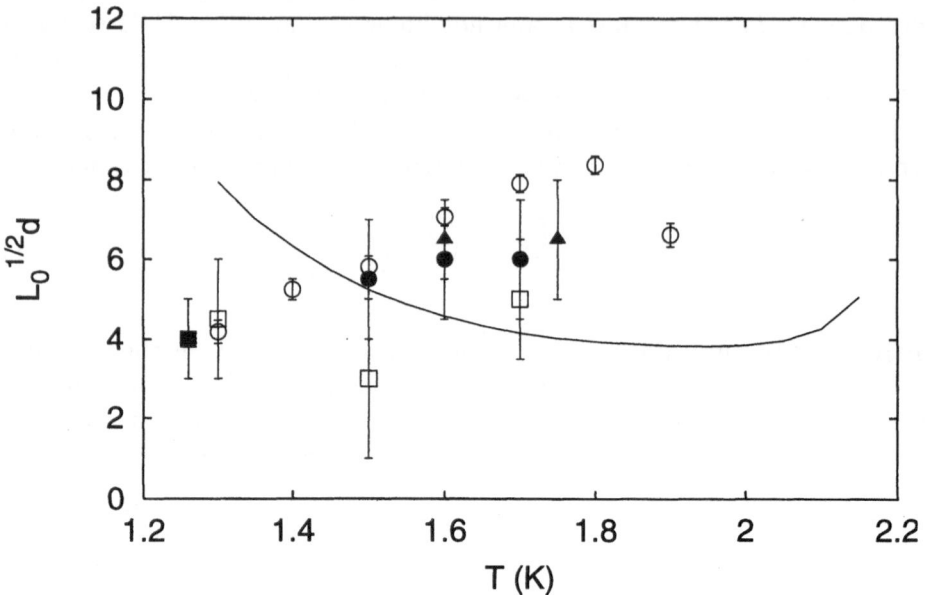

Figure 2. Transition between T-1 and T-2 states, comparison between theory and experiments.
Solid line: theoretical prediction.
White circles: Ladner et al. [4], $D = 0.0129$ cm, Re_{c2} ranging from 35 to 131.
Black circles: Martin & Tough [5], $D = 0.1$ cm, Re_{c2} from 122 to 186.
White squares: de Goeje & van Beelen [6], $D = 0.139$ cm, Re_{c2} from 26 to 272.
Black squares: Mares & van Beelen [7], $D = 0.0216$ cm, Re_{c2} from 33 to 61.
White triangle: Marees & van Beelen [7], $D = 0.0133$ cm, $Re_{c2} = 45$.
Black triangles: Griswold et al. [8], $D = 0.0132$ cm, Re_{c2} from 111 to 128.

References

1. R.J. Donnelly. *Quantized Vortices in Helium II*. Cambribge University Press, 1991.
2. J.T. Tough. Superfluid turbulence. In D.F.Brewer, editor, *Progress in Low Temperature Physics*, volume 8, chapter 3. North-Holland, 1982.
3. C.F. Barenghi, R.J. Donnelly, and W.F. Vinen. Friction on quantized vortices in helium II. a review. *Journal of Low Temperature Physics*, 52(3/4), 1983.
4. D.R. Ladner, R.K. Childers, and J.T. Tough. Helium II thermal counterflow at large heat currents. *Physical Review B*, 13(7):2918–2923, 1976.
5. K.P. Martin and J.T. Tough. Evolution of superfluid turbulence in thermal counterflow. *Physical Review B*, 27(5), 1983.
6. M.P. de Goeje and H. van Beelen. Turbulent counterflow of Helium II investigated with a sixfold Helmholtz resonator for second sound. *Physica B*, 133:109–128, 1985.
7. G. Marres and H. van Beelen. On the disappearance of the flow impedance of Helium II when mass and entropy flow in opposite directions. *Physica B*, pages 21–36, 1985.
8. D. Griswold, C.P. Lorenson, and J.T. Tough. Intrinsic fluctuations of the vortex-line density. *Physical Review B*, 35(7):3149–3161, 1987.

TRANSITION TO TURBULENCE OF FLUID FLOWING DOWN AN INCLINED PLANE: MODELING AND SIMULATION

C. RUYER-QUIL[1] AND P. MANNEVILLE

Laboratoire d'hydrodynamique, CNRS UMR 156,
École polytechnique F-91128 Palaiseau, France

Thin films flowing down an inclined plane exhibit a rich phenomenology (Chang, 1994). In particular, experiments (Liu *et al.*, 1995) have evidenced a well characterized sequence of nonlinear secondary instabilities (Eckhaus, sub-harmonic) to disordered wave dynamics and transitions from parallel to stream-wise and span-wise patterns which are similar to the Herbert- and Klebanov-type transitions encountered in transitional boundary layers. Contrasting with e.g. plane Couette flow, the system departs super-critically from the flat-film solution for which viscous stress and gravity equilibrate. Wavy films then exhibit a long-wave small-amplitude chaotic motion controlled by surface tension and well described by a Kuramoto-Sivashinsky equation (phase turbulence). Farther from criticality, the wave dynamics is dominated by large-amplitude, strongly-nonlinear localized structures such as solitary pulses.

It turns out that, as long as the flow rate is not too large, the interface remains smooth at the scale of the film thickness which suggests to approximate the equations by means of long-wavelength expansions in terms of a formal parameter ϵ expressing the smallness of space and time derivatives. Thus, keeping the leading contribution of surface tension as well as terms up to first order in ϵ, a system of equations resembling Prandtl equations can be obtained and consequently called boundary-layer equations (BL). Demekhin *et al.* (1983) showed that, in the case of BL equations, a similarity transformation can be found that shrinks the initial set of parameters, Reynolds number R and Kapitza number Γ, to a reduced Reynolds number δ. Assuming a parabolic velocity profile, Shkadov (1967) further applied the integral momentum method of von Kármán to the BL equations and was led to a simplified model in the form of two coupled evolution equations for the film thickness h and the local flow rate $q = \int_0^h u\, dy$ where u stands for the stream-wise velocity. Unfortunately this model cannot accurately pre-

U. Frisch (ed.), Advances in Turbulence VII, 93–96.
© *1998 Kluwer Academic Publishers.*

dict the onset of instability and does not take into account the dispersion introduced by viscous dissipation.

Yet, sticking to a strict gradient expansion of the primitive equations, one can eliminate all flow variables, supposed to be slowly varying and slaved to the local film thickness h. This led Benney (1966) to a systematic expression of the dynamics of the film in terms of h and its successive space-time derivatives. Stopping this approach at first order yields a single evolution equation for h that is accurate at the instability onset but can lead to non-physical catastrophic evolution farther from the critical point (Pumir et al., 1983).

Both Shkadov's model and Benney's equation fail to correctly predict the film behavior in the parameter range corresponding to current experiments. But, through a combination of the two approaches, we obtained a two-dimensional model enabling us to study the large amplitude waves free of span-wise modulations observed by Liu and Gollub (1994). Our derivation of the long-wave expansion of the Navier-Stokes equations keeps all terms up to order ϵ^2, which assures us that stream-wise viscous dispersion is correctly handled at least in a neighborhood of the critical point. This also accounts for corrections to Demekhin's similarity rule, which is valid at order ϵ only. Using velocity profiles appearing in Benney's expansion as a set of trial functions, we developed a method of weighted residuals combining an averaging method with collocation at boundaries that yields a model in terms of three coupled evolution equations for h, q and a new variable τ that measures the departure of the wall shear stress $\tau_w = (\partial u/\partial y)_{y=0}$ from the shear predicted by a parabolic velocity profile,

$$\partial_t h = -\partial_x q \,, \tag{1}$$

$$\partial_t q = h - \frac{3q}{h^2} - \tau + \partial_x \left[\frac{2}{35} h\tau q - \frac{6q^2}{5h} - \frac{6q\partial_x h}{h} + \frac{9\partial_x q}{2} \right] \tag{2}$$
$$- Bh\partial_x h + \Gamma h\partial_{x^3} h \,,$$

$$\partial_t \tau = \frac{7}{h} - \frac{21q}{h^4} - \frac{42\tau}{h^2} - \frac{18q^2\partial_x h}{5h^4} + \frac{6q\partial_x q}{5h^3} + \frac{2q\tau\partial_x h}{5h^2} + \frac{\tau\partial_x q}{15h}$$
$$- \frac{3q\partial_x \tau}{5h} + \frac{84q(\partial_x h)^2}{h^4} - \frac{63\partial_x q\partial_x h}{h^3} - \frac{7B\partial_x h}{h} + \frac{7\Gamma\partial_{x^3} h}{h} \,, \tag{3}$$

where $B = \cot\beta$ and β is the inclination angle. The variables have been made dimensionless using length and time scales based on the kinematic viscosity and the stream-wise gravitational acceleration. In our formulation, R equals the mean flow rate q_N (Ruyer-Quil and Manneville, 1998). Because of our choice of trial functions, Benney's equation at order ϵ^2 can be recovered by an asymptotic expansion of (1–3).

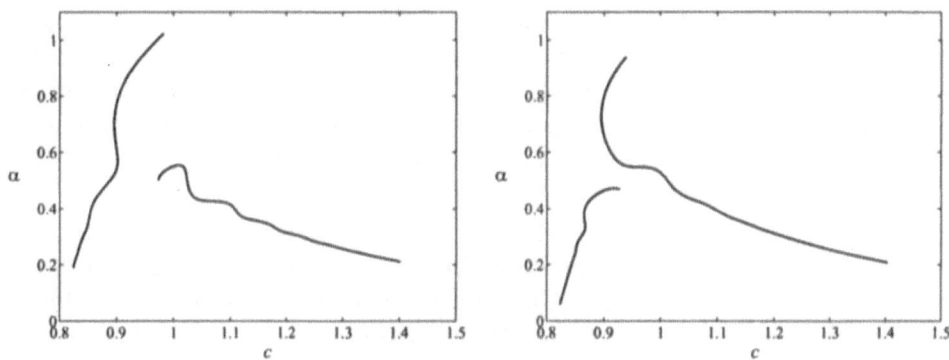

Figure 1. Celerity c versus wavenumber α for the stationary-wave families γ_1 and γ_2 bifurcating from neutral wavenumber α_c and $\alpha_c/2$. Left: $R = 7.59$, $\Gamma = 4374$. Right: $R = 3.40$, $\Gamma = 228.8$. Both sets of parameters correspond to the same reduced Reynolds number $\delta = 0.062$. Compare with fig. 15 and 17 of Salamon *et al.* (1994).

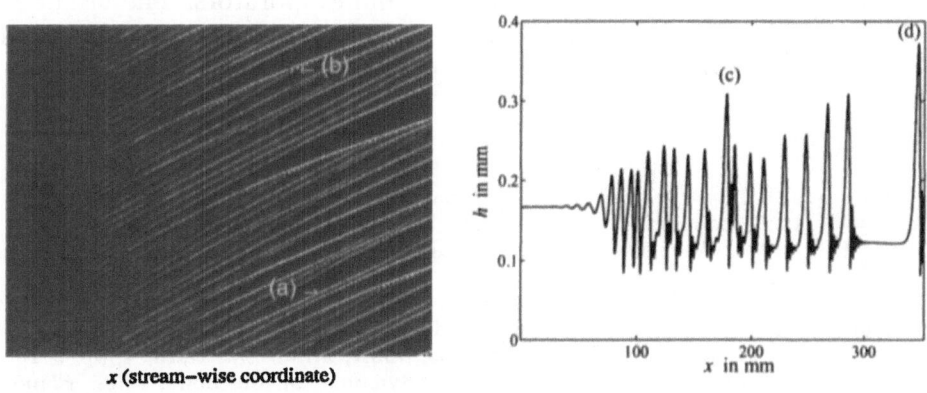

Figure 2. Simulation of noise-driven dynamics of a vertically falling water film ($R = 15.25$, $\Gamma = 2850$, Stainthorp and Allen, 1965). Left: spatio-temporal diagram showing the appearance of large amplitude solitary pulses from the coalescence of smaller saturated waves. Dark (light gray) regions correspond to thinner (thicker) film. Both repulsion (a) and coalescence (b) of fronts can be observed. Right: snapshot of film thickness at the end of the simulation. The merging of two waves is observed at (c). A large amplitude solitary wave is escaping the domain at (d).

Results of a preliminary study are in good agreement with both experiments (Liu and Gollub, 1994) and direct numerical simulations (Ramaswamy *et al.*, 1996) and show no non-physical finite-time singularity in the parameter range corresponding to current experiments. Moreover, our model accurately predicts the behavior of the families of stationary waves bifurcating from the neutral stability curve. In particular, the cusp catastrophe observed by Salamon *et al.* is recovered, cf. fig. 1. Ongoing numerical simulation of the noise-driven dynamics of (1–3) shows the development of spatio-temporal chaos dominated by two-dimensional solitary waves interaction as first noted by Chang *et al.* (1995), cf. fig. 2.

The model has been extended to include the span-wise coordinate z, in view of studying the secondary instabilities observed by Liu *et al.* (1995). Other modeling strategies, e.g. Galerkin method, have also been developed recently yielding a system of equations involving lower order nonlinearities than in (1–3) (Ruyer-Quil and Manneville, 1998). Just like the Swift–Hohenberg model for pattern formation in dissipative structures (Manneville, 1990), these models should offer good testing grounds for the study of the transition towards spatio-temporal chaos and more developed turbulence in free-surface open flows.

Thin films are encountered in a wide variety of industrial processes such as coating, galvanizing, in chemical engineering devices, e.g., wetted absorbers, chemical reactors, condensers and evaporators. The use of simplified models such as (1–3) providing good quantitative agreement with experiments could thus give rise to important applications.

References

[1] Work supported by a grant from the Délégation Générale à l'Armement.

Benney, J. (1966) Long waves on liquid film, *J. Math. Phys.*, **45**, pp. 150–155.

Chang, H.C. (1994) Wave evolution on a falling film, *Ann. Rev. Fluid Mech.*, **26**, pp. 103–136.

Chang, H.C., Demekhin, E. and Kalaidin, E. (1995) *J. Fluid Mech.*, **294**, pp. 123–154.

Demekhin, E.A., Demekhin, I.A. and Shkadov, V.Y. (1983) Solitons in viscous films flowing down a vertical wall, *Izv. Ak. Nauk SSSR, Mekh. Zhi. Gaza*, **4**, pp. 9–16.

Liu, J. and Gollub, J.P. (1994) Solitary wave dynamics of film flows, *Phys. Fluids*, **6**, pp. 1702–1712.

Liu, J., Schneider, B. and Gollub, J.P. (1995) Three-dimensional instabilities of film flows, *Phys. Fluids*, **7**, pp. 55–67.

Manneville, P. (1990) *Dissipative structures and weak turbulence*, Academic Press.

Pumir, A., Manneville, P. and Pomeau, Y. (1983) On solitary waves running down an inclined plane, *J. Fluid Mech.*, **135**, pp. 27–50.

Ramaswamy, B., Chippada, S. and Joo, S.W. (1996) A full-scale numerical study of interfacial instabilities in thin-film flows, *J. Fluid Mech.*, **325**, pp. 163–194.

Ruyer-Quil, C. and Manneville P. (1998) (a) Modeling film flows down inclined planes, *in preparation*; (b) A Galerkin approach to film flow dynamics, *in preparation*.

Salamon, T.R., Armstrong, R.C. and Brown, R.A. (1994) Traveling waves on vertical films, *Phys. Fluids*, **6**, pp. 2202–2020.

Shkadov, V.Y. (1967) Wave flow regimes of a thin layer of viscous fluid subject to gravity, *Izv. Ak. Nauk SSSR, Mekh. Zhi. Gaza*, **2**, pp. 43–51.

Stainthorp, F.P. and Allen, J.M. (1965) The development of ripples on the surface of a liquid film flowing inside a vertical tube, *Trans. Inst. Chem. Eng.*, **43**, pp. 85–91.

DNS OF THE GENERATION OF SECONDARY Λ-VORTICES IN A TRANSITIONAL BOUNDARY LAYER

D. MEYER, U. RIST AND S. WAGNER

Institut für Aero- und Gasdynamik, Universität Stuttgart
Pfaffenwaldring 21, D-70550 Stuttgart, Germany

1. Introduction

Especially for large-amplitude initial disturbances, so-called Λ-*vortices* develop during the non-linear late stages of laminar-turbulent transition. Once formed they may persist as so-called *hairpin vortices* well into the ensuing turbulent boundary layer. This makes their study important for understanding transition and turbulence in wall-bounded shear flows.

In Direct Numerical Simulations (DNS) as well as in controlled transition experiments, especially in K-type transition, Λ-vortices have been regularly observed by all authors studying this phenomenon. In such experiments or DNS, Λ-vortices may be generated periodically in time and in spanwise direction by the non-linear interaction of a large-amplitude Tollmien–Schlichting (TS-) wave with a spatially-periodic steady modulation in spanwise direction with a certain wavelength $\lambda_z = \lambda_{z0}$. In the present paper we shall demonstrate the influence of increasing λ_z with the idea of moving spanwise adjacent Λ-vortices further apart in order to identify their possible self interaction in the original case where $\lambda_z = \lambda_{z0}$.

Our DNS are a direct continuation of the work presented in [1]. A 2-D TS-wave of $u'_{max} \approx 4\%$ and a steady spanwise modulation are introduced within a narrow disturbance strip by suction and blowing at the wall. The flow-parameters are not changed compared to our earlier work. However, for the present simulations the steady 3-D disturbance is modified in such a way as to resemble the spanwise periodic disturbance in the vicinity of the so-called peak plane as close as possible and to move neighboring peaks further apart. Four different cases are studied: a case with spanwise periodic disturbance for reference and three cases with a local disturbance, as shown in fig. 1, with a respective peak plane distance of $2\lambda_{z0}$, $3\lambda_{z0}$, and $5\lambda_{z0}$.

U. Frisch (ed.), Advances in Turbulence VII, 97–100.
© 1998 *Kluwer Academic Publishers.*

2. Results

In fig. 2 the spanwise distribution of the disturbances in the boundary layer is shown. For the periodic case the amplitudes are larger than for the locally disturbed cases. For the case with $2\lambda_{z0}$ we can observe an overlapping and a mutual amplification of adjacent disturbances halfway between the peak planes. Especially for the cases with large distance between peak planes ($3\lambda_{z0}$ and $5\lambda_{z0}$) the boundary layer can be considered as locally disturbed, because the disturbance amplitudes in between two peaks go back to the level of the case with a pure 2-D TS-wave. When we look at fig. 3 we can notice a deferred development of the high-shear layers in the case with $5\lambda_{z0}$ compared to the periodic case. This means that the amplification rates are somewhat smaller in the latter case.

An explanation for this behavior can be found looking at the vortex structures in the boundary layer. Fig. 4 shows a top-view visualization of vortices within the numerical flow field using a technique described in [2]. The elongated parallel structures in fig. 4a are remains of the 2-D large-amplitude TS-wave for the case with $\lambda_z = 5\lambda_{z0}$. The primary Λ-vortices are centered around the peak plane at $z = 0$ while newly found secondary Λ's appear at the edges of the TS-part of the disturbance field. They seem to be generated by an inductive interaction of a Λ-vortex with the adjacent TS-wave. This assumption is supported by the fact that the generated secondary Λ's induce even more Λ's further downstream as can be seen in fig. 4b. In this way a cascade of Λ-vortices spreads out downstream until the TS-wave is "used up". In fig. 4c an overlapping and mutual amplification of the secondary structures can be observed, thus forming a new even stronger Λ-structure which develops downstream in much the same way as the Λ's in the original peak plane. We conclude that Λ-vortices are local, independent structures of the flow field, which can overlap and amplify each other. They have a strong inductive effect on their vicinity which leads to higher amplification rates and a subsequent earlier breakdown for the periodic case in which the Λ's are located very close to each other.

3. Conclusions

The results show that transition is somewhat delayed as soon as the spanwise distance of the vortices is increased. However, the qualitative nature of the transition process, i.e., the formation of Λ-vortices, high-shear layers, and their subsequent breakdown remains unchanged. The most important finding of the present study is, however, the observation of secondary Λ-vortices for $\lambda_z > \lambda_{z0}$ which are obviously formed by induction on each spanwise side of the Λ-vortices. It turned out that these secondary vortices are present in every simulation with $\lambda_z > \lambda_{z0}$ and that they can overlap forming a stronger secondary vortex as seen in the case with $\lambda_z = 2\lambda_{z0}$. Thus, we expect the newly found vortices to play a role in any flow involving Λ-vortices.

References

1. Rist, U. and Fasel, H. (1995): Direct numerical simulation of controlled transition in a flat-plate boundary layer, *J. Fluid Mech.* **298**, pp. 211–248
2. Jeong, J. and Hussain, F. (1995): On the identification of a vortex, *J. Fluid Mech.* **285**, pp. 69–94

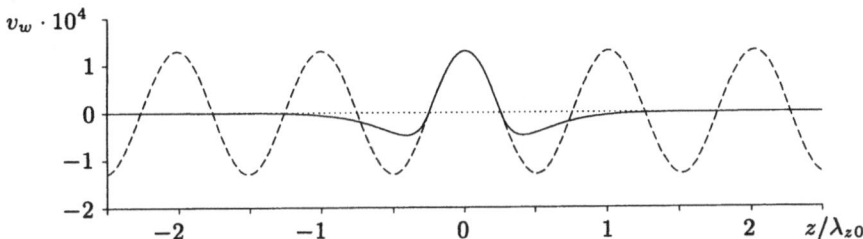

Figure 1. Comparison of the steady 3-D part of the wall-normal velocity amplitude v_w vs. spanwise coordinate of the periodic case (dashed line) with the locally disturbed cases (solid line).

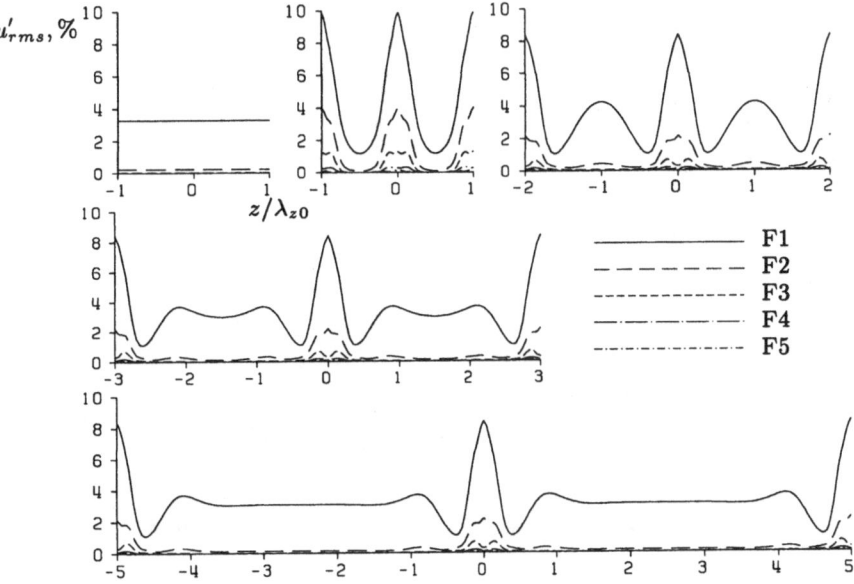

Figure 2. Maximum disturbance amplitudes of streamwise velocity component u_{rms} in % at downstream position x=400mm vs. normalized spanwise coordinate z/λ_{z0}. Plotted cases from left to right: pure 2-D TS-wave, spanwise periodic disturbance, local disturbance with $2\lambda_{z0}$, $3\lambda_{z0}$, and $5\lambda_{z0}$, respectively.

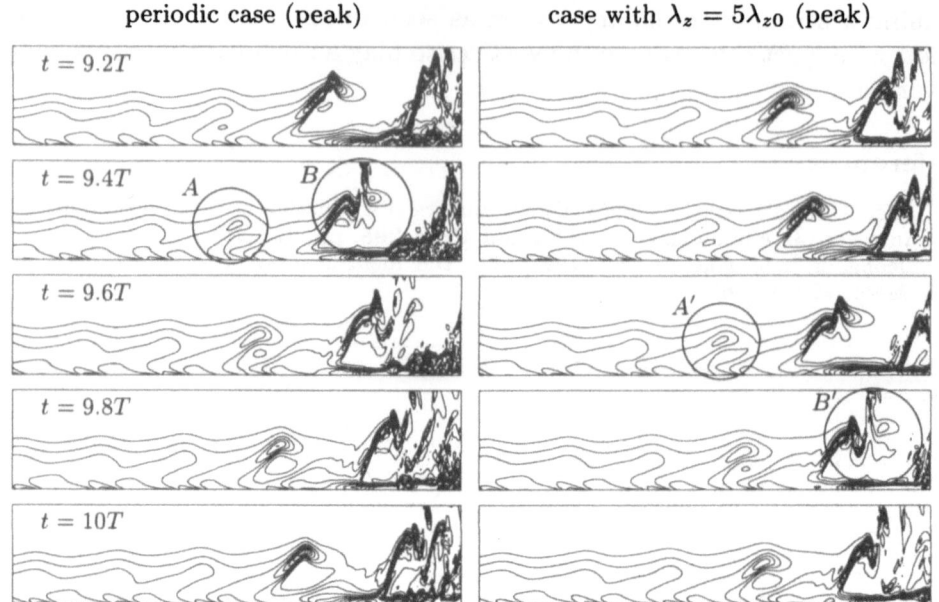

Figure 3. Comparison of isovalues of vorticity component ω_z at peak position. Downstream coordinate x between $x=283$mm and $x=475$mm; normal to wall coordinate y between $y=0$mm and $y=4.75$mm. Similar structures are marked with circles in both cases.

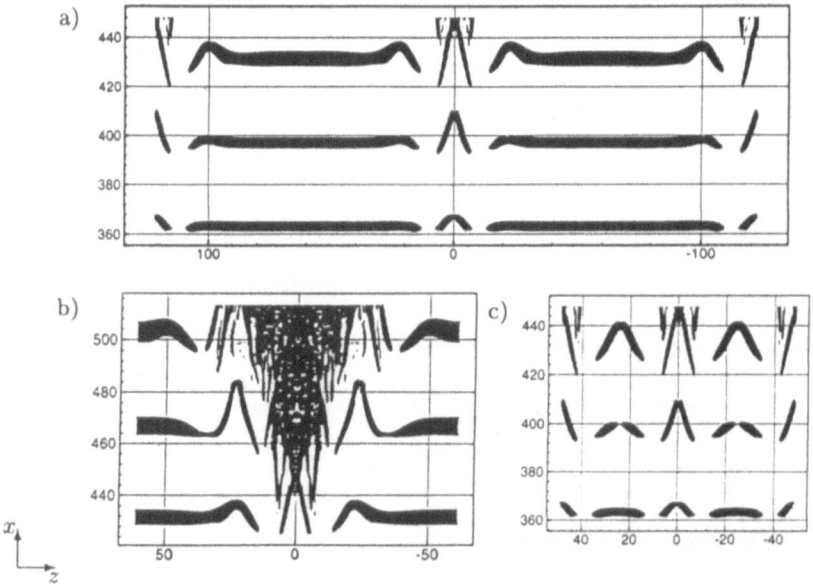

Figure 4. Top-view vortex visualization: a) case with $\lambda_z = 5\lambda_{z0}$, b) the same as in a) but further downstream, c) case with $\lambda_z = 2\lambda_{z0}$. All positions are given in mm.

ENERGY AMPLIFICATION OF STEADY DISTURBANCES IN GROWING BOUNDARY LAYERS

P. SCHMID
Department of Applied Mathematics, Box 352420
University of Washington, Seattle, WA 98105-2420, U.S.A.

The importance and necessity of linear growth mechanisms in transition to turbulence at subcritical Reynolds numbers has recently been recognized [1]. In the absence of exponentially growing linear modes, transiently growing disturbances can provide the amplification of disturbance energy that ultimately may result in turbulent fluid motion. This transient growth is due to the non-orthogonal structure of the linear eigenfunction which, in turn, is a consequence of the non-selfadjoint nature of the linearized Navier-Stokes equations. Many flows have been analyzed with regards to their transient amplification potential. Most of the analyses have concentrated on the temporal stability problem, and all of them have assumed a parallel mean flow which substantially simplifies the computations (see [2] for a review).

In this work, we will extend the established theory of transient energy growth to investigate the amplification of energy *in space* for a *non-parallel*, i.e. growing boundary layer.

This work has been motivated by the recent discovery of algebraically growing steady modes for a flat-plate boundary layer by Luchini [3] who extended the earlier work of Stewartson [4] and Libby & Fox [5] to include three-dimensional perturbations. The formulation is based on the boundary layer approximation of the Navier-Stokes equations and supports three-dimensional modal solutions of the form

$$
\begin{aligned}
u_j &= i\beta x^{-s_j+1} g_j(\eta) \\
v_j &= i\beta x^{-s_j+1/2} \left[\frac{\eta}{2} g_j(\eta) + (s_j - \frac{3}{2}) f_j(\eta) \right] \qquad \eta = \frac{y}{\sqrt{x}} \\
w_j &= x^{-s_j} h_j(\eta)
\end{aligned}
$$

where s_j are the eigenvalues and f_j, g_j, h_j are the eigenfunctions of the associated stability problem, and β denotes the spanwise wavenumber of

101

U. Frisch (ed.), Advances in Turbulence VII, 101–104.
© *1998 Kluwer Academic Publishers.*

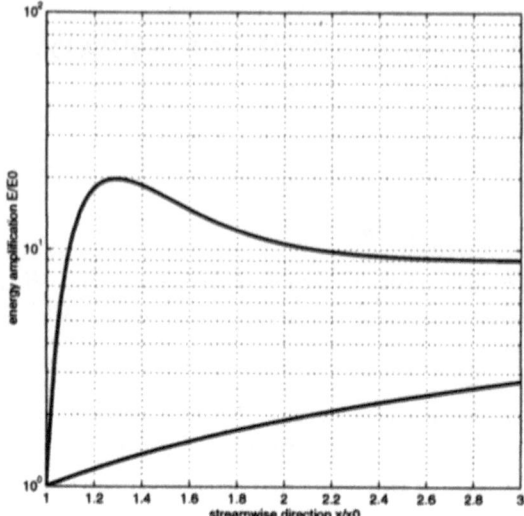

Figure 1.   Spatial amplification of disturbance energy. Upper curve: optimal superposition of modal solutions, lower curve: least stable modal solution.

the perturbation. All but one eigenvalue are greater than one which results
in algebraic growth of the streamwise velocity component as we follow the
disturbance in the streamwise direction once the associated eigenfunction
is excited.

Given these modal solutions, we will concentrate our efforts towards
optimizing the total disturbance energy gain between two streamwise locations, say x_0 and x with $x > x_0$, by superimposing a number of modal
solutions. Using a projection of the flow field onto the space spanned by the
above eigenvectors the disturbance energy density is given as the quadratic
form

$$\mathcal{E} = \Phi^H E(x)\Phi$$

where the vector Φ contains the weights of the eigenvectors and $E(x)$ is
related to the disturbance energy of the individual modal components. We
then formulate the optimization problem using a variational notation [6]
of the form

$$\mathcal{L} = \Phi^H E(x)\Phi + \lambda(\Phi^H E(x_0)\Phi - 1)$$

where the Lagrange multiplier λ enforces unit energy of the initial condition.
The resulting Euler-Lagrange equations lead to a generalized eigenvalue
problem given as

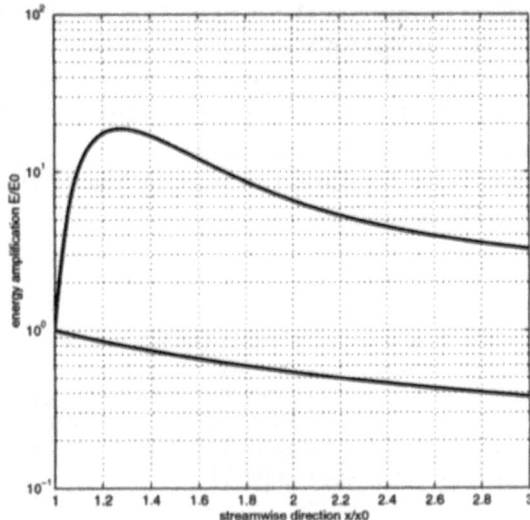

Figure 2. Spatial amplification of disturbance energy with least stable modal solution excluded in the optimization procedure. Upper curve: optimal superposition of modal solutions, lower curve: least stable modal solution.

$$E(x)\Phi = \lambda E(x_0)\Phi.$$

The largest eigenvalue $\lambda_{\max}$ represents the maximum energy amplification between location x_0 and x; the associated eigenvector $\Phi_{\max}$ provides the coefficients for the optimal linear combination of modes.

As can be seen from Figure 1, a substantial gain in energy can be achieved before the asymptotic behavior (given by the least stable - growing - mode) dominates the growth in energy. For Figure 2, similar computations were performed with the least stable (growing) mode excluded from the linear combination of modal solutions. Even in this case we observe a significant amplification of energy before the asymptotic behavior prevails. For comparison, the energy gain/loss obtained by only exciting the least stable eigenmode is added to both figures. Additional energy gain due to a superposition of many modes is clearly demonstrated in both cases.

Steady three-dimensional disturbances have been observed in experiments of unforced transition in a flat-plate boundary layer, and it is conjectured that they provide the finite-amplitude primary state of a secondary-instability based transition scenario. With the transient energy growth potential of steady boundary layer modes, as shown in the above figures, the spatially evolving perturbations investigated in this paper could play an important role in the natural transition of non-parallel boundary layer flow.

A more detailed parameter study exploring the dependence of the energy amplification on the initial location x_0 is currently underway. Furthermore, a connection between the above steady perturbations and unsteady perturbations which can be derived using an adjoint formulation as outlined in [7]) will be investigated in a future effort.

This work has been supported in part by the National Science Foundation under grant DMS-9406636.

References

1. Henningson, D.S. & Reddy, S.C. (1994) On the role of linear mechanisms in transition to turbulence, *Phys. Fluids* **6**, 1396.
2. Henningson, D.S. (1995) Bypass transition and linear growth mechanisms, In: *Advances in Turbulence V*, ed.: R. Benzi, Kluwer Academic Publishers, Dordrecht.
3. Luchini, P. (1997) Reynolds-number-independent instability of the boundary layer over a flat plate, *J. Fluid Mech.* **327**, 101.
4. Stewartson, K. (1957) On asymptotic expansion in the theory of boundary layer, *J. Math. Phys.* **36**, 137.
5. Libby, P.A. & Fox, H. (1964) Some perturbation solutions in laminar boundary-layer theory, *J. Fluid Mech.* **17**, 433.
6. Butler, K.M. & Farrell, B.F. (1992) Three-dimensional optimal perturbations in viscous shear flow, *Phys. Fluids A* **4** (8), 1637.
7. Andersson, P., Berggren, M. & Henningson, D.S (1997) Optimal disturbances in boundary layers, AFOSR Workshop on Optimal Design and Control, Sept. 1997, Arlington, Virginia.

NONLINEAR DYNAMICS OF LOW REYNOLDS NUMBER ROUND JETS: PERIODIC ATTRACTORS AND TRANSITION TO CHAOS

IONUT DANAILA[1,3], JAN DUŠEK[2] AND FABIEN ANSELMET[3]
[1] I.N.E.R.I.S., 60550 Verneuil en Halatte, France
[2] I.M.F., 2, rue Boussingault, 67000 Strasbourg, France
[3] I.R.P.H.E., 12, Av. Général Leclerc, 13003 Marseille, France

Direct numerical simulations have been shown [1, 2] to provide detailed information on the dynamics and coherent structures of the near field of a spatially developing axisymmetric jet. In [1] we demonstrated the shift from helical to axisymmetric structures with increasing diametral Reynolds number in the range [200; 500]. At the upper bound of this range, the *varicose* $m = 0$ mode is the most amplified (m is the azimuthal wave–number). The development of the unsteady flow is accompanied by the well known phenomena: 2D Kelvin–Helmholtz instability, roll–up and pairing, stream–wise filaments and side–jets. The onset of the asymptotic chaotic state is preceded by vortex rings reconnection and breakdown of the large structures due to strong stream–wise filaments. A similar transition process was observed in temporal simulations by Melander *et al.* [3].

In this paper, we investigate the mechanisms leading to instationarity and transition to chaos for Reynolds numbers close to the lower bound of the mentioned range. The numerical implementation was the same as that described in [1]. The 3-D Navier-Stokes (NS) solver Nekton based on the spectral element space discretization has been used to solve the incompressible NS equations in a cylindrical domain of stream–wise length equal to 15 nozzle diameters D and diameter roughly equal to 10 D. Numerical shear in a very small domain at the nozzle is responsible for the spontaneous onset of the instationarity.

For these low Reynolds numbers, the amplification of the helical mode is responsible for the jet symmetry breaking at the primary instability. Due to the axisymmetry of the base flow there exist two linearly independent, equally amplified, unstable modes, identified as the counter-rotating helical modes $m = \pm 1$. Their mutual interaction leads, close to the instationarity threshold, to a succession of 3 different regimes:

U. Frisch (ed.), Advances in Turbulence VII, 105–108.

Regime I, appearing for $0 < \epsilon = (Re - Re_{cr})/Re_{cr} < 1.2\%$, is characterized by a limit cycle dynamics with a single helical mode present in the flow; *Regime II* $(1.2 < \epsilon < 3\%)$ having a limit torus dynamics generated by the presence of both $m = \pm 1$ modes with unequal amplitudes; and finally, *Regime III* $(3 < \epsilon < 4.5\%)$ with, again, a limit cycle dynamics, resulting from the interaction of the two $m = \pm 1$ modes with equal amplitudes.

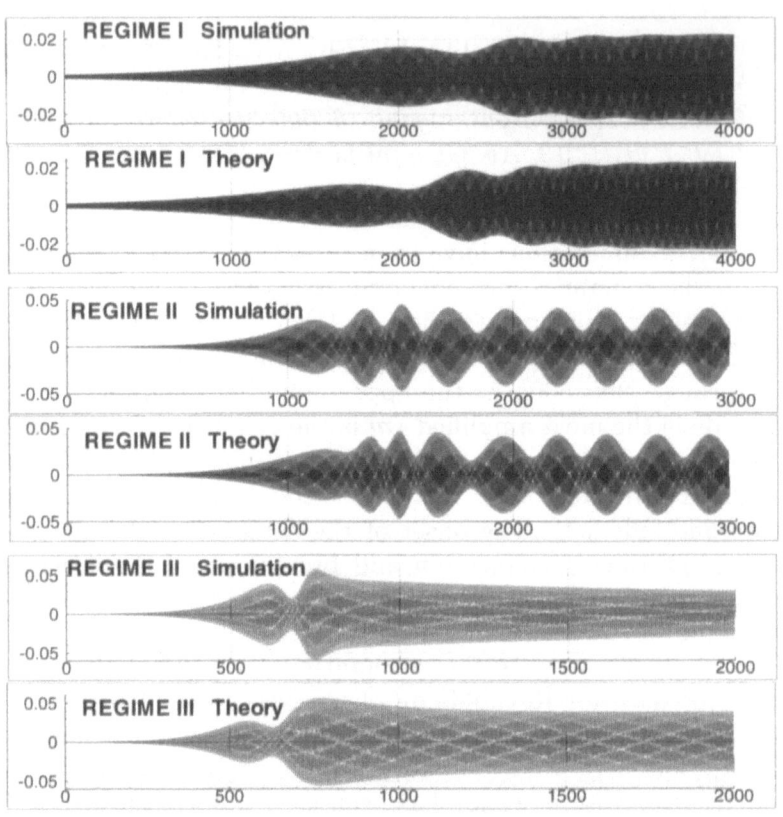

Figure 1. Azimuthal velocity at a point located in the jet mixing layer, at 2 nozzle diameters downstream. The period of oscillations is about 6 time units. Comparison between the simulated and theoretically predicted dynamics: limit cycle (regimes I and III) and limit torus (regime II).

All these stages could easily be predicted by a 5-th order weakly non-linear theory describing the interaction of the helical modes $m = \pm 1$ (see [2] for more details). The predictions of the theoretical model are in very good agreement with the results of direct numerical simulation, as illustrated in figure 1.

The chaotic state sets in about at $\epsilon = 5\%$ above the instationarity threshold. Figure 2a shows the very slow decay of the limit cycle resulting as an equilibrium of the two counter-rotating helical modes (Regime III). After very long transients, intermittent oscillations set in. The power spectra on the right side of the figure show that a new, about 50% higher, frequency appears.

a)

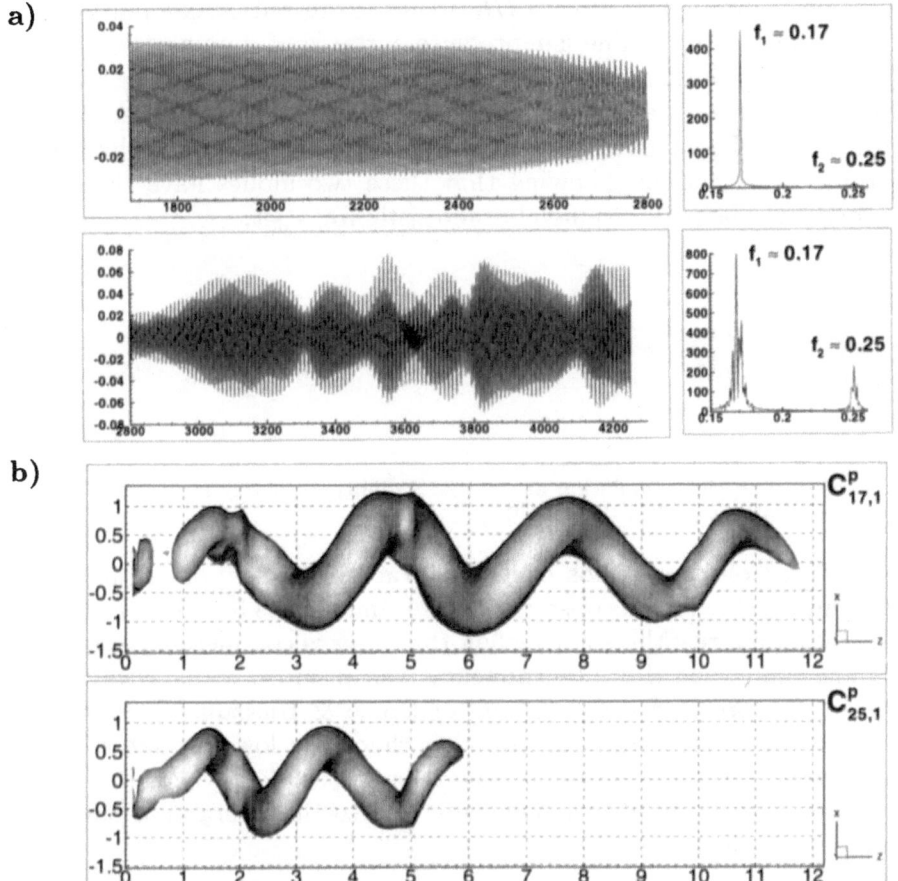

b)

Figure 2. Direct numerical simulation at $\epsilon \approx 5\%$. a) Azimuthal velocity signal and corresponding spectra for the direct simulation (same point as in Fig. 1). b) Iso–surfaces of low pressure characterizing the amplified modes; $c_{17,1}$ (up) corresponds to the helical mode of frequency f_1 and $c_{25,1}$ (down) to the helical mode of frequency f_2.

To detect the spatial structures responsible for this new frequency, we analyzed the flow–field by the Fourier analysis proposed in [4], and applied with success in [2], to characterize the unstable symmetry breaking modes. It consists in computing temporal Fourier modes through-out the flow-field in a sufficiently large time interval. In this case, we used a time interval corresponding to 17 periods of rapid oscillations visible in Fig. 2a.

It appeared that only temporal Fourier modes c_n with $n = 1, 17$ and 25 were really significant. The mode $n = 17$, corresponding to the basic frequency (denoted f_1), being the strongest and the mode $n = 25$ (corresponding to f_2) approaching progressively the level of mode $n = 17$. The obtained temporal modes (c_n) can be further decomposed into azimuthal Fourier modes with index m. The equilibrium of the $m = \pm 1$ helical modes, characterizing the decaying *Regime III*, is thus expressed by comparing the Fourier coefficients $c_{17,\pm 1}$. The iso–pressure surfaces of dominating modes $(2Real[c_{n,m} e^{-im\theta}]$, with $n = 17, 25$ and $m = 1)$ are shown in Fig. 2b. It clearly appears that the second frequency is associated with another helical mode. It is interesting to note that the wavelengths of both modes have the same ratio as their periods, showing that these two modes have the same phase velocity of about 0.5, value characteristic for a jet.

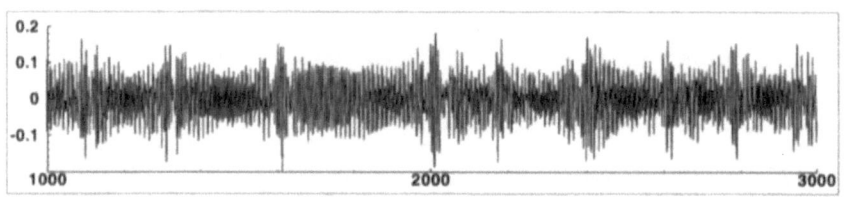

Figure 3. Direct numerical simulation at $\epsilon \approx 9\%$. Azimuthal velocity at the same point as in Fig. 1. Intermittent part of the signal.

The observed onset of chaos can be characterized as type II intermittency. A similar behavior was observed in the simulations for $5\% < \epsilon < 10\%$ above the instationarity threshold, when the chaos is completely developed (Fig. 3). The same type II of intermittency was observed at high Reynolds numbers, in the forced jet experiments of Broze & Hussain [5]. It is also interesting to note that the transition to chaos for low Reynolds numbers involves only the interaction of helical modes, while for high Reynolds numbers, the breakdown into turbulence is due to the interaction between axisymmetric and helical modes [3].

References

1. I. Danaila, J. Dušek and F. Anselmet, Coherent structures in a round, spatially evolving, unforced, homogeneous jet at low Reynolds numbers, *Phys. Fluids*, **9**, p. 3323, 1997.
2. I. Danaila, J. Dušek and F. Anselmet, Non–linear dynamics at a Hopf bifurcation with axisymmetry breaking in a jet, *Phys. Rev. E*, **57(4)**, April 1998.
3. M. V. Melander, F. Hussain and A. Basu, Breakdown of a circular jet into turbulence, in *Proceedings of T.S.F. 8*, München, 1991.
4. J. Dušek, Spatial structure of the Bénard - von Kármán instability, *European J. of Mechanics, B/Fluids*, **15**, p. 619, 1996.
5. G. Broze and F. Hussain, Transition to chaos in a forced jet: intermittency, tangent bifurcations and hysteresis, *J. Fluid Mech.*, **311**, p. 37, 1996.

OPTIMAL THREE-DIMENSIONAL PERTURBATIONS IN THE BLASIUS BOUNDARY LAYER

P. ANDERSSON, M. BERGGREN AND D.S. HENNINGSON
FFA, the Aeronautical Research Institute of Sweden,
Computational Aerodynamics Department,
P.O. Box 11021, S - 161 11 Bromma, Sweden.

Streamwise streaks are ubiquitous in transitional boundary layers, particularly when subjected to free-stream turbulence. Using the linearized, steady boundary-layer approximation, we numerically calculate the upstream disturbances experiencing maximum spatial energy growth, employing techniques commonly used when solving optimal-control problems for distributed parameter systems. The calculated optimal disturbances consist of streamwise vortices developing into streamwise streaks. We define the *disturbance energy* as the following, wall-normal integral of the velocity components

$$E(\mathbf{u}) = \int_0^\infty (\mathrm{Re}\, u^2 + v^2 + w^2)dy \equiv ||\mathbf{u}||^2 = (\mathbf{u}, \mathbf{u}), \qquad (1)$$

where Re is the Reynolds number based on the streamwise distance to the leading edge of the flat plate. The appearance of Re in the norm is a result of the boundary-layer scaling and ensures that the physical velocity components have equal weight.

We consider an upstream velocity perturbation $\mathbf{u}_{in}$ and its downstream response $\mathbf{u}_{out}(x)$, where x is their relative streamwise distance. Let $\mathcal{A}$ be the linear operator that maps $\mathbf{u}_{in}$ to $\mathbf{u}_{out}$, that is, $\mathbf{u}_{out}(x) = \mathcal{A}\mathbf{u}_{in}$. The action of $\mathcal{A}$ corresponds to solving a set of linear parabolic stability equations. We define the *maximum spatial transient growth*

$$G(x) = \max_{E(\mathbf{u}_{in})=1} E(\mathbf{u}_{out}(x)) = \max_{||\mathbf{u}_{in}||=1} (\mathcal{A}\mathbf{u}_{in}, \mathcal{A}\mathbf{u}_{in}) = \max_{||\mathbf{u}_{in}||=1} (\mathbf{u}_{in}, \mathcal{A}^*\mathcal{A}\mathbf{u}_{in})$$

and denote the corresponding maximizer $\tilde{\mathbf{u}}_{in}$ the *optimal perturbation*. The action of $\mathcal{A}^*$ is computed by marching the adjoint spatial stability equations upstream. Note that $G(x)$ is the maximum eigenvalue of the operator

U. Frisch (ed.), Advances in Turbulence VII, 109–112.

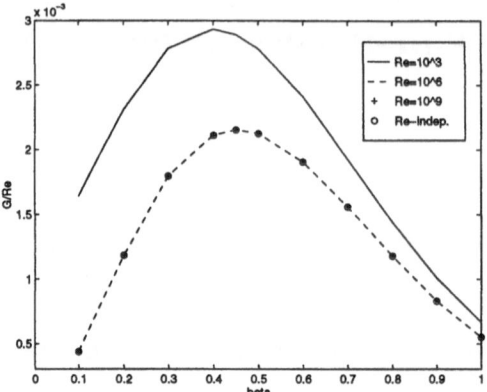

Figure 1. Maximum spatial transient growth divided by the Reynolds number vs. spanwise wavenumber

$\mathcal{A}^*\mathcal{A}$, which is computed using power iterations, $u_{in}^{n+1} = \mathcal{A}^*\mathcal{A}u_{in}^n$. We have $\lim_{n\to\infty} u_{in}^n/\|u_{in}^n\| = \tilde{u}_{in}$. From $\tilde{u}_{in}$ we then calculate $\tilde{u}_{out}(x) = \mathcal{A}\tilde{u}_{in}$ and the maximum energy growth, $G(x)$.

We study the scaling properties of the maximum spatial transient growth with respect to the Reynolds number. We integrate the equations a unit distance ($x = 1$) downstream, starting from the leading edge, and calculate G for several values of the spanwise wavenumber β. The calculations are repeated for three different Reynolds numbers Re $= 10^3$, 10^6 and 10^9, and once with the Reynolds-number-independent formulation used by Luchini [1]. Figure 1 depicts $G(x)/$Re versus β and shows that *the maximum spatial transient growth scales linearly with the distance from the leading edge for large Reynolds numbers.*

The v and w components of the optimal perturbation, for the spanwise wave number $\beta = 0.45$ and optimized with respect to downstream position $x = 1$, are given in figure 2a at the high-Reynolds-number limit. The corresponding u component of the response at the downstream position $x = 1$ caused by this optimal perturbation is given in figure 2b. For high Reynolds numbers, the u component almost completely vanishes in the optimal perturbation compared with the v and w components. Likewise, the v and w components vanish in comparison with the u component in the downstream response. This is a consequence of the appearance of the Reynolds number Re in the disturbance energy (1). Note that, because of the periodicity property in the spanwise direction, the upstream disturbance in figure 2a corresponds to streamwise vortices and the downstream response in figure 2b to streamwise streaks. Also plotted in figure 2b are the experimental data from Westin *et al.* [2]. All the streamwise velocity components have been normalized to unit maximum value. The presence of free-stream

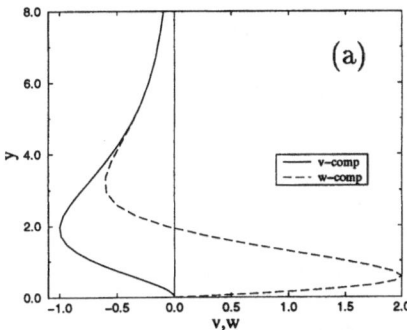
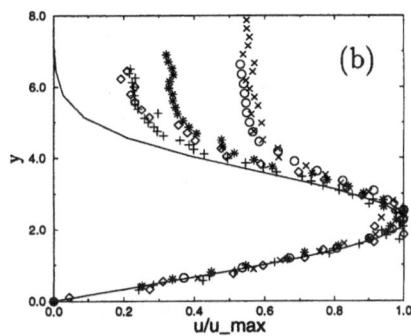

Figure 2. (a) The optimal perturbations at the leading edge maximized with respect to the downstream position $x = 1$. Here $\beta = 0.45$. The u component is zero. (b) Downstream response at $x = 1$ corresponding to the optimal perturbations in the left figure, that is $\beta = 0.45$. The v and w components are zero. Also a comparison with experimental data [2] measured in a flat-plate boundary layer at different downstream locations.

turbulence in the experiments prevents the root-mean-square streamwise velocity perturbations to vanish at infinity. The remarkably good agreement between the measured and calculated velocity profiles, and the fact that the calculations contained an optimization procedure while the experiments did not, indicate that the shown profile corresponds to some dominating, fundamental mode triggered in the flat plate boundary layer when subjected to high enough levels of free-stream turbulence. The fact that the power iterations converges quickly, also indicates the existence of a well-separated, dominating mode. The main conclusion is that *almost any initial disturbance will develop into a streamwise streak given a large enough Reynolds number.* A more complete version of the above material is given in Andersson *et al.* [3].

Based on these results, we propose a simple model for prediction of transition location, valid for situations where disturbances of this kind can be expected to dominate, that is, for free-stream turbulence levels at about 1–5 %. We make three assumptions in order to obtain our model.

1. We assume that the input energy $E(\mathbf{u}_{\mathrm{in}})$, as defined in (1), is proportional to the isotropic free-stream turbulence energy, Tu,

$$E(\mathbf{u}_{\mathrm{in}}) \propto \mathrm{Tu}^2. \tag{2}$$

This is an assumption about the receptivity process.

2. We assume that the initial disturbance grows with the optimal rate,

$$E(\mathbf{u}_{\mathrm{out}}) = G E(\mathbf{u}_{\mathrm{in}}) = \overline{G} \, \mathrm{Re} \, E(\mathbf{u}_{\mathrm{in}}), \tag{3}$$

where $\overline{G}$ is Reynolds-number independent. Recall that the last equality was found to hold for large enough Reynolds numbers.

TABLE 1.   Comparisons of different experimental
studies

	Tu(%)	Re$_T$	K
Roach& Brierley [4]			
T3AM	0.9	1,600,000	1138
T3A	3.0	144,000	1138
T3B	6.0	63,000	1506
Yang & Voke [5]	5.0	51,200	1131
Matsubara [6]			
grid A	2.0	400,000	1265
grid B	1.5	1,000,000	1500

3. We assume that transition occurs when the output energy reaches a specific value, E_T,

$$E(\mathbf{u}_{\mathrm{out}}) = E_T. \tag{4}$$

Combining assumptions 1–3, we obtain

$$\sqrt{\mathrm{Re}_T}\mathrm{Tu} = K, \tag{5}$$

where K is a universal constant for free-stream turbulence levels at 1–5 %. The experimental data used to verify this model are given in table 1. For these turbulence levels, the obtained model agrees well with a model by van Driest & Blumer [7] obtained from different physical arguments.

References

1. P. Luchini. Reynolds-number-independent instability of the boundary layer over a flat plate surface. part 2: Optimal perturbations. Submitted to *J. Fluid Mech*, 1997.
2. K. J. A. Westin, A. V. Boiko, B. G. B. Klingmann, V. V. Kozlov, and P. H. Alfredsson. Experiments in a boundary layer subject to free-stream turbulence. Part I: Boundary layer structure and receptivity. *J. Fluid Mech.*, 281:193–218, 1994.
3. P. Andersson, M. Berggren, and D. S. Henningson. Optimal Disturbances and Bypass Transition in Boundary Layers. submitted to *Phys. Fluids*, 1998.
4. P. E. Roach and D. H. Brierley. The influence of a turbulent free-stream on zero pressure gradient transitional boundary layer development. Part I: Test cases T3A and T3B. In O. Pironneau, W. Rodi, I.L. Ryhming, A.M. Savill, and T.V. Truong, editors, *Numerical simulation of unsteady flows and transition to turbulence*, pages 303–316. Cambridge University Press, 1992.
5. Z. Y. Yang and P. R. Voke. Numerical simulation of transition under turbulence. Technical Report ME-FD/91.01, Dept. Mech. Eng., University of Surrey, 1991.
6. M. Matsubara. Private Communications, 1997.
7. E. R. van Driest and C. B Blumer. Boundary layer transition: Freestream turbulence and pressure gradient effects. *AIAA J.*, 1:1303–1306, 1963.

EXPERIMENTAL STUDY OF THE STABILITY
OF A TRAVELLING ROLL SYSTEM IN A ROTATING DISK FLOW

L. SCHOUVEILER, P. LE GAL AND M.P. CHAUVE
Institut de Recherche sur les Phénomènes Hors Equilibre
UMR 6594, CNRS - Universités d'Aix-Marseille I & II
12, Avenue Général Leclerc,13003 Marseille, France

1. Introduction

Several studies have already adressed the problem of the stability of the two control parameters flow of a viscous fluid driven between a rotating and a stationary disk, when both the aspect ratio and the Reynolds number can be varied. The aspect ratio is defined as h/R where h is the axial distance between the two parallel disks and R their radius, and the Reynolds number is $Re = \Omega R^2/\nu$ where Ω is the angular velocity of the rotating disk and ν the kinematic viscosity of the working fluid. The experiments of San'kov and Smirnov (1984), Itoh (1988) or Sirivat (1991) revealed the existence of various bifurcation sequences for the transition to turbulence, depending on the considered aspect ratio, when the Reynolds number is increased. Recently, by exploring the two dimensionnal control parameter space (h/R, Re), we have determined the whole experimental transition diagram of this rotating disk flow (Schouveiler *et al.*, 1996, and Schouveiler, 1998) which completes the previous experimental studies and summarizes the regimes of the flow and the bifurcation sequences.

When the aspect ratio is large enough $(h/R > 7.1 \times 10^{-2})$, the stationary axisymetrical basic regime appears of Batchelor type with viscous effects confined in two boundary layers, one on each disk, separated by an inviscid rotating core. In this case, we have experimentaly reported (Schouveiler *et al.*, 1998) that the first stages of the transition are due to the developments of instabilities in the boundary layer of the stationary disk. This is in agreement with numerical results of Cousin-Rittemard (1996) and with the linear stability analyses of the similarity solutions for the infinite disk ideal case (San'kov and Smirnov, 1991, and Itoh, 1991). Thus, after a first bifurcation, we observe concentric circular rolls which appear near the periphery and propagate towards the disk axis. Then, above a second threshold, spiral rolls are formed at the periphery of the flow, they travel radially outwards and in the disk rotation direction. This spiral roll system coexists with the circular one and we have shown that the velocity perturbations induced by both instabilities are confined in the stationary disk boundary layer. We present in Figure 1 a visualization of the flow pattern resulting from the coexistence of the two travelling roll systems.

U. Frisch (ed.), Advances in Turbulence VII, 113–116.
© *1998 Kluwer Academic Publishers.*

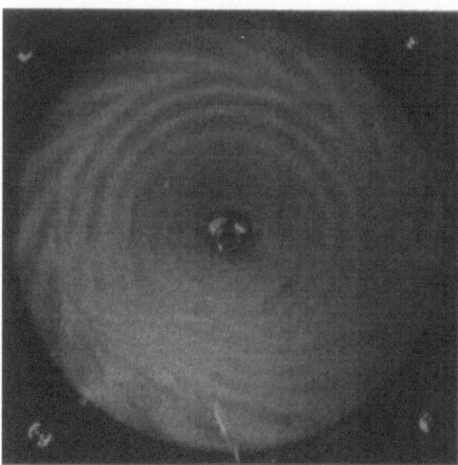

Figure 1. Visualization of the coexisting spiral and circular rolls for h/R = 11.43x10^{-2} and Re = 20.9x10^3.
Clockwise rotation.

We present in the following, the results of an experimental study of the stability of the travelling spiral roll system. First, let us define a mode of this system with the pair (m, f$_s$) where m is the integer number of rolls present on the circumference, and f$_s$ is the associated frequency measured in the laboratory frame.

2. Experimental details

The experimental setup is schematically shown in Figure 2. It consists of a horizontal rotating disk (of radius R = 140 mm) set in a vertical cylindrical housing whose top forms the stationary disk. The housing is fixed in the laboratory frame and completely filled with water at ambient temperature (v ≈ 1 mm^2/s). The radial distance between the edge of the disk and the vertical wall is less than 0.05 mm.

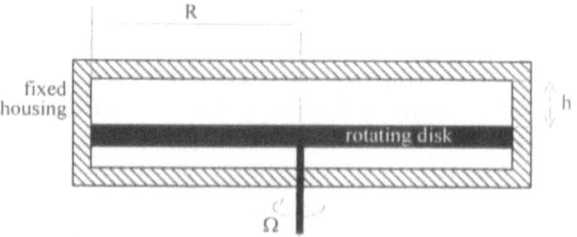

Figure 2. Experimental setup.

The angular velocity Ω of the rotating disk and the axial distance h between the two parallel disks are the two continuously adjustable experimental parameters. During the present study, h is fixed to 16 mm (h/R = 11.43x10^{-2}) and Ω (respectively the Reynolds number Re = ΩR^2/v) is varied in the range [0, 4π rad/s] (respectively [0, 240x10^3]).

The stability study is performed using two experimental techniques. The roll-like structures are first visualized by means of small anisotropic reflective particles in suspension in the flow (see Savas, 1985) and the flow is recorded through the stationary disk, made of Plexiglas, with a video camera. The amplitude of the structures is then obtained by measuring velocities of the pertubated flow with an ultrasonic Doppler anemometer. We measured, in this way, successive instantaneous profiles of the radial velocity. Spectral characteristics of the structures are computed by Fourier transform of these spatio-temporal velocity signals or of space-time images built by gathering of a pixel line, taken on a radius, extracted from the successive digitized video pictures.

3. Results and discussion

The diagram of Figure 3 presents all the (m, f_s) spiral roll modes which have been observed. It shows the existence of a band of stable modes above the threshold of the bifurcation, giving rise to the spiral roll regime, where even or odd m modes are possible. The system is thus constituted of corotating rolls, one roll corresponds to one wavelength. The linear behavior of the square amplitude evolution, of a given m mode, with the Reynolds number at the vicinity of the threshold permit to establish the supercritical nature of the bifurcation. When the spiral roll regime is reached by slowly increasing Re, the mode m = 18 is always selected and it appears then as the critical mode of the bifurcation. The other modes have been obtained either by different rates of variation of Re or because of a secondary instability as discussed below. For the different m modes, the points corresponding to the condition of marginal stability (zero amplitude) are deduced from the extrapolations of the linear evolutions of the square amplitude versus the Reynolds number. The marginal stability curve, noted (M) in Figure 3, is then obtained by the parabolic fit of these points.

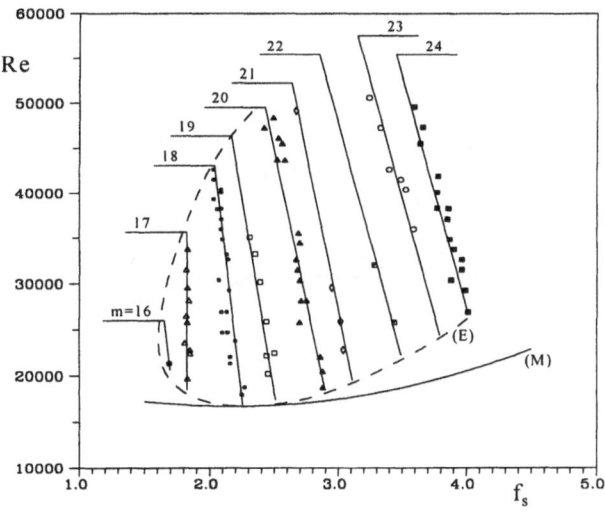

Figure 3. Stability domain, for $h/R = 11.43 \times 10^{-2}$, of the spiral roll modes (m, f_s).
——— : marginal stability curve (M), – – – : Eckhaus stability boundary (E).

Moreover, the simultaneous measurements of the f_s frequency and of the corresponding wavenumber (or of the number of rolls m) as presented in Figure 3 permit to establish the dispersion relation m = m(f_s) of the spiral roll system.

Finally, we have observed that the stability domain is limited by a boundary (dashed curve (E) in Figure 3) where the m modes, excepted the critical mode m = 18 at the threshold of the bifurcation, are destabilized with a non-zero amplitude. The band of stable modes appears thus limited by a secondary instability. To determine the nature of this instability, the boundary (E) has been approached from the stable mode side, by slowly changing the Reynolds number. When the stability limit (E) is reached, we observed a jump of the frequency f_s which is associated to a change of the selected mode m. For increasing (respectively decreasing) Reynolds number, this change occurs via the creation (respectively destruction) of one roll. Such a mechanism affecting a periodic pattern, as the spiral roll system, which induces a change of the wavelength by destruction or creation of one wavelength (here one roll) to return the system back in the stability domain is known as the Eckhaus instability. Although the Eckhaus instability is wellknown and has been exhaustively studied for stationary patterns, examples of such an instability on travelling waves are much more unusual.

References

Cousin-Rittemard N. (1996) Contribution à l'étude des instabilités des écoulements axisymétriques en cavité inter-disques de type rotor-stator, PhD thesis, Université Paris VI (in French).

Itoh M. (1988) Instability and transition of the flow around a rotating disk in a casing, Toyota Rep. 36, 28-36 (in Japanese).

Itoh M. (1991) On the instability of flow between coaxial rotating disks, *ASME Boundary Layer Stability and Transition to Turbulence, FED* **114**, 83-89.

Savas O. (1985) On flow visualization using reflective flakes, *J. Fluid Mech.* **152**, 235-248.

San'kov P. L. and Smirnov E. M. (1984) Bifurcation and transition to turbulence in the gap between rotating and stationary parallel disks, *Fluid Dyn.* **19** (5), 695-702.

San'kov P. L. and Smirnov E. M. (1991) Stability of viscous flow between rotating and stationary disks, *Fluid Dyn.* **26** (6), 3445-3448.

Schouveiler L., Le Gal P., Chauve M.-P. and Takeda Y. (1996) Experimental study of the stability of the flow between a rotating and a stationary disk, in S. Gavrilakis *et al.* (eds.), *Advances in Turbulence VI*, Kluwer Academic Publishers, Dordrecht, pp. 385-388.

Schouveiler L. (1998) Sur les instabilités des écoulements entre un disque fixe et un disque en rotation, PhD thesis, Université Aix-Marseille II (in French).

Schouveiler L., Le Gal P., Chauve M. P. & Takeda Y. (in press) Spiral and circular waves in the flow between a rotating and a stationary disk, *Exp. Fluids*.

Sirivat A. (1991) Stability experiment of flow between a stationary and a rotating disk, *Phys. Fluids A* **3** (11), 2664-2671.

DISCONTINUOUS TRANSITION TO SPATIO-TEMPORAL INTERMITTENCY IN THE PLANE COUETTE FLOW

S. BOTTIN[1], F. DAVIAUD[1], P. MANNEVILLE[2,1], O. DAUCHOT[1]
[1] *Groupe Instabilités et Turbulence, CEA Saclay,*
SPEC. F-91191 Gif-sur-Yvette, France.
[2] *Laboratoire d'Hydrodynamique, École Polytechnique,*
F-91128 Palaiseau, France

Our understanding of the transition to turbulence in shear flows still faces severe difficulties when the primary bifurcation is subcritical owing to our limited ability to identify nonlinear branching solutions. This is particularly the case for Poiseuille pipe flow and plane Couette flow (pCf) that are linearly stable for all Reynolds numbers and where the transition to turbulence strongly depends on the nature and amplitude of the perturbations brought to the flow. Contrasting with globally supercritical systems where the transition scenario involves long wavelength modulations of patterns, subcritical systems generically experience transitions where fronts develop between domains corresponding to different regimes. In this context, spatio-temporal intermittency (STI) [1] presents itself as a contamination scenario where laminar and turbulent domains coexist and fluctuate ([2], [3] and references therein). Owing to its similarity with directed percolation, STI can further be analyzed within the framework of the theory of phase transitions and critical phenomena: the average activity of the lattice (F_t) then plays the role of an order parameter.

In this study we present a preliminary report of the bifurcation diagram (figure 1, left) of pCf as experimentally determined from turbulent fraction measurements [4], suggesting a strong connection, at least at a phenomenological level, with the discontinuous transition *via* STI reported in [6].

Our pCf set-up is made of an endless transparent plastic ribbon guided by rollers achieving a shear flow without mean advection in a gap of width $2h = 7$mm between two walls moving in opposite directions (see [7] for details). The Reynolds number is defined as $Re = Uh/\nu$, where U is the linear speed of the ribbon and ν is the kinematic viscosity of the fluid (water). The experiment consists in generating spots by triggering the flow

U. Frisch (ed.), Advances in Turbulence VII, 117–120.

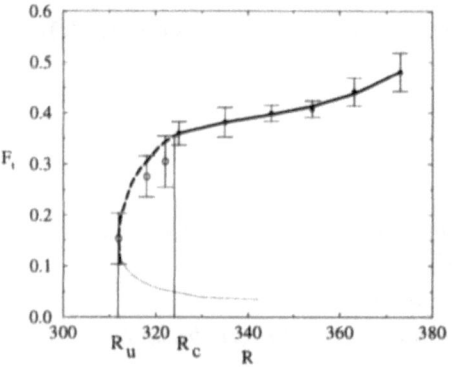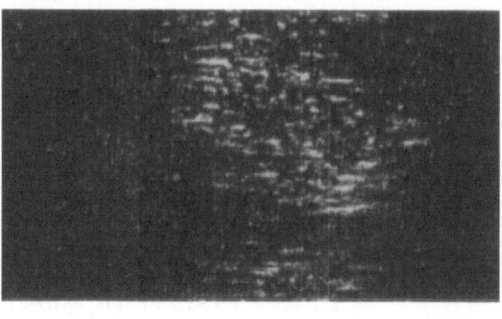

Figure 1. Left: Average turbulent fraction $\langle F_t \rangle$ of transitional pCf as a function of the Reynolds number Re and subsequent conjectured bifurcation diagram (see text for details). Right: Snapshot at time $t = 200$ of a turbulent domain for $Re = 325$.

with a small transverse jet at large Re $(Re \simeq 380)$ where they are left to mature during about 2 mn before Re is suddenly quenched down to the studied value Re_q.

A typical snapshot of a grown-up turbulent spot is shown in figure 1 (right). In the following, we determine the size of a spot from the ratio of the area of the turbulent domain to the total surface of the flow, denoted F_t. The turbulent area is evaluated from a snap-shot of the whole experimental surface by image processing.

While quenching the flow down to $Re_q > Re_c$ always yields a sustained turbulent state with an asymptotic average turbulent fraction $\langle F_t \rangle$, when $Re_q < Re_c$ turbulence always relaxes after a transient of variable duration. Figure 2 (left) displays the evolution of the turbulent fraction for three such experiments $(Re_q < Re_c)$. All the curves start by a short plateau at $F_t \simeq 0.45$ but, while F_t decreases rapidly to zero for $Re_q = 305$ (representative of all experiments with $Re_q < Re_u \simeq 312$), for $Re_q = 318$ a fluctuating intermediate stage of about 300 sec shows up before the ultimate decay stage. The case at $Re_q = 312$ appears to be somewhat in-between with a possibly significant shorter plateau at a lower average value. The lifetime of the transient is defined as the delay δt between quenching to the Re-value of interest and complete tranquilization. Statistics of lifetimes have been performed over between 50 and 120 independent runs, depending on the value of Re_q. Cumulated histograms of transient lifetimes δt longer than a given value τ, $N(\delta t > \tau)$, are displayed in figure 2 (right) as functions of τ for several values of Re. A clear exponential tail $N(\delta t > \tau) \propto \exp(-\tau/\tau_0)$ is visible, with τ_0 rapidly increasing as Re_q approaches 323 from below. According to the exponential-decay hypothesis, a good estimator for τ_0 is

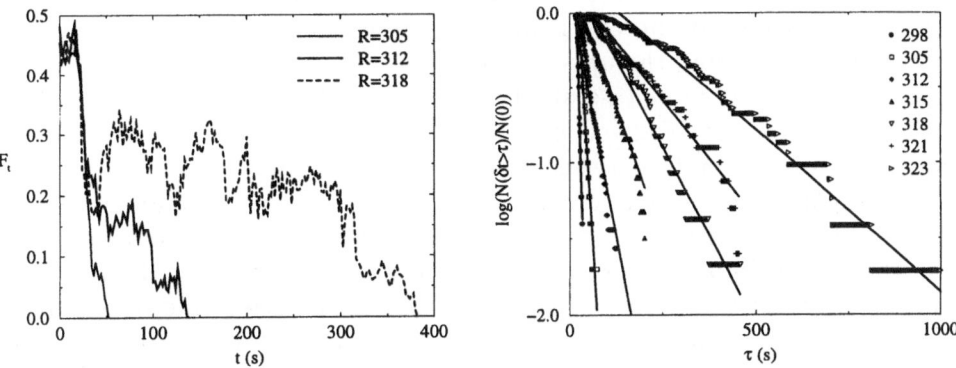

Figure 2. Left: Evolution in time of quenched spots for different values of Re. Right: cumulated lifetime distributions for turbulent transients at different values of the quench Reynolds number $Re_q < Re_c$ evidencing exponential decay (lin-log scales; solid lines are fits through the experimental data points).

obtained from the mean value of the lifetimes ($\langle \delta t \rangle \sim \tau_0$) at given Re that suggests a divergence in $1/(Re_c - Re_q)$, with $Re_c \simeq 323$.

The turbulent fraction $\langle F_t \rangle$ has been determined as a function of Re, yielding the bifurcation diagram of figure 1 (left). For $Re > Re_c$, a stationary regime of sustained turbulence is obtained, so that taking an average has a well defined statistical meaning. The mean of F_t is taken over 200 images taken every 20 sec in the asymptotic regime after the end of the growth stage and over 10 independent experiments at the same given Re. Corresponding data points are displayed as filled circles in figure 1 (left). Below Re_c but above Re_u, $\langle F_t \rangle$ no longer exists strictly speaking, but finite-time averages can be determined for long lasting turbulent patches. This cannot be done automatically as before since the terminal decay stage has to be eliminated before the mean is taken. Furthermore, as can be understood from Fig. 2 (left), visual inspection of individual records is required to decide, case by case, the time extension of the plateau over which the average is to be taken. Error bars affecting values obtained for $Re < Re_c$ now represent the extreme excursions of F_t before the ultimate decay. These finite-time average turbulent fractions are plotted as open circles in figure 1 (left).

The bifurcation diagram is in agreement with the expectation of a subcritical bifurcation, especially when taking into account the dotted line (lower branch). This dotted line is just a conjecture supported by the existence of aborted spots that never reach a F_t large enough to persist. It is suggestive of a separatrix in phase space within which deterministic return to laminar flow prevails (attraction basin of the laminar basic state) and beyond which other states become relevant. In this perspective, the upper branch (dashed and continuous line) represents the bifurcated state. Now, when considering the upper branch, one is led to consider the critical value Re_c as a limit of meta-stability for the turbulent state. In turns, this is reminiscent of the Maxwell plateau which, in the theory of first order transitions, discriminate between meta-stable and stable states.

Besides this thermodynamic interpretation, another one stems from the similarity between the decay of the turbulent state and transients escaping from the neighborhood of a chaotic repellor in the theory of dissipative dynamical systems: the exponential distribution of transient lifetimes arises from the fractal structure of the attraction basin of the basic state. For pCf this phenomenon has been studied numerically by Schmiegel and Eckhardt [8] who determined the map of lifetimes as a function of the perturbation amplitude and Reynolds number. However no sustained turbulent regime was obtained, which may be attributed to the fact that the size of the computational box is small whereas the flow is an extended geometry. The appeal to spatio-temporal intermittency as a transition scenario is one attempt to reconcile the two viewpoints by (i) giving a genuine spatio-temporal meaning to the decay of chaotic transients and (ii) allowing for a thermodynamic interpretation to the conversion of a repellor into an attractor in line with the statistical approach implied by directed percolation. In this respect, the existence of discontinuous transitions where, local transient chaos (possibly initiated by triggering) can be converted into sustained spatio-temporal chaos by spatial coupling [6] offers promising modeling perspectives.

References

1. Pomeau Y. *Physica D* **23**(1986) 3.
2. Kinzel W. *Ann. Israel Phys. Soc.* **5**, 425 (1983)
3. Chaté H. & Manneville P. "Spatiotemporal intermittency" in [5].
4. Bottin S., Daviaud F., Manneville P. & Dauchot O. *submitted to Europhys. Lett*
5. Tabeling P. & Cardoso O. Turbulence. A tentative dictionnary, Plenum Press(1995).
6. Chaté H. & Manneville P. *Europhys. Lett.* **6**(1988) 591.
7. Daviaud F., Hegseth J. & Bergé P. *Phys. Rev. Lett.* **69**(1992) 2511.
8. Schmiegel A. & Eckhardt B. *Phys. Rev. Lett.* **79**(1997) 5250.

ABSOLUTE/CONVECTIVE INSTABILITIES IN THE BATCHELOR VORTEX: VISCOUS CASE

C. OLENDRARU AND A. SELLIER
Laboratoire d'Hydrodynamique
Ecole Polytechnique, F-91128 Palaiseau Cedex, FRANCE

This study presents our recent results about the Absolute/convective instabilities in the Batchelor vortex. This basic flow is a good approximation of slowly evolving trailing vortices. It is entirely described in an axial (x, r, θ) cylindrical coordinate system by the axial and azimuthal distributions

$$U(r) = a + e^{-r^2}, \quad V(r) = 0, \quad W(r) = q[1 - e^{-r^2}]/r.$$

These nice properties of this two-parameter vortex explain why many authors studied temporal stability of such a flow both in the inviscid and viscous cases. However, it is worth locating in the $a - q$ plane the absolute/convective instability transition curves by resorting to a spatial approach. This task was recently achieved by Olendraru *et al.* [4] for an inviscid fluid and via a numerical treatment of the secular equation one can derives in these circumstances (see Ref. [3]) and by Delbende *et al.* [1], through a direct numerical simulation of the impulse response of the Batchelor vortex in the viscous case but for only one specific value of Reynolds number ($Re = 667$). In the same spirit of the work of Khorrami *et al.* [2] we employed a spectral method to solve the dispersion relation arising in the viscous case. This approach holds for any Reynods number and provides all the spatially amplified modes. The application of the Briggs-Bers criterion thereafter enables us to determine the absolute/convective instability transition curves in the plane $a - q$ for differents Reynolds numbers. The enclosed Fig.1 displays these curves for azimuthal wavenumber $m = -2$. Note that if the Reynolds number is large enough a transition takes place on the jet-like side without any co-flow ($a = 0$).

This work was supported by the Direction des Recherches, Études et Techniques (DRET) of the French Ministry of Defense under Grant #92-098 and by IDRIS under Grant #970936.

U. Frisch (ed.), Advances in Turbulence VII, 121–122.
© 1998 *Kluwer Academic Publishers.*

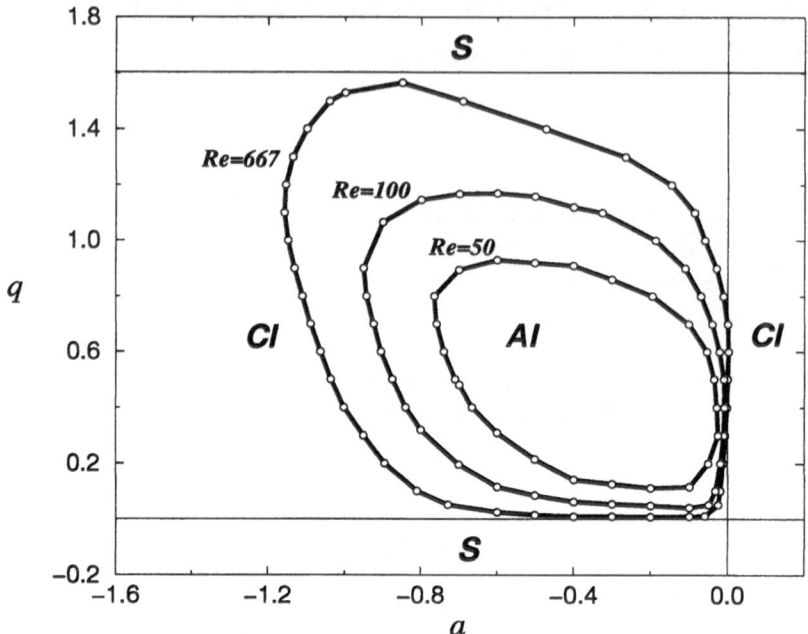

Figure 1. AI/CI transition curves for different Reynolds numbers for $m = -2$. The symbolds (S), (CI) and (AI) respectively indicate domains of stability, convective instability and absolute instability.

References

1. Delbende, I., Chomaz, J.M. and Huerre, P. (1998) Absolute/convective instabilities in the Batchelor vortex: a numerical study of the linear impulse reponse, *J. Fluid Mech.*, **355**, 229-254
2. Khorrami M. R., Malik M. R. et Ash R. L. (1989) Applications of Spectral Collocations Techniques to the Stability of Swirling Flows, *J. Comput. Phys.*, **81**, 206-229.
3. Lessen M., Singh P. J. and Paillet F. (1974) The stability of a trailing line vortex. Part 1: Inviscid theory, *J. Fluid Mech.*,**63**, 753-763.
4. Olendraru C., Sellier A., Rossi M. et Huerre P., (1996) Absolute/convective instability of the Batchelor vortex, *C. R. Acad. Sci. Paris*, t.**323**, Série II b, 153-159.

LINEAR AND NONLINEAR STABILITY ANALYSIS OF PERTURBED PLANE COUETTE FLOW

DWIGHT BARKLEY[1] AND LAURETTE S. TUCKERMAN[2]
[1] *Mathematics Institute, University of Warwick*
Coventry CV4 7AL, U.K.   barkley@maths.warwick.ac.uk
[2] *LIMSI, BP 133, 91403 Orsay Cedex, France, laurette@limsi.fr*

Despite being linearly stable for all Reynolds numbers, plane Couette flow becomes turbulent in laboratory experiments and in numerical simulations for $Re \approx 300$. In order to induce and to study the transition to three-dimensionality and then to turbulence, systematic perturbations are sometimes applied to the flow. In particular, plane Couette flow perturbed by a wire is currently the subject of laboratory experiments at CEA-Saclay [1, 2, 3].

We have performed a numerical linear and nonlinear stability analysis of plane Couette flow perturbed by the presence of a thin narrow ribbon midway between bounding plates. Both in the experiment and in the numerical study, the streamwise (x) dimension L is increased until length-independent results are obtained. In practice, this requires L between 20 and 40 times the cross-channel (y) dimension. In our numerical study the ribbon had formally infinite extent in the spanwise (z) direction. In experiments the perturbing wire has a large spanwise extent.

In our numerical approach we first solve the steady 2D Navier-Stokes equations subject to the boundary conditions:

$$\mathbf{u}(x - L, y) = \mathbf{u}(x + L, y)$$
$$\mathbf{u}(x, y = \pm 1) = \pm \hat{\mathbf{x}}$$
$$\mathbf{u}(x = 0, y) = 0, \qquad \text{for } -\rho \le y \le \rho,$$

where ρ is the half-height of the ribbon ($\rho = 0.086$ for all results reported). We use the spectral element code `Prism` [4], which permits grid refinement near the ribbon. Each velocity component is represented by $O(10^4)$ grid-points or basis functions. Figure 1 shows an example steady 2D flow.

Next we calculate the linear stability of the steady flows to three-dimensional z-periodic perturbations (eigenfunctions) of the form $(\mathbf{u}, p) =$

U. Frisch (ed.), Advances in Turbulence VII, 123–126.
© 1998 *Kluwer Academic Publishers.*

a)   b)

Figure 1. (a) Streamfunction near ribbon $(-3 < x < 3)$ of steady 2D flow $\mathbf{U}(x,y)$ at $Re = 250$. (b) Effect of ribbon is highlighted by subtracting unperturbed plane Couette flow $y\mathbf{e_x}$ from flow of (a). This flow displays centro-symmetry but not reflection symmetry in x and y. The effect of the ribbon is to set up a circulation opposing that of the plane Couette flow.

$(\hat{u}\cos\beta z, \hat{v}\cos\beta z, \hat{w}\sin\beta z, \hat{p}\cos\beta z)$. For fixed β, this is essentially a 2D calculation [5]. The vector $\hat{\mathbf{u}}(x,y,t) = (\hat{u}, \hat{v}, \hat{w})$ of Fourier coefficients evolves according to:

$$\frac{\partial\hat{\mathbf{u}}}{\partial t} = -(\hat{\mathbf{u}}\cdot\nabla)\mathbf{U} - (\mathbf{U}\cdot\nabla)\hat{\mathbf{u}} - (\nabla - \beta\hat{\mathbf{z}})\hat{p} + \frac{1}{Re}(\nabla^2 - \beta^2)\hat{\mathbf{u}}$$

$$(\nabla + \beta\hat{\mathbf{z}})\cdot\hat{\mathbf{u}} = 0$$

where ∇, etc. are two-dimensional differential operators. The boundary conditions on $\hat{\mathbf{u}}$ are the same as for the steady flow except that $\hat{\mathbf{u}}(x,y) = 0$ at $y = \pm 1$. The leading eigenvalues are calculated by the Arnoldi method [5]. Figure 2 shows the maximal growth rate of 3D eigenfunctions as a function of Re and β. At $Re_c \approx 230$ there is a linear instability with $\beta_c \approx 1.3$. The corresponding wavelength $\lambda_c \approx 4.8$ agrees well with experimental observation.

Finally, to determine the nonlinear characture of the instability, we carry out a single nonlinear 3D simulation starting from the initial condition $\mathbf{U} + \epsilon\mathbf{u}$, where $\mathbf{U}$ and $\mathbf{u}$ are the steady flow and eigenfunction at $Re = 250$ and ϵ is small. The evolution of the amplitude $A(t)$ with time is fit to the normal form $\dot{A} = \sigma A + \alpha A^3$. A positive (negative) value of α implies that the bifurcation is subcritical (supercritical). Figure 3 demonstrates that the growth is faster than exponential, and hence that the bifurcation is subcritical. We have further verified the subcritical nature of the instability by computing a steady 3D flow at $Re = 200$, below Re_c.

Figure 4 depicts the final steady 3D flow. Small streamwise vortices can be seen in each of the four corners of the $x = 0$ plane containing the ribbon. The lower $(y < 0)$ pair evolves with x into the strong pair of vortices at $x = 1$. The vortices at $x = 3$ are tilted, attesting to the nonlinear generation of the second spanwise harmonic 2β. Far from the ribbon, the 3D flow returns to plane Couette flow. This flow is similar to that observed in the Saclay experiments.

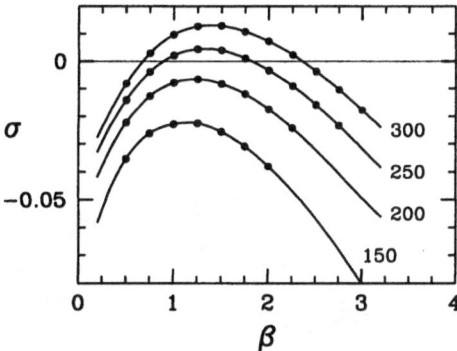

Figure 2. Growth rate σ of most unstable 3D eigenfunction as a function of spanwise wavenumber β for $Re = 150, 200, 250, 300$. Critical values for instability are $Re_c \approx 230$ and $\beta_c \approx 1.3$. Recall that in the absence of a ribbon or wire, plane Couette flow is linearly stable for all Re, β.

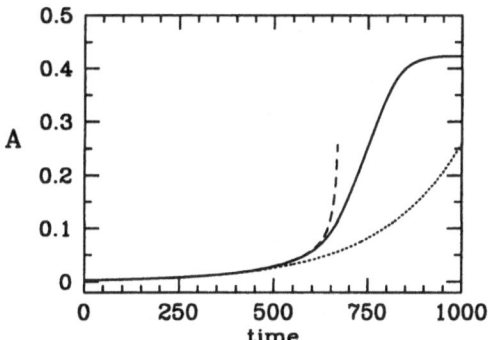

Figure 3. Nonlinear growth of the amplitude A of the 3D flow from simulation (solid) at $Re = 250$. First-order (dotted) and third-order (dashed) dynamics are shown with $\sigma = 0.0046$ and $\alpha = 1.4$.

In conclusion, we have accurately determined the extent to which the basic steady 2D profile is modified by the presence of a small spanwise-oriented ribbon. We have determined that a ribbon, comparable in size to the cylinders used in the Saclay experiments, is large enough to induce linear instability of the basic profile at Reynolds numbers of order a few hundred. We have found that the spanwise wavelength of the most unstable mode is in good agreement with the value seen experimentally. Finally, we have shown that the linear instability is subcritical and leads to a flow with streamwise vortices.

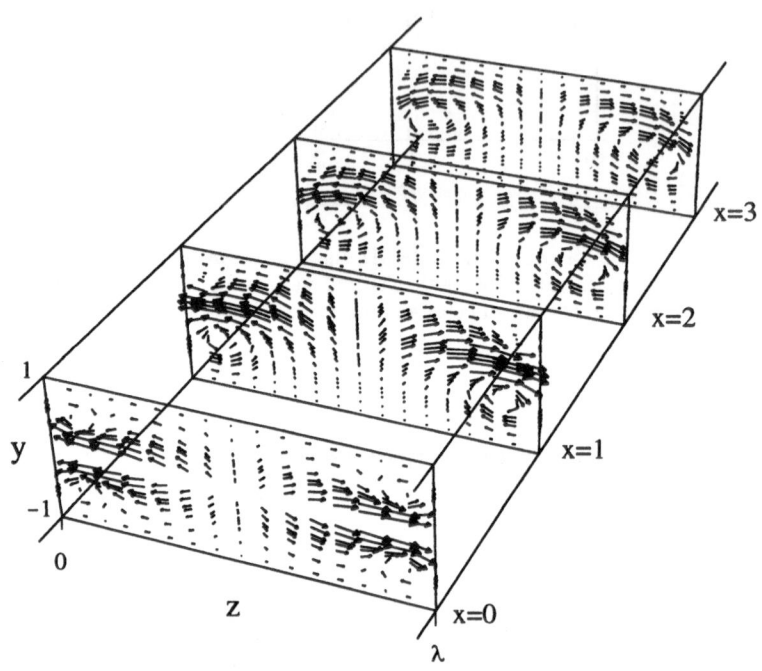

Figure 4.   Steady 3D velocity field at $Re = 250$. Shown are (v, w) velocity plots at four streamwise locations containing streamwise vortices.

References

1. O. Dauchot and F. Daviaud (1995) Streamwise vortices in plane Couette flow, *Phys. Fluids* **7**, 901–903.
2. S. Bottin, O. Dauchot and F. Daviaud (1997) Intermittency in a locally forced plane Couette flow, *Phys. Rev. Lett.* **79**, 4377–4380.
3. S. Bottin, O. Dauchot, F. Daviaud and P. Manneville (1998) Experimental evidence of streamwise vortices as finite amplitude solutions in transitional plane Couette flow, *Phys. Fluids*, submitted.
4. R.D. Henderson (1994), Unstructured spectral element methods; parallel algorithms and simulations, Ph.D Thesis, Princeton University.
5. D. Barkley and R. Henderson (1996) Floquet stability analysis of the periodic wake of a circular cylinder, *J. Fluid Mech.* **322**, 215–241.

EXPERIMENTAL STUDY OF THE TEMPERATURE AND VELOCITY BOUNDARY LAYERS IN THERMAL TURBULENCE AT LOW PRANDTL NUMBER.

T. SEGAWA[1], A. NAERT[2] AND M. SANO[1]

[1] *R.I.E.C., Tohoku University,*
2-1-1 Katahira, Aoba-Ku, Sendai 980-77, Japan.
[2] *Laboratoire de Physique,*
Ecole Normale Supérieure de Lyon,
46 Allée d'Italie, 69364 Lyon Cedex 07, France.

Despite recent progress, both from the theoretical and experimental side, the convective transport of heat in a turbulent flow is still far from being understood. For instance, the existence of an ultimate regime of convection at very high values of the Rayleigh number Ra (non-dimensional temperature difference) is only a conjecture. We report here on recent results obtained in a Rayleigh-Bénard convection experiment using mercury as a working fluid. Due to its low Prandtl number ($Pr = \kappa/\nu = 0.024$, κ is the thermal conductivity and ν is the kinematic viscosity), it is possible to realize in well controlled experimental conditions regimes of hard turbulence that would be difficult to reach in fluids of higher Pr, such as air, water or helium. The heat transport through a fluid layer between two horizontal plates at distance L, regulated at different temperatures ($T_{top} \leq T_{bottom}$) is characterized by the Nusselt number, ratio of the total heat flux over that transported by diffusion. The hard turbulence regime of thermal convection, is characterized by a scaling of Nu versus Ra, and a stable mean circulation in the cell. It appeared recently that, in this high Nu regime, the heat transport properties are affected by boundary effects, along the plates. Our experimental set-up consist in several cylindrical cells (diameter: 10 cm, height: 5, 10, 20 cm), covering a large range of Ra ($10^5 \leq Ra \leq 210^9$). We measured in these three cells the Nu-Ra relation. We evidenced a scaling exponent of 0.25 ± 0.02, distinct from that obtained in higher Pr fluids ($0.285 \simeq 2/7$), and that expected for the ultimate regime of convection ($1/2$) [1]. Because of the mean circulation in the cell, a velocity boundary layer develops along the the thermally active plates where temperature boundary layers also exists. The structure of the

U. Frisch (ed.), Advances in Turbulence VII, 127–128.

upper (cold) boundary layer has been investigated in the aspect ratio 1 cell ($10^6 \leq Ra \leq 10^8$), using a small thermistor, movable along the axis thanks to a stepping motor. The thickness λ_v of the velocity boundary layer is obtained from the high frequencies power spectra of the temperature fluctuations. The precision of this method has been recently improved giving a reliable scaling relation of λ_v versus Ra: $\lambda_v \propto Ra^{-0.20\pm0.02}$. The thickness of the thermal boundary layer is obtained from the temperature mean (or the standard-deviation) profile. As $\lambda_v \vert \lambda_T$ the velocity boundary layer is inbeded in the thermal one on the whole range of Ra accessible. The scaling relation is found to be $\lambda_T \propto Ra^{-0.20\pm0.02}$. Therefore, within the error bar, the temperature and velocity boundary layer thicknesses are in a constant ratio ($\lambda_v/\lambda_T = 0.63 \pm 0.05$): the velocity and temperature boundary layers are coupled for $10^6 \leq Ra \leq 10^8$). An ultimate regime of thermal convection was expected to result from such a matching, though with a distinct scaling exponent of Nu versus Ra. Note that in the regime we observe, the length scale characteristic of the temperature profile λ_T is distinct from the diffusive thermal length scale which scales as $1/Nu$, that is $\propto Ra^{-0.25}$. This is an indication that the boundary layer may be already turbulent. These features indicate that, if there is no further hydrodynamic transition, i.e. if the boundary layer is completely turbulent, we may have already reached in Hg the ultimate regime of thermal turbulence which is expected to occur for higher Pr fluids at very high Ra.

The fluctuations of the temperature are strongly non-Gaussian in the major part of the cell. In the center, they are symmetric with large intermittent wings, but they go asymmetric in the boundary layer region. We show that, in the whole range of Ra, different statistical quantities such as the mean, the standard-deviation, the skewness S, the flatness F and the skewness of the time derivative S' of the temperature have unique profiles, independent of Ra, if plotted versus the non-dimensional distance to the plate z/λ_T. The quantity $S \times S'$ is found negative in the bulk region (outside the velocity boundary layer). This feature implies that, except in a thin layer along the boundaries, temperature fluctuations are not buoyancy driven, but passively swept by the shear of the mean circulation [2].

References

1. Naert A., Segawa T., Sano M., *Phys. Rev.E* **56**, 2, (1997).
2. Segawa T., Naert A., Sano M., to appear in Phys. Rev. E

NONLINEAR FREQUENCY SELECTION
IN SPATIALLY DEVELOPING FLOWS

BENOÎT PIER

Laboratoire d'Hydrodynamique (LadHyX)
École polytechnique
F-91128 Palaiseau cedex, France

Spatially developing open flows such as bluff-body wakes, jets, or shear layers may, under certain conditions, sustain synchronized periodic oscillations over an extended streamwise domain [2]. Fluctuations saturate at a finite amplitude level in the unstable region and become tuned at an overall frequency. The intrinsic frequency and the associated spatial distribution of fluctuations define a *global mode* living on the underlying unstable basic state. Selection criteria for two distinct types of self-sustained oscillations, namely *soft* and *steep* global modes, may in general be derived, as outlined below.

A typical situation is a basic flow which is absolutely unstable (AU) in a central finite domain and stable far upstream and downstream. In the central unstable region perturbations grow, saturate at finite amplitude and become synchronized at an overall frequency whereas perturbations decay far downstream and upstream. The frequency selection mechanism can be expressed as a nonlinear eigenvalue problem: the matching of finite amplitude time-periodic oscillations in a central region to exponentially decaying tails in both upstream and downstream directions can only be achieved for a specific frequency.

The main assumption of the present study is the weak non-parallelism of the basic flow which is characterized by the slow streamwise coordinate $X = \epsilon x$, where $\epsilon \ll 1$ is the inhomogeneity parameter. Two nonlinear spatially extended systems have been or are currently under investigation: the one-dimensional complex Ginzburg–Landau (CGL) equation with slowly varying coefficients and the two-dimensional Navier–Stokes (NS) equations pertaining to slowly developing basic flows. The results have been firmly established in the case of CGL [3, 4] and work is in progress in the case of NS.

U. Frisch (ed.), Advances in Turbulence VII, 129–132.

Under the assumption of weak inhomogeneity of the basic flow, local stability properties are obtained within the parallel flow approximation by freezing the slow streamwise coordinate X. At the local level of description linear stability properties of streamwise propagating normal modes of the form $\psi \sim \exp i(\frac{1}{\epsilon} \int^X k(u)du - \omega t)$ are characterized by the local *linear* dispersion relation

$$\omega = \Omega^\ell(k; X), \tag{1}$$

where ω and k are the *complex* frequency and wavenumber. From this dispersion relation complex local linear spatial branches

$$k^{\ell\pm}(X; \omega)$$

are derived for each value of ω. In at most convectively unstable (CU) regions the $+$ and $-$ superscripts are unambiguously assigned to the downstream and upstream response to a monochromatic excitation of frequency ω applied at location X. A branch point $(k^{\ell+} = k^{\ell-})$ occurs when the frequency ω equals the local absolute frequency $\omega_0(X)$ defined as [2]

$$\omega_0(X) = \Omega^\ell(k_0, X) \quad \text{with} \quad \frac{\partial \Omega^\ell}{\partial k}(k_0, X) = 0.$$

Typically we consider flows with a finite AU domain ($\operatorname{Im} \omega_0(X) > 0$) which are stable far upstream and downstream.

In the unstable region, fully nonlinear periodic traveling wave solutions $\psi \sim \Psi(\frac{1}{\epsilon} \int^X k(u)du - \omega t)$ where Ψ is a 2π-periodic function, may be obtained, provided that the *real* frequency ω and *real* wavenumber k satisfy the local *nonlinear* dispersion relation

$$\omega = \Omega^{n\ell}(k; X). \tag{2}$$

Solving this equation for a given frequency yields the nonlinear wavenumber branches

$$k^{n\ell}(X; \omega).$$

In the context of the CGL equation with slowly spatially varying coefficients only two types of global solutions exist [4] in the form of soft or steep global modes and their resonance frequency is prescribed by distinct conditions on the dispersion relations, as summarized below.

Soft global modes [3] are characterized by a slowly varying spatial envelope and wavenumber over the entire extent of the flow (Fig. 1). Two wavenumber branches denoted by $k^{n\ell-}$ and $k^{n\ell+}$ meet within the nonlinear region at a saddle point X_s of the nonlinear dispersion relation (2). The soft global frequency ω_s is given by

$$\omega_s = \Omega^{n\ell}(k_s; X_s), \tag{3a}$$

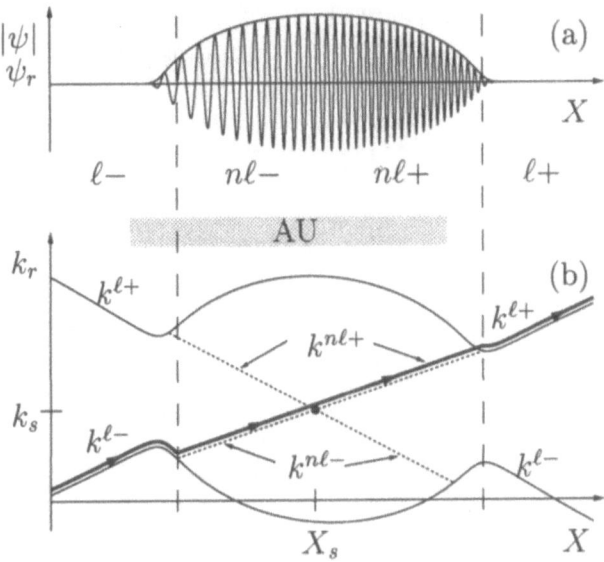

Figure 1. (a) Envelope $|\psi|$ and real part ψ_r of soft global mode of frequency ω_s. (b) Corresponding linear and nonlinear spatial branches in the X-k_r plane are represented by solid and dashed lines respectively. Local wavenumber follows path indicated by thick line in (b). Crossover between nonlinear branches occurs at X_s within the nonlinear region.

where the pair (k_s, X_s) satisfies the nonlinear saddle point condition

$$\frac{\partial \Omega^{nl}}{\partial k}(k_s; X_s) = \frac{\partial \Omega^{nl}}{\partial X}(k_s; X_s) = 0. \tag{3b}$$

As indicated in Fig. 1, note that the nonlinear ($nl-$ and $nl+$) regions and the AU region do not in general coincide.

Steep global modes [4] display a stationary front at the upstream boundary X_f of the AU region (Fig. 2). This front sharply links, within a few wavelengths, the upstream exponentially decaying part ($k^{\ell-}$ branch) to the downstream fully nonlinear saturated solution (k^{nl+} branch). In this case the global frequency ω_f is given by

$$\omega_f = \Omega^\ell(k_f; X_f) \tag{4a}$$

where the pair (k_f, X_f) satisfies the Dee–Langer [1] stationary front condition

$$\frac{\partial \Omega^\ell}{\partial k}(k_f; X_f) = 0 \quad \text{and} \quad \text{Im}\,\Omega^\ell(k_f; X_f) = 0. \tag{4b}$$

In other words, X_f is the location of *real* absolute frequency ω_f separating the CU and AU regions.

BENOÎT PIER

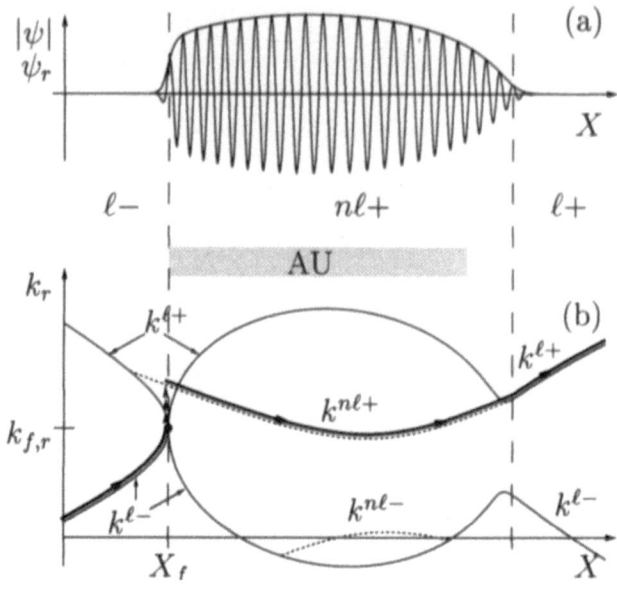

Figure 2. Same as Fig. 1 for a steep global mode of frequency ω_f with a sharp front located at the upstream boundary X_f of AU region (indicated in grey). The wavenumber jump at the front is indicated by repeated arrows.

The nature of the observed global mode is determined by formally computing the respective characteristic frequencies ω_s (3ab) and ω_f (4ab): the mode of largest frequency is selected and no other global mode may occur.

In all cases, frequency selection takes place at the downstream position where a − branch is linked to a + branch: $k^{n\ell-}$ and $k^{n\ell+}$ at X_s for a soft global mode, $k^{\ell-}$ and $k^{n\ell+}$ at X_f for a steep global mode. These stations effectively act as frequency generators for the entire flow. They may be interpreted as local oscillators inducing the upstream − and the downstream + branches, whether these are linear or nonlinear.

An essential difference between steep and soft global modes is that steep global modes only involve one nonlinear spatial branch. The sharp front allows an immediate crossover from the linear − to the nonlinear + branch. Generalization of the present theory to the general framework of the 2D Navier–Stokes equations is in progress.

References

1. G. Dee and J.S. Langer, Phys. Rev. Lett. **50**, 383 (1985).
2. P. Huerre and P.A. Monkewitz, Ann. Rev. Fluid Mech. **22**, 473 (1990).
3. B. Pier and P. Huerre, Physica D **97**, 206 (1996).
4. B. Pier, P. Huerre, J.-M. Chomaz, and A. Couairon, submitted to Phys. Fluids (1998).

SPATIAL DIRECT NUMERICAL SIMULATIONS OF TRANSITION IN ROTATING DEAN FLOW

T. RANDRIARIFARA[1] AND A. BOTTARO[2]
[1] *LMF-DGM, EPF-Lausanne, 1015 Lausanne, Switzerland*
[2] *IMFT, Université Paul Sabatier, 31062 Toulouse, France*

1. Introduction

The investigation of flows in situations where large scale streamwise vortices appear, such as in curved and/or rotating boundary layers and channels, is a task of interest in practical problems (i.e. water turbines) and at the same time it presents a fundamental appeal for its role of *model* configuration for the comprehension of the processes leading to turbulence. In this work, spatial direct numerical simulations are carried out to investigate the transition to turbulent flow in a curved channel rotating about its spanwise axis. Experiments [1, 2] show that this flow experiences transition via successive bifurcations: the first instability is characterized by large scale steady streamwise vortices; the secondary instability manifests itself as a travelling wave. The approach adopted here consists in first computing the base (vortical) flow for supercritical parameters. Then, linear secondary stability analyses (Floquet-type) are performed locally at different streamwise positions of the developing Dean flow. Finally, the steady vortical flow is excited, at the entrance of the computational domain, by an unsteady perturbation based on the secondary instability eigenmodes. Alternatively, a stochastic disturbance excitation is applied to let the dynamics of the flow amplify selectively the perturbation.

2. Numerical method

The three-dimensional Navier-Stokes equations are solved by a high accuracy Legendre spectral element method [3]. The technique consists in subdividing the physical domain onto non-overlapping subdomains on each of which the solution is approximated by Legendre polynomials. Pressure

U. Frisch (ed.), Advances in Turbulence VII, 133–136.
© 1998 *Kluwer Academic Publishers.*

grids are staggered to avoid spurious pressure modes. Time integration is third order accurate and it uses implicit backward differentiation formulas for the linear terms and an operator-integration-factor splitting method [4] for the non-linear terms. The discretized system of Navier-Stokes equations is decomposed in generalized block LU matrices [5] to handle the pressure-velocity decoupling. The Helmoltz and the pressure equations are iteratively solved by preconditioned conjugate gradient methods.

3. Linear secondary stability analyses

Firstly, the steady vortical flow is computed by a spatial approach over a domain of chosen spanwise periodicity [6]. The presence of the vortices generates inflectional velocity profiles which are Rayleigh-unstable. Inviscid, linear secondary stability analyses are then performed at different cross-sections using temporal normal mode analysis. Perturbations are taken in the form: $f'(t, x, y, z) = \hat{f}(y, z)e^{i\alpha(x-ct)}$, with f' representing the velocity or the pressure disturbances; x, y and z are, respectively, the streamwise, normal and spanwise coordinates; α is the (real) streamwise wavenumber and $c = c_r + ic_i$ is the (complex) phase speed. The inviscid stability equations with the locally parallel flow assumption yield (after elimination of the velocity components):

$$(\frac{\partial^2}{\partial y^2} + \frac{\partial^2}{\partial z^2} - \alpha^2)\hat{p} - \frac{2}{(U-c)}\frac{\partial U}{\partial y}\frac{\partial \hat{p}}{\partial y} - \frac{2}{(U-c)}\frac{\partial U}{\partial z}\frac{\partial \hat{p}}{\partial z} = 0, \quad (1)$$

$$\frac{\partial \hat{p}}{\partial y} = 0 \text{ at the walls}, \quad (2)$$

with $U(y, z)$ streamwise base flow. Equations (1) and (2) are solved by a spectral collocation method. Figure 1 shows the most unstable modes corresponding respectively to $\alpha = 7.2$, $c_r = 0.94$ (F = 103 Hz for U=1 m/s) for the sinuous mode and $\alpha = 7.2$, $c_r = 0.83$ (F = 91 Hz) for the varicose mode.

4. Results

The Dean flow is convectively unstable [7], which means that an impulse excitation introduced at some position propagates downstream and leaves the computational domain. The vortical flow is excited at a given streamwise position by an oscillating perturbation of assigned (small) amplitude and shape given by the eigenfunctions of the most unstable sinuous or varicose local mode of figure 1. Figures 2 and 3 show that these modes grow

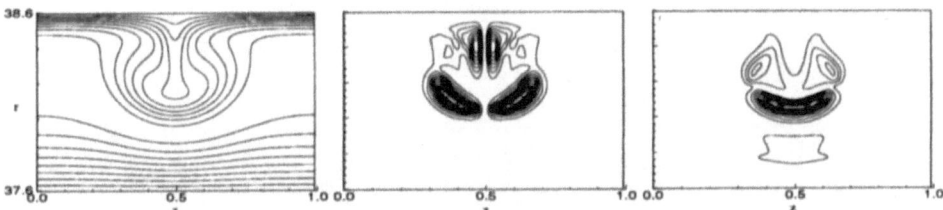

Figure 1. Left: streamwise velocity contours at one cross section (later taken to be the inflow for the transition computations); middle and right figures are the corresponding streamwise velocity eigenfunctions for the most unstable sinuous and varicose modes, respectively.

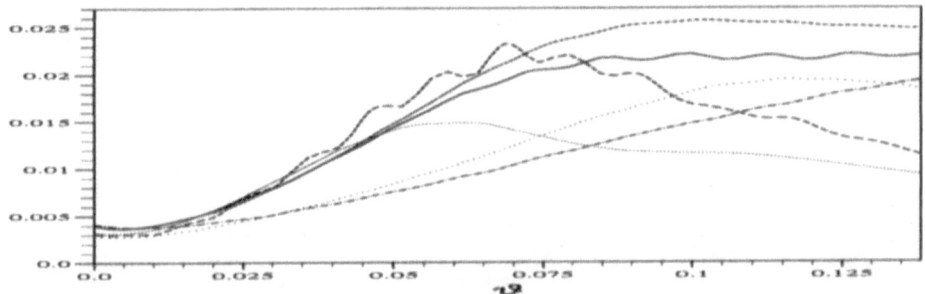

Figure 2. Rms energy as function of the angular distance along the curved channel, Re=1200. Continuous line: sinuous mode, Ro=0.05; dotted: varicose, Ro= 0.05; long-dashed: varicose, Ro=0; short-dashed: sinuous, Ro=0; dot-dashed: sinuous, Ro=-0.05; sparse dotted: varicose, Ro=-0.05. $Re = \frac{Ud}{\nu}$ and $Ro = \frac{\Omega d}{\nu}$; U: bulk speed, d: channel height, ν: kinematic viscosity.

exponentially downstream and saturate. Spatial oscillations of the fluctuations are observed. The rotation of the system affects the development of the secondary instability, and counter-rotation slows down the growth of the secondary waves, for both sinuous and varicose modes. A simulation initialized by a random (in time and space) disturbance velocity field exhibits the simultaneous development of the two types of modes (figure 4). The dominant frequencies of this latter simulation are in the range predicted by the linear theory. A kinetic energy balance analysis shows that the spanwise and the vertical shear of the main streamwise velocity are responsible for the energy production term, whereas the spanwise gradient of the velocity fluctuations contributes the most to the energy dissipation.

References

1. Matsson, O.J.E. & Alfredsson, P.H. (1992) Experiments on instabilities in curved channel flow, *Phys. Fluids* **A 4** (8), pp. 1666-1676.

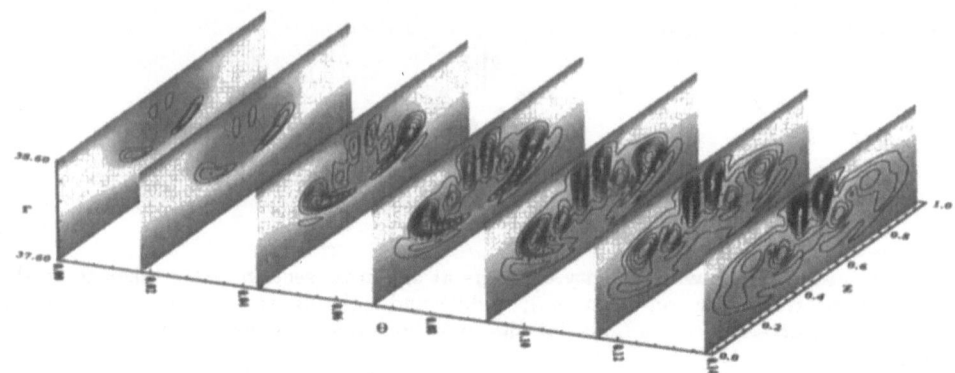

Figure 3.   Isolines of the rms values of the streamwise perturbation velocity for the sinuous mode, Ro=0.05, Re=1200 (the main streamwise velocity is shaded in the background). Maximum is 0.107*U.

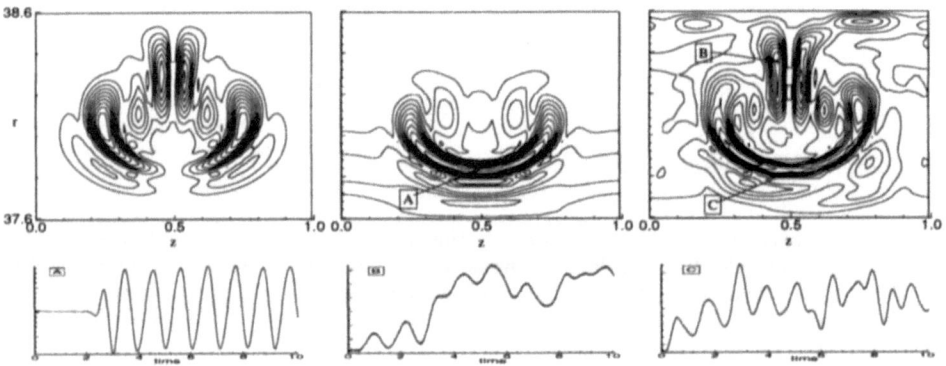

Figure 4.   Isolines of the streamwise perturbation velocity rms for θ=0.067 rad, Ro=0.05, Re=1200. Left: initial condition is the sinuous mode, middle: varicose mode, right: stochastic initial disturbance field. Bottom: time signals at the locations A (dominant frequency around 92 Hz), B (76 Hz) and C (86 Hz) indicated on the top figure.

2. Matsson, O.J.E. & Alfredsson, P.H. (1994) The effect of spanwise system rotation on Dean vortices, *JFM* **274**, pp. 243-265.
3. Couzy, W. (1995) Spectral element discretization of the unsteady Navier-Stokes equations and its iterative solution on parallel computers, *EPFL Thesis* **No 1380**.
4. Maday, Y., & Patera, A.T. & Ronquist, E.M. (1990) An operator-integration-factor splitting method for time-dependent problems: Application to incompressible flow, *J. Sci. Comp.* **Vol. 5**, No 4, pp. 263-292.
5. Blair, P. (1993) An analysis of the fractional step method, *J. Comp. Phys.* **Vol. 108**, pp. 51-58.
6. Randriarifara, T. & Bottaro, A. (1996) Effect of spanwise system rotation on spatially developing Dean vortices *Int. J. Num. Meth. Fluids* **Vol. 23**, pp. 1275-1287.
7. Huerre, P. & Monkewitz, P.A. (1990) Local and global instabilities in spatially developing flows, *Annu. Rev. Fluid Mech.* **22**, pp. 473-537.

SPATIAL NUMERICAL SIMULATION OF DISTURBANCE DEVELOPMENT IN A BOUNDARY-LAYER FLOW WITH COMPLIANT COATINGS

O. WIPLIER AND U. EHRENSTEIN

Université Lille I - Laboratoire de Mécanique de Lille,
Bd Paul Langevin, 59655 Villeneuve d'Ascq Cedex, France.

1. Introduction

The influence of compliant coatings on the transition in shear flows has extensively been studied since the pioneering experiments of Kramer [1]. There is now theoritical and numerical evidence that, although compliant walls may have a favorable influence on transition triggered by Tollmien-Schlichting waves (TSI), for highly compliant walls flow-induced surface instability (FISI) may become dangerous. A very recent survey of the various instability mechanisms in shear flows with compliant coatings can be found in Davies and Carpenter [2]. However, there is no clear answer about the capability of compliant coating to postpone transition.

To overcome the difficulty of non-parallelism in boundary layers some recent studies taking into account nonlinear effects focused on channel flow (which is truly parallel) with compliant coatings. The shortcomings of the locally parallel-flow assumption in boundary layers have been addressed by Ehrenstein and Rossi [3] who computed nonlinear travelling waves for the parallel Blasius flow with flexible walls. Hence, a numerical method taking into account both nonlinear and nonparallel effects would be a helpful tool for passive transition-control studies using compliant coatings. The present work focuses on such a numerical model capable of simulating the spatial disturbance growth in a coupled fluid-structure system.

2. Governing equations

As a prototype of compliant walls a spring backed elastic plate of Kramer's type is considered, the time-dependent vertical displacement $\eta(x,t)$ being

137

U. Frisch (ed.), Advances in Turbulence VII, 137–140.
© 1998 *Kluwer Academic Publishers.*

solution of

$$m\frac{\partial^2 \eta}{\partial t^2} + B\frac{\partial^4 \eta}{\partial x^4} + \kappa\eta = -p + \frac{2}{Re}\frac{\partial v}{\partial y}.$$

The quantities m, B and κ are respectively the mass density per unit length, the flexural rigidity and the spring stiffness, p is the perturbation fluid pressure and $\frac{2}{Re}\frac{\partial v}{\partial y}$ is the normal viscous stress of the perturbation fluid velocity. The kinematical condition at the moving boundary writes

$$u = 0, v = \frac{\partial \eta}{\partial t}$$

u, v being respectively the streamwise velocity and the wall-normal velocity. The basic state is a boundary-layer flow in the absence of a pressure gradient. Thus the interface remains flat and the displacement η is zero. The Navier-Stokes system for the disturbances in primitive variables is considered in a spatial domain $x_0 \le x \le x_1$ and $\eta(x,t) \le y < \infty$. A mapping is performed to transform the complex geometry into a cartesian one and the equations are discretized using 4-th order finite differences in x-direction and Chebyshev-collocation in wall-normal y-direction. At outflow a buffer-domain technique is used in order to relaminarize the flow and hence to avoid reflections at the outflow boundary [4].

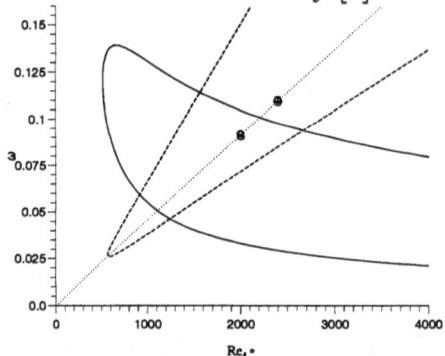

Figure 1. Neutral curves : — T.S.I. and - - F.I.S.I. with $m = 12.13$, $B = 379$ and $\kappa = 5.27 \times 10^{-2}$. The symbols $\otimes$ and $\oplus$ represent $Re = 2000$, $\omega = 0.09124$ and $Re = 2403$, $\omega = 0.1096$ respectively.

3. Numerical results

The linear stability of a boundary-layer flow with compliant coatings has extensively been studied for different classes of flexible surfaces since the work of Carpenter and Garrad [5]. Here, linear stability analyses (cf. Fig. 1), based on the parallel-flow assumption for the basic flow $\vec{U} = (U(y), 0)$, provide the inflow conditions (at the point marked $\otimes$) for the nonparallel spatial simulation of the disturbance development. For the numerical simulation of the nonlinear, nonparallel, spatial evolution, the time integration has been performed up to $18T$, T being the time period $2\pi/\omega$, and the length of the physically meaningful computational domain is about $13\lambda_{TS}$ with λ_{TS} the Tollmien-Schlichting wave length. The buffer domain used here has a length of $5\lambda_{TS}$. The spatial evolution of the wall displacement η is shown in Fig. 2. There is evidence of two wall-instability mechanisms, the dominant one being of FISI type, modulated by the TS-travelling-wave instability.

Fig. 3 depicts the maximum with respect to the wall-normal coordinate of the streamwise component of the disturbance flow. Clearly, the dominant flow instability is of Tollmien-Schlichting type. The amplitude of the disturbance first increases, then decreases after the perturbation has left the unstable region (cf. Fig.1, the end of the physical domain corresponds to the point marked $\oplus$).

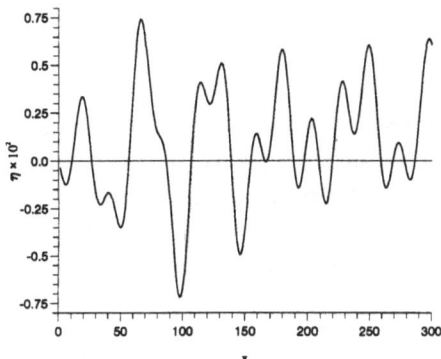

Figure 2.   Spatial evolution of the wall displacement.

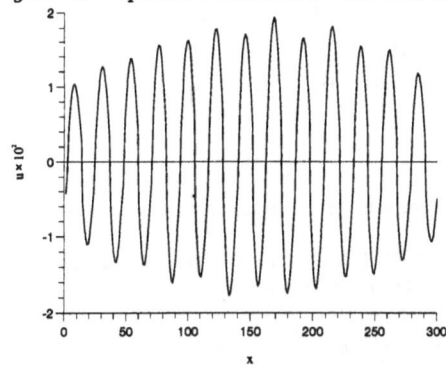

Figure 3.   Nonlinear spatial evolution of disturbance streamwise velocity

The power spectrum, by analyzing the instantaneous spatial distribution of the wall displacement, is shown in Fig. 4. Inspecting the maximum values one recovers the wavenumbers $\alpha = 0.111$ and $\alpha = 0.276$ corresponding respectively to the FISI and the TSI mechanism.

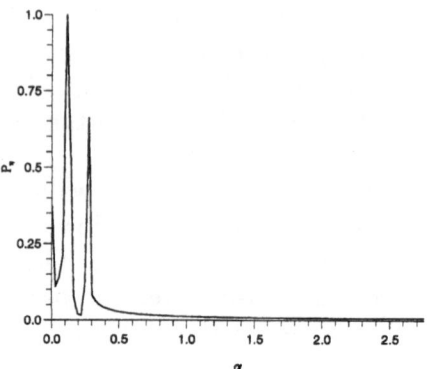

Figure 4. Power spectrum.

4. Discussion

In most of the models used to simulate the influence of compliant coatings on boundary-layer transition nonparallel and/or nonlinear effects are discarded. It has been shown that the present model, which takes into account the complete nonlinear fluid/structure interaction, is capable of reproducing several sources of instabilities, as illustrated in Fig. 1. Forthcoming numerical experiments will contribute to the understanding of passive transition control using compliant coatings.

Acknoledgements

This work has been supported by the D.G.A., Direction des Systèmes de Force et de la Prospective, France (Contrat ERS n97.1046A./DSP). Parts of the computations have been performed on the IBM/SP2 of the CNUSC, France.

References

1. Kramer, M.O. (1962) Boundary layer stabilization by distributed damping, *J. Amer. Soc. Nav. Eng.* **74**, 341.
2. Davies, C. and Carpenter, P.W. (1997) Instabilities in a plane channel flow between compliant panels, *J. Fluid Mech.*, **352**, 205.
3. Ehrenstein,. U. and Rossi, M. (1996) Nonlinear Tollmien-Schlichting waves for a Blasius flow over compliant coatings, *Phys. Fluids*, **8**, 1036.
4. Joslin, R.D., Street, C.L. and Chang C.L. (1993) Spatial direct numerical simulation of boundary-layer transition mechanims : validation of PSE Theory, *Theor. Comput. Fluid Dyn.*, **4**, 271.
5. Carpenter, P.W. and Garrad, A.D. (1985) The hydrodynamic stability of flow over Kramer-type compliant surfaces. Part 1 : Tollmien-Schlichting instabilities, *J. Fluid Mech.* **155**, 465.

A HEURISTIC ESTIMATION OF THE SMALLEST TRANSITON REYNOLDS NUMBER IN WALL BOUNDED FLOWS

CHRISTOPH VOIGT
Institut für nichtlineare Dynamik
Georg-August-Universität Göttingen
Bunsenstrasse 10
D-37073 Göttingen
Germany

Abstract: The aim of this paper is to present an estimation of the smallest Reynolds number in wall flows. A variational principle -which can be derived from large deviation principles or which can be seen as a variant of the recent introduced escape rate formalism in the nonequilibrium statistical mechanics context- is a means to an end. It is remarkably that the Kolmogorov and the Karman constants can be calculated by suitable assumptions and by a physical interpretation of the variational principle. These two constant determine the extend of the buffer layer and by it the smallest critical Reynolds number. These will be done for the pipe-, the Blasius- and the Couette-flow.

1. Introduction

When calculating critical Reynolds numbers the usual concept is to linearize the evolution equation which operates on laminar solutions of the Navier-Stokes equation and to analyse the spectrum of the problem. However, it is questionable what this concept has to do with the transition to turbulence. For example, there is the general accepted certainty that the laminar pipe flow is not linear instabil, but it is well known that there is a transition to turbulence. It is now suggested that the non-normality of the linearized operator may be the reason of transient growth which can trigger the transition to turbulence (Waleffe 1995 and quoted references there). This new concept cannot predict transition Reynolds numbers. The object of this paper is to show a new concept which may allow to estimate the smallest critical Reynolds numbers in wall flows. The new view is to ask what are the minimal conditions to have a self-sustained turbulent flow. The answer may be that turbulence is determined by inertia. This can be made quantitative by the claim that every turbulent flow have at least a rudimentary energy cascade. It is assumed that the log law of the wall is valid to make the things as easy as possible.

U. Frisch (ed.), Advances in Turbulence VII, 141–144.

2. Variational Principle

From the point of view of our theory turbulence is a large deviation from the thermodynamical equilibrium. The relaxation to equilibrium can be characterized by the so called rate function. This rate function is nothing else but the escape rate which is introduced to quantify nonequilibrium ensembles in statistical mechanics(Gaspard et.al.1995). The large deviation theory (Oono1989) describes the rate function as the relative Kolmogorov-Sinai entropy of the dynamics which produce the large fluctuation to the equilibrium dynamics. However, the required dynamics are the dynamics of the microscopic level. At first this seems to be curious. Nevertheless, the concept of local equilibrium which is fundamental to the Navier-Stokes dynamics and the relative entropy allow us to use macroscopic notions in the theory.

The large fluctuation from the equilibrium is characterized by the macroscopic kinetic energy of the macroscopic velocity field.The main idea is now to create a variational principle by using the contracting principle of the large deviation theory (Oono 1989).One has to minimize the rate function I with taking into account the condition of finite kinetic energy E of the macroscopic velocity field.

$$\delta\left[I+\beta E\right]=0, \quad \beta \text{ Lagrange parameter}$$

Assuming the velocity field is locally isotropic the rate function I and the kinetic energy E can be parametrized by the structure function of second order. As results we get that the structure function has the 2/3 scaling exponent. However, it can be also shown that the so called longitudinal Kolmogorov constant b is not universal but depend on the turbulence production mechanism. For two flows this dimensionless prefactor can be calculated explicitly. For homogeneous isotropis turbulence it has the value $b=\frac{3}{2}\left(\frac{4}{3}\right)^{\frac{2}{3}}=1.717$. The minimum of the variational principal is reached by $b=2.216$. Here, local isotropy is only fulfilled in an approximate sense. It can be shown that there must be an energy transfer from smale scales to large scales beside the usual energy cascade from large to small scales. There is every indication to believe that the minimum of the variational principle is reached in wall bounded flows (Härtel et al. 1998). With new interpretation of the summands I as turbulent energy production and βE as turbulent energy dissipation the inverse of the Karman constant κ can be caculated $\kappa^{-1}=2\left(\frac{2}{3}b^{\frac{3}{2}}-1\right)=2.4$. The log-law of the wall flow describes the inertial dominated part of the mean flow near walls. The transition from viscous - the linear flow profile - to inertial flow regime is about 11 y^{+}-units. This can be made plausible by remembering that the energy cascade has also a transition from inertial to viscous flow regimes near 11 Kolmogorov units, when $b = 2.216$, and near the wall the Kolmogorov and the y^{+}- units are identical. This allows to calculate the additive constant in the wall law. One gets : $u^{+}(y^{+})=2.4\ln\left(y^{+}\right)+5.23$

3. The Extension of the Buffer Layer

The velocity profil does not have a kink near $11\,y^{+}$-units. An interpolation profil is for example:

$$u^+(y^+) = y^+ \qquad\qquad y^+ \le y_0^+$$
$$u^+(y^+) = 2.4\ln([y^+ - y_0^+]\kappa + 1) + y_0^+ \qquad y^+ > y_0^+$$
$$y_0^+ = 5.23 + 2.4\ln 2.4 = 7.33$$

y_0^+ can be interpreted as the upper boundary of the viscous sublayer. For $y^+ > y_0^+$ the profil has not the fully developed log law. It is assumed that at a distance from the wall where the integral length of the turbulence is greater than the transition length at which the energy cascade is inertial dominated. This wall distance is called the upper boundary y_1^+ of the buffer layer. For distances greater than the buffer layer there are at least rudimentary energy cascade possible. As result one gets $y_1^+ = 3b\kappa^{-1} + y_0^+ = 65.8$.

4. Exampels

4.1. PIPE FLOW

Turbulence is only possible when there exists at least a rudimentary energy cascade. That means the dimensionless pipe radius must be greater than the upper boundary of the buffer layer. The laminar velocity profil $u(r)$ is given by $u(r) = u_c\left(1 - \left(\frac{r}{a}\right)^2\right)$, u_c is the centerline velocity, a the pipe radius. the friction velocity is given by $u_r^2 = \nu\frac{du}{dr}\Big|_{r=a}$, the friction velocity and the kinematic viscosity ν are used to make all objects dimensionless. The Reynolds number can now be written as $Re = \frac{u_c a}{\nu} = \frac{1}{2}a^{+2}$. Turbulence is possible when $a^+ > 65.8$. Therefore, Re must be greater than $Re_{kr} = \frac{1}{2}(65.8)^2 = 2165$.

4.2. BLASIUS FLOW

In this case the laminar profil cannot be given in closed form. It can be approximated in the following form $\frac{u}{U_0}(\frac{y}{\delta}) = 1.95(\frac{y}{\delta}) - 4.49(\frac{y}{\delta})^4 + 5.24(\frac{y}{\delta})^5 - 1.7(\frac{y}{\delta})^6$.

$\delta = 5.9_2\sqrt{\frac{\nu x}{U_0}}$ is the boundary thickness, y the wall distance. U_0 is the outer velocity.

With the same procedure as above one gets for the critcal Reynolds number with the displacement thickness $Re_{kr} = 590$.

4.3. COUETTE FLOW

The laminar profil is linear $u(y) = U_0\frac{y}{d}$, d is the distance between the two plates, U_0 is the difference velocity of the plates. To compare this with the pipe flow one has to take the half velocity difference and the half plate distance. The critcal Reynolds number is now $\mathrm{Re}_{kr} = \dfrac{U_0 d}{4\nu} = 1100$.

References

Gaspard, P.; Dorfmann, J.R. (1995) Chaotic scattering theory, thermodynamic formalism, and transport coefficients Phys.Rev.E**52**, 3525-3552

Härtel, C.; Kleiser, L. (1998) Analysis and modelling of subgrid-scale motions in near-wall turbulence J.Fluid Mech.**356**, 327-352

Oono, Y. (1989) Large deviation and statistical physics Prog.Theoret.Phys.Suppl **99**, 165-205

Waleffe, F. (1995) Transition in shear flows. Nonlinear normality versus non-normal linearity Phys.Fluids 7 , 3060-3066

III

Numerical Simulation

Simulation Numérique

III.

AN ADAPTIVE WAVELET METHOD COMPARED TO NON-LINEARLY FILTERED PSEUDO-SPECTRAL METHODS FOR TWO-DIMENSIONAL TURBULENCE

K. SCHNEIDER[1], N. K.-R. KEVLAHAN[2] AND M. FARGE[2]

[1] *ICT, Universität Karlsruhe (TH), Kaiserstraße 12,*
76128 Karlsruhe, Germany; and Centre de Physique Théorique,
CNRS - Luminy -, Case 907, 13288 Marseille, France.
[2] *LMD-CNRS, Ecole Normale Supérieure, 24 rue Lhomond,*
75231 Paris cedex 05, France.

1. Introduction

Wavelet methods have been associated with turbulence, almost since their invention, to analyze the structure and dynamics of the flow, for a recent review we refer the reader to [1]. These studies have shown that the strongest modes of the wavelet transform of a two-dimensional turbulent flow represent the coherent structures (e.g. vortices), while the weaker modes represent the unorganized background flow. Furthermore, it has been found that the coherent vortices can be well represented by only a very few wavelet modes. These observations suggest that wavelets could be an efficient basis for two-dimensional turbulent flows since the dynamics of such flows are largely controlled by their coherent vortices.

In the present paper we briefly sketch an adaptive wavelet method and compare it with filtered spectral methods for the generic case of a three vortex interaction. With respect to previous work [3] the nonlinear term is now also evaluated fully adaptively using a partial collocation technique.

2. Adaptive wavelet scheme

A two-dimensional incompressible viscous flow is described by the Navier–Stokes equations. In velocity-vorticity formulation these are

$$\partial_t \omega + \mathbf{v} \cdot \nabla \omega = \nu \nabla^2 \omega, \quad \nabla \cdot \mathbf{v} = 0, \tag{1}$$

where the velocity field $\mathbf{v} = (u, v)$, the vorticity $\omega = \nabla \times \mathbf{v}$, both with periodic boundary conditions and ν is the kinematic viscosity.

U. Frisch (ed.), Advances in Turbulence VII, 147–150.

For time discretization we use classical semi–implicit finite differences of second order composed of an Euler backwards step for the viscous term and an Adams–Bashforth extrapolation for the convection term. For ease of notation we write here the similar first order scheme

$$(I - \Delta t \nu \nabla^2) \, \omega^{n+1} = \omega^n - \Delta t \, \mathbf{v}^n \cdot \nabla \omega^n \qquad (2)$$

To solve (2) we develop ω^n into an orthonormal wavelet series

$$\omega^n(x,y) = c^n_{0,0,0} \, \phi_{0,0,0}(x,y) + \sum_{j=0}^{J-1} \sum_{i_x=0}^{2^j-1} \sum_{i_y=0}^{2^j-1} \sum_{\varepsilon=1}^{3} d^{\varepsilon,n}_{j,i_x,i_y} \, \psi^\varepsilon_{j,i_x,i_y}(x,y) \quad (3)$$

where ϕ and ψ are the 2π–periodic two–dimensional scaling function and the corresponding wavelets, respectively. Applying a Petrov-Galerkin scheme to (2) with test-functions $\theta^\varepsilon_{j,i_x,i_y}(x,y) = (I - \Delta t \nu \nabla^2)^{-1} \, \psi^\varepsilon_{j,i_x,i_y}(x,y)$ diagonalizes the resulting stiffness matrix and therefore avoids the solution of a linear system at each time step [2]. The solution of (2) then reduces to a change of basis

$$d^{\varepsilon,n+1}_{j,i_x,i_y} = < \omega^n - \Delta t \, \mathbf{v}^n \cdot \nabla \omega^n \,, \, \theta^\varepsilon_{j,i_x,i_y} > \qquad (4)$$

where $< \cdot, \cdot >$ denotes the L^2 scalar product. A reduction of the degrees of freedom is obtained by retaining only those coefficients with $|d^{\varepsilon,n+1}_{j,i_x,i_y}| > \epsilon$. The set of active coefficients in the next time step is then determined from the previous step by compression with the required tolerance and adding of neighbours. The nonlinear term $\mathbf{v}^n \cdot \nabla \omega^n$ is computed by partial collocation. This method (also called pseudo–wavelet scheme) can be sketched as follows: starting from the wavelet coefficients of ω^n we obtain the values of ω^n on a locally refined grid through an inverse wavelet transform. Solving the Poisson equation using a Petrov–Galerkin scheme with test–functions $(\nabla^2)^{-1} \psi^\varepsilon_{j,i_x,i_y}(x,y)$, we get the wavelet coefficients of the stream function Ψ^n. Applying an inverse adaptive wavelet transform, the stream function is reconstructed on a locally refined grid. Subsequently, the velocity $\mathbf{v}^n$ and $\nabla \omega^n$ are calculated using finite differences of 4th order on an adaptive grid. Then the scalar product $\mathbf{v}^n \cdot \nabla \omega^n$ can be calculated at the grid points. Finally, the right hand side of (4) is summed up on the adaptive grid in physical space and then the wavelet coefficients of the vorticity ω^{n+1} are calculated using the adaptive vaguelette decomposition [2].

3. Numerical results

In this section we compare a classical pseudo–spectral method, a nonlinearly filtered version, i.e. we retain only Fourier coefficients having absolute

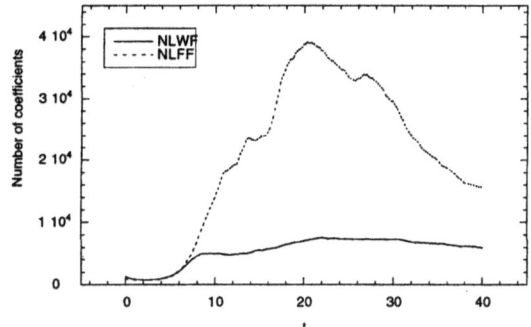

Figure 1. Evolution of the number of active modes for each method (NLFF = nonlinear Fourier filtering, NLWF = nonlinear wavelet filtering and adaptive wavelet simulation).

value above a given threshold and the adaptive wavelet method. For further details we refer the reader to [3]. The methods are each applied to a highly nonlinear flow typical of two-dimensional turbulence, the merger of two positive vortices pushed together by a weaker negative vortex. The change in the number of active modes as a function of time for the nonlinear Fourier filtering and wavelet methods is shown in Fig.1. The adaptive wavelet simulation uses 9.2 fewer modes than the reference pseudo-spectral simulation at the time of merging $t = 20$. The fact that there is no large peak in the number of wavelet modes at the time of merging (when the gradients of vorticity are the strongest) shows that the wavelet method is very efficient at representing the strong nonlinear interactions that characterize the dynamics of turbulence. This is because the nonlinear interactions are highly localized in physical space which implies de-localization in spectral space, and hence the need for a larger number of Fourier modes to represent the flow. In Fig.2 we show the vorticity field for the pseudo–spectral and the adaptive wavelet method with the corresponding computed wavelet coefficients. The comparison of the vorticity fields with the pseudo–spectral method, shows no significant difference. If we look at the energy spectra at $t = 40$ we can observe quantitatively that all relevant scales, in particular the small ones, are well resolved. However, the fine resolution is only required locally. These results suggest that the adaptive wavelet method combines a consistently high compression rate with high accuracy.

References

1. Farge, M., Kevlahan, N. K.-R., Perrier, V. and Schneider, K. (1998) Turbulence analysis, modelling and computing using wavelets. *Wavelets and Physics* (Ed. J.C. van den Berg), Cambridge University Press, in press.
2. Fröhlich, J. and Schneider, K. (1997) *J. Comput. Phys.*, **130**, pp. 174–190.
3. Schneider, K., Kevlahan, N. K.-R. and Farge, M. (1997) *Theoret. Comput. Fluid Dynamics* **9** (3/4), pp. 191–206.

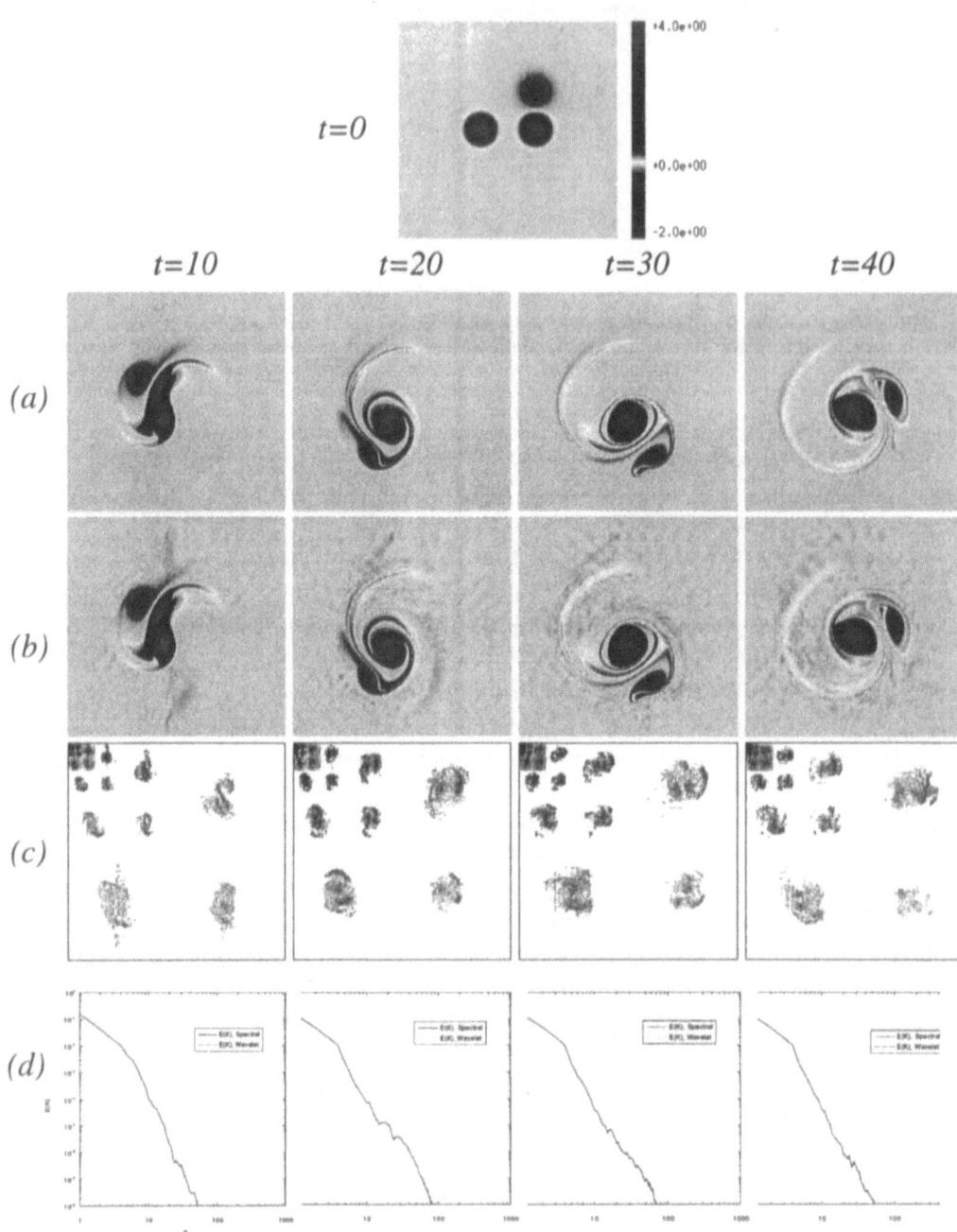

Figure 2. Evolution of the vorticity field as a function of time for the pseudo-spectral reference simulation (a) and adaptive wavelet method (b). Evolution of the active wavelet coefficients (c) used in the calculations of the adaptive wavelet method. Evolution of the energy spectrum for the two methods (d). For details we refer the reader to [3].

DYNAMICS OF COUNTERCURRENT SQUARE JETS

F.F. GRINSTEIN
Laboratory for Computational Physics & Fluid Dynamics
Code 6410, US Naval Research Laboratory, Washington, DC 20375-5344, USA

Extensive investigations have examined passive ways to manipulate the natural development of large scale vortices and their breakdown into turbulence to enhance the three-dimensionality of the flow, and thus entrainment and mixing. Compared to the axisymmetric jet, square jets are of special interest because they offer *passively* improved mixing at both ends: enhanced large scale entrainment due to axis-switching, and enhanced small scale mixing near corner regions and farther downstream -- due to the faster breakdown of vortex ring coherence. Although, passive control concepts enjoy reasonable success compared to active control scenarios -- which might demand an excessive energy input, practical performance requirements often demand the use of active control strategies. Coaxial jet systems with actuation based on imposing a co-annular secondary flow (U_s) around the primary jet (U_o), have been used as effective *active* control approaches to jet noise reduction, thrust vectoring, and mixing enhancement. Beyond the use of the more traditional coaxial jet systems with $r=U_s/U_o>0$, promising non-conventional techniques use *synthetic* secondary flows with $<U_s> = 0$ [1], or promote global instabilities and enhanced vortex strength based on the *countercurrent* jet shear layer enforced by a secondary suction flow with $r<0$ [2]. In the present work, we address the impact of combining active and passive control concepts using *countercurrent square* jets, seeking understanding of the instabilities and vortex dynamics underlying their entrainment-enhancing performance, and their dependence on initial conditions.

This research is based on the use of three-dimensional numerical simulations on structured grids. The numerical jet model solves the time-dependent compressible flow conservation equations for total mass, energy, and momentum, with appropriate inflow/ outflow open boundary conditions, and an ideal gas equation of state [3]. The explicit, finite-difference numerical model is *globally* second-order accurate in space and time. For the moderately-high Re regime investigated (Re>170,000 -- based on the primary jet velocity and circular-equivalent jet diameter), a monotonically-integrated LES (MILES) [3,4] approach based on the solution of the unfiltered Euler equations with Flux-Corrected Transport algorithms is used. Realistic countercurrent boundary conditions are implemented at the jet exit plane, involving: supersonic inflow conditions for the primary jet, subsonic outflow conditions for the annular suction region surrounding the primary jet, and reflecting wall conditions otherwise (Fig.1); stagnant-flow and subsonic outflow conditions are enforced at cross-stream and downstream boundaries, respectively; spurious reflections at open computational boundaries are minimized using one-dimensional characteristic analysis to define the additional boundary conditions needed for closure of the discretized system of equations [5]. The computational domain has a streamwise extent of 10D and extends up to ±5D away from the jet axis in the transverse directions.

U. Frisch (ed.), Advances in Turbulence VII, 151–154.
© 1998 *Kluwer Academic Publishers.*

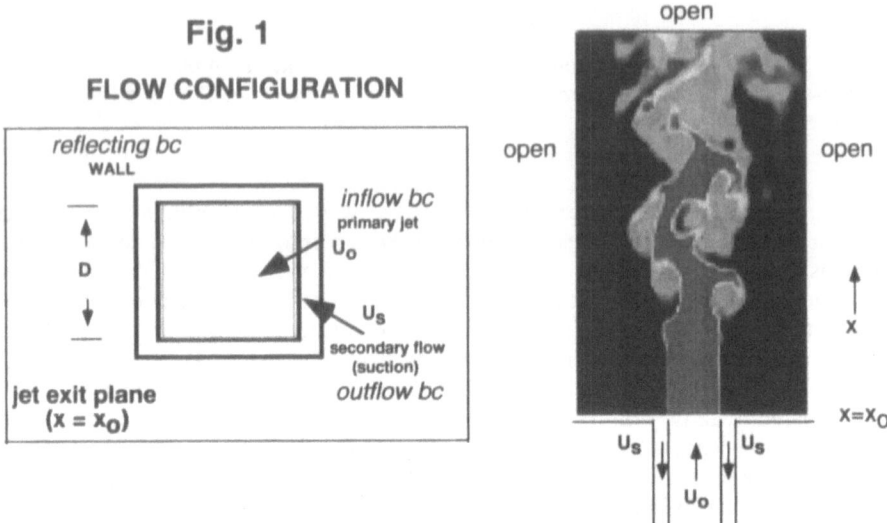

Fig. 1

FLOW CONFIGURATION

The cartesian grids use evenly spaced cells in the shear flow region of interest, and geometrical stretching in the cross-stream direction outside of the latter region to implement the open boundary conditions there.

In order to promote the absolutely unstable behavior in the jet shear-layer, the stability analysis reported by Ho & Huerre [6] and Huerre & Monkewitz [7], and the analysis and conditions in the countercurrent jet experiments by Strykowski et al. [2], were used as guidelines in choosing jet initial conditions. Stability analysis of the mixing layer in the incompressible regime indicates growth rates scaling with the velocity ratio $R=(1-r)/(1+r)$ [3], and its basic instability nature changing from convectively to absolutely unstable at a critical ratio $r_{cr}=-0.136$ [7], consistent with laboratory studies [8]. The stability analysis of the countercurrent shear layer was extended to the supersonic regime for the axisymmetric mixing layer, including also the effects of density ratio $s=\rho_2/\rho_1$ [2].

Countercurrent square jet simulations were performed in the supersonic regime, with $s=1$, jet Mach number $M=1.5$, and convective Mach number Mc in the range 0.75-0.85. The jet systems investigated consisted of air jets emerging into air surroundings at the same standard temperature and pressure. Comparisons were set between free jets ($r=0$) and countercurrent jets with $r = -0.15, -0.3, -0.4$, and otherwise identical initial and boundary conditions. The development in space and time of the initial transient from broad-band impulsively started jet systems is shown in Fig.2; the flow visualization is based on instantaneous distributions of the vorticity magnitude at three fixed times (selected to be the same for the three cases shown); by design, the initial transient has started to flow through the downstream boundary by the time of the left frames in the layout. The frames corresponding to the free jet case (Fig.2a), show convection of the transient through the downstream boundary, with very small leftover disturbances upstream – consistent with spatially growing disturbances of convectively unstable flow. On the other hand, the flow visualizations for the countercurrent jet cases (Fig.2b,c) indicate enhanced vortex strength and mixing of the jet with the surroundings and self-excited

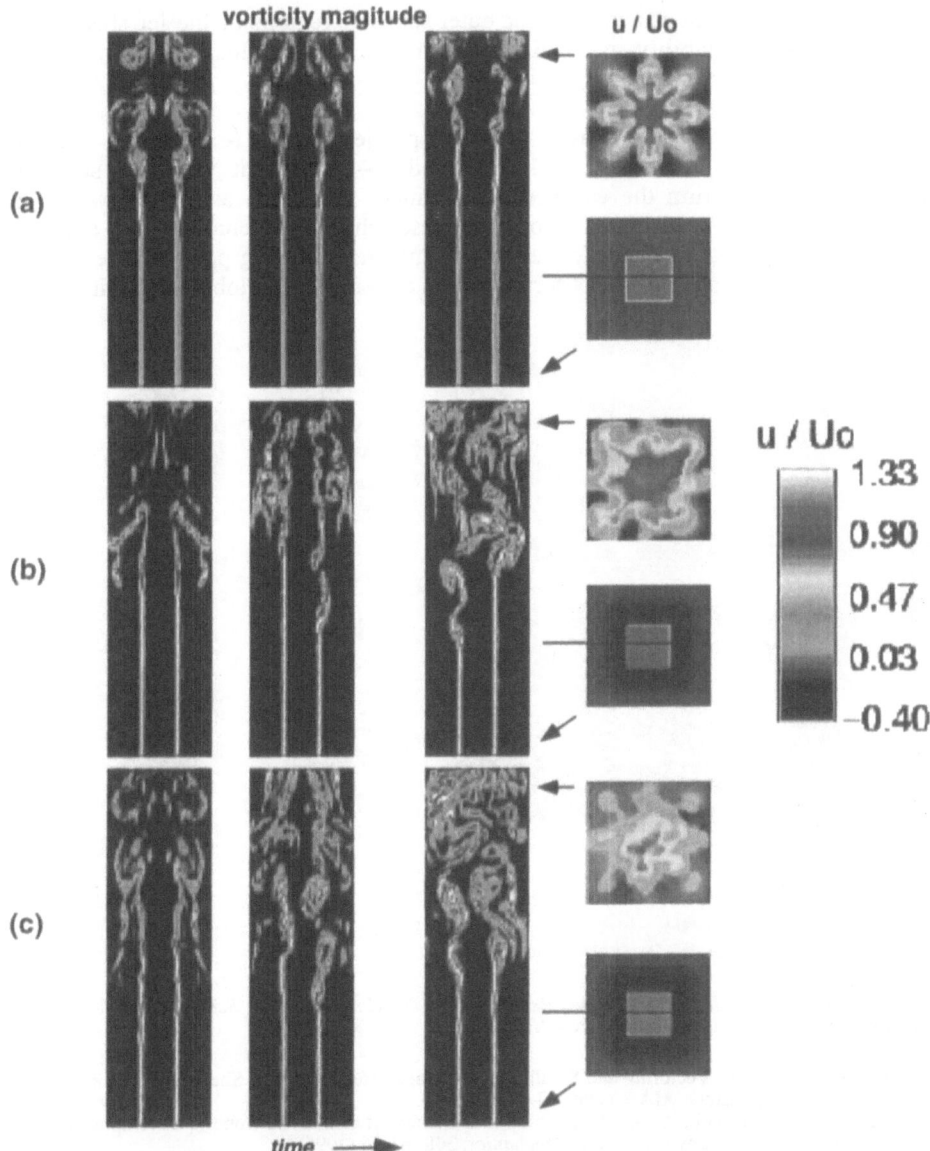

Fig. 2 INITIAL TRANSIENT DEVELOPMENT

instabilities depicting *global* absolutely unstable behavior (Fig.2). This absolutely unstable behavior is observed even for the case with weakest counterflow (r=-0.15), in contrast with significantly larger values of r_{cr} predicted by *local* stability theory (e.g., $r_{cr} \approx -0.35$ for the axisymmetric mode, and $r_{cr} \approx -0.31$ for the helical (m=6) mode)[1]. This discrepancy was also noticed in the laboratory experiments [2], where it was attributed to uncertainty in determining actual initial conditions.   Based on the results of the simulations, we speculate that these observations are due to feedback mechanisms involving pressure fluctuations acting on the outer subsonic regions of the jet shear layer -- such as the global instability mechanisms demonstrated in the (locally) convectively-unstable subsonic mixing layer [9].

The instantaneous flow visualizations of the developed jet regimes (e.g., Fig.3) indicate that the jet dynamics is dominated by helical modes -- consistent with the stability analysis in [2], and reaffirm the enhanced entrainment properties associated with the presence of co-annular suction. The ongoing research of countercurrent jet systems currently under way, seeks to quantify their entrainment-augmenting performance, and to further characterize the nature of the observed vortex dynamics and global instabilities.

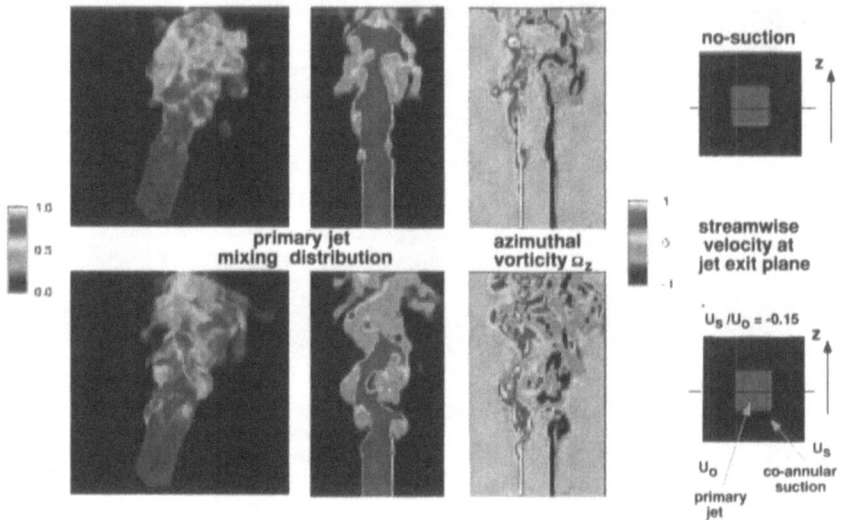

Fig. 3 DEVELOPED JETS
axially forced primary jet
St_D=0.2, forcing level 2.5% rms

This research is supported by the Office of Naval Research with Dr. Gabriel Roy as Scientific Officer, by the Naval Research Laboratory, and the DoD HPC-MP Program.

[1] Smith, B.L. and Glezer, A., Vectoring and Small-Scale Motions Effected in Free Shear Flows Using Synthetic Jet Actuators, AIAA Paper 97-0213 (1997).
[2] Strykowski, P.J. , Krothapalli, A., and Jendoubi, S., The Effect of Counterflow on the Development of Compressible Shear Layers, J. Fluid Mechanics, 308, 63-96 (1996).
[3] Grinstein, F.F. and DeVore, C.R., Dynamics of Coherent Structures and Transition to Turbulence in Free Square Jets, Physics of Fluids, 8, 1237-51 (1996).
[4] Grinstein, F.F. and Fureby, C., Monotonically Integrated Large Eddy Simulation of Free Shear Flows, AIAA Paper 98-0537 (1998), submitted to AIAA J.
[5] Grinstein, F.F., Open Boundary Conditions in the Simulation of Subsonic Turbulent Shear Flows, J. Comp. Phys., 115, 43-55 (1994).
[6] Ho, C.M. and Huerre, P., Perturbed Free Shear Layers, Annual Review of Fluid Mechanics, 16, 365-424 (1984).
[7] Huerre, P. and Monkewitz, P., Absolute and Convective Instabilities in Free Shear Layers, J. Fluid Mechanics, 159, 151 (1985).
[8] Strykowski, P.J. and Niccum, D.L., The Stability of Countercurrent Mixing Layers in Circular Jets, J. Fluid Mechanics, 227, 309 (1991).
[9] Grinstein, F.F., Oran, E.S., and Boris, J.P., Pressure Field, Feedback,and Global Instabilities of Subsonic Spatially Developing Mixing Layers, Physics of Fluids A, 3, 2401 (1991).

THE ROLE OF COHERENT STRUCTURE INTERACTIONS IN THE REGENERATION OF WALL TURBULENCE

JAVIER JIMÉNEZ AND ALFREDO PINELLI

School of Aeronautics, Universidad Politécnica de Madrid.
Plaza Cardenal Cisneros 3, 28040 Madrid, Spain

1. Introduction

The near-wall region is characterized by the presence of organized structures [1], but the way in which they interact to maintain the local dynamics and to extract energy from the mean flow is still controversial. The two dominant structures are the streamwise velocity streaks and the quasi-streamwise vortices. The former consist of long ($x^+ \sim 1000$) wavy arrays of alternating streamwise jets with an average spanwise wavelength of $z^+ \sim 100$. The latter are vortical structures almost aligned with the mean flow but slightly tilted away from the wall. While it is generally accepted that the vortices induce streaks by transferring mean streamwise momentum towards and away from the wall [2], there is less consensus on the generation mechanism of the streamwise vortices. Two conceptual models, widely discussed in the literature, are illustrated in Fig. 1: *A)* the streamwise vortices are a consequence of the breakdown of the streaks, probably due to inflectional instabilities of the local velocity profiles; *B)* the streamwise vortices are formed directly, or at least triggered, by the amplification of outer flow perturbations. Hypothesis *A* assumes that an autonomous wall cycle exists, while *B* predicts that wall turbulence depends on the existence of an outer flow.

In this paper we use numerical simulations to generate otherwise 'impossible' physical situations to enhance or weaken hypotheses *A* and *B* by introducing specific body forces. In this way we clarify their relative importance in natural wall turbulence. References [3, 4] use the same method to analyze other possible regeneration mechanisms.

U. Frisch (ed.), Advances in Turbulence VII, 155–158.

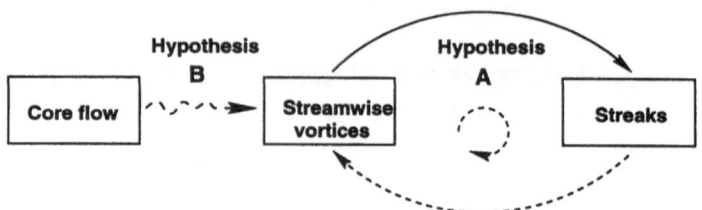

Figure 1.   Schematic representation of the hypotheses tested for the generation of near-wall turbulence. Solid lines, accepted mechanisms; dashed, uncertain ones.

2. The Role of the Outer Flow

We start by analyzing the possibility that no wall cycle exists and that the vortices are predominantly formed by the intensification of external perturbations. This is a plausible scenario since it is known from [5] that the near-wall scalings depend on the Reynolds number of the outer flow. The influence of this mechanism was established to be weak in a series of numerical experiments in [6], where the spanwise extent of a turbulent channel was shrunk until it could accommodate the basic wall structures ($L_z^+ \sim 100$) while inhibiting the larger structures in the core flow. Turbulence stayed active close to the wall, with statistical characteristics very similar to those obtained with a fully-developed core flow. That showed that a 'canonical' core was not crucial for the turbulent wall region, but it did not establish whether the wall could remain turbulent even in the absence of external disturbances. To check that possibility, we have run new simulations of turbulent channels in which the equations are modified to explicitly filter all the fluctuations above a certain distance from the wall. If periodicity is assumed in both streamwise and spanwise directions, the evolution equation for a generic Fourier mode can be written [7] as

$$\partial_t \hat{q}_{\alpha,\beta} = \hat{H}_{\alpha,\beta} + \nu(\partial_y^2 - \alpha^2 - \beta^2)\hat{q}_{\alpha,\beta}, \tag{1}$$

where $\hat{q}_{\alpha,\beta}$ is a Fourier mode of either the wall-normal vorticity component ω, or the Laplacian of the wall-normal velocity, and where $\hat{H}$ represents the nonlinear advection terms. In the simulations each numerical time step was modified to

$$\hat{q}_{\alpha,\beta}(y, t + \Delta t) = \left[\hat{q}_{\alpha,\beta}(y, t) + \Delta t \hat{H}_{\alpha,\beta}(y, t)\right] F(y), \tag{2}$$

where $F(y) = 1/2[1 - \tanh(y^2/\delta^2 - 1)]$ is a filter which acts as a dissipative body force, and damps all the fluctuations above $y \approx 0.65\,\delta$.

The result of the simulation was that as long as δ was chosen large enough to preserve the fluctuations below $y^+ \approx 75$, we still obtained a

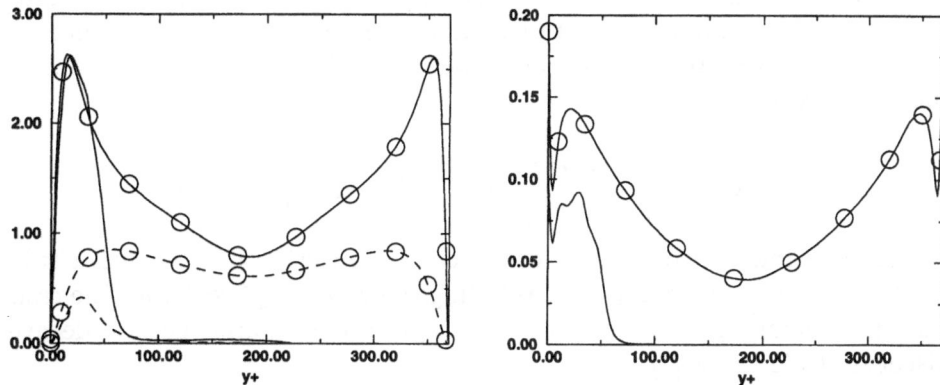

Figure 2. Left, velocity fluctuations: solid lines, streamwise; dashed, wall-normal. Right, streamwise vorticity fluctuations. Circles refer to the natural channel. All magnitudes are in wall units.

region close to the wall which shared many features with a natural fully turbulent channel flow, and which survived indefinitely. As shown in Fig. 2 the streamwise velocity and vorticity fluctuations (the indicators of streaks and quasi streamwise vortices in the wall region) were comparable to the ones of a natural channel. The fluctuations of the other components were lower than in a natural case, suggesting an influence of the external flow on the latter, as in [5]. The synthetic flow generated by (2) can be interpreted as a wall region where turbulence is continuously reproduced by an interaction cycle between wall streaks and streamwise vortices, without any influence from external disturbances. It actually constitutes an ideal environment in which to study that cycle.

3. The Role of the Wall Streaks

That the streaks are a necessary part of the cycle was tested next. Consider a numerical simulation of a channel in which the computational box is short with respect to the typical longitudinal extent of streaks. It was shown in [6] that turbulence could be sustained in boxes longer than about $L_x^+ \approx 400$, while natural streaks have longitudinal coherence lengths which are several times longer. The present experiments were run in doubly periodic boxes of size $L_x^+ \times L_z^+ \approx 500 \times 300$, wide enough to contain large core structures, at initial Reynolds numbers $Re_\tau \approx 200$. Each wall therefore contained two or three velocity streaks extending over the whole length of the box, each of which had associated on the average one streamwise vortex of each sign.

Since streaks span the length of the box and are roughly parallel to the mean flow, they can be approximately represented by that part of the streamwise velocity u which depends only on the transverse coordinates, $y - z$, but which is independent of x. Moreover, since we are only concerned with the spanwise modulation of u, we can define a streak component as the streamwise average

$$\Omega(y, z) = L_x^{-1} \int_0^{L_x} \partial u / \partial z \, \mathrm{d}x = L_x^{-1} \int_0^{L_x} \omega \, \mathrm{d}x. \qquad (3)$$

In our experiments we damped the streaks by filtering $\Omega(y, z)$ with $F(y) = 1/2 \tanh(y^2/\delta^2 - 1)$, as in (2), thus killing them *below* $y^+ \approx 1.5 \, \delta$. The result was that, as long as $\delta^+ > 60$, the cycle was destroyed and the wall decayed viscously to laminar [3].

4. Conclusions

These two experiments show that there is a self-sustained near-wall turbulence generation cycle, residing at $y^+ \approx 60 - 100$, and that it conforms essentially to hypothesis A in Fig. 1.

This work was supported the Spanish CICYT under contract PB95-0159. A.P. was supported by a HCM postdoctoral fellowship from the European Commission, and by CICYT.

References

1. Robinson, S.K. (1991) Coherent motions in the turbulent boundary layer, *Ann. Rev. Fluid Mech.* **23**, pp. 601–639.
2. Orlandi, P. & Jiménez, J. (1994) On the generation of turbulent wall friction, *Phys. Fluids* **6**, pp. 634–641.
3. Jiménez, J. & Pinelli, A. (1997) Wall turbulence: how it works and how to damp it, *AIAA Paper* **97-2122**.
4. Jiménez, J. & Pinelli, A. (1998) submitted to *J. Fluid Mech.*
5. Wei, T.& Willmarth, W. (1989) Reynolds number effects on the structure of a turbulent channel flow, *J. Fluid Mech.* **204**, pp. 57–95.
6. Jiménez, J.& Moin, P. (1991) The minimal flow unit in near wall turbulence, *J. Fluid Mech.* **225**, pp. 221–240.
7. Kim, J., Moser, R. & Moin, P. (1987) Turbulence statistics in fully developed channel flow at low Reynolds number, *J. Fluids Mech.* **177**, pp. 133–166.

SCALING EXPONENTS IN TURBULENT CHANNEL FLOW

G. AMATI[1,2], F. TOSCHI[3,4], S. SUCCI[5] AND R. PIVA[2]
[1] *CASPUR, P.le A. Moro 5, 00185, Roma, Italy.*
[2] *Dept. Meccanica & Aeronautica, Univ. "La Sapienza",*
Via Eudossiana 18, 00184, Roma, Italy.
[3] *Dept. Fisica, Univ. di Pisa, P. Torricelli, 56126, Pisa, Italy.*
[4] *INFM, Unitá di Tor Vergata, Roma.*
[5] *CNR-IAC, Viale del Policlinico 137, 00161, Roma, Italy.*

1. Intermittency and Scaling Exponents

The problem of understanding the statistical properties of turbulent flows is a very difficult one. In contrast to the intense efforts that have been devoted to the investigation of intermittency in homogeneous/isotropic flows, much less has been done for non-homogeneous and non-isotropic situations. In this paper we present some results about intermittency and scaling exponents in a wall bounded flow.

We study the scaling exponents of longitudinal streamwise structure functions defined as $S_p(r) \equiv < [(\vec{u}(\vec{x} + \vec{r}) - \vec{u}(\vec{x})) \cdot \hat{r}]^p >$. In the inertial range these scale as:

$$S_p(r) \simeq r^{\zeta_p} \tag{1}$$

Experimental and numerical investigations have shown that ζ_p are slightly, but definitely, different from Kolmogorov's dimensional prediction $\zeta_p = p/3$ (1941). In 1962 Kolmogorov suggested that $S_p(r) \simeq < \epsilon_r^{p/3} > r^{p/3}$, where ϵ_r is the energy dissipation field averaged over a volume of linear size r [1]. The exponents $\tau_p \equiv \zeta_p - p/3$ characterize the departure from the dimensional scaling $p/3$ and are related with the intermittency of dissipation ϵ_r. In fact, there is no theory able to compute the values of τ_p. A few year ago a model has been proposed by She and Levêque who suggested a relationship between structures in the flow (e.g. vortex tubes) and the anomalous exponents τ_p [2]. Actually this relationship could be of some interest since in wall turbulence the presence of structures is crucial for the dynamics of

U. Frisch (ed.), Advances in Turbulence VII, 159–162.
© *1998 Kluwer Academic Publishers.*

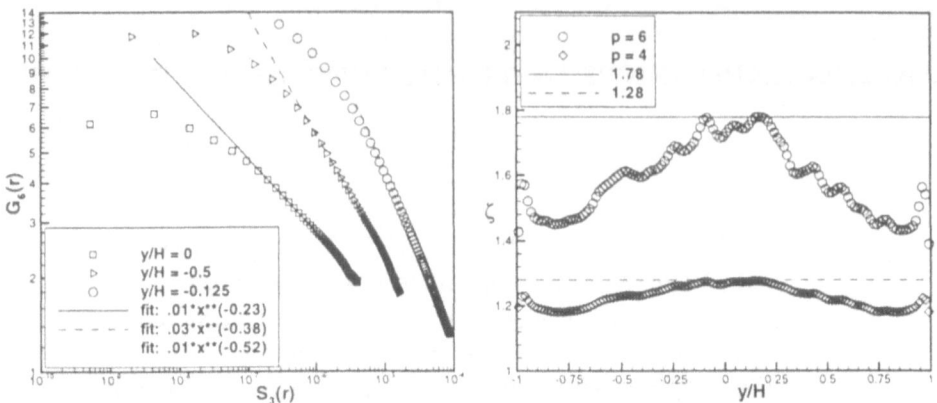

Figure 1.   Left picture: G_6 vs. S_3 for different distances from the wall. Right picture: scaling exponents ζ_p for $p = 4, 6$ as a function of y/H. Straight line represent homogeneous values.

the flow. Our main interest is to study the role of coherent structures for a deeper comprehension of intermittency in wall-bounded turbulence.

2. Scaling exponents in turbulent channel flow

To study the behaviour of scaling exponents in a wall bounded flow we have realized a direct numerical simulation of a channel flow, i.e. two no-slip wall and periodic boundary conditions along the other directions. The numerical scheme is a Lattice BGK on the parallel SIMD machine Quadrics [3]. This allowed us to obtain high statistics for each plane in the y direction, i.e. the normal-to-wall direction. The numerical grid was $256 \times 128 \times 128$ with a $Re \sim 3000$, measured at centerline. The quantity $2H$ will indicate, in the following, the channel width.

Scaling exponents are obtained using Extended Self Similarity (ESS) [4], which suggests a relation between structure functions of the form:

$$S_p(r) = S_3(r)^{\zeta_p/\zeta_3}. \tag{2}$$

ESS permitted us to extract exponents from our low Re dataset. Here we present and discuss results concerning the behaviour of scaling exponents as a function of the distance from the walls. Starting from eq. 2 plotting $G_p(r) \equiv S_p(r)/(S_3(r))^{p/3} \propto S_3(r)^{\tau_p}$, as a function of $S_3(r)$ we can infer directly τ_p, excluding the contribution of the dimensional exponent $p/3$. In Fig. 1 (left picture) we represent $G_6(r)$ vs. $S_3(r)$ for different elevations from the wall. It is evident that τ_p, hence ζ_p, change with the distance from the wall. This can also be seen in the figure of ζ_p vs. p in [5]. It is also clear that the scaling region varies from plane to plane. In particular the extension is wider at the center of the channel, and decrease towards the wall. Also we remark that u_{rms} depends on the elevation y, which means that each plane evolves with a different characteristic time scale. This implies that the dynamics is faster near the wall than near the center of the channel and explains the larger fluctuations in ζ_p observed in the central region of the channel, due to the fact we have more statistical independent samples near the wall then the center. More interesting is the dependence of τ_p on y/H. In Fig. 1 (right picture) we can see ζ_p as a function of y/H for $p = 4, 6$ compared with homogeneous values (see [6] for all the exponents). We notice that near the walls the flow presents a rather strong intermittency while in the center of the channel values of the exponents approach homogeneous/isotropic values. Scaling exponents ζ_p the region $y/H \in [0.1, 0.5]$ is $\sim 20\%$ lower with respect to homogeneous/isotropic values. Also it is in this region that the maximum number of coherent structures are present.

3. Intermittency and Coherent structures

In Fig. 2 (left) we observe an interesting relation between intermittency and a macroscopic quantity. The dots represent τ_6/τ_6^{hom} while the line represents the rms fluctuations of helicity density $h \equiv \vec{u} \cdot \vec{\omega}$, arbitrarily normalised. The good match between these different quantities suggests the existence of a direct relationship between intermittency and helicity. Let us consider the relation:

$$\frac{(\vec{u} \cdot \vec{\omega})^2 + (\vec{u} \wedge \vec{\omega})^2}{|\vec{u}|^2 |\vec{\omega}|^2} \equiv 1. \tag{3}$$

The first term is the square of the helicity while the second is a non-linear term related to energy transfer toward small scales. The regions with high helicity present low energy transfer. Hence the relation between helicity h and energy dissipation ϵ seems to be that helicity inhibits the energy cascade. This idea is supported by a clear anti-correlation between h and ϵ. In Fig. 2 (right) we show the joint PDF of squared helicity and dissipation measured in the buffer region (i.e. $|y/H| \in [0.95, 0.75]$, that is the region with higher helicity rms and intermittency). Regions with higher h are characterized by lower ϵ and vice-versa. Also, looking at a section of the channel (parallel to the wall) we detect a clear anti-correlation between structures characterized by high h and those characterized by high ϵ values. In this context structures aligned with mean velocity, like streamwise vortices, inhibit energy cascade (and so are long-lived structures). It follows a higher intermittency with respect to the center of the channel, where structures are not aligned, in a statistical sense, with the mean flow. The

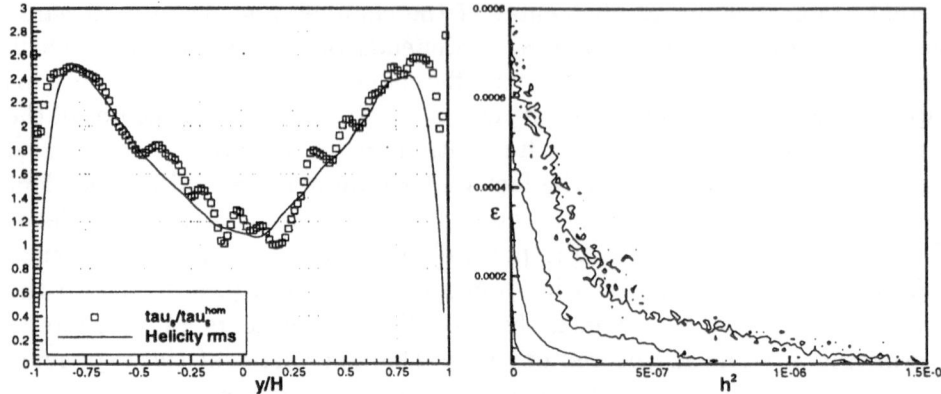

Figure 2. Left picture: comparison between h_{rms} (line) and τ_6/τ_6^{hom} (dot), as a function of y/H. Right picture: joint PDF of squared helicity h^2 and dissipation ϵ in the buffer region.

topological aspects of coherent structures and in particular of those with high helicity are presently under investigation [7]. Concluding we believe that a clear relationship between structures and intermittency would be a significant step to gain new insights into the physics of channel flow turbulence.

Acknowledgements

We would like to acknowledge ENEA, INFN and CASPUR where the numerical work was performed and Prof. R. Benzi for many useful discussions.

References

1. Kolmogorov, A.N. (1962) A refinement of previous hypothesis concerning the local structure of turbulence in a viscous incompressible fluid at high Reynolds number, *J. Fluid Mech.* **6**, pp. 82 - 85.
2. She, Z. S., Leveque, E. (1994) Universal Scaling Laws in Fully Developed Turbulence, *Phys. Rev. Lett.* **72 - 3**, pp. 336 - 339.
3. Amati G., Succi S., Piva R. (1997) Massively parallel Lattice Boltzmann Simulations of Turbulent Channel Flow, *Int J. of Mod. Phys. C* **8 − 4**, pp. 869 - 877.
4. Benzi, R., Biferale, L., Ciliberto, S., Struglia, M. V., Tripiccione, R. (1996) Generalized scaling in fully developed turbulence, *Physica D* **96**, pp. 162 - 181.
5. Amati G., Succi S., Piva R. (1998) Preliminary analysis of the scaling exponents in channel flow turbulence, *Fluid Dyn. Res.; in press.*
6. F. Toschi, G. Amati, S. Succi, R. Piva, R. Benzi (1998) Intermittency in a channel flow, *Proc. of II Monte Verita Colloquium on Fundamental Problematic Issue in Turbulence.*
7. F. Toschi, G. Amati, S. Succi, R. Benzi, R. Piva (1998) *In preparation.*

THREE-DIMENSIONAL SIMULATIONS OF
PLANAR TURBULENT JETS

D. K. BISSET and R. A. ANTONIA
Department of Mechanical Engineering
University of Newcastle, NSW 2308 Australia

1. Description and Methods

A uniform jet of fluid (velocity U_J) issues into quiescent surroundings from a slot of width H and much greater length in a large plate or wall. The origin is the centre of the slot; x,y,z are streamwise, transverse and spanwise coordinates; U,V,W and u,v,w are respective velocity components and fluctuations. Initially, for about $x/H<4$, two mixing layers surround a potential core. The interaction region follows; the mixing layers meet and strongly interact, causing rapid spreading of the jet and high turbulence levels. From $x/H = 20$ or more, the jet becomes self-preserving when the local mean centreline velocity U_0 and mean velocity halfwidth L are used for normalization. At Reynolds numbers ($\text{Re} = U_J H / v$) typical of laboratory experiments, say $6<\text{Re}\times10^{-3}<30$, the jet mixing layers are (a) strongly affected by boundary layers at the lip of the slot, and (b) susceptible to the influence of jet turbulence, acoustic pressure fluctuations, etc. This is one reason why published results differ quite widely. Even the self-preserving state is quite different in different facilities, as reviewed by Everitt & Robins (1978) for example. Simulation of flow with such a diverse character is not trivial. The authors are aware of only one other 3-D direct simulation of a planar jet (Stanley & Sarker 1997), though it was limited by its aspect ratio of only 3, and statistical results were not given.

Two spatial simulations of planar jets were carried out with the Advected Grid Explicit (AGE) method, Bisset (1998a). The first ($H = 24$ mm, $\text{Re} = 10,200$) covers a larger domain ($x/H<40$, aspect ratio 20) and is mainly concerned with the self-preserving region, while the second ($H = 25$ mm, $\text{Re} = 10,700$ and aspect ratio 9.6) concentrates on the initial and interaction regions ($x/H<14$) with better resolution. In the AGE method (Bisset 1998a) the computational (finite difference) grid moves at a fixed velocity (2.4 or 3.2 m/s in the present case) relative to the ground frame simulation domain, and grid points that leave the domain are 'recycled' to the beginning of the domain. The Navier-Stokes equations are computed in terms of variables U, V, W and P without assuming strict incompressibility, the advection terms being much smaller relative to the moving grid. For the jet simulation there are two important additions:

(i) The standard second-order central difference approximation for streamwise derivatives is equivalent to a parabola fitted through three points and differentiated at its midpoint. Here it is differentiated at the middle of the patch of fluid that will be advected through a gridpoint in the current timestep — see Bisset (1998b) for details.

(ii) Transverse velocity $\overline{V}$ at the edges of the domain must be non-zero to allow for entrainment [which was ignored by Bisset (1998a) in a two-stream mixing layer]. This is achieved automatically, for each of ten streamwise regions, by slowly adjusting

163

U. Frisch (ed.), Advances in Turbulence VII, 163–166.
© 1998 *Kluwer Academic Publishers.*

$\bar{V}$ such that mean pressure outside the turbulent zones is equal to ambient pressure.

Further details: The two simulations used $500 \times 250 \times 161$ ($376 \times 207 \times 161$) gridpoints in x, y and z respectively, with uniform spacing $\Delta x = \Delta y = 2$ mm (1 mm) and $\Delta z = 3$ mm (1.5 mm). The timestep was 80 μs (40 μs). Outlet and edge damping zones occupied 30 points in the x direction and 10 points at the ± limits in y. Damping was also applied to the first five points at the inlet (mainly to reduce pressure wave reflection) except in the vicinity of the jet inlet boundary layers. These were modelled in the small simulation as one grid plane thick with velocity $0.5U_J$ plus a random disturbance in U averaging ±10% of U_J. In the larger simulation the boundary layers were not identical, the values being $0.8U_J$, ±6% and $0.6U_J$, ±12%. The outlet condition for U was $\partial U / \partial x = 0$. The z boundary was periodic. After allowing startup transients to die down, the calculations were run for 32,768 (24,576) timesteps, or 2.62 s (0.98 s) of real time to obtain the results shown here (profiles are averaged over 16 spanwise positions). Calculations took 31.5 hrs (14.8 hrs) of CPU time on the Fujitsu VPP300 at the ANU Supercomputer Facility, and used 632 Mb (400 Mb) of memory.

2. Results

At $x = 32H$ the Kolmogorov lengthscale (based on the usual assumption of isotropy) was 0.23 mm and the Taylor microscale $\lambda = 8.1$ mm $(R_\lambda = 310)$. Therefore the simulations are somewhat under-resolved spatially for viscous scales, and could be considered (following Lesieur & Métais, 1996) as a form of automatic LES rather than DNS. Results are quite encouraging, however. Instantaneous velocity vectors from the larger simulation (with two out of three vectors omitted in each direction) in the plane $z = 0$ are shown in Figure 1. A few sectional streamlines are included to show the effects of entrainment. In spite of different conditions in the two inlet boundary layers, the initial mixing layer rollup was always fairly symmetrical and two-dimensional. However, normalized mean velocity profiles for $x/H = 17, 22, 27, 32$ and 37 showed minor effects of the inlet differences. Agreement was very good after small offsets were applied in y. The jet spreading rate dL/dx for the last three stations is 0.115 (experiments typically 0.09 to 0.11), and the corresponding maximum mean velocity develops as $(U_J^2 H)d(U_0^{-2})/dx = 0.187$ (experiments typically 0.14 to 0.22).

Figure 2(a) shows r.m.s. profiles of v for the five stations, indicating that the flow is not too far from self-similarity. Reynolds normal and shear stresses are shown in Figure 2(b); all fall within the fairly wide ranges of results given by Everitt & Robins (1978) and others. A few power spectra (including pressure) are given in Figure 2(c). There is some indication of an inertial range here, but not in the case of spectra from the centreplane (not shown). The normalized peak frequency in the v spectrum, $f_p U_0/L = 0.14$, is a little higher than the self-preserving value from experiments (about 0.11). Sectional streamlines for the same instant as Figure 1 are shown in Figure 3, calculated in a frame of reference moving at 1.2 m/s (about 51% of U_0 at $x/H = 32$). The symmetrical pattern of structures near the inlet is quite suddenly succeeded by much larger structures in an alternating pattern, the latter structures being responsible for the peak in the v (and w) spectra. Both these patterns are found in experiments, e.g. Antonia et al (1983), though the alternating pattern is sometimes found near the inlet too.

In the smaller simulation the mixing layers are thinner and better resolved, so the initial rollup is sufficiently rapid to allow vortex merging to occur within each mixing

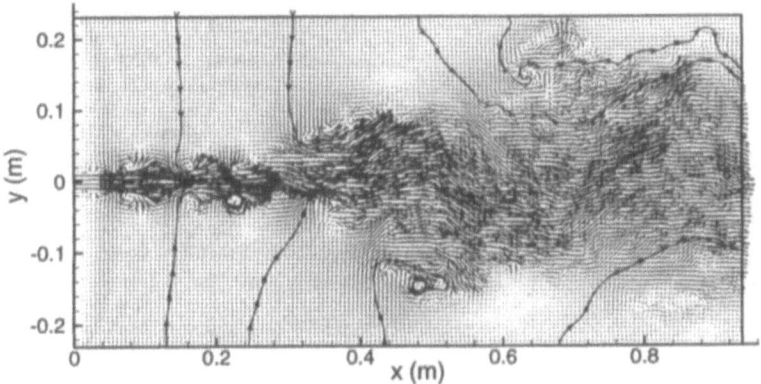

Figure 1. Instantaneous velocity vectors from the larger simulation

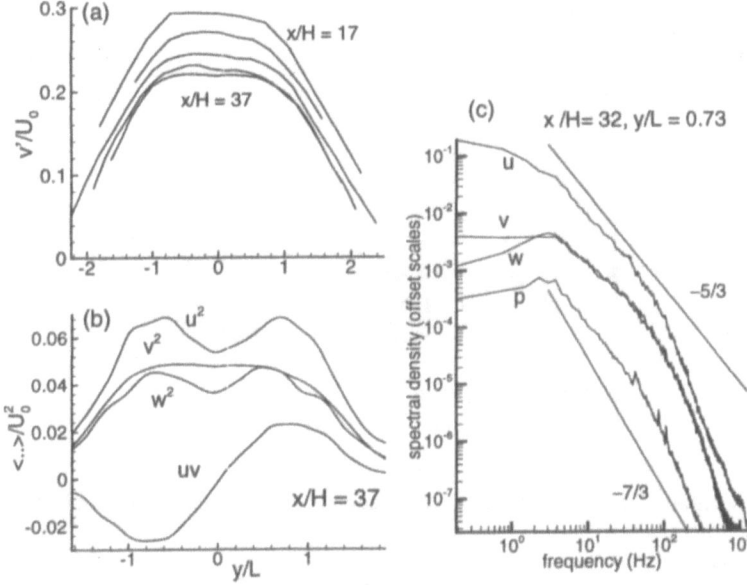

Figure 2. Selected profiles and power spectra for the larger simulation

layer before reaching the interaction region. A typical state is shown by contours of pressure fluctuations in Figure 4. The initial structures (delineated by their low-pressure cores) are always very long spanwise (though not exactly two-dimensional), but the large structures after interaction are highly three-dimensional. Long-time correlations from v-signals from above and below the centreplane change sign *twice* in x. They indicate that (a) the pattern of initial rollup in this simulation is mainly alternating, (b) the strucures after merging within the mixing layers ($x/H \approx 3$) tend to be placed symmetrically, and (c) a definite alternating pattern appears after interaction. Profiles at $x/H = 12$ are generally similar to, but weaker than, those of Figure 2(b).

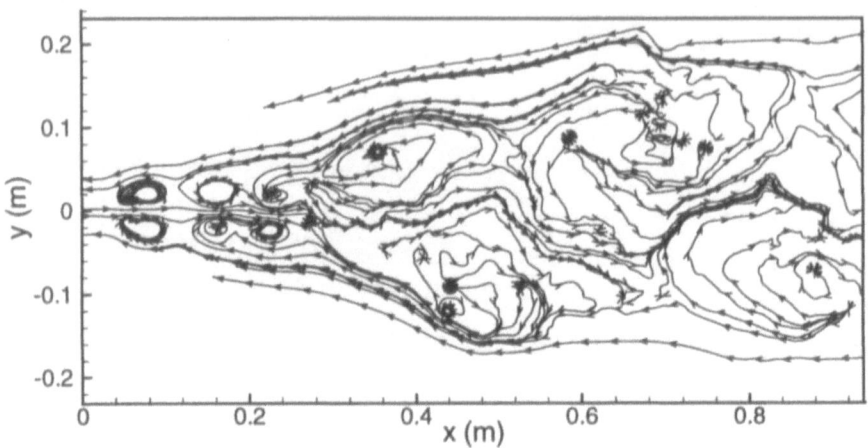

Figure 3. Sectional streamlines at $U_C = 1.2$ m/s (same data as Figure 1)

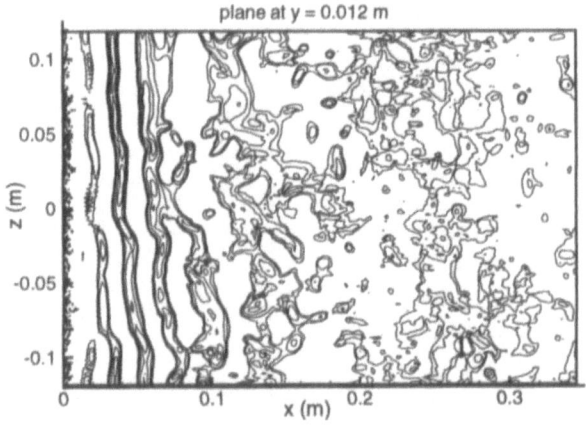

Figure 4. Contours of negative pressure fluctuations in a spanwise plane (smaller simulation)

3. References

Antonia, R.A., Browne, L.W.B., Rajagopalan, S. & Chambers, J. (1983) On the organized motion of a turbulent plane jet, *J Fluid Mech* **134**, 49-66

Bisset, D.K (1998a) The AGE method for direct numerical simulation of turbulent shear flow, *Int J Numerical Methods in Fluids* (to appear)

Bisset, D.K. (1998b) The AGE method: further developments and applications, *Conference on Numerical Methods for Fluid Dynamics*, Oxford UK, 31 March – 3rd April 1998

Everitt, K.W. & Robins, A.G. (1978) The development and structure of turbulent plane jets, *J Fluid Mech* **88**(3), 563-583

Lesieur, M., & Métais, O. (1996) New trends in large-eddy simulations of turbulence, *Ann Rev Fluid Mech* **28**, 45-82

Stanley, S. & Sarkar, S. (1997) Simulations of spatially developing two-dimensional shear layers and jets, *Theoret Comput Fluid Dyn* **9**, 121-147

THREE-DIMENSIONAL EFFECTS ON MIXING IN A COAXIA JET CONFIGURATION

M.V. SALVETTI[1] AND G. LOMBARDI[1]
[1] *Dipartimento di Ingegneria Aerospaziale, Università di Pisa, Pisa, Italy*

The flow originated by coaxial jets is of great interest in many applications, such as, for instance, the design of new industrial burners and spacecraft engines. In particular, the characterization and control of mixing between the streams is basic to obtain efficient combustion and low pollution. Even though the flow originated by coaxial jets has been studied extensively, the amount of information presently available on the mixing mechanisms is far from exhaustive.

Although limited to low Reynolds numbers, numerical simulation is useful because it provides simultaneously the time evolution of the vorticity field and of scalar concentrations, so that the effects of the dynamics of vortical structures on the mixing processes can be studied. Furthermore, in the numerical simulation the simultaneous "seeding" of both the internal and external jets is possible and the analysis of the effect of different flow parameters, such as Reynolds number or Schmidt number, is much easier than in experiments.

In a previous study, the mixing in a coaxial jet configuration has been investigated by means of axisymmetric direct numerical simulation [1] [2]. The effects on mixing of the mechanisms characterizing the vorticity dynamics in the nearfield, such as roll-up and pairing of vortices, have been analyzed. Comparisons with experiments have shown that, in spite of the low Reynolds numbers and of the axisymmetry assumption in the simulations, some of the basic mixing processes have been captured numerically. Nevertheless, it is known that 3D mechanisms play an important role even in the transitional region of coaxial jets.

The present study is the first step of an investigation of 3D effects on the mixing mechanisms in a coaxial jet configuration. To this end, 3D direct numerical simulations of the same coaxial jet configuration as in [1] and [2], characterized by internal to external radius and velocity ratios $R_i/R_e = 0.517$ and $U_i/U_e = 0.67$, have been carried out. The Reynolds

U. Frisch (ed.), Advances in Turbulence VII, 167–170.
© *1998 Kluwer Academic Publishers.*

number, based on the axis velocity U_i and the internal radius R_i, is equal
to 500. The Navier-Stokes equations in cylindrical coordinates and conservative variables have been solved numerically, together with the transport
equations for two passive scalars. The numerical method is based on a
finite-difference scheme, second-order accurate in space and time; details
can be found in [3] and [4]. At the inlet of the computational domain (corresponding to the jet outlet), the axial velocity is set to zero at $t = 0$, and
evolves to a prescribed profile $Q_z(r)$ in a time τ. The evolution law and
the value of τ are the same as in Ref. [4] and permit the reproduction of
the transient generated by a nearly impulsive start-up. Three-dimensional
instabilities are triggered by an azimuthal perturbation of the inlet velocity
profile. This perturbation is obtained by considering that the internal and
the external radii at the jet outlet vary with the azimuthal coordinate as
for a corrugated nozzle:

$$R_k(\theta) = R_k + \epsilon cos(n\theta) \quad k = i, e$$

where $\epsilon = 0.05R_i$, as in the experiments in [5]. By varying n, different
modes of azimuthal instability can be studied. The inlet concentration of
the first scalar is 1 in the internal jet and 0 elsewhere, while the other has
a concentration at the inlet equal to unity in the external jet and equal
to 0 elsewhere. At the outflow, radiative boundary conditions are used [4],
which permit the simulation of spatially evolving flows.

If the inlet velocity profile is perturbed as described above, 3D effects
can already be observed in the initial phase of the flow development, after
the impulsive start-up. In the present paper, because of space limitations,
we concentrate on the analysis of 3D effects on mixing only in the initial
stage of the flow development, which is interesting for combustion applications. As in the axisymmetric case [1] [4], this phase is dominated by
the evolution of the large start-up vortex and most of the mixing occurs in
correspondence to this large vortex. Since this vortex forms from the instability of the external jet, most of the mixing is between the scalar injected
in the external jet and the surrounding fluid. This can be seen, for instance,
from the comparison between the values of mixedness per unit axial length
obtained at $t=10$ for $n=5$, $n=7$ and in the axisymmetric case, computed
from the concentration of the internal and the external scalar respectively,
and shown in Fig. 1. The mixedness is a quantitative indicator of the mixing between each scalar and the surrounding fluids in a given domain. The
global mixedness is defined as follows:

$$f = \frac{4}{\pi L_r^2 L_z} \int_0^{2\pi} \int_0^{L_r} \int_0^{L_z} \xi_i(1 - \xi_i)r d\theta dr dz$$

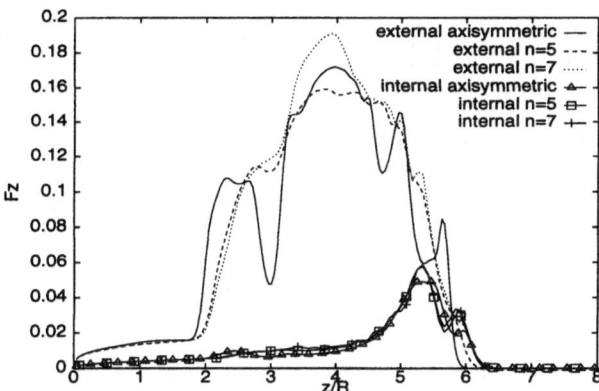

Figure 1. Mixedness per unit axial length obtained at $t=10$ for $n = 5$, $n = 7$ and in the axisymmetric simulations

where $L_r=6R_i$, $L_z=16R_i$ and ξ_i is the concentration of one of the scalars. The mixedness per unit axial length $F_z(z)$ is obtained by integrating only in the θ and r directions.

The three-dimensionality of the flow can clearly be seen from Fig. 2a, in which the flux of the absolute value of azimuthal vorticity through the different (z,r) planes is reported as a function of the azimuthal coordinate θ, at two different time instants, for $n = 5$ and $n = 7$. Only a sector $2\pi/n$ is shown, because of the periodicity of the flow. In particular, as a consequence of the azimuthal perturbation of the inlet velocity profile, vorticity concentrates in the central part of the sector, leading to a peak of the vorticity flux. In Fig. 2b the mixedness per unit azimuthal angle $F_t(\theta)$, computed from the concentration of the scalar seeded in the external jet is reported for $n = 5$ and $n = 7$ at the same instants as in Fig. 2a. The constant value obtained in the axisymmetric simulations is also shown. In the 3D simulations the mixedness is strongly increased in the central part of the sector, where vorticity concentrates, while it is lower than in the axisymmetric case on the external (z,r) planes. Thus, these azimuthal variations in the 3D cases, although significant, do not influence in a corresponding way the global mixedness f, as can be seen from Fig. 3. Nevertheless, even this global parameter shows that the rate of increase of mixing becomes larger with increasing three-dimensionality of the configuration.

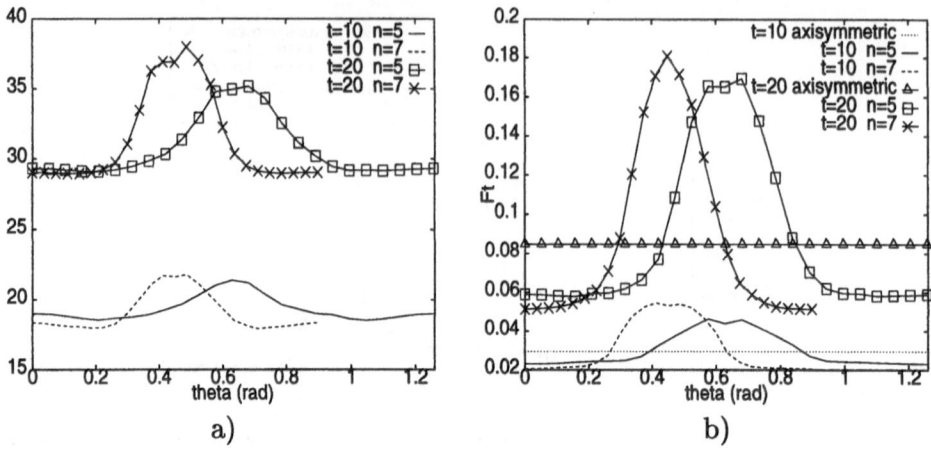

Figure 2. Flux of azimuthal vorticity (a) and mixedness per unit azimuthal angle (b) at $t = 10$ and $t = 20$

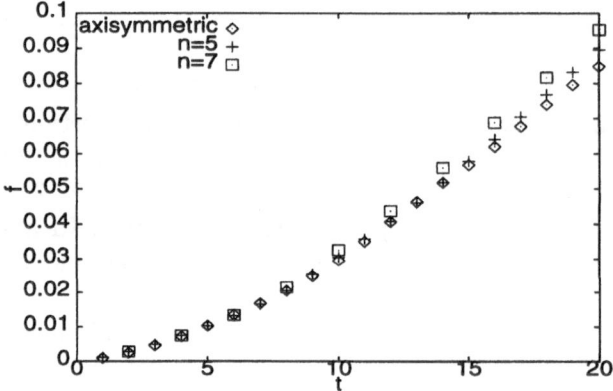

Figure 3. Time evolution of global mixedness for $n = 5$, $n = 7$ and the axisymmetric case

References

1. Salvetti, M.V., Lombardi, G., Talamelli, A. (1997) Investigation of mixing in a coaxial jet configuration, *Proceedings of Eleventh Symposium on Turbulent Shear Flows, September 8-11 1997, Grenoble (France)*.
2. Salvetti, M.V., Lombardi, G., Beux, F. (1998) A numerical investigation of mixing in a coaxial jet configuration through axisymmetric numerical simulations, *submitted for publication*.
3. Verzicco, R., Orlandi, P. (1996) A finite-difference scheme for the three-dimensional incompressible flows in cylindrical coordinates, *J. of Comput. Phys.* **123**, 402-414.
4. Salvetti, M.V., Orlandi, P., Verzicco, R. (1996) Numerical Simulations of Transitional Axisymmetric Coaxial Jets, *AIAA Journal* **34** (4), 736-743.
5. Prestridge, K., Lasheras, J.C. (1997) Entrainment and mixing patterns in coflowing forced jets subjected to axial and azimuthal forcing, *Proceedings of Eleventh Symposium on Turbulent Shear Flows, September 8-11 1997, Grenoble (France)*.

DIRECT NUMERICAL SIMULATION OF ADVERSE PRESSURE GRADIENT TURBULENT BOUNDARY LAYERS

M. SKOTE[1] AND D. S. HENNINGSON[1,2]
[1] *Department of Mechanics, KTH S-100 44 Stockholm Sweden.*
[2] *Aeronautical Research Institute of Sweden (FFA), Box 11021, S-161 11 Bromma, Sweden*

Direct numerical simulations (DNS) of the Navier-Stokes equations have been carried out with the objective of studying turbulent boundary layers with adverse pressure gradients (APG). The program uses spectral methods with Fourier discretization in the horizontal directions and Chebyshev discretization in the normal direction and utilizes the 'fringe method'. The turbulent flows have also been investigated using a differential Reynolds stress model (DRSM). Profiles for velocity and turbulence quantities obtained from the DNS were used as initial data.

The conditions needed for self-similarity have been investigated both experimentally and theoretically. Clauser [1] performed experiments where he adjusted the pressure gradient such that a self-similar turbulent boundary layer was obtained. A constant non-dimensional pressure gradient,

$$\beta = \frac{\delta^*}{\tau_w} \frac{dP}{dx},$$

was shown to be a condition for self-similarity. Here δ^* is the displacement thickness, τ_w is the shear stress at the wall and $\frac{dP}{dx}$ is the pressure gradient. Mellor & Gibbson [2] showed that self-similarity is obtained if the free stream variation is set to be a power law in the downstream direction, $U \sim x^m$, where U is the free stream velocity and x is the stream-wise coordinate.

Results from three direct numerical simulations of APG turbulent boundary layers as well as one zero pressure gradient case (ZPG) are presented. In the first APG case (APG1) the pressure gradient is close to that for which the corresponding laminar boundary layer would separate and in the second case (APG2) the pressure gradient is considerably higher. The pressure gradient is applied through the variation of the free stream velocity, which is described by a power law, $U \sim x^m$. For APG1 $m = -0.077$ and for APG2

U. Frisch (ed.), Advances in Turbulence VII, 171–174.

$m = -0.15$ which corresponds to $\beta \approx 0.24$ and $\beta \approx 0.65$ respectively. The simulations were performed at $Re_\Theta : 350 - 690$ and with a resolution in plus units of $\Delta X^+ = 13$ and $\Delta Z^+ = 3.6$. Some preliminary results from the third APG case (APG3) are included. In this case $m = -0.25$ and $\beta \approx 3.7$.

In this paper the near-wall behavior of turbulent statistics from our DNS is analyzed. A couple of examples of our investigations will be briefly described here. First the distribution of the shear stress and the production term in the turbulent energy budget are analyzed, and then the full mean energy budget is investigated.

When neglecting the non-linear, advective terms in the equations describing the mean flow of a two-dimensional, incompressible, turbulent boundary layer, the equation governing the inner part of the layer is obtained. This equation can, when using the inner length and velocity scales ν/u_τ and u_τ be written,

$$0 = -\frac{\beta}{Re_*} + \frac{\partial^2 u^+}{\partial y^{+2}} - \frac{\partial}{\partial y^+}\langle u'v'\rangle^+ \tag{1}$$

where $Re_* = \frac{u_\tau \delta^*}{\nu}$ and $\langle u'v'\rangle$ is the Reynolds shear stress. If the ratio $\frac{\beta}{Re_*}$ is smaller than the other terms, the equation reduces to the equation governing the inner part of a ZPG boundary layer. However, for the APG cases considered here this term can not be neglected. Equation (1) can be integrated to give an expression for the total shear stress,

$$\tau^+ \equiv \frac{\partial u^+}{\partial y^+} - \langle u'v'\rangle^+ = 1 + \frac{\beta}{Re_*}y^+. \tag{2}$$

The total shear stress, τ^+, from the DNS and the curves $\tau^+(y^+)$ represented by equation (2) are shown in Figure 1. The term

$$\frac{\beta}{Re_*} = \frac{\nu}{u_\tau^3}\frac{1}{\rho}\frac{dP}{dx} = \left(\frac{u_p}{u_\tau}\right)^3 \tag{3}$$

is evidently important for the shear stress distribution in the inner part of the boundary layer even though it does not violate the use of u_τ as the inner velocity scale. The velocity scale

$$u_p = \left(\nu\frac{1}{\rho}\frac{dP}{dx}\right)^{1/3} \tag{4}$$

has to be used if the term (3) becomes very large which happens if $u_\tau \ll u_p$, i.e. the boundary layer is close to separation.

The inner maximum of the production term in the turbulent energy budget,

$$p^+ = -\langle u'v'\rangle^+\frac{\partial u^+}{\partial y^+} \tag{5}$$

is 0.25 in the ZPG case (Figure 2b) as expected from the integrated inner equation, which in that case permits the expression (5) to be written

$$p^+ = \left(1 - \frac{\partial u^+}{\partial y^+}\right)\frac{\partial u^+}{\partial y^+}. \tag{6}$$

The maximum occurs at the position where $\frac{\partial u^+}{\partial y^+} = 0.5$. The inner maximum is increased due to the pressure gradient but the position is not changed as seen in Figure 2.

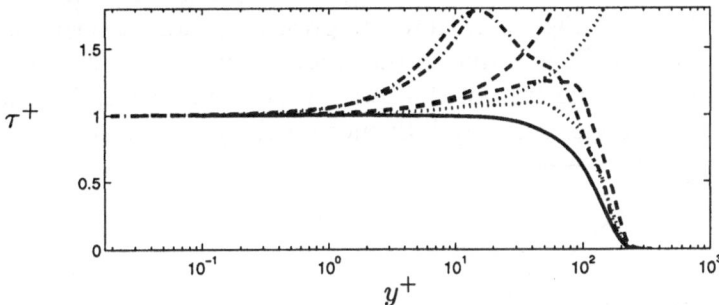

Figure 1.   Shear stress — ZPG; $\cdots$ APG1; - - APG2; - · - APG3. Shear stress evaluated from equation (2) diverges, whereas the total shear stress from DNS approaches zero for large y^+.

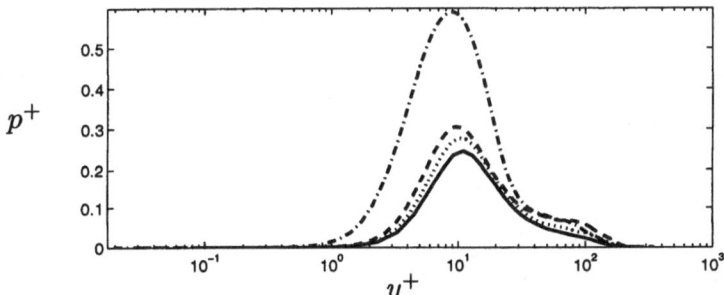

Figure 2.   Production — ZPG; $\cdots$ APG1; - - APG2; - · - APG3.

When multiplying equation (2) by $\frac{\partial u^+}{\partial y^+}$ we get mean energy budget.

$$\left(\frac{\partial u^+}{\partial y^+}\right)^2 - \langle u'v' \rangle^+ \frac{\partial u^+}{\partial y^+} = \frac{\partial u^+}{\partial y^+} + \frac{\beta}{Re_*}y^+\frac{\partial u^+}{\partial y^+}. \tag{7}$$

The same equation is obtained if equation (1) is multiplied by u^+. The terms are noted from left to right: direct dissipation, production, transport and pressure gradient term. The mean budget for the APG2 case in the inner region is shown in Figure 3. The largest contribution in the near wall region comes from the direct dissipation which is balanced by the transport term. At $y^+ = 5$ the pressure gradient term has reached its maximum and then stays constant. The production of turbulent energy has its maximum at $y^+ = 9$ where the production and direct dissipation are equal in magnitude. All the terms balance each other, though the total sum deviates from zero at large values of y^+. If the advective terms also are included in the total budget, the sum becomes zero but these terms are small compared to the others.

The same energy budget but for the DRSM prediction is almost identical. More interesting is to look at the budget for the high Reynolds number case, which is shown in Figure 4. Here the pressure gradient term has vanished from the budget. This is due to the term $\frac{1}{Re_*}$ in the pressure gradient term in equation (7). The sum of the terms is zero which implies that the advective terms no longer has any influence on the inner layer.

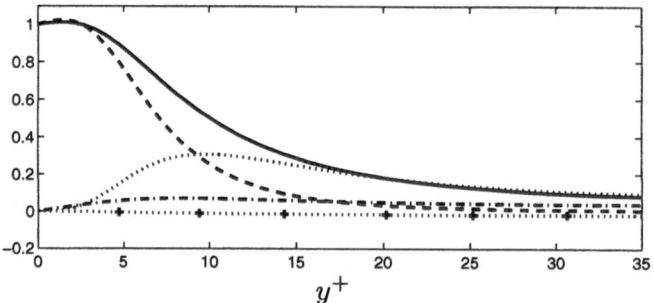

Figure 3. Energy budget for DNS: - - Dissipation; $\cdots$ Production; — Transport; - $\cdot$ - Pressure gradient term; $\cdot + \cdot$ Total.

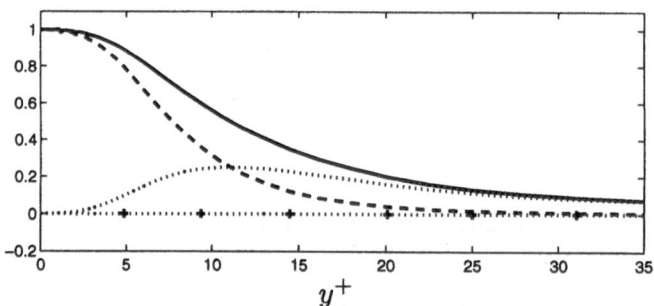

Figure 4. Energy budget for DRSM.

References

1. F. H. Clauser. Turbulent boundary layers in adverse pressure gradients. *J. Aero. Sci.*, 21:91–108, 1954.
2. G. L. Mellor and D. M. Gibson. Equilibrium turbulent boundary layers. *J. Fluid Mech.*, 24:225–253, 1966.

DNS OF SUBSONIC SPATIALLY DEVELOPING SHEAR FLOWS

B. WASISTHO, I. DE BRUIN, B. GEURTS AND H. KUERTEN

J.M. Burgers Centre - Faculty of Math. Sciences, University of Twente, P.O. Box 217, 7500 AE Enschede, The Netherlands

We perform direct numerical simulations (DNS) of transition in shear flows. The adequacy of the numerical method used is first validated in the linear regime of flow perturbations. In particular we consider a comparison study between DNS and the Linear Stability Theory (LST) in a spatially developing compressible mixing layer in two dimensions. The method is subsequently applied to a simulation of separation bubble induced transition in three dimensions. We use a fourth order central scheme for the spatial dicretization and four-stage second order Runge-Kutta for the time integration. The boundary conditions along the artificial boundaries are based on the non-reflecting characteristic method. A periodic boundary condition is employed in the spanwise direction. In addition, we apply a direct method of buffer domain technique to damp wave reflections at the outflow boundary. The description of the numerical method, the boundary conditions and the buffer domain technique can be found in [7].

The physical situation of a spatially developing mixing layer in 2D is used to illustrate the development in the laminar regime which can be compared with LST. We set the Mach number to 0.2, so that the effect of compressibility is negligible. The Reynolds number is 50, based on the upper stream velocity, viscosity and half the vorticity thickness at the inflow. Temperature is scaled by a reference temperature of 276 K. The dimensionless lower stream velocity is equal to 0.5, and both the upper and lower stream dimensionless temperatures are equal to one. In the normal direction the domain size is 60 units. In all three directions the grid is uniform. The extent of the domain in the streamwise direction is 6 wavelengths of the LST mode. Only one perturbation eigenfunction is imposed in order to be able to accurately compare the results with LST. We select the most unstable 2D mode from a linear stability analysis and prescribe this at the inflow boundary. In this particular case we have a streamwise wave number $\alpha = (\alpha_r, \alpha_i) = (0.329, -0.0391)$ and circular frequency $\omega = 0.247$. In Figures 1 respectively 2 the growth rates and Tollmien-Schlichting waves

U. Frisch (ed.), Advances in Turbulence VII, 175–178.
© 1998 *Kluwer Academic Publishers.*

are plotted. The results correspond well with Linear Stability Theory (solid line). Taking a finer resolution in the normal direction results in a much better correspondence.

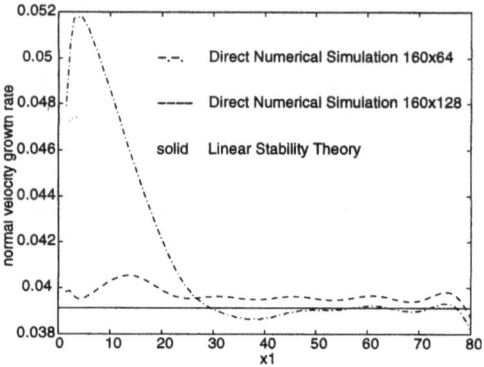

Figure 1.   Growth rates of the normal velocity.

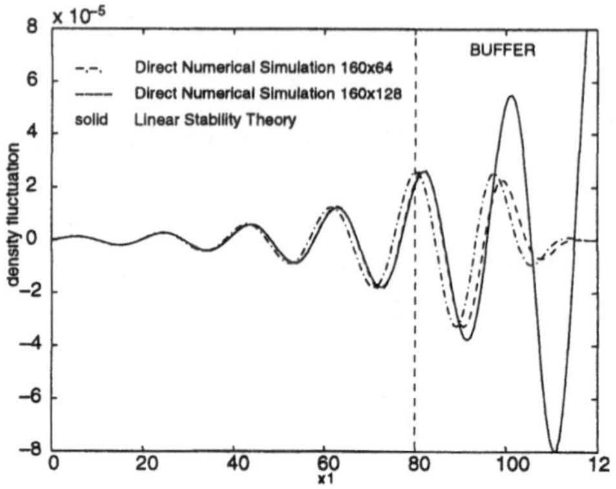

Figure 2.   Tollmien-Schlichting waves for the density in streamwise direction.

As an application of the method, we perform a laminar separation bubble simulation in 3D. For this purpose, a suction is prescribed along the upper boundary in order to produce an adverse pressure gradient. The wall is isothermal and the wall temperature is equal to its adiabatic value at the inflow boundary. The Mach number and reference temperature are respectively 0.2 and 276 K as before. Based on the inflow displacement thickness the Reynolds number is 330, and the inflow and outflow boundaries are located at $x_1 = 109.33$ and 348.14, respectively. This is equivalent to R^* (based on the laminar spreading of the displacement thickness) ranging from 330 to 488. The height of the domain is 20 and the spanwise extent is $[-\pi/|\beta|, \pi/|\beta|]$. Three- dimensional linear perturbation eigenfunctions are superimposed upon the inflow Blasius profiles. The circular frequency $\omega = 0.15$ ($F = 10^4 \omega/Re = 4.55$) and the wave numbers $\beta^\pm = \pm\alpha_r = \pm\Re(\alpha)$, where α is the streamwise wave number of the inflow perturbations. The eigenvalues resulting from this choice of parameters are:

$$\alpha_{2d} = 0.3455 + i0.00842,$$
$$\alpha_{3d} = 0.3170 + i0.01811 \ (\beta^\pm = \pm0.3170),$$

which are both damping modes. The streamwise extent of the computational domain corresponds to 13 waves of the 2D mode or 12 of the 3D mode.

A $402 \times 64 \times 48$ grid is used in the x_1, x_2 and x_3 direction, respectively, with a stretching in the normal direction. With these numbers of grid

points, the grid resolution is $(\Delta x_1, \Delta x_2, \Delta x_3)$=(0.597, 0.082, 0.413) based
on the inflow displacement thickness, or equivalently $(\Delta x_1^+, \Delta x_2^+, \Delta x_3^+)$=
(9.76, 1.34, 6.75), where $x_i^+ = x_i u_\tau Re$. Here, we use $u_\tau = 0.05 U_\infty$, which
is a typical value in the turbulent region in the present simulation.

In the recovery region, downstream of the reattachment point, the shape
factor decreases to 1.6, which is slightly higher than the value predicted by
the self-similarity relations ($H = 1.4$). However, this value is quite reasonable as in practice it depends on Re^* after transition (Ducros et al. in [1]).
Spalart reported $H = 1.43$ for $Re^* \approx 2000$ [5] and $H = 1.66$ for $Re^* = 498$
[6]. Our corresponding Re^* equals 580. The skin friction coefficient (Fig.
3) increases strongly just after reattachment indicating a high turbulence
intensity. The skin friction is in fair agreement with the empirical rela-

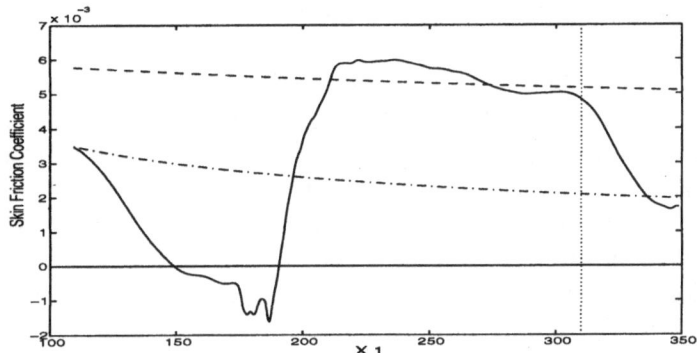

Figure 3. Skin friction of the mean flow; the present case (solid) is compared with the
empirical laws for zero pressure gradient (z.p.g.) laminar flow (dashed dotted) and for
the z.p.g. turbulent case (dashed).

tions in the turbulent regime, in the sense that it tends to the similarity
prediction.

An effect of the separation bubble on the turbulence downstream is that
the mean u_1 profile just after reattachment drops substantially below the
logarithmic law. The low values within the logarithmic layer resemble the
effect of surface roughness ([4]). Proceeding downstream, however, the profiles in the logarithmic regime increase gradually until the most downstream
profile recorded virtually coincides with the logarithmic law. This gradual
change of the streamwise velocity profiles clearly illustrate the recovery
process toward the zero pressure gradient state.

The Reynolds stresses $u_1' u_2'$ also show a recovery process in the downstream direction. The initially high level of Reynolds stress near the wall
decreases toward the level of zero pressure gradient turbulence. This self
similar turbulence is illustrated in the comparison of the Reynolds stress
$(u_1' u_2')$rms at the station $x_1 = 309.91 (Re_\tau = 240)$ with the DNS result of

Rai & Moin for $Re_\tau \approx 650$ (boundary layer, [3]) in Fig. 4 which shows an under-estimation in the whole range. The data from Eckelmann [2] (chan-

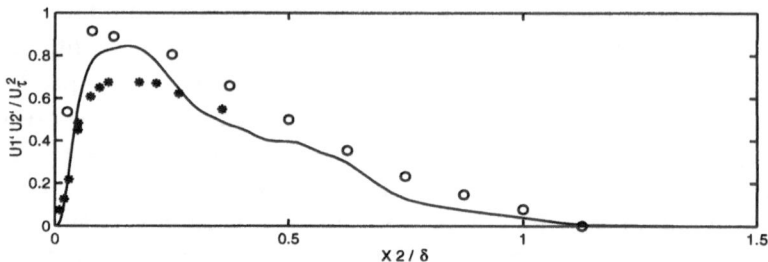

Figure 4. Dimensionless Reynolds stresses $-u_1'u_2'/u_\tau^2$ versus x_2/δ at the position $x_1 = 309.91$ compared to the data from Rai & Moin (1993) for $Re_\tau \approx 650$ (o) and the data from Eckelmann (1974) for $Re_\tau = 208$ (*).

nel flow) for $Re_\tau \approx 208$ are also incorporated in order to illustrate the effect of Reynolds number. Clearly, a lower Reynolds number results in a lower level of Reynolds stress. This may explain why our result is closer to the result of Eckelmann, at least in the region close to the wall.

The computational time for the 3D simulation is about 300 CPU hours on a Cray C90 computer. Concluding, the numerical method used is suitable for simulation of turbulent shear flows in a relatively simple geometry. It produces physically reliable results both in the free as wel as in the wall-bounded shear-layer case. In particular the flow of the mixing layer will form the basis of Large Eddy Simulations in the future.

References

1. F. Ducros, P. Comte and M. Lesieur (1995) Large-eddy simulation of transition to turbulence in a boundary layer spatially developing over a flat plate, *J. Fluid Mech.* **11**.
2. H. Eckelmann (1970) Experimentelle Untersuchungen in einer turbulenten Kanalströmung mit starken viskosen Wandschichten, *MPI f. Strömungsforschung und AVA Göttingen* Nr. 48.
3. M.M. Rai and P. Moin (1993) Direct numerical simulation of transition and turbulence in a spatially evolving boundary layer, *J. Comp. Phys.* **109** 169–192.
4. H. Schlichting (1979) *Boundary layer theory*, Mc. Graw Hill Book Company, 7th ed., New York.
5. P.R. Spalart (1988) Direct simulation of turbulent boundary layer up to $R_\theta = 1410$, *J. Fluid. Mech.* **187**, 61-98.
6. P.R. Spalart and G.N. Coleman (1997) Numerical study of a separation bubble with heat transfer, *Eur. J. Mech., B/Fluids* **16**, N° 2, 169–187.
7. B. Wasistho, B.J. Geurts and J.G.M. Kuerten (1997) Spatial simulation techniques for time-dependent compressible flow over a flat plate, *Computers and Fluids* **26**, 713-739.

NUMERICAL SIMULATION OF THE EVOLUTION OF NON-AXISYMMETRIC DISTURBANCES IN CYLINDRICAL LAMINAR PIPE FLOW

F.T.M. NIEUWSTADT[1], MA BING[2] AND ZHANG ZHAOSHUN[2]
[1] *J.M. Burgers Centre, 2628 AL Delft, the Netherlands.*
[2] *Dept. of Eng. Mech., Tsinghua Univ., Beijing, P.R. China*

1. INTRODUCTION

Since the well-known experiments of Reynolds in 1883 on transition of laminar to turbulent flow in a cylindrical pipe this problem has been studied without success. The main reason for this failure is the fact that laminar pipe flow is found to be stable for infinitesimal perturbations. Therefore transition can only follow from finite perturbations. This requires a non-linear theory which at the moment is not available.

In an experiment described in [1] the transition process has been considered for cylindrical laminar pipe flow subjected to a non-axisymmetric, divergence-free disturbance. In the present investigation we aim to simulate this transition experiment numerically with the objective to study the initial development of an imposed disturbance. This may shed some light on the mechanism which plays a role in the process of transition.

2. Numerical Model

To simulate the transition in cylindrical pipe flow we have developed a spectral element code in cylindrical coordinates. The elements consist of annular rings surrounding a central cylindrical element. In the axial and azimuthal direction an expansion in Fourier components is applied whereas in the radial direction a decomposition in Chebyshev polynomials is used in each annular ring. For further details on the computational method we refer to [2]. Note that in the following all variables have been made dimensionless with help of the pipe radius R and the bulk velocity U_B.

With this code we aim to study the streamwise development of disturbances, i.e. disturbances are generated in the beginning of the flow domain,

U. Frisch (ed.), Advances in Turbulence VII, 179–182.

subsequently they develop while they are transported downstream and they are finally convected out of the flow domain. Such a spatial development is not consistent with the periodic boundary conditions needed for the spectral decomposition in the axial direction. To resolve this problem we have applied the fringe technique proposed in [3]. With this method the disturbance is dampened at the end of the pipe section and the velocity profile is forced back to the laminar Poiseuille profile which is imposed at the entrance of the pipe. The effect of the fringe region on a developing disturbance as computed with our code, is shown in Fig. 1 below.

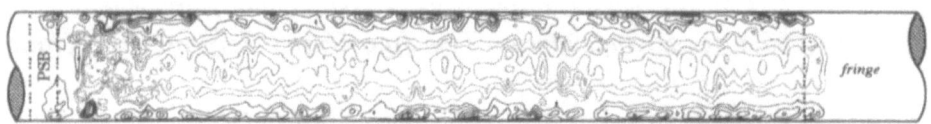

Figure 1.   Illustration of a disturbance which is generated in the flow domain on the left side called PSB and which is damped in the fringe region on the right side.

3. Results

For the numerical simulation we use the same disturbance as applied in the experiment described in [1], i.e. we apply periodic blowing/suction (PSB) radially out of a small slit in the pipe wall. This disturbance is given by

$$w(x,\theta,t) = A_w f\left(\frac{x - x_c}{\delta}\right) \sin(\Omega t) \sin(\theta)$$

where Ω the frequency. The azimuthal wave number is equal to one, i.e. while suction is applied on one part of the wall, blowing is applied on the opposite part. The function $f(x)$ gives the blowing/suction distribution across the slit which has a width equal to δ and which is centred around the location x_c. The amplitude A_w gives the strength of the disturbance. The subsequent development of the disturbance is then computed as a function of distance along the pipe.

The computations have been carried out for a Reynolds number (based on the pipe diameter $D = 2R$ and the U_B) equal to $Re_D = 3000$. The total length of the pipe is equal to 32π with the region between 28π and 32π as the fringe region. For the slit through which the disturbance is applied, $x_c = \pi$ and $\delta = \pi$. The number of collocation points in the axial, radial and tangential direction respectively are: $N_x = 128$, $N_r = 53$ and $N_\theta = 16$. The frequency of the disturbance is $\Omega = 5$.

For these parameters computations have been performed for various values of the A_w. The evolution of the disturbance is studied by means of E_{dis} which is defined as the kinetic energy of the disturbance velocity

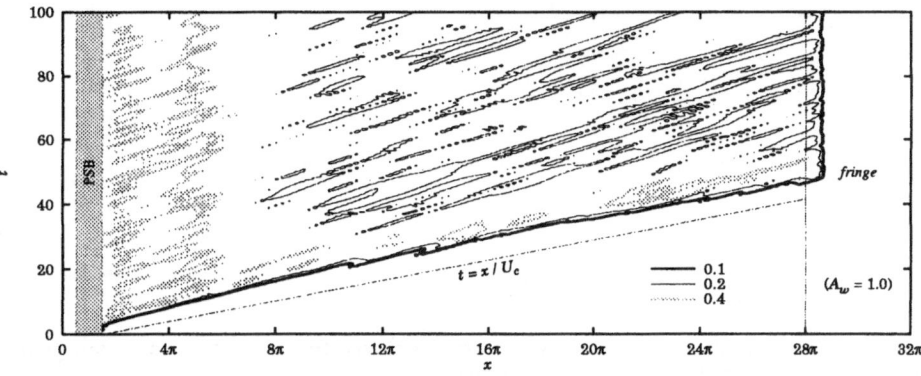

Figure 2. Contours of the disturbance energy $E_{dis}(x,t)$ for the amplitue $A_w = 1.0$.

u_{dis}. Here, u_{dis} follows by subtracting the laminar profile from the velocity computed at a given time t and at each position in the pipe. The E_{dis} is then computed as

$$E_{dis}(x,t) = \int_0^1 \int_0^{2\pi} \frac{1}{2}|u_{dis}|^2 \, r \, dr d\theta$$

so that for a laminar profile, the E_{dis} is equal to zero at each value of x.

In Fig. 2 we show the contour lines of E_{dis} as a function of x and t for the case that $A_w = 1.0$. The disturbance which is located in the region indicated in the figure by PSB, starts at $t = 0$ and then propagates towards the exit plane with a velocity close to the centreline velocity of the laminar flow profile $U_c = 2.0$. As soon as the disturbance reaches the fringe region it is quickly damped and the flow returns to the laminar profile which enters the entrance plane. When the disturbance has filled the whole pipe section between $x = \pi$ and $x = 28\pi$, a more or less statistically stationary flow exists.

To study the development of the disturbance energy in this statistically steady regime we introduce the time-averaged disturbance energy as $\overline{E}_{dis}(x)$. The $\overline{E}_{dis}$ as function of x is given in Fig. 3 for various values of A_w. Near the location where the disturbance is imposed, the $\overline{E}_{dis}$ is determined by the energy density of the blowing/suction velocity which is given by

$$E_{PSB} = \frac{\Omega}{2\pi\delta} \int_{x_c-\delta/2}^{x_c+\delta/2} \int_0^{2\pi} \int_0^{)2\pi/\Omega} \frac{1}{2}|u_{PSB}|^2(x,t,\theta) \, dx d\theta dt = \frac{3\pi}{32}A_w^2$$

and which is shown in the insert in Fig. 3. We see in this figure that for the amplitude $A_w = 1.0$, 0.6 and 0.5 the $\overline{E}_{dis}$ initially changes as a function of x between $\pi < x < 12\pi$ but then approaches a more or less constant

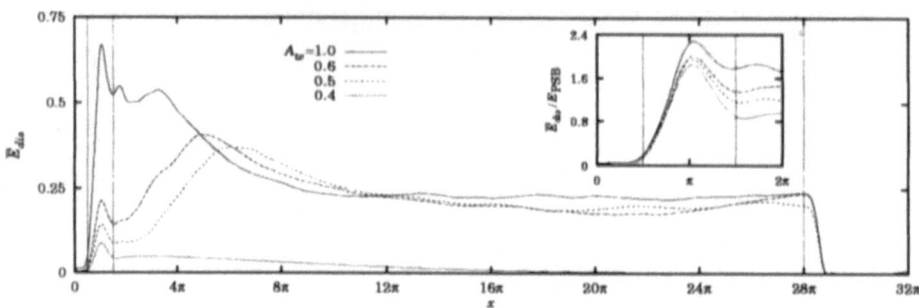

Figure 3. Time-averaged disturbance energy $\overline{E}(x)$ as a function of downstream distance for various values of the disturbance amplitude A_w.

value equal to 0.25. For $A_w = 0.4$ the development is quite different. Within the disturbance region a non-zero value of $\overline{E}_{dis}$ is found but $\overline{E}_{dis}$ decays to zero as soon as the disturbance is removed. Therefore for this value of the A_w the laminar flow can be considered as stable. This behaviour agrees qualitatively with the experimental results given in [1].

4. Conclusions

The spatial development of a localized disturbance introduced in a laminar pipe flow has been studied by means of numerical simulation. As a function of the disturbance amplitude we find a unstable and stable behaviour in agreement with the experiments discussed in [1]. For the unstable disturbances we find that they develop quickly and then all saturate at the same value for the disturbance energy.

References

1. A.A. Draad, G.D.C. Kuiken and F.T.M. Nieuwstadt (1998) Laminar-turbulent transition in pipe flow for Newtonian and non-Newtonian fluids. submitted to *J. Fluid Mech.*.
2. H. Shan, B. Ma, Z. Zhang and F.T.M. Nieuwstadt (1998) Direct Numerical Simulation of a Puff and Slug in Transitional Cylindrical Pipe Flow, *J. Fluid Mech.* under revision.
3. J. Nordström, N. Nordin and D. Henningson (1998) The fringe region technique and the Fourier method used in the Direct Numerical Simulation of spatially evolving viscous flows, to appear in *SIAM J. Sci. Comp.*

HOMOGENIZATION IN RANDOM FLOWS

E. VANDEN EIJNDEN, D. HANON AND J. P. BOON
Département de Physique, CP 231
Université Libre de Bruxelles
B-1050 Bruxelles, Belgium

The passive advection of a scalar quantity by a random velocity field is a problem encountered in a wide variety of situations like mass, charge and heat transport in turbulent fluids, dispersion in porous media, ... Of particular interest is the long-time large-scale dynamics of the scalar. In this work we use a combination of multiscale and projection techniques for analyzing the advection of a scalar by a random velocity field whose ensemble-average is periodic in space. General expressions relating the scalar transport to the properties of the flow are obtained. The method is applied to the analysis of scalar advection in a turbulent flow generated by a lattice gas automaton and whose ensemble-average is a two-dimensional ABC type flow.

We consider scalar advection by a d-dimensional random velocity field $v(r, t) = \{v_\alpha(r, t)\}$ ($\alpha = 1, \cdots, d$) taken to be solenoidal ($\partial v_\alpha / \partial r_\alpha = 0$), statistically stationary, with mean value

$$u_\alpha(r) = \langle v_\alpha(r, t) \rangle_{\text{ens}}, \tag{1}$$

and covariance

$$V_{\alpha\beta}(r, r', t) = \langle \delta v_\alpha(r, t + t') \delta v_\beta(r', t') \rangle_{\text{ens}}, \tag{2}$$

where $\delta v_\alpha(r, t) = v_\alpha(r, t) - u_\alpha(r)$ and $\langle \cdots \rangle_{\text{ens}}$ denotes the average over the appropriate ensemble of realizations of the velocity field. We also take $u_\alpha(r)$ to be periodic on some cell $\mathcal{C}$. A space-average over $\mathcal{C}$, denoted by $\langle \cdots \rangle_{\text{sp}}$, is introduced, such that, if $V_\mathcal{C}$ is the volume of the cell $\mathcal{C}$,

$$\langle u_\alpha(r) \rangle_{\text{sp}} = \frac{1}{V_\mathcal{C}} \int_\mathcal{C} dr \, u_\alpha(r) = 0. \tag{3}$$

Averaging over both ensemble and space will be denoted by brackets without subscript, i.e. $\langle \cdots \rangle \equiv \langle \langle \cdots \rangle_{\text{ens}} \rangle_{\text{sp}}$.

U. Frisch (ed.), Advances in Turbulence VII, 183–186.

Let $\rho(t) \equiv \rho(\boldsymbol{r}, t)$ be the ensemble-averaged concentration of a scalar advected by the random velocity field $v_\alpha(\boldsymbol{r}, t)$. Then $\rho(t) = \langle \tilde\rho(t) \rangle_{\text{ens}}$, where $\tilde\rho(t) \equiv \tilde\rho(\boldsymbol{r}, t)$ satisfies the stochastic Liouville equation

$$\frac{\partial}{\partial t}\tilde\rho(t) + v_\alpha \frac{\partial}{\partial r_\alpha}\tilde\rho(t) = 0. \tag{4}$$

Generally one cannot derive from (4) an equation for $\rho(t)$ without approximation (except if δv_α is a white-noise process, which we do not assume here). Furthermore, the evolution of $\rho(t)$ is presumably very complicated when considered over all scales. However, on scales much larger than the size of the cell $\mathcal{C}$, owing to the property that $\langle v_\alpha \rangle = 0$, the ensemble-averaged distribution $\rho(\boldsymbol{r}, t)$ is expected to be Gaussian with diffusive dynamics (*G-diffusive*), that is, it should asymptotically satisfy the equation

$$\frac{\partial}{\partial t}\rho(t) = D^\star_{\alpha\beta} \frac{\partial^2}{\partial r_\alpha \partial r_\beta}\rho(t), \tag{5}$$

for some constant *effective* diffusion tensor $D^\star_{\alpha\beta}$.

An expression for $D^\star_{\alpha\beta}$ can be obtained from the available statistical characteristics of the velocity field, $u_\alpha(\boldsymbol{r})$ and $V_{\alpha\beta}(\boldsymbol{r}, t)$, on the assumption that the motion is asymptotically G-diffusive and using the general procedure of multiple-scale asymptotics [1]. Similar calculations are presented e.g. in Ref. [2] and lead to the expression

$$D^\star_{\alpha\beta} = -\tfrac{1}{2} \lim_{t\to\infty} \left[\langle v_\alpha w_\beta(t) \rangle + \langle v_\beta w_\alpha(t) \rangle \right], \tag{6}$$

where $w_\alpha(t) \equiv w_\alpha(\boldsymbol{r}, t)$ is a random field with zero average ($\langle w_\alpha(t) \rangle = 0$) which satisfies the Liouville equation with a source term

$$\frac{\partial}{\partial t}w_\alpha(t) + v_\beta \frac{\partial}{\partial r_\beta}w_\alpha(t) = -v_\alpha, \tag{7}$$

with the initial condition $w_\alpha(\boldsymbol{r}, 0) = 0$. Of course, equation (7) is just as complicated as the original Liouville equation (4), but it offers a more convenient starting point for the evaluation of $D^\star_{\alpha\beta}$.

An equation for the *ensemble-average* $\langle w_\alpha(t) \rangle_{\text{ens}}$ can be derived[1] using standard projection techniques of statistical mechanics, and specifically within the quasilinear approximation which is valid if $\tilde\rho(t)$ evolves much more slowly than δv_α (see e.g. [3] and references therein). The quasilinear approximation leads to the following asymptotic expression for the

---

[1]The details of the calculation will be given elsewhere.

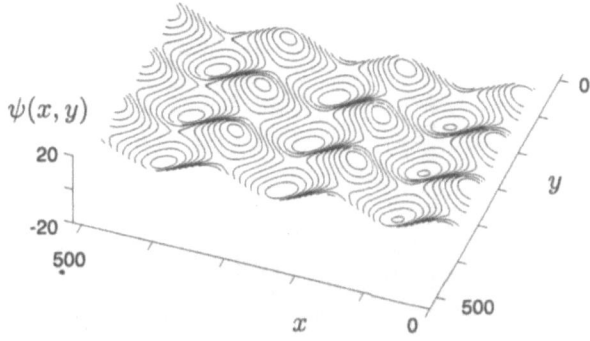

$$\psi(x, y)$$

Figure 1. The stream function $\psi(x,y)$ of the flow generated by the automaton.

ensemble-average of the stochastically non-linear term in (7) (thus allowing to close the ensemble-average of this equation)

$$\lim_{t\to\infty} \langle v_\alpha w_\beta(t)\rangle_{\text{ens}} = u_\alpha \langle w_\beta\rangle_{\text{ens}} - A_{\alpha\gamma}\frac{\partial}{\partial r_\gamma}\langle w_\beta\rangle_{\text{ens}} - B_{\alpha\beta}, \tag{8}$$

where $\langle w_\alpha\rangle_{\text{ens}} \equiv \langle w_\alpha(r)\rangle_{\text{ens}} = \lim_{t\to\infty}\langle w_\alpha(r,t)\rangle_{\text{ens}}$, and where the tensors $A_{\alpha\beta} \equiv A_{\alpha\beta}(r)$ and $B_{\alpha\beta} \equiv B_{\alpha\beta}(r)$ can be expressed in terms of $u_\alpha(r)$ and $V_{\alpha\beta}(r,r',t)$. Combining (6) and (8), one then obtains

$$D^\star_{\alpha\beta} = \tfrac{1}{2}\left\langle B_{\alpha\beta} - u_\alpha\langle w_\beta\rangle_{\text{ens}} + A_{\alpha\gamma}\frac{\partial}{\partial r_\gamma}\langle w_\beta\rangle_{\text{ens}} + \alpha \rightleftharpoons \beta \right\rangle_{\text{sp}}. \tag{9}$$

The ensemble average of (7) combined with (8) and (9) form a complete set of equations. It is important to notice that, without appeal to multiscale theory, the quasilinear treatment gives an expression like (8) for $\langle v_\alpha\tilde\rho(t)\rangle_{\text{ens}}$, with $\rho(t)$ instead of $\langle w_\alpha\rangle$ and without $B_{\alpha\beta}$ (source term). However, using this expression to close the ensemble-average of the Liouville equation (4) leads to an equation for $\rho(t)$ which is much more complicated than the equation for $\langle w_\alpha\rangle_{\text{ens}}$ obtained when (8) is used in the ensemble-average of (7) because, in contrast to the latter, the former equation for $\rho(t)$ has no stationary solution (since the evolution of $\rho(t)$ is asymptotically G-diffusive). Furthermore, it is difficult to extract from this equation the asymptotic form (5). Thus, the *combination* of multiscale and quasilinear treatments leads to essential simplification of the problem.

We apply the above results to the analysis of a turbulent flow generated by a two-dimensional lattice gas automaton [4], whose macroscopic dynamics is consistent with Navier-Stokes equations and which exhibits intrinsic spontaneous fluctuations which capture the essentials of actual fluctuations in real fluids [5]. We use the automaton to model a fluid subject to periodic

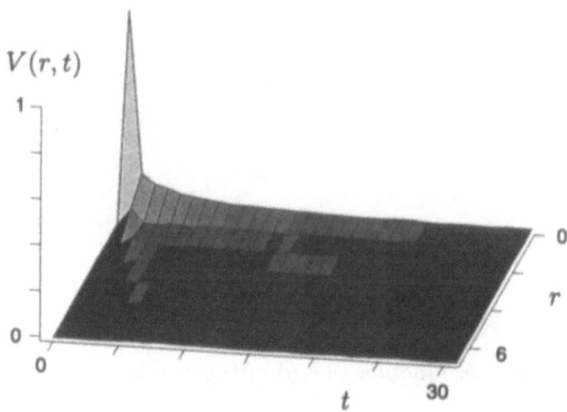

Figure 2.   The scalar covariance $V(r,t)$ of the flow generated by the automaton.

shear forcing along one direction. Beyond a critical value of the Reynolds number the ensemble-averaged part u_α of the flow evolves into a stationary ABC type flow characterized by the stream function $\psi(x,y)$ defined by

$$u_x = -\frac{\partial \psi}{\partial y}, \qquad u_y = \frac{\partial \psi}{\partial x}, \tag{10}$$

and shown in Fig. 1; the random part δv_α of the flow is statistically isotropic and may thus be characterized by the scalar covariance

$$V(r,t) = \langle \delta v_\alpha(\boldsymbol{r}+\boldsymbol{r}',t+t') \delta v_\alpha(\boldsymbol{r}',t') \rangle_{\text{ens}}. \tag{11}$$

As seen in Fig. 2, the structure of the function $V(r,t)$ shows that the automaton fluctuation cannot be approximated by a white-noise.

When a passive scalar (tracer) is implemented in the automaton [4], the tracer dynamics can be investigated, and the effective diffusion coefficient $D^\star_{\alpha\beta}$ can be measured and its value compared to the prediction of the theory outlined here. A detailed analysis will be presented elsewhere.

References

1. A. Bensoussan, J.-L. Lion and G. Papanicolaou, *Asymptotic Analysis for Periodic Structures* (North-Holland, Amsterdam, 1978).
2. L. Biferale, A. Crisanti, M. Vergassola, and A. Vulpiani, "Eddy diffusivity in scalar transport", Phys. Fluids **7**, 2725–2734 (1995).
3. N. G. Van Kampen, *Stochastic processes in physics and chemistry* (North-Holland, Amsterdam, 1992).
4. J. P. Boon, E. Vanden Eijnden, and D. Hanon, "A lattice gas automaton approach to *Turbulent Diffusion*", to be published in: Chaos, Solitons and Fractals (1998).
5. P. Grosfils, J. P. Boon, R. Brito, and M. H. Ernst, "Statistical hydrodynamics of lattice-gas automata", Phys. Rev. E **48**, 2655–2668 (1993).

RAPIDLY DISTORTED TURBULENT CHANNEL FLOW

M. HARTMANN AND D. RONNEBERGER
Drittes Physikalisches Institut der Universität Göttingen
Bürgerstraße 42-44, D-37073 Göttingen

While the turbulent 2D channel flow at time invariant boundary conditions is one of the most studied basic wall flows in turbulence research, far less is known about the same flow when subjected to unsteady boundary conditions. On the other hand, the effects of unsteadiness imposed on turbulent wall flows are encountered in quite a few cases of practical importance (e.g., sound propagation through flow ducts, heat transfer in pulsating flow, sediment transport by unsteady flow, etc). Several experimental investigations of the unsteady pipe and channel flow have been performed during the past decade (e.g. Mao & Hanratty [3], Tardu & Binder [4], Tardu et al [5], Beykirch et al [1]), however, these have been constricted to local measurements of the fluctuating velocity or wall shear stress in most cases. So our knowledge on the spatial and temporal structure of the flow is mainly based on the temporal structure of the measured signals. Consequently, the temporal modulation of the spatial structure of the flow that is caused by the imposed unsteadiness is far from being fully accessible so far.

Therefore the main purpose of the present work is to provide numerical data of the unsteady turbulent channel flow, and to analyze this flow with regard to its unsteady spatial structure. In the numerical simulation a finite difference scheme is used to solve the three-dimensional, time-dependent incompressible Navier-Stokes equations. As usually we make use of periodic boundary conditions in the streamwise and the spanwise direction, thus the flow is treated as a closed system in contrast to the open system regarded in the real world. Whether this is adequate in the study of the dynamics of the flow is still an open question, but the results obtained so far give no clues to significant differences between the (real) open system and the (simulated) closed system.

The unsteadiness is imposed on the mean flow – in the simulation as well as in the real experiments – either by harmonic oscillation of the longitudinal pressure gradient or, equivalently, by longitudinal oscillation of the walls of the 2D-channel: $u_{\mathrm{wall}}(t) = \hat{u} \cos(\omega_{\mathrm{w}} t)$, u = velocity component in

187

U. Frisch (ed.), Advances in Turbulence VII, 187–190.

the streamwise direction, ω_w = frequency of the wall. In both cases the imposed deformation of the phase averaged flow field consists of shear waves going out from the walls.

The simulations have been focused on rather high forcing frequencies so far. Basicly, it was expected that the response of the turbulence drops off when the forcing frequency is increased. This is so because the unsteady deformation of the flow is more and more restricted to the passive viscous sublayer, and additionally, because the structure of the turbulence needs some time to adjust to a change of the mean flow field and thus is expected to be unable to follow rapid oscillations. So the expectation was that only weak interactions between the turbulence and the imposed unsteadiness occur at high forcing frequencies leading to a better chance to find physical explanations of the results which possibly help to understand the more complex interaction at lower frequencies.

But even at the highest imposed frequency a not negligible modulation of the intensity of the streamwise velocity component $(\widetilde{u'u'})$ can be found even at the centerline of the channel. The propagation of $\widetilde{u'u'}$ at a speed in the order of u_τ has been found earlier but at lower frequencies (Tardu et al [5]). At the highest imposed frequency the propagation can be observed over more than two wave lengths (figure 1), and thus a diffusion process can be excluded. For comparison the amplitude (magnitude and phase) of the modulation of the streamwise velocity component $(\tilde{u})$ itself has been added to figure 1, and it exhibits the same spatial and frequency characteristics as $\widetilde{u'u'}$.

Within the first Stokes lengths $(y = \mathcal{O}(l_s),\ l_s = (2\nu/\omega_w)^{1/2})$ the shear wave $\tilde{u}$ shows excellent agreement compared to the Stokes solution. This indicates that according to the aforementioned expectation the interaction between the turbulent flow and the shear wave is too weak to cause any sensible effect on the propagation of the shear wave in the near wall region. But with increasing distance from the wall this interaction becomes significant, and the propagation speeds up.

An explanation for the reaction of the flow far away from the walls may be a viscoelastic property of the turbulence (Crow [2]). This becomes more obvious by direct inspection of the complex ratio between the modulated Reynoldsstress $-\widetilde{u'v'}$ and the modulated shear rate $\partial\tilde{u}/\partial y$: the phase of this eddy-viscoelasticity is between $0°$ and $-90°$ (figure 2) which is the intrinsic characteristic of a viscoelastic medium. The modulus of the local eddy-viscoelasticity equals the local eddy-viscosity of the mean flow outside the buffer layer and is comparatively small near the wall. Though the frequency has been changed by a factor of 11 in the simulations no significant dependency on the frequency has been found. This is quite unexpected result.

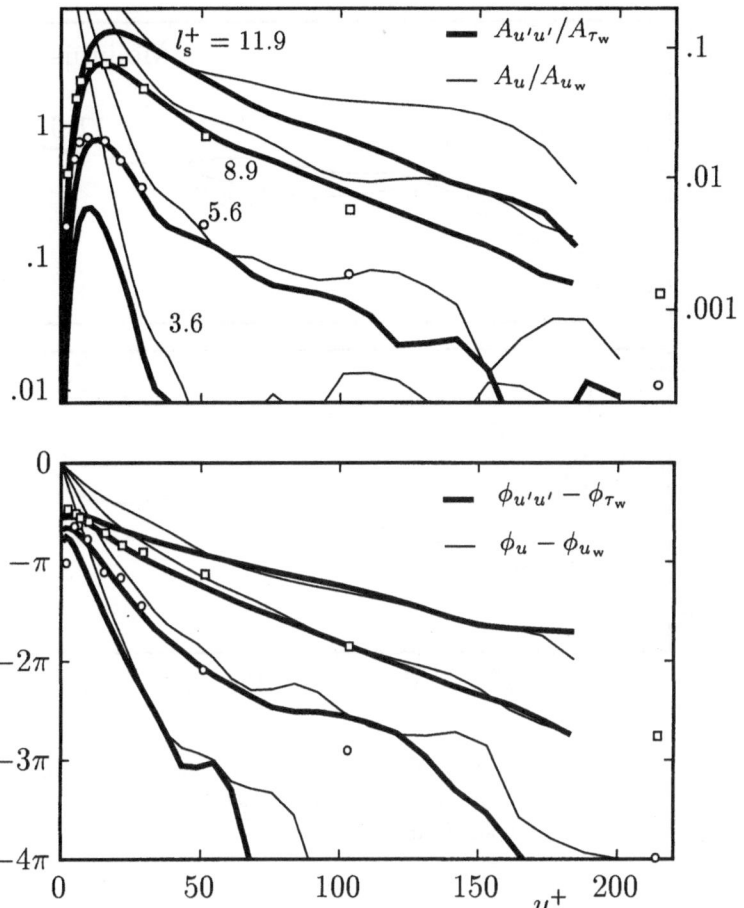

Figure 1. Propagation of shear waves through the turbulence: Magnitude $A_{u'u'}$ and phase $\phi_{u'u'}$ of the intensity $\widetilde{u'u'}$ of the streamwise velocity component (thick lines and symbols) as functions of the distance from the wall. The symbols pertain to real experiments. For comparison: Magnitude (right-hand ordinate in the upper diagram) and phase of $\widetilde{u}$ (thin lines). $\widetilde{u}(t) = A_u \cos(\omega_w t + \phi_u) = \langle u(t) \rangle - \overline{u}, \langle u(t) \rangle =$ phase average, $\widetilde{(u'u')}(t)$ correspondingly.

While this direct observation of the viscoelasticity of turbulence is interesting by itself it has also some implication in the search for stable turbulence structures. These structures are anticipated to restore themselves very rapidly after they have been distorted due to a perturbation of the surrounding flow field. This process which is expected to depend on the structure element under consideration has to be distinguished from the restoration of the flow field due to eddy-viscoelasticity which is inherent to any tangle of vortex lines.

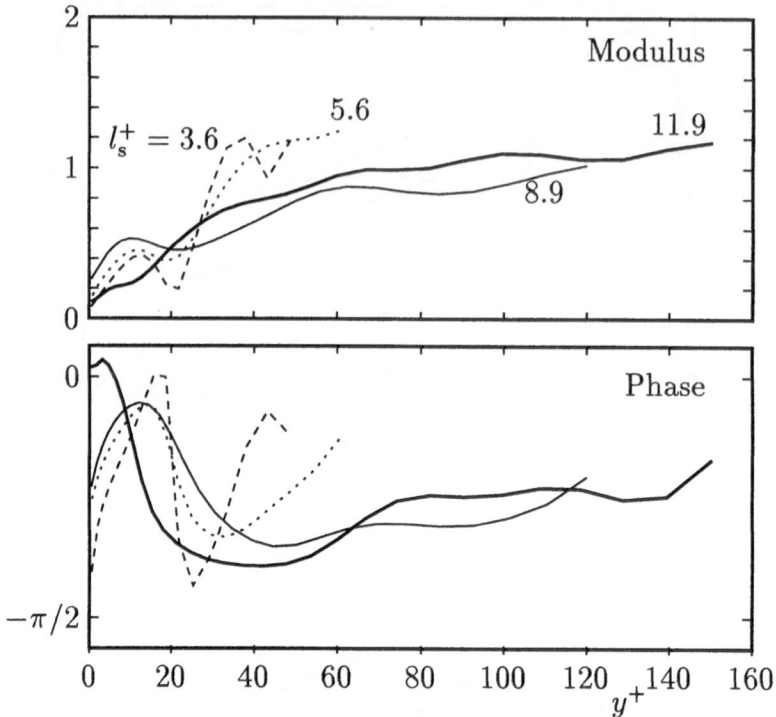

Figure 2. Eddy-viscoelasticity $-\widetilde{u'v'}/(\partial\widetilde{u}/\partial y)$ governing the propagation of a shear wave through the turbulent fluid (normalization to $-\overline{u'v'}/(d\overline{u}/dy)$). The large scatter of the results for small l_s^+ is due to the small amplitudes of $\widetilde{u'v'}$ and $\partial\widetilde{u}/\partial y$.

References

1. BEYKIRCH, M., HESSE, K., RONNEBERGER, D., 1996: The unsteady spectral properties of turbulent channel flow at time variant boundary conditions. *Advances in Turbulence VI*, S. Gavrilakis et al. (Ed.), Kluwer Academic Publishers, 513–514.

2. CROW, S. C., 1968: Viscoelastic properties of fine-grained incompressible turbulence. *J. Fluid Mech.* vol. 33, part 1, pp. 21–32.

3. MAO, Z.-X., HANRATTY, T. J., 1986: Studies of the wall shear stress in a turbulent pulsating pipe flow. *J. Fluid Mech.* vol. 170, pp. 545–564.

4. TARDU, S. F., BINDER, G., 1993: Wall shear stress modulation in an unsteady turbulent channel flow with high imposed frequencies. *Physics of Fluids A (Fluid Dynamics)* Vol. 5(8), pp. 2028–2037.

5. TARDU, S. F., BINDER, G., BLACKWELDER, R. F., 1994: Turbulent channel flow with large-amplitude velocity oscillations. *J. Fluid Mech.* vol. 267, pp. 109–151.

DIRECT NUMERICAL SIMULATION OF SUPERSONIC TURBULENT BOUNDARY LAYERS

T. MAEDER, N. A. ADAMS AND L. KLEISER
Swiss Federal Institute of Technology
Institute of Fluid Dynamics
ETH Zentrum, CH-8092 Zürich, Switzerland

Introduction

Morkovin's hypothesis asserts that the effect of compressibility on turbulence statistics is negligible in boundary layers up to about Mach 5. As a consequence, density behaves almost as a passive scalar, or as a variable flow property, and other dynamic effects associated with density or pressure fluctuations are small [1]. Experimental results confirm this hypothesis, although *a priori* assumptions about compressibility effects are usually employed for the processing of experimental data [2].

If the free-stream Mach number of a flow over an adiabatic wall is increased, the wall temperature rises. It follows that the mean temperature profile alters, resulting in a fluid property change within the boundary layer. Close to the wall, turbulence is characterized by a smaller Reynolds number than towards the boundary-layer edge. Besides the variable property effects, compressibility effects manifest themselves by additional correlations in the transport equations for Reynolds stresses or turbulent kinetic energy. Although data indicates that these terms are small for a zero-pressure-gradient boundary layer in the considered Mach number range ($M_\infty \leq 6$), other implications of compressibility are less obvious.

The strict Strong Reynolds Analogy (SRA), which can be derived from Morkovin's hypothesis under the assumption of small total-temperature fluctuations, is often employed when incompressible turbulence models are extended to compressible flow. It predicts a perfect correlation of temperature and velocity fluctuations. When the total-temperature variation across the boundary layer is no longer sufficiently small the strict assumptions do not apply and the validity of SRA, even in a less strict form, is unclear [1].

U. Frisch (ed.), Advances in Turbulence VII, 191–194.
© 1998 *Kluwer Academic Publishers.*

For relatively low Reynolds numbers the Direct Numerical Simulation (DNS) is a powerful tool to study turbulent flows. Recently, DNS was successfully applied to supersonic wall-bounded shear flows [3, 4]. The objective of the present work is to study the turbulence statistics of supersonic boundary layers with emphasis on compressibility effects. A set of DNS data at Mach numbers 3, 4.5 and 6 has been generated to cover a "critical" Mach number range around the validity-margin of Morkovin's hypothesis. The computed data is used to assess the relevance of modeling assumptions for compressible turbulence.

Simulation Method

The basic numerical approach follows the temporal simulation model as described in [5]. The computational domain has to satisfy two conditions. First, it needs to be short enough so that the mean-flow change in the streamwise direction within the domain can be neglected. Second, it must be large enough to allow turbulent fluctuations to decorrelate sufficiently.

A non-stationary mean-flow behavior is observed when streamwise mean-flow gradients are omitted due to streamwise periodicity. However, the mean-flow growth can be taken into account to some extent by imposing a given or quasi-simultaneously computed mean-flow through the addition of forcing terms to the basic equations [3]. Since the mean flow changes on a slow scale in the streamwise direction and the turbulent fluctuations change on a fast scale, the flow-field can be split into a slowly varying mean-flow part and a fast varying fluctuation part. The latter can be accurately represented in a streamwise periodic box with size of a few integral length-scales which is moved stepwise downstream during the calculation.

The mean flow is governed by the Reynolds-averaged parabolized Navier-Stokes equations, where the nonlinear interaction between mean flow and turbulent fluctuations is taken into account by a Reynolds-stress term. The effect of streamwise mean-flow variation on the turbulent fluctuations can be approximated by adding a forcing term to the Navier-Stokes equations, which are solved locally in a streamwise periodic domain to obtain the instantaneous flow field at the respective spatial station. The forcing is constructed such that the spatial average of the equations, which govern the full flow field locally, coincides with the equations governing the mean field. The correct mean flow is then obtained as the stationary limit of the mean solution computed at a fixed spatial station. Stationarity is reached if integral mean-flow quantities such as shape factor and skin-friction coefficient maintain a stationary value. When this is the case, the computational box can be marched another spatial step downstream.

Fourier collocation methods are employed for the spatial derivative calculations in the streamwise and spanwise directions. A compact central finite-difference scheme is used in the non-periodic wall-normal direction. The solution is advanced in time with a third-order low-storage explicit Runge-Kutta scheme. The derivatives on the slow streamwise scale are computed by a second order backward-differentiation formula, which uses mean-flow information of the two preceding spatial stations.

Results

We consider turbulent boundary-layer flows on an adiabatic flat plate at free-stream Mach numbers of $M_\infty = 3$, 4.5 and 6 with momentum-thickness Reynolds numbers $Re_\theta = 2988$, 3006 and 2656, respectively.

The mean streamwise-velocity profiles, transformed according to Van Driest (see [2]), are shown in figure 1a. Apparently, no pronounced logarithmic region is present. As pointed out in [2] the low Reynolds number region that dominates the inner layer at high Mach numbers makes the logarithmic region shrink dramatically.

The magnitude of the turbulent Mach number M_t indicates the significance of compressibility effects. For $M_t < 0.3$ compressibility effects are believed to be insignificant. Note that in our simulations M_t is at most 0.4, see figure 1b. Thus, we expect Morkovin's hypothesis to be valid for our data. As mentioned before, the SRA requires u' and T' to be perfectly correlated. Figure 1c shows that the magnitude of the correlation coefficient $|R_{uT}|$ is well below 1 across the boundary layer. After assuming its maximum of about 0.8 just outside of the viscous sublayer, it drops almost linearly towards the boundary-layer edge. We can conclude that the strict form of the SRA is not valid even for a $M_\infty = 3$ boundary layer.

A model for the dilatational dissipation ϵ_d has been derived from DNS data for homogeneous turbulence in [6], which uses the turbulent Mach number M_t as a parameter. According to this model the ratio ϵ_d/ϵ_s (where ϵ_s is the solenoidal dissipation) is proportional to M_t^2. In figure 1d the aforementioned ratio is plotted versus M_t^2. It is obvious that there is no "simple" functional relation between ϵ_d/ϵ_s and M_t. For the three investigated cases, the dilatational dissipation ϵ_d within the boundary layer is at least three orders of magnitude smaller than the solenoidal dissipation ϵ_s, so that ϵ_d can be considered negligible. It should be noted, however, that this is not the case for non-zero-pressure-gradient boundary layers, see [7].

References

1.   Lele S. K. (1994) Compressibility effects on turbulence, *Annu. Rev. Fluid Mech.* **26**, pp. 211–254.

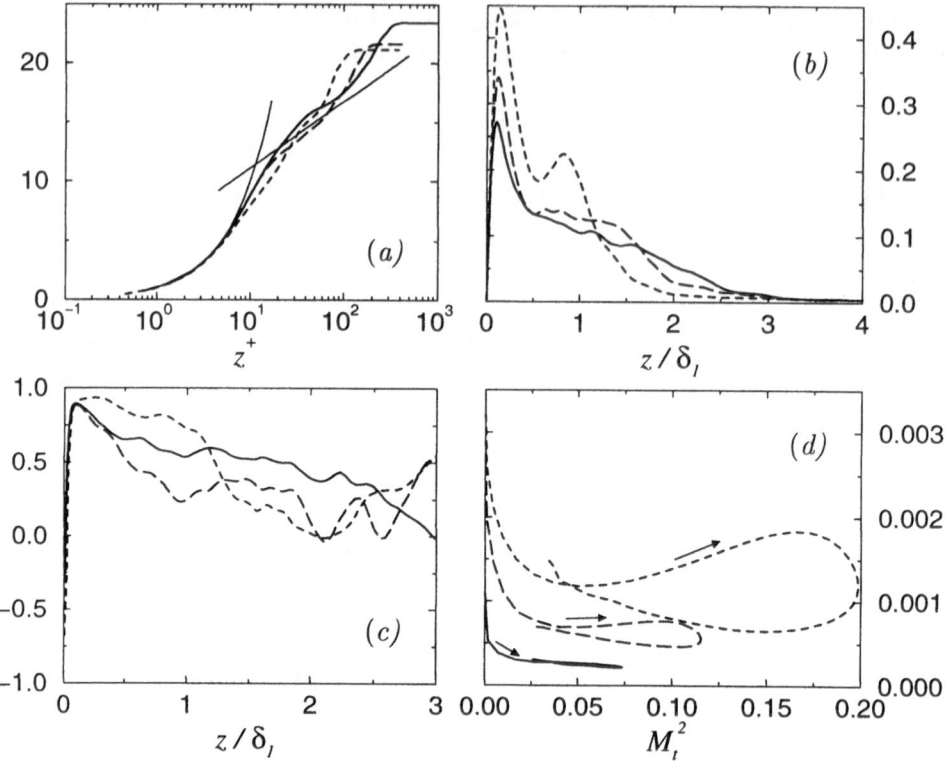

Figure 1. (*a*) Mean streamwise velocity profiles u_{VD}^+, Van Driest transformed (the thin solid lines represent the laws of the wall $u^+ = z^+$ and $u^+ = 2.44 \cdot \ln z^+ + 5.44$); (*b*) Turbulent Mach numbers M_t; (*c*) Distribution of $-R_{uT}$ in the wall-normal direction; (*d*) ratio ϵ_d/ϵ_s as a function of M_t^2 (arrows show direction of increasing z). Legend: —— $M_\infty = 3.0$, – – – $M_\infty = 4.5$, - - - $M_\infty = 6.0$, δ_1 : displacement thickness.

2. Smits A. J. and Dussauge J. -P. (1996) *Turbulent Shear Layers in Supersonic Flow*, AIP Press, Woodbury, New York.
3. Guo Y. and Adams N. A. (1994) Numerical investigation of supersonic turbulent boundary layers with high wall temperature, in *CTR Proceedings of the 1994 Summer Program*, Center for Turbulence Research, Stanford University and NASA Ames Research Center, Stanford, California, pp. 245–267.
4. Coleman G. N., Kim J. and Moser R. D. (1995) A numerical study of turbulent supersonic isothermal-wall channel flow, *J. Fluid Mech.* **305**, pp. 159–183.
5. Kleiser L. and Zang T. A. (1991) Numerical simulation of transition in wall-bounded shear flows, *Annu. Rev. Fluid Mech.* **23**, pp. 495–537.
6. Sarkar S., Erlebacher G., Hussaini M. Y. and Kreiss H. O. (1991) The analysis and modeling of dilatational terms in compressible turbulence, *J. Fluid Mech.* **227**, pp. 473–493.
7. Adams N. A. (1997) DNS of shock boundary-layer interaction – preliminary results for compression ramp flow, in *CTR Annual Research Briefs-1997*, Center for Turbulence Research, Stanford University and NASA Ames Research Center, Stanford, California, pp. 329–338.

SWIRLING PIPE FLOW SUBJECT TO AXIAL STRAIN

A.F. MOENE AND J.H. VOSKAMP

Fluid Dynamics Lab, Eindhoven University of Technology,
PO Box 513, 5600 MB Eindhoven, The Netherlands

AND

F.T.M. NIEUWSTADT

Laboratory for Aero and Hydrodynamics, Delft University of
Technology,
Rotterdamseweg 145, 2628 AL Delft, The Netherlands

1. Introduction

Swirling flow in combination with deformation still forms a major test case for many turbulence models. In order to better understand the relevant processes in such flows and to provide quantitative data for model validation, laboratory and numerical experiments are needed. Such data have been obtained on a simple flow: the passage of a swirling flow through a contraction. This flow can be considered as a prototype for flows that occur in cyclone separators and burners.

The only systematic investigation into the combined effect of strain and rotation on turbulence is the work by Leuchter and co-workers (Leuchter and Dupeuble, 1993; Leuchter, 1997). They restricted themselves to homogeneous turbulence, which made their results fit for theoretical analysis. Our approach is to study a flow that is closer to industrial practice.

2. Flow geometry

The contraction, used to generate the axisymmetric strain, has an inflow diameter D of $70mm$ and an outflow diameter of $40mm$. The semi angle of the contraction is 25°, giving a length of the contraction of about $35mm$ or $\frac{1}{2}D$. Note that the strain rate increases on passage through the contraction. The contraction is mounted in a test facility described in Steenbergen (1995). For the measurements without swirl, the contraction was located at $\approx 82D$ downstream of the pipe entrance. For the measurements *with* swirl, a swirl generator has been used at the pipe entrance. The

195

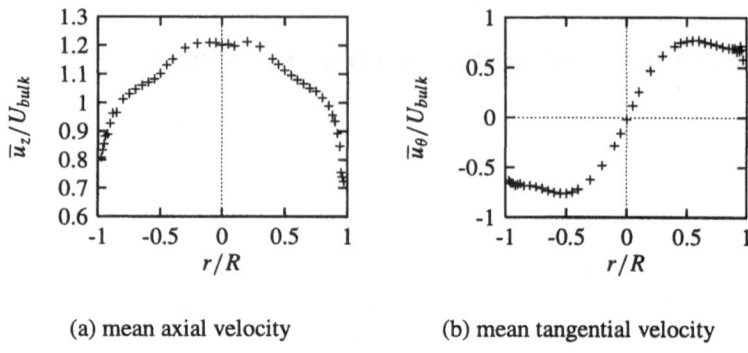

(a) mean axial velocity   (b) mean tangential velocity

Figure 1. Mean velocity profiles for $Re_D = 2 \cdot 10^4$ at $x/D = -1.8$.

distance between swirl generator and contraction is $\approx 14D$. Measurements have been performed at two Reynolds numbers: $Re_D = 2 \cdot 10^4$ and 10^5. Velocity measurements are made with a 2-D LDA system with water as the working fluid. The resulting mean velocity profiles for the flow with swirl are shown in figure 1.

3. Anisotropy of the stress tensor

In order to compare our results to those of Leuchter (1997), we first look at the data along the centerline of the pipe, where the flow can be assumed to be homogeneous (in the sense that $\frac{\partial}{\partial r}$ of $\bar{u}_z$ and the normal stresses equals 0, , $\frac{\partial \bar{u}_\theta}{\partial r} = constant$ and the shear stresses equal zero). In figure 2 the anisotropy of the normal stresses ($a_{\alpha\alpha}$) at the center line is shown, where $a_{\alpha\alpha}$ is defined as $a_{\alpha\alpha} = \overline{u'_\alpha u'_\alpha} - \frac{1}{3}k$, k is the turbulent kinetic energy and α is either z, r or θ. For the non-swirl case we see that the a_{zz} anisotropy becomes negative, as predicted by rapid distortion theory (RDT). For both Reynolds numbers, the results are in good agreement with an RDT analysis following Reynolds and Tucker (1975). In figure 2(b) it can be seen that the presence of swirl diminishes the effect of the axial strain on the anisotropy. The same result has been found by Leuchter (1997) who attributes this to an increased pressure-strain correlation.

Away from the pipe axis the flow can no longer be assumed to by homogeneous, nor axisymmetric. In figure 3 the profiles of all three anisotropies are shown, for both the non-swirl and swirl case. For the flow without swirl (3(a)), one can see that in a large part of the pipe cross section the anisotropy is reversed. Furthermore, the anisotropy is nearly constant for $-0.6 < r/R < 0.6$. In the swirling case, the region of the cross section over which the anisotropy is nearly constant does not change significantly between $x/D = 2.4$ and $x/D = 6.9$. This is due to a nearly complete suppression of the production of $\overline{u'_z u'_z}$ (since $\overline{u'_r u'_z}$, which occurs in the leading production term for $\overline{u'_z u'_z}$, is strongly reduced in the central part of the pipe). Furthermore, the anisotropies relax to their pre-strain values far more slowly than

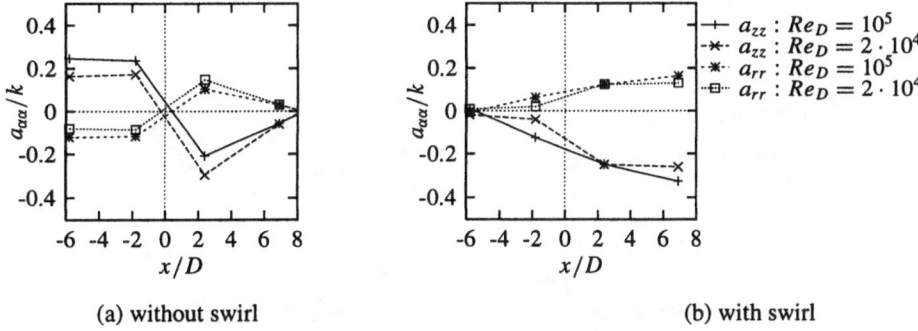

(a) without swirl   (b) with swirl

Figure 2. Development of the anisotropy $(a_{\alpha\alpha}/k)$ on passage through the contraction, where $a_{\alpha\alpha} = \overline{u'_\alpha u'_\alpha} - \frac{1}{3}k$. The origin of the contraction (where the radius starts to diminish) is at $x = 0$. The symbols represent measurements, the lines only serve to connect the points.

in the case without swirl.

4. Anisotropy of the length scales

In figure 4 the axial development of the integral length scales [1] of u_z and u_θ at the pipe axis is shown. In figure 4(a) we see that the effect of the axial strain on the length scales is to increase $L_{r,z}$ and to decrease $L_{z,z}$, which is consistent with the idea that the strain stretches the vortices with axes parallel to the pipe axis and shrinks those with axes normal to the pipe axis. The effect of rotation is to increase the length scales, indicating a tendency towards two-dimensionality. Comparison of figures 4(a) and 4(b) shows that the swirl enhances the growth of $L_{r,z}$ and suppresses the decline of $L_{z,z}$.

References

Leuchter, O. (1997). Rotation effects on strained homogeneous turbulence. *ERCOFTAC Bulletin*, (32), 52–57.

Leuchter, O. and Dupeuble, A. (1993). Rotating homogeneous turbulence subjected to axisymmetric contraction. Presented at the 9th Symposium on Turbulent Shear Flow, Kyoto (Japan), August 16-18, 1993.

Reynolds, A. and Tucker, H. (1975). The distortion of turbulence by general uniform irrotational strain. *J. Fluid Mech.*, **68**, 674–693.

Steenbergen, W. (1995). *Turbulent pipe flow with swirl*. Ph.D. thesis, Eindhoven University of Technology.

---

[1] $L_{\alpha,x} = \int_0^\infty \rho_\alpha(r_x)dr_x$, where ρ_α is the autocorrelation of component α and r_x is the lag in direction x. The integral *length* scale of component α, $L_{\alpha,z}$, is derived from the integral *time* scale of component α with Taylor's hypothesis.

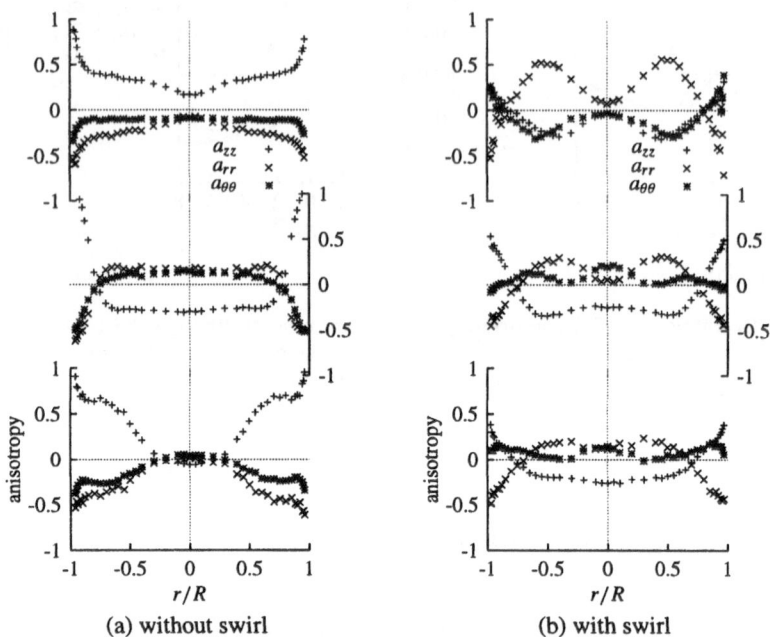

Figure 3. Anisotropy of the normals stresses throughout the pipe cross section for $Re_D = 2 \cdot 10^4$. 3(a) gives the data for the flow without swirl, 3(b) refers to the swirl case. Axial positions are, from top to bottom: $x/D = -1.8$, 2.4 and 6.9. Note that the radial coordinate is scaled with the *local* pipe radius.

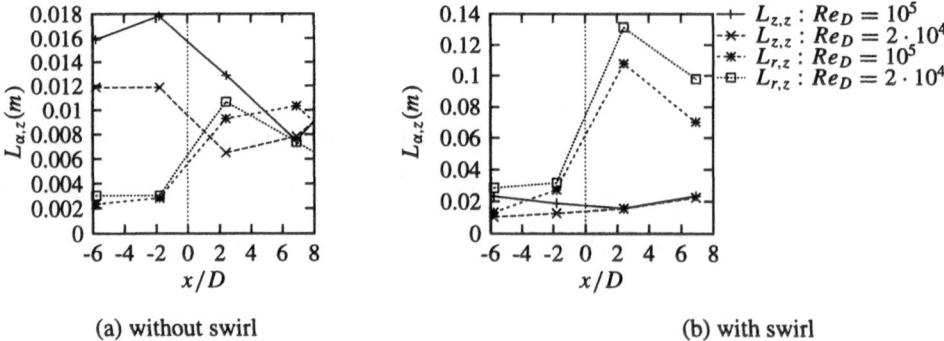

Figure 4. Development of the integral length scales on passage through the contraction. The length scales are in the axial direction: $L_{z,z}$ and $L_{r,z}$ (see footnote 1). 4(a) shows the results for the case *without* swirl and figure 4(b) gives the development for the case *with* swirl. Note the difference in vertical scale.

IV

High Reynolds Numbers and Intermittency

Grands Nombres de Reynolds et Intermittence

VI

A CASCADE MODEL OF TURBULENCE

LEO P. KADANOFF
The James Franck Institute
The University of Chicago
5640 South Ellis Ave
Chicago Illinois 60637 USA

e-mail : LeoP@UChicago.edu

This note describes a model of turbulent energy-cascades called the GOY model. This model contains a view of the basic processes of turbulence, within the context of simplification in which the k-space (k is the wave vector) is split into shells of different magnitudes of k. The dynamics moves energy from shell to shell, and conserves energy and "helicity" through the inertial range.

The GOY model was invented by Ohkitani and Yamada (1987, 1988, 1989) and by Gledzer (1973). Some of the recent interest in this model is due to the paper by Jensen, Paladin, and Vulpiani (1991). The model contains two important parameters: the shell width (λ) and a number (ϵ) which controls the motion of energy toward smaller of larger scales.

After describing some of the relation between this model and the Navier–Stokes equation, I point out that the GOY model includes some of the main physical ideas of the Kolmogorov theory of turbulence. I then describe some our own work carried on at the University of Chicago by Jane Wang, Detlef Lohse, Roberto Benzi, Norbert Schörghofer and myself. In different regions of the parameter space, the model agrees with both K41 and K62, and even agrees quite well with experimental data. This agreement is based upon a physical identification of helicity within the model. A detailed analysis of information transfer through the inertial subrange is given.

In one region of the parameter space the model is in a laminar state, in another the laminar state becomes unstable to chaotic behavior, in a third there is only the chaotic behavior. The phase diagram is described and the stability of the laminar state is treated in detail.

The linear stability problem is defined by a linear response matrix, and the eigenvalues and eigenstates of that matrix. For a system with N-shells,

U. Frisch (ed.), Advances in Turbulence VII, 201–202.
© *1998 Kluwer Academic Publishers.*

there are $2N$ eigenvalues. The scaling properties of these quantitites are treated in detail. For small values of ϵ, all eigenvalues are stable. Then as ϵ increases, one begins to find a few unstable eigenvalues. For still-larger ϵ, there is a kind of phase transition into a state in which a finite proportion of the eigenvalues become unstable. Apparently this transition is connected with a change in which the helicity goes from being dominated by large scale effects, to one in which it is dominated by small-scale behavior. Biferale (1995) predicted that there would be a change in behavior at this value ϵ

This talk is based upon two recent articles by Kadanoff (1995, 1997). This publication is supported in part by the U.S. National Science Foundation.

References

Biferale, L., private communication (1995).

Gledzer, E.B. *Sov. Phys. Dokl.* 18, 216 (1973).

Jensen, M.H., Paladin, G. and Vulpiani, A. *Phys. Rev.* A 43, 798 (1991).

Kadanoff, L.P. A model of turbulence, *Physics Today*, (September 1995) p. 11; Cascade Models of Turbulence and Mixing, *Tr. J. of Physics*, 21, 1-14 (1997).

Ohkitani, K. and Yamada, M. *J. Phys. Soc. Jpn.* 56, 4210 (1987); *Prog. Theor. Phys.* 79, 1265 (1988); *Prog. Theor. Phys.* 81, 329 (1989).

SPACE AND LEVEL DEPENDENCE IN
RANDOM MULTIPLICATIVE CASCADES

M. BLANK

CNRS, Observatoire de la Cote d'Azur, BP 4229, F-06304
Nice Cedex 4, France, e-mail: blank@obs-nice.fr [†]

Multiplicative cascade models in problems related to turbulence were introduced firstly by Novikov and Stewart and by Yaglom to describe stochastic transfer of energy along the inertial range. In spite of a great deal of results about such models (see, e.g. [4]), it turns out that still there are a lot of problems in the field. The aim of this paper is to discuss several mathematical questions that appear very naturally in the study of cascade models.

Consider a binary tree structure, obtained by hierarchically partitioning the original volume of size l_0 in sub-volumes of size $l_n = 2^{-n}l_0$, corresponding to fluctuations at different scales. The energy dissipation, associated to a cube at scale n, is multiplicatively linked to the energy dissipation, at the larger scale $n-1$. From the formal point of view this procedure can be defined as follows. Let $\{\xi_{ij}\}$ be a sequence of nonnegative random values with two indices. The pair of indices (i, j) defines the position of the node in our binary tree:

$$
\begin{array}{ccccc}
& & 1 & & \\
& \xi_{11} & & \xi_{12} & \\
\xi_{21} & \xi_{22} & & \xi_{23} & \xi_{24} \\
\xi_{31} & \xi_{32} & \cdots & \xi_{37} & \xi_{38} \\
\end{array}
\tag{1}
$$

$$\cdots\cdots\cdots\cdots\cdots\cdots\cdots\cdots$$

On the n-th level of the tree (1) there are 2^n nodes. These nodes correspond to the elements of the hierarchical partition of the initial volume Δ into

---

[†]On leave from Russian Academy of Sciences, Inst. for Information Transmission Problems, B.Karetnij Per. 19, 101447, Moscow, Russia.

U. Frisch (ed.), Advances in Turbulence VII, 203–206.

2^n disjoint sub-volumes $\Delta_k^{(n)}$ of equal size. One may think about these elements as elements of the partition of the unit interval into 2^n equal subintervals. We associate the vertical direction (levels of the tree) with the time coordinate, and the horizontal direction with the space coordinate. In distinction to the traditional approach we do not assume here that our random values are mutually independent.

There are two qualitatively different ways to look at models of this type. The first one, which we shall call "local", corresponds to the statistics, obtained by the multiplication of values ξ_{n,k_n} along only one branch of the tree. This is the point of view of the local observer, who can observe only events taking place in some small region (but on a number of different scales). The second way, which we shall call "global", is the following. Consider a random measure μ_n on our phase space, such that for any k the value $\mu_n(\Delta_k^{(n)})$ is equal to the multiplication of all random values on the nodes along the branch from the root of the tree to the considered node.

$$\mu_n(\Delta_k^{(n)}) := 1 \cdot \xi_{1,k_1} \cdot \xi_{2,k_2} \cdot \ldots \cdot \xi_{n,k_n},$$

where the sequence of positive integers $\{k_m\}$ corresponds to the nodes along the considered branch. Therefore for each n the measure μ_n is a piecewise constant distribution on our phase space. Under some relatively weak assumptions these random measures μ_n converge (in the weak topology) to a deterministic limit measure μ_∞. This construction is known in the mathematical literature as a *multiplicative chaos model* [3]. The difference between these two points of view is of the same nature as the difference between the averaging in time and in space in usual dynamical system theory, which clearly may lead to very different results.

In terms of above random measures one can define a structure function of the cascade as

$$\Phi_n(q) := \frac{\log(\sum_i \mu_n^q(\Delta_i))}{\log(2^n)}.$$

The index n here corresponds to the hierarchical level up to which we calculate the structure function. It is well known that in the traditional model of cascades with independent identically distributed (iid) random values the structure function can demonstrate phase transitions [3]. We restrict ourselves in this paper to the phase transitions in the weak sense, i.e. the divergence of higher order moments for the series for $\Phi_n(q)$, while in the strong sense the latter means that the corresponding limit function differs from $\Phi_\infty(q)$ and thus the statistics of the random measures μ_n does not converge to the corresponding statistics of the limit measure μ_∞. This question may be considered as a problem of statistical mechanics, where the measure μ_∞ plays the role of the *free energy* and the parameter q corresponds to the *inverse temperature*. The absence of phase transitions yields the so called self-averaging property of the free energy, which characterizes the unique random state.

The first question that we study is the condition under which the described (weak) phase transition does not take place. To do it consider a *conservative* cascade. This means that for any n the measure μ_n is probabilistic, i.e. $\mu_n(\Delta) = 1$, which is not the case for independent random values (when one can assume only that $\mathbf{E}\left[\mu_n(\Delta)\right] = 1$). Here $\mathbf{E}\left[\cdot\right]$ means mathematical expectation. To construct such a conservative cascade suppose that $\xi_{i,2k+2} = 1 - \xi_{i,2k+1}$ for any k, and random values $\xi_{i,2k+1} \in (0,1)$ are iid for all i and k. Our assumptions correspond to the nature of the cascade process, because subtrees of the binary tree (1) of the same volume should have the same statistical properties. Under these assumptions we prove the convergence of $\Phi_n(q)$ to its mathematical expectation and thus the absence of phase transitions. See details in [2].

Recently there was a long discussion in the literature about the adequacy of various cascade models for turbulence predictions (see, e.g. [4, 5] and further references therein). It is of interest that practically only random cascades with *independent* random values ξ were taken into account. A possible corollary to this discussion is that to make random cascade models more realistic one should consider some dependency between the components of the cascade. Above we have discussed the space dependence, resulting from the energy conservation, which immediately helps to avoid some unpleasant properties of the independent cascades. However, practically nothing was done to study the time (level) dependence. Probably the first result in the field was obtained in our paper with U. Frisch and L. Biferale [1], where we have constructed pure deterministic cascades. The intermediate case with Markov dependent (in time) cascades is practically not studied up to now. Clearly, by the nature of a computer experiment (or statistics obtained from a real experiment) one can observe a cascade only up to some level, say n. Therefore we nether deal with the limit structure function, but only with its finite time approximation (for a finite value of n). Due to this reason the dependence of $\Phi_n(q)$ on the time coordinate becomes very important. As it is well known if $\xi_{i,j}$ in different generations are iid, then the logarithm of the structure function grows linearly with time (see, e.g. [4]). Now we are in a position to study this question for a more general case.

Consider the simplest possible time dependent random model – the case when the dependence between random values ξ_{i,j_i} and $\xi_{i+1,j_{i+1}}$ belonging to the same branch of the tree (1) is defined by an ergodic Markov chain with a finite number of states corresponding to N different values $a_1, \ldots, a_N$. Denote by $P = (p_{ij})_{i,j=1}^N$ the transition probability matrix of the chain and by $p = (p_i)_{i=1}^N$ its unique stationary distribution, i.e. $P^*p = p$. Clearly the type of dependence of the structure function on the parameter n during the transient part of the process can be more or less arbitrary. Therefore we restrict our analysis to the stationary part. It is of interest that our construction is similar to that of mathematical theory of fast dynamos, i.e. the phenomenon by which rapid magnetic field growth can be sustained in the presence of a prescribed velocity field.

Let $S_n(q) := \sum_i \mu_n^q(\Delta_i)$. Simple calculations show that

$$S_n(q) = \sum_{i_1} \xi_{1,i_1}^q \sum_{i_2} \xi_{2,i_2}^q \cdots \sum_{i_n} \xi_{n,i_n}^q,$$

where the summation is going over all n-tuples $i_1, i_2, \ldots, i_n$ belonging to the same branch of our tree. Denote by A and $\mathbf{p}$ two diagonal $N \times N$ matrices, whose i-th diagonal elements are a_i and p_i respectively, and let $\mathbf{1}$ be a vector with all unit components. One can prove that in this notation

$$\mathbf{E}\,[S_n(q)] \;=\; 2^n \cdot p^*(PA^q)^n \mathbf{1} = 2^n \cdot \mathbf{1}^*(A^q P^*)^n \mathbf{p}\, \mathbf{1}$$

$$\mathbf{E}\left[S_n^2(q)\right] \;=\; 2^n \cdot \sum_{i=0}^{n} \mathbf{1}^*(A^{2q} P^*)^{n-i} \mathbf{p}((PA^q)^i \mathbf{1})^2$$

for any $n \geq 1$, which gives an exact representation of the first two moments of the main part of the structure function $\Phi_n(q) = \log(S_n(q))/\log(2^n)$.

Observe that for typical values $(a_i)_{i=1}^N$ the behavior of $\mathbf{E}\,[S_n(q)]$ as a function on the variable n asymptotically for large n is completely defined by the largest eigenvalue of the matrix (PA^q), while the "second largest" eigenvalue gives the correction. Since the latter in other terms can be described as an weighted (multiplied by a scalar function a_i) transfer operator, (PA^q) plays a role of a discrete analogue of the well known fast dynamo operator. It is of interest that typically the centered second order moment of $S_n(q)$ does not vanish when $n \to \infty$. Notice also that our calculations can be easily generalized for the case of k-ternary trees ($k \geq 2$) instead of the binary one ($k = 2$).

References

1. Biferale L., Blank M., Frisch U. Chaotic cascades with Kolmogorov 1941 scaling, *J. Stat. Phys.*, **75**:5-6(1994), 781-795.
2. Blank M. Multiplicative cascade models and multifractality, in: *Small-Scale Structures in Three-Dimensional Hydro and Magnetohydronamic Turbulence*, eds M.Meneguzzi, A.Pouquet, P.L.Sulem. Lecture Notes in Physics, 462(1995), 159-165, Springer-Verlag.
3. Collet P. and Koukiou F. Large deviations for multiplicative chaos, *Commun. Math. Phys.*, **147** (1992), 329-342.
4. Frisch U., Turbulence, 1995, Camb. Univ. Press.
5. Molchan G.M. Turbulent cascades: limitations and a statistical test of the lognormal hypothesis, *Phys. Fluids*, **9**:8 (1997), 2387-2396.

INTERMITTENT DISSIPATION IN DEVELOPED TURBULENCE

G. FALKOVICH
Physics of Complex Systems, Weizmann Institute of Science, Rehovot, 76100, Israel

I briefly describe here emerging methods designed for the description of probability density function (PDF) of the gradients of turbulent fields. The formalism is applied to Burgers' turbulence and to the scalar advected by both compressible and incompressible turbulence.

Developed turbulence is intermittent in space and time. Rare high peaks are responsible for the tails of dissipation PDF while quiet regions contribute PDF near zero; this prompts an attempt to describe intermittency at the level of dissipation PDF by two complementary approaches. The first (instanton) approach describes rare strong fluctuations as optimal fluctuations realizing probability extrema, the formalism is based upon a path-integral representation of conditional probability with optimal fluctuations being saddle points. A complementary description of gradient's PDF near zero is an adiabatics when high-order spatial derivatives are neglected. Following [1], let us demonstrate how both methods applied together give a consistent description of the PDF of dissipation by using, probably, the simplest (yet nontrivial) turbulent problem of a passive scalar advected by one-dimensional random flow which is smooth in space and white in time. Velocity field is thus a random laminar flow while the scalar field will be considered fully turbulent and multiscale (Batchelor-Kraichnan problem). That requires Prandtl number (viscosity-to-diffusivity ratio) to be large. The scalar field θ is subject to advection, diffusion and pumping:

$$\partial_t \theta + v\nabla\theta = \kappa\Delta\theta + \phi. \tag{1}$$

Both the velocity v and the source function ϕ are supposed homogeneous, Gaussian and δ-correlated in time: $\langle \phi(x,t)\,\phi(x',t')\rangle = \chi(|x-x'|)\delta(t-t')$ where χ decays on a scale L, $\chi(0) = P$ is the flux of θ^2. The correlation function of the velocity has the correlation length $L_v \gg L$:

$$\langle v(x,t)v(x',t')\rangle = [VL_v - VL_v^{-1}(x-x')^2]\delta(t-t') \tag{2}$$

U. Frisch (ed.), Advances in Turbulence VII, 207–210.
© *1998 Kluwer Academic Publishers.*

Let us first implement a simple adiabatic approach neglecting diffusion term. Then for the single-point PDF $\mathcal{P}(\omega, t) = \langle \delta[\theta_x(x, t) - \omega] \rangle$ one obtains a closed Fokker-Planck equation

$$\frac{\partial \mathcal{P}}{\partial t} = (D\omega^2 + T)\frac{\partial^2 \mathcal{P}}{\partial \omega^2} + 4D\omega \frac{\partial \mathcal{P}}{\partial \omega} + 2D\mathcal{P} \,, \qquad (3)$$

where we denote $T = \chi''(0)$ and $D = VL_v^{-1}$, the variances of ϕ_x and v_x respectively. That equation has an equilibrium steady solution

$$\mathcal{P}(\omega) \propto (T + D\omega^2)^{-1}, \quad \mathcal{P}(\epsilon) \propto (\kappa T/D + \epsilon)^{-1}, \qquad (4)$$

applicable for $\epsilon = \kappa\omega^2 \ll P = \langle \epsilon \rangle$. Since $T \simeq P/L^2$ and the Peclet number $Pe^2 = DL^2/4\kappa$ is assumed to be large then (4) has a wide interval of validity. Note that T/D is a square gradient produced by the pumping during the typical stretching time D^{-1}. For $Pe \gg 1$, $T/D \ll P/\kappa$.

Let us now describe the tail of $\mathcal{P}(\epsilon)$ at $\epsilon \gg \langle \epsilon \rangle$. It is clear from (2) that the correlation functions of the strain $\sigma = v_x$ are x-independent that is σ can be treated as a random function of time t only. The Martin-Siggia-Rose action $\mathcal{I}$ for the n-th order moment of the gradient θ_x is given by

$$i\mathcal{I} = \int E dt - \int dt dx \, p\partial_t\theta + \frac{n}{2}\log(\theta_x)^2$$

$$E = \int dx \, p(-\sigma x\partial_x\theta + \kappa\partial_x^2\theta) + \frac{1}{2}\int dx_1 dx_2 \, p_1\chi(x_{12})p_2 - \frac{(\sigma - D)^2}{4D} \,.$$

Assuming $n \gg 1$ we calculate $\langle \theta_x^n \rangle$ in the saddle-point approximation. The flow configuration (optimal fluctuation or instanton) that gives the main contribution into the moment corresponds to the action extremum:

$$\partial_t\theta + \sigma x\partial_x\theta - \kappa\partial_x^2\theta = -i\int dx' p(t, x')\chi(x - x') \,, \qquad (5)$$

$$\partial_t p + \sigma\partial_x(xp) + \kappa\partial_x^2 p = n\delta(t)\delta'(x)/\theta_x(0, 0) \,, \qquad (6)$$

where $\sigma = D + 2D\int dx \, px\partial_x\theta$. Then $\langle \theta_x^n \rangle = \theta_x^n(0, 0)\exp I_{extr}$. The solution of (5,6) that satisfies the boundary conditions $\theta(x, -\infty) = 0$ and $p(x, +0) = 0$ is found in [1], it corresponds to some optimal process that produces large gradient with largest possible probability. At a distant past, the strain $\sigma \approx D$ stretches small initial perturbation up to the width of order L and the force pumps θ up to the amplitude of order $\sqrt{P/D}$. At the same time, the strain σ decreases due to the negative contribution of the second term. Then second stage starts when $\sigma \ll D$, the width of θ does not change and only the amplitude grows due to the force. Of all realizations of the force the constant one is preferred, since it gives the fastest growth, θ increases as ϕt. The weight of such a process is $\exp(-\phi^2 t - t/2)$. The second term in the exponent is the probability to have small σ during the time t. Then $\phi \sim 1$ and $t \propto n$. By the end of this stage $\theta_x \propto n$. And finally, during the last stage the profile having the amplitude n and width L is compressed by the large σ which can be estimated as $\sigma \sim -Dn$. The duration of that stage (and therefore the final width) is determined by diffusion: $\sigma x \partial_x \theta \sim \kappa \partial_x^2 \theta$ at the end, so the width of θ shrinks to $L\sqrt{\kappa/n}$. Now, θ has amplitude n and width $n^{-1/2}$, therefore the final answer for θ_x^n is $\propto n^{3n/2}$, which corresponds to the stretched exponential tail

$$\ln \mathcal{P}(\epsilon) \propto (\epsilon/\langle\epsilon\rangle)^{1/3} . \tag{7}$$

Let us describe an alternative formalism (which exploits the smallness of $1/Pe$) considering after [2] a scalar advection in a 2d incompressible random flow. We consider (1) and replace (2) by

$$\langle v_\alpha(t_1, \mathbf{r}_1) v_\beta(t_2, \mathbf{r}_2)\rangle = \delta(t_1 - t_2)[V_0 \delta_{\alpha\beta} - D(3\delta^{\alpha\beta} r^2/2 - r^\alpha r^\beta)] \tag{8}$$

One cannot treat diffusion perturbatively because $\mathcal{P}(\epsilon)$ is a nonperturbative object with respect to κ: it is zero at $\epsilon \neq 0$ and zero diffusivity yet have nonzero limits as $\kappa \to +0$. Still, the presence of a small parameter calls for finding a proper way to simplify the description, which is provided by the dynamical Lagrangian formalism. Indeed, our goal is to express unknown (statistics of dissipation) via known (statistics of pumping). Since Peclet number is the ratio between pumping and diffusion scales, then any piece of scalar has a long way to go between birth and death, our goal is to describe how statistics is modified along the way. Dynamical formalism explicitly reveals the presence of two different time scales, a short one related to stretching and a long one related to diffusion (which eventually restricts the process of stretching). Time scale of stretching fluctuations is of order of inverse Lyapunov exponent while the whole time of stretching is $\ln Pe$ times larger. At $Pe \to \infty$, we are able to calculate PDF $\mathcal{P}(\epsilon)$ rigorously, exploiting time separation and executing explicitly separate averaging over slow and fast degrees of freedom. Introducing $\sigma_{\alpha\beta} \equiv \nabla_\beta v_\alpha$ and passing into the comoving reference frame we write (1) as $\partial_t \theta + r_j \sigma_{jl}(t)\partial_l \theta - \kappa\Delta\theta = \phi(t, \mathbf{r})$. Making spatial Fourier transform and introducing $\hat{W} = \hat{\sigma}W$, one can write $\theta_k(t)$ as an integral over Lagrangian trajectories

$$\int_0^t dt' \phi\left(t - t', \hat{W}^{-1,T}(t')\mathbf{k}\right) \exp\left[-\kappa k_\mu \int_0^{t'} \left[\hat{W}^{-1}(\tau)\hat{W}^{-1,T}(\tau)\right]_{\mu\nu} d\tau k_\nu\right] .$$

Averaging over ϕ a product of $2n$ replicas of the inverse Fourier of $k\theta_k$, we get the n-th moment of $\epsilon = \kappa (\nabla\theta)^2$ and restore the PDF

$$\mathcal{P}(\epsilon) = \int_{0^+ - i\infty}^{0^+ + i\infty} \frac{ds\, e^{s\epsilon}}{2\pi^2 i} \int dm\, e^{-m^2} \left\langle e^{-s \int_0^\infty dt\, Q} \right\rangle_\sigma , \tag{9}$$

$$Q = \int d\mathbf{q} \chi_q \left[\frac{\mathbf{q}\hat{W}(t)\mathbf{m}}{2\pi Pe} \right]^2 \exp\left[-\frac{\mathbf{q}\hat{\Lambda}(t)\mathbf{q}}{Pe^2} \right], \tag{10}$$

where $\hat{\Lambda}(t) \equiv D\hat{W}(t) \int^t dt' \hat{W}^{-1}(t')\hat{W}^{-1,T}(t')\hat{W}^T(t)$ and $\mathbf{q} = \mathbf{k}L$, an extra integration over the auxiliary vector $\mathbf{m}$ takes care of combinatorics. The next step is Kolokolov transformation to replace time-ordered exponent by a regular function, the natural price to pay is measure nonlocality in the path integral. Then we implement time separation which makes reduction to two subsequent path integrations both local in time: one describing the long evolution (since the stretching effectively goes along one direction it is reduced to ODE that can be solved for any pumping χ) and another describing the fast fluctuations which can be integrated exactly due to strain Gaussianity [2]. With that method, one can re-derive the PDF for 1d case and find $\mathcal{P}(\epsilon)$ in two dimensions. The tail of distribution at $\epsilon \gg \langle\epsilon\rangle$ is shown to be the same stretched exponent (7). That tail owes it's form to the fluctuations of the dissipation scale r_d. Indeed, r_d^{-1} is proportional to the (Gaussian) strain; assuming that the fluctuations of $\delta\theta$ and r_d are independent one gets 2/3-tail for $w = \delta\theta/r_d$ integrating together exponential $\mathcal{P}(\delta\theta)$ and Gaussian $\mathcal{P}(1/r_d)$: $\mathcal{P}(w) = \int \mathcal{P}(\delta\theta)\mathcal{P}(1/r_d)\delta(w - \delta\theta/r_d)d\delta\theta dr_d^{-1}$. It is likely that the same law is correct for the Prandtl number of order unity as well because the tail is determined by the fluctuations with r_d smaller than the mean value (and the viscous scale), velocity can be considered smooth at such scales. At $\epsilon \ll \langle\epsilon\rangle$, we find $\mathcal{P} \propto 1/\sqrt{\epsilon}$, note that this is different from what one may get from adiabatics at 2d. The reason is geometrical: there are large regions of almost smooth scalar field that are unaffected by diffusion in one dimensional compressible flow, while at higher dimensions a piece of scalar having small gradient in some direction necessarily has large gradient in another direction due to incompressibility. That is why adiabatics works only in 1d.

We also apply instanton formalism to Burgers equation $u_t + uu_x - \nu u_{xx} = f$ driven by a Gaussian force f and for a dissipation $\epsilon = \nu u_x^2$ we find $\ln \mathcal{P}(\epsilon) \propto -(\epsilon/\bar{\epsilon})^{3/4}$ at $\epsilon \gg \bar{\epsilon} = \langle\epsilon\rangle \max\{Re, Re^{-1}\}$ [3]. Note that this formula is valid for arbitrary Reynolds number (taken on pumping scale) while (7) is for large Peclet number only.

References

1. Balkovsky E. and Falkovich G. (1998) Two complementary descriptions of intermittency, *Phys. Rev. E* **57**, pp. R1231–R1234.
2. Chertkov M., Falkovich G. & Kolokolov I. (1998) Intermittent dissipation of a passive scalar in turbulence, *Phys. Rev. Lett.* **79**.
3. Balkovsky E., Falkovich G., Kolokolov I. & Lebedev V. (1997) Intermittency of Burgers' turbulence, *Phys. Rev. Lett.* **78** pp.1452–1455.

INHOMOGENEOUS TURBULENCE IN THE CLOSED VON KÁRMÁN FLOW

F. CHILLA AND J.F. PINTON

École Normale Supérieure de Lyon
CNRS URA 1325, 46 allée d'Italie, F-69364 Lyon France

Many studies in turbulence have been devoted to the homogeneous and isotropic (HI) case, and certain universal feature have emerged, as the Kolmogorov $k^{-5/3}$ spectrum or the values of the structure function exponents [1,2]. In particular, most of the turbulence phenomenology is based on the existence of a "universal" properties in an inertial range of scales whose dynamics is independant of the characteritics of the large scale motion and of the dissipative zone. However, real flows at high Reynolds numbers are strongly inhomogeneous, anisotropic and it is worthwhile to study the relevance and usefulness of HI concepts in these situations.

We analyze here the turbulence features in the neighborhood of a strong vortex which may be stable or unstable. One motivation is that in inhomogenenous flows coherent structures in the form of vortex tubes are commonly observed [3].

The flow is produced in the gap between two coaxial disks of radius R rotating in the same direction. Air is the working fluid; the flow is enclosed in a cylindrical vessel. The rotation rates of each disk is set independantly and controled with a feedback loop. In this situation one observes a large scale dynamics that depends on the disks rotation rates [4,5,6], superimposed to turbulent fluctuations (the integral Reynolds number of the flow is $Re > 5 \ 10^4$). For equal rotation rates ($f_1 = f_2 = 30$Hz), the vertical vorticity concentrates to build a strong stable axial vortex. The flow has a large core in solid body rotation. When the disks rotate at quite different rates ($f_1 = 12$Hz, $f_2 = 40$Hz) the axial vortex is unstable. One observes an intermittent sequences of formation and breakdown of a strong and thin axial vortex. We consider here the influence of the large scales dynamics on the small scale turbulence features. We use local hot-wire anemometry to study the velocity field characteristics as a function of the distance r to the rotation axis. The component $v_{r\theta} = \sqrt{v_r^2 + v_\theta^2}$ is recorded; in the region

U. Frisch (ed.), Advances in Turbulence VII, 211–214.

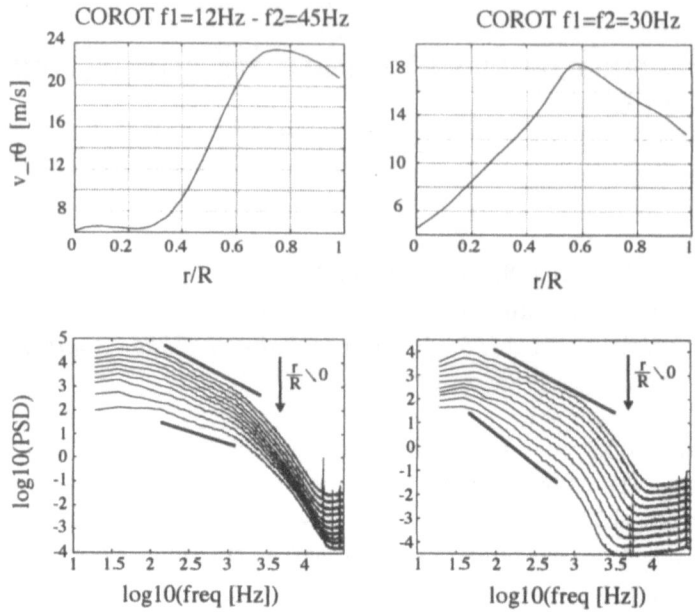

Figure 1. (Top): mean velocity profile; (Bottom): power spectrum. In each case the left column pertains to the unstable vortex dynamics and the arrow indicates the direction of variation as one gets closer to the rotation axis.

discussed here the azimuthal component dominates.

One issue is the shape of the velocity spectrum. We observe that, at each distance r, it displays a power law behavior indicating the existence of a self similar region. Sufficiently far away from the rotation axis one gets a Kolmogorov-like $k^{-5/3}$ spetrum. It is no longer the case as one gets nearer to the vortex structure; the slopes of velocity spectra are not constant and change continuously:
- in the case of equal rotation rates (stable vortex) the slope of the spectrum goes from -1.5 at $r/R \sim 0.5$ to -2.2 at the center.
- in the case of different rotation rates (unstable vortex), the spectrum slope varies from -1.65 to -1.1 as r/R again ranges from 0.5 to 0.
As can been seen the large scale dynamics influences the energy transfers in the neighborhood of a coherent structure. In the case of a stable vortex, the spectra exhibit a steeper slope; this may indicate a tendency to bidimensionalization. On the opposite the spectra have a slower decrease as one approaches the core of an intense, unstable vortex – this again may be expected: sharp discontinuous 'spikes' would yield a k^{-1} spectrum. We observe that this behavior is at odds compared to our study of the open flow [6] (i.e. with the lateral walls removed). In that situation the vortex structure is weaker and has a slow precession motion; the velocity spectra

at every distance from the vortex have a Kolmogorov scaling region.

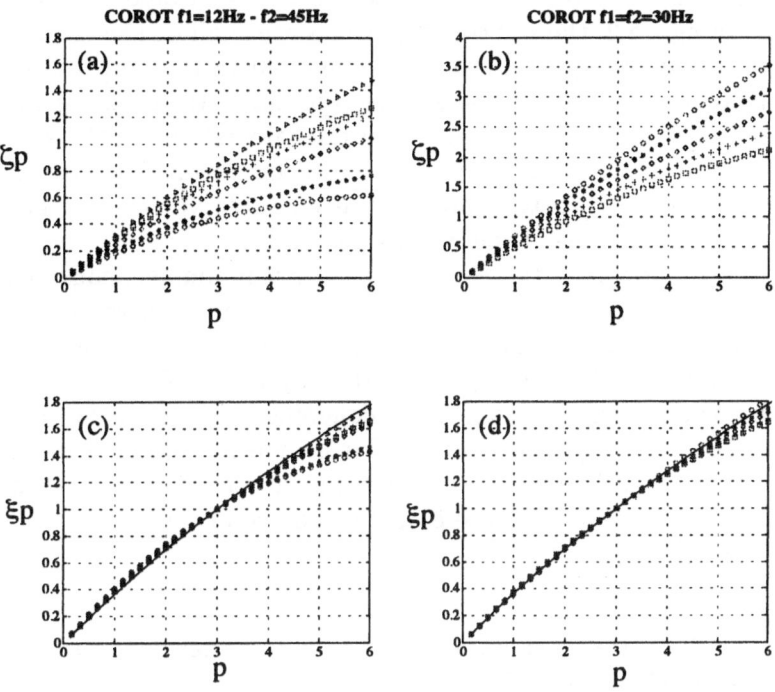

Figure 2. Structure function exponents at several distances near the vortex core : unstable vortex case on the left (a&c) {(△3.5cm), (□3cm), (+2.5cm), (◇2cm), (*1cm), (o0cm)} and stable vortex case on the right (b&d) {(□2cm), (+1.5cm), (◇1cm), (*0.5cm), (o0cm)}.

Another issue is the intermittency behavior, i.e. the change in the shape of the PDFs of the velocity increments through the scales. It can be measured *via* the structure function exponents:

$$\zeta(p) \equiv \frac{d \ln < \delta|v(\ell)|^p >}{d \ln \ell} \ ,$$

calculated in the inertial domain evidenced in the spectra. The results are displayed in figure 2 – top curves. Since the 2nd order exponent is related to the scaling of the velocity spectrum, the $\zeta(p)$ functions at each location are ordered as the spetra slopes. In particular, the Kármán-Howarth value $\zeta(3) = 1$ is no longer a fixed point. However, in each flow geometry and for every location of the probe, the $\zeta(p)$ functions are non linear and the magnitude of intermittency effects is related to their curvature. To study this and to offset the variations of $\zeta(3)$, we calculate the relative exponents:

$$\xi(p) \equiv \frac{d \ln < \delta|v(\ell)|^p >}{d \ln < \delta|v(\ell)|^3 >} \ ,$$

again in the 'inertial range' detected on the spectra – although the scaling range is somewhat extended using this scheme. As can be observed in the bottom part of figure 2, the relative exponents are similar. Deviations from the isotropic ('She-Lévèque'[7]) values are marked in the neighborhood of the instable vortex and otherwise smaller (possibly within the error bars –this present work uses 10^6 data points in the calculations of the structure functions). However we note a trend: the $\xi(p)$ curvature increases as one gets closer to the axis (vortex core) in the unstable vortex configuration whereas it decreases in the stable vortex geometry.

These results show that the velocity cascade is inhomogeneous in space. Its characteristics depend on the distance to the coherent vortex and are quite different from that of homogeneous, isotropic turbulence. The flow is also anisotropic at all scales, as indicated by analysis of other velocity components. The influence of large scale coherent motions on the turbulent cascade has also been observed in a similar geometry [8] and in the neighborhood of large shearing zone [9]. It remains a goal of further studies to relate the observed changes to the global structure of the flow.

References

1. Nelkin M. (1994), Universality and scaling in fully developed turbulence, *Adv. in Physics*, **43**(2), 143-181.
2. Arnéodo A. et al., (1996), *Europhys. Lett.*, **34**(6),411-416.
3. Bonnet J.-P., Glauser M.N. eds., (1995), Eddy structure identification in free turbulent shear flows, Kluwer Acad. Press.
4. Pinton J.-F., Chillà F., Mordant N., (1998), Intermittency in the closed flow between coaxial corotating disks, to appear in *European J. Mech./B Fluids*.
5. Chillà F., Pinton J.-F.,Labbé R.(1996) On the influence of a large-scale coherent vortex on the turbulent cascade,*Europhys. Lett.*, **35**, 271.
6. Labbé R.,Pinton J.-F., Fauve S. (1996) Study of the von Kármán flow between coaxial co-rotating disks, *Phys. Fluids*, **8**(4), 914-922 .
7. She Z.S., Lévèque (1994) Universal scaling laws in fully developed turbulence,*Phys. Rev. Lett.*, **72**, 336 .
8. Andreotti B., Douady S., Couder Y., (1998), Experimental investigation of turbulence near a large scale vortex, to appear in *European J. Mech./B Fluids*.
9. Toschi F., Benzi R., (1997) private communication.

UNCOVERING A LOG-NORMAL CASCADE PROCESS IN HIGH REYNOLDS NUMBER TURBULENCE FROM WAVELET ANALYSIS

A. ARNEODO, S. MANNEVILLE AND J.F. MUZY
Centre de Recherche Paul Pascal, Univ. Bordeaux I
avenue Schweitzer, 33600 Pessac, France

We present a generalization of the Castaing *et al.* [1] approach of velocity intermittency using the wavelet transform (WT). This description consists in looking for a multiplicative cascade process directly on the velocity field assuming that the probability density function (pdf) of the modulus maxima of the WT (WTMM) at a given scale a, $P_a(T)$, can be expressed as a weighted sum of dilated pdf's at a larger scale $a' > a$:

$$P_a(T) = \int G_{aa'}(u) \, P_{a'}(\mathrm{e}^{-u}T) \, \mathrm{e}^{-u} \mathrm{d}u, \quad \text{for } a' > a. \tag{1}$$

The reader is referred to Ref. [2] for the computation of the WT of the velocity field and the restriction of the analysis to the modulus maxima of the WT (WTMM method). Our numerical method for estimating the Fourier transform $\widehat{G}_{aa'}$ of the kernel $G_{aa'}$ is described in Refs. [3, 4]. This wavelet-based method is applied to a turbulent velocity signal recorded in the Modane wind tunnel and kindly provided by Y. Gagne and Y. Malecot. The Taylor scale based Reynolds number is $R_\lambda \simeq 2000$ and the sample is $2.5 \cdot 10^8$ points long, with a resolution of roughly 3η (where η is the Kolmogorov scale), corresponding to 25000 integral scales. We then propose a two-point statistical analysis based on space-scale correlations [5].

1. Experimental results of the one-point statistical approach

The computation of $\widehat{G}_{aa'}$ for various pairs of inertial scales $a < a'$ reveals the existence of a function $s(a, a')$ and of a single kernel $\widehat{G}$ such that $\widehat{G}_{aa'} = \widehat{G}^{s(a,a')}$ (*i.e.* formally $G_{aa'} = G^{\otimes s(a,a')}$). According to the definitions given in Ref. [1], this means that the underlying cascade process is self-similar

U. Frisch (ed.), Advances in Turbulence VII, 215–218.

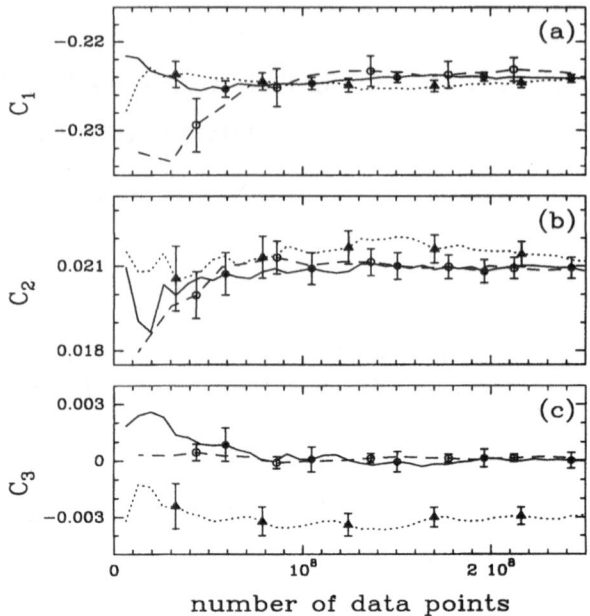

Figure 1.   The first three cumulants of $G_{aa'}$ vs the sample length. Turbulent velocity signal for $a = 770\eta$ and $a' = 1540\eta$ ($\circ$ and dashed line), log-normal numerical process with $m = 0.32$ and $\sigma^2 = 0.03$ ($\bullet$ and solid line) and log-Poisson numerical process with $\lambda = 2$, $\delta = 0.89$ and $\gamma = -0.082$ (triangles and dots) for the two corresponding scales $a = 2^8$ and $a' = 2^9$. The analyzing wavelet is a first order compactly-supported wavelet [4].

and that $s(a, a')$ accounts for the number of cascade steps from scale a' to scale a. Such a cascade is said scale-similar (or scale-invariant) if $s(a, a') = \ln(a'/a)$. In the present case, $s(a, a')$ turns out to be very well fitted by the functional form $s(a, a') = (a^{-\beta} - a'^{-\beta})/\beta$, where $\beta \simeq 0.095$ quantifies the departure from scale-similarity (scale-invariance is restored for $\beta \to 0$) [6]. Thus, the cascade process is self-similar but not scale-invariant.

To analyze precisely the shape of G, we use the Taylor series expansion: $\widehat{G}(p) = \exp\left(\sum_{k=1}^{\infty} c_k \frac{(ip)^k}{k!}\right)$, where the (real valued) coefficients c_k are the cumulants of G. Figure 1 shows the first three cumulants $C_k = s(a, a') c_k$ (for a given pair of inertial scales $a < a'$) for the turbulence data and for both a log-normal (mean m and variance σ^2) and a log-Poisson (with the λ, δ and γ parameters consistent with the ones proposed by She and Leveque [7]) synthetic numerical processes. Even though both numerical processes perfectly fit the first two cumulants, the log-Poisson model yields a third order cumulant that is more than one order of magnitude higher than the experimental one, whereas the very small (theoretically zero) third order cumulant of the log-normal numerical process still remains within the

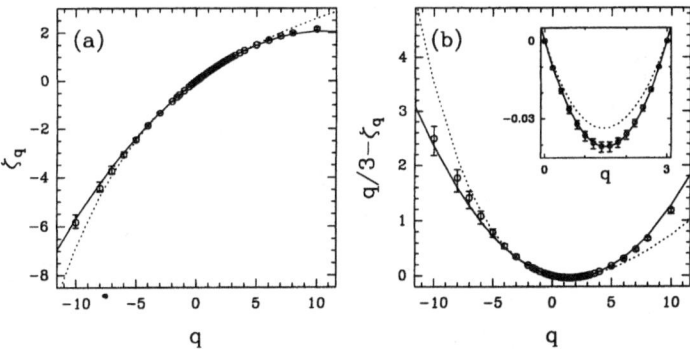

Figure 2. WTMM estimation of the ζ_q spectrum for the turbulent velocity signal [4]. (a) ζ_q vs q. (b) Deviation of the experimental spectrum from $q/3$. The experimental measurements (o) are compared to the theoretical quadratic ESS spectrum of a log-normal process with $m = 0.32$ and $\sigma^2 = 0.03$ (solid line) and to the She and Leveque [7] log-Poisson prediction (dots).

error bars of the experimental one. We thus exclude log-Poisson statistics as proposed in Ref. [7] and conclude that the statistics underlying the velocity fluctuations are log-normal, as soon as $R_\lambda > 1000$.

Testing the convolution formula (1) on the WTMM pdf's using a Gaussian kernel yields results in very good agreement with the log-normal cascade picture [4]. A second test of log-normality lies in the scaling exponents ζ_q of the velocity structure functions. As shown in Fig. 2(a), the experimental spectrum, obtained using extended self-similarity (ESS) [8] and extended to negative values of q thanks to the WTMM method, remarkably coincides with the quadratic log-normal prediction. The log-Poisson prediction [7] provides a good approximation of ζ_q for $q \in [-6, 6]$. However, plotting the deviation of the ζ_q's from the Kolmogorov (1941) linear $q/3$ spectrum (Fig. 2(b)), reveals a systematic departure of the log-Poisson prediction from the experimental spectrum, whereas the log-normal model still perfectly fits the experimental data.

2. Space-scale correlations of the WT: a two-point statistical analysis

To get a deeper insight into the nature of the statistics, we study the space-scale correlations of the "magnitude" $\omega(x, a)$ of the velocity field at point x and scale a [5]. $\omega(x, a)$ is defined as the logarithm of a local average over a size a of the wavelet coefficients at scale a around the point x. Here, we use the velocity increments to compute $\omega(x, a)$. Figure 3 shows various correlation functions $C(\Delta x, a, a') = \overline{\tilde{\omega}(x, a)\,\tilde{\omega}(x + \Delta x, a')}$, where the overline stands for ensemble average and $\tilde{\omega}$ for the centered process

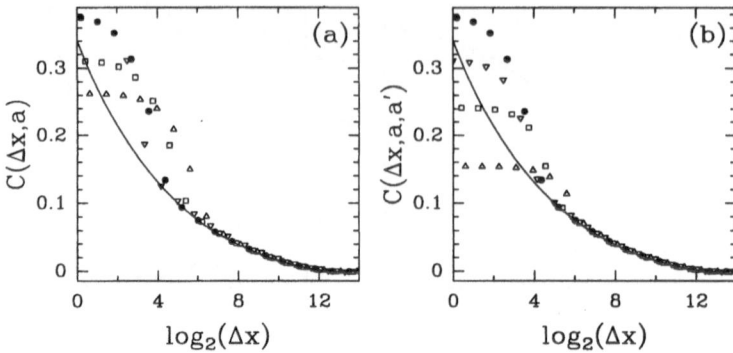

Figure 3. Increment magnitude correlation functions of the turbulent velocity signal. (a) "One-scale" correlation functions $C(\Delta x, a) = C(\Delta x, a, a)$ at scale $a = 8$ (down triangles), 16 ($\bullet$), 32 (squares) and 64 (up triangles). (b) "Two-scale" correlations functions at scale $a = 8, a' = 16$ (down triangles), $a = 16, a' = 16$ ($\bullet$), $a = 16, a' = 32$ (squares) and $a = 16, a' = 64$ (up triangles). The solid line corresponds to a non scale-invariant, log-normal cascade model with $\beta = 0.3$, $\sigma^2 = 0.27$. a, a', and Δx are expressed in mesh size ($\simeq 3\eta$) units.

$\omega - \overline{\omega}$. Once again, the experimental results are in very good agreement with a log-normal, non scale-invariant cascade. However, the cross-over from the value $C(\Delta x = 0, a, a)$ down to the fitted curve and the fact that the scale-invariance breaking exponent β takes a different value when computed from the space-scale correlations of the increment magnitude ($\beta = 0.3$) and from the WTMM estimation of the kernel G ($\beta = 0.095$) suggest that simple (even non scale-invariant) self-similar cascades are not sufficient to account for the space-scale structure of the velocity field.

References

1. Castaing B., Gagne Y. and Hopfinger E. J., *Physica D* **46** (1990) 177; Chabaud B., Naert A., Peinke J., Chillà F., Castaing B. and Hebral B., *Phys. Rev. Lett.* **73** (1994) 3227; Castaing B. and Dubrulle B., *J. Phys. II France* **5** (1995) 895.
2. Muzy J.F., Bacry E. and Arneodo A., *Phys. Rev. Lett.* **67** (1991) 3515; *Int. J. Bifurcation and Chaos* **4** (1994) 245; Arneodo A., Bacry E. and Muzy J.F., *Physica A* **213** (1995) 232.
3. Arneodo A., Muzy J.F. and Roux S.G., *J. Phys. II France* **7** (1997) 363.
4. Arneodo A., Manneville S. and Muzy J.F., *Eur. Phys. J. B* **1** (1998) 129.
5. A. Arneodo, E. Bacry, S. Manneville and J.F. Muzy, *Phys. Rev. Lett.* **80** (1998) 708.
6. The reader may find more quantitative results on the breaking of scale-invariance in different flow configurations in Refs. [4]. In particular, the exponent β is found to decrease with R_λ, strongly indicating that the non scale-similar multiplicative process asymptotically converges towards a scale-similar cascade.
7. She Z.S. and Leveque E., *Phys. Rev. Lett.* **72** (1994) 336.
8. Benzi R., Ciliberto S., Trippiccione R., Baudet C., Massaioli F. and Succi S., *Phys. Rev. E* **48** (1993) R29; Benzi R., Ciliberto S., Baudet C., Ruiz-Chavarria G. and Tripiccione R., *Europhys. Lett.* **24** (1993) 275.

A TWO-FLUID PICTURE OF INTERMITTENCY IN SHELL-MODELS OF TURBULENCE.

J.L. GILSON, I. DAUMONT AND T. DOMBRE
*Centre de Recherches sur les Très Basses Températures-CNRS,
BP166, 38042 Grenoble Cedex 9, France.*

Shell-models were introduced a long time ago as useful toy-models for addressing the problem of anomalous scaling in fully developed turbulence [1]. It was noticed very soon [2] that elementary bricks of intermittency in such 1D-cascade models could be coherent pulses growing in a self-similar way as they move from large to small scales. Such ideal objects have presumably no direct counterpart in Euler dynamics.

It was shown in [3] that genuine self-similar solutions of the equations of motion in the inviscid limit display an unique scaling exponent (to be denoted below as z_0), provided they are localized in k-space. The exponent z_0, giving the logarithmic slope of the velocity gradient spectrum left behind the pulse, was found to be quite close to the Kolmogorov value 2/3 in the case of the Gledzer-Ohkitani-Yamada (GOY) model, in the range of parameters where it reproduces very well the multiscaling properties of real turbulent flow. Taken literally, this result would suggest weak intermittency effects, in disagreement with statistical data.

In Ref.[4] the role played by interaction of pulses with the rest of the flow in producing more singular events was elucidated, and a two-fluid picture was introduced, where coherent structures form in and propagate into a featureless random background. The physical soundness of this approach is best demonstrated by the Fig.1 depicting the evolution of a pulse at a Reynolds number $Re = 10^6$. The left side showing the amplitude of velocity gradient against the shell index n, reveals that the coherent structure emerges from a disorganized K41-ramp. On the right side one sees how phases tend to order in the trail of the pulse, while a phase-slip of order π (able to induce a local change of sign in the energy flux), settles at its leading edge. For a cascade of infinite length, finite departure of the scaling exponent from z_0 requires that the corresponding singular event finds,

U. Frisch (ed.), Advances in Turbulence VII, 219–222.

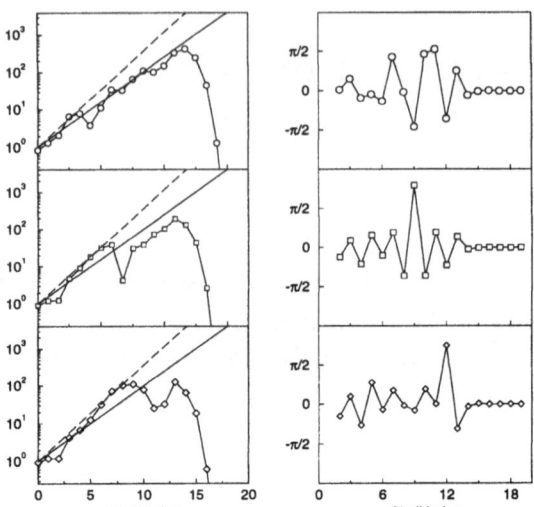

Figure 1. Three successive snapshots of the propagation of a pulse for $Re = 10^6$. Amplitudes b_n are depicted on the left side, together with power laws Q^{nz} for $z = 2/3$ (solid line) and $z = 0.85$ (dashed line), while phases $\Psi_n = \theta_{n-2} + \theta_{n-1} + \theta_n$ entering into the expression of the energy flux are shown on the right.

all along the way, incoherent fluctuations of the right amplitude to feed its anomalous growth. If by contrast the random background obey K41 scaling, as suggested by Fig.1, then nontrivial multiscaling properties can appear only as an intermediate asymptotics. Many dynamical facts seem to speak in favour of this situation. First, it is difficult to imagine how instabilities at small scales, involving shells far downstream the center of the pulse and time scales much shorter than its turnover time, may conspire to increase either the slope or the absolute level of the K41-ramp. Second, the inversion of the energy flux just in front of the pulse, revealed by Fig.1, makes rather unlikely a continuous adaptation of the level of the ramp to the incoming pulse. In many circumstances, we observed in contrary that this level tends to decrease.

In order to test possible breakdown of multifractality upon increasing the cascade length, we studied the statistics of effective growth exponents associated to absolute maxima of the energy flux ϵ_n, defined on each time interval between two successive maxima of ϵ_2 at large scales. We were led to this choice of observable by the fact that predictions on its density of probability can be made within the semi-analytic instanton approach to be presented below. Results are summarized on Fig.2. We observe that the statistics of z is almost Gaussian at all scales, with a maximum z^* around 0.82 significantly higher than $z_0 = 0.72$. However, there is indeed a slow decrease of z^* with n, as well as a nonlinear behaviour of the curvature

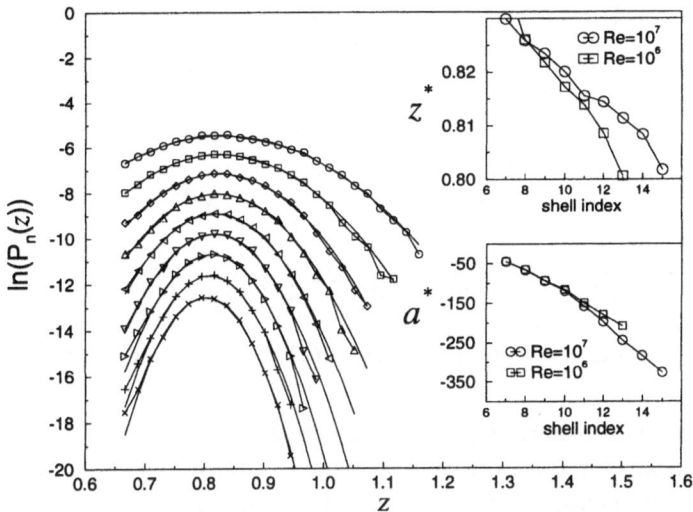

Figure 2. Histograms of exponents z defined according to $\epsilon_n/\epsilon_2 \equiv 2^{3(n-2)(z-2/3)}$ where ϵ_n and ϵ_2 are defined in the text. Statistics was run over 3.10^4 large-scale-time-units, with $Re = 10^7$ (the Kolmogorov dissipative scale is reached on the 17th shell). Histograms for shells $n = 7$ (top) to $n = 15$ (bottom) are shown together with a quadratic fit. Curves where displaced relatively to each other along the vertical axis for the sake of clarity. In the top-right (resp. bottom-right) inset, evolution with n of the most probable value z^* (resp. curvature of the parabola a^*) is reported for $Re = 10^6$ and $Re = 10^7$.

a^* which in a true asymptotic multifractal regime should decrease linearly with n. Intrinsic effects of the length of the cascade seem to be present in the GOY model but it is fair to say that they show up in a weak form. More extensive computations at still higher Reynolds numbers are needed to clarify this issue from a numerical side.

Analytic approach may provide indirectly interesting insights into these matters. We have recently investigated in great details stochastic dynamical systems related to the original GOY model, wherein the turbulent medium is described as a random force, acting on the coherent field. One is then dealing with equations of the following type :

$$\frac{d\mathbf{b}}{dt} = \mathbf{N}[\mathbf{b}] + \boldsymbol{\eta}\,, \tag{1}$$

where $\mathbf{b}$ embodies the coherent part of the velocity field, $\mathbf{N}[\mathbf{b}]$ is the nonlinear kernel of the GOY model and η is a Gaussian random force, delta-correlated in time and in shell space such that : $\langle \eta_n^*(t)\eta_{n'}(t') \rangle = \Gamma\, D_n\, \delta_{nn'}\delta(t - t')$. In our study, we assumed the noise to be relevant at all scales (restoring thereby scale-invariance in the problem). This implies the presence of a scale factor $(\mathbf{b}, \mathbf{b})^{3/2}$ in the variance D_n, multiplying some arbitrary form factor Δ_n which physically should make noise more active

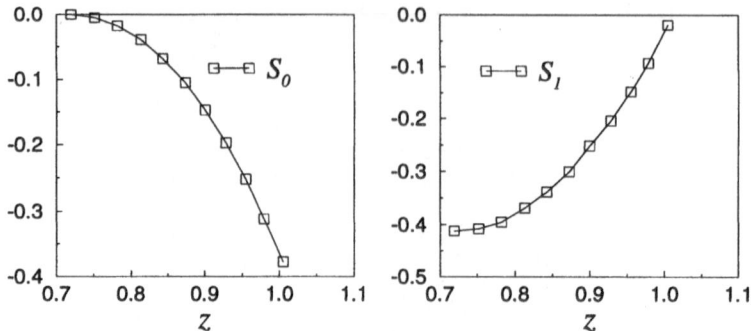

Figure 3. Plots of the two quantities $S_0(z)$ and $S_1(z)$ defined in the text.

just in front of the coherent pulse, than far away. The overall factor Γ, supposed to be small for the sake of consistency, completes the parametrization of the turbulent background. The presence of noise allows now a whole continuum of scaling exponents and we expect, in the semiclassical limit $\Gamma \ll 1$, the density of probability of developing an effective growth exponent z after $n \gg 1$ cascade steps to take the form :

$$P_n(z) \sim \sqrt{n} \, \exp\left[n \left(\frac{S_0(z)}{\Gamma} + S_1(z) \right) \right]. \tag{2}$$

where $S_0(z)$ is the action per unit cascade step of the self-similar extremal solution (or instanton) of scaling exponent z and $S_1(z)$ measures its weight in phase space, as estimated from a summation over quadratic fluctuations around it. We were able to compute both quantities (details will be presented elsewhere) and came to the following conclusions : provided the action of noise is local in shell space, $S_0(z)$ takes the form of a parabola centered naturally around z_0. Interestingly enough, $S_1(z)$ takes also (seemingly in a generic manner) the shape of a parabola, inverted with respect to the previous one and centered around an exponent smaller than z_0 (Fig.3 shows the results obtained with $\Delta_n = C_{n-1}^2 |C_{n-1}|$, and $\mathbf{C} = \mathbf{b}/(\mathbf{b}, \mathbf{b})^{1/2}$). Thanks to this nice effect of fluctuations, it is therefore possible to displace the maximum of $P_n(z)$ towards larger values of exponents, in accord with the histograms shown on Fig.2. Unfortunately, the minimum of $S_1(z)$ seems to be too close to z_0 to account quantitatively for both the positions and curvatures.

References

1. M.H. Jensen *et al.* Phys. Rev. A **43**, 798 (1991); and references therein.
2. E.D. Siggia, Phys. Rev. A **17**, 1166 (1978).
3. T. Dombre and J.-L. Gilson, Physica D (1998) **111**, 265-287.
4. J.-L. Gilson and T. Dombre, Phys. Rev. Lett (1997) **79**, 5002.

ON THE GENERATION OF INTERMITTENT GRADIENTS IN A DETERMINISTICALLY FORCED BURGERS' EQUATION

RAFAEL GÓMEZ-BLANCO AND JAVIER JIMÉNEZ

School of Aeronautics. Universidad Politécnica de Madrid
28040 Madrid, Spain

A deterministically forced Burgers' (FB) equation is considered in which the forcing is adjusted so that the energy spectrum remains $E(k, t) = k^\alpha$. The phases of the individual Fourier modes are not modified. The system is studied as a function of the logarithmic slope α. A similar procedure was used by [1, 2] in 3-D isotropic turbulence, with α fixed to the Kolmogorov value $-5/3$.

We study this system as a model for the turbulent 3-D Navier–Stokes equations, and in particular for the appearance of strong intermittent velocity gradients. It is hoped that some of the results can be extended to other chaotic systems with many degrees of freedom. We show that, at least in this case, intermittency is not a property of the chaotic evolution itself, and in fact decreases as the system moves away from the original bifurcation to chaos. It is associated with the long times spent by the system in neighbourhoods of the phase space associated with the formation of localized negative gradients.

The 'inertial' spectrum for the classical viscous Burgers' (CB) equation, $\partial_t u + u\partial_x u = \nu_2\, \partial_x^2 u$, has a slope $\alpha = -2$, due to the formation of compressive shock waves. The steepening of negative (compressive) velocity gradients is the basic mechanism of the Burgers' nonlinear operator [3, pg. 19–36], and would lead to discontinuities of u in finite time, which are smoothed by viscosity. Positive gradients, in contrast, are destroyed by the nonlinearity. Since the final state of the equation is a single decaying shock, its gradients are spatially intermittent in a trivial sense.

A slightly more general system is the hyperviscous Burgers' (HB) equation, in which the viscous dissipation is substituted by an iterated derivative $(-)^{n+1}\nu_n\partial_x^{2n}$. The shocks of the CB are monotonic in x, and are known to be stable [3, pg. 96–112], but those for $n > 1$ are not, and contain leading and trailing trains of wiggles which become longer and stronger as $n \gg 1$.

U. Frisch (ed.), Advances in Turbulence VII, 223–226.

It is unlikely that these wiggly shocks are stable in the same sense as in the CB, but no definite results are available.

We study here systems in which the Fourier transform of the velocity is truncated to $2N$ discrete Fourier modes, with zero spatial mean. Such systems are sometimes used to mimic the viscous minimum length by the spectral cutoff although, in the absence of forcing or damping, they conserve energy. They are roughly equivalent to the $n \to \infty$ limit of the HB and, instead of forming shock waves, they reach at long times an equipartition spectrum with $\alpha = 0$. Our forced model avoids this by fixing the spectrum, and is therefore defined in the N-dimensional space of the phases of the complex Fourier coefficients (the amplitudes are fixed by the forcing). Note that by moving from $\alpha = -2$ to $\alpha = 0$ the model approximates increasing hyperviscosity orders, and that we should therefore expect a transition from stable steady shocks to more complex behaviours.

The phases enter the nonlinear term as triads whose wavenumbers verify that $k + p + q = 0$. There are approximately $(N/2)^2$ triads, but only N of them are independent. One of those corresponds to a spatial shift and it is not dynamically significant, and the effective number of degrees of freedom is only $N - 1$. In our analysis we assume a fixed value for the shift, but all the features described below correspond to one-parameter families in which the physical structures could be located anywhere along the axis.

The system is studied numerically in triad space, with N in the range 2–60. The $N - 1$ independent triads are selected so that low-order ones represent as much as possible the lowest wavenumbers, which are the most energetic when α is negative. Chaos is defined by the presence of positive Lyapunov exponents, and intermittency is characterized by super-Gaussian values for the normalized moments of the p.d.f. of the velocity derivative, $g = \partial_x u$. Note that the p.d.f. is accumulated over x (spatial intermittency) and time. Thus, even steady solutions have a well defined p.d.f. for g.

For any value of α, the system has at least 2^{N-1} fixed points, most of which are unstable. Spatial intermittency is always maximum in the neighbourhood of a 'shock' fixed point, where all the triads are $\pi/2$, and which represents a compression shock in physical space when $\alpha = -2$ and $N \to \infty$. The spatial intermittency of this point increases with N, roughly corresponding to the Reynolds number, as the shock becomes thinner. The shock is a global attractor for the CB equation in the subspace of the phases, corresponding to the steepening process mentioned above. It is not an equilibrium point in that case because the amplitudes decay as the energy is dissipated, but the decay is prevented here by the forcing.

In our ordering, the first triad w_1 is the most important in determining the velocity. If $w_1 = \pi/2$, the strongest gradient tends to be negative (compressive shock), while $w_1 = -\pi/2$ approximates an expansion. The basic bifurcation is already present in the one-dimensional case, in which the evolution equation is simply $w_{1t} = A(\alpha) \cos w_1$, and the coefficient $A(\alpha)$ changes sign near $\alpha = -2$. Below this threshold the compressive shock is stable and the expansive one unstable, while above it the stabilities switch. Note that, even in this one-dimensional case, the velocity is described by two complex Fourier modes, because of the passive spatial shift.

At higher dimensions the bifurcation becomes more complicated. For α below a critical value α_3, the shock point still behaves like a global attractor, but for $\alpha_2 < \alpha < \alpha_3$ the system presents transient chaos. When $\alpha > \alpha_3$, the shock becomes linearly unstable and the system becomes chaotic if N is high enough ($N > 10$). If N is low, both simple unsteady behaviour and other steady solutions are possible.

In the transient chaos regime, but close to α_2, a narrow chaotic region appears near $w_1 = -\pi/2$, reminiscent of the stable expansion shocks of the one-dimensional case. There is however no global attractor in this region, and the system eventually reaches a state with strong localized negative gradients (an incipient compression shock). This marks the neighbourhood of the stable manifold of the compressive shock, and the system evolves quickly towards it. Outside the narrow chaotic region, the system moves quickly towards the shock without previous chaotic motion. In this regime, the steady compression shock is trivially spatially intermittent.

As α is increased and the physical character of the shock becomes more complex, the separatrix of the chaotic region moves closer to the shock. The time that the trajectories spend in the chaotic region increases with α. At $\alpha = \alpha_3$ it becomes infinitely long, and the chaotic region includes the shock. In a typical high-dimensional case ($N = 20$), the eigenvalue that first looses stability is real, and the bifurcation is a 'subcritical' pitchfork in which two high-dimensional saddles merge into the shock, which then also becomes a saddle. Just before the bifurcation the system approaches the shock along the stable manifold of one of the saddles, and is rejected along the one-dimensional heteroclinic orbit connecting the saddle to the shock. It can be shown that this orbit is the locus of locally highest spatial intermittency, and corresponds to the final steepening of a single shock wave.

The regime immediately beyond this bifurcation is characterized by spacio-temporal intermittency. The merged stable manifolds of the two saddles are intact, and the system approaches the shock along it, only to by repelled along the, now unstable, one-dimensional orbit. As long as the eigenvalues along the stable manifold are $O(1)$, as they were before the bifurcation, but the newly unstable one is small, the shock forms quickly (inertially) but disintegrates slowly. The system intermittently spends long periods in its neighbourhood, where the gradients are large. As α is in-

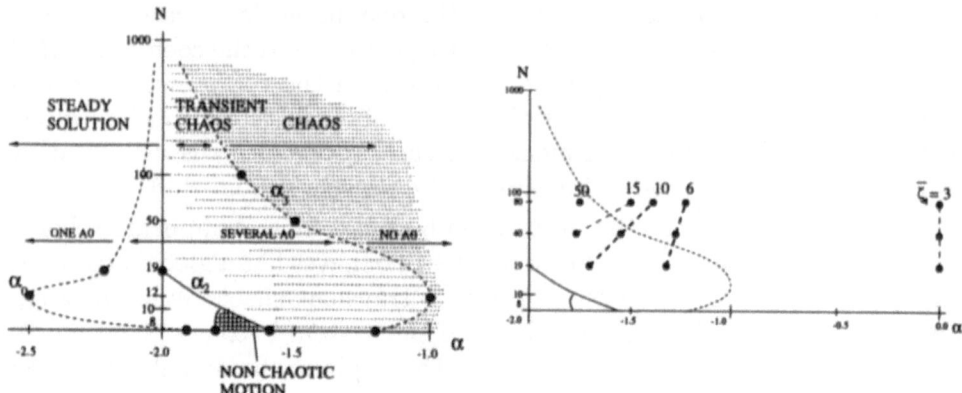

Figure 1. Left: Transition scenario for the system. Right: Isolines of the fourth-order flatness, as a measure of intermittency. The value of $\overline{\zeta_4} = 3$ for the equipartition slope $\alpha = 0$ corresponds to Gaussian statistics

creased and the unstable eigenvalue becomes $O(1)$ the residence times become small, and intermittency disappears.

The main conclusion of the present investigation is that intermittency is associated with the existence of a particular fixed point of the equation, which is a global attractor for steep spectra, and whose properties do not disappear completely immediately after its looses its stability. Intermittency is due to the continued tendency of the system to create compressive shocks, and occurs as long as the instability of these shocks is weak with respect to the time scales of shock formation.

The process is reminiscent of the vortical structures (weakly unstable equilibrium solutions) which are known to dominate the intermittent gradients of three-dimensional turbulence. It is interesting that, also in that case, hyperviscous simulations, which are known to generate wiggly unstable vortices [4], become less intermittent as the hyperviscosity exponent increases [5].

This work was supported the Spanish CICYT under contract PB95-0159. R.G.B. was supported in part by a fellowship from Iberdrola.

References

1. L. Shtilman and J.R. Chasnov, CTR Summer School 137 (1992).
2. Z.S. She and E. Jackson, Phys. Rev. Lett. **70** 1255 (1993)
3. G.B. Whitham, Linear and Nonlinear Waves. Wiley (1974)
4. J. Jiménez, J. Fluid Mech. **279** 169 (1994)
5. J. Jiménez and A.A. Wray, CTR Res. Briefs, 287, Stanford, Ca. (1994).

REYNOLDS NUMBER DEPENDENCE OF TRANSVERSE SCALING EXPONENTS IN GRID TURBULENCE

R. A. ANTONIA AND T. ZHOU

Department of Mechanical Engineering, University of Newcastle,
N.S.W., 2308, Australia

Abstract.

Experiments over a relatively small range of the Taylor microscale Reynolds number R_λ suggest that the inequality between ζ^L and ζ^T, the scaling exponents of the longitudinal and transverse velocity increments, may disappear at sufficiently high values of R_λ. The scaling exponents of locally averaged values of the energy dissipation rate ϵ and enstrophy ω^2 imply a considerably smaller value of $(\zeta^L - \zeta^T)$ than that estimated directly.

1. Introduction

Significant evidence has emerged over the last few years concerning the anomalous inertial range scaling of the transverse velocity increments and how it may differ from that exhibited by the longitudinal velocity increments. With some exceptions, the available data (e.g.[1, 2, 3, 4]) indicate that $\zeta^T < \zeta^L$. Most of the estimates of ζ^T and ζ^L have been made with the ESS (extended self-similarity) method[5] although the same scaling range has not always been used in each case.

The magnitude of the inequality $\zeta^T < \zeta^L$ will almost certainly depend on several factors, e.g. the Reynolds number, the presence of shear, the level of structural organization of the flow and the anisotropy. Some of these parameters are no doubt interrelated. Various explanations have been offered for the inequality, perhaps the most popular one (e.g.[3, 4, 6]) tending to associate the longitudinal velocity increment with either the strain rate or the energy dissipation rate ϵ and the transverse increment (here identified with δu_2, where u_2 is the velocity fluctuation in the x_2 direction) with the mean square vorticity (or enstrophy) ω^2 respectively.

227

U. Frisch (ed.), Advances in Turbulence VII, 227–230.
© *1998 Kluwer Academic Publishers.*

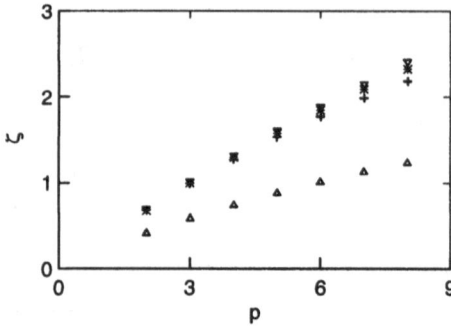
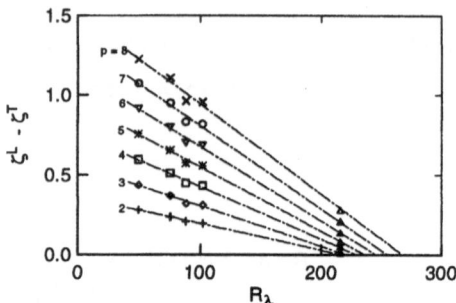

Figure 1. Dependence of scaling exponents on the order p of the moments of
either velocity increments or of ϵ_r and ω_r^2.
∇, ζ^L; $\triangle$, ζ^T; $*$, ζ^d; $+$, ζ^v. ($R_\lambda = 75$).

Figure 2. Dependence of the difference
$(\zeta^L - \zeta^T)$ on R_λ. — · —, least squares linear fits to the present data for $p = 2 - 8$;
$\triangle$, Chen et al. [3] ($p = 2 - 8$).

The main purpose of this paper is to test this association in shearless
grid turbulence; a particular feature of the experiment is that reasonably
reliable measurements are made of ϵ and ω^2. We also consider the dependence of $\zeta^L - \zeta^T$ on the Taylor microscale Reynolds number R_λ.

2. Experimental Details

Grid turbulence was chosen partly to avoid the complications of the effect
of mean shear and also because it provides a simple means of calibrating a
vorticity probe, since $\langle \epsilon \rangle$ and $\langle \omega^2 \rangle$ are known with relatively good accuracy.
Measurements were made[7] using a 3-component vorticity probe behind a
biplane square mesh grid ($M = 24.76$ mm, circular rods $d = 4.76$ mm) at
several stations downstream of the grid ($R_\lambda = 55$). X-wire measurements
were also made at $x_1/M = 70$ downstream of a biplane square mesh ($M = 24.76$ mm, square rods $d = 4.76$ mm) at four values of R_λ (50 to 103). More
experimental details can be found in [7].

3. Experimental Results

ESS estimates of $\zeta^L(p)$ and $\zeta^T(p)$ are shown in Figure 1 for $2 \leq p \leq 8$. Here
$\langle (\delta u_1)^p \rangle \sim r^{\zeta^L(p)}$, and $\langle (\delta u_2)^p \rangle \sim r^{\zeta^T(p)}$. Taylor's hypothesis was used to
convert temporal to spatial increments. $\zeta^T(p)$ is appreciably smaller than
$\zeta^L(p)$. The difference between these two exponents increases with p. Similar
results were also reported by Herweijer and van de Water[8] for grid turbulence ($R_\lambda = 300$) and by Chen et al.[3] for a direct numerical simulation
of isotropic turbulence ($R_\lambda = 216$). For turbulence downstream of a screen
($R_\lambda = 37 - 82$), $\zeta^L(p)$ and $\zeta^T(p)$, were approximately equal [1] for $p \leq 6$.

But as noted in [7], the ESS estimate of $\zeta^T(p)$ in [1] was based on the region where $\langle|(\delta u_2)^3|\rangle$ exhibits a linear dependence on r. The use of $\langle|(\delta u_2)^3|\rangle$ is difficult to justify, especially when $\langle(\delta u_2)^3\rangle$ should be zero by isotropy. To assess the effect R_λ may exert on $\zeta^L(p) - \zeta^T(p)$, X-probe measurements at four different Reynolds numbers ($R_\lambda = 50 - 103$) have been made. The exponent $\zeta^L(p)$ remains approximately constant while $\zeta^T(p)$ increases as R_λ increases. As shown in Figure 2, the difference $\zeta^L(p) - \zeta^T(p)$ may disappear at sufficiently large R_λ. Least squares linear regressions to the present data suggest that, after linearly extrapolating to zero, the difference may disappear at $R_\lambda = R_\lambda^*$ say. For $p = 8$, $R_\lambda^* = 270$ while for $p = 2$, $R_\lambda^* = 220$. Note the true values of R_λ^* may be somewhat larger since the linear extrapolation is relatively crude. Values of $\zeta^L(p) - \zeta^T(p)$ from [4] ($R_\lambda = 82$, not shown here) are significantly smaller than the present values, possibly reflecting the good level of isotropy achieved in this simulation. Data from [3] for larger p appear to lie on the extrapolation of our data. Our data, when considered in conjunction with the higher Reynolds number data of Mydlarski and Warhaft[10] suggest that the scaling range anisotropy is unlikely to disappear much before $R_\lambda \simeq 10^3$. To account for the possibly different roles played by ϵ and ω^2 on the scaling of the longitudinal and transverse increments, [3] proposed a modification to the refined similarity hypothesis (RSH) [9], which they called "refined similarity hypothesis for the transverse velocity increments" or RSHT. RSH and RSHT imply that $\delta u_1 \sim (r\epsilon_r)^{1/3}$ and $\delta u_2 \sim (r\omega_r^2)^{1/3}$ respectively. Assuming $\langle(\delta u_1)^p\rangle$ and $\langle(\delta u_2)^p\rangle$ scale as $r^{\zeta^d(p)}$ and $r^{\zeta^v(p)}$ in the IR, while $\langle\epsilon_r^p\rangle$ and $\langle(\omega_r^2)^p\rangle$ scale as $r^{\tau^d(p)}$ and $r^{\tau^v(p)}$

$$\zeta^d(p) = p/3 + \tau^d(p)/3$$
$$\zeta^v(p) = p/3 + \tau^v(p)/3 \ .$$

The exponents τ^d and τ^v can be inferred from the variations of $\langle(\epsilon_r)^p\rangle$ and $\langle(\omega_r^2)^p\rangle$ with r over the same range as that used for ζ^L and ζ^T. A necessary condition for RSHT is that $\langle(\epsilon_r)^p\rangle$ should scale differently from $\langle(\omega_r^2)^p\rangle$ and that the difference between their scaling exponents should be compatible in sign with that of ζ^L and ζ^T. ζ^d and ζ^v are also shown in Figure 1. The inequality $\zeta^v < \zeta^d$ is much smaller, regardless of p, than the inequality $\zeta^T < \zeta^L$. This result suggests that RSHT does not fully explain the inequality $\zeta^T < \zeta^L$. A more likely source of the inequality may be that, when R_λ is small as in the present experiment, any anisotropy in the scaling range will tend to influence δu_2 more than δu_1. As R_λ increases, the scaling range anisotropy will be improved and the difference between ζ^L and ζ^T should become smaller as illustrated in Figure 2. One might expect that, in the presence of a mean shear, R_λ^* could be significantly larger than without shear. The atmospheric data of Dhruva et al.[11] indicate that $\zeta^L - \zeta^T$ may not entirely disappear at $R_\lambda \simeq 10^4$.

4. Conclusions

RSHT cannot account for the inequality of the longitudinal and transverse velocity increment scaling exponents found in this experiment since the measured moments of ϵ_r and ω_r^2 exhibit only slightly different power-law exponents in the scaling range. A more likely source for the inequality $\zeta^T < \zeta^L$ is the strong anisotropy which reflects the small R_λ of the present flow. Results at different R_λ, when considered together with the higher R_λ grid data of [10], suggest that the inequality will not disappear until R_λ reaches about 10^3.

Acknowledgements

The support of the Australian Research Council is gratefully acknowledged.

References

1. Camussi, R. and Benzi, R. 1997. Hierarchy of transverse structure functions, *Phys. Fluids*, **9**, 257-259.
2. Pearson, B. R. and Antonia, R. A. 1997. Velocity structure functions in a turbulent plane jet, *Proc. Eleventh Turbulent Shear Flow Conference*, Grenoble, 3-117 to 3-121.
3. Chen, S., Sreenivasan, K. R., Nelkin, M. and Cao, N. 1997. A refined similarity hypothesis for transverse structure functions, *Phys. Rev. Lett.*, **79**, 1253.
4. Boratav, O. N. and Plez, R. B. 1997. Structure and structure functions in the inertial range turbulence, *Phys. Fluids*, **9**, 1400-1415.
5. Benzi, R., Ciliberto, S. Tripiccione, R. Baudet, C, Massaioli, F. and Succi, S. 1993. Extended self-similarity in turbulent flows, *Phys. Rev. E*, **48**, R29-32.
6. Kahalerras, H., Malecot, Y. and Gagne, Y. 1996. In S. Gavrilakis, L. Machiels and P. A. Monkewitz (eds.) *Advances in Turbulence VI*, Dordrecht, Kluwer Academic Pub., 235-238.
7. Antonia R. A., Zhou, T. and Zhu, Y. 1997. Three-component vorticity measurements in a turbulent grid flow, *J. Fluid Mech.* [submitted]
8. Herweijer, J. A. and van de Water, W. 1995. Transverse structure functions of turbulence, in R. Benzi (ed.) *Advances in Turbulence V*, Dordrecht, Kluwer Academic Pub., 210-216.
9. Kolmogrov, A. N. 1962. A refinement of previous hypothesis concerning the local structure of turbulence in a viscous incompressible fluid at high Reynolds number, *J. Fluid Mech.*, **13**, 82-85.
10. Mydlarski, L. and Warhaft, Z. 1996. On the onset of high-Reynolds number grid-generated wind tunnel turbulence, *J. Fluid Mech.*, **320**, 331-368.
11. Dhruva, B., Tsuji, Y. and Sreenivasan, K. R. 1997. Transverse structure functions in high-Reynolds-number turbulence, *Phys. Rev. E*, **56**, R4928-R4930.

INTERMITTENCY AND EDDY VISCOSITIES IN DYNAMICAL MODELS OF TURBULENCE

F. TOSCHI[1,2], L. BIFERALE[2,3], S. SUCCI[4] AND R. BENZI[5]
[1] *Dipartimento di Fisica, Università di Pisa,*
Piazza Torricelli 2, I-56126, Pisa, Italy
[2] *INFM, Unità di Tor Vergata.*
[3] *Dipartimento di Fisica, Università "Tor Vergata",*
Via della Ricerca Scientifica 1, I-00133, Roma, Italy.
[4] *Istituto Applicazioni Calcolo, CNR,*
Viale Policlinico 137, I-00161, Roma, Italy
[5] *AIPA, Via Po 14, I-00100, Roma, Italy.*

Abstract. The dependence of intermittent inertial properties on ultraviolet eddy viscosity closures is examined within the framework of shell-models for turbulent flows. Inertial intermittent exponents turn out to be fairly independent on the way energy is dissipated at small scales.

One of the most striking features of fully developed turbulence is its intermittent spatial and temporal behavior. The structure functions $S_p = \langle \delta_r v^p \rangle$ are usually used as indicators of intermittency in the inertial range.

$$S_p(r) \sim \left(\frac{r}{L}\right)^{\zeta_p} \tag{1}$$

In the presence of intermittency their scaling exponents, ζ_p, differ from the dimensional estimate (Kolmogorov 1941), $\zeta_p = p/3$ (see [1]).

As is well known, for most turbulent flows of practical interest, the dissipative scale η is far too short to be resolved by any foreseeable computer. A simple estimate gives that the computational effort grows like the fourth power of the Reynolds number. Under this state of affairs, it is clear the interest and relevance of Large Eddy Simulations (LES): simulations where only "large" scale are resolved. In LES the effect of the "small" scale (Sub-Grid Scale, SGS) is taken in account by some SGS model. Generally speaking, the common aim of these SGS models is to incorporate the effects of unresolved scales ($r < \Delta$, Δ being a typical mesh size) on the resolved

U. Frisch (ed.), Advances in Turbulence VII, 231–234.
© 1998 *Kluwer Academic Publishers.*

p	$\zeta^D(p)$ $N = 16$	$\zeta^S(p)$ $N = 16$	$\zeta^S(p)$ $N = 20$	$\zeta^S(p)$ $N = 24$
1	0.368 ± 0.007	0.367 ± 0.002	0.367 ± 0.002	0.367 ± 0.001
2	0.700 ± 0.005	0.699 ± 0.002	0.699 ± 0.002	0.699 ± 0.001
3	1.0 ± 0.0	1.0 ± 0.0	1.0 ± 0.0	1.0 ± 0.0
4	1.271 ± 0.007	1.273 ± 0.004	1.268 ± 0.007	1.272 ± 0.003
5	1.52 ± 0.01	1.522 ± 0.007	1.50 ± 0.02	1.518 ± 0.008
6	1.74 ± 0.02	1.75 ± 0.01	1.71 ± 0.04	1.74 ± 0.02
7	1.94 ± 0.04	1.97 ± 0.01	1.90 ± 0.07	1.96 ± 0.02
8	2.12 ± 0.05	2.17 ± 0.02	2.08 ± 0.09	2.16 ± 0.03
9	2.29 ± 0.08	2.37 ± 0.02	2.3 ± 0.1	2.36 ± 0.04
10	2.5 ± 0.1	2.57 ± 0.03	2.4 ± 0.1	2.56 ± 0.04
11	2.6 ± 0.1	2.76 ± 0.04	2.6 ± 0.1	2.76 ± 0.05
12	2.8 ± 0.2	2.96 ± 0.06	2.8 ± 0.2	2.96 ± 0.06

TABLE 1. Scaling exponents for the 0-dimensional (chain) model with eddy viscosity, $\zeta^S(p)$, for $N = 16, 20, 24$ and with normal viscosity, $\zeta^D(p)$, with $N = 16$. Exponents has been extracted by mean of ESS for precision requirements.

ones ('Large Scales', $r > \Delta$). One of the guiding ideas in Sub-Grid Scale (SGS) modeling is the concept of eddy viscosity. This was introduced over a century ago by Boussinesq and later developed further by G. Taylor and L. Prandtl [2] and it builds upon a direct analogy with the kinetic theory of gas. According to this analogy, the effect of short 'microscopic' scales on large 'macroscopic' scales can be likened to a sort of diffusion process characterized by an effective viscosity much larger than the molecular one. Strictly speaking, this is justified only when a sharp separation between fast and slow modes exists, but it turns out that the analogy proves useful in practice also in situations where such an assumption would not hold in principle.

One of the simplest and most popular SGS models is the one by Smagorinski [4], the eddy viscosity is given in term of $C_S \Delta^2 |S| S_{ij}$, where S_{ij} is the large (resolved) scale strain tensor.

One of the outstanding questions is whether the intermittent scaling exponents are affected by the specific mechanism by which a turbulent flow manages to dissipate sub-Kolmogorov scales. Besides its theoretical interest, this question may provide some practical clues on the direction of turbulence modeling and Large Eddy Simulation.

In this proceeding we report some results presented in [3]. Because of the statistical accuracy required, we have completed our study using a class of 0

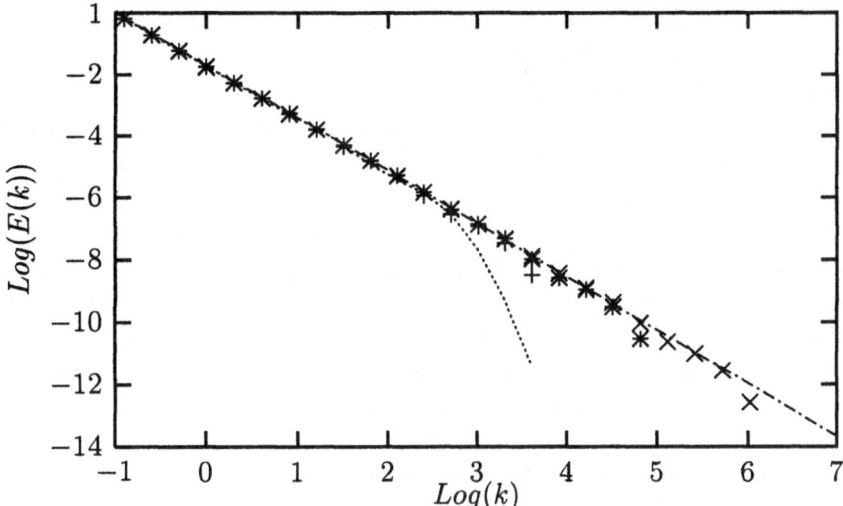

Figure 1. Log-log plot of the energy spectra, $E(k)$, versus the wavenumber, k, for the 0-dimensional (chain) model with eddy-viscosity at three different resolutions $N = 16$ (pluses), $N = 20$ (stars), $N = 24$ (crosses). For the sake of comparison the case with normal viscosity is also reported for $N = 16$ (dotted line). The straight (dash-dotted) line has slope $-1 - \zeta_2$.

and 1-dimensional helical shell models [7]. We have run simulations with the usual dissipation $\nu k_n^2 u_n$ and with a shell-model surrogate of Smagorinski eddy viscosity. When using eddy viscosity we have put the normal viscosity to zero ($\nu = 0$) and added a term $\nu_S |u_n| k_n$ to the equation acting only on the last and last-but-one shells. The constant ν_s has been taken equal to unity. In the case of the 1-dimensional model the variables u_n are replaced by $u_{n,j}$ which also depend upon the spatial position (index j).

Our results confirm that, within error bars, the values of the intermittency exponents are unaffected from the kind of dissipation used, whether the normal one or the eddy one. In the case of the 0-dimensional model, we have run several simulations with eddy viscosity and different numbers of shell (16, 20, 24) in order to ascertain the independence of the value of intermittent exponents from the length of the chain (see Table 1). In the case of 1-dimensional model [7] we have also checked the validity of RKSH (Refined Kolmogorov Similarity Hypothesis) which connects the intermittent value of structure functions scaling exponents with the scaling exponents of the powers of coarse grained energy dissipation (ϵ_r). In particular we have validated the relation:

$$S_p(r) = \left\langle \epsilon_r^{p/3} \right\rangle \cdot S_3(r)^{p/3}. \tag{2}$$

We have found that the RKSH is well verified. Furthermore, we have verified directly that the inertial range, in the case of eddy viscosity, extends up to the last-but-one shell, a great enhancement with respect to the normal dissipation case (see Figure 1).

The eddy-viscosity closure that we have adopted may also be regarded as a multiplicative closure of the small-scales equations of motion, i.e. it is tantamount to assuming that $u_{n+1} \sim a_{n+1,n} \cdot u_n$ with an appropriate multiplicative random coefficient $a_{n+1,n}$. The fact that intermittency is not affected by the details of the eddy-viscosity models indicates that fine-tuning of the coefficients in front of the eddy-viscosity term is probably not demanded. Nevertheless, oversimplified eddy-viscosity models based only on dimensional analysis would probably fail on the same goal, due to their inability to dissipate violent intermittent bursts.

In conclusion we suggest that intermittency is largely unaffected by the particular kind of dissipation. In concrete applications, like LES of complex flows, we suggest that far away from boundaries, eddy viscosity should be able to correctly reproduce the dynamics of large scales. Near to the boundaries a different physics could be present (see [8] for a preliminary study) and the eddy viscosity models should incorporate it somehow, otherwise they will not be able to reproduce the dynamic and the intermittency properly.

References

1. Frisch, U. (1995) *Turbulence: The legacy of A. N. Kolmogorov*, Cambridge University Press, Cambridge, UK.
2. Frisch, U., Orszag, S., (1990) Turbulence: challenge for theory and experiments, *Phys. Today*, **23**.
3. Benzi, R., Biferale, L., Succi, S., Toschi, F. (1998) Intermittency and eddy viscosity in dynamical models of turbulence *Phys. of Fluids*, submitted. Preprint at *chao-dyn/9802020*.
4. Smagorinski, J., (1963) General circulation experiments with the primitive equations. I. The basic experiment, *Mon. Weather Rev.*, **91**, 99-164.
5. Benzi, R., Biferale, L., Kerr, R., Trovatore, E. (1996) Helical shell models for 3-dimensional turbulence, *Phys. Rev. E*, **53**, 3541.
6. Benzi, R., Biferale, L., Trovatore, E. (1996) Universal energy transfer statistics in turbulent dynamical models, *Phys. Rev. Lett.*, **77**, 3114.
7. Benzi, R., Biferale, L., Tripiccione, R., Trovatore, E. (1997) (1+1)-dimensional turbulence, *Phys. of Fluids*, **9**, 2355.
8. Amati, G., Toschi, F., Succi, S., Piva, R. (1998) Scaling exponents in turbulent channel flows, *Contribution to ETC7*.

DYNAMICAL ORGANIZATION AROUND TURBULENT BURSTS

F. OKKELS[1,2] AND M. H. JENSEN[1]
[1] *Niels Bohr Institute and Center for Chaos and Turbulence Studies, Blegdamsvej 17, DK-2100 Ø, Denmark*
[2] *Optics and Fluid Mec. Dep., Risø Research Center, Denmark.*

It is well-known from experiments and numerical simulations that intermittency in turbulent flows produces corrections both to the classical Kolmogorov $-5/3$ scaling law of the energy spectrum but also for other moments in the inertial range [1, 2]. Still very little is known about the structure of the intermittent bursts which arise in turbulent flows. This work presents an analysis of the intermittent bursts in the turbulent GOY shell model, inspired by work of Dombre and Gilson [3]. Shell models are formed by various truncation techniques of the Naiver-Stokes equations and have become paradigm models for the study of turbulence at very high Reynolds numbers [4, 5, 6, 7, 8, 9, 10]. In the GOY model wave-number space is divided it into N separate shells with characteristic wavenumber $k_n \sim \lambda^n$ $(\lambda = 2)$ where $n = 1, \ldots, N$. Each shell is assigned a complex amplitude u_n which can be imagined as the velocity gradient on a scale $\ell_n = 1/k_n$. By assuming conservation of phase space, energy and helicity and interactions among the nearest and next nearest neighbor shells, one arrives at the following evolution equations [6, 8, 9]

$$\left(\frac{d}{dt} + \nu k_n^2\right) u_n = ik_n\left(u_{n+1}^* u_{n+2}^* - \frac{1}{4}u_{n-1}^* u_{n+1}^* - \frac{1}{8}u_{n-2}^* u_{n-1}^*\right) + f\delta_{n,4} \quad (1)$$

with boundary conditions $a_{N-1} = a_N = b_1 = b_N = c_1 = c_2 = 0$, and constant forcing f on the fourth shell.

The set (1) of N coupled ordinary differential equations can be numerically integrated by standard techniques. In the simulations, we use the following values: $\delta = 1/2, N = 19, \nu = 10^{-6}, k_0 = 2^{-5}, f = (1+i) * 0.005$.

Shell models were constructed to simulate the statistical properties of turbulence, but the detailed dynamics of the amplitudes in the 2N-dimensional phase space can also be investigated.

U. Frisch (ed.), Advances in Turbulence VII, 235–238.

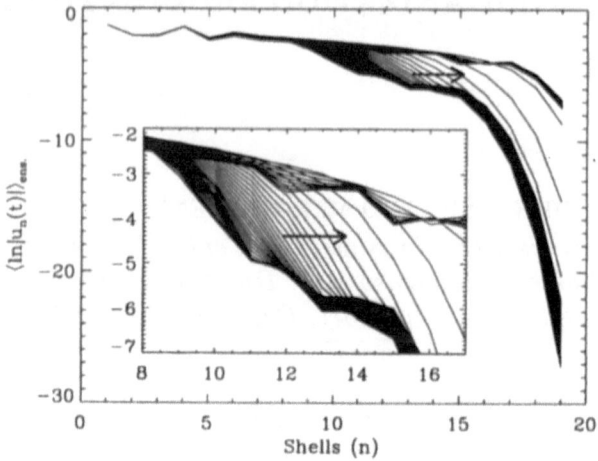

Figure 1. $\langle \ln |u_n(t)| \rangle$ ensemble averaged over 71 bursts is plotted as a function of n during the burst in time intervals of 3/50 n.u. Inserted is a enlargement of the propagating front. The arrows show the propagation-direction of the front.

It turns out that the intermittent bursts consist of a collection of different organizations of the amplitudes u_n which travel with exponential increasing speed from the lower up towards the higher shells where they are damped away by viscosity [3]. Every burst follows a common pattern where the most prominent characteristics is that the energy cascades down the scales by a front having a constant slope in a $\ln |u_n(t)|$ vs. n plot. This is seen in Fig.1 showing the time-evolution of the ensemble average of $\ln |u_n(t)|$ over 71 bursts, each translated such that they occur at the same time. During the burst $\langle \ln |u_n(t)| \rangle$ is plotted as a function of n in time intervals of 3/50 n.u. Inserted is a enlargement of the propagating front. The arrows show the propagation-direction of the front.

To explain the reason for bursts, we must distinguish between the behaviour at the Inertial Sub Range (ISR) where the coupling terms dominate and the Viscous Sub Range (VSR) where the viscosity dominates.

1. Propagation of bursts at the Viscous Sub Range (VSR)

An analysis of the model shows that the bursts are roughly unaffected by the phases of the complex amplitudes in the VSR [11], and therefore the phases will be neglected in the following description, introducing the real-valued amplitudes (r_n). The vanishing amplitudes is signalled by an attraction to and a repulsion from $r_n = 0$, which is a trivial but important fixed point of the model. This dynamics appears as a result of a balance

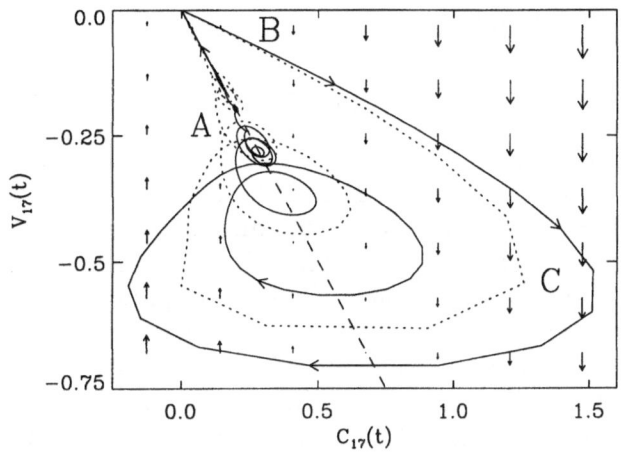

Figure 2.   Trajectories of V_n vs. C_n during a burst for the the real valued model (solid) and complex model (dotted). On the same graph is shown the flow field of the viscosity V_n visualized by arrows. The dashed line correspond to $\dot{r}_n = 0$. The labels represent the attraction to the fixed point $r_n = 0$ (A), the repulsion away from the fixed point (B), and the end of the repulsion (C). The arrows on the trajectories indicate the direction of the temporal evolution.

between the viscosity term and the coupling term. To see this the GOY model, in terms of the real variable r_n, can be written in its simplest form as

$$\dot{r}_n = V_n + C_n \tag{2}$$

where $V_n = -\nu k_n^2 r_n$ and $C_n = k_n(r_{n+1}r_{n+2} - \frac{1}{4}r_{n-1}r_{n+1} - \frac{1}{8}r_{n-2}r_{n-1})$. The forcing is neglected because we focus on the dynamics of the high wavenumbers. Since V_n and r_n are proportional with opposite sign, Eq. 2 can be written as $\dot{V}_n = -\nu k_n^2(V_n + C_n)$. The values of V_n is shown as a grid of arrows in Fig.2, since it determines the attraction (repulsion) to (from) the fixed point depending on the values of (C_n, V_n). Also shown in Fig.2 are the trajectories of (C_{17}, V_{17}) during a burst.

From Fig.2 it is possible to explain the propagation of bursts in the GOY model. During the attraction (A) the viscosity term is slightly dominating the coupling term so the amplitude approaches zero slowly, a bit below the diagonal $(V_n = -C_n)$. The only thing that can alter this attraction towards zero is a change of the coupling term, and this happens just when a burst approaches from the lower shells. Then the amplitude is forced into a region of negative $\dot{V}_n$ where it diverge away from zero and then participating in the further motion of the burst. At the end of the repulsion, the viscosity term begins to dominate again and the dynamics is repeated from (A). *The intermittent burst propagates by a "domino" effect through the shells.* The qualitative similarity between bursts in the real valued and the complex model, seen in Fig. 2, indicating the weak effect of the phases on bursts.

2. Creation of bursts in the Inertial Sub Range (ISR)

Recent work show that in the ISR different parts of the coupling now play a similar role as the viscosity term and the coupling term in the VSR. It can be shown that the dominance of the different coupling terms depends on the local slope of $\ln |u_n(t)|$ vs. n. Thus the laminar state between bursts having a small negative slope makes the first coupling term in Eq.1 (which we call C_1) dominate while the third coupling term in Eq.1 (called C_3) dominate at the steep negative slope of the front, as seen in Fig.1. The assumption that C_1 plays the role of the viscosity in the ISR and C_3 plays the role of the total coupling in the ISR appears to be correct because there is a general attraction of the amplitudes towards zero during the laminar stage, while C_3 dominate during the burst.

The reason for the creation and propagation of bursts is that during the laminar state, the domination of C_1 prevents bursts from propagating since C_1 only depends on the two higher lying neighboring shells and therefore cannot transport energy towards the higher shells. When the forcing causes the negative slope of the lower shells to increase, C_3 begins to dominate, making the burst propagate because C_3 depends on the lower lying neighboring shells. In that way the increasing negative slope of the next shells leads to a propagation of the burst.

The conclusion is that bursts are formed and transported by a chain-reaction caused by the balance of the viscosity and the coupling terms in the VSR and by the first and third coupling term in the ISR.

References

1. U. Frisch, "Turbulence: The legacy of A.N. Kolmogorov", Cambridge University Press (1995).
2. A.N. Kolmogorov, C.R. Acad. Sci. USSR **30**, 301; ibid 32, 16 (1941).
3. T. Dombre and J.-L. Gilson, "Intermittency, chaos and singular fluctuations in the mixed Obukhov-Novikov shell model of turbulence", Preprint (1996).
4. T. Bohr, M. H. Jensen, G. Paladin, and A. Vulpiani, Dynamical systems approach to turbulence (Cambridge University Press, Cambridge), in press.
5. E. B. Gledzer, Sov. Phys. Dokl. **18**, 216 (1973).
6. M. Yamada and K. Ohkitani, J. Phys. Soc. Japan **56**, 4210 (1987)
   Prog. Theor. Phys. **79**,1265 (1988).
7. M. H. Jensen, G. Paladin, and A. Vulpiani, Phys. Rev. A **43**, 798 (1991).
8. L. Kadano,, D. Lohse, J. Wang, and R. Benzi, Phys. Fluids **7**, 617 (1995).
9. L. Biferale, A. Lambert, R. Lima, and G. Paladin. Physica D **80**, 105 (1995).
10. N. Schörghofer, L. Kadanoff, and D. Lohse, Physica D **88**, 40 (1995).
11. F. Okkels, Master thesis, CATS, University of Copenhagen, Denmark (1997).
    F. Okkels and M. H. Jensen, Phys. Rev. E, in press (1998)

SCALE RATIOS, STATISTICAL SYMMETRIES
AND INTERMITTENCY

A. POCHEAU
IRPHE, UMR 6594 CNRS, Universités Aix-Marseille I & II
Centre de Saint-Jérôme, S.252, 13397 Marseille, France

Existence (or lack) of preferred scale has largely been used as a useful guide to investigate statistics of stochastic extended systems. In particular, widespread use of power laws in literature actually comes from the fact that these are the only *single* scale functions pertaining to systems relying on *no* preferred scale. However, the analogous question of existence (or lack) of preferred scale *ratio* has surprisingly been mostly overlooked in literature, despite its possible relevance to stochastic systems. The goal of this work is to determine, from a general point of view, its various implications.

Let us first emphasize that existence of preferred scale or scale ratios are actually two distinct issues. In particular, deterministic fractals or hierarchical structures single out no scale but a specific scale ratio: 3 in a Koch curve; 2 in the famous Richardson's cascade since energy is transfered from vortices at a *scale* to vortices at the *half-scale*. On a physical ground, note that preferred scale ratios could be generated by intrinsic mechanisms, as proposed for laminar growth processes [1], but this seems doubtful in stirred and stretched systems since scale ratios are then not dynamically invariant. Also, preferred scale ratio might be given by the ratio of high to low cut-off scales, e.g. $Re^{3/4}$ in hydrodynamics, but the fully-developed-turbulence limit $Re \to \infty$ would then be singular and, thus, non-universal. On the opposite, absence of preferred scale ratio agrees with presently known phenomenology since dissipative structures [2], although studied at a definite scale in experiments, extend to all scales, in a way expressed with power laws [3], i.e without use of preferred scale nor scale ratio. Moreover, no preferred scale ratio is singled out by Kolmogorov K41 theory [4] because of its self-similar nature, and none might even be by intermittent statistics since absence of preferred scale ratios goes beyond self-similarity.

For these reasons, absence of preferred scale ratios in stochastic systems stands as a reasonnable assumption. It is supported by the fact that no physical mechanism pointing to preferred scale ratios has been identified in

U. Frisch (ed.), Advances in Turbulence VII, 239–242.

the inertial range of turbulence [5]. What may then be the possible statistics
in this case ? Do they include intermittency or not and are they linked
to other statistical properties ? On the other hand, many authors using
an explicit scale ratio nevertheless claim that its value may be considered
as arbitrary [5]. Is this legitimate ? Finally, how can we identify hidden
preferred scale ratios in models or derivations ? These open questions show
the need for clarifying the statistical consequences of the status ascribed to
scale ratios in stochastic extended systems.

Hereafter, absence of preferred scale (scale ratio) will be referred to as
global (local) scale-invariance. To derive general implications of these symmetries, it is useful returning to their basic meaning: the physical equivalence of all length (length ratio). According to it, physics is unable to
provide absolute standards to measure length (length ratio) so that the
question "how long (longer) ?" loses physical relevance. Note that the same
undeterminacy must also apply to any variable related to length, so as to
deny the possibility of indirectly distinguishing between length (length ratio) by using it. Altogether, these physical equivalences imply that suitable
changes of units (unit ratios) are physically undetectable. In particular, for
any change of length (length ratio), it must be possible to *preserve* physical laws by suitably *adjusting* units (unit ratios) pertaining to remaining
variables: physical laws must be *scale-covariant*.

A natural procedure to express scale-covariance follows from its definition: take a family of scales to observe a system, ($l_i = \rho^i l_s, i \in \Re$), change
length $l_s \rightarrow l'_s$ (length ratio $\rho \rightarrow \rho'$) and look for the same physical law
again by changing scales (scale ratios) of remaining variables. If this can be
achieved for *any* change of length (length ratio), that law singles out *none*:
it then pertains to a system scale-invariant in a global (local) sense.

Consider now a stochastic field $u(l,t)$, e.g. increment of velocity over
distance l in a turbulent fluid and denote u_s, l_s the standards of velocity
and length. Assuming global scale-covariance implies scale power laws for
structure functions: $\forall i, < (u/u_s)^i >_{(l)}=< (u/u_s)^i >_{(l_s)} (l/l_s)^{e_i}$. Here, scale
exponents e_i define an intermittency function $\zeta(.)$, $e_i = \zeta(i)$, which, in case
of absence of preferred scale ratio, should be scale-covariant in a local sense.
To express this symmetry, we must implement gauge changes simulating
change of unit ratios, determine the *new* $\zeta(.)$ function observed with the
new standards and ask for covariance for *any* change of length ratios [6, 7].
As $\zeta(.)$ relates scaling exponents e_i to moment orders, hereafter labelled o_i,
this requires not only velocity and length standards to measure e_i but also
moment standards to measure o_i.

Change of length ratio $\rho \rightarrow \rho' = \rho^\lambda$ corresponds to data change $l/l'_s =
(l/l_s)^\lambda$ and thus to standard change (1). Being a power law, change (1)
favours no length. In a similar way, change of u-standard $u_s \rightarrow u'_s(u, l)$

simulating change of scale ratios for u must favour no scale, either directly by preferred u value or indirectly by preferred l value. It can therefore only correspond to data change in power laws of u and l, $u/u'_s = (u/u_s)^\alpha (l/l_s)^\beta$, and thus to standard change (2). Moment orders refer to a parametrization of the series of moments $\mathcal{M} = [m^i = (u/u_s)^i, i \in \Re]$. Owing to its multiplicative structure, this set is actually derived from the measurement of two moments only, say m_0 and m_1: $\mathcal{M} = [m^i = m_0(m_1/m_0)^{o_i}, o_i \in \Re]$, $m_0 \in \mathcal{M}$, $m_1 \in \mathcal{M}$. Here, o_i labels the order of moment m^i and m_0 and m_1 the standards used for this measure. If, as usually assumed, $m_0 = 1$ and $m_1 = u/u_s$ then $o_i = i$. For another choice m'_0, m'_1 (3), the new moment order o'_i is related to o_i by $o'_i = \epsilon i + \omega$ with $\epsilon = (\mu - \kappa)^{-1}$ and $\omega = -\kappa\epsilon$.

$$l_s \to l'_s = l_s(l/l_s)^{1-\lambda} \tag{1}$$
$$u_s \to u'_s = u_s(u/u_s)^{1-\alpha}(l/l_s)^{-\beta} \tag{2}$$
$$m_0 \to m'_0 = (u/u_s)^\kappa \quad ; \quad m_1 \to m'_1 = (u/u_s)^\mu \tag{3}$$

Standard changes (1) (2) (3) make scaling exponents $e_i \equiv \zeta(o_i)$ and moment orders $o_i \equiv i$ change to e'_i and o'_i, so that they should be usually related by a different intermittency function $\zeta'(.)$: $e'_i = \zeta'(o'_i)$. However, invoking scale-covariance for *any* change of length ratio, we shall require that, for *any* λ, there exists at least a set of gauge parameter $\mathcal{G} = (\alpha, \beta, \lambda, \epsilon, \omega)$ for which the intermittency function is the same: $\zeta'(.) = \zeta(.)$.

Changes of standards (1) (2) (3) cause a change of scaling exponents $e_i \to e'_i = (e_{\alpha i} + \beta i)/\lambda$ and moment orders $o_i = i \to o'_i = \epsilon i + \omega$. Altogether, they yield the criterion for scale-covariance, $e'_i = \zeta(o'_i)$, i.e. :

$$\lambda\zeta(\epsilon i + \omega) = \zeta(\alpha i) + \beta i \tag{4}$$

Solutions of criterion (4) with $\zeta(0) = 0$ are *necessarily* of the forms (5) (log-Poisson statistics [8]) or (6) (log-Levy statistics [9]) [6, 7]:

$$\zeta(i) = ai + b[1 - exp(-ci)] \tag{5}$$
$$\zeta(i) = ai + b[(i + d)^c - d^c] \tag{6}$$

All do not satisfy criterion (4) however (Fig.1) [7]. Regarding solutions (5):

1. $bc = 0$: *All* scale ratio are physically *equivalent*. This is K41 [4].
2. $abc < 0$ and $a + bc \neq 0$: There exists *discrete families* of physically *equivalent* scale ratios ($\rho_i = \rho^{\lambda^i}, i \in N$) i.e., for instance, *discrete* family of cascades characterized by scale ratios ρ_i.
3. $abc > 0$ or $a + bc = 0$ or $a = 0$: *All* scale ratios are *distinguished*. This regime refers either to a *single* cascade process ($abc > 0$), to uniformity ($a + bc = 0$) or to statistical homogeneity ($a = 0$). In the latter case, further analysis shows that scale ratios are distinguished *not by fluctuations* but by mean state [6, 7].

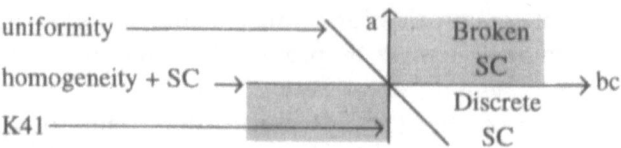

Figure 1. Symmetry diagram for log-Poisson statistics. Parameters a,b,c refer to (5). Dashed (light) regions refer to broken (discrete) scale-covariance (SC).

Following our gauge analysis, the only intermittent statistics relying on *no* preferred scale ratio are log-Poisson [6, 10] or log-Levy statistics [6, 11] in a necessarily homogeneous statistical state [6, 7]. Apart from linear $\zeta(.)$, this implies exponential or monomial $\zeta(.)$ with *no* additional linear part.

On a more general ground, our study provides a link between $\zeta(.)$ functions and the status of scale ratios in systems, models or theories. At first, the many models [5] yielding $\zeta(.)$ functions which are not solutions of criterion (4) necessarily include, explicitly or implicitly, preferred scale ratio. Moreover, the She-Lévêque model [3] satisfies (5) but for $a > 0, bc > 0$ owing to incompressibility $(d\zeta/di \geq 0)$ and Hölder inequalities $(d^2\zeta/di^2 \leq 0)$ [5]. Following our analysis, it thus relies on hidden preferred scale ratio.

Although being a useful tool for studying statistics, this work will remain incomplete until definite answers are given as to the existence (or lack) of preferred scale ratios in stochastic systems. Meanwhile, it points to a striking alternative: if no physical mechanism singles out scale ratio, only few intermittent statistics are selected by arguments independent of differential equation, phenomenological description or probabilistic analysis; if some do, determination of preferred scale ratio stands as a major issue for statistics, to therefore address lucidly.

References

1. D. Sornette, A. Johansen, A. Arneodo, J.F. Muzy and H. Saleur (1996) *Phys. Rev. Lett.* **76**, pp. 251–254.
2. S. Douady, Y. Couder and M.E. Brachet (1991) *Phys. Rev. Lett.* **67**, pp. 983–986
3. She, Z.S. & Lévêque, E. (1994) *Phys. Rev. Lett.* **72**, pp. 336–339.
4. A. N. Kolmogorov (1941) *C.R.Acad.Sci.* URSS **30**, pp. 9-13.
5. Frisch, U. (1995) *Turbulence*, Cambridge University Press.
6. Pocheau, A. (1996) *Europhys.Lett.* **35**, pp. 183–188.
7. Pocheau, A. submitted to *Phys. Rev.Lett.*
8. B. Dubrulle (1994) *Phys. Rev. Lett.* **73**, pp. 959-962; Z.-S. She and E.C. Waymire (1995) *Phys. Rev. Lett.* **74**, pp. 262-265.
9. J-P. Kahane (1995), in *Levy Flights and Related Phenomena in Physics*, eds M. Shlesinger, G. Zaslavsky and U. Frisch, *Lecture notes in Physics* **450**.
10. Dubrulle, B. & Graner, F. (1996) *J.Phys.II(France)* **6**, pp. 797–816.
11. B. Dubrulle, FM. Bréon, F. Graner and A. Pocheau, submitted to Eur. J.Phys.

"MEASUREMENT" OF TURBULENCE INTERMITTENCY IN PHYSICAL AND NUMERICAL EXPERIMENTS

Y. MALECOT, C. AURIAULT AND Y. GAGNE
LEGI/IMG-CNRS, B.P.53X, 38041 Grenoble, France.

AND

O. CHANAL AND B. CASTAING
CRTBT/CNRS/UJF, B.P.166X, 38042 Grenoble, France.

1. Introduction

In order to improve the modelling of turbulent flows, a better knowledge of the small scale intermittency is still needed. The main feature of the intermittency is that the velocity field has actually no scale invariance at finite Reynolds [1]. A few statistical models take into account a Reynolds number dependence [2], [3].

In this paper, we use a model introduced by Castaing [4], which permits to easily "measure" the small scale intermittency from the experimental velocity increment probability density functions (hereafter pdfs). This analysis exhibits a β exponent, which universaly depends on the Reynolds number of the flow [5].

In section 2, we experimentally show how the lack of scale invariance of the velocity field suggests to study the cumulants of the multiplier distribution in the cascade process. In section 3, it is shown that measurements of longitudinal and transverse velocity increments obtained in several different physical experiments lead to a β exponent as predicted in [4]. In section 4, the same analysis is applied on some numerical data (D.N.S) and is used to estimate the Reynolds number in the L.E.S.

2. No scale invariance

Figure 1 shows the behaviour of the logarithmic derivative $\zeta_3 = \frac{d\ln<|\delta u_r|^3>}{d\ln r}$ versus r/l_0. Considering the absolute value of $\delta u(r)$ we let apart the problem of the skewness of the longitudinal difference and stress on the comparison

243

U. Frisch (ed.), Advances in Turbulence VII, 243–246.
© 1998 *Kluwer Academic Publishers.*

between longitudinal and transverse increments. Complete scale invariance would result in a plateau ($\zeta_3 = 1$ according to K41). Such an analysis would thus concerns only $R_\lambda > 500$ and a very limited range of scales. On the contrary, the analysis we present below concerns all the studied R_λ and all the scales for which ζ_3 lies between the two horizontal dashed lines (i.e. the scales where ESS is checked).

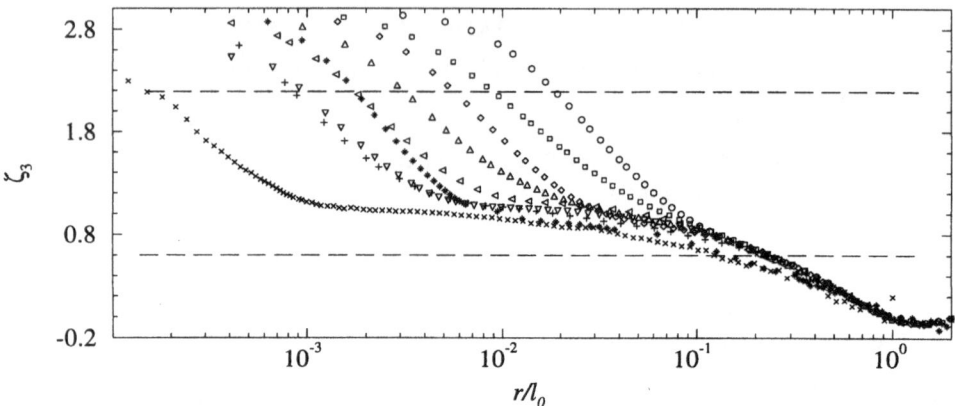

Figure 1. $\zeta_3 = \frac{d \ln <|\delta u_r|^3>}{d \ln r}$; $R_\lambda = 89(\circ)$; $R_\lambda = 124(\square)$; $R_\lambda = 208(\diamond)$; $R_\lambda = 352(\triangle)$; $R_\lambda = 463(\triangleleft)$; $R_\lambda = 703(\triangledown)$; $R_\lambda = 800(*)$; $R_\lambda = 985(+)$; $R_\lambda = 2000(\times)$.

3. Experimental evidence of the cumulant scaling

The model introduced by Castaing in 1990 [4] is based on two main hypothesis. The first hypotheses consists in expressing any velocity increments δu_r with an independent multiplier random variable α_r such as:

$$\delta u(r) = \alpha_r \, \delta u(l_0) \quad \forall \, r \leq l_0$$

where l_0 is the integral scale,

with,
$$\bar{P}_r(\ln |\delta u|) = \int G_r(\ln \alpha) \, \bar{P}_{l_0}(\ln |\delta u| - \ln \alpha) \, d \ln \alpha$$
$$\Leftrightarrow \qquad \bar{P}_r = G_r \otimes \bar{P}_{l_0}$$

where $\bar{P}_r$ (resp. G_r) is the pdf of $\ln |\delta u(r)|$ (resp. the α_r multiplier). The second hypothesis assumes that the G_r distribution is infinitely divisible which is equivalent to the ESS (cf. [6]).

From an experimental point of view, we can measure the structure fonction of the velocity increments, $< |\delta u(l)|^p >$, which are also directly linked to the cumulant generating function ψ_r of the G_r distribution:

By definition,
$$\ln < \alpha_r^p > = \ln \frac{<|\delta u(r)|^p>}{<|\delta u(l_0)|^p>}$$

and,
$$\psi_r(p) = \ln < \alpha_r^p > = C_{1_r} p + C_{2_r} \frac{p^2}{2} + ... + C_{i_r} \frac{p^i}{i!} + ...$$

In the Kolmogorov 41 theory, the shape of the velocity increment pdfs is assumed to be the same at each scale. So, the multiplier α_r is not a random variable anymore. Except the first one $(C_1(r) = \frac{1}{3}\ln(\frac{r}{l_0}))$, all the cumulants are equal to zero.

The Kolmogorov-Obukov 62 theory assumes a lognormal distribution for the multiplier α. So, all the cumulants beyond the second one are equal to zero. Scaling invariance leads to:

$$C_1(r) \propto C_2(r) = -\mu\ln(\frac{r}{l_0}) \quad \text{with } \mu = constant$$

The variational model [2] takes into account both the integral and the viscous scales. Its main predictions are:

$$\frac{\partial C_i(r)}{\partial \ln(r)} \propto r^{-\beta} \quad \Leftrightarrow \quad C_i(r) = k_i((\frac{l_0}{r})^\beta - 1) \tag{1}$$

and,

$$\beta \propto \frac{1}{\ln(l_0/\eta)} \quad \Leftrightarrow \quad \frac{1}{\beta} = \frac{1}{\beta_0}\ln(\frac{R_\lambda}{R_*}) \tag{2}$$

Measurements have been performed in several types of flow at different Reynolds number (see [5] and [7] for experimental details). Both longitudinal and transverse velocity increments have been analysed. The main results are:

(i) Beyond the third cumulant $C_3(r)$ which is very small, all the other cumulants are neglictible, whatever the scale r is.

(ii) The three non zero cumulants are nearly proportional according to the infinite divisibility hypothesis.

(iii) These cumulants $C_i(r)$ behave like the prediction (1) with the same β exponent for transverse and longitudinal velocity increment at a given Reynolds number (cf. Fig.2a).

(iv) Additional transverse experimental data are in good agreement with the prediction (2) (cf. Fig.2b)

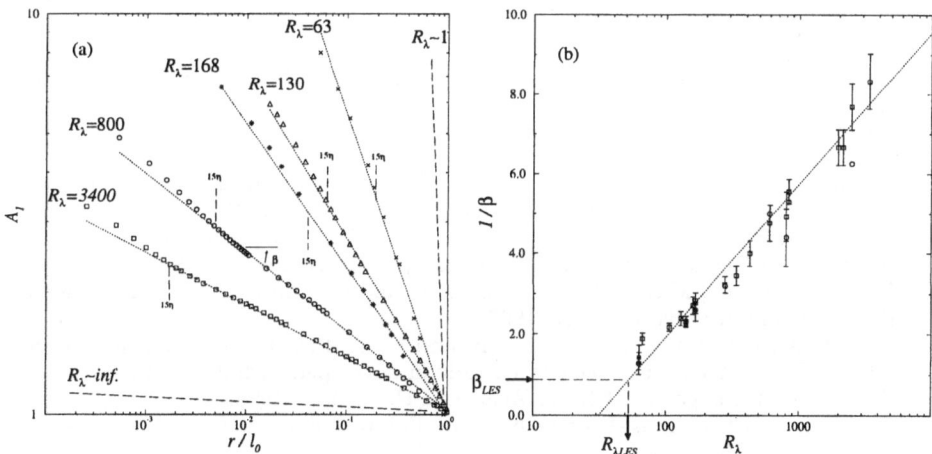

Figure 2. (a) $A_1(r) = \frac{C_1(r)}{k_1} + 1$; $R_\lambda = 3400$ ($\square$); $R_\lambda = 800$ (o); DNS 512^3 at $R_\lambda = 168$ (∗); $R_\lambda = 130$ ($\triangle$); DNS 128^3 at $R_\lambda = 63$ (×). (b) Transverse (square) and longitudinal (circle) increments; Experiments ($\square$, o), DNS ($\blacksquare$, •).

4. An application to the L.E.S.

We have also applied the previous analysis on several DNS (128^3, 384^3 and 512^3) performed by Jimenez [8] corresponding to a Reynolds number of $R_\lambda = 63$, $R_\lambda = 142$ and $R_\lambda = 168$ (cf. Fig.2a). On Figure 2b, we notice that all the physical and numerical data collapse on a same universal curve which means that β is not dependent on the type increment. That result is not so surprising, since our analysis is based on the absolute value of $|\delta u_r|$, avoiding any skewness effect.

Inversely, it is possible to estimate the Reynolds number from the knowledge of β. This can be useful, particularly in the case of Large Eddy Simulation (LES). The subgrid scale modelling prevents from knowing the effective dissipative scale, and therefore from calculating R_λ. Applying our cumulant method to the calculated scales of the LES of a periodic box of 128^3 points, performed by Metais [9], we can estimate a β exponent only with the velocity increment pdfs corresponding to these scales.

Using the universal curve, $\beta(R_\lambda)$, of the Fig.2b, we deduce a Taylor Reynolds number of $R_{\lambda LES} = 53$. Therefore, its seems that with a given numerical resolution (namely 128^3), the LES leads to a lower value than this obtained with the DNS. In fact, a detailed analysis (cf. [10]) shows that:
$$R_{\lambda LES} \simeq 3 R_{\lambda DNS}$$

Acknowledgements: We are grateful to O. Metais and J. Jimenez for their numerical data. This work was supported by the DSP contract 97/1045.

References

1. FRISH, U. *Turbulence: the legacy of A.N. Kolmogorov.* Cambridge University Press, 1995.
2. CASTAING, B. Conséquences d'un principe d'extrèmum en turbulence. *J. Phys.*, 50:147, 1989.
3. FRISH, U., and VERGASSOLA, M. A prediction of the multifractal model: the intermediate dissipation range. *Eur. Phys. Lett.*, 14:439–444, 1991.
4. CASTAING, B., GAGNE, Y. and HOPFINGER, E.J. Velocity probability density functions of high Reynolds number turbulence. *Physica D*, 46:177, 1990.
5. KAHALERRAS, H., MALÉCOT, Y., GAGNE, Y. and CASTAING, B. Intermittency and Reynolds number. *Phys. Fluids*, 1998.
6. NAERT, A., CHABAUD, B., HÉBRAL, B. and CASTAING, B. Experimental check of infinite divisibility for the velocity cascade in developed turbulence. In *Advances in Turbulence VI*, page 251. Kluwer Academic Publishers, 1996.
7. CHANAL, O. PhD thesis, Institut National Polytechnique de Grenoble, 1998.
8. JIMÉNEZ, J., WRAY, A.A., SAFFMAN, P. and ROGALLO, R. The structure of intense vorticity in isotropic turbulence. *J. Fluid Mech.*, 255:65–90, 1993.
9. MÉTAIS, O. and LESIEUR, M. Large-Eddy Simulation of isotropic and stably-stratified turbulence. In *Advances in Turbulence II*. Springer-Verlag, 1989.
10. MALÉCOT, Y. and GAGNE, Y. Intermittency and Reynolds number in LES of turbulent flows. In *Dynamics and Statistics of Concentrated Vortices in Turbulent Flows*, page 51. Euromech Colloquium 364, 1997.

MULTISCALE VELOCITY CORRELATIONS IN TURBULENCE

L. BIFERALE[1,2], F. TOSCHI[2,3], R. BENZI[4]

[1] *Dipartimento di Fisica, Università di Tor Vergata,*
Via della Ricerca Scientifica 1, I-00133 Roma, Italy.
[2] *INFM, unità di Tor Vergata.*
[3] *Dipartimento di Fisica, Università di Pisa,*
Piazza Torricelli 2, I-56126, Pisa, Italy.
[4] *AIPA, Via Po 14, 00100 Roma, Italy.*

Abstract. We review some results [1] on Multiscale correlation functions in high Reynolds number experimental turbulence and synthetic signals. Fusion Rules predictions as they arise from multiplicative, almost uncorrelated, random processes for the energy cascade are tested. Leading and sub-leading contribution, are well captured by assuming a simple multiplicative random process for the energy transfer mechanisms.

Intermittency in the inertial range is usually analyzed by means of the statistical properties of velocity differences, $\delta_r v(x) = v(x) - v(x + r)$. In particular, in the last twenty years [2], overwhelming experimental and theoretical works focused on structure functions: $S_p(r) = \langle(\delta_r v(x))^p\rangle$. A wide agreement exists on the fact that structure functions show a scaling behaviour, $S_p(r) \sim \left(\frac{r}{L}\right)^{\zeta(p)}$ in the limit of very high Reynolds numbers. The velocity fluctuations are anomalous in the sense that $\zeta(p)$ exponents do not follows the celebrated dimensional prediction made by Kolmogorov, $\zeta(p) = p/3$. In order to better characterize the energy transfer mechanism, it is natural to look also at correlations among velocity fluctuations at different scales. Multiscale correlations functions should play in turbulence the same rôle played by correlation functions in critical statistical phenomena.

Recently, some theoretical work [3, 4, 5] and an exploratory experimental investigation [6] have been devoted to the behavior of multiscale velocity correlations:

$$F_{p,q}(r, R) \equiv \langle(\delta_r v(x))^p (\delta_R v(x))^q\rangle \qquad (1)$$

247

U. Frisch (ed.), Advances in Turbulence VII, 247–250.

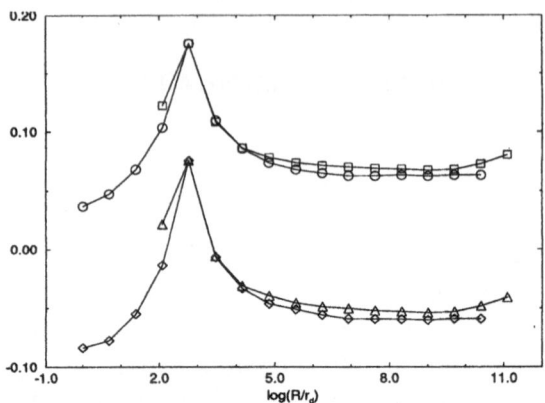

Figure 1. Experimental and numerical $F_{p,q}(r,R)/S_{p+q}(R) \cdot S_p(r)$ at fixed r and changing the large scale R. Circles correspond to $p = 2, q = 2$, diamonds to $p = 4, q = 2$ for the experimental data. Squares correspond to $p = 2, q = 2$ and triangles to $p = 4, q = 2$ for the synthetic signal. Small scale r is fixed to $r = 16$ in units of the Kolmogorov scale. The data for $p = 4, q = 2$ have been shifted along the vertical axis for the sake of presentation.

In the following, in order to simplify our discussion, we will confine our analysis for the case of longitudinal velocity differences.

Stochastic cascade processes are simple and well known useful tools to describe the leading phenomenology of the intermittent energy transfer in the inertial range. The main finding reported here is that multiscale correlations (1) are in *quantitative* agreement, for any separation of scale r/R, with the prediction one obtains by using a pure uncorrelated multiplicative process for the energy cascade. The main idea turns around the hypothesis that small scale statistics is fully determined by a cascade process conditioned to some large scale configuration:

$$\delta_r v(x) = W(r, R) \cdot \delta_R v(x) \qquad (2)$$

where, requiring homogeneity along the cascade process, the random function W should depend only on the ratio r/R. Structure functions are then described in terms of the W process: $S_p(r) = C_p \langle [W(r/L)]^p \rangle$. Pure power laws arise in the high Reynolds regime: in this limit we must have $\langle [W(\frac{r}{R})]^p \rangle \sim (\frac{r}{R})^{\zeta(p)}$. In the same framework, it is straightforward to give, under the hypothesis of weak correlation among multipliers, the leading prediction for the multiscale correlation functions (1):

$$F_{p,q}(r, R) = C_{p,q} \left\langle \left[W\left(\frac{r}{R}\right) \right]^p \right\rangle \left\langle \left[W\left(\frac{R}{L}\right) \right]^{p+q} \right\rangle \sim \frac{S_p(r)}{S_p(R)} \cdot S_{p+q}(R) \quad (3)$$

This expression was for the first time proposed in [3] and analyzed in great details in [4, 5] where it was considered to rigorously express the leading behavior of (1) when $r/R \to 0$. Let us notice that, beside any rigorous claim, expression (3) is also the zero-*th* order prediction starting from any multi-plicative uncorrelated random cascade satisfying $\langle[W(\frac{r}{R})]^p\rangle \equiv S_p(r)/S_p(R)$. In this proceedings we want to report about the results presented in [1] by analyzing high Reynolds number experiments and synthetic signals. We discuss three points: (i) whether the prediction (3) gives the correct lead-ing behavior in the limit of large separation of scales $r/R \sim 0$, (ii) if this is the case, what one can say about sub-leading behavior for separation $r/R \sim O(1)$, (iii) what happens to those observables for which the "multi-plicative prediction" (3) is incorrect because of symmetry reasons. Indeed, let us notice that for correlation like: $F_{1,q}(r,R) = \langle(\delta_r v)(\delta_R v)^q\rangle$ the multi-plicative prediction gives: $F_{1,q}(r,R) = \frac{S_1(r)}{S_1(R)} \cdot S_{1+q}(R)$. Such a prediction is wrong because, if homogeneity can be assumed, $S_1(r) = 0$ for all scales r. In this case prediction (3) does not represent the leading contribution.

Synthetic signals are built in terms of wavelet decomposition with coef-ficients defined by a pure uncorrelated random multiplicative process [7]. Such a signal should therefore show the strong fusion rules prediction (3) and it will turn out to be an useful tool for testing how much deviations from (3), observed in experiments or numerical simulations, are due to im-portant dynamical effects or only to unavoidable geometrical corrections.

First of all, let us notice that for any 1-dimensional string of number (such as the typical outcome of laboratory experiments in turbulence) the multiscale correlations (1) feel strong geometrical constraints. In particular we may always write down "Ward-Identities" (WI):

$$S_p(R - r) \equiv \langle[\delta_R v - \delta_r v]^p\rangle = \sum_{k=0,p} b(k,p)(-)^k F_{k,p-k}(r,R), \qquad (4)$$

where $b(k,p) = p!/[k!(p-k)!]$. The "Ward-Identities" will turn out to be useful for understanding sub-leading predictions to the multiplicative cas-cade process. One may argue that in geometrical set-up different from the one specified in (1) the same kind of constraint will appear with eventually different weights among different terms.

Let us check the strong fusion rules prediction (3) for moments with $p > 1, q > 1$. In Figure 1 we have checked the large scale dependency by plotting $F_{p,q}(r,R)/S_{p+q}(R) \cdot S_p(R)$ as a function of R at fixed, r, for different values of p, q. The expression (3) predicts the existence of a plateau (inde-pendent of R) at all scales R where the leading multiplicative description is correct. From Figure 1 one can see that, in the limit of large separation $R \to L$ at fixed r, $F_{p,q}(r,R)/S_{p+q}(R) \cdot S_p(R)$ shows a tendency toward a plateau. On the other hand, there are clear deviations for $r/R \sim O(1)$. Such deviations show a very slow decay as a function of the scale separation. In order to understand the physical meaning of the observed deviations to the fusion rules (3), we compare, in Figure 1, the experimental data against the equivalent quantities measured by using the synthetic signal. We notice an almost perfect superposition of the two data sets, indicating that the deviations observed in real data can hardly be considered a "dynamical effect". Using the WI plus our multiplicative receipt for $p = 4$ we quickly read that the leading contribution to $F_{2,2}$ is $O(r^{\zeta(2)}) \cdot O(R^{\zeta(4)-\zeta(2)})$, while sub-leading terms scale as $O(r^{\zeta(4)})$, and as $O(r^{\zeta(3)}) \cdot O(R^{\zeta(4)-\zeta(3)})$.

This superposition of power laws is responsible for the slowly-decaying correlations in Fig. 1. The result so far obtained, i.e. that both the experimental data and the synthetic signal show the same quantitative behaviour, is a strong indication that multiscale correlation functions, at least for $p > 1, q > 1$, are in good agreement with the random multiplicative model for the energy transfer.

For multiscale correlations where the direct application of the random-cascade prediction is useless, like $F_{1,q}(r, R)$, we use the WI plus the multiplicative prediction applied to all terms, except the $F_{1,q}$. One obtains the expansion: $F_{1,q}(r, R) \sim \left[O\left(\frac{r}{R}\right)^{\zeta(2)} + \cdots + O\left(\frac{r}{R}\right)^{\zeta(q+1)} \right] \cdot S_{q+1}(R)$,

Let us summarize what is the framework we have presented until now. Whenever the simple scaling ansatz based on the uncorrelated multiplicative process is not prevented by symmetry arguments, the multi-scale correlations are in good asymptotic agreement with the fusion rules prediction even if strong corrections due to sub-leading terms are seen for small-scale separation $r/R \sim O(1)$. Subleading terms are strongly connected to the WI previously discussed, i.e. to geometrical constraints. In the other cases (i.e. $F_{1,q}(r, R)$) the geometry fully determines both leading and sub-leading scaling. All this findings, led us to the conclusions that multiscale correlations functions measured in turbulence are fully consistent with a multiplicative, almost uncorrelated, process.

We acknowledge useful discussions with A.L. Fairhall, V. L'vov and I. Procaccia. We are indebted to S. Ciliberto, R. Chavarria and Y. Gagne for having allowed us the access to the experimental data. L.B. and F.T have been supported by INFM (PRA TURBO).

References

1. R. Benzi, L. Biferale, F. Toschi, *Phys. Rev. Lett.*, (1998), In press.
2. U. Frisch, *Turbulence. The legacy of A.N. Kolmogorov* (Cambridge University Press, Cambridge, 1995).
3. G. Eyink, *Phys. Lett. A*, **172**, 355, (1993).
4. V.S. L'vov, I. Procaccia, *Phys. Rev. Lett.*, **76**, 2896, 1996.
5. V.S. L'vov, I. Procaccia, *Phys. Rev. E*, **54**, 6268, 1996.
6. A.L. Fairhall, B. Druva, V.S. L'vov, I. Procaccia, K.S. Sreenivasan, *Phys. Rev. Lett.*, (1997), To appear.
7. R. Benzi, L. Biferale, A. Crisanti, G. Paladin, M. Vergassola, A. Vulpiani, *Physica D*, **65**, 352, (1993).

DISCRETE SCALE INVARIANCE IN TURBULENCE?

D. SORNETTE[1,2]
[1] *LPMC, CNRS UMR6622 and Université de Nice-Sophia Antipolis*
B.P. 71, 06108 NICE Cedex 2, France
[2] *IGPP and Department of ESS, UCLA*
Los Angeles, CA 90095-1567

Scaling laws provide a quantification of the complexity of turbulent flows. For instance, the second order longitudinal structure function has the form $\langle (v_r)^2 \rangle = C_K (\bar{\epsilon} r)^{2/3}$, according to Kolmogorov 1941 [1] where r, which lies in the inertial range, is the scale at which velocity differences are measured, and $\bar{\epsilon}$ is the mean rate of energy dissipation per unit mass. Dimensional analysis shows that $\langle (v_r)^2 \rangle = (\bar{\epsilon} r)^{2/3} F(\Re, r/L)$, where $F(x, y)$ is a universal function to be determined, L is the external or integral scale. Kolmogorov's assumption is that, for the Reynolds number $\Re \to \infty$ and $r/L \to 0$, $F(x, y)$ goes to a constant C_K. This is the so-called complete similarity of the first kind [2] with respect to the variables $\Re$ and r/L.

The existence of the limit of $F(\Re, r/L \to 0)$ has first been questioned by L.D. Landau and A.M. Obukhov, on the basis of the existence of intermittency – fluctuations of the energy dissipation rate about its mean value $\bar{\epsilon}$. Indeed, Barenblatt's classification leads to the possibility of an *incomplete similarity* in the variable r/L. This would require the absence of a finite limit for $F(\Re, r/L)$ as $r/L \to 0$, and leads in the simplest case to the form $\langle (v_r)^2 \rangle = C_K (\bar{\epsilon} r)^{2/3} \left(\frac{r}{L} \right)^{\alpha}$, where α is the so-called intermittency exponent, believed to be small and positive. If α is real, this corresponds to a similarity of the second kind [2]. Incomplete self-similarity [3, 4] may stem from a possible $\Re$-dependence of the exponents. The case where α is complex, leading to $\langle (v_r)^2 \rangle = C_K (\bar{\epsilon} r)^{2/3} \left(\frac{r}{L} \right)^{\alpha_R} \cos[\alpha_I \log(r/L)]$, could be termed a similarity of the third kind, characterized by the absence of limit for $F(\Re, r/L)$ and accelerated (log-periodic) oscillations. To our knowledge, Novikov has been the first to point out in 1966 that structure functions in turbulence should contain log-periodic oscillations [5]. His argument was that if an unstable eddy in a turbulent flow typically breaks up into two or three smaller eddies, but not into 10 or 20 eddies, then one can suspect the existence of a prefered scale factor, hence the log-periodic oscillations. They

U. Frisch (ed.), Advances in Turbulence VII, 251–254.

have been repeatedly observed but do not seem to be stable and depend on the nature of the global geometry of the flow and recirculation [1, 6] as well as the analyzing procedure.

The theory of complex exponents and log-periodicity has advanced significantly [7] in the last few years. Complex exponents reflect a discrete scale invariance (DSI), i.e. the fact that dilational symmetry occurs only under magnification under special factors, which are arbitrary powers λ^n of a prefered scaling ratio λ. Complex exponents have been studied in the eighties in relation to various problems of physics embedded in hierarchical systems. In the context of turbulence, shell models construct explicitly a discrete scale invariant set of equations whose solutions are marred by unwanted log-periodicities. Only recently has it been realized that discrete scale invariance and its associated complex exponents may appear "spontaneously" in euclidean systems, i.e. without the need for a pre-existing hierarchy. Systems that have been found to exhibit self-organized DSI are Laplacian growth models [8], rupture in heterogeneous systems [9], earthquakes [10], animals [11] (a generalization of percolation) among many other systems. In addition, general field theoretical arguments [11] indicate that complex exponents are to be expected generically for out-of-equilibrium and/or quenched disordered systems. This together with Novikov's argument suggest to revisit log-periodicity in turbulent signals. Demonstrating unambiguously the presence of log-periodicity and thus of DSI in turbulent time-series would provide an important step towards a direct demonstration of the Kolmogorov cascade or at least of its hierarchical imprint. For partial indications of log-periodicity in turbulent data, we refer the reader to fig. 5.1 p.58 and fig. 8.6 p.128 of Ref. [1], fig.3.16 p. 76 of Ref.[12], fig.1b of Ref.[13] and fig. 2b of Ref. [14].

It is a common observation that the oscillations, if any, "move" when changing the length of the signal over which the averaging is carried out. They thus have the aspect of noise. However, previous numerical simulations on Laplacian growth models [8, 15] and renormalization group calculations [11] have taught us that the presence of noise modifies the phase in the log-periodic oscillations in a sample specific way leading to a "destructive interference" upon averaging. In the turbulence context, we propose that one realization corresponds approximately to a signal measured over one turn-over time scale L/v_L. In contrast to this sample specific phase dependence, we stress that the prefered scaling ratio λ has universal properties.

It is thus important to carry out an analysis on each sample realization separately, without averaging, as has been demonstrated to work for other systems [8]. An enticing alternative is to introduce a new averaging scheme that does not destroy the oscillations. The standard averaging procedure, that we could term "Grand canonical" [16], is known to introduce spurious sample-to-sample fluctuations of relative amplitude proportional to $L^{-d/2}$ in d dimensions. In contrast, the concept of "canonical" averaging [16] consists in identifying, for each realization, the corresponding specific value of the critical control parameter K_c^R. The natural control parameter then becomes $\Delta = (K - K_c^R)/K_c^R$ and the averaging can then be performed over the different samples with the same Δ. This should then lead in principle to a "rephasing" of the log-oscillations. This "canonical averaging" has been demonstrated for log-periodic signatures of the acoustic emission precursors prior to rupture and in Laplacian growth models [15].

We propose to adapt this "canonical" averaging scheme to the analysis of structure functions of turbulent flows. There are probably several possible schemes to implement it. Let us suggest here one based on the energy dissipation rate. The strategy is to look for a reference quantity that is specific to a given turn-over time realization. For critical phenomena, a natural candidate is the susceptibility whose maximum determines the sample specific critical point location K_c^R [16, 15]. For turbulence, we suggest to determine the scale r_c at which the dissipation rate is the largest in a given turn-over time series. This can be derived by a direct measurement of the velocity gradient at small scales or from the third-order structure function $S_3(r) = -\frac{4}{5}\bar{\epsilon}r$, which obeys the exact four-fifth Kolmogorov law under a set of assumptions [1]. This specific scale r_c translates into a specific "phase" in the logarithm of the scale which, when used as the origin, allows one to phase up the different measurements of a structure function $S_p(r) = A_p(\bar{\epsilon}r)^{p/3}$ in different turn-over time realizations. We expect, as in Laplacian growth models and in rupture, that the log-periodic oscillations will be reinforced by this canonical averaging.

What could be the mechanism that creates these characteristic scales? There are undoubtly the integral length L and the scales associated to the dissipation range. But what could produce an approximate geometrical series of scales in the inertial range? We conjecture two possible routes. The first one is inspired from a recent discovery that the continuous nonlinear Einstein partial differential equations of general relativity in the presence of a scalar field self-interacting through gravitation may generate a log-periodic spectrum of black hole masses with develop according to a log-periodic self-similar time dynamics [17]. The mechanism might result from the existence of a limit cycle in the renormalization group description of a field close to the negative density limit (in turbulence, could this be obtained from a negative effective viscosity?). The other route is that scale invariant equations that present an instability at finite wavevector k decreasing with the field amplitude may generate naturally a spectrum of internal scales. An example is $\frac{\partial v}{\partial t} = -2v\frac{\partial^2 v}{\partial x^2} - v^2\frac{\partial^4 v}{\partial x^4}$. This equation is scale invariant in the sense that if $v(t, x)$ is a solution, then $\gamma^2 v(t, \gamma x)$ is also a solution for arbitrary γ. A linear stability analysis shows that a mode $v_0 e^{\sigma t} e^{ikx}$ grows with $\sigma = 2v_0 k^2 - v_0^2 k^4$, i.e. the most unstable mode occurs at finite $k_{m.u.} = \frac{1}{\sqrt{v_0}}$. Thus, a finite characteristic scale appears that is completely controlled by the amplitude of the field. Starting from an approximate homogeneous level v_0, the instability produces a large scale $\frac{2\pi}{k_{m.u.}} = 2\pi\sqrt{v_0}$. As dips in the field develop, the amplitude there decreases and the corresponding instabilities will create smaller length scales, and so on. Preliminary simulations [18] confirm this intuitive picture : the resulting DSI field is seen to result from a cascade of instabilities with characteristic wavelengths controlled by the amplitude.

I hope that these conjectural ideas will stimulate further works on these fascinating log-periodic structures in turbulent signals. I am grateful to U. Frisch, L. Gil, N. Goldenfeld, A. Johansen, A. Noullez and G. Simms for discussions.

References

1. Frisch, U. (1995) *Turbulence, the legacy of A.N. Kolmogorov*, Cambridge University Press.
2. Barenblatt, G. I. (1996) *Scaling, self-similarity, and intermediate asymptotics*, Cambridge University Press.
3. Barenblatt, G.I. & Goldenfeld, N. (1995) *Phys. Fluids* **7**, pp. 3078-3082.
4. Dubrulle, B., (1996) *J. Phys. France II* **6**, pp. 1825-1840.
5. Novikov, E.A. (1966) *Dokl.Akad.Nauk SSSR* **168/6**, pp. 1279; (1990) *Phys.Fluids A* **2**, pp. 814–820.
6. Anselmet, F., Gagne, Y., Hopfinger, E.J. & Antonia, R.A. (1984) *J. Fluid Mech.* **140**, pp. 63
7. Sornette, D. (1998) Discrete scale invariance and complex dimensions, *Physics Reports*, in press (april 1998) (http://xxx.lanl.gov/abs/cond-mat/9707012).
8. Sornette, D., Johansen, A., Arnéodo, A., Muzy, J.-F. & Saleur, H. (1996) *Phys. Rev. Lett.* **76**, pp. 251–254; Huang, Y., Ouillon, G., Saleur, H. & Sornette, D. (1997) *Phys. Rev. E* **55**, pp. 6433–6447.
9. Anifrani, J.-C., Le Floc'h, C., Sornette, D. & Souillard, B. (1995) *J.Phys.I France* **5**, pp. 631–638.
10. Sornette, D. & Sammis, C.G. (1995) *J.Phys.I France* **5**, pp. 607–619; Johansen, A., Sornette, D., Wakita, H., Tsunogai, U., Newman, W.I. & Saleur, H. (1996) *J.Phys.I France* **6**, pp. 1391–1402.
11. Saleur, H. & Sornette, D. (1996) *J.Phys.I France* **6**, pp. 327–355.
12. Arnéodo, A., Argoul, F., Bacry, E., Elezgaray, J. & Muzy, J.-F. (1995) *Ondelettess, multifractales et turbulences*, Diderot Editeur, Arts et Sciences
13. Tchéou, J.-M. & Brachet, M.E. (1996) *J.Phys.II France* **6**, pp. 937–943.
14. Castaing, B. (1997) in *Scale invariance and beyond*, eds. Dubrulle, B., Graner, F. & Sornette, D., EDP Sciences and Springer, pp. 225–234.
15. A. Johansen & D. Sornette (1998) Evidence for discrete scale invariance by canonical averaging, preprint.
16. Pazmandi, F., Scalettar, R.T. & Zimanyi, G.T. (1997) *Phys. Rev. Lett.* **79**, pp. 5130–5133 (1997).
17. Choptuik, M.W. (1993) *Phys. Rev. Lett.* **70**, pp. 9–12.
18. Gil, L. & Sornette, D. (1998), in preparation.

INTERMITTENCY IN COMPRESSIBLE FLOWS

D. PORTER[1], A. POUQUET[2] AND P. WOODWARD[1]

[1] *LCSE & Department of Astronomy, University of Minnesota,
116 Church Street, Minneapolis, Minn 55455, USA.*
[2] *UMR 6529, OCA, BP 4229, 06304 Nice Cedex 4, France.*

We determine scaling exponents of structure functions for a computation
with the Piecewise Parabolic Method [1] (or PPM) of a compressible flow at
a *r.m.s.* Mach number of unity, using periodic boundary conditions and 512^3
grid points, with δx the uniform grid increment. In [2] were analyzed similar
computations with a one–dimensional shear wave forcing, resulting in a
sizable departure from isotropy; hence we choose here a forcing consisting
of a superposition of shear waves at $k_0 = 1$ in three orthogonal directions.
Contrary to [3], we maintain a constant temperature on average by adding a
Stefan–like σT^4 cooling law to the energy equation appropriate for optically
thin media. In the quasi–steady state, kinetic energy $E_K \sim 0.7$ and heat
energy $E_H \sim 1.5$; the *r.m.s.* Mach number locally has excursions up to
4, with density contrasts of up to 30 and density fluctuations of order
unity. The Taylor Reynolds number is close to 100, with however minimal
dissipation in the large scales because of the nature of the PPM method.
As in similar computations but for the decay problem [4], the flow consists
of a superposition of strong planar shocks, vortex filaments and sheets as
well as spirals [2], with on average a weak baroclinic term.

Taking 14 samples in the statistically steady state each 0.2 turn–over
time apart leads to a set of $\sim 10^9$ data points. The scaling of longitudinal
and transverse structure functions of order p, $S_p^{L,T} \sim r^{\zeta_p^{L,T}}$, is computed
using absolute values of velocity increments. In absence of a theorem for
third–order structure functions of compressible flows equivalent to that of
Kolmogorov [5] except for the Burgers equation, these scaling laws are
evaluated as a function of displacement r. Nevertheless, we checked that
the Extended Self–Similarity (or ESS) methodology [6] whereby $\langle \delta u_L^3(\mathbf{r}) \rangle$
can replace r gives comparable results, but with a reduction in the scatter
in local slopes, as can be seen in Figure 1; the horizontal extent of the lines

U. Frisch (ed.), Advances in Turbulence VII, 255–258.

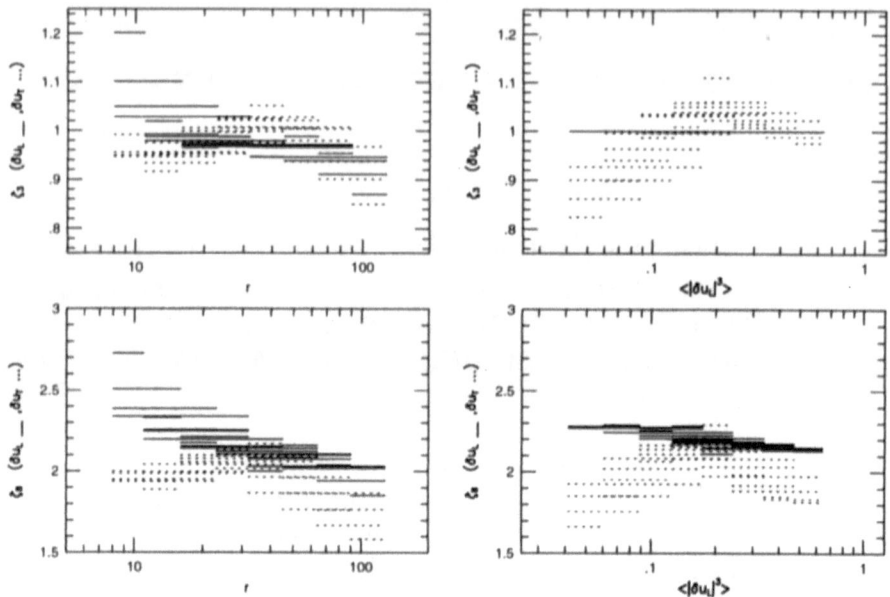

Figure 1. Local slope of structure functions of u_L (solid line) and u_T (dotted) for $p = 3$ (top) and $p = 8$ (bottom) without (left) or with (right) the ESS methodology. The extent of the lines indicates the range of scales over which the fits are performed.

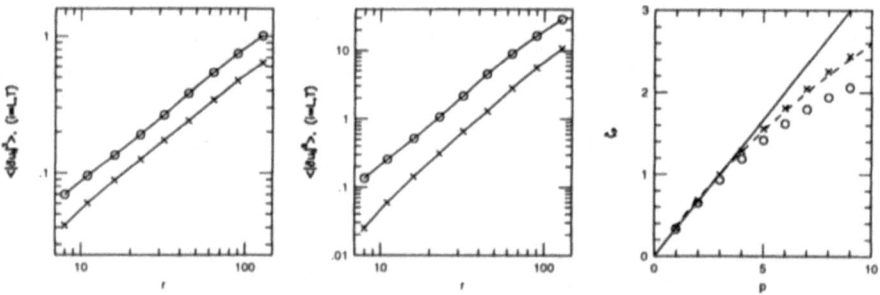

Figure 2. $p = 3$ (left) and $p = 8$ (middle) scaling for u_L (cross) and u_T (circle); $\zeta_p^{L,T} = f(p)$ (right); the solid (*resp.* dash) line follows the Kolmogorov (*resp.* SL) law.

indicates the range of scales over which the exponent is computed. The error in the determination of $\zeta_p^{L,T}$ increases with p, but it is less drastic when restricting their evaluation to a neighborhood of the inertial range where linear scaling obtains for $S_3^L(r)$, *i.e.* $r \in [11, 23]\delta x$.

We find $\zeta_3^{L,T} \sim 1$; scaling holds for a factor ten in scale at low order, and ~ 4 at high p. Figure 2 displays the moments of the velocity for $p = 3$ (left) and $p = 8$ (middle), with a circle for u_T and a cross for u_L; finally, at right is shown $\zeta_p^{L,T} = f(p)$, the solid line following the $p/3$ Kolmogorov law

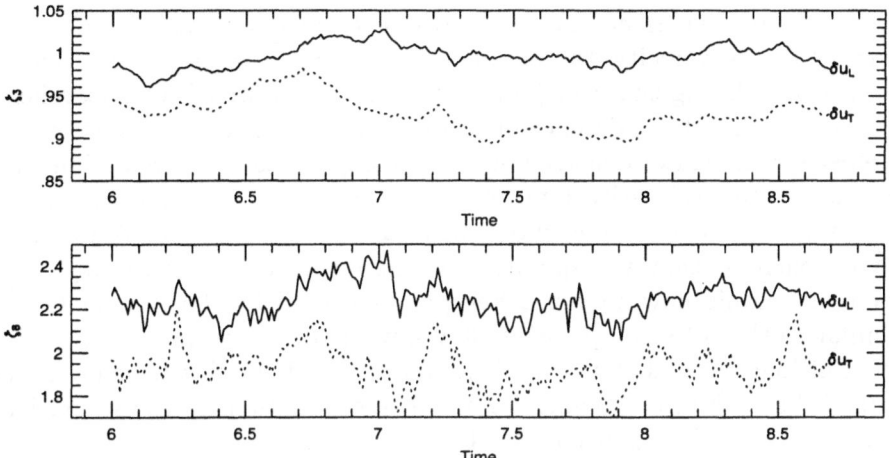

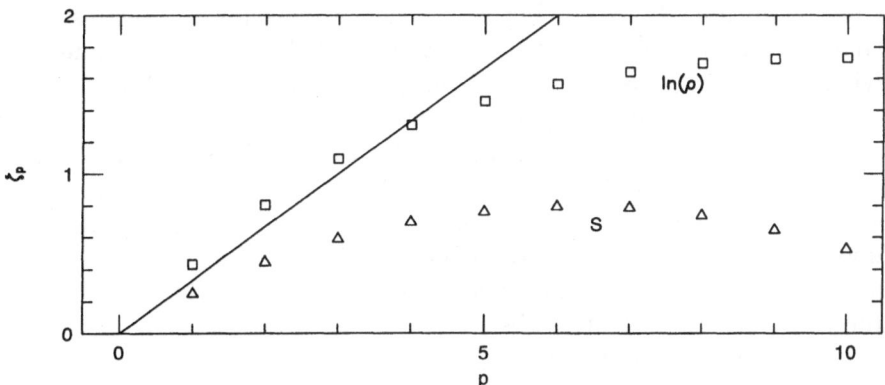

Figure 3.   Temporal variation of $\zeta_3^{L,T}$ and $\zeta_8^{L,T}$ over 2.7 turnover times.

Figure 4.   Intermittency exponents of the $\log_e$ density (squares) and entropy (triangles).

from which a clear departure is observed. Little variation in $\zeta_p^{L,T}$ obtains when increasing the range of fit to $64\delta x$. The scaling exponents of u_L are seen to agree with the She–Leveque (SL) model [7] shown with a dash line, and the scaling exponents of u_T begin to depart from those of u_L for $p > 4$.

To examine the temporal variation of these scaling exponents, 270 snapshots of the flow – evenly spaced over 2.7 turnover times – were taken at full 512^3 resolution, which amounted to 337 GBytes of data. The instantaneous scaling exponents $\zeta_3^{L,T}$ and $\zeta_8^{L,T}$ shown in Figure 3 as functions of time display large variations, with the transverse structure functions exponents (dotted line) being lower than the longitudinal ones (solid line) at all times.

Finally, we show in Figure 4 the scaling exponents for the density and entropy, computed as $\delta\ell_\rho(\mathbf{r}) \sim r^{\zeta_p^\rho}$ where $\ell_\rho = log_e\rho$, and $\delta S(\mathbf{r}) \sim r^{\zeta_p^S}$. Contrary to the velocity, the exponents decrease at high p. Note that the assertion stipulating that the ζ_p curve for a given variable is convex relies on the assumption of the existence of a bound for that field; such a bound may not exist for the log of density nor the entropy, since $e.g.$ rarefaction waves can lead to an arbitrarily large density contrast ρ_{max}/ρ_{min}, and spatial variations in the dissipation may lead to arbitrarily large entropy jumps.

We conclude that the quantitative signature of intermittency of the driven supersonic flow analyzed here, as measured by the exponents $\zeta_p^{L,T}$, is similar to that of incompressible flows, with good agreement between the scaling of longitudinal structure functions S_p^L and the She-Leveque model, whereas the scaling exponents for transverse functions are systematically lower than those for S_p^L when $p > 4$.

The ratio $\chi = E^C/E^V$ measuring the relative shock–to–vortex energy is as significant a parameter a $priori$ for determining the compressibility of a flow as the Mach number measuring the relative intensity of turbulent to acoustic motions. For decaying 2D Navier–Stokes flows, χ settles around 10% irrespective of the initial state [8]. Similar observations are made for 3D decaying flows using the PPM algorithm (Porter, Pouquet & Woodward, in $preparation$): strong vorticity production occurs first by shock–shock interactions and then by vortex stretching, and proves sufficient to overcome the energy in the longitudinal modes $\forall\chi(t = 0)$, with E^C decaying consistently throughout the runs. But the driven case is unclear; 3D computations of supersonic flows which include rapid heating and cooling for modeling the interstellar medium at the kpc scale [9] show at least at low resolution no production of vorticity in the absence of either rotation or magnetic field, and different intermittency exponents might emerge in such cases.

This work was supported at the University of Minnesota by the National Science Foundation, through a grand challenge grant ASC-9217394, and by the Department of Energy's Office of Energy Research, through grant DE-FG02-87ER25035. In Nice, AP received support from CNRS grant PNST.

References

1. Woodward P. & Colella P. 1984 $J.$ $Comput.$ $Phys.$ **54**, 115.
2. Porter D., Pouquet A. & P.R. Woodward 1995 $Lecture$ $Notes$ in $Physics$ **462**, 51, M. Meneguzzi, A. Pouquet & P.L. Sulem Eds., Springer–Verlag.
3. Kida S. & S. Orszag 1990 $J.$ $Scientific$ $Comp.$ **5**, 85.
4. Porter D., Woodward P.R. & Pouquet A. 1998 $Phys.$ $Fluids$ **10**, 237.
5. Kolmogorov, A. 1941 $Dokl.$ $Akad.$ $Nauk$ $SSSR$ **32**, 16.
6. Benzi R., Ciliberto S., Tripicciona R., Baudet C., Massaioli F. & Succi S. 1993 $Phys.$ $Rev.$ E, **48**, R29.
7. Z.S. She Z.S. & Leveque E. 1994 $Phys.$ $Rev.$ $Lett.$ **72**, 336.
8. Passot, T. & A. Pouquet 1987 $J.$ $Fluid$ $Mech.$, **181**, 441.
9. Vazquez, E., Passot, T. & Pouquet, A. 1996 $Astrophys.$ $J.$, **473**, 881.

STATISTICS OF LONGITUDINAL AND TRANSVERSE VELOCITY INCREMENTS

WILLEM VAN DE WATER

Physics Department, Eindhoven University of Technology, P.O.Box 513, 5600 MB Eindhoven, the Netherlands

The possible existence of universal scaling properties of velocity fluctuations is an exciting aspect of strong turbulence. The prime instrument to quantify scaling is the velocity structure structure function that is defined as $G_p(r) = \langle (\Delta u(r))^p \rangle$, where the velocity increment $\Delta u(r)$ is measured over a distance r, $\Delta u(r) = u(x+r) - u(x)$ and the average is done over x. There are two distinct ways of measuring Δu. In the *longitudinal* case, the measured component of $\Delta \boldsymbol{u}$ points in the same direction as $\boldsymbol{r}$, whereas these vectors are perpendicular in the *transverse* case.

The structure function has scaling behavior $G_p(r) \sim r^{\zeta(p)}$. According to Kolmogorov (1941) theory, turbulent velocity fluctuations are self-similar with $\zeta(p) = p/3$. It has become clear that measured scaling exponents differ from this prediction. There is now concensus about the experimental value of the scaling exponents $\zeta(p)$, especially so for moments of the absolute values $\langle |\Delta u(r)|^p \rangle$ of longitudinal velocity increments. As high-order moments are an average over increasingly large and increasingly rare velocity increments, the statistical accuracy is a problem.

A surprising discovery by us was [1] that the transverse scaling exponents showed a larger deviation from self-similarity than the longitudinal ones. Since then this was confirmed in direct numerical simulations of turbulent flows.[4, 5] Figure 1 shows the scaling exponents of longitudinal and transverse structure functions. The transverse velocity increments were measured in grid-generated turbulence, the statistics of longitudinal velocity increments were measured in jet- and grid turbulence. As has become customary in the recent past, for both cases we compute the statistical moments of the

U. Frisch (ed.), Advances in Turbulence VII, 259–262.
© 1998 *Kluwer Academic Publishers.*

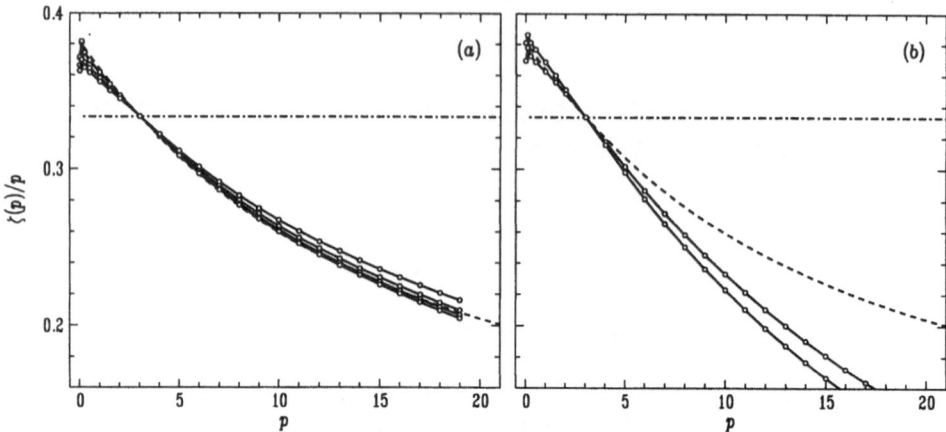

Figure 1. (a) Scaling exponents $\zeta^L(p)/p$ of *longitudinal* structure functions that were measured in jet- and grid turbulence. Dash-dotted line: Kolmogorov's prediction $\zeta(p) = p/3$; dashed line: prediction of the She-Leveque model [3]. (b) Scaling exponents $\zeta^T(p)/p$ of *transverse* structure functions that were measured in grid turbulence.

absolute value of the velocity increments $\widetilde{G}_p(r) = \langle |\Delta u|^p \rangle$. Such practise, however, ignores the skewness of the longitudinal velocity fluctuations that is intimately tied to the turbulent energy cascade. We have found that important new information is contained in the proper longitudinal moments. Of course, as the probability distribution function (PDF) of transverse velocity differences is symmetrical, only moments of absolute transverse velocity differences exist. Longitudinal velocity increments were measured using a single stationary probe. In this case time delays t are interpreted as spatial separations r usings Taylor's hypothesis $r = Ut$, with U the mean flow velocity. Transverse velocity increments were measured with an array of probes that is oriented perpendicularly to the mean flow direction. In this case no recourse to Taylor's hypothesis is needed. In all cases the scaling exponents were determined directly from the measured $G_p(r)$. "Extended self-similarity"[6], which is another way of extracting scaling exponents, works poorly for the proper moments $G_p(r) = \langle (\Delta u(r))^p \rangle$ [2].

In Fig. 1 the results have been plotted in a manner that most clearly shows the effect of intermittency. The longitudinal exponents $\zeta^L(p)$ are very nicely parametrized by the model of She and Leveque [3]. As absolute values $|\Delta u|$ are used, also moments with $-1 < p \leq 1$ are meaningful, and Fig. 1 shows that also the small moments $p < 3$ exhibit intermittency. The exponents were normalized to $\zeta(p = 3) = 1$, which holds in the longitudinal case. However, in the experiments small but significant deviations were found $\zeta^L(3) = 0.96 \cdots 1.03$ and $\zeta^T(p) = 1.08$, which we ascribe to experimental artifacts.

It is a striking observation that the transverse exponents $\zeta^T(p)$ show a larger deviation from self-similarity than the longitudinal ones, and for $p > 3$ the model of Ref. [3] disagrees with the transverse exponents. Several explanations are possible. For example, it has been suggested that the strongest events in turbulence are line vortices. In our experimental arrangement a longitudinal measurement cuts a line through the turbulent velocity field whereas the transverse measurement cuts a plane. In the latter case, the chances to detect line vortices are bigger than in the longitudinal arrangement.

The moments G_p follow from the distribution functions $P(\Delta u)$ as $G_p(r) = \int x^p P(x)\,\mathrm{d}x$ and there have been several suggestions to deduce scaling information *directly* from the distribution functions [7]. This is a tantalizing suggestion as it may be a way to circumvent the statistics problem. The idea is to express $P(\Delta u)$ at inertial-range separations r (which are affected by intermittency) into those at large separations that are close to Gaussian. A typical outcome of such a procedure is illustrated in Fig. 2 that shows a measured transverse $P^T(\Delta u)$ together with a model. While model and experiment can be matched well at large separations r, the figure shows marked discrepancies at smaller r/η. It appears that this is especially so for the *transverse* $P^T(\Delta u)$.

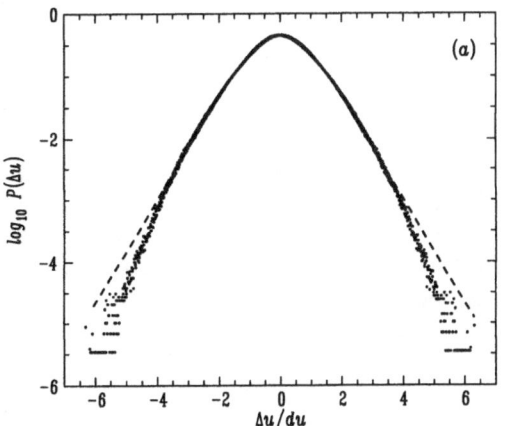

Figure 2. Comparision of probability distribution functions $P^T(\Delta u(r))$ of velocity differences with a scaling model. The velocity fluctuations are *transverse* with the separation vector r oriented perpendicularly to the measured component of u. The fit between model and experiments is good when the separation r/η is large. When the separation r is in the inertial range, $r/\eta = 250$ (as shown), there are marked discrepancies.

A more heuristic representation of $P(\Delta u)$ is through stretched exponentials, $P(\Delta u(r)) \sim \exp(-\alpha|\Delta u(r)|^\beta)$, where the constants α and β may be different for the negative $\Delta u(r) < 0$ and positive $\Delta u(r) > 0$ velocity differences, and, of course depend on r. The PDF has exponential tails at small separations and evolves to Gaussian at large r. Accordingly the stretching exponent β evolves from 1 to 2. This representation appears to be statistically faithful; with our statistical accuracy discrepancies are barely significant.

In our experiments we employ realtime data processing, and long integration times (up to 1.5×10^9 samples) are possible. Still, moments with $p > 12$ remain uncertain. Figure 3 shows the odd-order $G_{19}(r) = |\langle(\Delta u(r))^{19}\rangle|^{1/19}$. The directly measured structure functon shows oscillations which make it difficult to determine the scaling exponent. These oscillations are caused by insufficient statistics; they change and often disappear when averaging over longer integration times. As Fig. 3 illustrates, they also disappear when extrapolating the tails of the PDF's using stretched exponentials.

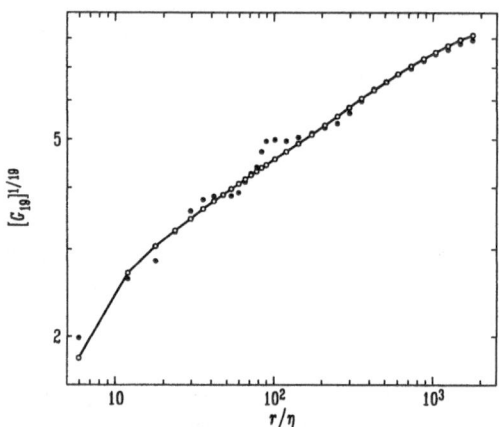

Figure 3. Dots: longitudinal structure function $|G_{19}(r)|^{1/19}$ as computed directly from measured PDFs. The structure function has oscillations that make it difficult to determine a scaling exponent. Open circles $|G_{19}(r)|^{1/19}$ as computed from a stretched- exponential extrapolation from measured PDF's. After extrapolation, the structure function shows clear scaling behavior.

A stretched exponential representation of the tails of PDF's is a good way to reduce statistical fluctuations of high-order structure functions, but it can never be a substitute for long integration times. On the other hand, in this way true large moments ($p > 20$) will always remain elusive. From a conceptual point of view, the question about the limiting behavior $\zeta(p \to \infty)$ is extremely interesting as it points to the nature of the most intermittent structures in turbulence. We thus may never be able to answer it.

References

1. W. van de Water and J.A. Herweijer, Phys. Scripta **T67**, 136-140 (1996).
2. W. van de Water and J.A. Herweijer, Phys. Rev. E. **51** 2669 (1995).
3. Z-S She and E. Leveque, Phys. Rev. Lett. **72**, 336 (1994).
4. O.N. Boratav and R.B. Pelz, Phys. Fluids **9**, 1400 (1997).
5. S. Chen, K.R. Sreenivasan, M. Nelkin and N. Cao, Phys. Rev. Lett. **79**, 2253 (1997).
6. R. Benzi, S. Ciliberto, R. Tripiccione, C. Baudet, F. Massaioli, S. Succi, Phys. Rev. E. **48** , R29 (1993).
7. B. Chabaud, A. Naert, J. Peinke, F. Chilla, B. Castaing and B. Hébral, Phys. Rev. Lett. **73**, 3227 (1994); R. Friedrich and J. Peinke, Phys. Rev. Lett. **78**, 863 (1997).

AN ALTERNATIVE TO SHELL-MODELS:

MORE COMPLETE AND YET SIMPLE MODELS OF INTERMITTENCY

Y. CHIGIRINSKAYA* , D. SCHERTZER
LMM, Université P. & M. Curie, Paris, France
S. LOVEJOY
Physics Dept., McGill University, Montreal, Canada

1. Introduction

The complexity and unsolvability of the Navier-Stokes equations have lead to the consideration of some simplified caricatures of them, which nevertheless preserve some fundamental properties of the original ones. Well-known examples are the Burgers equation, which has unfortunately the drawback of introducing compressibility, and the so-called shell-models (Gledzer et al., 1981) . More complete models are indeed needed since the spatial dimension is absent in shell-models, whereas it is crucial for the development of intermittency. We developed these models by keeping only certain type of interactions of Bernoulli's form of the Navier-Stokes equations.

We show that for 3-D turbulence as well as for 2-D turbulence, the equations of evolution due to direct interactions between eddies and sub-eddies are analogous to the Euler equations of a gyroscope. One may note that the recognition of the similarities (Arnold, 1966; Obukhov, 1971) between the Navier-Stokes equations of hydrodynamic turbulence and the Euler equations of a gyroscope can be traced back to Lamb (1963). This profound analogy in both cases had the advantage to yield two quadratic invariants in a straightforward manner. The corresponding indirect interactions are obtained by coupling an infinite hierarchy of gyroscopes. Overall we derive from rather abstract considerations on the structure of the Navier-Stokes equations (its Lie structure) dynamical space-time models which can be called Scaling Gyroscope Cascade (SGC) models. For simplicity sake of numerical simulations, we restricted our attention to one dimensional cuts of 2-D or 3-D turbulence.

2. Scaling Gyroscopes Cascades (SGC) models

Consider Bernoulli's form of the Navier-Stokes equations, for the velocity field $\underline{u}(\underline{x},t)$ with $\underline{P}(\nabla)$ (resp. $\hat{\underline{P}}(\underline{k})$ in Fourier space) denoting the projector on divergence-free vector fields :

---

* On leave to EE&S Dept., Clemson University-Clemson Research Park, Anderson, SC 29634-0919, USA, E-mail: IOULIAT@clemson.edu.

U. Frisch (ed.), Advances in Turbulence VII, 263–266.
© 1998 *Kluwer Academic Publishers.*

$$\left(\frac{\partial}{\partial t} - v\Delta\right)\underline{u}(x,t) = \underline{P}(\nabla)\cdot\underline{L}(x,t) \tag{1}$$

$$P_{i,j}(\nabla) = \delta_{i,j} - \nabla_i\nabla_j\Delta^{-1}; \quad \hat{P}_{i,j}(\underline{k}) = \delta_{i,j} - k_ik_j/k^2 \tag{2}$$

and $\underline{L}(x,t) = \underline{u}(x,t) \wedge \underline{\omega}(x,t)$ is the Lamb vector ($\underline{\omega}$ is the vorticity field). Therefore, in a general manner, the Navier-Stokes equations (in Fourier space) corresponds to an infinite hierarchy of gyroscope-type equations:

$$\left(\frac{\partial}{\partial t} + vk^2\right)\hat{\underline{u}}^*(\underline{k}) = \underline{\hat{P}}(\underline{k})\cdot \int_{\underline{k}+\underline{p}+\underline{q}=0}\hat{\underline{u}}(\underline{p}) \wedge \hat{\underline{\omega}}(\underline{q})\,d\underline{p} \tag{3}$$

the (complex) analogues of the momentum and rotation being respectively the triplet $[\hat{\underline{u}}(\underline{k}),\hat{\underline{u}}(p),\hat{\underline{u}}(q)]$ and $[\hat{\underline{\omega}}(\underline{k}),\hat{\underline{\omega}}(p),\hat{\underline{\omega}}(q)]$ for a triad $(\underline{k}+\underline{p}+\underline{q}=0)$ of direct interaction, the Lie bracket being the vector product modulated by the projector $\underline{\hat{P}}(\underline{k})$.

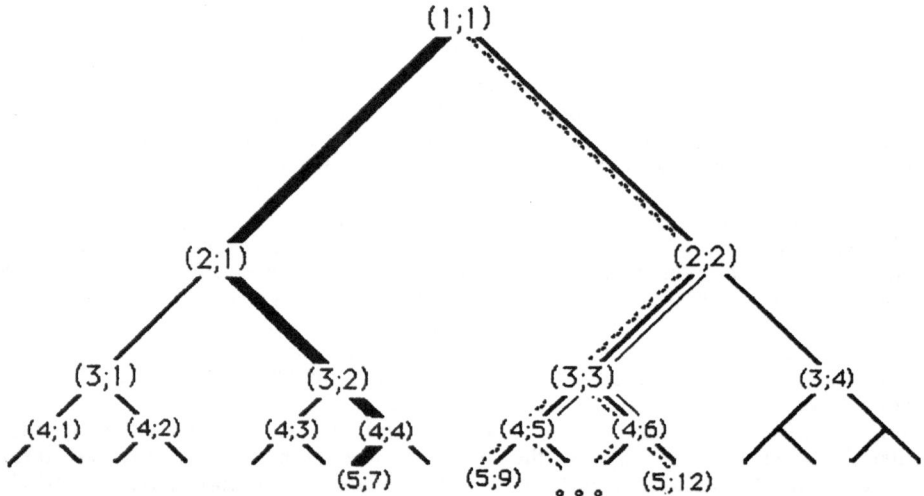

Figure 1. Schematic diagram of discrete SGC models. The interactions for eddy (3;3) in 3-D (the light thin line) and 2-D (the dashed line) turbulence. The thick line points out one of the possible reduction to a shell-model.

However, the projector $\underline{P}(\nabla)$ is non longer relevant as soon as the Lamb vector is divergence free. This is obtained at first order (Chigirinskaya and Schertzer, 1996) for nonlocal interactions $(\max(\underline{k},\,\underline{p},\,\underline{q}) \geq \lambda \,\min(\underline{k},\,\underline{p},\,\underline{q})$, λ being the arbitrary nonlocalness parameter) which satisfy some orthogonal conditions $(\{|\underline{k}| << |\underline{p}| \sim |\underline{q}| \;\; and \;\; \underline{p}\perp\underline{k}\}$ and $|\underline{p}| << |\underline{k}| \sim |\underline{q}|$ and $\underline{u}(\underline{p})\|\underline{k})$. This furthermore reduces the convolution (in the Fourier space) to a quite simpler integral.

3. Discretization of Scaling Gyroscopes Cascades

In order to take into account the spatial dimension, while keeping an exponential discretization of scales (which is not manageable with fast Fourier transforms), we introduce a tree-structure of eddies (Figure 1): each eddy having $N(\lambda)=\lambda^D$ sub-eddies whose location is labeled by (i) (in correspondence to its center $\underline{x}_n^i$, the distance between two neighboring centers being of the order of $l_n = \frac{L}{\lambda^n}$, L being the outer scale). This type of space and scale analysis has been widely used for phenomenological cascade models and is indeed a precursor of orthogonal wavelet decompositions (Zimin, 1991). To the eddy of size l_n and a location $\underline{x}_n^i$ correspond a velocity field $\hat{\underline{u}}_n^i$ and vorticity field $\hat{\underline{\omega}}_n^i = i\underline{k}_n^i \wedge \hat{\underline{u}}_n^i$. Fourier/wavelet components, as well as the wave-vector $|\underline{k}_n^i| = k_n$ (the wave-number k_n being the inverse of scale of the corresponding eddies).

Log K(q, η)

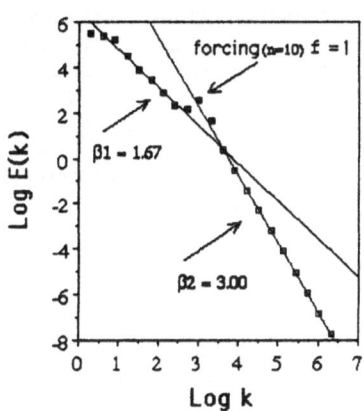
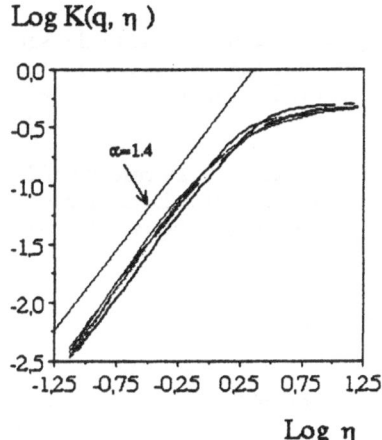

Figure 2. The energy spectrum (averaged over 1024 realizations) of the SGC for 2-D turbulence displays an inverse energy cascade (n<10) and a direct cascade of enstrophy (n>10).

Figure 3. Log-Log plot of the moment scaling function K(q=1.3,η) of the energy spatial flux (n=5;6 and 7) of 2-D and 3-D SGC. This yields α=1.4 (Levy index of multifractality), C_1=0.25 (mean singularity).

For 3-D turbulence the tree-structure of interactions is based on the fundamental triads of (direct) interactions $(\underline{k}_{n-1}^i, \underline{k}_n^{2i-1}, \underline{k}_n^{2i})$, between a mother and two daughter eddies $(i = \overline{1, 2^{n-1}})$. Therefore, one obtains (Chigirinskaya and Schertzer, 1996) the following (in general complex) scalar equation of evolution for the velocity amplitude $\hat{u}_n^i$ $(a(i) = E(\frac{i+1}{2})$ being the location index of its "ancestor"; E(x) being the integer part of the real x $(E(x) \leq x < E(x)+1))$:

$$\left(\frac{\partial}{\partial t} + vk_n^2\right)\hat{u}_n^i = k_{n+1}\left[\left|\hat{u}_{n+1}^{2i-1}\right|^2 - \left|\hat{u}_{n+1}^{2i}\right|^2\right] + (-1)^i k_n \hat{u}_n^i \hat{u}_{n-1}^{a(i)} \tag{4}$$

For 2-D turbulence one has to consider (Chigirinskaya et $al.$, 1997) the interactions between three successive levels (mother, daughter and grand-daughter),- a second order approximation. This yields an algebra more involved than for the case of 3-D turbulence (Eq.1) and which is generated by commutators of $\hat{\Psi}$ and $\hat{\omega}$ (Ψ being the potential vector; ($\underline{k},\underline{p},\underline{q}$) being the triads: $\underline{k}+\underline{p}+\underline{q}=\underline{0}$):

$$C_{n,n'}^{i,i'} = \left(\underline{q}_{n'}^{i'},\underline{p}_{n}^{i},z\right)\left[\Psi\left(\underline{p}_{n}^{i}\right),\omega\left(\underline{q}_{n'}^{i'}\right)\right]$$

$$\left(\frac{\partial}{\partial t}+\nu k_{n}^{2}\right)\hat{\omega}_{n}^{i} = C_{n-1,n-2}^{a(i),a(a(i))} + \sum_{d(i)=2^{2i-1},2^{2i}}\left(C_{n+1,n-1}^{d(i),a(i)} + \sum_{d^{2}(i)=2^{2i-2},2^{4}} C_{n+1,n+2}^{d^{2}(i),d(i)}\right)$$

(5)

4. Some Results and Perspectives

The analogues of the energy and of the square of angular momentum are indeed invariant and the SGC yield the inverse energy cascade sub-range as well as the direct enstrophy sub-range for the 2-D turbulence (Figure 2). Not only do the SGC yields concrete models which can be used to investigate fundamental questions of turbulence (Schertzer et $al.$, 1997), in particular its intermittency, but the multifractal characteristics of the numerical simulations of inverse energy cascade of 2-D turbulence are extremely close to those of the direct energy cascade of 3-D turbulence (Figure 3). We also find a surprisingly close agreement with various empirical studies of atmospheric turbulence.

5. Acknowledgments

We heartily thank A. Avez, U. Frisch, M. Larcheveque, H.K. Moffatt and A.M. Yaglom for enlightening discussions. Partial support by INTAS grant 93-1194 is acknowledged.

6. References

Gledzer, E.B., Dolzhansky, E.V., and Obukhov, A.M. (1981) $Systems$ of $Fluid$ $Mechanical$ $Type$ and $Their$ $Application$, Nauka, Moscow (in Russian).

Zimin, V.D. (1991) in R.Z. Sagdeev et al. (eds.), $Nonlinear$ $Dynamics$ of $Structures$, World Scientific, Singapore.

Arnold, V.I. (1966) Sur la géometrie différentielle des groupes de Lie de dimension infinie et ses applications a l'hydrodynamique des fluides parfaits, $Ann.$ $Inst.$ $Fourier$ $(Grenoble)$ **16**, 1, 319-361.

Obukhov, A.M. (1971) On invariant characteristics of systems of fluid mechanical type, $Fluid$ $Dynam.$ $Trans.$ **5**, 2, 193-199.

Lamb, H. (1963) $Hydrodynamics$, 6th Ed., Cambridge University Press, Cambridge.

Chigirinskaya, Y. and Schertzer, D. (1996) Cascade of scaling gyroscopes: Lie stucture, universal multifractals and self-organized criticality in turbulence, in A. Molchanov and W.A. Woyczynski (eds.),$Stochastic$ $Models$ in $Geophysics$, Springer-Verlag, New York, pp. 57-81.

Chigirinskaya, Y., Schertzer, D., and Lovejoy, S. (1997) Scaling gyroscopes cascade: universal multifractal features of 2-D and 3-D turbulence, in M. Giona and G. Biardi (eds.), $Fractals$ and $Chaos$ in $Chemical$ $Engineering$, World Scientific, Singapore, pp. 371-384.

Schertzer D., Lovejoy, S., Schmitt, F., Chigirinskaya, Y., and Marsan, D. (1997) Multifractal cascade dynamics and turbulent intermittency, $Fractals$ **5**, 3, 427-471.

V

Industrial Applications and Modelling

Applications Industrielles et Modélisation

TURBULENCE MODELLING IN COMPLEX INDUSTRIAL FLOWS

P.-L. VIOLLET AND D. LAURENCE
Électricité de France, Direction des Études et Recherches
Service ER, 1 av. Général de Gaulle, 92141 Clamart, France
Service AEE, LNH, 6 quai Watier, 78401 Chatou, France

1. Trust in CFD

In the 1980's at EDF DER, much expectation was placed in Computational Fluid Dynamics (CFD), starting directly with 3D Navier-Stokes solvers and k-epsilon models. The codes ESTET and N3S became fully operational to replace models ten years later, too late for design studies, and yet are used today for an unexpectedly wide range of studies. In todays economy-driven world, the objective is scarsely to design completely new reactors, but rather to ensure safety and economy of the operation of the current ones and to optimize minor changes for ones currently built. The good behavior of the plants constructed many years ago, and designed to last thirty years, allows to expect their life-span to be extended further, yielding substantial savings. This of course has entailed the reconsidering of many safety studies, which at the same time have been extended to a much larger number components and flow conditions. In the same time, it has appeared necessary to undertake a number of new studies on both old and new types of fossil fuel-fired power plants. Tackling this broader range of problems, with economical constraints, and shorter delays has become possible through CFD.

The delay for evolving from scale models to CFD has not been only a matter of coding and testing, but also a matter of assessing how much trust we can put in CFD, and in particular its turbulence model. The reliability of a numerical study simply cannot be transferred from the scientist to the numerical tool. However, when used by well experienced engineers, CFD can be a very powerful tool, a means for them to elaborate complex yet trustworthy designs and studies. This necessary "know-how" is gained by an awareness of the modelling assumptions pertaining to each category of model, and the experience of running many models through many test

269

U. Frisch (ed.), Advances in Turbulence VII, 269–272.
© 1998 *Kluwer Academic Publishers.*

cases, such as the series of IAHR and ERCOFTAC workshops which will be illustrated in the presentation.

2. Complex geometry

One major application of CFD at EDF is the study of the thermal hydraulics of the flow inside a PWR nuclear reactor. The geometry is very complex and the generation of a detailed mesh has required 2–3 months, even with unstructured finite element meshes. It has proved nevertheless fruitful since many flow conditions have been computed over the past four years. The water is pressurised at 155 bars and 300 C. The height of the reactor is $h = 12$ m. Scale model experiments cannot accurately represent all operating conditions, especially when Reynolds number and buoyancy effects have to be considered (and they are in any case one order of magnitude more costly than the several hundred hours of computations on a CRAY C90 required here). A temperature step is introduced in one of the three inlets, and its mixing prior to entering the reactor core is studied. The simple k-epsilon model used yielded however good comparisons with available data (Alvarez and Martin 1994).

3. Complex Flow

This first application serves to illustrate that "complex geometry" does not imply "difficulty to model". Indeed, the second example is a simple circular pipe with hot fluid flow at Re $= 900\,000$ with a T junction to a smaller pipe leaking cold fluid into to the circuit. The flow rate of the leakage is only 0 to 1% of the main pipe flow, still in this range, hot fluid creeps up, into the "dead leg". This motion is helical and unsteady, leading to potentially severe unsteady thermal stresses. It has been observed on actual plants and scale models, but is fairly difficult to reproduce in a numerical simulation. Refined computations, ranging from 10^5 to 5×10^5 nodes (Deutsch *et al.* 1996) have clearly shown that the SMC performs significantly better in reproducing the unsteady vortex penetration in the dead leg. Only the SMC simulation on the finer 1.5 million nodes mesh was able to reproduce the experimentally observed penetration $H/D = 8$. This shows that although the problem can seem a priori simple, it actually requires considerable modelling and computational efforts.

4. Complex Physics

In this last example, a flame holder, the geometry is a simple rectangular cylinder. Without combustion, the second moment closure (SMC) exhibits the well known vortex shedding which the k-epsilon model fails to reproduce. With combustion however, the flame actually suppresses the unsteadiness of the wake, but still the k-epsilon model predicts a much thicker flame than the SMC. The case of a backstep flame holder is more severe, although the k-epsilon model is usually trusted for such a flow. In this case, the streamwise component of mass flux plays a major role. The flame predicted by the SMC is significantly narrower and nearly separates at the reattachment point. On the other hand the k-epsilon model only predicts a mass flux collinear to the gradient of mean concentration, i.e. in the cross stream direction. The conclusion of Bailly and Garreton (1996) is that before combustion can be seriously modelled, it is essential to start with a SMC, not for the sake of better mean flow predictions, but because the full Reynolds stress tensor is needed to accurately predict the turbulent fluxes of species. The same can be said of all industrial flows involving complex physics such as two-phase flows, sediment transport, chemistry, electrical arcs .... Hence the motivations for EDF to develop SMC.

5. Perspectives

Benchmark exercises have shown that realizable SMC's, developed using appealing theoretical constraints, actually bring visible but small improvement in complex flows (Rodi *et al.* 1998). Most of the progress in recent years is due to increase of computing power (University capacities nearly outdating that of Industry), and this enabled finer grids and "low Re" versions of the models in place of "wall functions". While test cases are chosen from well documented laboratory experiments at moderate Reynolds numbers, actual industrial studies are still out of reach of such "low Re" approaches and it may be worthwhile to revisit the 20-year-old "wall functions". In the "low Re" framework, the idea of elliptic relaxation (Durbin 1993, Wizman *et al.* 1996) to represent the kinematic blocking of normal fluctuations by a solid wall is a novel approach which has shown much improvement on a variety of complex flows with impingement, separation, heat transfer, etc. A clear outcome of recent workshops (Hanjalic, Obi and Hadzic 1997), is that Large Eddy Simulation brings outstandingly superior results compared to RANSE for several bluff body flows, for example the surface mounted cube. At EDF, Rollet-Miet (1997) applied the N3S–LES code to the crossflow in a staggered tube bundle, and produced results in excellent agreement with experiments, with little effect from the specific choice of sub-grid scale model (see Fig. 1). This came as a total surprise since for many years we had tested a variety of RANSE models (RNG, SMC, realizable SMC ...) with moderate improvement. LES does seem to be the ultimate solution of turbulence modeling, but still a mesh of the order of a million nodes enables to perform LES of only a small detail of the

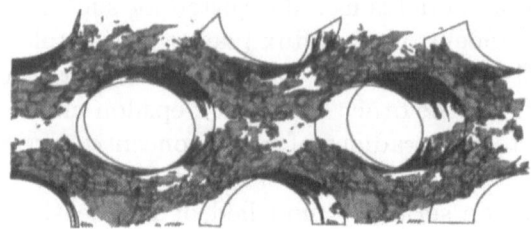

Figure 1. LES of the crossfow in a tube bundle (N3S-LES by Rollet-Miet), iso-surface of high instantaneous velocity (= bulk velocity). Fluctuations are highly three-dimensional with no sign of coherent structures. The wake clearly extends to the stagnation point of the next cylinder. Flow from right to left.

PWR presented in Section 2. A current joint CEA–EDF project is aiming at LES of the order of 50 million nodes by use of massively parallel machines. Simultaneously, the RANSE approach is continued to answer the current demand for conjugate fluid–solid problems involving complex physics.

References

Alvarez D., Martin A. and Schneider J-P., "Boron mixing in a 900Mw PWR vessel", 4th Int. Meeting on Nuclear Reactor Thermal Hydraulics, Taiwan, April 1994.

Bailly P. and Garreton D., "Experimental and numerical study of a combustion zone stabilized by a square cylinder", 26th Symp. on Combustion, The combustion Institute 1996. (Also EDF–DER report 96NB00131.)

Boudjemadi R., Maupu V., Laurence D. and Le Quéré P., "Budgets of Turbulent Stresses and Fluxes in a Vertical Slot Natural Convection Flow at Rayleigh Ra $= 10^5$ and 5.4×10^5". J. Heat & Fluid Flow, vol. 18, n. 1, pp. 70–79, Feb. 1997.

Deutsch E., Mechitoua N. and Mattei, J.D., "Flow simulations in piping system dead legs using SMC and k-epsilon models", 6th symp. on Flow Modelling and Turbulence Measurements, Florida State University, Sept. 1996.

Durbin P. A., "A Reynolds stress model for near-wall turbulence", J. Fluid Mech., vol. 249, pp. 465–498, 1993.

Hanjalic K., Obi S. and Hadzic I., 6th ERCOFTAC/IAHR/COST Workshop on Refined Flow Modelling, Delft University. Technology, Heat Transfer Section, June 1997.

Rodi W., Bonnin J.C., Buchal T. and Laurence D., "Testing of calculation methods for turbulent flows: Workshop results for 5 test cases". Collection de Notes de la DER EDF, 98NB0004, 1998.

Rollet-Miet P., "Simulation des Grandes Échelles sur maillages non-structurés pour géométries complexes". Thèse Ecole Centrale de Lyon, décembre 1997.

Wizman V., Laurence D., Durbin P., Demuren A. and Kanniche M., "Modeling near wall effects in second moment closures by elliptic relaxation", J. Heat & Fluid Flow vol. 17, pp. 255–266, 1996.

EFFECT OF PARTICLE PROPERTIES ON TURBULENCE INTENSITY OF A SUSPENSION

G. OOMS[1] AND W.M.M. SCHINKEL[2]
[1] *Technological University Delft, Laboratory for Aero- and Hydrodynamics, Rotterdamseweg 145, 2628 AL Delft, The Netherlands.*
[2] *Shell International Exploration and Production, Research and Technical Services, P.O.Box 60, 2280 AB Rijswijk, The Netherlands.*

The behaviour of particles in a fluid suspension is of great interest to many branches of technology. Therefore,in an earlier paper by Felderhof and Ooms [1] the influence of particles suspended in a fluid on the effective mass density of the suspension was studied. To simplify the calculations the equations of fluid motion were linearized and the presence of the particles was represented by a point-force coupling. It was shown how at high concentration the low-concentration, single-particle behaviour is modified by retarded hydrodynamic interactions between the particles. It was found that the effective mass density depends on the frequency.

In a subsequent paper by Felderhof and Ooms [2] this study was extended to investigate the diffusion of particles in a turbulently flowing suspension, taking into account particle inertia and friction, as well as hydrodynamic interactions. A time-dependent effective diffusion coefficient which describes the average dispersion of particles, was derived. It was found that the turbulent diffusion is strongly affected by the particle lag and by the hydrodynamic interactions. In addition there is a marked dependence on the ratio of the integral length scale of turbulence and the particle diameter.

Some typical results based on this study are given in Figs. 1 and 2. Fig.1 shows the ratio of the turbulence spectra of the fluid in the presence and in the absence of particles as function of the dimensionless wavenumber (qa) and the dimensionless frequency $(\omega a^2/\nu)$ for a ratio of the fluid density and the particle density (ρ/ρ_p) of 10^{-2}, a ratio of the integral lengtscale of turbulence and the particle diameter (Λ/a) of 10^4, a particle volume fraction (ϕ) of 10^{-1} and a turbulent Reynolds number $(\langle v_0^2 \rangle^{1/2}\Lambda/\nu)$ of 10^3. At small wavenumbers and low frequencies the particles follow the motion of the fluid and the suspension has an increased effective mass density. This

U. Frisch (ed.), Advances in Turbulence VII, 273–276.
© 1998 *Kluwer Academic Publishers.*

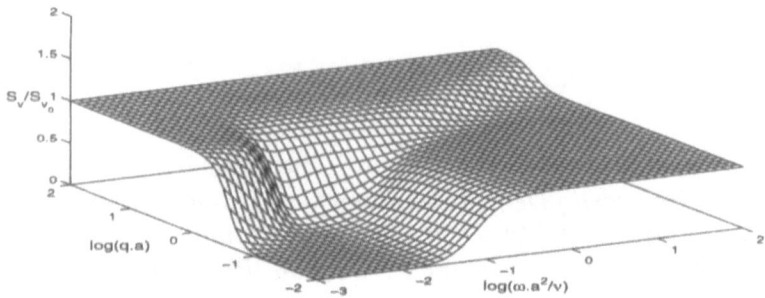

Figure 1.   Ratio of turbulence spectra of fluid in presence and absence of particles; $\rho/\rho_p = 10^{-2}$, $\Lambda/a = 10^4$, $\phi = 10^{-1}$, $\langle v_0^2 \rangle^{1/2} \Lambda/\nu = 10^3$

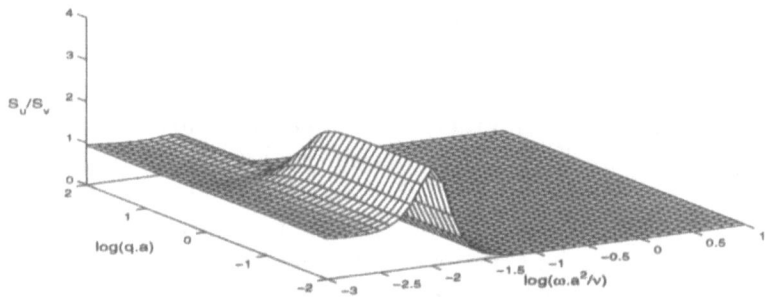

Figure 2.   Ratio of turbulence spectra of particles and of fluid in presence of particles; $\rho/\rho_p = 10^{-2}$, $\Lambda/a = 10^4$, $\phi = 10^{-1}$, $\langle v_0^2 \rangle^{1/2} \Lambda/\nu = 10^3$

strongly reduces the spectral energy of the fluid. In Fig.2 the ratio of the turbulence spectrum of the particles and the turbulence spectrum of the fluid in the presence of particles is given as function of the dimensionless wavenumber and frequency. As can be seen there is a region of wavenumbers and frequencies where the turbulence energy of the particles is considerably larger than that of the fluid. From the calculated spectra the effect of the particles on relevant quantities, such as turbulence intensity and turbulent diffusion coefficient, can be derived.

Two extensions have been made of the study of Felderhof and Ooms. The first one concerns the friction law for the particles in the flow field of the fluid. Felderhof and Ooms assumed that the friction was given by Stokes' law. This restricted the validity of their calculations to small values of the particle Reynolds number. Therefore a friction law has now been included, which is also valid at larger values of the particle Reynolds number. Results of new calculations show a significant influence of the type of friction law on the turbulence spectrum, and hence on the turbulence intensity and turbulent diffusion.

The second extension deals with the Green's function for the hydro-dynamic interaction between two particles. In the papers of Felderhof and Ooms the Green's function depends on the distance between the particles, the frequency of the flow field, the mass density of the fluid and its molecular viscosity. However, in a turbulent flow field the hydrodynamic interaction is also influenced by turbulence eddies with length scales smaller than the distance between the particles. This influence is now taken into account by replacing the molecular viscosity by the sum of the molecular viscosity and the turbulent viscosity due to eddies with a length scale smaller than the particle distance. This is descibed in detail by Ooms and Schinkel [3].

The extended calculation method has been used to study the effect of the particle properties (diameter, mass density) on the turbulence spectra of the fluid and of the particles, and hence on other quantities like turbulence intensity and turbulent particle diffusion. To that purpose the values of the two dimensionless groups (ρ/ρ_p) and (Λ/a) were varied over a wide range and the influence on the spectra calculated. (ρ/ρ_p) was varied between 10^{-3} and 10^{+3} and (Λ/a) between 10^0 and 10^{+3}. The value of (ϕ) was kept constant at 10^{-1}, the value of $\langle v_0^2 \rangle^{1/2}\Lambda/\nu$ at 10^{+4}. As the allowed number of pages for this publication is limited, only some first results (presented in Fig.3 and Fig.4) are given for the turbulence intensity of the fluid and of the particles. Fig.3 shows the ratio of the turbulence intensity of the fluid in the presence and absence of particles $(\langle v^2 \rangle/\langle v_0^2 \rangle)$ as function of the ratio of the integral length scale and the the the particle diameter (Λ/a) and as function of the ratio of the fluid density and the particle density (ρ/ρ_p). Fig.4 gives the ratio of the particle turbulence intensity and the turbulence intensity of the fluid in the absence of particles $(\langle u^2 \rangle/\langle v_0^2 \rangle)$ as function of (Λ/a) and of (ρ/ρ_p). As is clear from Fig.3 the turbulence suppression of the fluid increases with decreasing particle diameter and increasing particle mass density. So for a certain particle volume concentration small particles are much more effective than large particles in damping the fluid turbulence. Also heavy particles are more effective in this respect than light ones. Fig.4 shows that the turbulence intensity of relatively large particles with a low mass density can be considerably larger than the turbulence intensity of the fluid. Experiments are considered to check the results of this study.

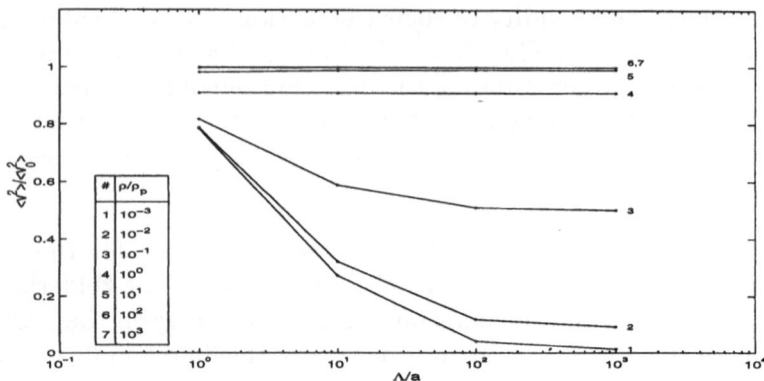

Figure 3. Ratio of turbulence intensity of fluid in the presence and absence of particles as function of (Λ/a) and (ρ/ρ_p); $\phi = 10^{-1}$, $\langle v_0^2 \rangle^{1/2} \Lambda/\nu = 10^4$

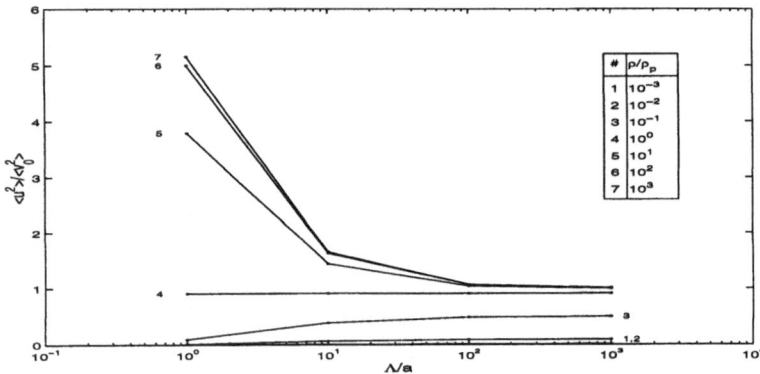

Figure 4. Ratio of turbulence intensity of particles and of fluid in the absence of particles as function of (Λ/a) and (ρ/ρ_p); $\phi = 10^{-1}$, $\langle v_0^2 \rangle^{1/2} \Lambda/\nu = 10^4$

References

1. Felderhof, B.U. & Ooms, G. (1989) Effective mass density of fluid suspensions, *Phys. Fluids* A1,pp.1091–1097.
2. Felderhof, B.U. & Ooms, G. (1990) Effect of inertia, friction and hydrodynamic interactions on turbulent diffusion, *Eur. J. Mech.* B9,pp.349–368.
3. Ooms, G. & Schinkel, W.M.M. (1998) Effect of turbulence on hydrodynamic interaction and diffusion in a suspension, *Third International Conference on Multiphase Flow, Lyon, France, June 8-12, 1998.*

ON THE TWO–WAY INTERACTION BETWEEN A TURBULENT MIXING LAYER AND DISPERSED SOLID PARTICLES

E. ORY, I. A. JOIA AND R. J. PERKINS

LMFA/ECL, 36 avenue Guy de Collongue, B.P. 163, 69131 Ecully Cedex, France

1. Introduction

Particle–laden two–phase flows can have a considerable influence on the safety or economic viability in many environmental and industrial processes. Investigations, over the past decade, provided convincing evidence that large–scale instantaneous structure of the turbulence, rather than gradient diffusion, plays an important role in determining particle dispersion (see Eaton & Fessler 1994 for a review on preferential concentration of particles by turbulence). Even when the average mass loading is low, there will be regions in the flow where the instantaneous mass loading is high enough for the particles to have a significant effect on the flow. In order to study two–way interaction between a turbulent mixing layer and dispersed solid particles we have developed an adaptive numerical scheme, based on vortex methods, which is used to integrate the vorticity–velocity formulation of the Navier–Stokes equations. Following Tang *et al.* (1990), we study, using a slightly different algorithm which conserves the total circulation, the effects of particles on the flow vorticity through the forces acting on the particles.

2. Method

A Lagrangian numerical method, based on vortex methods, has been developed for the direct numerical simulations of viscous incompressible free shear flows following Koumoutsakos' (1993) formulation. The classical vortex method (Leonard, 1980) has been enhanced to properly account for diffusion effects. The velocity–vorticity formulation helps in eliminating the pressure from the unknowns of the equations. In vortex methods, the vorticity field is considered as a discrete sum of an irrotationnal background

U. Frisch (ed.), Advances in Turbulence VII, 277–280.

velocity field and the individual vorticity fields of the vortex. Vortex methods have several inherent computational advantages, namely their adaptivity and the ability to obtain physical insight by directly dealing with the vorticity field. The Lagrangian representation of the convective terms avoids many difficulties associated with its discretization on an Eulerian mesh such as excess numerical diffusion.

Once the flow has reached a statistically stationary state, particles are released in the computational domain. Their trajectories are calculated by integrating their equation of motion, using the fluid velocities obtained from the vortex method. At very low particle concentrations the force exerted by the particle on the flow is negligible, as the particle concentration increases the forces exerted by the particle on the fluid can no longer be neglected. In such cases it is important to understand how the particle motion will modify that of the carrier fluid. Our model is similar to that of Joia et al. (1998), the detail is therefore omitted here, suffice to say that, if the particles exert a force $(-\mathbf{F}_f)$ on the flow, then this can be expressed as an additional source term for the vorticity ($\frac{D\boldsymbol{\omega}}{Dt} = -\boldsymbol{\nabla} \times \mathbf{F}_f/\rho_f$), and can therefore be treated in a way analogous to the vorticity introduced by the vortices shed at the origin of the flow.

3. Conditions

We have used the vortex method to construct a two–dimensional model of a mixing layer. The irrotational background velocity field is represented by a uniform flow of velocity $\overline{U}$ together with a semi–infinite bound vortex, with a circulation per unit length ΔU. Vortices are shed from the splitter plate at intervals of Δt, with circulation $\overline{U}\Delta U\Delta t$ and are advected at the local fluid velocity. The influence of vortices which have left the simulation domain at the downstream boundary $(x = L)$ is included through the use of a second semi–infinite bound vortex. Spherical glass particles of diameters ϕ and relaxation times τ_p were then, loaded at the origin of the mixing layer. The range of St values covered $(0.5 < St < 2.5)$ turned out to involve three different stages of particle dispersion. These simulations were performed for the same conditions as those in the experiments of Ishima et al. (1992) and Hishida et al. (1993):

Table 1. Experimental and computational conditions

$\overline{U}$ [m.s^{-1}]	ΔU [m.s^{-1}]	L [m]	ρ_f [kg.m^{-3}]	ϕ [μm]	ρ_p [kg.m^{-3}]	τ_p [s]
				42		0.014
8.5	9	0.45	1.23	72	2590	0.041
				135		0.144

For each particle type, various mass loadings are investigated. The mass loading, here, is defined as the ratio of the particle mass flux to the fluid mass flux at half distance of the simulation domain ($x = L/2$). The characteristic length scales, needed to calculate a fluid mass flux, are defined by the sizes of the large–scale structures. When studying different mass loadings, the number of particles released into the simulation is not changed. However, the number of 'real' particles which are represented by each 'computational' particle is varied. In these conditions, two–way coupling simulations have been made for mass loadings of 0.1, 0.5, 1 and 5. A grid, needed to calculate the forces exerted on the flow, is laid on the domain and contains 70×50 cells. Preliminary tests have shown that small differences on the cell sizes could change, qualitatively, the instantaneous structure of the flow, but will not change the conclusions of the results.

4. Results and Discussion

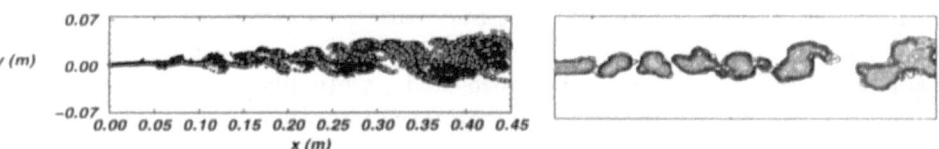

Figure 1.   Location of the vortices • and the $72\mu m$–particles o and the relative vorticity contours – mass loading $= 0$

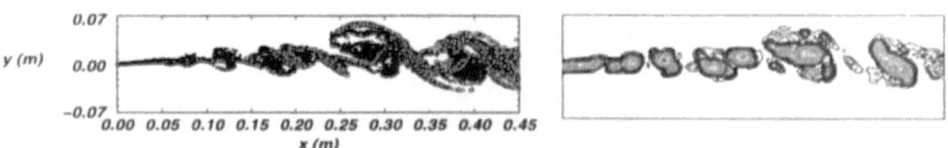

Figure 2.   Location of the vortices • and the $72\mu m$–particles o and the relative vorticity contours – mass loading $= 1$

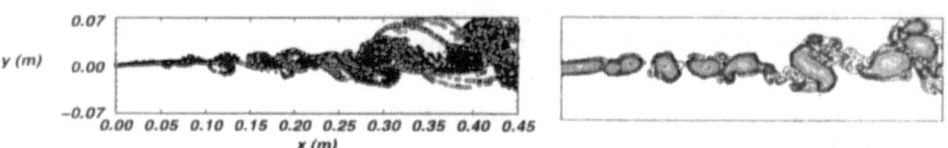

Figure 3.   Location of the vortices • and the $72\mu m$–particles o and the relative vorticity contours – mass loading $= 5$

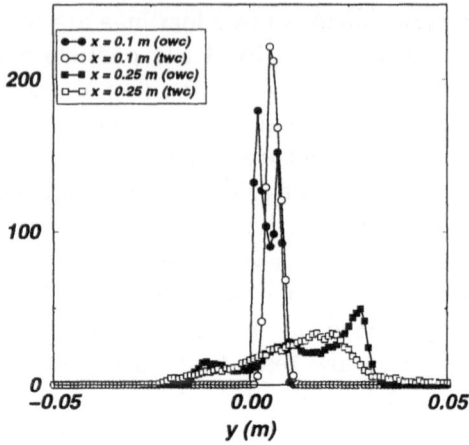

Figure 4. $72\mu m$–particle pdf – mass loading $= 0$ and 5

Fig. 1 to Fig. 3 show $72\mu m$–particle and vortex locations at the same instant with the relative vorticity contours for different mass loadings. For a mass loading different of zero the large–scale, eddy–like structures tend to be modified and we can see owing to the vorticity contours that the vorticity field become more and more continuous as the mass loading increases. Qualitatively, we can also, note that the lateral dispersion of the particles increases with the mass loading. This is confirmed by Fig. 4 which shows a comparison between the pdf of the distribution of particle number density of the $72\mu m$–particle size in a one–way coupling and a two–way coupling configurations.

References

Eaton, J. K. & Fessler, J. R. 1994. Preferential concentration of particles by turbulence. *Int. J. Multiphase Flow*, **20 - Suppl.**, 169–209.

Hishida, K., Ando, A. & Maeda, M. 1993. Experiments on particle dispersion in a turbulent mixing layer. *Int. J. Multiphase Flow*, **18 - 2**, 181–194.

Ishima, T., Hishida, K. & Maeda, M. 1992. Effect of particle residence time on particle dispersion in a plane mixing layer. *J. Fluids Eng.*, **115**, 751–759.

Joia, I. A., Ory, E. & Perkins, R. J. 1998. A Discrete Vortex Model of Particle Laden Jets. *In: 3rd International Conference on Multiphase Flow*. June 8–12, Lyon, France.

Koumoutsakos, P. 1993. *Direct numerical simulations of unsteady separated flows using vortex methods*. PhD. Thesis, California Institute of Technology.

Leonard, A. 1980. Review – Vortex methods for flow simulation. *J. Comput. Phys.*, **37**, 289–335.

Tang, L., Crowe, C. T., Chung, J. N. & Troutt, T. R. 1990. A numerical model for droplets dispersing in a developing plane shear layer including coupling effects. *Pages 27–33 of: Numerical methods for multiphase flows*, vol. 91. ASME.

EXPLICIT ALGEBRAIC MODELLING OF PASSIVE SCALAR FLUX

P. M. WIKSTRÖM[1], S. WALLIN[2] AND A. V. JOHANSSON[1]
[1] Department of Mechanics, KTH S-100 44 Stockholm Sweden.
[2] Aeronautical Research Institute of Sweden (FFA), Box 11021,
S-161 11 Bromma, Sweden

Modelling of the passive scalar flux is considered. An algebraic relation for the scalar flux in terms of mean flow quantities may be formed by applying some equilibrium condition on the transport equations for the scalar flux. This modelling approach is analogous to explicit algebraic Reynolds stress modelling (EARSM) for the Reynolds stress anisotropies. The resulting set of algebraic equations are implicit and non-linear even with linear modelling of the pressure-scalar gradient correlation. A method to solve this implicit relation in a fully explicit form is proposed where the non-linearity in the scalar production to dissipation ratio has been considered and solved. This method is analogous to that used in the EARSM by Wallin et al. [5]. The solution is exact in two-dimensional mean flows. In three-dimensional mean flows the tensorial form is exact but an approximation for the scalar production to dissipation ratio is introduced. For a special choice of two of the model parameters a drastically simplified model results. This also gives a simple formulation even for three-dimensional mean flows. Comparisons are made with DNS data of a turbulent channel flow, with a passive scalar, by Wikström et al. [7].

In nearly homogeneous steady flows the advection and diffusion of the non-dimensional scalar flux, $\xi_i \equiv \overline{u_i \theta}/\sqrt{k k_\theta}$, may be neglected, see e.g. Adumitroaie et al. [2], Abe et al. [1] and Girimaji et al. [3]. This is in many engineering flows a reasonable approximation especially if the driving forces (the velocity and scalar gradients) are large. This gives the following implicit algebraic equation system

$$\frac{1}{2}\xi_i \left(\frac{P_\theta - \varepsilon_\theta}{k_\theta} + \frac{P_k - \varepsilon}{k} \right) = \frac{P_{\theta i} - \varepsilon_{\theta i} + \Pi_{\theta i}}{\sqrt{k k_\theta}} \tag{1}$$

which together with the modelling of $-\varepsilon_{\theta i} + \Pi_{\theta i}$ can be seen as a system of equations for the non-dimensional scalar flux, ξ_i. A model form that has

U. Frisch (ed.), Advances in Turbulence VII, 281–284.

been widely used, see e.g. Shabany & Durbin [4], can be written as

$$-\varepsilon_{\theta i} + \Pi_{\theta i} = -c_{\theta 1}\frac{\varepsilon}{k}\overline{u_i\theta} + c_{\theta 2}\overline{u_j\theta}\frac{\partial U_i}{\partial x_j} + c_{\theta 3}\overline{u_j\theta}\frac{\partial U_j}{\partial x_i} + c_{\theta 4}\overline{u_iu_j}\frac{\partial\Theta}{\partial x_j} \qquad (2)$$

This is the most general form, conserving the superposition principle for passive scalars. Since the model is tensorially linear in $\overline{u_i\theta}$, also (1) is tensorially linear. The scalar flux is, however, included in the expression for the production of the scalar variance, $P_\theta \equiv -\overline{u_i\theta}\frac{\partial\Theta}{\partial x_i}$, and the algebraic equation system thus has a scalar non-linearity in $\overline{u_i\theta}$ which needs to be considered in a solution that is fully self-consistent, see Girimaji et al. [3].

Adumitroaie et al. [2] have proposed a solution method for the algebraic scalar flux equation system (1) which is quite simple and compact also for three-dimensional flows. That model is, however, not fully explicit since the production terms, P_k and P_θ, are left implicit. In this paper we will extend the method to a fully explicit model.

The formal solution of the system (1) reads

$$\xi_i = -(1 - c_{\theta 4})A_{ij}^{-1}\left(a_{jk} + \frac{2}{3}\delta_{jk}\right)\Theta_k \qquad (3)$$

and is obtained using the inverse of the matrix $\mathbf{A}$. One interesting observation is that the influences from the mean velocity gradient through $\mathbf{A}^{-1}$, the mean scalar gradient and the Reynolds stress anisotropy are tensorially separated. The formal solution in terms of the inverse of the matrix $\mathbf{A}$ which will be inverted analytically follows the solution method proposed by Adumitroaie et al. [2].

The matrix $\mathbf{A}$ is a function of the mean flow strain- and rotation-rate tensors and depend on the specific choice of the model of $-\varepsilon_{\theta i} + \Pi_{\theta i}$. $\mathbf{A}$ is also dependent on the $P_\theta/\varepsilon_\theta$ ratio and relation (3) is thus not fully explicit. A fully explicit and self-consistent solution is then obtained by solving $P_\theta/\varepsilon_\theta = z\xi_i\Theta_i$ where z is the time scale ratio, $\frac{\frac{1}{2}\overline{\theta^2}/\varepsilon_\theta}{k/\varepsilon}$, and Θ_i is a suitably normalized mean temperature gradient. The resulting equation for the $P_\theta/\varepsilon_\theta$ ratio is a fourth order polynomial equation for general three-dimensional mean flows, without a useful closed solution. For two-dimensional mean flows the equation for the $P_\theta/\varepsilon_\theta$ ratio is cubic and has a reasonably simple closed solution. The expression for the $P_\theta/\varepsilon_\theta$ ratio together with the relation (3) thus forms a fully explicit and self-consistent solution of the algebraic scalar flux equation system (1).

The explicit relation (3) is not valid if the determinant of $\mathbf{A}$ is zero. In that case no solution of the implicit relation (1) exists either and it is reasonable to suspect that there would be numerical problems associated with the solution of the corresponding transport model. The determinant can indeed be zero in some parameter regimes which have been identified. This is, however, not very likely to happen in not too extreme flow situations, but could have importance for the convergence of a numerical solution and must be treated with care.

For a special choice of two of the model parameters, $c_{\theta 2} = 1$ and $c_{\theta 3} = 0$, a drastically simplified model results. The matrix $\mathbf{A}$ is then proportional to the identity matrix and the equation for the $P_\theta / \varepsilon_\theta$ ratio is a quadratic polynomial equation, with a simple solution, see Wallin et al. [6]. This is the case in both two- and three-dimensional mean flows. The simplified model is equivalent to a modelled dispersion tensor that is proportional to the Reynolds stress tensor. This has similarities with the Daly & Harlow model for turbulent diffusion of the Reynolds stresses.

A direct numerical simulation (DNS) of a turbulent channel flow with a passive scalar has been performed using a Prandtl number of 0.71. The Reynolds number based on the centerline mean velocity is 4800 and the Reynolds number based on the wall friction velocity, u_τ, and the channel half width, δ, is 265. The simulation code uses spectral methods, with Fourier representation in the streamwise(x) and spanwise(z) directions, and Chebyshev polynomials in the wall-normal(y) direction. The computational domain is 12.56δ, 2δ, 5.5δ in the streamwise, wall-normal and spanwise directions respectively and the number of grid points are 256×193×192. This gives a resolution in, e. g., the x–direction of 13.0 wall units.

The temperature at each wall is kept constant with a higher temperature on the upper wall. This boundary condition, which represents a case where the passive scalar is introduced at the upper wall and removed from the lower wall, results in an antisymmetric mean temperature profile in the channel.

In Figs. 1–2 model predictions of the present EARFM, in a turbulent channel flow, are investigated using the DNS data. All quantities are normalized by $u_\tau \theta_\tau = \alpha \left(\frac{\partial \Theta}{\partial y} \right)_{wall}$. In Fig. 1 the simplified model is used. For this model the scalar flux ratio $\overline{u\theta}/\overline{v\theta}$ and the Reynolds stress ratio $\overline{uv}/\overline{v^2}$ becomes equal in a turbulent channel flow. According to the DNS data $\overline{uv}/\overline{v^2}$ is smaller than $\overline{u\theta}/\overline{v\theta}$. Since the model then is not able to predict both components of the flux vector simultaneously such an assumption either underpredicts the $\overline{u\theta}$ component or overpredicts the $\overline{v\theta}$ component. Still this simplified model gives a significantly more realistic description of the heat fluxes than a standard eddy-diffusivity model.

The model parameters used in Fig. 2 give a very good agreement for both components in the center of the channel, whereas $\overline{u\theta}$ is underpredicted for $y/\delta > 0.6$. The ability to obtain good agreement for $y/\delta < 0.6$ may be expected, since the equilibrium assumption is most appropriate in this region.

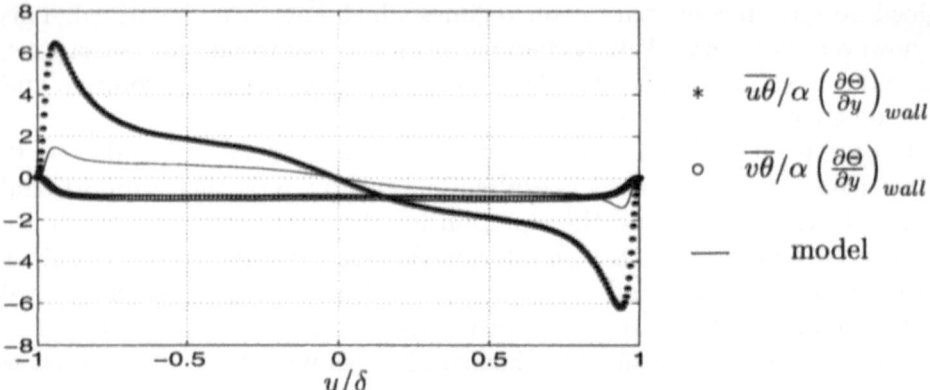

Figure 1.   Model comparison with $c_{\theta 1} = 4.0$, $c_{\theta 2} = 1$, $c_{\theta 3} = 0$ and $c_{\theta 4} = 0$.

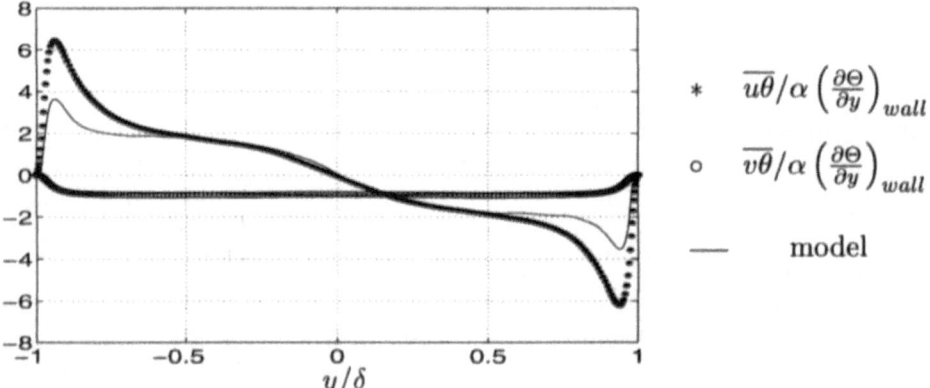

Figure 2.   Model comparison with $c_{\theta 1} = 2.5$, $c_{\theta 2} = 0$, $c_{\theta 3} = 0$ and $c_{\theta 4} = 0.35$.

References

1. Abe, K., Kondoh, T. & Nagano, Y. (1996) A two-equation heat transfer model reflecting second-moment closures for wall and free turbulent flows. *Int. J. Heat and Fluid Flow.*, 17:228–237.
2. Adumitroaie, A., Taulbee, D. B. & Givi, P. (1997) Explicit Algebraic Scalar-Flux Models for Turbulent Reacting Flows. *A. i. ch. e. J.*, 8.
3. Girimaji, S. S. & Balachandar, S. (1997) Analysis and modeling of buoyancy generated turbulence using numerical data. *NASA CR-201736, ICASE Report*, No. 97-42.
4. Shabany,Y. & Durbin, P. A. (1997) Explicit Algebraic Scalar Flux Approximation. *Int. AIAA J.*, 35:985–989.
5. Wallin, S. & Johansson, A. V. (1996) A new explicit algebraic Reynolds stress turbulence model including an improved near-wall treatment. *Flow Modeling and Turbulence Measurements VI, Tallahassee FL.*, Chen, Shih, Lienau & Kung (eds), 399–406.
6. Wallin, S. Wikström, P. M. & Johansson, A. V. (1997) Explicit algebraic modelling of passive scalar flux. FFA TN 1997-52.
7. Wikström, P. M. & Johansson, A. V. (1998) DNS and heat-flux transport modelling in a turbulent channel flow. *Sent to Turbulent Heat Transfer–2, Manchester, UK.*.

SPECTRAL MODELLING AND DIRECT NUMERICAL SIMULATION OF ELLIPTIC FLOW

O. LEUCHTER[1] AND J.P. BERTOGLIO[2]
[1] *ONERA, 92190 Meudon, France*
[2] *ECL/LMFA, 69130 Ecully, France*

1. Introduction

Elliptic flows belong to a family of plane flows with uniform mean velocity gradients, combining the effects of rotation and strain. The elliptic domain is defined by the predominance of rotational gradients, providing elliptically shaped streamlines in the distortion plane. Elliptic flows represent attractive reference flows in the context of the stability of strained vortex flows (see e.g. Ref.[1]), and in the domain of turbulence modelling, where they constitute challenging test cases for the assessment of single-point closure models assigned to predict rotation-dominated turbulent flows.

Elliptic flows have been investigated experimentally in an original facility where grid generated turbulence in solid-body rotation is subjected to plane strain by means of a specifically shaped distorting duct, Ref. [2]. The experimental setup allows one to achieve a whole period of elliptic flow, with a rotation rate twice the strain rate.

Rapid distortion theory (RDT) in the (inviscid) spectral formulation was extensively used to provide a reliable theoretical background for interpreting the pertinent features observed in the experiments, [2],[3]. The undulatory nature of the variation of the anisotropy components and the rapid pressure terms have been clearly identified as key features of the linear rotation effects in elliptic flows.

The objective of the present work is to investigate elliptic flows by comparing a **nonlinear spectral approach** using the EDQNM closure and direct numerical simulations (DNS). Both approaches are aimed at providing new data sets for elliptic flows, with the long-term objective to develop Reynolds stress models (RSM) for flows with dominant rotation.

U. Frisch (ed.), Advances in Turbulence VII, 285–288.

2. Flow Characteristics

The flow under consideration is generated by superimposing plane strain (with rate D) and solid-body rotation (with rate $\Omega > D$), according to the velocity gradient matrix:

$$\left(\frac{\partial U_i}{\partial x_j}\right) = \begin{pmatrix} 0 & 0 & 0 \\ 0 & 0 & D-\Omega \\ 0 & D+\Omega & 0 \end{pmatrix} \tag{1}$$

The principal directions of strain are inclined at ± 45 deg, so that for the limiting case of pure shear ($\Omega = D$), the shear rate reduces to $\partial U_3/\partial x_2$. The basic parameter of the flow is the ratio $\omega = \Omega/D$ (> 1 for elliptic flow) which is related to the aspect ratio A of the elliptic streamlines by $A = \sqrt{(\omega + 1)/(\omega - 1)}$. One single period of elliptic flow, $T = 2\pi/\sqrt{\Omega^2 - D^2}$, is considered with $\omega = 2$ (or $A = \sqrt{3}$), in accordance with the experimental values.

The main parameters of the flow with regard to turbulence are the normalized strain rate $S_0 = Dq_0^2/2\epsilon_0$ and the Taylor micro-scale Reynolds number $Re_\lambda = \sqrt{5/3}\, q_0^2/\sqrt{\nu\epsilon_0}$. The quantities q_0^2 and ϵ_0 stand for (twice) the kinetic energy and the dissipation rate, respectively, at $t = 0$. The initial state of the turbulence is assumed to be isotropic. The parameter S_0 represents the relative strength of the distortion ($S_0 = \infty$ corresponds to the RDT assumption). The following values have been retained in the calculations: $S_0 = 0.5$, 1 and 2, and $Re_\lambda = 14$, 26 and 41. The highest Reynolds number corresponds to the experiment, the lowest and the medium values represent the maximum Reynolds numbers considered for the 64^3 and 128^3 simulations, respectively.

3. Numerical Description of the Flow

The nonlinear spectral approach consists of numerically solving the non-linear Craya equation for the spectral tensor Φ_{ij}, defined as the Fourier transform of the two-point velocity correlation tensor in physical space, see Ref.[4]. The closure approach is of the EDQNM-1 type, in the sense that the effects of the mean velocity gradients are not explicitly accounted for in the spectral equations for the third-order moments. The inhibiting effect of rotation on the spectral transfer, however, is taken into account by introducing a heuristic modification of the usual (isotropic) expression for the eddy-damping term, see Ref. [5].

The DNS calculations were made with a pseudo-spectral method using distorted Fourier modes and periodic boundary conditions. They were done, like in Ref.[6], for significantly lower Reynolds numbers than those corresponding to the experiments (see above). The initial velocity field for the

DNS is generated with a preliminary calculation of isotropic decay starting from a 3D random field. The calculation is pursued until the velocity-gradient skewness reaches its asymptotic level. More details about the DNS are given in Ref. [7].

4. Results

Figure 1 shows for the nominal conditions of the experiments ($Re_\lambda = 41$, $S_0 = 0.5$), the evolution of the normalized turbulent energy and the second invariant II of the anisotropy tensor $b_{ij} = \overline{u_i u_j}/q^2 - 1/3$. Both quantities are plotted against the normalized time t/T, where T is the period of the elliptic flow. The results for the rotation-free reference case of pure

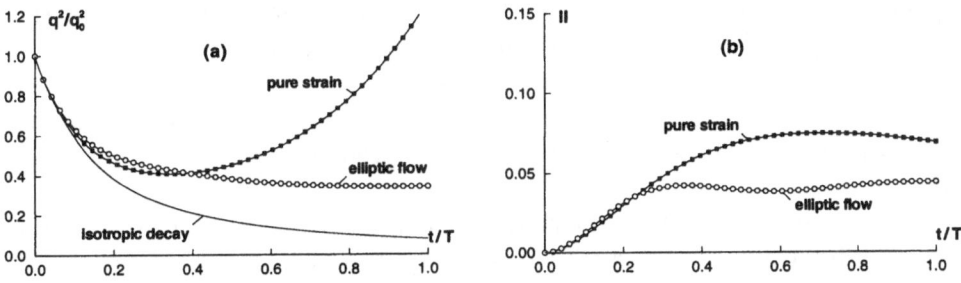

Figure 1. EDQNM results ($Re_\lambda = 41$, $S_0 = 0.5$): (a) turbulent energy, (b) anisotropy

strain are shown for comparison (as well as those for isotropic decay). The figures show a significant reduction of the energy and the anisotropy for elliptic flow, a consequence of the linear rotation effects acting through the production and the rapid pressure terms.

Standard RSM would predict smaller reductions than those shown in Fig.1, due to the inadequacy of the models regarding the rapid pressure terms in rotation-dominated flows. In particular, the blocking effect of rotation on the anisotropy level and its wavy evolution which are typical features of elliptic flows (see below) can not be reproduced by RSM.

A few DNS calculations were made to compare with the EDQNM calculations, starting from the same initial conditions. A comparison for isotropic decay at $Re_\lambda = 14$ (not shown here) revealed satisfactory agreement for the evolution of the turbulent energy and the dissipation rate. Figure 2 shows a comparison between EDQNM and DNS results for elliptic flow at $Re_\lambda = 26$ and $S_0 = 0.5$. The agreement is somewhat less perfect than one would expect, in the sense that the DNS predicts an inhibition of the energy decay in the second half of the period, whereas the energy predicted by EDQNM continues to decay. Moreover, the DNS seems to develop additional undu-

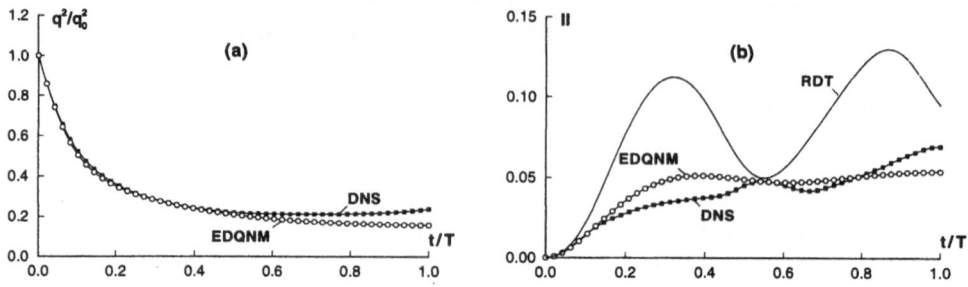

Figure 2. Comparison between EDQNM and DNS ($Re_\lambda = 26$, $S_0 = 0.5$): (a) turbulent energy, (b) anisotropy

lations of the anisotropy, the origin of which is not well elucidated. This is probably due to the occurence of instabilities of Fourier modes in the low-wavenumber range, in relation with the intrinsically unstable character of elliptic flows. This difficulty was also pointed out in Ref.[6]. Further effort in the context of the present work will deal with the identification of the origin of the differences between the two predictive methods for elliptic flow.

The limiting case of RDT is shown for comparison in Fig.2b. It exhibits strong undulatory variations of growing amplitude which appear as specific features of elliptic flows. Both, the EDQNM and DNS results clearly show, in reasonable qualitative agreement, the strong damping of the undulations by the nonlinear mechanisms neglected in the RDT.

References

1. Waleffe, F. (1990) The three-dimeensional instability of strained vortices, *Physics of Fluids A*, vol. **2**, pp. 76–80.
2. Leuchter, O., Benoit, J.P. & Cambon, C. (1992) Homogeneous turbulence subjected to rotation-dominated plane distortions, *Advances in Turbulence IV*, ed. F.T.M. Nieuwstadt, pp. 197–202, Kluwer Academic Publishers.
3. Leuchter, O. (1993) Effects of rotation-dominated distortions on homogeneous turbulence, *Proceedings of the 5th International Symposium on Refined Flow Modelling and turbulence Measurements*, Paris, pp. 97–105.
4. Bertoglio, J.P. (1981) A model of three-dimensional transfer in non-isotropic homogeneous turbulence, *Third Symposium on Turbulent Shear Flows*, Davies.
5. Cambon, C., Bertoglio, J.P. & Jeandel, D. (1982) Spectral closures for homogeneous turbulence, *Proceedings of the 1980-1981 AFOSR-HTTM-Stanford Conference on Complex Turbulent Flows*, eds. S.J.Kline, B.J. Cantwell & G.M. Lilley, vol. III, pp. 1307–1311.
6. Blaisdell, G. A. & Shariff, K. (1996) Simulation and modeling of the elliptic streamline flow, *Proceedings of the Summer Program 1996, Center for Turbulence Research*, Stanford University, pp. 433–446.
7. Leuchter, O. & Cambon, C. (1997) EDQNM and DNS predictions of rotation effects in strained axisymmetric turbulence, *Proceedings of the 11th Symposium on Turbulent Shear Flows*, Grenoble.

VORTEX CORE PRECESSION IN A GAS CYCLONE

A.J. HOEKSTRA, E. VAN VLIET, J.J. DERKSEN

AND H.E.A. VAN DEN AKKER

Kramers Laboratorium voor Fysische Technologie,
Delft University of Technology,
Prins Bernhardlaan 6, 2628 BW Delft, the Netherlands

1. Introduction

A prominent industrial application of confined vortex flow is the gas cyclone separator. In the vortex core, the flow is dominated by coherent low frequency oscillations, which have also been observed in vortex breakdown cases [1, 2]. This precession of the vortex core (PVC) tends to affect velocity measurements obtained with experimental one-point techniques like laser-Doppler anemometry (LDA) and contributes to the measured Reynolds stresses as pseudo-turbulence. In this way, a proper evaluation of the performance of Reynolds stress transport models is complicated.

In this paper, we present some measurement results and propose a procedure to estimate velocity profiles, which are resolved for the precessing motion of the vortex core.

2. Velocity measurements

The experimental facility consisted of a gas cyclone separator of industrial dimensions (diameter cyclone chamber $D_c = 0.29m$, total height $L = 1.2m$) and a two-component Laser-Doppler anemometer. Swirl is generated by a single tangential inlet, and the fluid is discharged through a concentric exit pipe or vortex finder. Strongly swirling flow is usually characterised by both the Reynolds number, Re, and a swirl parameter, Ω, which is some measure of the ratio of angular to axial momentum. For gas cyclones, Re is usually defined by the inlet velocity (U_{in}) and the inlet width (t), while Ω is a geometric swirl number, proportional to the vortex finder diameter (D_e).

U. Frisch (ed.), Advances in Turbulence VII, 289–292.
© *1998 Kluwer Academic Publishers.*

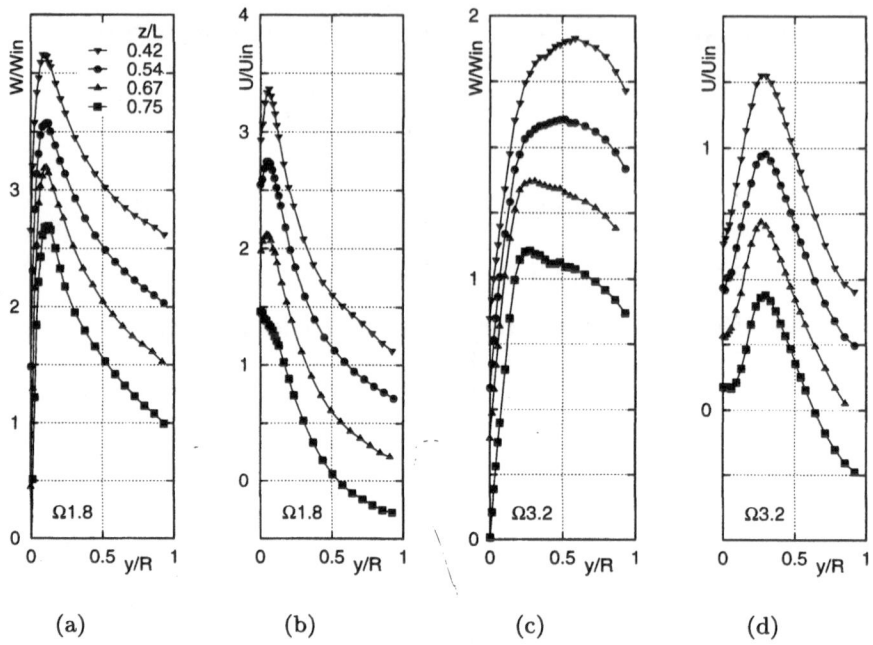

(a)   (b)   (c)   (d)

Figure 1. Mean normalised tangential and axial velocity profiles in the separation section of a gas cyclone for two different geometric swirl numbers, where for (a,b) $\Omega = 1.8$ and (c,d) $\Omega = 3.2$. Four measurement stations are presented (z/L); for each curve the origin is shifted upwards by (a,b) 0.5 and (c,d) 0.25 velocity units, respectively.

In figure 1, mean velocity measurements in the separation section of the cyclone are presented for $Re \doteq U_{in}t/\nu = 2.5\,10^4$ and $\Omega \doteq \pi D_e D_c / 4 A_{in} = 1.8$ and 3.2. Both the axial and swirl profiles exhibit a strong effect of the vortex finder diameter on the flow, as was also observed by Escudier *et al.* [3]. From the shape of the velocity profiles the flow can be separated into three regions, viz. a core region with a forced-vortex type of swirl distribution, an up-flow annular region, and a down-flow annular or wall region. For the annular regions the swirl distribution is free-vortex like. The size of the core region strongly depends on Ω. For $\Omega = 1.8$, the data from various axial locations exhibit a strong similarity, whereas for $\Omega = 3.2$ the vortex core is widening near the vortex finder.

In the vortex core the axial velocity exhibits a minimum at the axis, which results in a reverse flow region for $\Omega = 3.2$. This can be explained by the presence of an adverse pressure gradient at the centreline, owing to a distinct swirl decay in downstream direction as a result of exit pipe wall friction [4]. The core region is dominated by large velocity gradients in the radial direction.

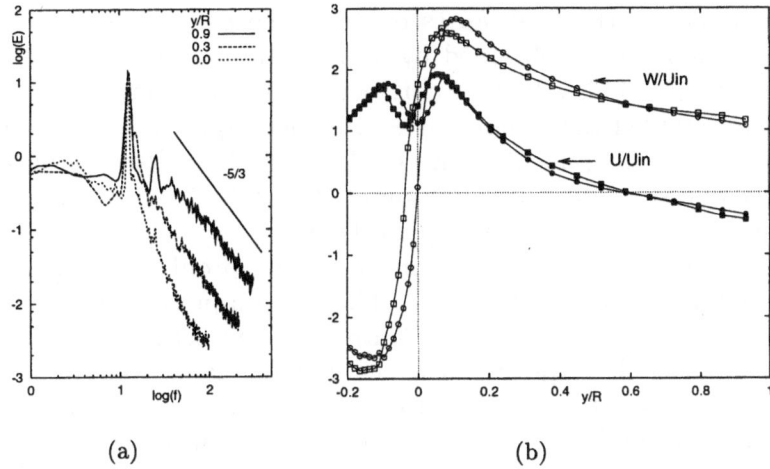

Figure 2. At measurement station $z/L = 0.42$ for $\Omega = 1.8$ and $Re = 2.5\,10^4$: (a) Spectral estimates of the tangential velocity fluctuations at the wall, annulus, and core region. (b) Phase resolved mean tangential and axial velocity profile for $\Omega = 1.8$ obtained with adaptive noise cancelling, where ($\square$) and ($\bigcirc$) are in opposite PVC phase.

3. Vortex core precession

The presence of a precessing vortex core is revealed by a frequency analysis of the LDA time series. In figure 2a, power spectral estimates are presented for the tangential velocity fluctuations measured in the wall, annular and core region. In all spectra, a distinct low-frequency peak is observed, which can be attributed to the precessing motion of the vortex core. In the core region, a low-frequency motion prevails, while in the outer regions the fluctuations include turbulent, high-frequency motions. This has also been observed by Kitoh [4] for swirling pipe flow. The power spectra of the axial velocity fluctuations reflect similar characteristics.

Although the PVC frequency peak is present in all spectra, its energy content does change with radial position at a given axial station. An explanation could be found in the LDA reference frame, which does not coincide with the rotating vortex centre. Provided that the PVC moves in a circular orbit about the geometric centreline, the measured velocity component can be shown to depend on the local mean velocity gradient in the measurement volume and on the orientation of the measurement volume with respect to the rotating vortex origin. In this way, extrema in the mean velocity profile are smoothed by spatial averaging, in particular near the vortex centre.

To determine conditional or phase resolved mean velocity profiles, a trigger method is required, which is complicated by the quasi-periodic character of the PVC. Provided that the PVC is uncorrelated with the turbulent time scales of the flow, an estimate of the periodical interference in the LDA signal can be obtained by applying an adaptive signal processing technique [5]. This adaptive noise cancelling configuration applies the delayed, re-sampled LDA signal as a reference input to form an estimate of the periodical component owing to the PVC. The filter coefficients are iteratively adjusted during evaluation of the LDA signal by use of a least mean squares stochastic gradient algorithm.

An example of the conditional mean velocity profiles obtained with the adaptive filter is shown in figure 2b. The profiles presented correspond to the extreme positions of the vortex centre with respect to the measurement volume. Only at these positions is the genuine tangential velocity component measured, without any contribution of the radial component. The forced vortex region is hardly affected by the PVC, whereas the annular region exhibits a strong effect of the wall boundary on the tangential velocity distribution. The region of axial velocity deficit coincides with the size of the vortex core, and its minimum with the vortex centre (see also [6]).

4. Conclusion

We reported on the flow field in a gas cyclone and its dependency on geometry. It was found that adaptive noise filtering on LDA data is capable of recovering details of the flow field that were obscured by precession of the vortex core.

Acknowledgements

The authors gratefully acknowledge the support by the Netherlands Foundation for Chemical Research (SON) and the financial aid from the Netherlands Organisation for Scientific Research (NWO).

References

1. Chanaud, R.C. (1965) Observations of oscillatory motion in certain swirling flows, *J. Fluid Mech.* **21**, pp. 111–127.
2. Garg, A.K. and Leibovich, S. (1979) Spectral characteristics of vortex breakdown flow fields, *Phys. Fluids* **22** (11), pp. 2053–2064.
3. Escudier, M.P., Bornstein, J. and Zehnder, N. (1980) Observations and LDA measurements of confined turbulent vortex flow, *J. Fluid Mech.* **98**, pp. 49–63.
4. Kitoh, O. (1991) Experimental study of turbulent swirling flow in a straight pipe, *J. Fluid Mech.* **225**, pp. 445–479.
5. Widrow, B. and Stearns, S. (1985) *Adaptive signal processing*, Prentice-Hall, New Jersey.
6. Yazdabadi, P.A., Griffiths, A.J. and Syred, N. (1994) 0 of the PVC phenomena in the exhaust of a cyclone dust separator, *Exp. Fluids* **17**, pp. 84–95.

SUPPRESSION OF COMBUSTION INSTABILITIES BY ACOUSTIC CONTROL OF SHEAR LAYER PROPERTIES

C. O. PASCHEREIT[1], E. GUTMARK[2] AND W. WEISENSTEIN[1]

[1] *ABB Corporate Research Ltd, 5405 Baden, Switzerland.*
[2] *Louisiana State University, Mech. Eng. Department, Baton Rouge, LA 70803-6413, USA*

Introduction

Flow instabilities and related large scale structures play an important role in combustion and heat release processes by influencing the mixing between fuel and air in diffusion flame configurations and the mixing between the fresh fuel/air mixture and hot combustion products and fresh air in premixed combustors. The temporal and spatial evolution of the instability waves in nonreacting flows was extensively studied in mixing layers [1, 2], jets [3, 4] and flows over backward facing steps [5]. However, the evolution of flow instability waves in more complex flows such as swirling flows and their interaction with combustion is not as well understood. Unlike large scale structures in nonswirling flows which are predominantly axisymmetric, swirl enhances the role of azimuthal unstable modes. Interaction between large scale structures related to flow instabilities, acoustic resonant modes in the combustion chamber and the heat release process can potentially give rise to undesired thermo-acoustic instabilities in the combustor. The effect of swirl on the longitudinal and azimuthal instability modes and the way it modifies the combustion process leading to thermo-acoustic instabilities is not well understood and requires further investigation. To control swirl induced instability, and to enhance combustion performance, controlling the large scale vortices is of critical importance [6, 7]. However, flow control has been demonstrated primarily for non-swirling flows, where the large-scale instabilities are well understood, and the coherence of the vortices can be enhanced by flow excitation. To attempt to do the same in swirling flows, the coherence of large scale structures in this type of flow have to be established, and the effect of forcing has to be explored.

293

U. Frisch (ed.), Advances in Turbulence VII, 293–296.

In the present work, instability modes were investigated and acoustically controlled in an experimental low-emission swirl stabilized burner.

Experimental Setup

The structure of the instability modes was studied in a water tunnel as well as in a combustor test-rig [7]. The combustion chamber consists of an air cooled double wall quartz glass to provide full visual access to the flame. The exhaust system is an externally air cooled tube with the same cross section as the combustion chamber to avoid acoustic reflections at area discontinuities. The acoustic boundary conditions of the exhaust system could be adjusted from almost anechoic (reflection coefficient $|r| < 0.15$) to open end reflection. The modal structure of the flow instabilities and its interaction with the acoustic modes was determined from pressure fluctuations and OH-chemiluminescence radiation. Pressure fluctuations were measured using B&K water-cooled microphones. Time varying heat release was recorded with two filtered fiber optic probes to detect OH radiation. The probes were distributed in azimuthal as well as in axial direction.

Results and Discussion

Several modes of instability where forced to occur by adjusting the acoustic boundary conditions at the combustor exit (Fig. 1). The instability associated with the premixed burner operation were at an axisymmetric mode at a normalized frequency $St = fD/\overline{U} = 0.58$ and a helical mode at $St = 1.16$, where f is the instability frequency, D the burner diameter and $\overline{U}$ the burner exit velocity. The $St = 0.58$ instability was the predominant mode at 29 dB above the background noise level. An additional helical mode was observed for non-premixed operation at a frequency of $St = 2.0$. All the instability modes were related to combustion within large scale structures which were excited in the combustion chamber due to interaction between various flow instabilities and the acoustic resonant modes of the combustor as determined by the combustion chamber acoustic boundary conditions.

The flow structure downstream of the burner was assessed in a water tunnel facility. The normalized axial velocity and the RMS values of the turbulent fluctuations were measured in a plane perpendicular to the burner axis (Fig 2). The negative mean velocity region near the burner axis shows the recirculation region which provides one of the mechanisms for flame stabilization. The flow instability associated with this region was predominantly helical and resembled a wake instability. This instability had a Strouhal number $St = 1.2$ equivalent to the helical mode observed in premixed combustion operation (Fig 3). The other region receptive to flow instabilities was the sudden expansion area at the burner exit plane. A

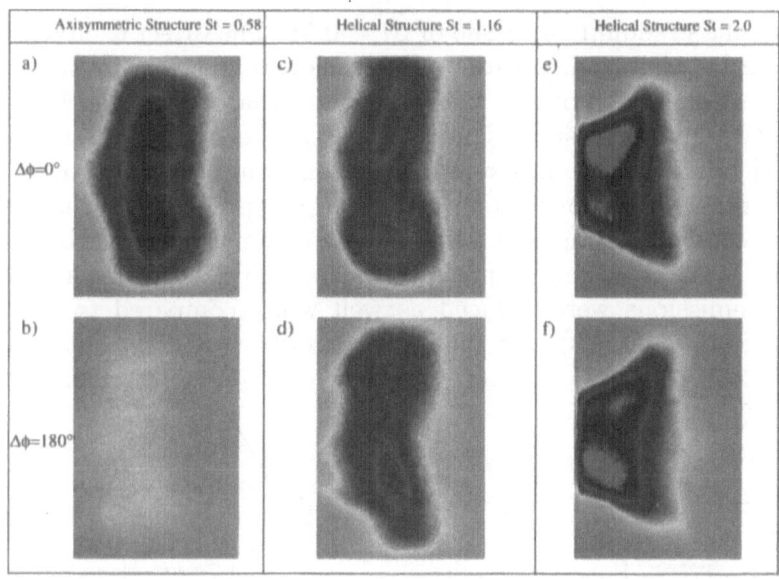

Figure 1. Visualization of phase averaged OH-images at two phase angles of 0 and 180 degrees. **(a,b)** Axisymmetric structure (premixed, $St = 0.58$). **(c,d)** Helical structure (premixed, $St = 1.16$). **(e,f)** Helical structure (diffusion flame, $St = 2.0$). The flow direction is from left to right.

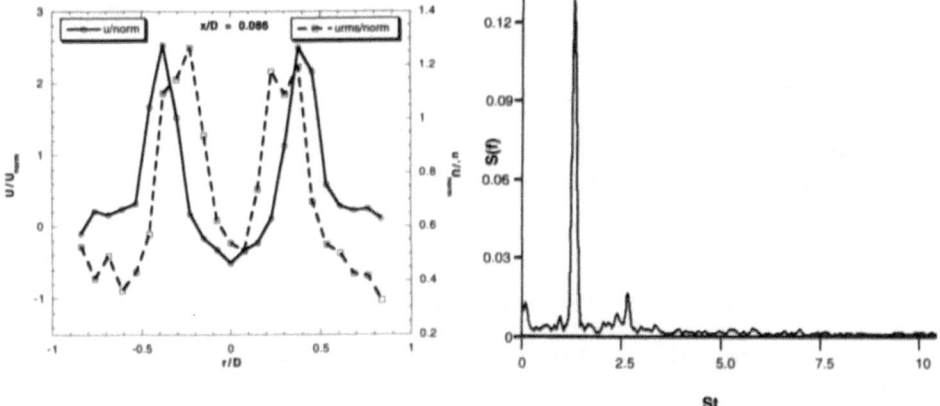

Figure 2. Axial mean and *RMS* velocity profile in a plane parallel to the burner axis, $x/D = 0.086$.

Figure 3. Helical instability in the shear layer of the recirculation region near the burner axis, r/D=-0.15, $x/D = 0.086$.

closed-loop feedback control system using upstream loudspeakers for shear layer excitation and water-cooled microphones to monitor pressure fluctuations was designed to reduce the coherence of the large scale structures thus to suppress the level of pressure oscillations due to combustion instabilities at the predominant mode of $St = 0.58$. The direct excitation of the shear layer benefits from natural amplification of flow instabilities and thus requires less energy to obtain the same effect than noise cancellation. The phase between the pressure signal and the speakers driving signal was varied and the effect on the pressure fluctuations amplitude recorded (Fig. 4). This controlled behavior is compared with the pressure fluctuations level when the controller is not operated. At the optimal phase angle of 270 degrees, the instability was suppressed by nearly 5 dB. For optimal control, the NOx emissions were reduced as well when compared to the uncontrolled situation (Fig. 5). At the optimal control conditions it was shown

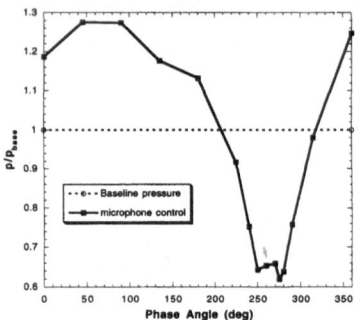 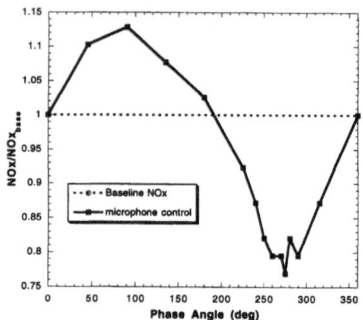

Figure 4.   Amplitude if the pressure fluctuations at the instability frequency $St = 0.58$ for various phase shifts in a closed-loop control system.

Figure 5. NOx emissions for various phase shifts in a closed-loop control system.

that the major effect of the control system was to reduce the coherence of the vortical structures which gave rise to the thermoacoustic instability.

References

1.   D. Oster and I. Wygnanski. The forced mixing layer between parallel streams. *Journal of Fluid Mechanics*, **123**:91–130, 1982.
2.   C. Ho and Patrick Huerre. Perturbed free shear layers. *Ann. Rev. Fluid Mech.*, **16**:365–424, 1984.
3.   S.C. Crow and F.H. Champagne. Orderly structure in jet turbulence. *Journal of Fluid Mechanics*, **48**:567, 1971.
4.   C. O. Paschereit, I. Wygnanski, and H. E. Fiedler. Experimental investigation of subharmonic resonance in an axisymmetric jet. *Journal of Fluid Mechanics*, **283**:365–407, 1995.
5.   M. A. Z. Hasan. The flow over a backward facing step under controlled perturbation: Laminar separation. *Journal of Fluid Mechanics*, **238**:73–96, 1992.
6.   E. Gutmark, T. P. Parr, K. J. Wilson, D. M. Hanson-Parr, and K. C. Schadow. Use of chemiluminescence and neural networks in active combustion control. In *Twenty third Symp. (Intl.) on Comb.*, pages 1101–1106. The Comb. Inst., 1990.
7.   C. O. Paschereit, E. Gutmark, and W. Weisenstein. Structure and control of thermoacoustic instabilities in a gas-turbine combustor. In *36th AIAA Aerospace Science Meeting and Exhibit*, Reno, Nevada, January 12-15 1998.

COMPRESSIBILITY EFFECTS AND SUCCESSIVE STAGES OF TRANSITION TO TURBULENCE IN THE TRANSONIC FLOW AROUND A WING BY DNS

A. BOUHADJI AND M. BRAZA

UMR CNRS-INPT 5502 , Institut de Mécanique des Fluides de Toulouse, Av. Camille Soula, 31400 Toulouse Cedex, France

In this paper, the study of successive changes of the unsteady transonic flow around a wing in transition to turbulence is performed by Direct Navier-Stokes Simulation in two-dimensional and three-dimensional approach. The flow configuration is a NACA0012 section, constant along the spanwise direction, at zero incidence. A detailed presentation of the methodology used may be found in ref. [1]

The behaviour of the flow at Mach=0.2 and Re=10^4 has been already studied by Mehta [2], Steger [3] and Rhie [4]. A symmetrical separation zone has been observed in these studies. We perform this test case and we find the same position of the separation point ($x/c \simeq 0.8$). Indeed, more than 6.10^5 time iterations had been done by our study and no noticeable unsteadiness has been detected, on the contrary, a perfectly steady state persist and two symmetrical recirculation zones are present downstream of the airfoil. The lift coefficient is almost nil (6.10^{-4}) and the drag coefficient is equal to 0.03792. By increasing the Mach number to 0.3, a very weak oscillation of the lift coefficient is perceptible, but the flow stay almost steady. Beyond M=0.35, the dynamic system reaches a symmetry breaking state under the action of truncation and round-off errors, because the externally supplied energy ($Re = 10^4$) is sufficiently high. It displays the onset of a von Kármán instability due to the two symmetrical located inflexion points in the near wake mean velocity profile. This mode (that we call mode I), exhibits an alternating vortex pattern which seems to persist until M=0.9. For higher Mach numbers, M=0.75 to M=0.8, a low frequency mode II appears. The visualisations of the iso-Mach number contours within one period, show the formation of weakly supersonic alternating zones at the lower and upper side of the profil. This phenomenon corresponds to a cycle of buffeting. These two kind of processes interact non-linearly and provide a lift coefficient oscillating in respect to both frequencies and having

U. Frisch (ed.), Advances in Turbulence VII, 297–300.

considerably increased its amplitude (fig.1 c-d). The low frequency mode II is attenuated in Mach range (0.8-0.85) and it is found to disappear at Mach=0.85, where only mode I persists. At this Mach number the local supersonic zones are more stonger due to the curvature of the profil accelerating the flow, these zones are terminated by a shock wave in the two sides of the airfoil and are enslaved only by the von Kàrmàn instability. With increasing the Mach number from 0.85 to 0.9, the unsteady behaviour still persists, but the amplitudes of the lift coefficient become weaker (fig.1 e). This corresponds to the displacement of the shock wave dowstream to the trailing edge. Due to the hyperbolic character of the flow, it seems that the shock wave prevents the perturbation from travelling upstream. This is verified at M=0.95 and M=0.98, where a strong shock appears in the near wake. Upstream, the flow is perfectly steady (fig.2 g-h).

The 3D flow at Reynolds number 7000 and Mach number 0.85, show the appearance of significant 3D effects. These effects are in association with the clear development of mode I, which seems to persist in the 3D flow. Figure 3 shows the iso-contours of the longitudinal vorticity component ω_y in the (x,z) plan at $y/c=0$. The most dark kernels correspond to counterclockwise filaments and the less dark ones to clockwise vortex structures. The wavelength of each kind of streamwise vortices is found to be $\lambda/c=0.45$ This pattern is found to persist at a farther downstream section, where the vortices are enlarged under the diffusion effect. The spanwise evolution of the longitudinal vorticity component ω_x in (y,z) plan shows also the organization of the flow pattern according to contra-rotating vortex kernels as it is discussed for the previous figure. The time-dependent evolution of w component at selected spanwise positions (fig.4) allows assessment of the growth of the 3D secondary instability through the regularly oscillatory fluctuating behavior of this velocity component. This is a main effect in interaction with the primary instability (mode I) developped downstream the shock wave, as discussed in the 2D case. It is also obtained that the mean values of the different physical quantities for the 3D case are lower than the ones of the 2D case. Indeed, the frequency of the 3D oscillations is found to be lower than the 2D case. This frequency decrease in the 3D case may be attributed to a decrease of the energy amount devoted to sustain the alternating pattern, on the profit of the simultaneous development of the three-dimensional motion.

References

1.   Bouhadji, A., Braza, M., (1997) 3D direct simulation of the shock-boundary layer interaction and Von-Kármán instability in transonic flow around a wing *Proceedings, 11th symposium on turbulent shear flows*, Grenoble, September 8-11.
2.   Mehta U.B., (1977) Dynamic Stall of an Oscillating Airfoil *Proceedings of AGARD*

Conference on Unsteady Aerodynamics
3. Steger, J.L., (1978) Implicit Finite Difference Simulation of Flow about Arbitrary Two-Dimensional Geometries, *AIAA Journal*, Vol.**16**, pp. 679-686.
4. Rhie, C.M., (1981) A Numerical Study of The Flow Past An Isolated Airfoil with Separation, *PhD Thesis*, University of Illinois

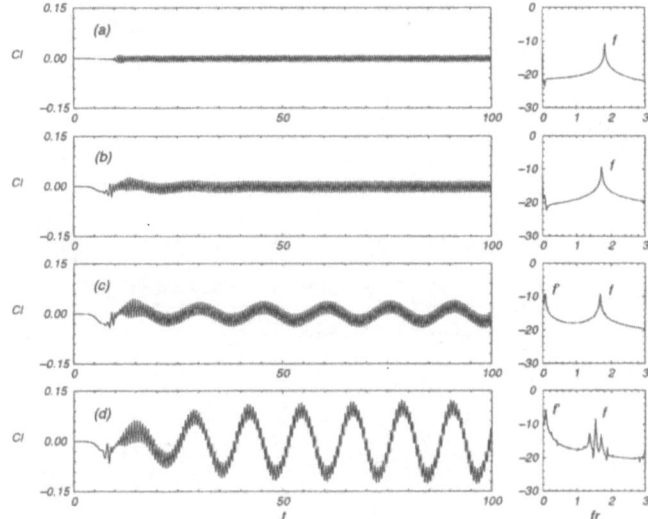

Figure 1.   Time dependent evolution of the lift coefficient C_l for the Mach numbers (a)-M=0.5, (b)-M=0.7, (c)-M=0.75, (d)-M=0.8

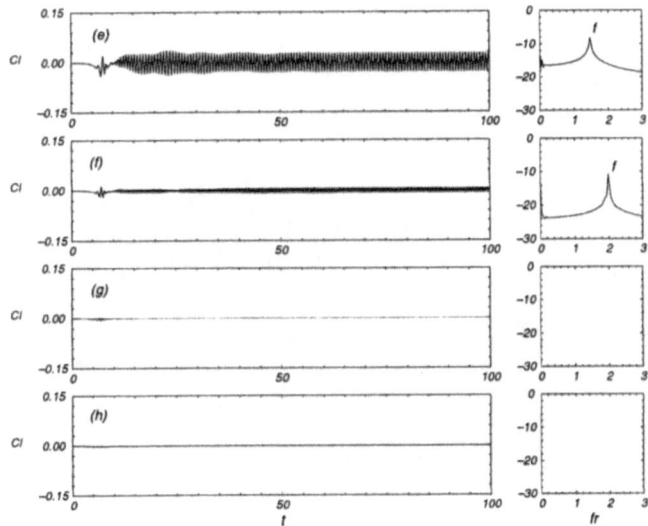

Figure 2.   Time dependent evolution of the lift coefficient C_l for the Mach numbers (e)-M=0.85, (f)-M=0.9, (g)-M=0.95, (h)-M=0.98

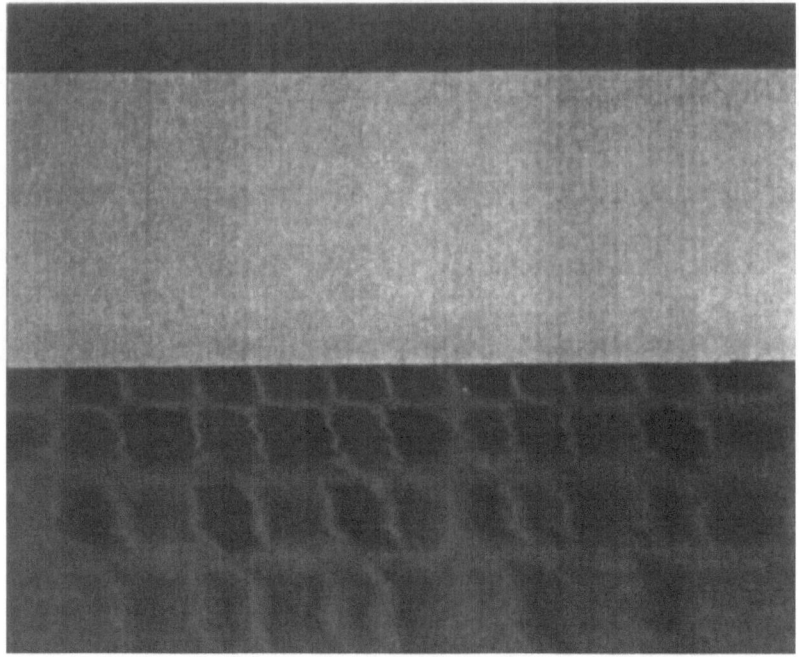

Figure 3.   Longitudinal vorticity ω_y component contours in (x, z) plan at $y/c = 0$

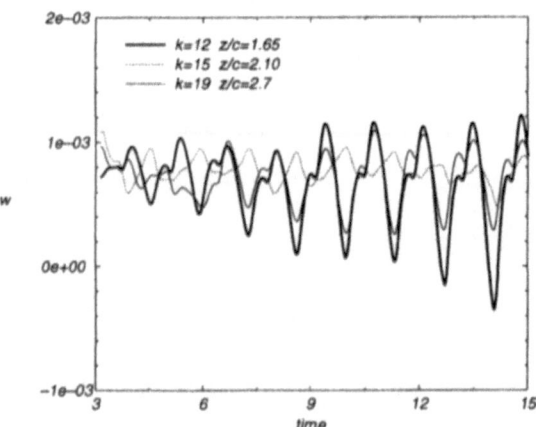

Figure 4.   Time-dependent evolution of w component velocity at $x/c=1.24$ $y/c=0.054$ in three spanwise sections

SECONDARY FLOW IN ROTATING TURBULENT PLANE COUETTE FLOW: DIRECT SIMULATION AND SECOND - MOMENT MODELLING

H.I. ANDERSSON, B.A. PETTERSSON REIF & K.H. BECH
Division of Applied Mechanics
Norwegian University of Science and Technology
N-7034 Trondheim, Norway

Abstract. The purpose of this study was twofold. First, we compiled turbulence statistics from a direct numerical simulation of plane Couette flow subjected to moderate anticyclonic system rotation. Thereafter, the resulting three-componental mean flow was mimicked by solving the Reynolds-averaged Navier-Stokes equations in combination with a second-moment closure in which near-wall effects were taken into account by elliptic relaxation. The model predictions reproduced not only the rotational-induced roll-cell pattern but also the associated damping of the turbulence.

1. Background

When *laminar* Poiseuille and Couette flow in plane channels are rotated about a spanwise axis, i.e. in an orthogonal mode, the flow may be susceptible to a roll-cell instability analogous to the centrifugal instability due to streamline curvature. The existence of an array of regularly spaced pairs of counter-rotating vortices or roll-cells aligned with the primary flow even in the *turbulent* flow regime was first observed in Poiseuille flow by Johnston *et al.* (1972) and in Couette flow by Bech & Andersson (1996 a,b).

The tendency of such secondary flows to develop in rotating channel flows makes it important to distinguish between these organized large-scale flow structures and the real turbulence. At the outset conventional one-point turbulence models were not intended to account for persistent flow structures embedded in the instantaneous flow field. Indeed, Pettersson & Andersson (1997) observed that even sophisticated second-moment closures (SMC) failed to properly accommodate the influence of a rotational-induced secondary flow field.

2. Rotating Turbulent Plane Couette Flow

The shear-driven turbulent Couette flow responds differently to system rotation than does the pressure-driven Poiseuille flow. The mean velocity increases monotonically from one wall to the other in the Couette flow and the mean vorticity therefore attains the same sign throughout the flow. Thus, the entire Couette flow will be exposed either to cyclonic or

U. Frisch (ed.), Advances in Turbulence VII, 301–304.

anticyclonic rotation. This attractive feature makes the Couette flow particularly amenable to in-depth exploration of the intricate influence of system rotation.

To this end we have considered the statistically steady Couette flow bounded by two infinite parallel planes separated a distance 2h, the fluid motion being induced solely by the velocity difference $2U_w$ between the planes. This two-parameter problem is characterized by the Reynolds number $Re = U_w h/\nu$ and the rotation number $Ro = 2\Omega h/U_w$, where Ω denotes the constant angular velocity of the reference frame about a spanwise axis of rotation. Counter-rotating pairs of streamwise-oriented vortices may arise from an imbalance between the Coriolis force and the pressure-gradient in the wall-normal direction when the imposed background vorticity 2Ω is antiparallel to the mean flow vorticity -dU/dy. It is noteworthy that the persistent roll-cell patterns observed by Bech & Andersson (1996a, 1996b, 1997) at low (Ro=0.01) and moderate (Ro=0.10 and 0.20) anticyclonic rotation vanished at the high rotation number Ro=0.50. In order to explore the role of the rotational-induced secondary flow, the present study focuses on plane Couette flow subjected to moderate anticyclonic rotation.

3. Direct Numerical Simulation

The complete Navier-Stokes equations were integrated numerically in time and space over a 10 π h x 2h x 4π h computational domain and system rotation was accounted for by Coriolis force terms in the momentum equations. An adapted version of the finite-difference code ECCLES developed by Gavrilakis *et al.* (1986) was used for the direct numerical simulations (DNS). The Reynolds number Re was set to 1300 and results for a wide range of rotation numbers have been published elsewhere (Bech & Andersson 1996a, 1996b, 1997), after first having assessed the realism of the numerically generated Couette flow in the absence of rotation (Bech *et al.* 1995). Here, we focus on intermediate anticyclonic rotation at which persistent roll-cells were present. At Ro=0.20 these energetic rolls contained 2.24 times the volume-averaged mean turbulent kinetic energy $<<k_o>>$ at Ro=0, while the turbulence energy was reduced to 0.48 $<<k_o>>$. In this regime the flow was homogeneous only in the streamwise x-direction and a triple decomposition was used in order to distinguish between the 2D three-componental secondary flow and the 3D turbulence.

4. Second - Moment Modelling

Within the framework of the Reynolds - averaged Navier - Stokes (RANS) equations a first approach would be to treat the secondary flow field as a part of the turbulence, which then is subjected to conventional modelling. Following this strategy, Pettersson & Andersson (1997) observed some significant discrepancies at intermediate rotation numbers. Here, a more refined and physically appealing approach is adopted, cf. Andersson (1998), namely to treat the secondary flow as an integral part of a 2D three-componental mean flow governed by the RANS equations and let the turbulence model account only for the real turbulence. For this purpose we used a slightly modified version of Durbin's (1993) SMC model, in which near-wall effects are accounted for by the novel elliptic relaxation approach and the common use of non-linear wall-damping functions is thereby avoided.

System rotation was naturally included not only in the rotational production terms but also in the mean intrinsic vorticity in the non-linear variable-coefficient pressure-strain model due to Ristorcelli *et al.* (1995), which is consistent with the principle of material frame indifference in the limit of two-dimensional turbulence. Unlike Pettersson & Andersson (1997), who modified the dissipation rate equation to become explicitly dependent on the imposed system rotation, no attempts were made in the present study to sensitizing the SMC equations for rotation by means of ad hoc modifications.

5. Results and Discussion

Two-dimensional three-componential SMC predictions were accomplished for low and intermediate rotation numbers, but the computations failed to converge for Ro above 0.25. The variation of $Re_\tau = u_\tau h/\nu$ with Ro, where $u_\tau = (\tau_w/\rho)^{1/2}$ denotes the friction velocity, broadly followed the same trend as the DNS data. In fact, Re_τ increased from 82.2 to 107.2 when the rotation number varied from 0 to 0.2 in the DNSs, while the corresponding variation predicted by the SMC was from 84.6 to 112.0. The energetic counter-rotating vortical flow was faithfully reproduced by the RANS and the turbulence level was reduced to 0.42 $<<k_o>>$ at Ro = 0.2, i.e. in accordance with the DNS. The distribution of the streamwise mean velocity component $U_1(y,z)$ in Figure 1a reveals a substantial inhomogeneity in the spanwise z-direction caused by the roll-cells. The spanwise-averaged profiles $<U_1>$ in Figure 1b compare more favourably with the DNS data than did the 1D one-componential calculations by Pettersson & Andersson (1997). In that study the mean shear rate dU_1/dy was somewhat overpredicted in the central core region in which U_1 exhibits a nearly linear variation.

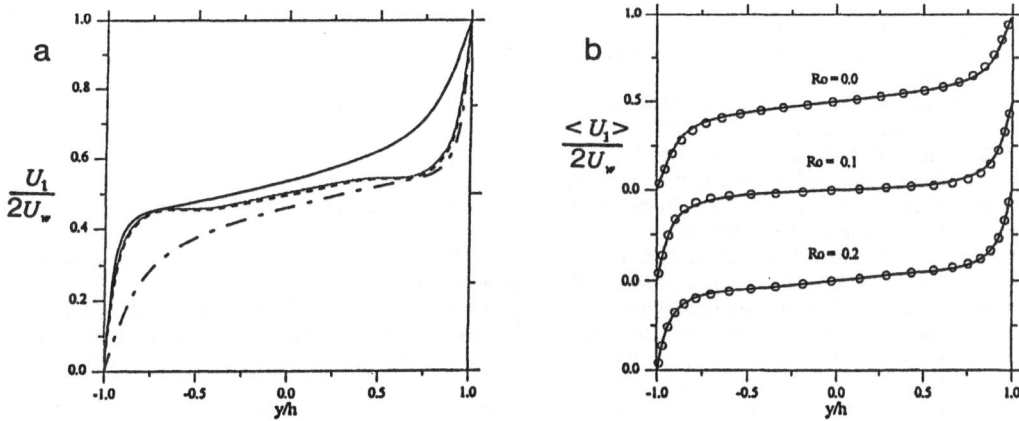

Figure 1. Predicted streamwise mean velocity. a) Profiles of $U_1(y,z)$ at four different spanwise positions at Ro = 0.2; b) Spanwise-averaged mean velocity profiles $<U_1>$ compared with DNS data (symbols).

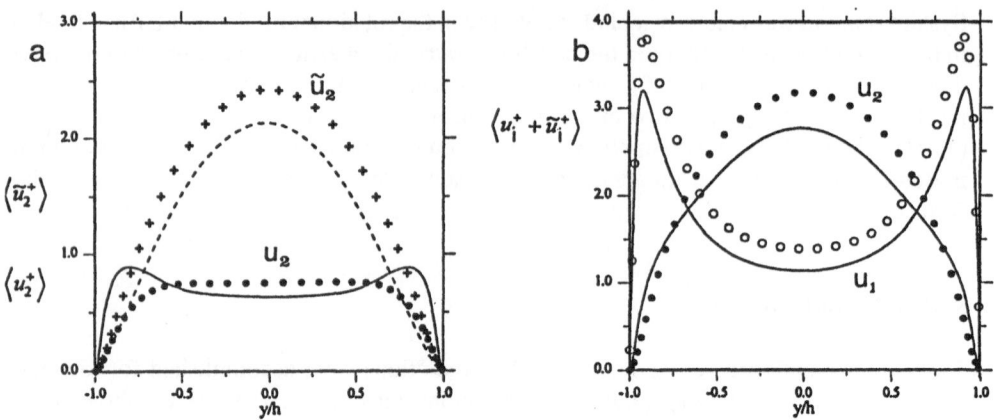

Figure 2. Spanwise-averaged directional energies at Ro = 0.2. SMC predictions (lines) and DNS data (symbols). a) Partition between roll-cell energy $\tilde{u}_2$ and turbulence u_2 in the wall-normal direction; b) Total energy $\tilde{u}_i + u_i$ in the streamwise and wall-normal directions.

In the present approach the secondary flow is part of the resolved mean flow field governed by the RANS. The partition of the overall kinetic energy between the roll-cells $\tilde{u}_i$ and the turbulence u_i is seen in Fig.2a. Here, the computed results have been averaged in the z-direction prior to presentation. Likewise, the total energy in the streamwise and wall-normal velocity components is shown in Fig. 2b. The model calculations compare surprisingly well with DNS data.

The DNSs were supported by The Research Council of Norway (Programme for Supercomputing) through a grant of computing time. Ms. I. Wiggen prepared the camera-ready manuscript with appreciated care.

References

Andersson, H.I. (1998) Organized structures in rotating channel flow, in J.N. Sørensen and E.J. Hopfinger (eds.), *Proc. IUTAM Symposium on Simulation and Identification of Organized Structures in Flows*, Kluwer, in print.

Bech, K.H. & Andersson, H.I. (1996a) Secondary flow in weakly rotating turbulent plane Couette flow, *Journal of Fluid Mechanics* **317**, 195-214.

Bech, K.H. & Andersson, H.I. (1996b) Growth and decay of longitudinal roll cells in rotating turbulent plane Couette flow, in S. Gavrilakis et al. (eds.), *Advances in Turbulence VI*, Kluwer, pp. 91-94.

Bech, K.H. & Andersson, H.I. (1997) Turbulent plane Couette flow subject to strong system rotation, *Journal of Fluid Mechanics* **347**, 289-314.

Bech, K.H., Tillmark, N., Alfredsson, P.H. & Andersson, H.I. (1995) An investigation of turbulent plane Couette flow at low Reynolds numbers, *Journal of Fluid Mechanics* **286**, 291-325.

Durbin, P.A. (1993) A Reynolds stress model for near-wall turbulence, *Journal of Fluid Mechanics* **249**, 465-498.

Gavrilakis, S., Tsai, H.M., Voke, P.R. & Leslie, D.C. (1986) Large-eddy simulation of low Reynolds number channel flow by spectral and finite difference methods, in U. Schumann and R. Friedrich (eds.), *Direct and Large Eddy Simulation of Turbulence*, Notes on Numerical Fluid Mechanics Vol. 15, Vieweg, pp. 105-118.

Johnston, J.P., Halleen, R.M. & Lezius, D.K. (1972) Effects of spanwise rotation on the structure of two-dimensional fully developed turbulent channel flow, *Journal of Fluid Mechanics* **56**, 533-557.

Pettersson, B.A. & Andersson, H.I. (1997) Near-wall Reynolds-stress modelling in noninertial frames of reference, *Fluid Dynamics Research* **19**, 251-276.

Ristorcelli, J.R., Lumley, J.L. & Abid, R. (1995) A rapid-pressure covariance representation consistent with the Taylor-Proudman theorem materially frame indifferent in the two-dimensional limit, *Journal of Fluid Mechanics* **292**, 111-152.

BUBBLE DISPERSION IN A TURBULENT BOUNDARY LAYER ALONG A VERTICAL FLAT PLATE

S. TRAN-CONG, J. L. MARIÉ AND R. J. PERKINS
LMFA/ECL, 36 avenue Guy de Collongue,
B.P. 163, 69131 Ecully Cedex, France

1. Introduction

Many experimental investigations of upward bubbly flows (Serizawa *et al.*, 1975; Liu, 1993; Zun *et al.*, 1992) have demonstrated that there can be a peak in the void fraction profile, close to the wall, but that this depends on bubble size. In downward pipe flows, the peak in void fraction profile occurs in the centre of the pipe, which has led to suggestions that the lift force is the primary mechanism for bubble migration (because it depends on the relative velocity of the bubble – which does not change sign – and the vorticity, which changes sign depending on whether the flow is upwards or downwards). Recent investigations (Moursali *et al.*, 1995; Tran-Cong *et al.*, 1997) have shown that the bubble migrations are random, are characterized by a very short timescale, and are very sensitive to bubble size. The migration cannot be explained simply by the action of a lift force. In this paper, we present a statistical analysis of measured bubble trajectories, and some simple models, to suggest that the bubble migration towards the wall is determined principally by the large scale unsteady structures in the turbulent boundary layer.

2. Experimental results

The experimental apparatus is shown in Figures 1a,b; the techniques are described in detail in Tran-Cong *et al.* (1997). The experiments were performed in a vertical upflowing boundary layer on a flat plate, with velocities up to 1m/s. Bubbles were released from a nozzle, at various distances from the plate, and their trajectories were filmed using high-speed video camera. The images were then analysed (Perkins & Hunt, 1989; Lunde & Perkins, 1995) to yield velocities and deformations. The "trapping effi-

U. Frisch (ed.), Advances in Turbulence VII, 305–308.
© 1998 *Kluwer Academic Publishers.*

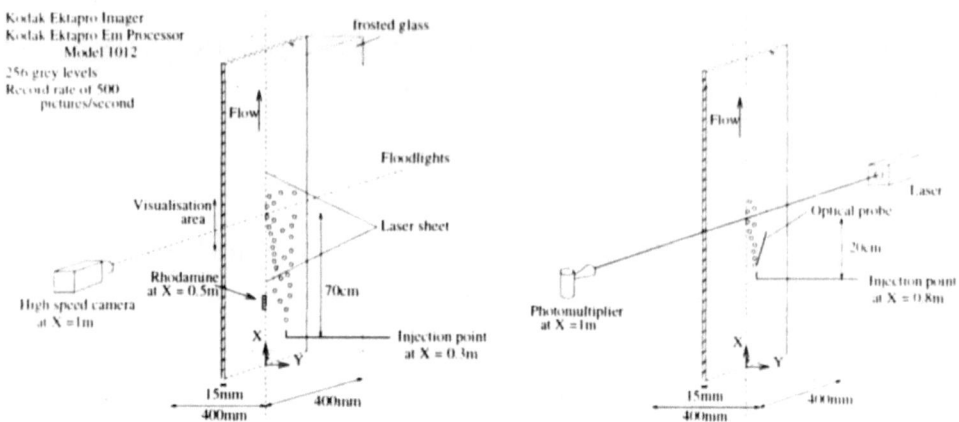

Figure 1.　Experimental apparatus and techniques. (a) Visualisation. (b) Method for counting the "trapping efficiency".

ciency" of the boundary layer was measured, as a function of bubble size and distance from the wall. The measurements show that bubble migration is a maximum for bubble diameters of about 3.5mm; Figures 2a,b shows typical trajectories for bubbles smaller than critical size of 3.5mm ($d_b \sim 2mm$) and larger ($d_b \sim 6mm$). Analysis of many such trajectories shows that bubbles of all sizes are deflected towards the wall, but only the smaller bubbles remain trapped at the wall. (The severe deformation of the large bubbles close to the wall probably helps them to escape.) Small bubbles approach the wall at an angle of about 13° and their trajectories show much less variation than those of the large bubbles. The start of the migration is marked by a very sudden change in direction.

We have also investigated the interaction between the bubbles and large scale turbulent structures in the boundary layer, by injecting Rhodamine dye directly into the boundary layer through a slot in the plate. The flow was illuminated by a light sheet from an Argon-Ion laser, in a plane normal to the wall (see Figure 1a), and the flow was filmed using a standard video camera. The analysis of these films is still in progress; the major difficulty is that the bubble trajectories are 3–D (spirals or zig–zags) so they do not remain in the light sheet for very long. Nevertheless, the films appear to show a correlation between the instantaneous structure of the boundary layer and the motion of bubbles towards the wall.

3. Simple models

In order to investigate the role of the unsteady structure of the boundary layer, we have also developed some simple models, and we can compare the

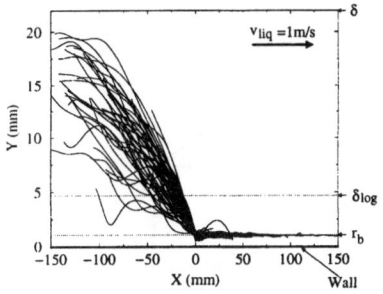 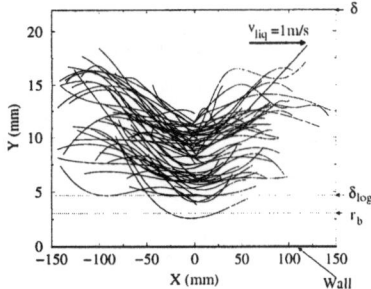

Figure 2.   Bubble trajectories close to the wall, for different diameters. (a) $d_b = 2mm$. (b) $d_b = 6mm$. The impingement points are shifted to the same origin $X = 0$

results from these models with the measured bubble trajectories.

(i) Mean velocity profile

We consider the standard logarithmic velocity profile for a turbulent boundary layer on a flat plate, and we compute bubble trajectories by integrating an equation of motion which includes drag, buoyancy, added mass and lift forces (Auton *et al.*, 1988). Some sample trajectories are shown in Figure 3a, together with the corresponding measured trajectories (with the same bubble initial positions and velocities). Here, migration is caused by the lift force; the trajectories show that the computed migration is much slower than that measured in the experiments, and cannot explain the observed sudden change in direction. Close to the wall, where the shear is much stronger, the computed trajectories are closer to the measured ones.

(ii) Burgers vortex

We model the unsteady large scale structure of the boundary layer as a Burgers vortex moving at constant velocity, with an image vortex to reproduce the presence of the wall. The vortex parameters are chosen to represent approximately the large scales of the turbulent boundary layer. As before, we compute bubble trajectories, by integrating the appropriate equation of motion. Some sample trajectories are shown in Figure 3b, together with the corresponding measured trajectories. The bubbles undergo a very sudden change in direction, far from the wall, and the computed trajectories agree well with measured trajectories. However, closer to the wall, the observed migration is much more rapid than predicted by the model. We therefore conjecture that bubble migration is initiated by the large scale structure in the boundary layer, but that the lift force (and possibly smaller scale structure?) becomes more important close to the wall. We intend to examine this further, using another simple model, based on a POD analysis of the turbulent boundary layer (Joia *et al.*, 1997).

This is only a 2–D model for the velocity field and the motion of the

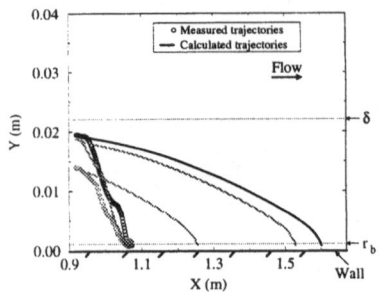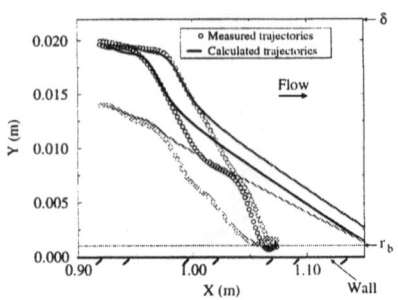

Figure 3. Computed trajectories, for 2mm diameter bubbles. (a) In a time-averaged. (b) In a Burgers vortex.

bubbles, but it appears from our experiments that the 3–D motion of the bubbles is only important in the outer flow. Once the bubbles begin to migrate, they move towards the wall very rapidly. The main effect of the zig–zag or spiral motion of the bubbles appears to be to bring them close enough to the wall to begin the migration. The model also neglects the deformation of the bubbles, particularly in the strong shear close to the wall; this appears to be important in determining why the large bubbles that are deflected towards the wall do not get trapped there.

References

Auton, T. J. **Hunt**, J. C. R. & **Prud'Homme**, M. 1988. The force exerted on a body in inviscid unsteady non-uniform rotational flow. *J. Fluid. Mech.*, **197**, 241–257.

Joia, I. A., **Ushijima**, T., & **Perkins**, R. J. 1997. Numerical study of bubble motion in a turbulent boundary layer using Proper Orthogonal Decomposition. *paper FEDSM97-3534 of: Seventh International Symposium on Gas–Liquid Two–Phase Flows.* ASME Fluids Engineering Division Summer Meeting.

Liu, T. J. 1993. Bubble size and entrance length effects on void development in a vertical channel. *Int. J. Multiphase Flow*, **19**, 99–113.

Lunde, K. & **Perkins**, R. J. 1995. A method for the detailed study of bubble motion and deformation. *pp 395–405 of: Advances in Multiphase Flow 1995.* Eds. Serizawa, A., Fukano, T. & Bataille, J. – Elsevier Science B. V.

Moursali, E., **Marié**, J. L., & **Bataille**, J. 1995. An upward turbulent bubbly boundary layer along a vertical flat plate. *Int. J. Multiphase Flow*, **21 - 1**, 107–117.

Perkins, R. J. & **Hunt**, J. C. R. 1989. Particle tracking in turbulent flows. *pp 286–291 of: Advances in Turbulence 2.* Eds. Fernholz, H. H. & Fielder, H. E. – Springer-Verlag.

Serizawa, A., **Kataoka**, I., & **Michiyoshi**, I. 1975. Turbulence structure of air-water bubbly flows – Part I, II & III. *Int. J. Multiphase Flow*, **2**, 221–259.

Tran-Cong, S., **Marié**, J. L. & **Perkins**, R. J. 1997. Experimental study of the bubble dispersion in a turbulent boundary layer. *paper FEDSM97-3533 of: Seventh International Symposium on Gas–Liquid Two–Phase Flows.* ASME Fluids Engineering Division Summer Meeting.

Zun, I., **Kljenak**, I. & **Serizawa**, A. 1992. Bubble coalescence and transition from wall void peaking to core void peaking in turbulent bubbly flow. *pp 233–249 of: Dynamics of Two-Phase Flows.* Eds. Jones, O.C. and Michiyoshi, I. – Boca Raton.

TURBULENT PLANE MIXING LAYER PERTURBED BY THE WAKE OF A CIRCULAR CYLINDER

D. HEITZ, G. ARROYO AND P. MARCHAL

Cemagref, 17, avenue de Cucillé 35044 Rennes cedex , France.

AND

J. DELVILLE, J.-H GAREM AND J.-P. BONNET

LEA (UMR CNRS 6609), 43, route de l'aérodrome 86036 Poitiers Cedex, France.

1. Introduction

The evolution of the wake of a cylinder embedded in a plane mixing layer, although rarely investigated [1],[2],[3], corresponds to a generic problem which can be encountered in numerous practical configurations as protection of food products against contamination by airborne particles in the Food Industry, heat exchangers, off-shore structures, wind-turbines, smokestacks....

In this study, we analyse the turbulent field downstream of a cylinder placed perpendicular to the splitting plate of a turbulent plane mixing layer. This configuration is a simplified version of a typical flow over a manipulating arm approaching the food element in an ultraclean automatic production line. Our attention is focused on the three dimensional dynamics of the flow past the cylinder. Particular interest is paid to the Reynolds stress re-organization and to the spectral characteristics of velocity fluctuations.

2. Experimental details

The experimental investigation has been carried out in a low speed wind-tunnel. A circular cylinder was placed in a turbulent mixing layer with velocities of 6 and 9 m/s. The Reynolds numbers, based on the cylinder diameter, are respectively 6000 and 9000 for each side of the mixing layer. At the location of the cylinder, the vorticity thickness of the mixing layer is equal to the cylinder diameter, i.e. $\delta_\omega = D = 15$ mm. Measurements

309

U. Frisch (ed.), Advances in Turbulence VII, 309–312.

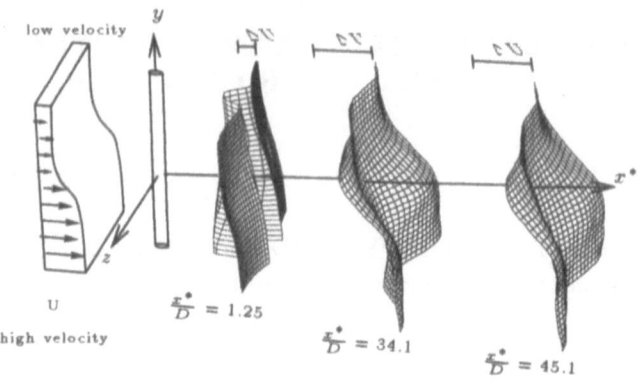

Figure 1.   Streamwise evolution of mean velocity profiles.

are performed by means of a miniature four hot-wires probe which allows to measure simultaneously the instantaneous three velocity components. From these experiments, mean velocity profiles and turbulent quantities were obtained and a spectral analysis was undertaken.

3. Results and discussion

The development of the wake submitted to the shear layer conditions was analyzed through the near velocity profiles. Fig. 1 presents the streamwise evolution of the mean velocity profiles. The wake formation zone follows an unusual behaviour. It appears longer on the high velocity side and a secondary flow takes place from the low to the high velocity side of the mixing layer [3].

The highly 3D feature of the flow, was studied through a detailed analysis of the Reynolds stress. Fig. 2 gives an example of structure functions measured at 5 diameters (in the near wake region) and 65 diameters (in the far wake). The $< u'v' >$ stress can be considered as characteristic of the mixing layer while $< u'w' >$ is characteristic of the wake. It can be observed that, in the near wake region, the wake structure looks dominant and is hardly modified by the mixing layer interaction. As the flow evolves, the mixing layer recovers gradually its own characteristics while the wake turbulent energy vanishes.

These modifications are linked with a complex spectral behaviour. Indeed, the typical frequencies of the wake and of the mixing layer are quite different. Fig. 3 shows the results obtained in the near wake ($x^*/D = 5$) and in the far wake ($x^*/D = 65$). Two typical frequencies of the spanwise component are observed in the near wake region: about 90 Hz (region 1

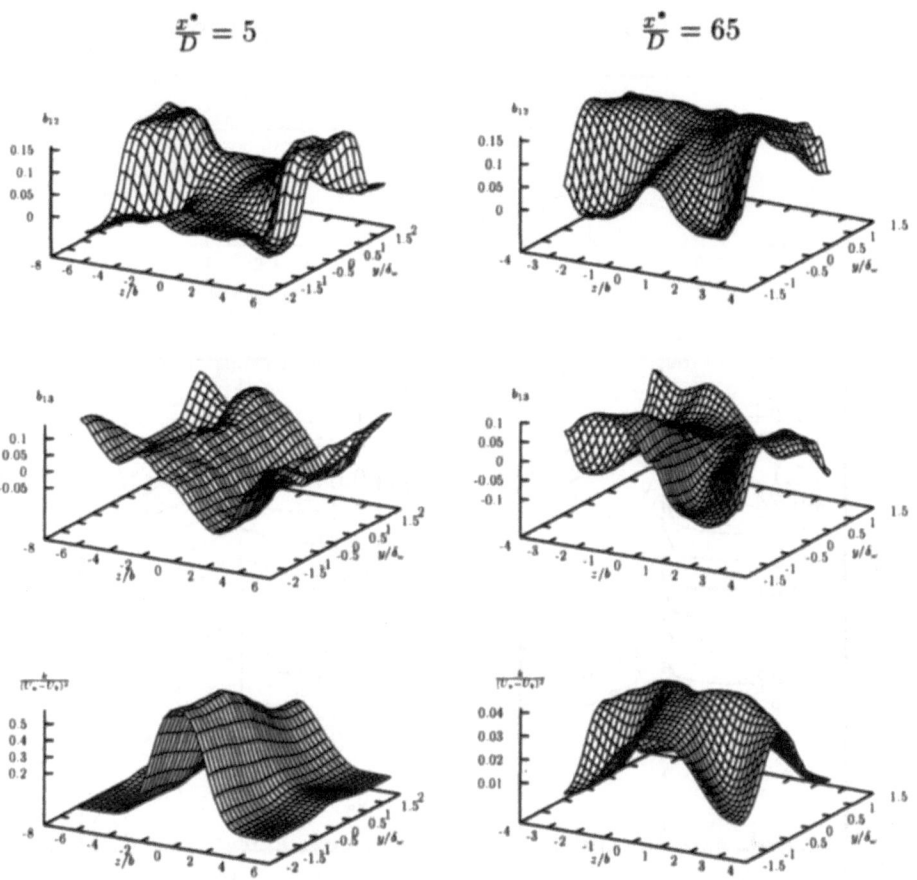

Figure 2. From top to bottom, Reynolds stress invariants ($b_{12} = < u'v' > /2k$ and $b_{13} = < u'w' > /2k$) and turbulent energy measured at 5 and 65 diameters downstream of the cylinder.

Fig. 3) for the low velocity wake and about 120 Hz (region **2** Fig. 3) for the high velocity wake. The typical frequencies of the wake appear here also quite insensitive to the mixing layer interaction. This interaction evolves downstream and in the far wake the interaction region is dominated by the mixing layer frequency (about 60 Hz at $x^*/D = 65$). From all the available data, a possible scenario of the vortical system can be proposed especially for the intermediate wake. The upper (low velocity side) part of the wake can be associated with a vortex reconnection, while wake eddies of the lower part (high velocity side) can be directly connected to the major vorticity of the mixing layer.

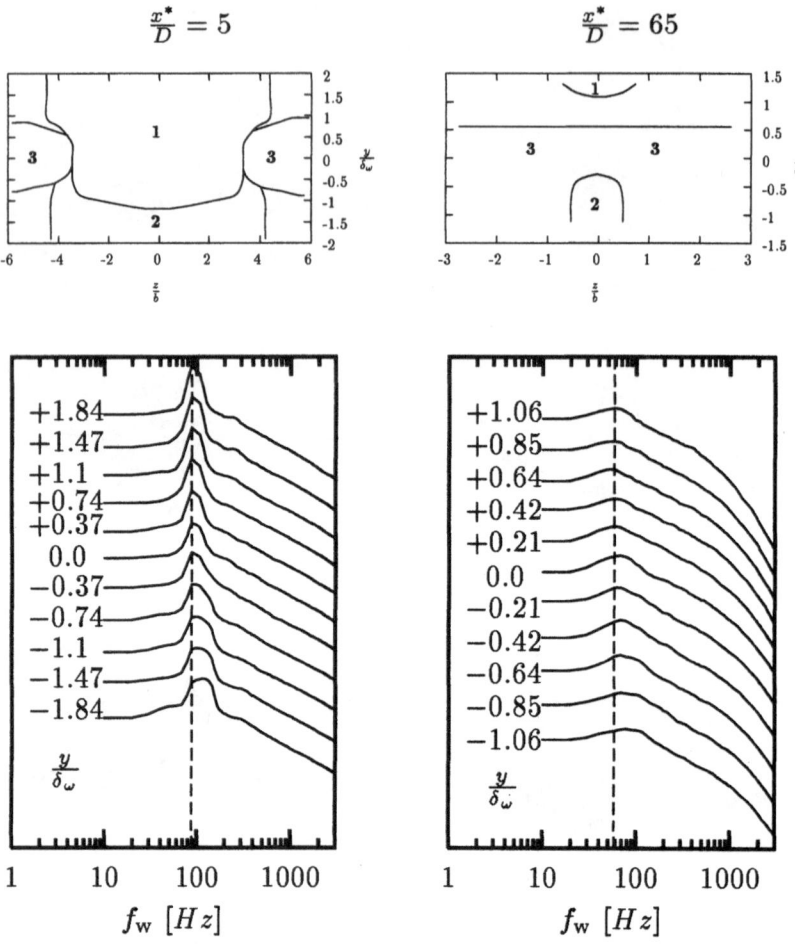

Figure 3. Top: schematic of the regions influenced by **1** low velocity wake, **2** high velocity wake and **3** mixing layer, typical frequencies at 5 and 65 diameters downstream of the cylinder. Bottom: spectra of the spanwise velocity component on wake centerline $z/b = 0$ (with b characteristic wake half width) at 5 and 65 diameters downstream of the cylinder.

References

1. S. Tavoularis, H. Stapountzis & U. Karnik (1987) Vortex shedding from bluff cylinders in strongly sheared turbulent streams, *J Wind Eng. and Industrial Aerodynamics.* **26**, pp. 165–178.
2. H.G.C. Woo, J.E. Cermak & J.A. Peterka (1989) Secondary flows and vortes formation around a circular cylinder in a constant-shear flow *J. Fluid Mech.* **204**, pp. 523–542.
3. D. Heitz, J. Delville, G. Arroyo, J.-H. Garem, J.-P. Bonnet & Ph. Marchal (1997) Interaction of the wake of a circular cylinder and a plane mixing layer, *TSF 11'th. Grenoble – France*, vol.1, pp. 5–1,5–6.

THE ROLE OF ENTRAPMENT PHENOMENA IN THE MODIFICATION OF A PLANE MIXING LAYER BY BUBBLES

E. CLIMENT [1] AND J. MAGNAUDET [2]

[1] *Institut de Mécanique des Fluides de Strasbourg. UMR CNRS/ULP 7507*
   2 Rue Boussingault, 67000 Strasbourg, France.
   E-mail : climent@imf.u-strasbg.fr

[2] *Institut de Mécanique des Fluides de Toulouse. UMR CNRS/INPT-UPS 5502*
   2 Avenue Camille Soula, 31400 Toulouse, France.
   E-mail : magnau@imft.fr

1 - Introduction

Understanding the interactions occurring between clouds of bubbles (or solid particles) and turbulent flows represents a key step in the development of models required for predicting turbulent dispersed flows encountered in engineering or in oceanography. Such interactions are currently divided into two sets. The first one is simply the dispersion of bubbles due to the motion of the continuous phase. This kind of interactions (one-way coupling) has been widely studied in various flow configurations. The second sort of interactions (two-way coupling) is related to the modifications induced by the dispersed phase on the structure of the surrounding turbulent flow. Such interactions are much more difficult to study because they result basically from an inverse-cascade process, since bubbles of small size may affect the whole range of scales of the flow and may even in certain cases drive the largest scales. In some recent attempts, two-way coupling effects were studied by forcing the Navier-Stokes equations by source terms representing the influence of the dispersed phase. In the present study we use the same kind of methodology in order to examine how a plane upflowing mixing layer is modified by a large number of bubbles and how the dispersion of these bubbles is affected by the interaction process.

At the present time it is strictly impossible to compute a turbulent dispersed flow up to the smallest scales. Thus a spatial filtering is required for computing only the flow scales which are much larger than the bubble size. In practice this filtering is produced by the computational grid. As a consequence, source terms modeling the net effect of the subgrid scales on the resolved scales appear in the filtered Navier-Stokes equations. Therefore, in order to describe a two-phase dispersed flow, we solve the "forced" Navier-Stokes equations governing the large scales of the carrying flow together with the Lagrangian equation governing the motion of each bubble of the dispersed phase.

U. Frisch (ed.), Advances in Turbulence VII, 313–316.
© 1998 *Kluwer Academic Publishers.*

Neglecting direct hydrodynamic interactions between bubbles, this Lagrangian equation takes into account the major hydrodynamic forces acting on a clean spherical bubble moving at moderate to high Reynolds number in a time-dependent non-uniform flow namely, buoyancy, viscous drag, added-mass, and lift. The forcing corresponding to the effects of the bubbles on the continuous phase is introduced through a spatial distribution of momentum composed of three source terms described in detail in Ref. [1]. Briefly speaking the first of these terms is a point-force representing the momentum transferred by each bubble to the surrounding fluid (obviously in most cases the dominant contribution in this term comes from buoyancy effects due to density difference between the continuous phase and the dispersed one). The two other source terms model crudely the potential effects due to the finite size of the bubbles and the rotational effects occurring in their wakes, respectively. It was shown in Ref. [1] that this model is for example able to reproduce the large-scale velocity perturbations induced by the rise of a single bubble in a quiescent liquid. Here this model is applied without any modification to a two-dimensional vortical flow configuration. Since the computations reported below require the simultaneous tracking of $O(10^4)$ to $O(10^5)$ bubbles, parallel computing is used to solve the motion of the dispersed phase.

2 - Entrapment phenomena

A preliminary study was first carried out in order to determine the possibility of bubble entrapment in the coherent vortices of this flow. In line with previous analytical studies carried out in simpler flows [4], this study revealed that when the bubble diameter is small enough, bubbles can be trapped by the stable fixed point existing near the center of each vortex. The entrapment criterion is entirely determined by the rise velocity V_L of the bubble in a still liquid and by the velocity difference ΔU across the mixing layer: entrapment occurs if $V_L < \Delta U/2$. Under such conditions, bubbles can accumulate in the vicinity of the fixed points and strong modifications of the flow in these regions can be expected. On the other hand, when the bubble diameter is larger than a critical value satisfying $V_L > \Delta U/2$, the fixed points disappear and bubbles can cross the vortices without experiencing major deviations. A detailed analysis of the role of each hydrodynamic force acting on a bubble reveals that the added mass effect and the lift force are the main centripetal forces contributing to the entrapment while the viscous drag has always a centrifugal effect.

3 - Modifications of a single vortex by bubbles

Before looking at the effects of bubbles on a mixing layer, the case of a single vortex interacting with a cloud of bubbles was considered. According to the process described above, small bubbles are first trapped near the stable fixed point where they transfer momentum, mainly through the point-force term. They communicate an upward momentum to the fluid, the effect of which is to decrease velocity gradients in the core of the vortex. At a certain time, the added mass force becomes unable to trap the bubbles anymore. Under such conditions, the fixed point becomes unstable, so that bubbles are released from the vortex. The capture process can then start again. Overall this mechanism produces a nearly periodic phenomenon in which the fixed point is alternately stable and unstable.

4 - Effects of two-way coupling in a mixing layer

On the basis of these preliminary studies, two different bubble diameters were selected in order to analyse two-way interactions in a vertical, two-dimensional, spatially evolving mixing layer. A first interesting feature displayed by the computations is that in all cases, the roll-up of the vortices in the central region of the mixing layer occurs earlier than in the single-phase situation, owing to the stronger perturbations produced by the presence of bubbles. It follows that the linear evolution of the spreading rate starts earlier when two-way coupling effects are taken into account, as clearly seen in Figure 1. When the bubbles injected in the flow are small enough to satisfy the entrapment criterion, spectacular modifications of the flow occur. First of all a very significant increase of the spreading rate is observed. As could be expected, this increase grows with the injection rate and reaches up to 30% for the highest void fraction considered in the computations (Figure 1). Furthermore, one notices that bubbles tend to tilt significantly the centerline of the mixing layer towards the low-velocity side of the flow. These two results reproduce the main experimental findings described in Ref. [3]. Moreover, compared to the results obtained under one-way coupling assumptions, a strong reduction of the lateral dispersion of the bubbles is observed (Figure 2).

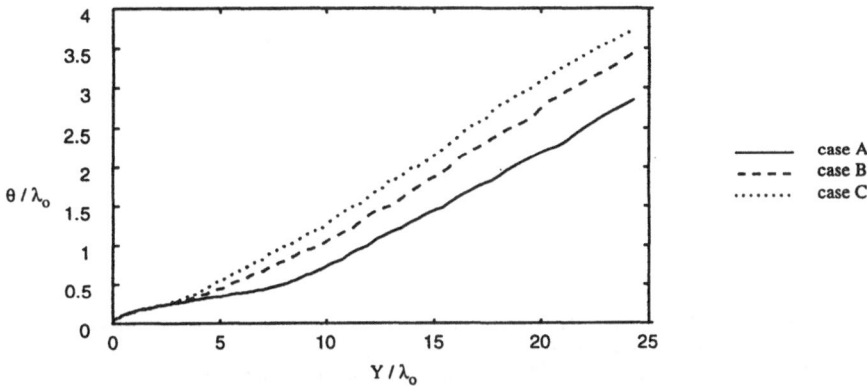

Figure 1. Spreading rate of the mixing layer.
Case A: one-way coupling, Case B: two-way coupling (average injection rate 0.25%),
Case C: two-way coupling (average injection rate 0.50%)

This effect is obviously related to the increase of the spreading rate and can be explained by looking at the number of stable fixed points present in the flow. Since this number decreases for the reasons explained above when the influence of the dispersed phase on the flow is taken into account, less bubbles are able to follow the motion of the vortices in the two-way coupling situation. Bubbles released out of the vortices tend to accumulate in the central region of the mixing layer, thus decreasing the average lateral dispersion (Figure 2). In contrast with the above findings, the results obtained with large bubbles reveal little change of the flow field compared to the single-phase situation; in particular, the spreading rate of the mixing layer is nearly unaffected by the presence of the bubbles (its variation is less than 5%). The reason is that large bubbles passing through the vortices have a very short interaction time with the vortical structures, so that only weak velocity perturbations result from the momentum transfer. Also the lateral dispersion of bubbles is only slightly modified in that case. These features observed with large bubbles are in agreement with the experimental results obtained by Loth and Cebrzynski [2]. Hence the present study reconciles the trends found by Roig *et al.* [3] and those found by the latter authors and emphasizes the critical role played by the entrapment process in the modifications produced by bubbles in a vortical flow.

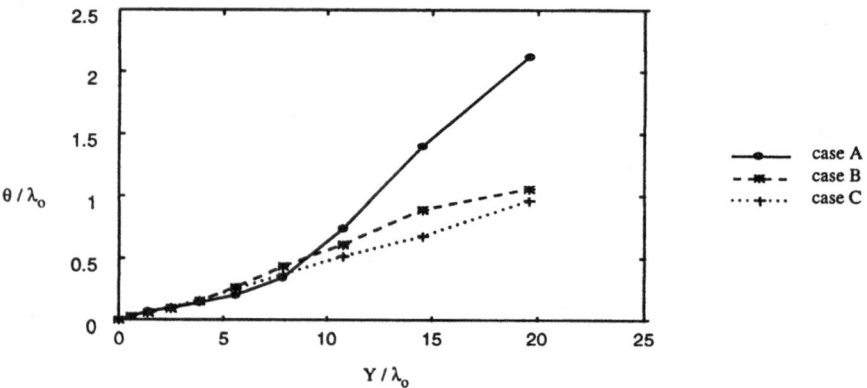

Figure 2. Lateral dispersion of the bubbles.
Case A: one-way coupling, Case B: two-way coupling (average injection rate 0.25%),
Case C: two-way coupling (average injection rate 0.50%)

References

1. Climent, E. & Magnaudet, J. (1997) Simulation d'écoulements induits par des bulles dans un liquide initialement au repos. *C. R. Acad. Sci. Paris* **324** (II-b) pp. 91-98.
2. Loth, E. & Cebrzynski, M.S. (1995) Modulation of shear layer thickness due to large bubbles. *Int. J. Multiphase Flow* **21**, pp. 919-927.
3. Roig, V., Suzanne, C. & Masbernat, L. (1998) Experimental investigation of a bubbly mixing layer. *Int. J. Multiphase Flow* **24**, pp. 35-54.
4. Tio, K.K., Lasheras, J.C., Ganan-Calvo, A.M. & Linan, A. (1993) The dynamics of bubbles in periodic vortex flows. *Appl. Sci. Res.* **51**, pp. 285-290.

AN ANALYTICAL EXPRESSION FOR THE SPECTRUM OF COMPRESSIBLE TURBULENCE IN THE LOW MACH NUMBER LIMIT

G. FAUCHET AND J.P. BERTOGLIO

CNRS, LMFA, UMR 5509, ECL, BP 163, 69131 Ecully Cedex,
France.

In a recent paper [1], a statistical closure for weakly compressible isotropic turbulence was proposed. The model is a one-time two-point closure sharing strong similarities with the previously derived extension to compressible turbulence of the E.D.Q.N.M. model [2]. Both models were obtained from a D.I.A. formulation. They differ in the assumptions introduced to express the two-time correlations in terms of one-time correlations. The improved model was found to compare satisfactorily with Direct and Large Eddy Simulations ([3] and [4]).

The aim of the present study is to perform an asymptotical analysis of the statistical closure in the limit of a low Mach number (and high Reynolds number), and to derive an analytical expression for the spectrum of the purely compressible part of the velocity field.

The purely compressible (or dilatational) part of the velocity fluctuation is defined, in Fourier space, as:

$$u_i^C(\vec{K}, t) = \frac{K_i K_j}{K^2} u_j(\vec{K}, t) \tag{1}$$

and the solenoïdal field is defined as:

$$u_i^S(\vec{K}, t) = u_i(\vec{K}, t) - u_i^C(\vec{K}, t) \tag{2}$$

For statistically steady isotropic turbulence, the closed set of equations leads to:

$$2\nu' K^2 E^{CC}(K) = T^{CC}(K) \tag{3}$$

where $E^{CC}(K)$ is the spectrum associated $u_i^C(\vec{K}, t)$ and $\nu' = 4/3\,\nu$ is the viscosity acting on the compressible mode. $T^{CC}(K)$ is a transfer term involving different contributions reflecting the various types of interactions between the solenoidal and compressible parts of the field. In the low turbulent Mach number limit, $M_t = \frac{u'}{C_o} \ll 1$ (C_o being the sound speed), and

317

U. Frisch (ed.), Advances in Turbulence VII, 317–320.

within the range of the turbulent length scales, the leading contribution to $T^{CC}(K)$ is:

$$T_1^{CC}(K) = \int_{P,Q \in \Delta} \frac{K^3}{PQ} (x + yz)^2 \, \Theta_{KPQ}^{CC-SS-SS} \, E^{SS}(Q) E^{SS}(P) dPdQ \quad (4)$$

in which the integration domain is the part of the {P,Q} plane where $\vec{K}$, $\vec{P}$ and $\vec{Q}$ form a triangle, and x, y and z denote the cosines of the angles in this triangle. E^{SS} is the spectrum of the solenoidal field $u_i^S(\vec{K}, t)$. The other contributions to T^{CC} are terms having a similar structure but involving $E^{CC} E^{CC}$ or $E^{SS} E^{CC}$ products.

One of the results in [1] was that, a low M_t, the equation for E^{SS} was not affected by compressibility. Therefore the approach followed here is to model E^{SS} as if the turbulence was incompressible, then to evaluate T_1^{CC} in order to finally deduce an analytical expression for E^{CC}. For the present analysis, use is made of a classical spectral shape for E^{SS} : $E^{SS} = C_s K^s$, for $K \le K_l$; $E^{SS} = C_K \epsilon_s^{2/3} K^{-5/3}$, for $K_\eta \ge K \ge K_l$; and $E^{SS} = 0$ for $K > K_\eta$, in which C_K is the Kolmogorov constant, ϵ_s is the dissipation rate (associated with the solenoidal mode), K_η is the Kolmogorov wavenumber, and K_l is a wave-number related to the integral length scale (C_s being determined by continuity of the spectrum at K_l).

An expansion of the characteristic time $\Theta_{KPQ}^{CC-SS-SS}$ is performed in terms of ν' and C_o. At the leading order, it is found

$$\Theta_{KPQ}^{CC-SS-SS} = \frac{\sqrt{\pi}}{2\sqrt{\alpha(K)}} e^{-\frac{K^2 C_o^2}{4\alpha(K)}} + 2\nu' K^2 \frac{\alpha(K)}{K^4 C_o^4} \quad (5)$$

in which

$$\alpha = \alpha(K) + \alpha(P) + \alpha(Q) \quad (6)$$

where $\alpha(K)$ is a damping factor associated with the two-time correlations

$$\alpha(K) = aK^3 E^{SS}(K) \text{ with } a \simeq 0.2 \quad (7)$$

The methodology then consists in performing a non-local expansion of T_1^{CC} when $K < K_l$ (assuming that $K \ll P,Q$). This leads to an analytical expression for T_1^{CC} at small K. For $K > K_l$, that is to say in the inertial range of the solenoidal spectrum, it is not possible to use the same technique. However assuming locality of the interactions, it is found

$$T_1^{CC}(K) = \frac{4\alpha(K)\nu'}{C_o^2} E_{inc}^{PP}(K) \quad (8)$$

in which $E_{inc}^{PP}(K)$ is the spectrum of the so-called "incompressible" pressure (i.e. the solution of a Poisson equation applied to u^S). This spectrum is expressed using the form proposed by George [5]:

$$E_{inc}^{PP}(K) = \frac{1}{2C_o^2} C_G C_K^2 \epsilon_s^{4/3} K^{-7/3} \text{ (with } C_G = 1.32) \quad (9)$$

It is finally found that, in the inertial range, $E^{cC}(K)$ scales as $M_t^4 \, Re_L^0 \, K^{-3}$ (in agreement with Ristorcelli's pseudo-sound theory [6]). For smaller wave-numbers, a plateau appears, where $E^{CC}(K)$ is proportional to $M_t^4 \, Re_L^0 \, K^0$. At very small K, that is to say outside of the turbulent length scale range, T_1^{CC} is found to increase strongly. A bump in T_1^{CC} is detected, corresponding to the acoustic emission by the turbulent motion. This behavior of T_1^{CC} creates a strong increase of $E^{CC}(K)$. Then another contribution to T^{CC} begins to play a significant role: an absorption term, and the balance equation for $E^{CC}(K)$ can be re-written as

$$2 \, (\nu' + \nu_t^{ac}) K^2 E^{CC}(K) = T_1^{CC}(K) \tag{10}$$

with

$$\nu_t^{ac}(K) = \int_{P,Q \in \Delta} \frac{P^2}{Q \, K^2} \, z \left(1 - z^2\right) \Theta_{PKQ}^{SS-CC-SS} \, E^{SS}(Q) dP dQ \tag{11}$$

Applying a non-local expansion to (11) in the low M_t limit, an analytical expression for the eddy viscosity, $\nu_t^{ac}(K)$, is found:

$$\nu_t^{ac}(K) = \frac{1}{30} \sqrt{\frac{2 \, \pi \, C_K}{a}} \, \frac{\epsilon_s^{1/3}}{K_l^{4/3}} \, F \left(\frac{K c_b}{2 \sqrt{2} \, a \, M_t \, K_l} \right) \tag{12}$$

$$\text{with } F(x) = 5 \, \frac{1 - \exp(-x^2)}{x^2} - \exp(-x^2) \tag{13}$$

This expression shows that the eddy viscosity is constant over a large plateau at small K in conformity with the usual concept of what an eddy viscosity should be. It decreases rapidly when K becomes larger than the wave-number corresponding to the peak of the acoustic emission term.

The complete analytical expression for $E^{CC}(K)$ is a function of the Reynolds number (built on the integral length scale) Re_L, of the turbulent Mach number M_t, as well as of the Kolmogorov constant and of the solenoidal energy and dissipation. It also involves the exponent s of the solenoidal energy spectrum at low K. In Fig. 1, where the analytical expression is compared to the numerical prediction of the closure (for s=4), it can be observed that the agreement is fairly good. From the expression of the spectrum, analytical expressions for different one-point quantities can be evaluated by wave-number integration. For the turbulent kinetic energy k_c and the dissipation ϵ_c related to the compressible mode, it is found:

$$k_c = 0.25 \, c_b^{-6} \, k_s \, Mt^4 \tag{14}$$

$$\epsilon_c = 1.65 \, c_a \, c_b^{-5} \, \epsilon_s \, M_t^4 \ln(Re_L) \, Re_L^{-1} \tag{15}$$

in which k_s is the turbulent kinetic energy associated to the solenoïdal part of the velocity field and c_a and c_b are coefficients depending on the slope

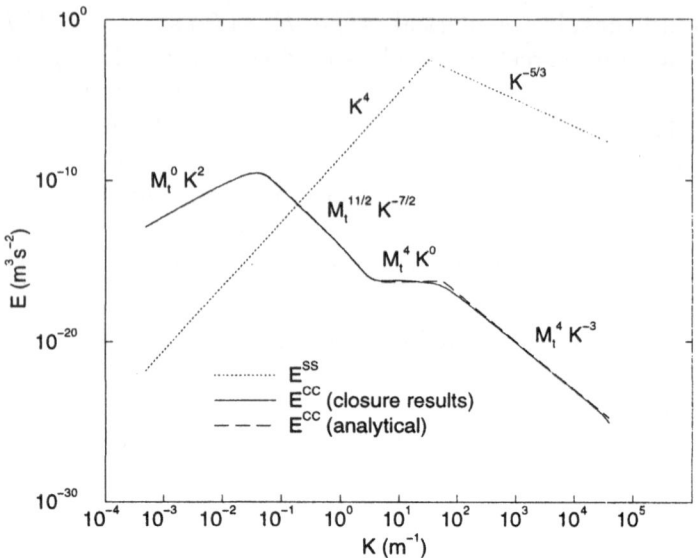

Figure 1. Comparison between analytical model and numerical results of the closure; Re_L=6000, M_t=6 10^{-4}, s=4.

of E^{SS} at large scales [1]. In the acoustic range, the production P and the energy k_{ac} are found to be:

$$P = 4.2\, c_b^{-5}\, \epsilon_s\, M_t^5 \tag{16}$$

$$k_{ac} = 0.52\, c_b^{-5}\, M_t^3\, k_s \tag{17}$$

These scalings are in agreement with classical theories in acoustics.

References

1. Fauchet G., Shao L., Wunenburger R. & Bertoglio J.P., 1997, An improved two-point closure for weakly compressible turbulence and comparisons with Large Eddy Simulation, 11th Symp. on Turb. Shear Flows, Grenoble, Sept. 8-11, 97.
2. Bataille F. & Bertoglio J.P., 1993, Longtime behaviour study and decay of a compressible turbulence, 9th Symp. on Turb. Shear Flows, Kyoto, Aug. 16-18, 93.
3. Ristorcelli J.R & Blaisdell G.A., 1997, Validation of a pseudo-sound theory for the pressure-dilatation in DNS of compressible, 11th Symp. on Turb. Shear Flows, Grenoble, Sept. 8-11, 97.
4. Shao L. & Bertoglio J.P., 1996, Large Eddy Simulations of weakly compressible isotropic turbulence, 6th European Turb. Conference, Lausanne, July 96.
5. George W., Beuther P. & Arndt R., 1984, Pressure spectra in turbulent shear flows, Journ. of Fluid Mech., vol. 148.
6. Ristorcelli J.R., 1995, A pseudo-sound constitutive relationship for the dilatational covariances in compressible turbulence, ICASE Report 95-22.

---

[1] $c_a = \frac{3\,s+5}{3\,s}$ et $c_b = \sqrt{\frac{3\,s+5}{3\,(s+1)}}$

THE EFFECT OF HEATED BUILDING ON ATMOSPHERIC DIFFUSION IN URBAN STREET CANYONS

T. KANZAKI AND Y. ICHIKAWA
Department of Atmospheric Science
Central Research Institute of Electric Power Industry
Komae, Tokyo, 201-8511 Japan

The effect of heated building walls on gas dispersion in urban street canyons was experimentally studied. Measurements of gas concentration, wind velocity and air temperature were conducted around cubical model buildings in a wind tunnel. The concentration, velocity and temperature statistics were estimated. The results show that the effect of heated building walls prevents the diffusion of air pollutants to leeward street canyons.

1. Introduction

The diffusive-advective process of air pollutants emitted from industrial smokestacks and vehicles along the wayside is affected by the turbulent motion and buoyancy force in the atmospheric boundary layer. In order to predict the dispersion of air pollutants in the urban boundary layer, it is important to estimate the effects of complicated roughness and heated building surfaces on the concentration field. The purpose of this study is to investigate the effect of heated surfaces of buildings on gas dispersion in urban street canyons.

2. Experiments

The experiments were conducted in a wind tunnel under neutral stratification conditions. Figure 1 shows the experimental setup and measuring system. The test section of the tunnel was 20 m long, 3.0 m wide and 1.5 m high. The turbulent boundary layer was produced using roughness elements such as spires and angles placed on the floor at the entrance of the test section. In order to form street canyons, nine cubical model buildings were set downstream in the fully-developed boundary layer, as shown in figure 2. The height of the model buildings was $Hb=0.16$ m. A point source was installed in front of the model buildings ($x/Hb=-1$) at the height of $z/Hb=1.0$. To

321

investigate the effects of heated walls on gas dispersion, four side walls and the roof of the center model building were heated to a constant temperature of 90℃. Tracer gas consisting of ethylene and dried air was emitted from the upstream point source. The instantaneous plume concentration was measured using a modified flame ionization hydrocarbon analyzer. Wind velocity and air temperature were measured using a 2-D laser-Doppler velocimeter and a cold-wire thermometer, respectively. The mean wind velocity at the boundary-layer edge was set to U_L=1.5 m/s, the initial concentration of the tracer gas was C_i= 2.95×10^5 ppm, the Reynolds number based on the height of the boundary layer L, Re=$(U_L L / \nu)$, was 40000, and the Richardson number Ri= $(g \Delta T / T_L)(L/U_L^2)$ was 0.74, where ΔT was the temperature difference between the surface of the heated model building and the ambient flow. This flow geometry simulated the urban boundary layer at a scale of 1: 300. The measurements were conducted at the centerline in the downstream region of $-1 \leq x/Hb \leq 8.0$ and $0 \leq z/Hb \leq 2.5$.

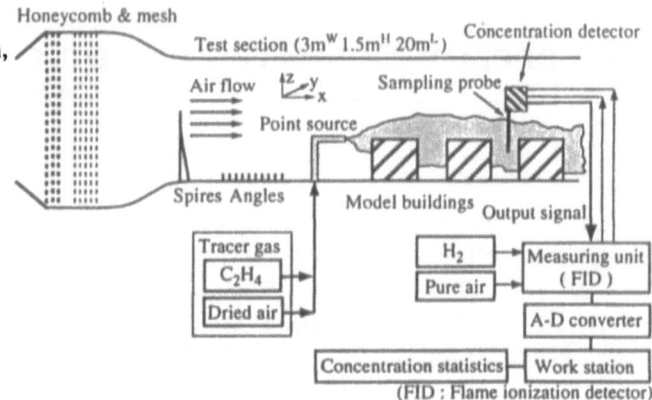

Figure 1. The experimental setup and the measuring system.

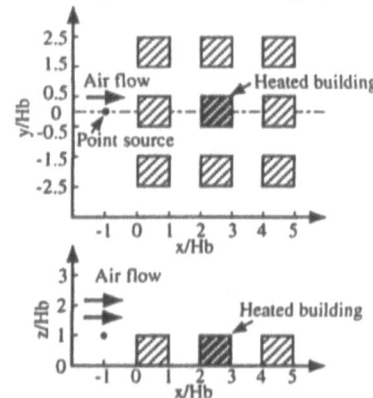

Figure 2. The configuration of cubical model buildings.

3. Results and Discussion

Figure 3 shows the vertical profiles of mean temperature difference between the interior of the street canyon and the ambient flow at x/Hb=1.5, 3.5, 5.5. In the windward canyon of the heated building, the air temperature rises to less than 3 ℃ near the ground surface. On the other hands, the heated building strongly affects the temperature field in the leeward canyon. The air temperature rises up to 6℃ at x/Hb=3.5 and a higher

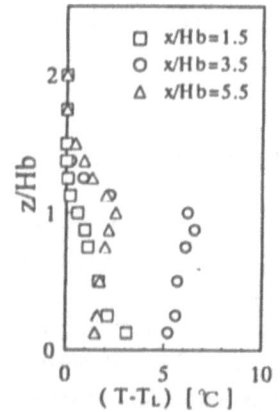

Figure 3. Profiles of the temperature rise in street canyons.

temperature region is formed there. The heating effects have been retained in the downstream region.

Figure 4 shows the profiles of turbulent kinetic energy around the model buildings along the x-y centerline for the nonheated case and the heated case. The turbulent kinetic energy, $k=(1/2)(u'^2+v'^2+w'^2)$, was nondimensionalized by the mean wind velocity at the boundary-layer edge, U_L, where u', v' and w' are the rms values of longitudinal, transverse and vertical velocity fluctuations, respectively. The values in the figures are (k/U_L^2). For the nonheated case, the profile shows that turbulent kinetic energy is mainly produced near the windward corner of the upstream building. A comparison between the profiles for the nonheated case and for the heated case reveals that the behaviors of both contours are almost the same in the windward region of $-1 \leq x/Hb \leq 2$. However, for the heated case, turbulent kinetic energy increases in the leeward canyon of $3 \leq x/Hb \leq 4$. The increase is about twice at the roof level compared to that in the non-heated case. This means that turbulent kinetic energy is vigorously generated under a thermally unstable condition which is caused by the temperature difference between the heated walls and the ambient flow.

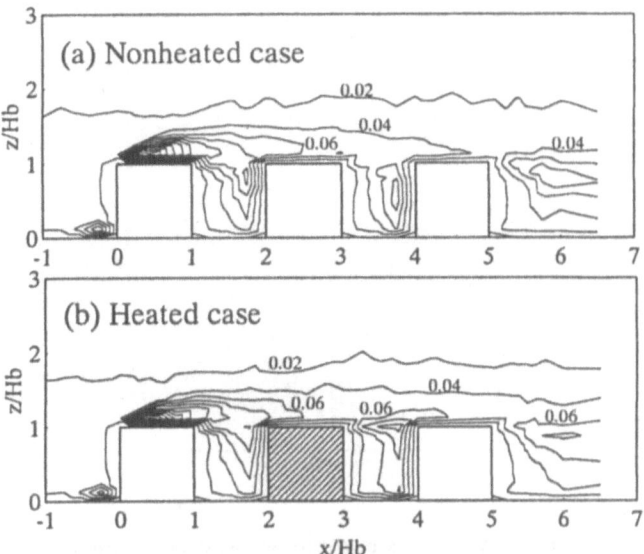

Figure 4. Profiles of the turbulent kinetic energy along the x-y centerline.

Figure 5 shows the profiles of mean gas concentration around the model buildings along the x-y centerline for the nonheated case and the heated case. The mean concentration was nondimensionalized by the initial concentration Ci. In the nonheated case, tracer gas was transported to the ground level in street canyons. In the leeward canyon, the mean concentration was $5 \times 10^{-5} \leq C/Ci \leq 9 \times 10^{-5}$ in the region of x/Hb=1.5 and z/Hb≤0.75. On the other hand, in the heated case, the mean concentration was $1 \times 10^{-5} \leq C/Ci \leq 4 \times 10^{-5}$ in the region of x/Hb=1.5 and z/Hb≤0.75. This means that the downflow passing the building to the leeward canyon is prevented by the buoyancy force owing to the effect of the heated building. Therefore, gas dispersion at the ground level in the leeward canyon is suppressed. Thus, the heated building walls evidently affect the gas dispersion process in urban street canyon.

Figure 6 shows the profiles of the peak-to-mean concentration ratio in street canyons at x/Hb=1.5, 3.5 and 5.5. Profiles of the ratio in the heated case are very similar to those

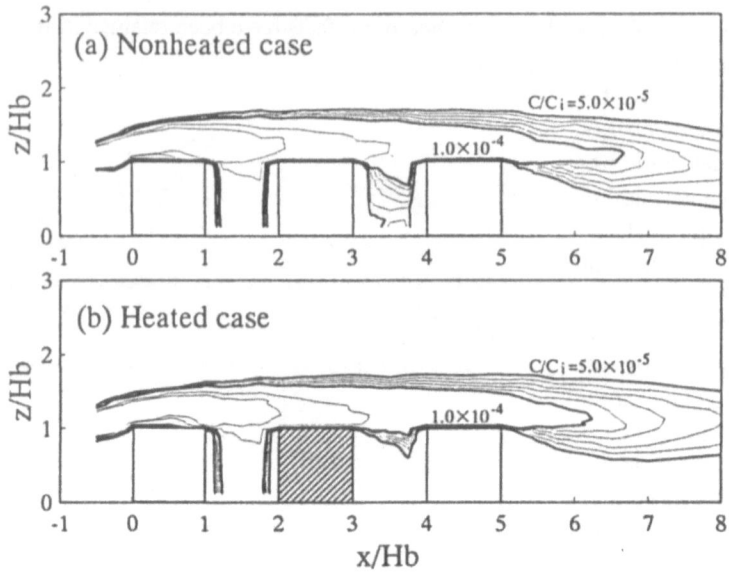

Figure 5. Profiles of the mean gas concentration along the x-y centerline.

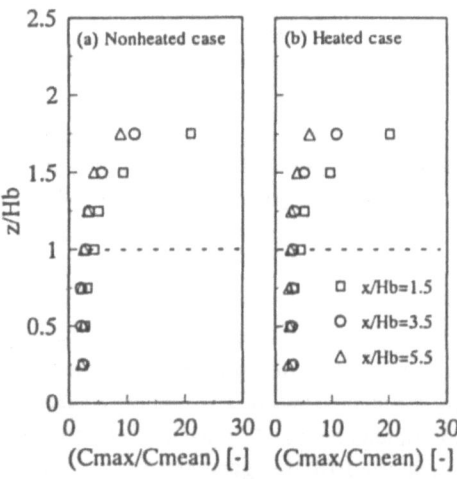

Figure 6. Profiles of the peak-to-mean gas concentration ratio in street canyons.

in the nonheated case. The ratio is $4 \leq (C_{max}/C_{mean}) \leq 5$ at the height of $z/Hb \leq 0.75$. The profiles of fluctuating concentration in the non-heated and heated cases are also very similar, though they are not shown here. These results suggest that the heated building affects the mean concentration field in the leeward canyon, but the effect is not strong enough to change the concentration fluctuation field. It is concluded that the instantaneous maximum concentration in street canyons can be estimated from the peak-to-mean concentration ratio in the nonheated case.

4. Conclusions

To estimate the effect of heated building walls on gas dispersion in urban street canyons, experiments of atmospheric diffusion were precisely conducted in the wind tunnel. The experimental results show that gas dispersion at the ground level in the leeward canyon is suppressed by the effect of the heated building. Furthermore, it is suggested that the instantaneous maximum concentration in street canyons can be estimated from the peak-to-mean concentration ratio in the nonheated case.

VI

Vortex Dynamics

Dynamique du Tourbillon

FORMATION AND DISRUPTION OF CONCENTRATED VORTICES IN TURBULENCE

H.K. MOFFATT
Isaac Newton Institute for Mathematical Sciences
20 Clarkson Road
Cambridge CB3 0EH, UK

The Burgers vortex (Burgers 1948) is described by an exact solution of the Navier-Stokes equations in which the effects of uniform axisymmetric stretching are in equilibrium with viscous diffusion. Burgers introduced this vortex as 'a mathematical model illustrating the theory of turbulence', and he noted particularly that the vortex had the property that the rate of viscous dissipation per unit length of vortex was independent of viscosity in the limit of vanishing viscosity (i.e. high Reynolds number). This is of course an attractive feature in the light of Kolmogorov's theory of turbulence, in which the rate of dissipation of energy per unit volume ϵ and the kinematic viscosity ν are regarded as independent variables.

The idea that the small-scale structures of turbulence might be representable in terms of a random distribution of vortex sheets or tubes was taken up by Townsend (1951). Townsend showed that a random distribution of vortex sheets would give rise to an energy spectrum proportional to k^{-2} (multiplied by an exponential viscous cut-off factor) this power-law reflecting the fact that on any straight line through the field of turbulence, intersecting a finite number of vortex sheets, there will be (in the limit of vanishing viscosity) a finite number of discontinuities of velocity per unit length. A random distribution of vortex tubes gave rise to a power-law k^{-1} (again modified by an exponential cut-off), this slower fall-off with k being associated with the more singular behaviour in physical space associated with a line vortex.

The Kolmogorov spectrum $k^{-5/3}$ lies tantalisingly between k^{-1} and k^{-2}, suggesting that the typical (or generic) structures in **x**-space which may be responsible for such a spectrum should involve some compromise between tubes and sheets, for example spiral structures (Lundgren 1982, Gilbert 1988), these possibly arising through the interaction of tubes and sheets (Krasny 1986, Moffatt 1993).

U. Frisch (ed.), Advances in Turbulence VII, 327–330.

During the last fifteen years, evidence from direct numerical simulation (DNS) of turbulence has accumulated indicating the presence of concentrated tube-like structures in the vorticity field (Siggia 1981, Kerr 1985, Yamamoto & Hosokawa 1988, Vincent & Meneguzzi 1991, and others). This has led to a great revival of interest in the primitive theories of Burgers and Townsend, and a re-evaluation of possible models of turbulence in terms of simple vortex structures.

It was noticed by Kida & Ohkitani (1992) that the dissipation structure in the concentrated vortices of 3D turbulence exhibit two maxima, off-set from the centre of the vortex, and they suggested that this might be explained in terms of the action of non-axisymmetric strain acting on the vortex. This suggestion was taken up by Moffatt, Kida & Ohkitani (1994) who developed a high Reynolds number asymptotic theory of a vortex subjected to non-axisymmetric strain. Determination of the dissipation structure involved pursuing the analysis to third order in the small parameter Re^{-1}, and this analysis did indeed reveal the two peaks in the dissipation structure, arising from a symmetry-breaking splitting of the circle of maximum dissipation that occurs for the Burgers vortex. Remarkably, at leading order in the asymptotic analysis, the axisymmetric Burgers vortex emerges in spite of the non-axisymmetric character of the strain. This is because, at high vortex Reynolds number, the vortex spins rapidly in the strain field, and experiences the θ-averaged strain, which is axisymmetric. A similar behaviour had been previously recognised by Ting & Tung (1965), and by Neu (1984). Even more remarkably, the solution of Moffatt, Kida & Ohkitani indicates that the vortex can survive for an exponentially long time even when one of the rates of strain in the plane of cross-section of the vortex is positive. Again, this is because it is only the θ-averaged strain that is relevant at leading order.

The characteristic dissipation structures identified by Moffatt, Kida & Ohkitani are present also, although for rather different reasons, in two-dimensional freely decaying turbulence, at the stage when identifiable vortices emerge from a random initial state (McWilliams 1984, 1990, Jiménez, Moffatt & Vasco 1996). Each vortex in such a field moves with the local velocity induced by all the other vortices, and is also subject to the two-dimensional strain field associated with the presence of all the other vortices. At high Reynolds number, the effect of this strain field is to distort each vortex cross-section to slightly elliptical form; the associated dissipation field has precisely the same structure as that determined in the earlier work of Moffatt, Kida & Ohkitani. This remarkable result is a consequence of the analogy between steady stretched three-dimensional vortices and unsteady unstretched two-dimensional vortices, as described by Lundgren (1982).

The elliptic deformation of vortices in both two- and three-dimensional turbulence makes them prone to the type of three-dimensional resonant instability identified by Bayly (1986) and Pierrehumbert (1986). As shown, however, by Le Dizès, Rossi & Moffatt (1996), stretching carries the wave-number of sinusoidal perturbations through the unstable wave-number band in a finite time, so that infinitesimal disturbances are always asymptotically stable. This mechanism is not present for two-dimensional (unstretched) vortices, and one may reasonably conjecture that 2D turbulence is always unstable to 3D disturbances (in the absence of stabilising mechanisms such as stratification or magnetic field).

A serious limitation of the Burgers model in the context of three-dimensional turbulence lies in the assumption of the uniformity of the strain field, and the associated infinite length of the stretched vortices. In fact, the region of concentration is of finite length, the vortex lines diverging more or less rapidly at the ends of these regions of concentration. This finite length is apparent also in experiments (Douady, Couder & Brachet 1991) designed to detect intense vortex filaments in turbulent flow of liquids seeded with small gas bubbles.

Variation of the strain field arises through the non-uniform action of the 'other vortices' near to the parent vortex whose structure is considered. In so far as these other vortices may be treated as point vortices, the non-uniform strain field acting on the parent vortex is a strain field associated with a non-uniform potential flow. In this lecture, a simple model will be developed involving the action of distributed vortices on a two-dimensional stretched vortex sheet (of Burgers type). The problem is treated by using the potential ϕ and stream function ψ of the non-uniform straining flow as independent coordinates. The advection-diffusion equation for the parent vortex sheet has universal form in terms of these coordinates, and a wide family of exact solutions of the Navier-Stokes equations is thus generated. A variety of solutions will be described, which provide a good indication of the manner in which such vortex sheets may disrupt in regions of strong non-uniformity of strain.

A similar technique runs into difficulties for the analogous axisymmetric problem, because in this case the advection-diffusion equation, expressed in terms of the relevant ϕ and ψ, is not universal. Nevertheless, the approach does indicate one mechanism by which vortex disruption can occur.

References

Bayly, B. (1986) Three-dimensional instability of elliptical flow, *Phys. Rev. Lett.* **57**, p. 2160.

Burgers, J.M. (1948) A mathematical model illustrating the theory of turbulence, *Adv. Appl. Mech.* **1**, pp. 171–199.

Douady, S., Couder, Y. & Brachet, M.E. (1991) Direct observation of the intermittency of intense vorticity filaments in turbulence, *Phys. Rev. Lett.* **67**, 983–986.

Gilbert, A.D. (1988) Spiral structures and spectra in two-dimensional turbulence, *J. Fluid Mech.* **193**, pp. 475–497.

Jiménez, J., Moffatt, H.K. and Vasco, C. (1996) The structure of the vortices in freely decaying two-dimensional turbulence, *J. Fluid Mech.* **313**, pp. 209–222.

Kerr, R.M. (1985) Higher-order derivative correlations and the alignment of small-scale structures in isotropic numerical turbulence, *J. Fluid Mech.* **153**, pp. 31–58.

Kida, S. and Ohkitani, K. (1992) Spatio-temporal intermittency and instability of a forced turbulence, *Phys. Fluids* A4, pp. 1018–1027.

Krasny, R. (1986) Desingularisation of periodic vortex sheet roll-up, *J. Computational Physics* **65**, pp. 292–313.

Le Dizès, S., Rossi, M. & Moffatt, H.K. (1996) On the three-dimensional instability of elliptical vortex subjected to stretching, *Phys. Fluids* 8(8), pp. 2084–2090.

Lundgren, T. (1982) Strained spiral vortex model for turbulent fine structure, *Phys. Fluids* **25**, pp. 2193–2203.

McWilliams, J.C. (1984) The emergence of isolated coherent vortices in turbulent flow, *J. Fluid Mech.* **146**, pp. 21–43.

McWilliams, J.C. (1990) The vortices of two-dimensional turbulence, *J. Fluid Mech.* **219**, pp. 361–385.

Moffatt, H.K. (1993) Spiral structures in turbulent flow, *New Approaches and Concepts in Turbulence, Monte Verità*, Birkhäuser Verlag Basel, pp. 121–129.

Moffatt, H.K., Kida, S. and Ohkitani, K. (1994) Stretched vortices – the sinews of turbulence; large-Reynolds-number asymptotics, *J. Fluid Mech.* **259**, pp. 241–264.

Neu, J.C. (1984) The dynamics of stretched vortices, *J. Fluid Mech.* **143**, pp. 253–276.

Pierrehumbert, R.T. (1986) Universal short-wave instability of two-dimensional eddies in an inviscid fluid, *Phys. Rev. Lett.* **57**, p. 2157.

Siggia, E.D. (1981) Numerical study of small scale intermittency in three-dimensional turbulence, *J. Fluid Mech.* **107**, pp. 375–406.

Ting, L. and Tung, C. (1965) Motion and decay of a vortex in a nonuniform stream, *Phys. Fluids* **8**, pp. 1039–1051.

Townsend, A.A. (1951) On the fine scale structure of turbulence, *Proc. R. Soc. Lond.* A **208**, pp. 534–542.

Vincent, A. and Meneguzzi, M. (1991) The spatial structure and statistical properties of homogeneous turbulence, *J. Fluid Mech.* **225**, pp. 1–25.

Yamamoto, K. and Hosokawa, I. (1988) A decaying isotropic turbulence pursued by the spectral method, *J. Phys. Soc. Japan* **57**, pp. 1532–1535.

VORTEX STRUCTURES IN TURBULENT SUPERFLUID FLOW

C.F. BARENGHI[1], D.C. SAMUELS[1], G.H. BAUER[2]
[1] *Mathematics Department, University of Newcastle, Newcastle upon Tyne, NE1 7RU England,*
[2] *Physics Department, University of Illinois, Champaign, Illinois 61801 USA,*

AND

R.J. DONNELLY[3]
[3] *Physics Department, University of Oregon, Eugene, Oregon 97403 USA*

There is current experimental interest [1] in exploiting the very low kinematic viscosity of gaseous and liquid Helium (about 100 times smaller than water's and 1000 times smaller than air's) to study very intense turbulence at ultra high Reynolds numbers or Rayleigh numbers. Our concern here is turbulence in Helium II, the quantum phase of liquid Helium, and the calculation which we describe is motivated by the construction of a superfluid wind tunnel at the University of Oregon.

We have known since the times of Landau that Helium II is the intimate mixture of two fluid components, the *superfluid* and the *normal fluid*. The former corresponds to the quantum ground state and has zero viscosity like a classical Euler fluid; the latter corresponds to the thermal excitations and is viscous like a classical Navier Stokes fluid. The superfluid contains vortex lines of quantised circulation, which scatter the thermal excitations and provide a coupling mechanism (mutual friction) between the superfluid and normal fluid components. The hydrodynamics of Helium II is therefore richer than the hydrodynamics of an ordinary classical liquid, but it can be well described using the methods and the ideas of classical fluid mechanics because it consists simply of an Euler fluid coupled to a Navier Stokes fluid via vortex filaments [2].

Because of the presence of two fluid components, the flow of Helium II is typically very different from the flow of a classical fluid, and many early experiments which initially seemed to give paradoxical results were easily

U. Frisch (ed.), Advances in Turbulence VII, 331–334.

explained using Landau's two - fluid theory. But the difference between the flow of Helium II and the flow of a classical fluid appears to exist only as long as we consider flows at relatively small Reynolds numbers. Recent experiments have showed that isothermal, intense turbulence in Helium II at very large Reynolds number is surprisingly similar to classical turbulence [3]. The evidence comes from experiments on turbulent Taylor - Couette flow between rotating concentric cylinders, turbulence in pipe and channel flows, turbulent vortex rings and turbulence created by moving a grid in a stationary sample of Helium II. To explain this evidence the idea has been put forward that at high Reynolds numbers the very high density of vortex lines which is present in the superfluid is able to *lock* together the two fluid components into a seemingly classical behaviour.

Motivated by these experiments, we have performed numerical simulations of the motion of vortex lines in an idealized model of normal fluid turbulence. The idea behind our model is that the *vortex tubes* which must be present in the (classical) normal fluid turbulence are responsible for the locking mechanism. The vortex tubes in turbulence appear with finite length and in group of several and have lifetimes of the order of few large eddy turnover times. They have been observed in the numerical calculations of classical turbulence and, on the experimental side, they have been seen in turbulent water and turbulent Helium I (the classical phase of liquid Helium). To model the vortex tubes we have chosen an Arnold - Beltrami - Childress (ABC) flow. Using cartesian coordinates x, y, z the normal fluid velocity $\mathbf{v}^n$ has components

$$v_x^n = A \sin\left(2\pi z/\lambda\right) + C \cos\left(2\pi y/\lambda\right), \tag{1}$$

$$v_y^n = B \sin\left(2\pi x/\lambda\right) + A \cos\left(2\pi z/\lambda\right), \tag{2}$$

$$v_z^n = C \sin\left(2\pi y/\lambda\right) + B \cos\left(2\pi x/\lambda\right), \tag{3}$$

where λ is a length scale and A, B and C are parameters. ABC flows are solutions of the steady Euler equation and of the time dependent, forced Navier Stokes equation. Despite the apparent simplicity, the streamlines have a complex lagrangian pattern which includes chaotic particle path at certain values of parameters. ABC flow have also been used to study turbulent processes of dynamo action in magneto - hydrodynamics. Finally, ABC flows have nonzero helicity, a property which has been associated with turbulence structures both in experiments and numerical simulations.

Our numerical simulations calculated the time evolution of an arbitrary initial vortex configuration in the presence of a driving ABC flow. The vortex lines were discretized [4] into a variable number of points, and the position $\mathbf{s}$ of each point was integrated in time according to the equation

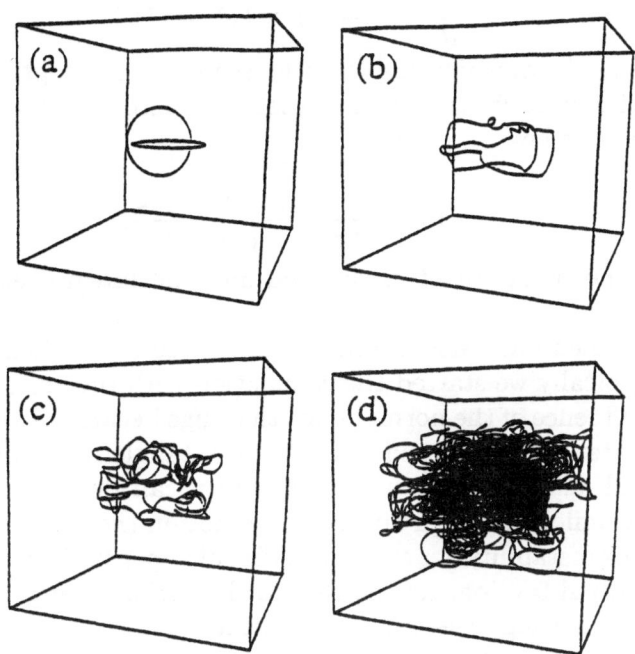

Figure 1.   Time evolution of two initial vortex rings into a turbulent tangle of vortex lines

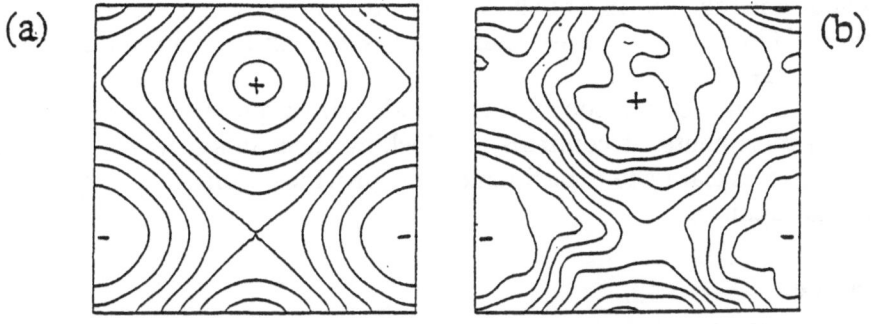

Figure 2.   Contour plots of one component of the normal fluid vorticity (a) and of the superfluid vorticity (b). Note that they match. The superfluid vorticity is obtained by dividing the computational box into small regions, and counting how many vortex lines go through each region in a given direction. The normal fluid vorticity is the vorticity of the driving ABC flow.

$$\frac{ds}{dt} = \mathbf{v}^i + \alpha \mathbf{s}' \times (\mathbf{v}^n - \mathbf{v}^i), \tag{4}$$

where a prime denotes derivative with respect to arclength, α is a known coefficient of mutual friction, and the self - induced velocity $\mathbf{v}^i$ results from the Biot - Savart law

$$\mathbf{v}^i(\mathbf{s}) = \frac{\Gamma}{4\pi} \int \frac{(\mathbf{z} - \mathbf{s}) \times d\mathbf{z}}{|\mathbf{z} - \mathbf{s}|^3}, \tag{5}$$

where Γ is the quantum of circulation and the integral extends over all vortex lines.

We performed our calculations [5] in a three dimensional periodic box of size λ. Typically we started the calculation with two initial vortex rings. Under the influence of the normal flow, the rings became distorted, the total length of vortex line increased and a vortex tangle developed (see Figure 1). We found that a vortex wave instability can destabilize vortex lines parallel to a sufficiently high normal fluid motion (hence the importance of helicity). This instability generates bundles of superfluid vortex lines which, driven by mutual friction, concentrate in the regions where the vorticity of the normal fluid is high (see Figure 2). Although the *microscopic* superfluid velocity pattern in the bundles is very complicated, its *macroscopic* average over a region larger than the intervortex separation has the same spatial pattern as the vorticity field of the normal fluid. This *vorticity matching* has been observed in the experiments. Investigation of the growth time scale for the vortex lines showed that it is of the same order of the ABC flow time scale; since the lifetime of the vortex tubes observed in turbulence is of the order of few turnover times, then there is enough time for the matching process to take place.

Although our model is too simple to make direct quantitaive comparison with the experiments, it confirms the locking mechanism which has been postulated to explain the experiments and provides a physical explanation for this mechanism.

References

1. Donnelly, R.J. (1991) *High Reynolds Number Flows Using Gaseous and Liquid Helium*, Springer Verlag.
2. Barenghi, C.F. (1997) Vortex lines and transitions in superfluid hydrodynamics, *Proc. Roy. Soc. Lond. A* **355**, 2025–2034.
3. Barenghi, C.F., Swanson, C.J. & Donnelly, R.J. (1995) Emerging issues in helium turbulence, *J. Low Temp. Phys.* **100** 385–413.
4. Samuels, D.C. (1992) Velocity matching and Poiseuille pipe flow properties of He II, *Phys. Rev. B* **50**, 11714–11724.
5. Barenghi, C.F., Samuels, D.C., Bauer, G.H. & Donnelly, R.J. (1997) Vortex lines in a model of turbulent flow, *Phys. Fluids* **9**, 2631–2643.

ON GEOMETRICAL PROPERTIES OF VELOCITY DERIVATIVES IN NUMERICAL TURBULENCE.

A. TSINOBER, M. ORTENBERG AND L. SHTILMAN
Faculty of Engineering, Tel-Aviv University
Tel-Aviv 69978, Israel

1. Introductory notes and motivation

One of one most basic phenomena and distinctive features of three-dimensional turbulence is the predominant vortex stretching, which is manifested in positive net enstrophy generation $\sigma \equiv \omega_i \omega_j s_{ij}$, $\langle \sigma \rangle > 0$. This process consists both of vortex stretching ($\sigma > 0$) and of vortex compressing ($\sigma < 0$)[1] and cannot occur without its concomitants - vortex tilting and folding with large curvature of vortex lines. The ultimate clarification of relations between curvature and dynamically relevant quantities such as enstrophy ω^2, strain $s^2 \equiv s_{ij}s_{ij}$, enstrophy generation σ and it's rate $\alpha \equiv \sigma/\omega^2$ can be obtained from looking at global properties. The hope is that some insights can be gained from local analysis. This what is mostly done in this work on the basis of a DNS data set of Navier-Stokes equations without forcing in a box with periodic boundary conditions and random Gaussian initial conditions. The results below correspond to the time moment right after the total enstrophy has reached its maximum at $Re_\lambda \approx 75$ [3].

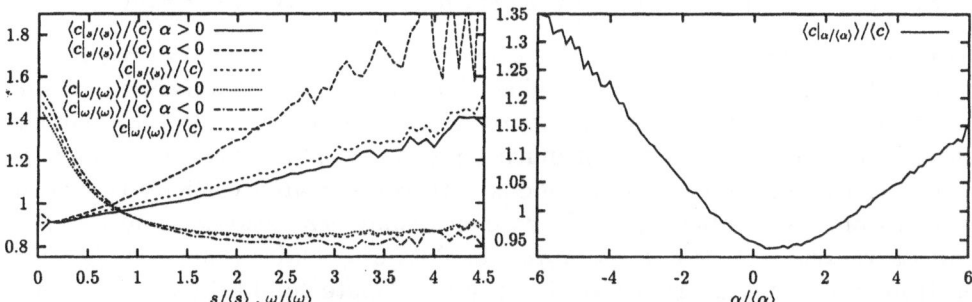

Figure 1. Conditional averages of curvature C of vortex lines in slots of ω and s for the whole field and for $\alpha > 0, < 0$ (left) and in slots of α (right).

---

[1]A turbulent flow field consists of about 2/3 points with $\sigma > 0$ and 1/3 - with $\sigma < 0$.

U. Frisch (ed.), Advances in Turbulence VII, 335–338.

2. Curvature and vortex tilting

The dependence of curvature on ω is practically the same for the whole field
[2], for positive and for *negative* rate of enstrophy generation α (fig. 1, left).
Contrary to common expectation the curvature is *increasing* with strain s
both for negative and *positive* enstrophy generation and/or its strain (fig. 1,
left). Such behavior of curvature C undermines the simple analogy with the
behavior of material lines (ML) in turbulent flows, i.e. along with correlation
between curvature and regions of vortex compressing ($\alpha < 0$) there exists also
a significant *positive* correlation between curvature C and regions of vortex
stretching ($\alpha > 0$) (fig. 1, right) in contrast with the case of material lines, in
which the analogous correlation is always *negative*.
Similar behavior is observed in closely related process - vortex tilting, which is
characterized by the rate of change of direction of vorticity $\eta_i = s_{ij}\varpi_j - \alpha\varpi_i$,
$\varpi_i = \omega_i/\omega$ [1] (fig. 2).

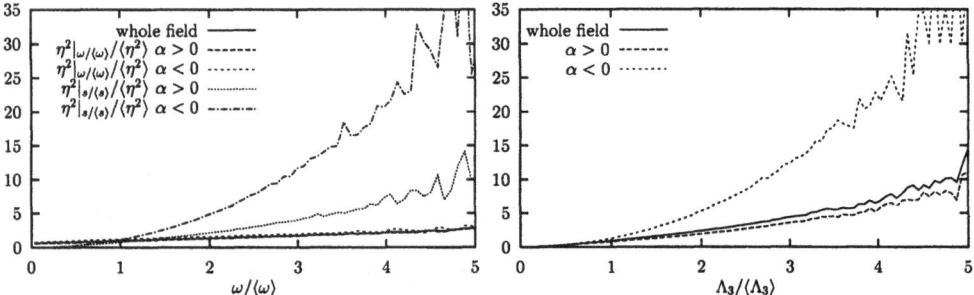

Figure 2.   Conditional averages of the magnitude of the rate of change of vorticity direction
η^2. Left - in slots of ω and s – the direction of vorticity is changing much stronger in strain
dominated regions. Right - in slots of the smallest (negative) eigenvalue Λ_3 of the rate of
strain tensor s_{ij} –the rate of change is largest in (sub)regions of vortex compression with
large magnitude of Λ_3. Note that η^2 is increasing in slots of Λ_1 and Λ_2 too but at slower
rates.

3. Regions of strongest interaction of vorticity and strain

Regions with finite curvature and vortex stretching ($\alpha > 0$) are distinct from
those with concentrated vorticity and small curvature (fig. 3) and correspond to
large strain s^2 and strongest interaction between vorticity and strain. Among
other things this is exhibited in much larger enstrophy generation σ and its
rate α in these regions than in regions with concentrated vorticity (fig 4). It
is seen from the comparison of the rate enstrophy generation $\alpha \equiv \omega_i \omega_k s_{ik}/\omega^2$
and its viscous reduction $\nu\omega_i \nabla^2 \omega_i/\omega^2$ in slots of ω and s (figure 4) that the
enstrophy dominated regions are in an approximate equilibrium in the sense
that their fairly large (but not largest!) enstrophy generation (and its rate) is
approximately balanced by the viscous reduction. In this sense they are less
active than the strain dominated regions possessing much larger (apparently
maximal) enstrophy generation and its rate which is much larger than its viscous reduction. Similar behavior is observed for magnitude of vortex stretching
vector W^2; $W_i = \omega_j s_{ij}$ (see also 2).

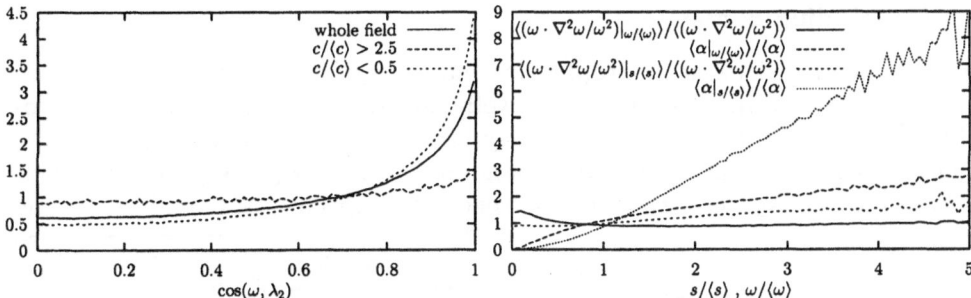

Figure 3. PDFs of $\cos(\omega, \lambda_2)$ conditioned on curvature C of vortex lines.

Figure 4. Rate of enstrophy generation $\alpha \equiv \omega_i \omega_k s_{ik}/\omega^2$ and its viscous reduction $\nu \omega_i \nabla^2 \omega_i/\omega^2$ in slots of ω and s.

The above results are consistent with those of table 1 showing that the largest contribution to the rate of enstrophy generation comes from the regions associated with the *largest* eigenvalue Λ_1 of the rate of strain tensor s_{ij} and not from the ones associated with the *intermediate* eigenvalue Λ_2 to which belong the regions of concentrated vorticity.[2]

	$\langle\lambda_1 \cos(\omega, \lambda_1)\rangle$	$\langle\lambda_2 \cos(\omega, \lambda_2)\rangle$	$\langle\lambda_3 \cos(\omega, \lambda_3)\rangle$
DNS	1.47	0.49	-0.97
Grid	1.17	0.46	- 0.63

TABLE 1. Contribution to the total mean of enstrophy generation rate $\langle\alpha\rangle \equiv \langle\Lambda_i \cos^2(\omega, \lambda_i)\rangle$ from the terms corresponding to the eigenvalues Λ_i of the rate of strain tensor s_{ij}. Grid turbulence and DNS; $Re_\lambda \approx 75$.

4. Rate of change of enstrophy generation

This is governed by $W^2 - \omega_i \omega_j \frac{\partial^2 p}{\partial x_i \partial x_j}$ [4]. The last term can be interpreted as interaction of vorticity and pressure Hessian. Its behavior is *qualitatively* different in enstrophy (negative and decreasing) and strain (positive and increasing) dominated regions, which results in considerable reduction of $W^2 - \omega_i \omega_j \frac{\partial^2 p}{\partial x_i \partial x_j}$ in the enstrophy dominated regions and its strong increase in the strain dominated ones (fig. 5). The difference in behavior of quantities characterizing various aspects of nonlinearities such as shown in figures 2, 4, 5 and some others in strain and enstrophy dominated regions can be interpreted as reduction of nonlinearities in the latter. Nevertheless, enstrophy dominated regions possess, e.g. rather large enstrophy generation.

[2]Similar results are true for the contributions to the mean enstrophy generation $\langle\sigma\rangle \equiv \langle\omega^2 \Lambda_i \cos^2(\omega, \lambda_i)\rangle$ [3]

5. Concluding remarks

It is common to look at enstrophy dominated regions as contrasted to those
dominated by strain. The former are well defined by, say, high enough enstrophy ω^2 and are tube-like objects. They form a subset of much larger *locally*
quasi-two-dimensional regions corresponding to large $\cos(\omega, \lambda_2)$, i.e. alignment
between ω and λ_2. However, it is not enough to specify the magnitude of the

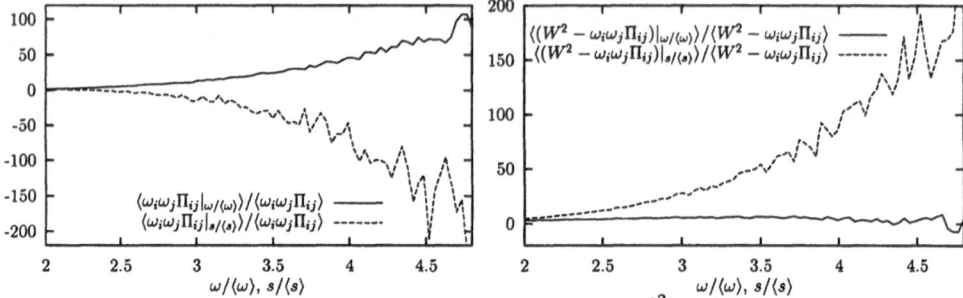

Figure 5.   Conditional averages of the nonlocal term $\omega_i \omega_j \frac{\partial^2 p}{\partial x_i \partial x_j}$ (left) and of the inviscid
rate of change of enstrophy generation $W^2 - \omega_i \omega_j \frac{\partial^2 p}{\partial x_i \partial x_j}$ (right) in slots of ω and s.

total strain s^2 in order to 'visualize' in a unique way the strain dominated
regions. They contain a number of *qualitatively* different regions, e.g. the following dynamically significant subsets [3]: i - The largest enstrophy generation
occurs in regions of strongest vorticity/strain interaction, which are different
from those with concentrated vorticity. These regions contribute most to the
enstrophy generation and its rate and at least in this respect are dynamically
more important than those of concentrated vorticity. They are associated with
alignment between ω and λ_1, large values of Λ_1, and fairly large curvature of
vortex lines; ii - Regions with large magnitude of Λ_3, alignment between ω and
λ_3 large curvature of vortex lines, vortex compressing, tilting and folding. Considerable part of curvature of vortex lines is generated in these regions along
with the other part produced in the regions of largest enstrophy generation via
self-induction.

References

1. Constantin, P.: Geometrical statistics in turbulence, *SIAM Rev.*, **36** (1994), 73 - 98.
2. Galanti, B., Procaccia, I. and Segel, D.: Dynamics of vortex lines in turbulent flows, *Phys.
    Rev.*, **E 54** (1996), 5122 - 5133.
3. Tsinober, A., Shtilman, L. and H. Vaisburd H.: A study of properties of vortex stretching
    and enstrophy generation in numerical and laboratory turbulence, *Fluid Dyn. Res.*, **21**
    (1997), 477 - 494.
4. Tsinober, A., Shtilman, L. and Vaisburd, H.: Vortex stretching and enstrophy generation
    in numerical and laboratory turbulence, *Lect. Notes Phys.*, **462** (1995), 9 - 16.

[3]In addition to regions wrapped around the enstrophy dominated regions and associated
with alignment between ω and λ_2 (just like in the enstrophy dominated regions) and mostly
positive Λ_2.

STATISTICAL PROPERTIES OF RANDOM DISTRIBUTION OF STRAINED VORTICES IN TURBULENCE

N. HATAKEYAMA AND T. KAMBE

Department of Physics, University of Tokyo, Hongo, Bunkyo-ku, Tokyo 113-0033, Japan.

At high Reynolds numbers, a number of elongated intense vortex structures are observed to distribute randomly in space, which has been reported in recent computer simulations [1, 2] and experiments [3] of turbulence. These are called the *worm* structures. Each worm is found to be the *Burgers vortex* under local straining [2] and to be responsible for the signal usually referred to as *intermittency* [3]. Bearing these in mind, statistical properties of a model field associated with a random distribution of the isolated Burgers vortices are investigated and found to be in good agreement with those obtained in the real turbulence field [4].

The turbulent velocity field is denoted as $v(x)$. Defining the *longitudinal* difference in the separation direction s by $\Delta v_\ell(x, s) = [v(x + s) - v(x)] \cdot s/s$ where $s = |s|$, the pth-order structure function is given by $S_p = \langle [\Delta v_\ell(x, s)]^p \rangle$ where $\langle \cdot \rangle$ is an ensemble average. In the homogeneous isotropic turbulence, the structure function S_p is independent of the location r and the direction s/s, and follows a power law in the *inertial range* as $S_p(s) \sim s^{\zeta_p}$ where ζ_p is the *scaling exponent* of the pth order. In the limit of very high Reynolds number, the third-order structure function in the inertial-range is described exactly by the Kolmogorov's *four-fifths law*:

$$S_3(s) = \langle (\Delta v_\ell)^3 \rangle = -\frac{4}{5} \varepsilon s, \qquad (1)$$

where ε is the mean rate of the energy dissipation [5]. The Kolmogorov theory also implied the scaling exponents of the other-order by dimensional arguments, $\zeta_p = p/3$ (referred to as K41 below), but it is known that the exponents in real turbulence deviate from K41.

We consider the velocity field of the strained vortex (Burgers vortex) which is a superposition of a velocity $v_\omega = (0, v_\theta(r), 0)$ induced by an axisymmetric vortex of Gaussian core and an irrotational solenoidal straining

U. Frisch (ed.), Advances in Turbulence VII, 339–342.

velocity $\boldsymbol{v}_e = (-ar, 0, 2az)$ where $a > 0$, in the cylindrical coordinate system (r, θ, z). Then the total flow field is $\boldsymbol{v}(\boldsymbol{x}) = (-ar, v_\theta(r), 2az)$ and the exact steady solution of the incompressible Navier-Stokes equation is derived as $v_\theta = (\Gamma/2\pi r_b)[1 - \exp(-\hat{r}^2)]/\hat{r}$ where $\hat{r} = r/r_b$, $r_b = (2\nu/a)^{1/2}$ and Γ is the total circulation.

The structure function of this velocity field is given by the following procedures. Firstly a spherical average is taken with respect to a running point $\boldsymbol{r} + \boldsymbol{s}$ centered at a fixed $\boldsymbol{r}$:

$$\langle (\Delta v_\ell)^p \rangle_{\text{sp}} (r, s) = \frac{1}{4\pi} \int_{-\pi}^{\pi} d\phi \int_0^{\pi} (\Delta v_\ell)^p \sin \zeta \, d\zeta, \tag{2}$$

in the spherical polar coordinates $\boldsymbol{s} = (s, \zeta, \phi)$. Here all directional distribution of the vortices is considered to be averaged statistically.

Secondly a volume average is taken with respect to the point $\boldsymbol{r}$:

$$\langle (\Delta v_\ell)^p \rangle_{\text{vol}} (s) = \frac{1}{\pi r_0^2 z_0} \int_{-z_0/2}^{z_0/2} dz \int_0^{r_0} \langle (\Delta v_\ell)^p \rangle_{\text{sp}} 2\pi r \, dr, \tag{3}$$

where the z-integral can be dropped because the longitudinal velocity increment given by the Burgers vortex is independent of z. At this stage the structure functions are estimated for various strengths of the Burgers vortices with the fixed radius r_b. Choosing suitable range of the volume integration with respect to r, the scaling region corresponding to the inertial range is obtained. Especially, using $r_0 = 2.5 r_b$, the third-order structure function is found to follow the four-fifths law (1) (with ε averaged over the same volume) independent of the *vortex Reynolds number* $R_\Gamma = \Gamma/\nu$, while the scaling exponents of the other-order structure functions deviate increasingly below K41 as R_Γ becomes larger [4].

Thirdly, the average with respect to R_Γ is taken as

$$\langle (\Delta v_\ell)^p \rangle = \int_0^\infty \langle (\Delta v_\ell)^p \rangle_{\text{vol}} P(R_\Gamma) \, dR_\Gamma, \tag{4}$$

where $P(R_\Gamma)$ is a probability density of R_Γ. Here we propose $P(R_\Gamma) = (C^3/2) R_\Gamma^2 \exp(-C R_\Gamma)$ where $C = (3/4\pi) R_\lambda^{-1/2}$, so that $P(R_\Gamma)$ is approximately the same form especially in the limit of $R_\Gamma \to 0, \infty$ as those obtained in DNS [2] and the experiment [3]. The resulting structure functions for $R_\lambda = 2000$ shown in figure 1 actually lead to the power-laws. In figure 2, the scaling exponents ζ_p of the structure functions are found to be in good agreement with the real turbulence date [1, 3]. From these results, the anomalous scaling of the high-order structure functions of the turbulence field is regarded as controlled by the distribution of the strengths of the worm structures.

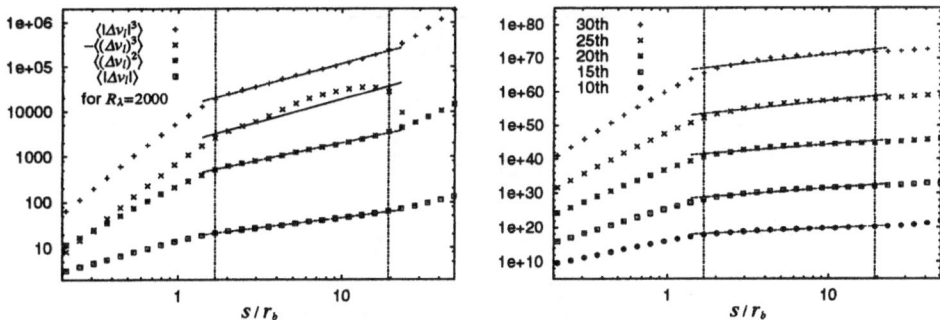

Figure 1. The first- to third-order structure functions (left) and the higher-order structure functions (right) for $R_\lambda = 2000$. A solid line represents the four-fifths law (1) and dashed lines are log-linear least-square fits within the inertial range (between two vertical dot-dashed lines).

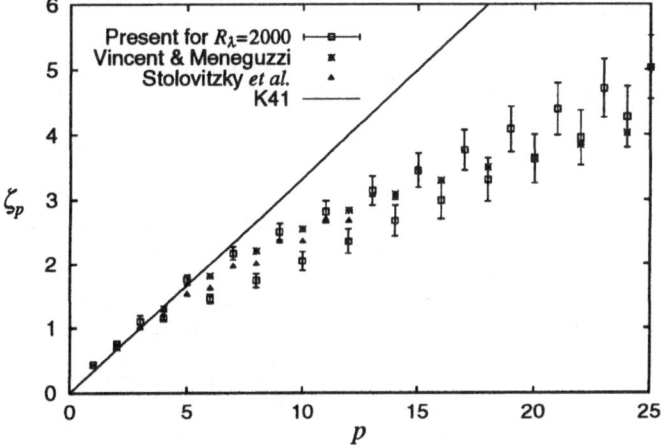

Figure 2. The scaling exponents ζ_p of the structure functions for $R_\lambda = 2000$ are compared with those of DNS [1] and the experiment [6].

It is considered that there is an invariant measure of the Navier-Stokes turbulence in the limit of large Reynolds numbers, for which a probability distribution function (PDF) $P(s, \Delta v_\ell)$ is defined and the pth-order structure function is expressed as an integral $S_p(s) = \int (\Delta v_\ell)^p P(s, \Delta v_\ell)\, d\Delta v_\ell$. This is modeled by the strained vortex in the present study. This PDFs for three values are shown in figure 3. It is found that the PDF of the longitudinal velocity increment with the separation scaling to the inertial and viscous range has exponential tail, and deviates increasingly away from the Gaussian distribution as the separation decreases. These are consistent with those observed in real turbulence.

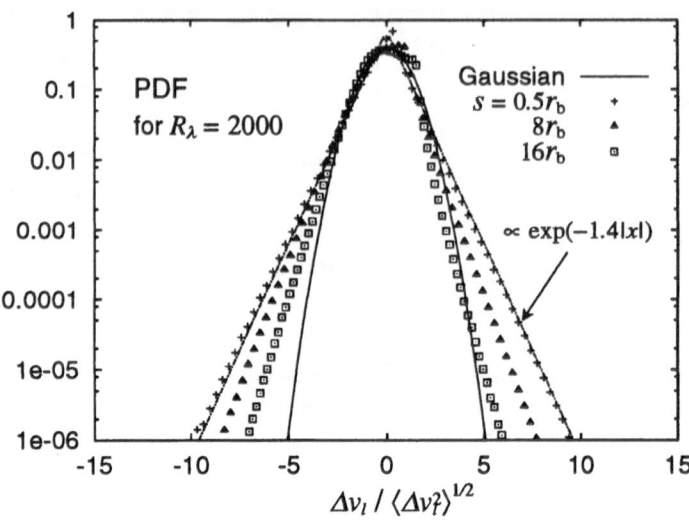

Figure 3. The PDFs of the longitudinal velocity increment for three different separations, where $s = 0.5r_b$ is the scale within the viscous range and $s = 8r_b, 16r_b$ are the scales within the inertial range.

References

1. Vincent, A. & Meneguzzi, M. (1991) The spatial structure and statistical properties of homogeneous turbulence, *J. Fluid Mech.* **225**, pp. 1–25.
2. Jimenéz, J., Wray, A.A., Saffman, P.G. & Rogallo, R.S. (1993) The structure of intense vorticity in isotropic turbulence, *J. Fluid Mech.* **255**, pp. 65–90; Jimenéz, J. & Wray, A.A. (1994) On the dynamics of small-scale vorticity in isotropic turbulence, *CTR Annual Res. Briefs*, pp. 287–312.
3. Belin, F., Maurer, J., Tabeling, P. & Willaime, H. (1996) Observation of intense filaments in fully developed turbulence, *J. Phys. II France* **6**, pp. 573–583.
4. Hatakeyama, N. & Kambe, T. (1997) Statistical laws of random strained vortices in turbulence, *Phys. Rev. Lett.* **79**, pp. 1257–1260.
5. Kolmogorov, A.N. (1941) Dissipation of energy in locally isotropic turbulence, *C. R. Acad. Sci. URSS* **32**, pp. 16–18.
6. Stolovitzky, G., Sreenivasan, K.R. & Juneja, A. (1993) Scaling functions of scaling exponents in turbulence *Phys. Rev. E* **48**, pp. 3217–3220.

STRUCTURE OF A STRETCHED VORTEX

P. PETITJEANS
Laboratoire de Physique et Mécanique des Milieux Hétérogènes,
Ecole Supérieure de Physique et de Chimie Industrielles
10, rue Vauquelin, 75005 Paris, France

Abstract

Experimental measurements performed under conditions which reproduce most of the dynamical characteristics of the natural evolution of vorticity filaments in turbulent flows are presented here. Strong deviations from Burgers vortex (which is a non-confined stretched vortex model) are observed and analyzed. A new model of non-uniform stretched vortex is proposed.

1 Introduction

Recent recognition of the role of small scale filaments and layered vortex sheets in turbulence have led to interest in the dynamical behavior of vortex structures with concentrated vorticity. Indeed, the instability, reconnection, merging and breakdown of these vortical structures are at the origin of intermittence in turbulent energy dissipation. Numerical simulations (Siggia, 1981; Jimenez, 1991) and experiments (Cadot et al., 1995) show clearly that the dynamics of stretched filaments are responsible for the deviation from the Gaussian behavior observed in turbulence statistics. Structures with concentrated vorticity arise in many different situations. For instance, concentrated vorticity can appear as a result of instabilities such as the Bénard - Von Karman vortex street, and in shear flows. If the vorticity field is stretched especially in the case of structures resulting from instabilities, the vorticity may be strongly enhanced. This process occurs naturally in the complex field of a turbulent flow. In these flows, vorticity can be locally stretched by the local velocity gradients. This is the reason a Burgers model (where the stretching is constant and uniform) may not be the relevant model for these structures. In order to better understand the detailed mechanisms of the evolution of a stretched vortex, we have performed a new experiment where isolated stretched vortices are produced. Most of the dynamical characteristics of the evolution of vorticity filaments in turbulent flows is reproduced in our channel, in a well-controlled experimental situation.

U. Frisch (ed.), Advances in Turbulence VII, 343–346.

2 Experiment and results

The experiments are performed in a water channel where the first section generates a laminar flow, and where the second is the working section which consists of a straight section 60 cm long. In the middle of this section, a slot is made in the bottom of each lateral side which is used to create a controlled suction of the main flow. The initial vorticity coming from the boundary layer on the lower wall is enhanced by the stretching and produces a vortex on the lower wall. Below a certain flow rate at the exit of the channel, the vortex remains stationary at this location (Petitjeans et al., 1997), sometimes with a low-amplitude precession. Above this flowrate, the vortex is elongated in the direction of the flow and may break up, in which case.

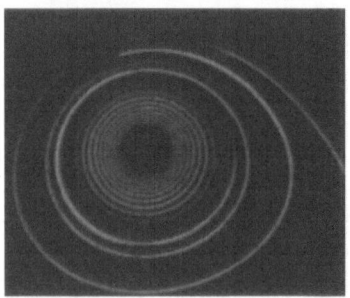

Figure 1

A cross-section of the visualization at the middle of the channel is shown Fig. 1. From visualizations, it is possible to measure the components $U_r(r)$ and $U_\theta(r)$ of the vortex, by following the front of the dye in time, taking a frame each 0.04 s. After three or four turns around the vortex, the curve tends to a logarithmic law defining a logarithmic spiral motion $r \sim \exp(-\alpha\theta)$ where α is a constant (Vassilicos and Brasseur, 1996). Note that a logarithmic spiral has a Kolmogorov dimension $D_K = 1$, i. e. it has a trivial self-similar structure. Only an algebraic spiral $(r \sim \theta^\alpha)$ has a non-trivial self-similar structure and a Kolmogorov dimension $D_K > 1$.

The stretching velocity $U_s(r, s)$ has also been measured by laser Doppler velocimetry. Figure 2 shows the radial dependence of the stretching velocity $U_s(r)_{s\ fixed}$ for two different spanwise positions $z = 2.5$ cm and $z = 4.5$ cm from the center of the channel. We observe that the stretching essentially occurs in a small region around the vortex axis. Figure 3 shows the axial dependence of the stretching velocity $U_s(s)_{r=0}$ along the axis of the vortex. From the center of the channel (in the spanwise direction) to about half way from the side, the stretching, i. e. the gradient of the stretching velocity along the vortex axis, is constant.

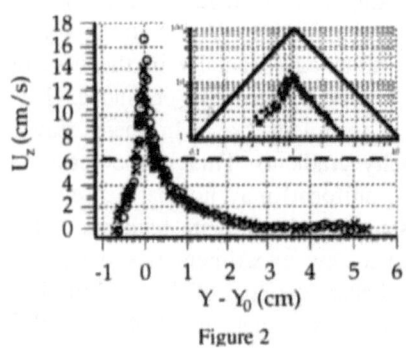

Figure 2

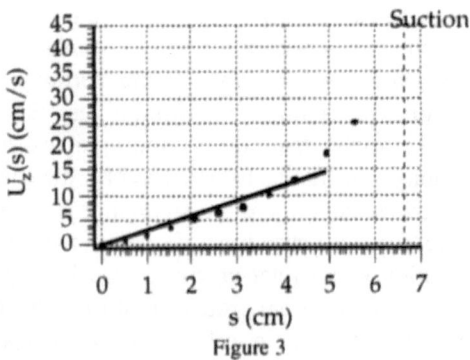

Figure 3

The figure 2 illustrate shows clearly that the stretching is not transversally uniform (i. e. it does depend on the distance r from the axis). Such a result was to be expected since in our experiment the suction is located at a particular position on the lateral walls. We claim that this is also the case for most naturally occurring stretched vortices, and is particularly the case for filaments of vorticity in turbulent flows, where local stretching strongly enhance the vorticity (parallel to the stretching) of the turbulent flow. A model often used to describe these vortices is the Burgers vortex, which is the stationary vortex created by constant and uniform stretching (viscous diffusion of vorticity exactly balances vorticity amplification by stretching):

$$U_z = Az \tag{1}$$

$$U_r = -\frac{1}{2}Ar \tag{2}$$

$$U_\theta = \frac{\Gamma}{2\pi r}\left\{1 - \exp\left(-\frac{A}{4v}r^2\right)\right\} \tag{3}$$

$$\omega_z = \frac{\Gamma A}{2\pi}\exp\left(-\frac{A}{4v}r^2\right) \tag{4}$$

where the constant A is the stretching rate, Γ is the circulation, v is the kinematic viscosity, and ω is the vorticity. This model cannot represent our kind of stretched vortex, which is not stretched uniformly. Figure 4, which represents the radial velocity $U_r(r)$ shows that this velocity scales like $-r^{1/2}$. From other measurements of $U_r(r)$, we observe that it does not depend on z, at least far enough from the walls. From figure 2 and 4, we observe that the stretching velocity $U_z(r, z)$ scales as $r^{-1/2}z$. The stretching velocity scales linearly with the curvilinear axis s, at least far enough from the lateral walls, i. e. in about 3/4 of the width. These measurements are independent and coherent with the continuity equation. So, we obtain:

$$U_r = -\frac{2}{3}Ar^{1/2} \tag{5}$$

$$U_z = Azr^{-1/2} \tag{6}$$

where A is a constant.

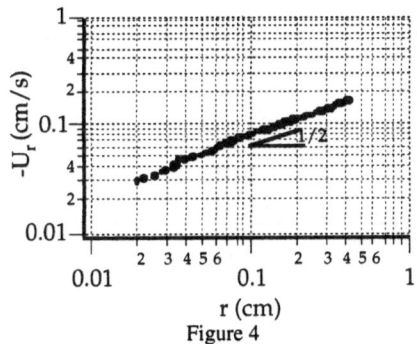

Figure 4

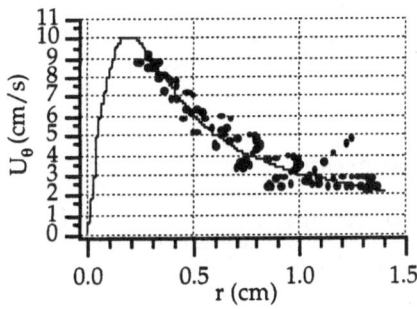

Figure 5

The important point is that the stretching here is confined around the axis of the vortex. These expressions are now used to solve the vorticity equation where we assume, at the order zero, that the vorticity is stationary, is axial, and depends only on r. The vorticity obtained with U_r and U_z is

$$\omega_z = \omega_0 \exp\left(\frac{-4A}{9v} r^{3/2}\right) \tag{7}$$

Then, from Kelvin's circulation theorem, the azimuthal velocity is:

$$U_\theta = \frac{c}{r} \int_0^r x.\exp\left(\frac{-4A}{9v} x^{3/2}\right) dx \tag{8}$$

where c is a constant. This model of a stretched vortex is the simplest possible model consistent with the measured velocity profiles. It is not an exact solution of the Navier Stokes equations because of the curved form of the vortex, because the vortex is slightly non-stationary, and because of the influence of the wall which deforms the vortex. A formal perturbative development is required to quantify the approximation and work is in progress to complete this calculation.

3 Conclusion

Experimental observations show that most stretched vortices in nature in general and in turbulence in particular are non-uniformly stretched, which is inconsistent with the basic assumptions of the Burgers model. On the basis of our results, we propose a new vortex description which can better model these vortices. The vortex we have proposed takes into account the non-uniformity of the stretching. This simple model clearly show that the non-uniform stretching significantly alters the vorticity profile. Work is still in progress in order to take into account of the deformation and non-stationarity of our vortex, and another experiment is planned to eliminate these problems and to create a truly stationary, isolated, locally stretched vortex.

Acknowledgments:
I would like to thank Jean-Henry Robres, Nicholas Kevlahan, Alain Pumir, Eduardo Wesfreid for helpful discussions.

References:

Cadot O., Douady S., and Couder Y., 1995, Characterization of the low-pressure filaments in a three-dimensional turbulent shear flow, Phys. Fluids, **7**, 630-646
Jimenez J., 1991, On small scale vortices in turbulent flows, Phys. Fluids A, **4**, 652-654
Petitjeans P., Wesfreid J. E., and Attiach J.C., 1997, Experiments in Fluids, **22**, 351-353
Petitjeans P., and Wesfreid J. E., 1997, Applied Scient. Research, **57**, 279-290
Siggia E. D., 1991, Numerical study of small-scale intermittency in three-dimensional turbulence, J. Fluid Mech., **107**, 375-406
Vassilicos J. C. and Brasseur J. G., 1996, Self-similar spiral flow structure in low Reynolds number isotropic and decaying turbulence, Phys. Rev. E, **54** (1), 467-485

DYNAMICS OF LOW-PRESSURE SWIRLING VORTICES IN TURBULENCE

S. KIDA AND H. MIURA

Theory and Computer Simulation Center, National Institute for Fusion Science, Gifu 509-52, Japan

The dynamics of slender swirling vortices in a fully developed turbulence is investigated numerically. The vortices are educed by the use of our new method, the sectional-swirl-and-pressure-minimum scheme (Refs. [1] and [2]). In this scheme the central lines of vortices are traced by searching the sectionally local minimum of the pressure under the swirl condition that the velocity field projected on a plane normal to the third eigen-vector of the pressure hessian is elliptical. The vortex core is defined as those regions in which the above swirl condition is satisfied. The core radius estimated by this method is $1.58r_*$ for an axi-symmetric Burgers vortex tube with vorticity distribution $\omega(r) = \omega_0 \exp[-r^2/2r_*^2]$. The vortex identified by this scheme may be conveniently called the low-pressure swirling vortex.

We present here some of the physical characteristics of the low-pressure vortices and their dynamical behavior in a decaying numerical Navier-Stokes turbulence in a periodic cube of side 2π. The initial velocity field is given by a prescribed energy spectrum $E(k) = (k/k_0)^4 \exp\left[-2(k/k_0)^2\right]$ ($k_0 = 4$) with randomized phases. The dealiased spectral code with resolution $N^3 = 128^3$ is employed. The viscosity is set at $\nu = 5 \times 10^{-5}$.

In Fig. 1(a) we plot the temporal evolution of the enstrophy. It takes a maximum at $t \approx 87$. The distribution of the core-radius R of the tubular vortices is shown by the thick solid line in Fig. 1(b) at $t = 144$ by which the smallest scales of motion have been fully excited and the tubular vortices well developed. The Taylor micro-scale length λ and the Kolmogorov length $l_K = \epsilon^{-\frac{1}{4}}\nu^{\frac{3}{4}}$ at this time are 0.31 and 0.028, respectively, which may be compared with the effective mesh-size of the simulation $\Delta x = 2\pi/128 = 0.049$. The micro-scale Reynolds number R_λ is 46. The thin (or dashed) line in Fig. 1(b) indicates the PDF of the core-radius of stronger (or weaker) vortices with vorticity greater (or smaller) than the rms value (cf. Jiménez

U. Frisch (ed.), Advances in Turbulence VII, 347–348.

et al. 1993 [3] and by Belin et al. 1996 [4]). The mean core-radius is $5.6l_K$
both for the total and the stronger vortices.

Three snaps of the vortex axes are shown in Fig. 2. It is interesting to
see that those two vortices highlighted are approaching with each other in
anti-parallel Ref. [5]. Such an anti-parallel approach is considered one of the
fundamental dynamical processes of tubular vortices as well as the vortex
reconnection.

References

1. Miura, H. & Kida, S. (1997) Identification of Tubular Vortices in Turbulence, *J. Phys. Soc.Japan* **66**, pp. 1331–1334
2. Kida, S. & Miura, H. (1997) Identification and analysis of vortical structures, *Eur. J. Mech./B* (submitted)
3. Jiménez, J., Wray, A.A., Saffman, P.G. & Rogallo, R.S. (1993) The structure of intense vorticity in isotropic turbulence. *J. Fluid Mech.* **255**, 65-90.
4. Belin, F., Maurer, J. Tabeling, P. & Willaime, H. (1996) Observation of intense filaments in fully developed turbulence. *J. Phys. II France* **6**, 573-583.
5. Siggia, E.D. (1985) Collapse and amplification of a vortex filament. *Phys. Fluids* **28**, 794-805.

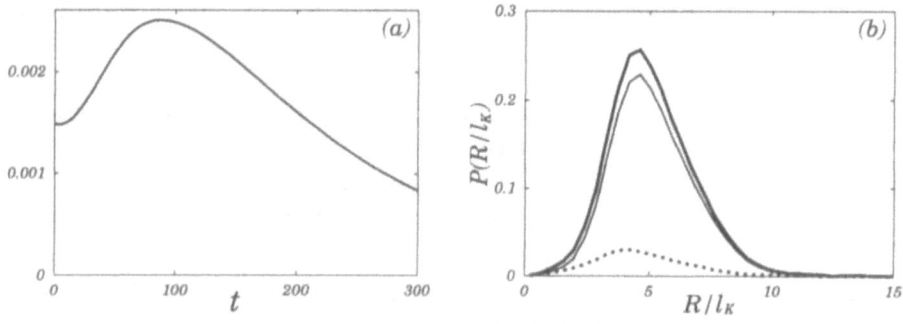

Figure 1.   (a) Evolution of enstrophy. (b) PDF of core-radius of tubular vortices.

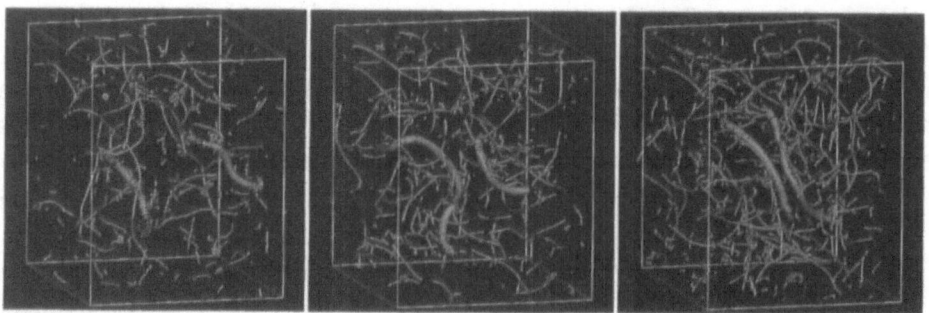

Figure 2.   Anti-parallel approach of two vortices. $t = 0$, 30 and 80.

EXPERIMENTAL INVESTIGATION OF THE INTERACTION BETWEEN STRETCHING AND VORTICITY IN CONSTRAINING GEOMETRIES

B. ANDREOTTI, S. DOUADY AND Y. COUDER
Laboratoire de Physique Statistique, Ecole Normale Supérieure
24 rue Lhomond, 75231 Paris Cedex 5, France.

1. Introduction

It is possible, using well chosen geometries, to generate flows which constrain specific global distributions of vorticity and strain. We report the investigation of two such flows. The first is initially dominated by strain and the second by vorticity. In each case, we both study directly the interaction between stretching and vorticity at moderate Reynolds numbers and some properties of the turbulent regime.

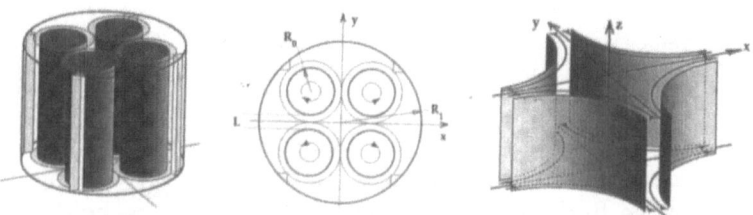

Figure 1. (a) Sketch of the four rollers apparatus, (b) its cross section, (c) the corresponding hyperbolic flow.

2. The four rollers experiment

We investigated experimentally the quasi hyperbolic flow obtained in the four rollers experiment introduced by Taylor [1] (Fig. 1). Observation of the central region of the cell at low Reynolds number (Fig. 2 a) shows that a good approximation of the hyperbolic flow ($v_x = \gamma x, v_y = -\gamma y, v_z = 0$) is obtained (pure strain). We measured the velocity gradient in the central zone and found it proportional to the angular velocity of the cylinders: $\gamma = k\Omega$, the coefficient of proportionality k being a characteristic of the

U. Frisch (ed.), Advances in Turbulence VII, 349–352.

overall geometry of the system.   The stability of the ideal hyperbolic flow

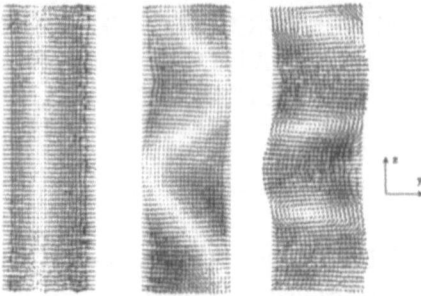

Figure 2.   The velocity field measured by a PIV technique in the (Oy,Oz) plane below the instability threshold (a), just above it (b) and well above it (c). Spanwise only a part of the field is shown.

was first investigated theoretically by Aryshev [2] and by Lagnado et al [3, 4] who concluded to its instability.   Kerr and Dold [5] also visited this

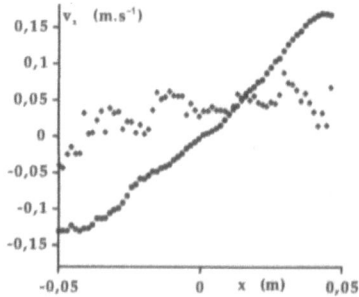

Figure 3.   The velocity profile in the stretching direction (Ox). The steepest curve is an averaged profile $V_x(x)$. The other one corresponds to a single profile taken along the axis of a vortex.

problem and found a class of Burgers like steady solutions, characterised by a periodic array of alternate vortices aligned in the stretching direction. Experimentally we find that over a certain velocity threshold the central line destabilizes into a sinusoid (Fig. 2 b). The instability threshold can be defined by a Reynolds number $Re = \frac{\gamma L^2}{\nu}$ based on the stretching γ and on L the gap between two adjacent cylinders. With this definition we find [6] the threshold value to be $Re_c = 19$. Above this transition the formation of an array of steady alternate vortices is observed (Fig. 2 c). The distance separating two successive vortices in the vertical direction is of the order of L the gap between the cylinders. In our geometry we observe two pairs of opposite vortices and two disturbed structures near the ends. At approximately $Re \sim 45$ the vortices become chaotic. Well defined vortices are observed in all the explored range of Reynolds numbers (up to $Re \sim 10000$). By means of an ultrasonic probe, we measure instantaneously the whole longitudinal velocity along the stretching direction Ox at high Reynolds numbers. Though fluctuating most of these profiles are similar and their average is shown on Fig. 3. They exhibit a strong gradient in the cell's centre, forced by the rollers. It is possible to trigger single measurements precisely at the time when the ultrasonic beam coincides with the axis of a vortex. These profiles *always* show a drastic reduction of the stretching as seen on Fig. 3.

3. The central vortex geometry

This set up is formed of a cylindrical tank with a rotating bottom, a system well known to create a central vortex [7, 8]. In our system (Fig. 4), an axial pumping is added along the vortex axis. When the pumping is switched

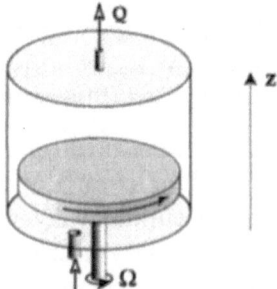

Figure 4.   Sketch of the second experiment showing the cylindrical tank, the rotating disk and the pumping hole.

on it generates an axial velocity gradient which stretches the vortex. We measure this longitudinal $V_z(z)$ velocity profile using again the ultrasonic technique. Figure 5 shows six profiles obtained after the pump has been switched off. The longitudinal gradient first increases (Fig. 5 a), reaches a

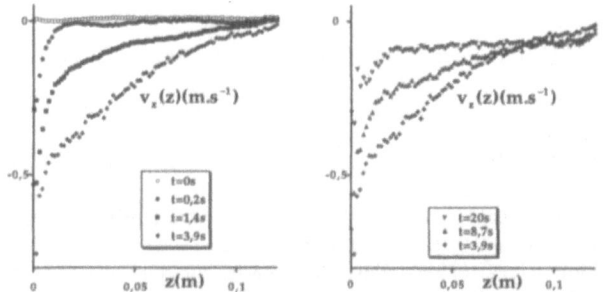

Figure 5.   The temporal evolution of the longitudinal velocity profile in the core of the vortex after the pump has been switched on. (a) Profiles at times $0s$, $0.2s$, $1.4s$, $3.9s$. (b) Profiles at times $3.9s$, $8.7s$, $20s$.

large value which stretches and amplifies the vortex but decreases at later times (Fig 5 b) and stabilises at a weak value in the steady regime [6]. This effect is very similar to that described in part 2. In both experiments there exists an amplification of the vorticity by the stretching. In turn, the vortex rotation retroacts on the stretching itself and reduces it. This is a kind of

Lenz law which has not considered before. This effect can be ascribed to
the bidimensionalisation induced by the vortex rotation. In collaboration
with J. Maurer we used a similar geometry to obtain a large Reynolds
number vortex in low temperature helium gas [9]. The tangential velocity

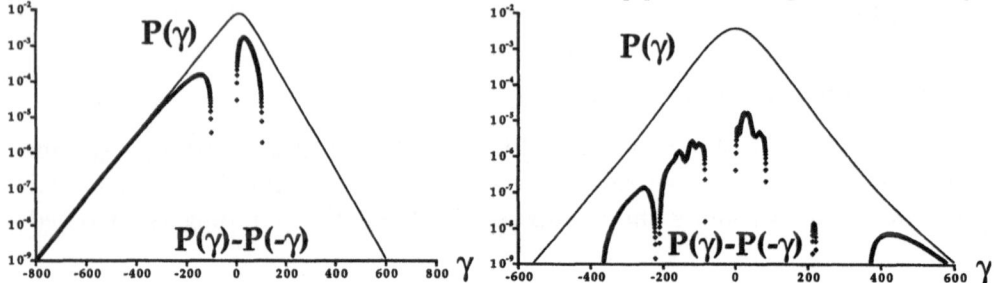

Figure 6. The PDFs of the longitudinal velocity derivative (fine lines) and their odd
parts (diamonds), (a) far from the vortex and (b) near the vortex core.

fluctuations are recorded for different positions of a probe which can be
set at a tunable radial position. Far from the vortex, the fluctuations are
similar to those of a turbulent boundary layer. As usual, the stretching
probability distribution function is assymetric (Fig. 6 a), indicating both
a statistical favouring of stretching over compression and the existence of
an energy flux from large to small scales. In contrast, in the vortex central
region some drastic changes are observed [9]. In particular, the longitudinal
velocity derivative PDF (stretching) becomes nearly symetric. This result
clearly indicates that the large scale rotation reduces the trend to (small
scale) vortex stretching usually observed for such large Reynolds numbers.

References

1. Taylor, G. I. (1934) The formation of emulsions in defibnable field of flow, *Proc. Roy. Soc. A* **146**, 107–125.
2. Aryshev, Y. A., Golovin, V. A. and Ershin, S. A. (1982) Stability of colliding flows, *Fluid Dyn.* **1** 755–759.
3. Lagnado, R. R., Phan Thien, N. and Leal, L. G. (1984), The stability of two dimensional linear flows, *Phys. Fluids* **27**, 1094–1101.
4. Lagnado, R. R. and Leal, L. G. V (1990) Visualization of three-dimensonnal flow in a four-roll mill experiment, *Exp. Fluids* **9**, 25–32.
5. Kerr, O. and Dold, J.W. (1994) Periodic steady vortices in a stagnation point flow, *Fluid Mech.* **276**, 307–325.
6. Andreotti B., Douady S. and Couder Y. (1997) About the interaction between vorticity and stretching in coherent structures, in *Turbulence modelling and vortex dynamics*, edited by O. Boratav, A. Eden and A. Erzan, Springer, 92–107.
7. Turner, J. (1966) The constraint imposed on tornado like vortices by the top and bottom boundary conditions, *J. Fluid Mech* **25**, 377–386.
8. Escudier, M. P. (1984) Observations of the flow produced in a cylindrical container by a rotating endwall, *Exp. Fluids* **2**, 189–196.
9. Andreotti B. Maurer, J. Couder, Y. and Douady S. (1998) Experimental investigation of turbulence near a large scale vortex, to appear in *Eur. Jour. of Mech. B Fluids*.

STRETCHING OF VORTICITY AND PASSIVE VECTORS IN TURBULENCE

K. OHKITANI

RIMS, Kyoto University, Kyoto 606-8502, Japan

Vortex stretching process in turbulence is a nonlinear process. To characterize the underlying vorticity-strain correlation, we numerically compare vortex stretching with a linear stretching process of material line elements with diffusion, hereafter called passive vectors. We point out crucial differences between them.

We consider the 3D Navier-Stokes equations in the vorticity form

$$\frac{\partial \omega}{\partial t} + (u \cdot \nabla)\omega = (\omega \cdot \nabla)u + \nu \nabla^2 \omega, \tag{1}$$

$$\nabla \cdot u = \nabla \cdot \omega = 0, \quad \omega = \nabla \times u$$

and the equations for the passive vectors δl

$$\frac{\partial \delta l}{\partial t} + (u \cdot \nabla)\delta l = (\delta l \cdot \nabla)u + \nu \nabla^2 \delta l, \tag{2}$$

$$\nabla \cdot \delta l = 0$$

with standard notations.

If the initial conditions $\omega(x,0)$ and $\delta l(x,0)$ are the same, then the solutions are the same by the uniqueness argument for the solutions of (1). The question we raise here is : *If the initial conditions are not the same but they have the same Fourier spectrum of energy, how can their subsequent time evolution differ ?*

Numerical simulations under periodic boundary conditions by the use of the Fourier pseudospectral method are performed with 128^3 grid points. In spite of limited Reynolds numbers a significant difference is found in their evolutions. The initial conditions for ω and δl have the same spectrum with excitation only at a few lowest wavenumber modes but are different in their phases.

U. Frisch (ed.), Advances in Turbulence VII, 353–356.

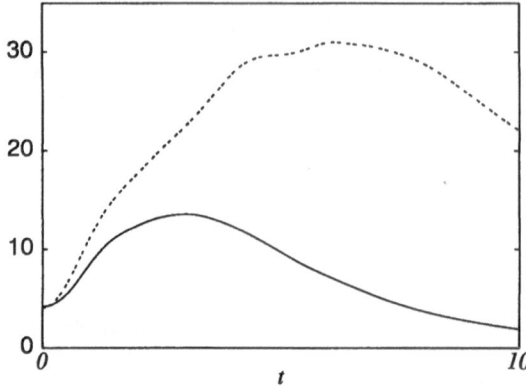

Figure 1.   Time evolutions of the enstrophy (solid) and the passive vector variance (dashed).

Time development of the enstrophy $\langle |\omega|^2/2 \rangle$ and the passive vector variance $\langle |\delta l|^2/2 \rangle$ is shown in Fig.1 for $\nu = 0.005$ ($\langle\ \rangle$ =spatial average). The norm increases more rapidly in the linear process than in the nonlinear vortex stretching process, by a factor of 2.7 at their peak values. Below we investigate how this difference comes about. We note that the normalized correlation coefficient between ω and δl remains small (a few percents) throughout the evolution.

Statistics on their alignment with the rate-of-strain tensor S is examined. At the developed stage $t = 4$, PDFs of the cosines of angles between ω (and δl) and each eigenvector are given in Figs.2. Here we have denoted by e_1, e_2 & e_3 the normalized eigenvectors of the rate-of-strain tensor S associated with the largest, intermediate and smallest eigenvalues. (See [2] for more detailed analyses on the statistics of ω.)

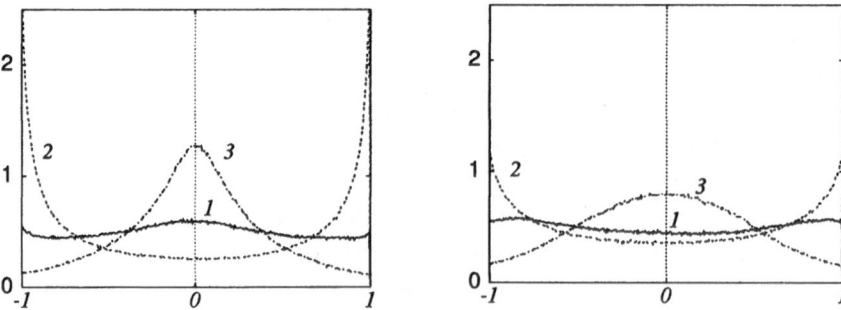

Figure 2.   PDFs of cosines of the angle between ω (left) [δl (right)] and e_1 (1:solid), e_2 (2:dashed) & e_3(3:dotted).

From these figures we can gather the followings.

1. The preference to e_2 is stronger in ω than in δl.
2. The vorticity ω shows more tendency to be perpendicular to the most compressing e_3 than δl.
3. There is little difference in the preference to the most expanding e_1.

Features 2 and 3 suggest that ω would grow larger than δl, which is not the case. What matters is the feature 1, that is, the alignment with the intermediate eigenvector of S. The PDF results in Figs.2 are consistent elaborations of [1], who found by interpolating passively advected scalars that the stretching rate of material lines is larger than that of vortex lines.

We will discuss the scaling property of the global quantities, such as the total enstrophy. We increased the Reynolds number $(= 1/\nu)$ by a factor of 2. In Figs.3 we show time evolution of the energy dissipation rate $\nu \langle |\omega|^2 \rangle$ and a similar quantity $\nu \langle |\delta l|^2 \rangle$ for the identical initial condition but with different values of ν. It is important to observe that the difference in $\nu \langle |\delta l|^2 \rangle$ is more pronounced at larger Reynolds numbers. The former one appears to behave fairly independent of ν, consistent with a finite amount of energy dissipation rate in developed turbulence, while the latter at its peak appears to increase for smaller values ν, suggesting that $\nu \langle |\delta l|^2 \rangle$ increases with the Reynolds number. (The latter does *not* represent a dissipation rate of some physical quantity, because of lack of an energy-like conserved quantity for δl. Still, its dependence on ν is worth being studied.)

We recall a basic characteristic of decaying turbulence starting from smooth initial conditions, that is, in the early stage when the vortex layers dominate the energy dissipation rate vanishes with ν, while in the latter fully developed stage it becomes in dependent of it. The increase of the peak value of $\nu \langle |\delta l|^2 \rangle$ as $\nu \to 0$ implies that δl contain richer small-scale structures than ω; the structure of δl corresponds to another extreme. Indeed, a PDF of δl has a longer tail than an approximately exponential tail of ω (figure omitted).

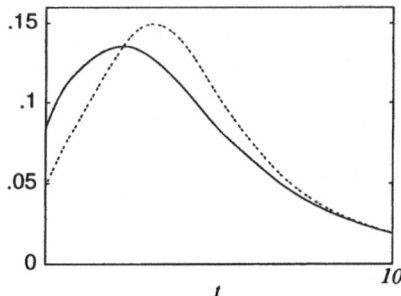
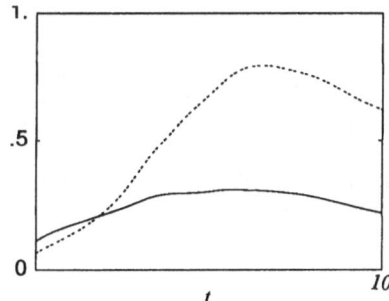

Figure 3.   Energy dissipation rate $\nu \langle |\omega|^2 \rangle$ (left) and the corresponding $\nu \langle |\delta l|^2 \rangle$ (right); $\nu = 0.005$ (solid) and 0.0025(dashed).

We have learned about the most salient feature of turbulence. *In order to render the energy dissipation rate independent of the Reynolds number, not only vortex stretching but also vortex compression is essential.* To simply stretch line elements, a linear process would do a much better job. In view of the above PDF analysis (feature 1), this depletion of nonlinearity can be attributed to the alignment of ω with the intermediate eigenvector of S because this eigenvalue can be negative.

Explanation of this finding is lacking but would be useful for understanding turbulence. Theoretically, the tensor S can be represented by a linear functional of ω (see, for example, [3]). This constraint, the origin of nonlinearity, makes the stretching term in (1) effectively weaker than the linear one in (2), in such a way that the energy dissipation rate takes an appropriate finite value. It is necessary to estimate precisely $S \cdot \omega$ to settle the problem (see [4] for relevant matters).

Finally, at the cost of giving up incompressibility we consider yet another linear equation

$$\frac{\partial \delta m}{\partial t} + (u \cdot \nabla)\delta m = S \cdot \delta m + \nu \nabla^2 \delta m \tag{3}$$

by switching off the second term in the identity

$$(\delta m \cdot \nabla)u = S \cdot \delta m + \frac{1}{2}\omega \times \delta m.$$

The equation (3) is closer to the vorticity equation (1) since the effect of rotation due to vorticity is absent from the RHS. Numerical simulations show that $\langle|\delta m|^2\rangle$ grows much more rapidly than $\langle|\delta l|^2\rangle$. Therefore it is the linearity that causes the rapid growth of the norm in (2).

References

1. Huang, M.-J. (1996) Correlations of vorticity and material line elements with strain in decaying turbulence, *Phys. Fluids*, **8**, pp. 2203-2214.
2. Tsinober, A., Shtilman L., & Vaisburd, H. (1997) A study of properties of vortex stretching and enstrophy generation in numerical and laboratory turbulence, *Fluid Dynamics Research*, **21**, pp. 477-494.
3. Constantin, P. (1994) Geometric statistics in turbulence, *SIAM Review* , **36**, pp. 73-98.
4. Ohkitani, K. (1998) Asymptotic formula for stretching of vorticity and its alignment with rate-of-strain tensor, preprint.

INSTABILITY OF NON-AXISYMMETRIC VORTEX FLOWS

C. ELOY AND S. LE DIZÈS
Institut de Recherche sur les Phénomènes Hors Équilibre
UMR 6594, 12 av. du Gal Leclerc, 13003 Marseille, France

It is known that a vortex constrained in an elliptic cavity is subject to a three dimensional instability. This elliptic instability has been studied both analytically (see for instance [1, 2, 3]) and experimentally [4, 5]. The subject of the present work is to extend the analysis to cavities deformed with a different azimuthal wavenumber.

1. Principle

Let us consider the two-dimensional, incompressible flow given by the following streamfunction (in cylindrical coordinates) :

$$\psi = -\frac{1}{2}r^2 + \frac{\varepsilon}{n}r^n \sin n\theta, \tag{1}$$

where n is an integer and ε is real. This solution of the Navier-Stokes equations describes a rotating flow with constant vorticity in a deformed rotating cylinder whose border is the streamsurface $\psi = -1/2$ (see Fig. 1). It also mimics a vortex core in a strain field with a symmetry of order n.

The goal of this work is to determine the linear stability properties of this flow in the limit of high Reynolds number and small deformation ε. For elliptic streamlines ($n = 2$), the instability has been described by Tsai

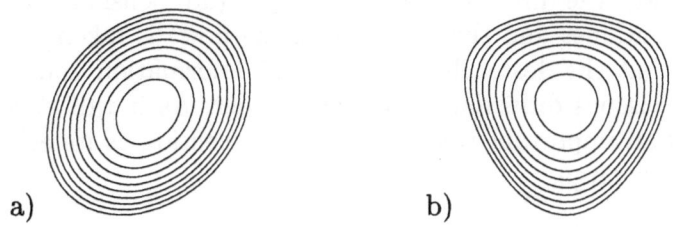

a)   b)

Figure 1. Streamlines of the flow given by (1), with $\varepsilon = \frac{1}{4}$ and a) $n = 2$, b) $n = 3$. The flow is rotating counterclockwise.

U. Frisch (ed.), Advances in Turbulence VII, 357–360.

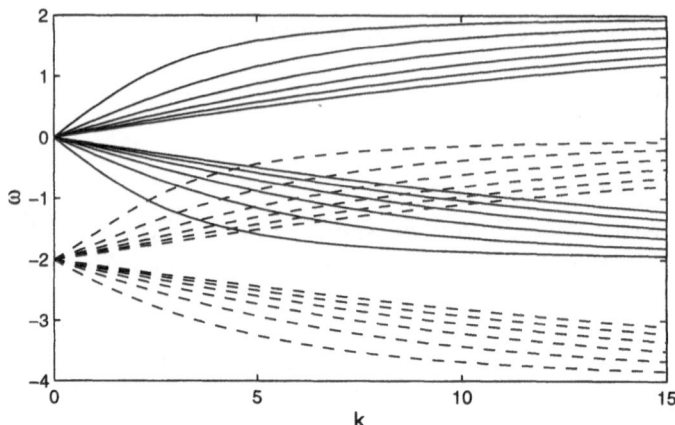

Figure 2. Dispersion relation of the Kelvin waves in the plane (k, ω) for $l_1 = 0$ (solid line) and $l_2 = 2$ (dashed line). Only the first six branches are represented. Every intersection corresponds to a point of resonance.

& Widnall [1] and Moore and Saffman [2]. Their instability mechanism can be extended to larger n : two Kelvin waves, given by

$$\mathbf{U}(r, \theta, z, t) = \mathbf{U}_l(r)e^{i(kz+\omega t+l\theta)}, \tag{2}$$

with same axial wavenumber k, same frequency ω and with azimuthal wavenumbers l_1 and l_2 satisfying $l_2 - l_1 = n$, resonate and are amplified by the strain field with a growth rate proportional to ε. For each Kelvin wave, the dispersion relation between k, ω and l has the following property :$-l - 2 < \omega < -l + 2$ (see Fig. 2). It follows that for $n \geq 5$, the resonance cannot occur and consequently there is no possible instability by this mechanism. This paper focuses on the cases $n = 2$ and $n = 3$. The limit case $n = 4$ deserves a particular analysis, for the sake of brevity, it will not be treated here.

In this study, the dimensionless viscosity ν (adimensioned using the radius of the container and the vorticity of the flow) is taken to be $O(\varepsilon)$ in such a way that the destabilizing effects of deformation and the stabilizing effect of viscous diffusion are of same order. As it will be seen below, viscosity has an important role in the mode selection as its effect is larger when k or l are large.

2. Results and discussion

For a given n, there is an infinity of possible couples (l_1, l_2) since the only imposed condition is $l_2 - l_1 = n$ (by symmetry, the study can be restricted

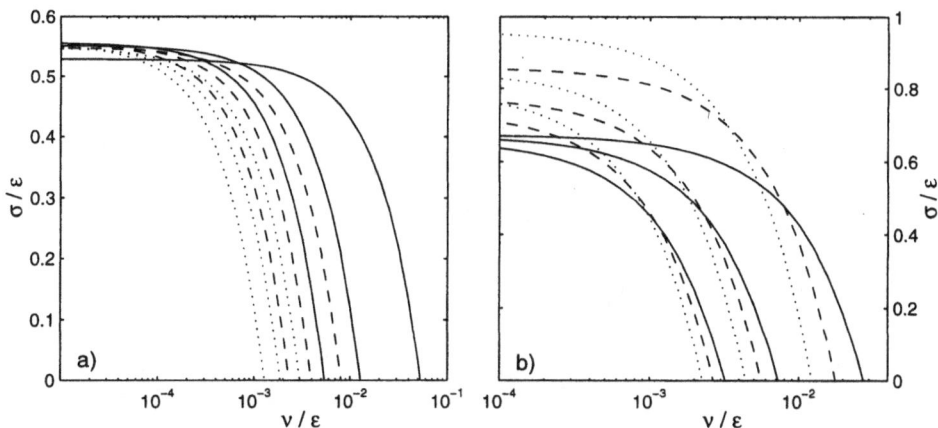

Figure 3.   Normalized growth rate as a function of ν/ε for resonant wave combinations $(l_1, l_1 + n, i)$ a) $n = 2$ (elliptic case) and b) $n = 3$. Solid lines : $l_1 = -1$, dashed lines : $l_1 = 0$ and dotted lines : $l_1 = 1$.

to $l_1 + l_2 \geq 0$). Given a couple (l_1, l_2), there is also an infinity of points in the plane (k, ω) corresponding to resonance (see Fig. 2). At each of these points an inviscid growth rate can be computed : it appears that all resonant points are unstable. However, the growth rate is significantly higher when the crossing point corresponds to symmetric branches of the dispersion relation. These special "symmetric" modes are noted (l_1, l_2, i), where i numbers the branch such that k grows with i.

For the elliptic case ($n = 2$) the effect of viscous selection is summarized on Fig. 3a. which shows the evolution of the growth rate σ with respect to the viscosity ν (both normalized by ε). When $\nu > 0.053\varepsilon$, viscosity stabilizes any perturbation and the vortex filament is stable. When ν is reduced, unstable modes appear but the most unstable mode is always a combination of two helical Kelvin waves of the form $(-1, 1, i)$ where i increases as ν/ε decreases. In other words, the vortex filament is destabilized by an undulation whose wavelength grows with ν/ε (see Fig. 4a). Note however that other modes such as $(0, 2, i)$ (leading to vortex spliting) have a close growth rate for small ν/ε (see Fig. 3a).

The results for $n = 3$ are displayed on Fig. 3b. The vortex filament is found to be stable for $\nu > 0.028\varepsilon$. Contrarily to the elliptic case, the most unstable mode is of the form $(l, l + 3, 0)$ where l increases as ν/ε decreases. This means that when viscous effects are weaker, the azimuthal structure of the most unstable mode becomes more complex. Two of these modes are pictured on Fig. 4b,c.

The limit value of the maximum growth rate for large wavenumbers and

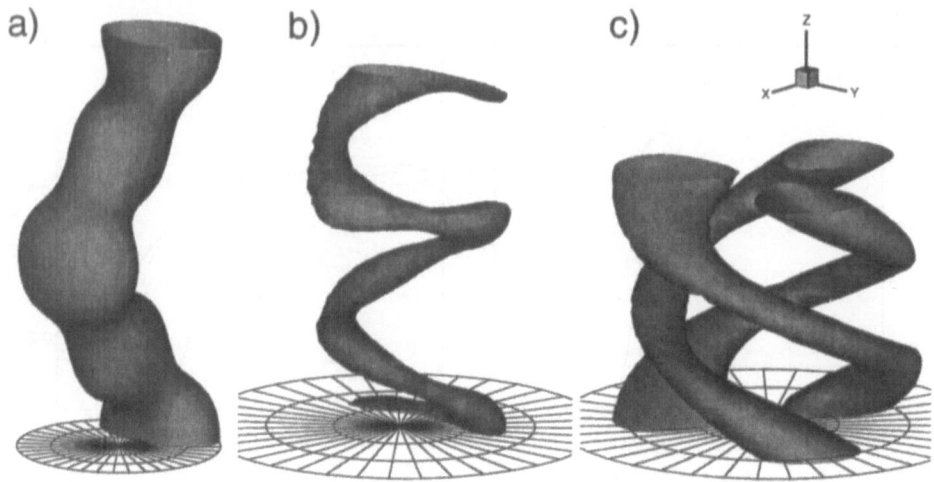

Figure 4.   Vorticity contour of a superposition of the basic flow and the perturbation for the modes a) $(-1,1,0)$, b) $(-1,2,0)$ and c) $(0,3,0)$. The bottom disc pictures the size of the container.

small viscosity have also been calculated. We have obtained $\sigma_{\max} = 9\varepsilon/16$ for $n = 2$ (in agreement with Waleffe [3]) and $\sigma_{\max} = 49\varepsilon/32$ for $n = 3$.

This work shows how a non-axisymmetric environment destabilizes a vortex filament. It also provides an explanation in terms of instability modes for several dynamical behaviors of filaments observed in turbulence [6]. In particular, the spontaneous undulation, splitting or stranding of the filaments could be viewed as modes of the present instability. This interpretation is reinforced by recent results on the Burgers vortex [7] which suggest that the instability is weakly dependent on the vortex profile.

References

1. Tsai, C.-Y. and Widnall, S.E. (1976) The stability of short waves on a straight vortex filament in a weak externally imposed strain field, *J. Fluid Mech.* **73**(4), 721–733.
2. Moore, D.W. and Saffman, P.G. (1975) The instability of a straight vortex filament in a strain field, *Proc. R. Soc. Lond.* A **346**, 413–425.
3. Waleffe, F. (1989) *The 3D instability of a strained vortex and its relation to turbulence*, PhD thesis, Massachusetts Institute of Technology.
4. Malkus, W.V.R. (1989) An experimental study of global instabilities due to tidal (elliptical) distortion of a rotating elastic cylinder, *Geophys. Astrophys. Fluid Dynamics* **48**, 123–134.
5. Chernous'ko, Y.L. (1978) An experimental study of secondary multi-eddy flows in elliptical cylinders, *Izv. Akad. nauk SSSR FAO*, **14**(2), 151–153.
6. Cadot, O., Douady, S. and Couder, Y. (1995) Characterization of the low pressure filaments in three-dimensional turbulent shear flow, *Phys. Fluids*, **7**(3), 630–646.
7. Eloy, C. and Le Dizès, S. (1997) Instability of the Burgers and Lamb-Oseen vortices in a strain field, *Submitted to J. Fluid Mech.*.

WAVES AND INSTABILITIES IN ROTATING FLOWS

F.S. GODEFERD[1,2], S. LEBLANC[1], L. LOLLINI[1] & C. CAMBON[1]
[1] *LMFA UMR 5509, École Centrale de Lyon, France.*
[2] *also at Dip. di Ing. Aerospaziale, Poli. di Milano, Italy.*

1. Context

Solid body rotation is an external effect that can modify greatly the structure of turbulent flows, as well as the instability mechanisms of shear flows. Hydrodynamic stability analyses may explain the persistence of existing coherent structures in turbulent flows by showing that some of the vortices submitted to rotation are unstable, whereas others are stable, parameterized by the combination of external rotation and local vorticity and strain in the flow. However, these analyses cannot explain the *creation* of these coherent structures *per se*. It is thus an important matter to study rotating flows from both these two previous points of view, by means of existing stability theories, and of numerical experiments of turbulent flows. Previous works on homogeneous turbulence, with or without initial coherent vortices, have already drawn a precise pictures of how these flows may be affected by the Coriolis force (Cambon *et al.*, 1997, Cambon *et al.*, 1994 for example).

The common feature of all these flows seems that inertial waves can propagate through the fluid, as a result of the Coriolis force effect. Although these waves can hardly be isolated in the fully nonlinear regime, they can be identified easily in the purely linear one. They are therefore shown to be playing an important structuring role in turbulent flows, since they are essentially dispersive, with priviledged directions of propagation. The effect of these waves on coherent structures can also be identified in the weakly nonlinear case, by analysing the interaction of wave packets with large scale vortices.

In the following section 2, we show results of direct numerical simulations of a rotating flow in the linear regime, that allows a direct observation of the propagation of inertial waves between two infinitely extending plane walls. On the other hand, in section 3, we present results of DNS of Taylor-

U. Frisch (ed.), Advances in Turbulence VII, 361–364.

Green cells perturbed three-dimensionally to show the effects of the Coriolis force on the stability of stagnation points, according with the criterion proposed by Leblanc & Cambon, 1997 extending previous results by Lifschitz & Hameiri, 1991.

2. Evidence of inertial waves in rotating flows

Without mean flow, the Euler equations in a rotating frame may be written as $\partial_t u + 2\Omega \times u = -\nabla\pi$, with the incompressibility condition $\nabla \cdot u = 0$, where $\pi = p/\rho - (\Omega \times r)^2/2$. Real valued solutions, sought under the shape of plane waves, lead to eigen solutions that follow the dispersion relation $\sigma = \pm 2\Omega \cos\theta$. In the above equation, θ is the angle between the rotation vector and the wavevector. This dispersion relation for inertial waves shows the anisotropic character of these waves.

In order to reproduce numerically the behaviour of these waves in a fluid between two infinitely extending plane walls, we use a pseudo-spectral numerical code which has been developped by Pascal & Buffat, 1996. The inertial waves are generated by introducing a point force in the above momentum equation, under the shape of a sine acceleration, with a given time frequency ($n = 1$ here).

The computational domain in the horizontal direction is chosen such that reflection at the walls can occur before the wave reaches the periodic boundary, leading to an aspect ratio of 4 for the channel. Each simulation's duration corresponds to at least twenty forcing periods, while ensuring again that waves do not reach the periodic boundaries, a situation which would lead to unphysical results, since periodicity would artificially modify the kinematics of the waves. The resolution is shown to be possibly relatively low (here $32 \times 64 \times 64$ points) without impairing the accuracy of the results. The resulting flow field is visualised at the end of the simulations, showing the lines of constant phase, or wavecrests. The angle between these and the horizontal direction is therefore θ. Details of these simulations are reported by Lollini & Cambon, 1996.

Figure 1 shows lines of constant phase for the inertial waves at three different values of the rotation rate Ω. The patterns on these three visualisations strikingly resemble those that are visible on a photograph in the experiment by McEwan, 1970. Moreover, the propagation angles of the waves predicted by linear theory and our simulations are in close agreement, with an error of less than one percent.

3. Instabilities

In the following, we consider the fact that solid body rotation, with an axis perpendicular to a plane flow, play either a destabilizing or a stabilizing role upon *three-dimensional* perturbations. The topology of such a base flow is very important, in that a unified stability criterion, in the inviscid case, may be derived from its local characteristics, in terms of quadratic stagnation points: hyperbolic, elliptic or parallel mean shear. Leblanc & Cambon, 1997 have thus proposed a criterion for short-wave instability localized along the streamlines. Of course, such a modification of the growth of wave packets in rotating fluids, is not without relation with the presence of the so-called "inertial waves", observed in the previous section. We illustrate in this brief account the fact that the growth of a perturbation to the vorticity field, through a vortex stretching mechanism, may be inhibited and eventually killed by the Coriolis force. For the sake of brevity, we shall not detail the linearized equations for the perturbation evolution, but proceed directly to writing the stability criterion (see Leblanc & Cambon, 1997 for a more complete account of the analysis). The general discriminant expressed in the rotating frame writes $\Phi(x,y) = (\Psi_{xx} - 2\Omega)(\Psi_{yy} - 2\Omega) - (\Psi_{xy})^2$, in which the streamfunction in the plane (x,y) of the flow is denoted Ψ. From the theory, a sufficient condition for instability is that $\Phi(x_0, y_0) < 0$ at a stagnation point x_0 in the flow. In that case, the inviscid growth rate of both velocity and vorticity perturbations is given by: $s = (-\Phi(x_0, y_0))^{1/2}$. Different bands of instability are therefore readily identified for such planar flows that exhibit either elliptical or hyperbolic stagnation points. If both are present, the hyperbolic and elliptic instabilities are concurrential, and the fastest growing ones determines the evolution of the flow. As an illustration of a flow containing hyperbolic stagnation points, we perform Direct Numerical Simulations of square Taylor-Green cells with axes aligned with the rotation vector Ω. The DNS method is a fully de-aliased pseudo-spectral one, using 128^3 points. Details of the numerics are explained in *e.g.* Staquet & Godeferd, 1998. In the non-rotating case, clearly, the symmetry in the flow is not broken, as shown on figure 2(a), where transverse vortices—also called "ribs"— appear in between the Taylor-Green ones. This is the result of the instability arising from the presence of the hyperbolic point between the cells. In the rotating case (figure 2(b)), however, only strong cyclonic vortices remain, since the anti-cyclonic ones are unstable. In addition to this, the pattern of the ribs is itself modified, for only those between the cyclonic vortices are conserved.

References

CAMBON, C. , BENOÎT, J.-P. , SHAO, L. , & JACQUIN, L. 1994. Stability analysis and LES of rotating turbulence with organized eddies. *J. Fluid Mech.*, 278, 175.

CAMBON, C. , MANSOUR, N. N. , & GODEFERD, F. S. 1997. Energy transfer in rotating turbulence. *J. Fluid Mech.*, 337, 303–332.

LEBLANC, S. & CAMBON, C. 1997. On the three-dimensional instabilities of plane flows subjected to Coriolis force. *Phys. Fluids*, 9, 1307–1316.

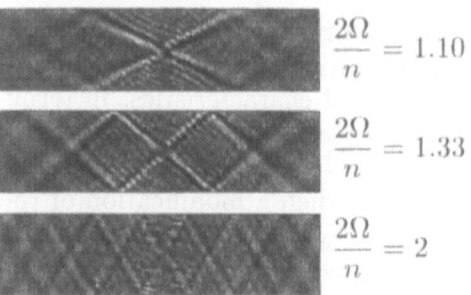

Figure 1. Two-dimensional plot of the lines of constant phase for the inertial waves (the wavevector is orthogonal to these lines). One sees the reflexion on the two horizontal walls.

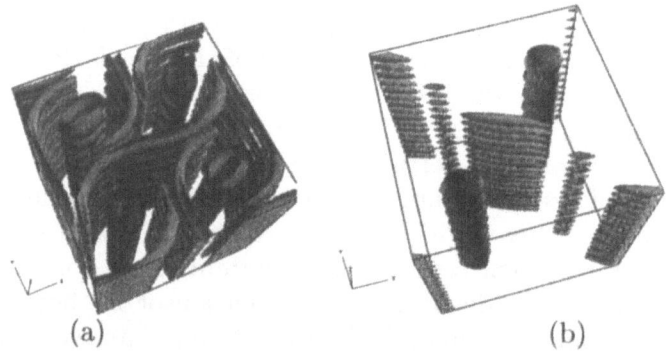

Figure 2. Iso-surface of the vorticity magnitude of the perturbed flow in the nonlinear regime for $k = 8$. (a) Non rotating case, at $t = 10$. (b) Rotating case with $\Omega = 0.3$, at $t = 30$.

LIFSCHITZ, A. & HAMEIRI, E. 1991. Local stability conditions in fluid dynamics. *Phys. Fluids*, A(3), 2644.

LOLLINI, L. & CAMBON, C. 1996. Numerical simulations of inhomogeneous turbulence generated by an oscillating grid and subjected to solid-body rotation. *Advances in Turbulence VI, Kluwer.*

McEWAN, A.D. 1970. Inertial oscillations in a rotating fluid cylinder. *J. Fluid Mech.*, 40, 603–639.

PASCAL, H. & BUFFAT, M. 1996. LES. of turb. flows compressed and/or sheared between two walls on parallel computer using a "Divergence-free spectral Galerkin method. *Comp. Fluid Dyn.*, Wiley & Sons Ltd, 884–891.

STAQUET, C. & GODEFERD, F.S. 1998. Statistical modelling and DNS of decaying stably-stratified turbulence : Part 1. Energetics. *J. Fluid. Mech.*, to appear.

THE FORMATION OF VORTEX STRUCTURES IN SUPERFLUID TURBULENCE

D.C. SAMUELS
Dept. of Mathematics, Univ. of Newcastle,
Newcastle upon Tyne, NE1 7RU, UK.

The turbulent flow of helium II is complicated by the fact that one must actually consider two different, but interacting, turbulent velocity fields: that of the normal fluid component and that of the superfluid component. Aside from the forcing due to the interaction with the superfluid, the normal fluid obeys a Navier-Stokes equation, and the turbulence of this component is expected to be classical. The superfluid, however, is an Eulerian fluid with the additional restriction that the vorticity in the superfluid is confined to quantized vortex filaments. These superfluid vortex filaments have an extremely small core radius (approximately one Angstrom) and are thus true vortex filaments. The circulation of these filaments is quantized, and due to energy considerations only vortices with one quantum of circulation need to be considered. At high flow rates these quantized vortex filaments can form dense tangles which are referred to as *quantum turbulence.*

The particles that make up the normal fluid will scatter off the superfluid vortex filaments, thus exchanging energy and momentum between the two fluids. This interaction between the two fluids is called *mutual friction,* and is modeled as a force that is proportional to the local relative velocity $V_N - V_S$ where V_N is the local velocity of the normal fluid and V_S is the local superfluid velocity. We consider how the superfluid vorticity reacts to the driving force (through mutual friction) from various different types of normal fluid flows.

Early simulations considered only uniform normal fluid flows. These studies found that a homogeneous, random superfluid vortex tangle was formed when forced by a uniform normal fluid flow [1]. But in turbulent flows the normal fluid will be far from uniform. It is well known that in classical turbulence the regions of highest vorticity are organized into concentrated vortex tubes, or possibly vortex sheets. Through simulations, we are examining the effect of these normal fluid vorticity structures on the superfluid vortex filaments.

U. Frisch (ed.), Advances in Turbulence VII, 365–368.
© 1998 *Kluwer Academic Publishers.*

There are some simple rules which describe the dynamics of superfluid vortex filaments in non-uniform normal fluid flows. Since the mutual friction force is proportional to the local velocity difference between the two fluids, $\mathbf{V_N} - \mathbf{V_S}$, vortex filaments will remain at rest on any line or plane within the fluid where the two velocities $\mathbf{V_N}$ and $\mathbf{V_S}$ are equal. Simulations have shown [3] that these regions of initially matched velocities act as attractors for superfluid vortex filaments with vorticity vectors which are parallel to the local normal fluid vorticity and that superfluid vortex filaments with vorticity vectors *antiparallel* to the local normal fluid vorticity are repelled from these regions. So any regions within the flow which have $\mathbf{V_N} = \mathbf{V_S}$ initially will tend to build up a population of superfluid vortex filamants that, if a local spatial average is performed, will match the local normal fluid vorticity. Mutual friction tends to equalize the vorticity fields of the two fluid components.

By the arguement given above, the cylindrical vorticity structures in classical turbulence should attract superfluid vortex filaments. Such a vortex structure in the turbulent velocity field of the normal fluid will have a line along its core at which $\mathbf{V_N} = 0$. With the superfluid initially at rest, then $\mathbf{V_N} = \mathbf{V_S}$ along this line, and superfluid vortex filaments should build up within the normal fluid vortex structure. But there is still the question of the speed of the growth of the superfluid vorticity field. Experiments seem to indicate [2] that the superfluid and normal fluid components are locked together in turbulent flows, so that they effectively move as a single fluid. If this locking together of the two fluid components is to occur by the processes described above, then the time scales needed for the growth of large concentrations of superfluid vorticity within the normal fluid vortex structures must be shorter than the lifetimes of these structures.

The fast growth of the superfluid vorticity is provided by a process known as the Glaberson instability [4]. This is an instability of the superfluid vortex filaments to the growth of helical vortex waves. This instability is driven by a normal fluid velocity along the direction of the vortex filament. As the amplitude of these helical waves grows they form new loops of vortex filaments, which can themselves undergo this instability again, in turn forming yet another generation of vortex loops, leading to an exponential growth in the length of the superfluid vortex filament. The time scale of this exponential growth can be measured in simulations [4] but there is currently no theoretical derivation of this growth time scale.

As a test of this process we consider a single vorticity concentration in the normal fluid, modelled as a gaussian distribution of vorticity. To limit the size of the calculation we linearly increase the core size of this gaussian distribution for positions outside the range $-1 \leq Z \leq +1$. We want to avoid the use of periodic boundary conditions since that would allow a single vortex filament to loop indefinitely many times along the normal fluid vortex core simply by moving continuously in one direction and through the periodic boundaries, leading to an artificially rapid growth rate of the superfluid vortex filaments.

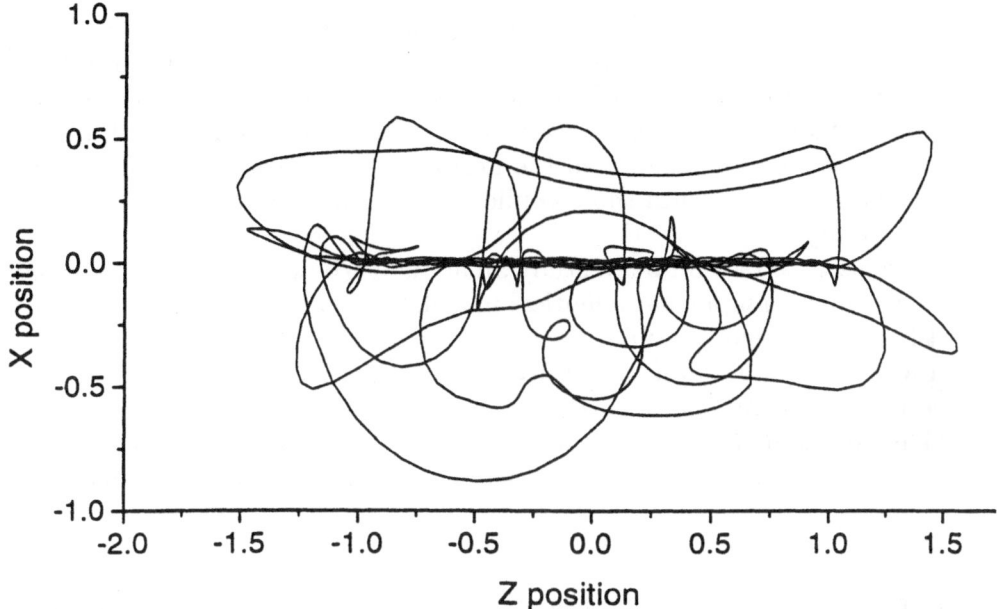

Figure 1. A concentration of superfluid vortex filaments developing at the location of a normal fluid vortex. The lines represent the superfluid vortex filaments. The normal fluid vortex is cylindrical with a radial gaussian distribution of vorticity. The normal fluid vortex extends from $-1 \leq Z \leq +1$, and is not shown in the figure for clarity. The superfluid vortices are concentrated in the core of the normal fluid vortex, with a diffuse cloud of vortex filaments outside the structure. This diffuse cloud of filaments provides new filament length by the Glaberson instability, and this new length is captured by the normal fluid vortex core. This is an early stage of the growth. Eventually the vortex filaments become much thicker, but are always concentrated mainly in the core of the normal fluid vortex.

In this simulation we find that, starting with a very small seed of vorticity in the superfluid, a spatially organized superfluid vortex structure does form in the same position as the normal fluid vortex. An example of this is shown in figure 1. The formation of these structures is however limited by the need to trigger the Glaberson instability for the formation of new superfluid vortex filament. We have found that this requires that the circulation of the normal fluid vortex structure must be above a minimum critical level. While normal fluid vortices of lower circulations will still attract superfluid vortex filaments, they will not cause the Glaberson instability and thus not have the rapid growth in filament line length.

This study showed that isolated vortex structures in the normal fluid would form corresponding vortex structures in the superfluid. We mean this to be a first step in understanding the behavior of superfluid vortex filaments in turbulent flows, and the concentrated vortex structures in classical turbulence are not isolated and are strongly interacting with background flows, particularly with strain. To take a step closer to real turbulence, we have considered a model flow with many closely spaced vortex structures oriented in all three directions, the ABC flow. In this case [5] we see that the added complexity of the flow does not disturb the growth of the superfluid filaments, and once again we get an exponential growth of the superfluid vortex filaments, and the locally averaged vorticity distribution in the superfluid matches the complicated spatial structure of the normal fluid vorticity.

The next step would be to study an actual turbulent flow in the normal fluid, not just a simple mathematical model. To do this we also need to convert from a kinematic simulation, where the normal fluid flow is prescribed and forces the superfluid, to a fully dynamical simulation where both the superfluid flow and the normal fluid flow are calculated. At the moment we have developed an understanding of how the superfluid reacts to the normal fluid. How the normal fluid will react in turn is still unknown.

References

1. Schwarz, K.W. (1988), Three dimensional vortex dynamics in superfluid He4: Homogeneous superfluid turbulence, *Phys. Rev. B* **38**, pp. 2398–2417.
2. Donnelly, R.J. (1991), *High Reynolds Number Flows using Liquid and Gaseous Helium*, ed. R.J. Donnelly. pp. 3–49, Springer Verlag.
3. Samuels, D.C. (1992) Velocity matching and poiseuille pipe flow of superfluid helium, *Phys. Rev. B.* **46**, pp. 11714–11724.
4. Samuels, D.C. (1993) Response of superfluid vortex filaments to concentrated normal-fluid vorticity, *Phys. Rev. B.* **47**, pp. 1107–1110.
5. Barenghi, C.F., Samuels, D.C., Bauer, G.H. & Donnelly, R.J. (1997), Superfluid vortex lines in a model of turbulent flow, *Phys. Fluids 9*, pp. 2631–2643.

VORTEX KNOTS

R.L. RICCA[1], D.C. SAMUELS[2] AND C.F. BARENGHI[2]
[1] *Mathematics Department, University College London, London WC1E 6BT, UK*
[2] *Mathematics Department, University of Newcastle, Newcastle upon Tyne NE1 7RU, UK*

In this paper we present new results concerning the evolution and stability of vortex knots in the context of the Euler equations. For the first time, since Lord Kelvin's original conjecture of 1875, we have direct numerical evidence of stability of vortex filaments in the shape of torus knots. The results are based on the analytical solutions of Ricca [1] for thin vortex filaments and numerical integration of the Biot-Savart induction law. Moreover, a comparative study of vortex knot evolution under the so-called Localized Induction Approximation (LIA), which is a low-order approximation to the Biot-Savart law, confirms the stability results predicted by the LIA analysis. In particular, we show that thin vortex knots which are unstable under LIA have a greatly extended lifetime when the Biot-Savart law is used, but thick vortex knots have the same stability behaviour for both equations of motion.

Applications of ideas from modern topology to fluid mechanics have been pioneered by Moffatt [2] and co-workers [3], whose results clearly demonstrate the importance of the new techniques in the study of knotted and linked structures in fluid flows. The use of geometric and topological methods in fluid mechanics has indeed proven to be very useful in the analysis of the entanglement of filamentary vortex structures as observed in direct numerical simulations of homogeneous turbulence (see, for example [4] and [5]). The most advanced visiometrics of streamlines and vorticity lines associated with the formation of coherent structures reveal that a high degree of braiding, re-connection and formation of new linkings of field lines is a generic feature of turbulent flows. Moreover, the study of complex flow patterns using topological techniques finds useful applications in the study of filament structures present in a wide spectrum of physical scales, from magnetic flux tubes in solar physics to quantized vortex lines in superfluidity [6]. Yet, from a theoretical viewpoint very little is known about

U. Frisch (ed.), Advances in Turbulence VII, 369–372.

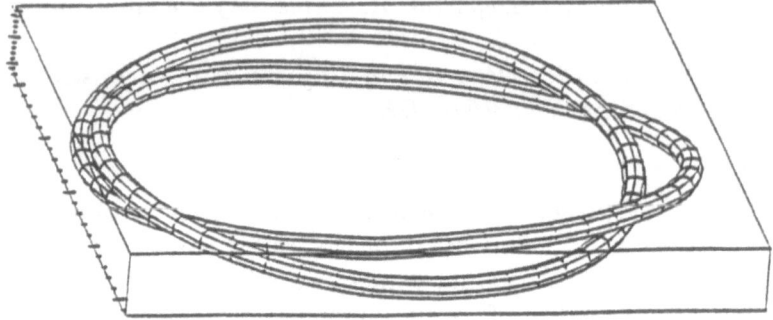

Figure 1.   Evolution of torus knot $\mathcal{T}_{2,3}$ under LIA. The knot is found to be stable as predicted by the LIA analysis of Ricca. The knot is visualized by centering a thin tube on the knot axis, as shown in figure. Hence, the tube is a virtual object and its thickness is not measured by a_0.

the effects of topology on the evolution of complex structures, and there is therefore a call for more information about these processes and their mathematical modelling.

The aim of the present work is twofold: to investigate the relationship between geometry, topology and dynamics of topologically complex vortex structures; to model the topological entanglement of vortex structures using knotted vortex lines as elementary constituents. We have concentrated our attention to *torus knots* $\mathcal{T}_{p,q}$ which are thin vortex filaments wrapped around a mathematical torus, where p and q are relatively prime integers. A torus knot is characterized by its *winding number* $w = q/p$ which is a topological invariant and measures the number of wraps of the knot along the small circle of the torus per number of wraps along the large circle of the torus. In principle, vortex motion in the context of the Euler equations is governed by the Biot-Savart law which determines the self-induced velocity $\mathbf{u}(\mathbf{X})$ of a vortex line C of strength Γ in the following way:

$$\mathbf{u}(\mathbf{X}) = \frac{\Gamma}{4\pi} \int_C \frac{\hat{\mathbf{t}} \times (\mathbf{X} - \mathbf{R}(s))}{|\mathbf{X} - \mathbf{R}(s)|^3} \, ds \ . \tag{1}$$

Here $\mathbf{X}$ is the position vector, $\mathbf{X} = \mathbf{R}(s)$ the vector equation for $\mathcal{C}$, s the arc-length and $\hat{\mathbf{t}} = d\mathbf{R}/ds$ the unit tangent along $\mathcal{C}$. Note that the integral of eq. (1) is a global geometric functional and retains all the induction effects associated with the geometry of $\mathcal{C}$, while preserving topology. Unfortunately, explicit analytic solutions to the Biot-Savart law are only known for very simple geometries and in general are very difficult to obtain. Moreover, numerical simulations based on (1) are rather expensive to run, because the motion of each single vortex point depends on the motion of *all* the other points in which the vortex line is discretized. A standard way to overcome these difficulties is to use a cut-off technique based on the Localized Induction Approximation to (1). Under LIA the filament motion is governed essentially by local curvature effects and in the limit of very thin vortex filaments (1) is replaced by

$$\mathbf{u}(\mathbf{X}) = \frac{\Gamma}{4\pi} \ln\left(\frac{R_{\text{eff}}}{a_0}\right) \mathbf{R}' \times \mathbf{R}'' \,, \tag{2}$$

where R_{eff} is some length-scale, which we choose equal to $8c$ (c local radius of curvature) in order to reproduce the correct velocity for vortex rings; a_0 represents a very small cut-off parameter, and typically $a_0 \approx 10^{-8}$.

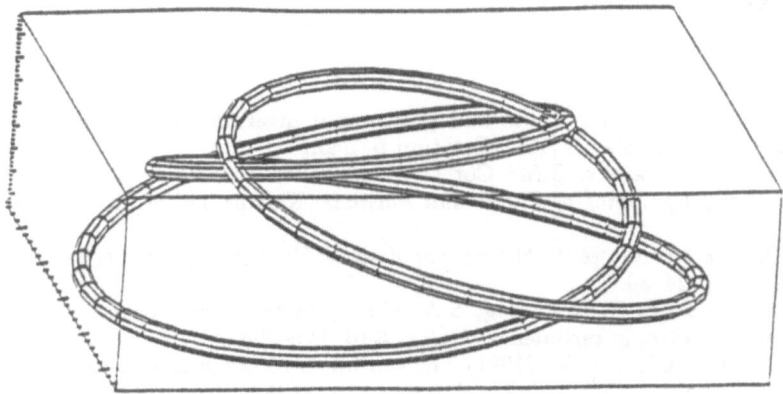

Figure 2. Evolution of torus knot $\mathcal{T}_{3,2}$ under LIA. The knot is found to be unstable as predicted by the LIA analysis of Ricca. The knot is however stabilized when its evolution is governed by the Biot-Savart law.

The numerical calculations which we have performed [7] confirm the validity of Ricca's [1] stability criterion under LIA evolution, i.e. that torus knots are stable if $w > 1$. Figure 1 shows the stable knot $T_{2,3}$ and Figure 2 shows the knot $T_{3,2}$ as it becomes unstable and unfolds. These results provide useful information for studying more sophisticate models of vortex structures under LIA. Another interesting result that we have found is the discovery of a strong stabilizing effect due to the Biot-Savart law. Take for example the knot $T_{3,2}$: this knot becomes immediately unstable under LIA, whereas it remains stable under Biot-Savart, travelling a considerable distance. Although we find that these knots eventually de-stabilize (remember that numerical noise is always present), the time which elapses and the distance over which the knots travel before breaking-up is very large and has physical significance. Moreover, there are cases (for relatively thin knots) in which the evolution under Biot-Savart is almost identical to that given by LIA, an unexpected result worth investigating.

Finally, let us point out that unstable vortex knots evolve under Biot-Savart towards a reconnection event. This is another interesting feature of vortex knot evolution. In view of the great interest in the formation of singularities in the Euler equations, an issue that represents an outstanding problem in the mathematics of ideal fluid mechanics, unstable vortex torus knots prove to be a simple and effective means of investigation. No doubt that these results will stimulate more numerical work and will certainly give new impetus to the mathematical search for the existence of steady and stable vortex knot solutions to the Euler equations.

Acknowledgements. One of us (R.L.R.) wishes to thank The Leverhulme Trust for financial support.

References

1. Ricca, R.L. (1993) Torus knots and polynomial invariants for a class of soliton equations. *Chaos* **3**, 83–91. [1995 Erratum **5**, 346.]
2. Moffatt, H.K., Zaslavsky, G.M., Comte, P. & Tabor, M. (Ed.) (1992) *Topological Aspects of the Dynamics of Fluids and Plasmas*. Kluwer, Dordrecht, The Netherlands.
3. Ricca, R.L. & Berger, M.A. (1996) Topological ideas and fluid mechanics. *Phys. Today* **49** (12), 24–30.
4. She, Z.-S., Jackson, E. & Orszag, S.A. (1990) Intermittent vortex structures in homogeneous isotropic turbulence. *Nature* **344**, 226–228.
5. Vincent, A. & Meneguzzi, M. (1991) The spatial structure and statistical properties of homogeneous turbulence. *J. Fluid Mech.* **225**, 1–25.
6. Barenghi, C.F., Samuels, D.C., Bauer, G.H. & Donnelly, R.J. (1997) Vortex lines in a model of turbulent flow, *Phys. Fluids* **9**, 2631–2643.
7. Samuels, D.C., Barenghi, C. F. & Ricca, R.L. (1998) Quantized vortex knots *J. Low Temp. Physics* **100**, 509–514.

TURBULENCE STRUCTURE AROUND A COLUMNAR VORTEX - RDT AND VORTEX WAVE EXCITATION

TAKESHI MIYAZAKI[1,*] AND JULIAN C.R. HUNT[1]

[1] *Department of Applied Mathematics and Theoretical Physics
University of Cambridge, Silver Street, Cambridge, CB3 9EW,
UK*
*Permanent address: Department of Mechanical and Control
Engineering
University of Electro-Communications, Chofu, Tokyo 182, Japan*

In this paper, the structure of initially isotropic homogeneous turbulence around solid and fluid columnar vortices (with circulation Γ and radius σ) is investigated using rapid distortion theory (RDT). We assume that the turbulence velocity u_0 is small compared with Γ/σ, that the length scale L is such that $(L/u_o) \gg \sigma/\Gamma$ and that the Reynolds number is very large. The initial condition is a columnar vortex embedded in a field of initially homogeneous isotropic turbulence (Fig 1a). In order to investigate the different elements of the phenomena, namely vortex stretching by differential rotation, blocking of external fluctuations by the vortex and wave generation on the vortex, we consider three idealisations of the columnar vortex. For the first "solid cylinder model: (SC)", the core of a columnar vortex is replaced by a solid cylinder rotating with a constant angular velocity $\Omega = \Gamma/2\pi\sigma^2$. The fluid motion outside the core is irrotational with the circulation Γ. The turbulent eddies are stretched by the differential rotation induced around the columnar vortex and the azimuthal component of turbulent vorticity grows algebraically, in proportion to time t, while the other components do not grow. At the same time the eddies are blocked by the presence of the vortex. Thus, RDT also enables us to simulate the change in the random velocity fields around the vortex. It is found that the turbulent eddies are wrapped around the columnar vortex to form fine-scale structures like vortex rings, with similar properties to those computed by [1], who conducted direct numerical simulations at moderately high Reynolds number using a spectral method (128^3) (Fig 2). What they observed are 1) azimuthal alignment of the external vorticity, 2) excitation of bending

373

U. Frisch (ed.), Advances in Turbulence VII, 373–376.

waves on the columnar vortex and 3) the tendency of the external vorticity to form vortex rings which become axisymmetric and interact with each other. Although the third of these manifestations is the result of nonlinear interactions between eddies of the external turbulence under the action of the straining motions, the first two are consistent with the RDT results.

The analysis shows that the velocity field near the vortex core becomes approximately axisymmetric quite rapidly, within two or three revolutions of the columnar vortex. The radius of influence of the columnar vortex r_c increases with time, as given by $r_c = \sqrt{\Gamma t}$. Because of the boundary condition the dominant turbulent velocity component is axial near the vortex core surface, whereas the radial velocity is the largest component away from the vortex core, by a distance greater than the integral scale. This enhances the momentum and scalar transport in the radial direction. The turbulence energy decays like r^{-5} radially and grows in proportion to t^2.

The analysis can be extended to analyse the generation and interactions with a wave excited on a "hollow core vortex: (HC)". It has only two vortex wave modes for each axial and azimuthal wave number (k_z, m). On the other hand if the vortex has a continuous radial vorticity distribution (such as a Rankine's combined vortex), for each (k_z, m) there is a countably infinite number of modes. The algebraic manipulations for HC are simple and provide clear understanding of the mechanisms of vortex wave generation. In order to consider "realistic" core-dynamics, we proceed to "Rankine's combined vortex: (RCV)" that has a solidly rotating fluid core, although the essential mechanisms of vortex wave excitation remain the same.

Although the dominant and most rapidly growing *turbulent* velocity fields are axisymmetric, at the level of a linear approximation the *waves* excited on the surface of the vortex by the turbulence are not axisymmetric. This is because the largest pressure disturbances induced by turbulent eddies are asymmetric at linear order (Fig 1a). The amplitudes of these asymmetric waves saturate within two or three revolutions, again. They are similar to the bending waves simulated by [1]. The wave amplitude is larger for smaller axial and azimuthal wave numbers. It is shown that the influence of vortex waves on turbulence (at linear order) is weak, even at the core surface, and that it decays exponentially in the radial direction.

If nonlinear effects are considered, however, the "vortex rings" with circulation $\gamma \sim u_o \sigma$ induce an image vorticity within the vortex tube and thence self-induced motion parallel to the vortex axis with velocities v of order $u_0 t \Gamma / \sigma^2 \sim t \frac{\Gamma \gamma}{\sigma^3}$ (Fig 1b). These motions can excite large amplitude axisymmetric vortex waves resonantly. It is found that the growing radial deformation $\Delta \sigma$ of the vortex becomes comparable with its initial radius σ on a time scale of the order of L/u_o – independently of the ratio $\Gamma/(\sigma u_o)$, provided the ratio is large (Fig 3). This is a possible 'suicidal' mechanism for the breakdown of vortices in turbulent flows. See also [3]. At the same time these distorted eddies near the vortex may also be advected by larger scale eddies, leading to a similar and possibly more effective mechanism for wave excitation, which is analogous to turbulent eddies inducing waves on stable inversion layers, e.g. [2].

These model calculations should improve our understanding both of how laminar vortices behave in turbulent flows and of how turbulence **within** a slender vortex interacts with its mean velocity field and influences its overall dynamics (e.g. [4]). Professor Kawai has noted that streamwise vortices from the corners of buildings waves grow and sometimes burst, leading to large pressure fluctuations on the surface of the building.

References

1. Melander, M.V. & Hussain, F. (1993) *Physical Review* **E48** (4), 2669-2689.
2. Carruthers, D.J. & Hunt, J.C.R. (1986) *J. Fluid Mech.* **165**, 475-501.
3. Marshall, J.S. (1997) *J. Fluid Mech.* **345**,1-30.
4. Govindaraju, S.P. & Saffman, P.G. (1971) *Phys. Fluids* **14**, 2074-2080.

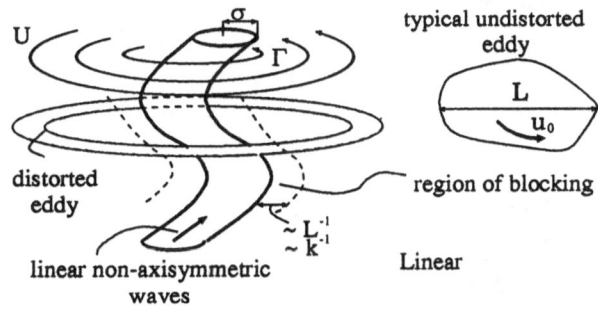

Figure 1a.   Linear rapid distortion process

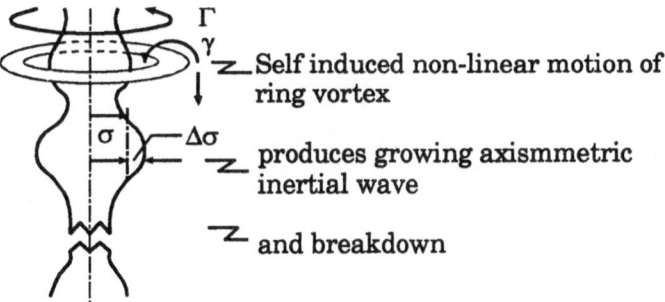

Figure 1b.   Non linear breakdown process

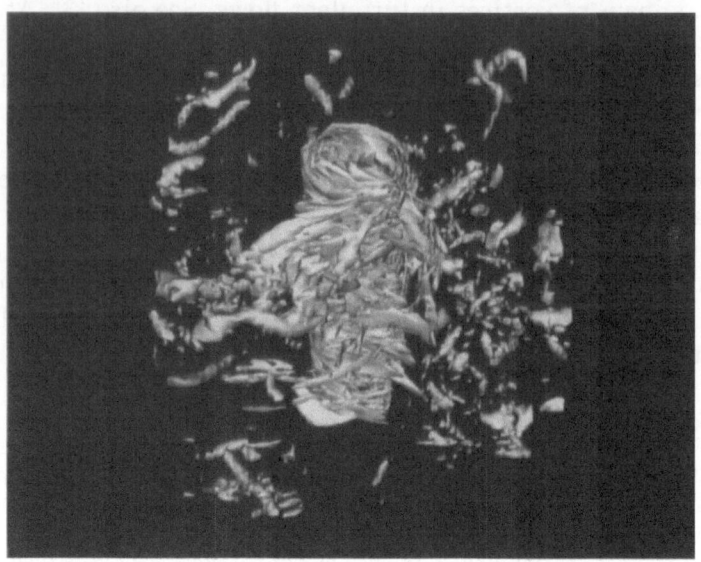

Fig 2. Simulation of the linear distortion of an initally random velocity field showing the formation of ring vortices

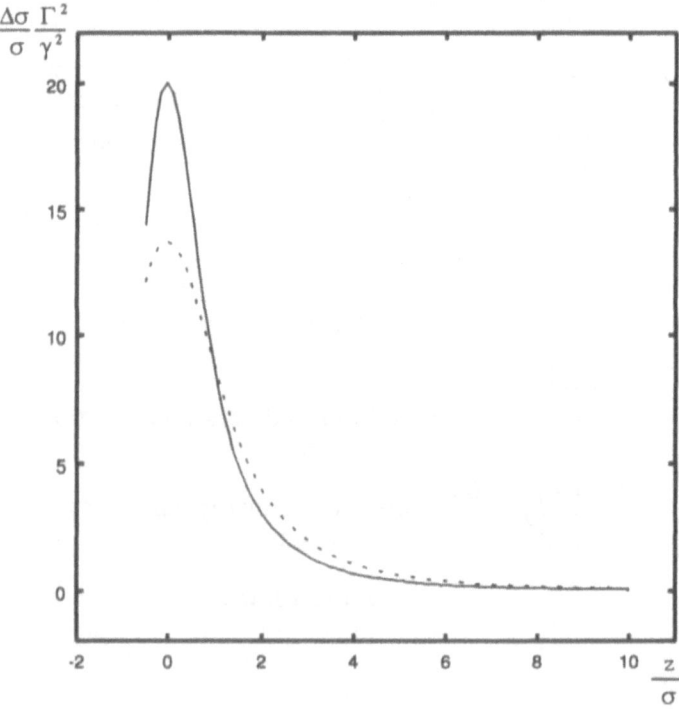

Fig 3. Amplitude $\Delta\sigma$ of the non-linear wave on the vortex; the variation with axial positions z (relative to the center of the wave), normalized on the circulations of the vortex (Γ) and of the distorted eddy (γ) and the radius.

LOW–TEMPERATURE SUPERFLUID TURBULENCE: EXPERIMENTAL AND NUMERICAL RESULTS

M.E. BRACHET[1], C. NORE[2], M. ABID[3], J. MAURER[1] AND P. TABELI
[1] LPS, CNRS, Universités Paris 6 et 7,
24 Rue Lhomond, 75231 Paris Cedex 5, France.
[2] LIMSI, CNRS, BP 133 91403 ORSAY Cedex France.
[3] IRPHE, UMR CNRS, Université d'Aix-Marseille I,
13397 Marseille Cedex 20, France.

Superfluid turbulence is studied experimentally in a low–temperature Helium flow and numerically with the Gross-Pitaevskii equation in the Taylor-Green (TG) vortex flow. It is known that [1, 2] the kinetic energy transfer is of the same magnitude in both the superfluid and viscous TG vortex. Furthermore, the energy spectrum of the superflow is compatible with Kolmogorov's scaling. Many vortex reconnection events take place throughout both the superfluid and viscous flow. Experimentally, pressure fluctuation spectra show little difference above and far below the super-fluid transition temperature, where the normal-fluid component of Helium is less than 5% in mass (at $T = 1.2K$).

1. Summary of numerical results

One of the main quantitative results obtained in [1, 2] is the remarkable agreement of the energy dissipation rate with the corresponding data in the incompressible viscous TG flow. Both the moment $t_{max} \sim 10$ of maximum energy dissipation and its value $\epsilon(t_{max}) \sim 10^{-2}$ at that moment are in quantitative agreement with the viscous data [3] (see Fig. 1, left). Furthermore, it can be checked on this figure that the dissipation is rather insensitive to the value of the coherence length ξ.

Another important quantity studied in turbulence is the scaling of the kinetic energy spectrum. It was found in [1, 2] that, in the interval $2 \leq k \leq 16$, the spectrum of incompressible kinetic energy is compatible with the $k^{-5/3}$ Kolmogorov scaling. The wavenumber $k = 16$ corresponds to the averaged distance between vortex lines and the wavenumber $k = 2$ to the integral scale.

U. Frisch (ed.), Advances in Turbulence VII, 377–380.

The vortex lines of the superfluid TG vortex are visualized in physical space in Fig. 1, right, at $t = 9$ where a complex vortex tangle is present. Note that this vortex tangle was formed by stretching and reconnection of the vortex lines that were present at $t = 0$.

These numerical results have lead us to conjecture the existence of a Kolmogorov regime in experimental low temperature superflows [1, 2].

2. Experimental results

The experimental set-up is similar to the one described in [4]. Some modifications have been made to work down to $1.2K$. The flow is produced in a cylinder, 8 cm in diameter and 12 cm high, limited axially by two counter-rotating disks. One disk is flat and, on the other one, are fixed 8 radial blades, forming an angle of 45° between each other. A stator is mounted at half the total height of the cell in order to stabilize the turbulent shear region. The two disks are driven by two DC motors rotating from 1 to 30 Hz. The whole system is enclosed directly in a liquid Helium bath which is used as the experimental fluid, the main difference with the set-up described in [4]. The temperature of the fluid is fixed by the pressure above the liquid bath, which is itself controlled by the pumping system.

Local pressure fluctuations are measured by using small total-head pressure tubes, immersed in the flow. The pressure sensors are hollow metallic tubes, connected to a quartz pressure transducer WHM 112 A22 from PCB. Details are given in [5].

In normal fluids, the pressure measured at the tip of the total-head tube can be related to the upstream flow $U(t)$ and the local pressure $P(t)$ using Bernoulli theorem:

$$P_{\text{meas}}(t) = P(t) + \rho U^2(t)/2 \qquad (1)$$

In the flow region where the probe is immersed, a well established axial mean flow U exists so that, after removing the mean parts of Equation (1), one gets:

$$p_{\text{meas}}(t) = p(t) + \rho U u(t) \qquad (2)$$

where p_{meas}, p and u are the fluctuations of the measured pressure, the actual pressure, and the local velocity respectively. It is currently admitted that, in ordinary turbulent situations, and at low fluctuation rates, Equation (2) is dominated by the dynamic term, so that, by measuring the pressure fluctuations at the total head tube, one has a direct access to the velocity fluctuations $u(t)$.

The situation is less clear when the probe is immersed in the superfluid. It is however possible to write an equation similar to (2). Details can be found in [5, 6].

The analysis of the pressure fluctuations obtained with the total head tube placed at 2 cm above the mid plane and 2 cm from the cylinder axis, yields interesting informations. Figure 2 shows the spectra of the pressure fluctuations above and below T_λ (i.e respectively at $2.3K$ and $1.4K$). Fig. 2, left, clearly shows, as expected, that such fluctuations follow a Kolmogorov regime between the injection scale (signalled by the peak at 25 Hz) and the largest resolved frequency, i.e 900 Hz. The spectrum obtained at $1.4K$ is similar to that obtained at $T = 2.3K$ (see Fig. 2, right). A clear Kolmogorov like regime exists for the same range of frequencies. The corresponding Kolmogorov constant turns out also to be indistinguishable from the classical value. We have further analyzed the deviations from Kolmogorov in the superfluid regime. The striking result is that they have the same magnitude as in classical turbulence. More details are given in [6].

These observations - both on global and local quantities - agree pretty well the theoretical approach developed in the first part of the paper. In particular, it seems rather clear that Kolmogorov cascade survives in the superfluid regime.

References

1. Nore C., Abid M. and Brachet M. (1997a) Decaying Kolmogorov turbulence in a model of superflow, *Phys. Fluids*, 9, 2644-2669.
2. Nore C., Abid M. and Brachet M. (1997b) Kolmogorov turbulence in low-temperature superflows, *Phys. Rev. Lett.*, 78, 3896-3899.
3. Brachet M. E., Meiron D. I., Orszag S. A., Nickel G., Morf R. H. and Frisch U. (1983) Small-scale structure of the Taylor-Green vortex, *J. Fluid Mech.*, 130, 411-452.
4. Maurer J., Tabeling P. and Zocchi G. (1994) Statistics of Turbulence between two counterrotating Disks in Low-temperature Helium Gas, *Europhys. Lett.*, 26, 31-36.
5. M. Abid, M.E. Brachet, J. Maurer, C. Nore and P. Tabeling (1998) Experimental and numerical investigations of low–temperature superfluid turbulence, to appear in *Eur. J. Mech. B Fluids*.
6. Maurer J. and Tabeling P. (1998) *Local investigation of superfluid turbulence*, submitted to Europhys. Lett.

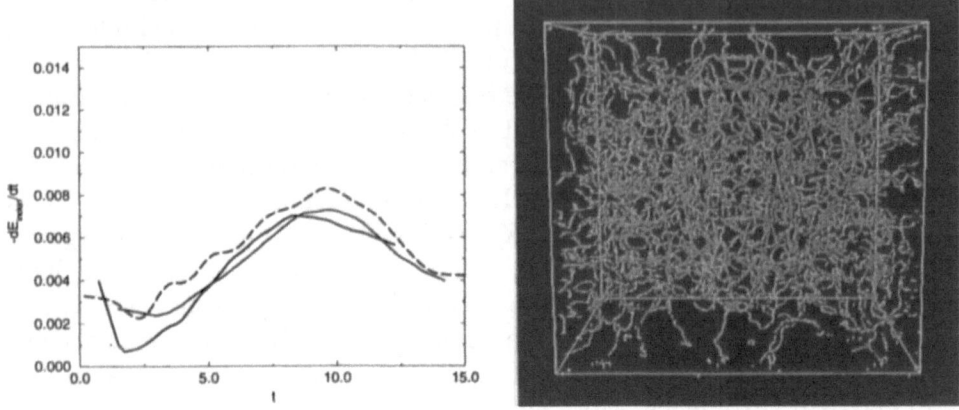

Figure 1. Left: Time evolution of the total incompressible kinetic energy dissipation $-dE^i_{kin}/dt$ for: $\xi = 0.1/(4\sqrt{2})$ and $N = 256$ (long-dashed), $\xi = 0.1/(6.25\sqrt{2})$ and $N = 400$ (dotted) and $\xi = 0.1/(8\sqrt{2})$ and $N = 512$ (solid), where ξ is the coherence length and N the grid resolution. All runs are realised with sound velocity $c = 2$. Right: Three-dimensional visualization of the vorticity field for the Taylor-Green flow at time $t = 9$. The visualization is obtained by drawing the 30000 vectors of highest norm in the $[0, \pi] \times [0, \pi] \times [0, \pi]$ box.

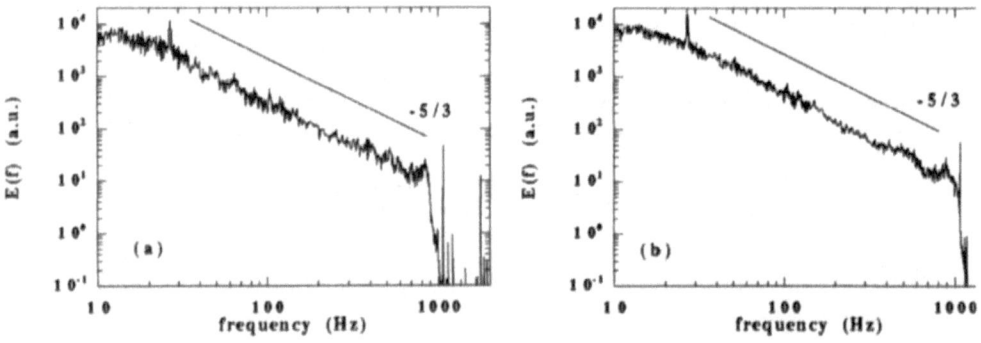

Figure 2. Experimental pressure fluctuation spectrum (in non-dimensional units) measured with a total head pressure tube immersed in the flow at $T = 2.3\ K$ (left) and $T = 1.4\ K$ (right).

STATIONARY DISSIPATIVE STRUCTURES DRIVEN BY LARGE–SCALE MHD FIELDS

K.N. BERONOV AND V.P. BERONOVA

Research Institute for Mathematical Sciences,
Kyoto University,   Sakyo-ku, Kyoto, Japan

A family of steady solutions to the incompressible, viscous and resistive MHD equations is derived. They are proposed as models useful in understanding the structure of large-scale three-dimensional MHD flows observed in various contexts. Although the explanation of these observations remains only qualitative, the range of effects addressed suggests the importance of the common mechanism captured by the solutions. These represent equilibria, at small scales in turbulent MHD flows, between the straining by large-scale fields, and the dissipation. The large-scale fields are considered slowly varying and locally modeled as steady, linear, and curl-free; the small-scale structures are modeled by steady solutions with two-dimensional structure and translational or rotational symmetry.

Dissipative MHD solutions driven by a large-scale *flow* have been considered before: A *current* layer model is introduced in [3]; the model of stretched *magnetic flux* tube in axisymmetric strain seems common [5, 14]. These are MHD analogs of the Burgers vortex layer and tube solutions, but the additional driven field is different for each geometry. Two missing elements are now added: a *magnetic field* is allowed for in the large-scale driving model, and *current*-and-vortex tubes are constructed, which describe, to leading order, the stable structures resulting from (Kelvin-Helmholtz or resistive) layer instabilities. (How axisymmetric cores dominate strong vortex tubes in shear fields is shown in [10].)

Magnetic driving couples the driven fields. When there is no linear magnetic field, we pursue further the analogy between neutral and MHD dynamics: All known MHD layer solutions are included into a larger family, in parallel with the classification [13] of a set of exact vorticity layer solutions. The Burgers layer has a time-dependent self-similar variant; self-similar versions exist for the whole MHD layer family.

U. Frisch (ed.), Advances in Turbulence VII, 381–384.

Strong current layers are commonly observed in large-scale MHD flows. Their location usually overlaps with that of strong vorticity layers [9]. The initial exponentially fast shrinking of layers at X-points in two-dimensional [4] and three-dimensional MHD flow eventually saturates due to dissipation. As in hydrodynamic turbulence, the *creation* of current sheets and their collocation with vorticity layers can be described in terms of self-similar, time-dependent layer solutions (driven only by a large-scale *velocity*).

Approximate statistical alignment of turbulent magnetic and velocity fields [1] motivates a restriction of the model large-scale magnetic field to one aligned with the driving velocity $(-Ux, -Vy, (U+V)z)$, i.e. to $(-Ax, -By, (A+B)z)$, where $A, B, U, V =$ const., $AV = BU$ provide steady forcing. Quasi two-dimensional steady states are obtained if the driving flow has one direction of stretching: $U > 0$, $V \geq 0$. The driven fields are fully determined by the scalar vorticity $\omega(x, y)$ and current $j(x, y)$ obeying

$$
\begin{aligned}
\nu_m \triangle j &= \mathcal{J}(\psi; \omega) - \mathcal{J}(\phi; j) + x\partial_x(A\omega - Uj) + y\partial_y(B\omega - Vj) \\
&\quad + (U-V)\,\square\psi - (A-B)\,\square\phi + 2(\psi_{,xy}\,\square\phi - \phi_{,xy}\,\square\psi)\,, \\
\nu \triangle \omega &= \mathcal{J}(\psi; j) - \mathcal{J}(\phi; \omega) + x\partial_x(Aj - U\omega) + y\partial_y(Bj - V\omega) \\
&\quad - (U+V)\omega + (A+B)j\,, \\
\triangle\phi &= -\omega\,, \qquad \triangle\psi = -j\,, \qquad\qquad \triangle = \partial_x^2 + \partial_y^2\,, \\
\mathcal{J}(f; g) &= f_{,x}\,g_{,y} - f_{,y}\,g_{,x}\,, \qquad\qquad\quad \square = \partial_x^2 - \partial_y^2\,.
\end{aligned}
$$

Translational or circular symmetry of the solutions annihilates the nonlinear terms. The localized solutions of the two resulting ODEs are classified.

Layer solutions $(\partial_y = 0)$ with collocated current and vorticity show coupling of their amplitudes when $A \neq 0$. When $B = V = 0$, layers have a Gaussian profile, and the number of solutions depends on the ratio A/U: When $A > U$, only a single solution exists, but when $|A| < U$, two linearly independent solutions coexist, characterized by layer thicknesses $\sim \sigma_\pm^{-1/2}$,

$$
\sigma_\pm = \frac{U(\nu_m + \nu) \pm \sqrt{U^2(\nu_m - \nu)^2 + 4A^2\nu_m\nu}}{2\nu_m\nu}\,, \qquad \nu_m, \nu > 0\,.
$$

Whenever some $\sigma > 0$, a solution exists, unique up to a freely chosen amplitude a, but the *relative* amplitudes of the current and vorticity are fixed: $j_{\max} = a(U + A - \nu\sigma)$ and $\omega_{\max} = a(U - A - \nu_m\sigma)$.

Axisymmetric tube solutions are found only in the case $U = V$, $A = B$, and then only in convergent series form. The asymptotics at large radii show that a localized solution is only possible when $|A| < U$. Moreover, at $A = 0$ there is no current in the steady tube; perturbations of current layers will cause dissipation without creating stable current tubes. A steady current tube requires a large-scale field $A \neq 0$ and a vortex to roll it up.

MHD turbulence simulations [9] have shown that, in the presence of substantial magnetic field, layers are stable; in analogous flows without magnetic field, vorticity layers decompose into tubes. We believe such layer stability will persist at higher Reynolds numbers: When $A > U$, there is no steady tube-like state which could attract a growing perturbation. Even if the magnetic strength is smaller (so that localized tube states are formally allowed) but is still *comparable* with the velocity strain rate, the effective radius of the tube must be so large that the solution cannot be realized. When U is prevailing over A, the qualitative picture is analogous to the hydrodynamic case: Current-vorticity tubes emerge from the instability of current-vorticity layers [6], which is of Kelvin-Helmholtz or tearing nature, depending on the ratio of the magnetic to the dynamic Reynolds number [8]. The new solutions provide steady profiles for the stability analyses of stretched layers and tubes, and show the need to include magnetic flux driving in the analysis (for the pure flow-driving case see [7]), as the type of instability depends not only on the magnetic Prandtl number ν_m/ν [7, 8], but also on the relative magnitude of large-scale fields $|A|/U$.

Observations and numerical simulations show [11] that a *thin* current sheet forms *inside* the near-Earth plasma sheet, with possible subsequent instability and tube formation, resulting in a substorm. An "inner layer" appears in our model, as well, when $|A| < U$: Instability may lead not to tube formation, but only to *redistribution* of energy between two independent layer solutions. Although total vorticity is preserved by the time-dependent variant of the above equations (under periodic or decay boundary conditions), total current is not, and perturbations of one steady layer state can lead to another, given by a different linear combination of the two layer solutions of different thicknesses, wherein the current is partially dissipated. Advective concentration into sheets dominates diffusion on larger scales, enhancing the narrower layer. If two layer solutions are allowed, so are tubes; the thinner and stronger layer is more unstable and rolls up.

A recent book [2] emphasizes that simulations of MHD turbulence starting from different initial conditions saturate to *only two* statistically steady states, one with approximately constant magnetic $\beta \leq 1$, and another with $\beta \gg 1$. This dichotomy occurs in 2D periodic DNS, in the distinction of fast/slow turbulent dynamos, during nonlinear evolution of the magnetorotational instability, in observations of optically thin/dense accretion disks (emitting hard-and-intermittent/soft-and-homogeneous X-radiation). It is remarked [2, Section 4.2] that (1) optically thin, low-β disks correspond to higher values (~ 0.1) of the "magnetic viscosity parameter", which is proportional to magnetic fluctuation correlations, than the high-β disks, and (2) that "numerous X-point like structures in the nonlinear stage of magnetorotational instability" are found in DNS of 3D compressible MHD with rotation [12]. Our view is that the dynamics of energy transfer controls the overall properties of those turbulent plasmas, and the dissipative, linear stagnation flow/flux model is relevant. Then, the dichotomy in the number of model solutions, with respect to the ratio U/A, can be translated into a β dichotomy: In initially high-β flows $|A| \ll U$, current/vortex tubes form, and in turn create strain fields orthogonal to the larger ones they are supported by. It can be shown that when $U/|A|$ is large, driven velocity and magnetic fields at X-points between tubes are in a proportionally large ratio. Thus, cascading is possible. In low-β flows, cascading is inhibited at large scales, except for X-points where, contrary to the average, $U/|A| > 1$. Energy is dissipated effectively only at these locations, hence the strong intermittency. In other words, while vortices mix high-β fields, layers separate regions with different field strengths, thus increasing the "magnetic viscosity parameter" of low-β fields.

References

1. Biskamp, D. (1993), Nonlinear Magnetohydrodynamics, *Cambridge Monographs on Plasma Physics*, Vol. 1, Cambridge Univ. Press.
2. Tajima, T. and Shibata, K. (1998), Plasma Astrophysics, *Frontiers in Physics*, Vol. 98, Addison-Wesley.
3. Sonnerup, B.U.Ö. and Priest, E.R. (1975), Resistive MHD stagnation-point flows at a current sheet, *J. Plasma Phys.* **14**, p.525.
4. Sulem, P.L., Frisch, U., Pouquet, A., and Meneguzzi, M. (1985), On the exponential flattening of current sheets near neutral X-points in two-dimensional ideal MHD flow, *J. Plasma Phys.* **33**, p.191.
5. Hughes, D.W., and Proctor, M.R.I. (1988), Magnetic fields inthe scalar convection zone, *Ann. Rev. Fluid Mech.* **20**.
6. Einaudi and Rubini (1989), Resistive instabilities in a flowing plasma. II. Effects of viscosity, *Phys. Fluids* **B 124**.
7. Phan, T., and Sonnerup, B.U.Ö. (1991), Resistive tearing-mode instability in a current sheet with equilibrium viscous stagnation-point flow, *J. Plasma Phys.* **46**.
8. Ofman, L., Chen, X.L., Morrison, P.J., and Steinolfson, R.S. (1991), Resistive tearing mode instability with shear flow and viscosity, *Phys. Fluids* **B 3**, p.1364.
9. Politano, H., Pouquet, A., and Sulem, P.L. (1995), Current and vorticity dynamics in three-dimensional magnetohydrodynamic turbulence, *Phys. Plasmas* **2**(8), p.2931.
10. Moffatt, H.K., Kida, S., Ohkitani, K. (1994), Stretched vortices — the sinews of turbulence: large-Reynolds-number asymptotics, *J. Fluid Mech.* **259**, p.241.
11. Schindler, K. (1995), Formation of thin current sheets in magnetospheres, in *"Small-scale structures in three-dimensional hydrodynamic and magnetohydrodynamic turbulence"*, eds. Meneguzzi, M., Pouquet, A., Sulem, P.L., *Lect. Notes in Phys.* **462**.
12. Hawley, J.F., Gammie, C.F., and Balbus, S.A. (1995), Local three-dimensional magnetohydrodynamic simulations of accretion disks, *Astrophys. J.* **440**, p.742.
13. Beronov, K.N. (1997), Vorticity layers in unbounded viscous incompressible flow with uniform strain, *Fluid Dyn. Res.* **21**, p.285.
14. Bajer, K. and Moffatt, H.K. (1997), On the effect of a neutral vortex on a stretched magnetic flux tube, *J. Fluid Mech.* **339**, p.121.

VII

Astro/Geophysical Flow and Convection

Ecoulements Astro/Géophysiques et Convection

VII

ON SELF-SIMILAR EVOLUTION FOR MULTI-DIMENSIONAL BURGERS TURBULENCE

S.N. GURBATOV[1] AND U. FRISCH[2]
[1] *Radiophysics Dept., University of Nizhny Novgorod, 23, Gagarin Ave., Nizhny Novgorod 603600, Russia.*
[2] *Observatoire de la Côte d'Azur, Lab. G.D. Cassini, B.P. 4229, F-06304 Nice Cedex 4, France*

This work is devoted to the evolution of random solutions of the unforced Burgers equation in d dimensions ($\mathbf{v}$ is the velocity and ψ the potential)

$$\partial_t \mathbf{v} + (\mathbf{v} \cdot \nabla)\mathbf{v} = \nu \nabla^2 \mathbf{v}, \qquad \mathbf{v} = -\nabla \psi \tag{1}$$

in the limit of vanishing viscosity ν. The one-dimensional "nonlinear diffusion equation" was originally introduced in the thirties by Jan M. Burgers as a model for turbulence. The three-dimensional Burgers equation has also received attention as an approximate model for the formation of large-scale structure of the Universe when pressure is negligible; it describes then the statistical properties of gravitational turbulence, that is, the nonlinear stage of the gravitational instability developing from random initial perturbations [1]-[3]. Other problems leading to multi-dimensional Burgers equations or variants include surface growth under deposition of dust and flame front motion. In such instances, the potential corresponds to the shape of the surface or of the front.

We are interested here in the solution in the limit of vanishing viscosity ($\nu \to 0$), when the potential and velocity have the following "maximum representation" [2], [3]

$$\psi(\mathbf{x}, t) = \max_{\mathbf{y}} \Phi(\mathbf{x}, \mathbf{y}, t), \quad \mathbf{v}(\mathbf{x}, t) = \frac{\mathbf{x} - \mathbf{y}(\mathbf{x}, t)}{t}, \tag{2}$$

$$\Phi(\mathbf{x}, \mathbf{y}, t) = \psi_0(\mathbf{y}) - \frac{(\mathbf{x} - \mathbf{y})^2}{2t}. \tag{3}$$

In eq. (2) $\mathbf{y}(\mathbf{x}, t)$ is the Lagrangian coordinate at which Φ achieves its global (absolute) maximum for given $\mathbf{x}$ and t.

U. Frisch (ed.), Advances in Turbulence VII, 387–390.
© *1998 Kluwer Academic Publishers.*

At late times in the evolution from random initial conditions, the velocity field has a cellular structure with "local" self-similarity. The Lagrangian coordinates $\mathbf{y}(\mathbf{x}, t)$ becomes then a discontinuous function of $\mathbf{x}$, constant throughout cell regions, and jumping at the boundaries of these cells. In each cell, fluid particles fly away from a small region near the cell center. The velocity field has discontinuities (shocks) at the cell boundaries; these shock surfaces or walls form a connected structure. The longitudinal component of the velocity vector $\mathbf{v}(\mathbf{x}, t)$ consists of a sequence of sawtooth pulses, just as in one dimensions. The transverse components consist of sequences of rectangular pulses with random amplitudes. Wall motion results in a continuous change of cell shape with cells often swallowing their neighbors and thereby inducing growth of the external scale $L(t)$ of the Burgers turbulence.

In the limit of the vanishing viscosity, as the time t tends to infinity, the solutions become statistically self-similar. First, we consider the "cellular" model of initial condition, in which we assume, that the space is divided into cells of volume L_0^d and we assign an initial value for ψ_0 chosen independently within each cell. We show that the asymptotic behavior of the turbulence at large times when $L(t) \gg L_0$ is then determined by the tail of the cumulative distribution $F(H) = \text{Prob}(\psi_0 < H) = 1 - f(H)$. In the one-dimensional case, the mean value $\langle (\partial_x \mathbf{v})^2 \rangle$ is the energy of the turbulence and is a decreasing function of time. However, when $d > 1$ the energy is not a quantity conserved by the nonlinear term. For the deposition problem this means that the roughness of the (hyper)-surface, measured by its mean-square gradient $E(t) = \langle |\mathbf{v}|^2 \rangle$, may either decrease or increase with time. The mean energy $E(t)$ is given by the time derivative of the mean potential $E(t) = 2\partial_t \langle \psi(\mathbf{x}, t) \rangle$. Thus, we can infer the mean energy from the probability distribution of the absolute maximum of Φ [2,3]:

$$P(H) = \partial_H Q(H, t), \quad Q(H, t) = \text{Prob}(\psi < H). \quad (4)$$

At large times the cumulative distribution has the form

$$Q(H, t) = e^{-N(H, t)}, \quad (5)$$

$$N(H, t) = \frac{1}{L_0^d} \int f\left(H + \frac{\mathbf{y}^2}{2t}\right) d^d \mathbf{y}, \quad (6)$$

where $N(H, t)$ is the mean number of events such that $\psi > H$. From (2) it follows that the probability distribution function of the velocity field is determined by the probability distribution function of the coordinate of absolute maximum, for which we have

$$P(\mathbf{y}, \mathbf{x}, t) = \frac{1}{L_0^d} \int f\left(H + \frac{(\mathbf{x} - \mathbf{y})^2}{2t}\right) \frac{\partial Q(H, t)}{\partial H} dH. \quad (7)$$

It is well known in probability theory that extreme value statistics can be classified into three different universality classes. Thus, in the Burgers turbulence also three classes of initial distributions leading to self-similar evolution are identified: a) distributions with a power-law tail $f(H) \propto H^{-\gamma}$, b) compactly supported potentials $f(H) \propto (H_M - H)^{-\gamma}\, \gamma \leq 0$, c) stretched exponential tails $f(H) \propto H^{\alpha} \exp(-H^{\beta})$. In case c), which includes Gaussian initial conditions, we find that the (mean) "energy" $E(t) = \langle |\mathbf{v}|^2 \rangle$ decays $\propto t^{-1}$ with a logarithmic correction, just as it does in one dimension (Kida's law).[1] In cases a) and b) we find that the mean potential increases and that the energy and external scale have a power-law time dependence:

$$E(t) \propto t^{-p}, \quad p = \frac{2(\gamma - d)}{2\gamma - d}; \quad L(t) \propto t^m, \quad m = \frac{\gamma}{(2\gamma - d)}. \qquad (8)$$

Note, that the probability distribution function in case a) exists only for $\gamma > 1 + d/2$. More surprisingly, for case a) when $1 + d/2 < \gamma < d$, we find self-similar evolution with the energy *increasing* in time. This increasing takes place only if $d \geq 3$. When $\gamma \gg d$ the energy decays as t^{-1}, almost as for case (c). For a power-law tail of the initial potential the probability distribution function of the velocity also has a power-law tail $\propto |\mathbf{v}|^{-2\gamma}$.

For compactly supported initial potentials (γ is now negative) the probability distribution function of the velocity has a stretched exponential tail $\propto \exp\{-|\mathbf{v}|^{d+2|\gamma|}\}$. When $\gamma \to \infty$ we recover a t^{-1} law for the mean energy. If $\gamma = -1$, so that the distribution is nearly uniform near the edge of $H = H_M$ of ths suppport, we have $p = \dfrac{2(1 + d)}{2 + d}$. In the limit $\gamma \to 0$ the energy decays as t^{-2} in all dimension d, and the structure of the velocity field is "frozen", with a probability distribution function $\propto \exp\{-|\mathbf{v}t|^d\}$ for all $\mathbf{v}$. We remark also that this "frozen" behavior takes place for arbitrary γ when the dimension of the space $d \to \infty$. We also show that when the initial velocity is Gaussian, homogeneous, and *anisotropic* the one- and two-point distribution of the velocity can be obtained and become asymptotically self-similar and isotropic.

In the one-dimensional case we also consider the situation when the initial velocity has long correlations. Specifically, we assume, that the initial velocity is homogeneous and Gaussian with a spectrum proportional to k^n at small wavenumbers k and falling off quickly at large wavenumbers. When $1 < n < 2$ the spectrum, at long times, has three scaling regions as illustrated on the Figure: first, a $|k|^n$ region at very small k's with a time-independent constant, stemming from an "outer region" in which the initial conditions are essentially frozen; second, a k^2 region at intermediate wavenumbers, and, finally, the usual k^{-2} region, associated to the

---

[1] A result also reported by J.P. Bouchaud and M. Mézard (cond-mat/970747).

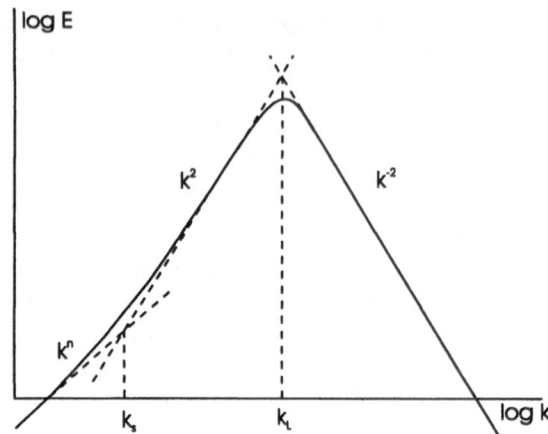

shocks. The spectrum does not evolve in a globally self-similar way. Indeed, the switching from the $|k|^n$ to the k^2 region occurs around a wavenumber $k_s(t) \propto t^{-1/[2(2-n)]}$, while the switching from k^2 to k^{-2} occurs around $k_L(t) \propto t^{-1/2}$. For details, see Ref. [4].

This work was supported by grants RFBR No 96-02-19303 and INTAS-RFBR No 95-IN-RU-0723. S. Gurbatov thanks the French Ministry of Education, Research and Technology for support during his visits to the Observatoire de la Côte d'Azur.

References

1. Shandarin S.F. & Zeldovich Ya.B. *Rev. Mod. Phys.*, **61**, (1989), 185.
2. Gurbatov S.N., Malakhov A.N., & Saichev A.I., *Nonlinear Random Waves and Turbulence in Nondispersive Media: Waves, Rays and Particles.* Manchester University Press, 1991.
3. Vergassola M., Dubrulle B., Frisch U.& Noullez A. *Astr. Astrophys.***289** (1994), 325.
4. Gurbatov S, Simdyankin S., Aurell E., Frisch U. & Tóth G. *J. Fluid Mech.*, **344** (1997), 339.

A NEW CONCEPT OF THE THIRD-ORDER TRANSPORT IN NON-LOCAL TURBULENCE CLOSURE FOR CONVECTIVE BOUNDARY LAYERS

S. ZILITINKEVICH[1,2], V. GRYANIK[3,4], V.N. LYKOSSOV[5] AND
D.V. MIRONOV[3]
[1] Meteorology Programme, Uppsala University,
Villavägen 16S-752 36, Uppsala, Sweden.
[2] RISØ National Laboratory,
P. O. Box 49, DK-4000 Roskilde, Denmark.
[3] Alfred Wegener Institute for Polar and Marine Research,
Am Handelshafen 12, 27570 Bremerhaven, Germany.
[4] A. M. Obukhov Institute of Atmospheric Physics, RAS,
Pyzhevsky 3, 109017 Moscow, Russia.
[5] Institute for Numerical Mathematics, RAS,
Gubkina str. 8, "D" 117312 Moscow, Russia.

The turbulence closure problem for the convective boundary layer (CBL) is considered with the main aim to advance in understanding the nature and developing an improved parameterisation of non-local transport due to large-scale semi-organised structures. The problem is treated by the example of the vertical turbulent flux of potential temperature, $\langle w\theta \rangle$. Hereafter the angle brackets denote the ensemble mean, small letters denote turbulent fluctuations, and Θ stands for mean potential temperature. An overview is given of various closure schemes ranging from comparatively simple counter-gradient correction to the Boussinesq approximation, to sophisticated closures based on a set of budget equations for the second-order moments (fluxes and variances). It is emphasised that the key role in the non-local transport is played by the third-order moments.

As an alternative to conventional turbulent diffusion parameterisation for the flux of flux of potential temperature (e.g. Canuto et al. 1994), a turbulent advection + diffusion parameterisation is developed,

$$\left\langle w^2\theta \right\rangle = w_a \left(C_\theta \left\langle w\theta \right\rangle - C_k K_{w\theta} \frac{\partial \Theta}{\partial z} \right) - K_{w\theta} \frac{\partial \left\langle w\theta \right\rangle}{\partial z}. \tag{1}$$

U. Frisch (ed.), Advances in Turbulence VII, 391–394.

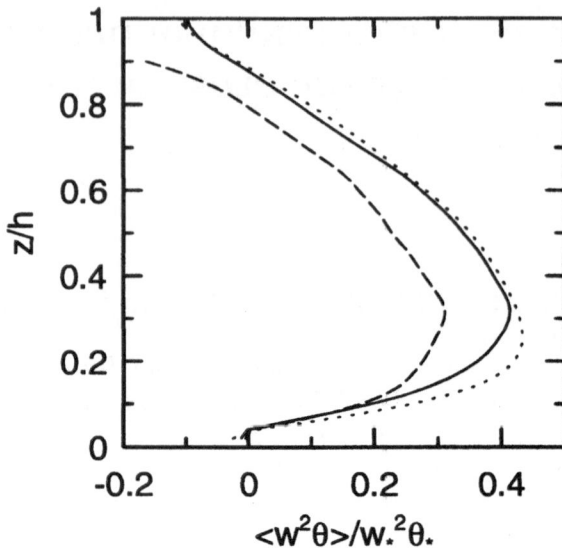

Figure 1. Vertical profile of the flux of flux of potential temperature. Dotted line shows LES data, solid line is computed from the proposed parameterisation, and dashed line, from the quasi-normal approximation. Curves are made dimensionless with the Deardorff convective scales, $w_* = (\beta Q_s h)^{1/3}$ and $\theta_* = Q_s/w_*$, where β is the buoyancy parameter, Q_s is the surface temperature flux, and h is the CBL depth.

Here, $C_\theta = 1$ and $C_k = 0.1$ are dimensionless constants, $K_{w\theta} = 0.2\tau \left\langle w^2 \right\rangle$ is the turbulent diffusivity, $\tau = e/\epsilon$ is the turbulence time scale, e is the turbulence kinetic energy, ϵ is its dissipation rate, and w_a is the "large-eddy skewed-turbulence advection velocity" given by

$$w_a = \left\langle w^3 \right\rangle / \left\langle w^2 \right\rangle \equiv S_w \left\langle w^2 \right\rangle^{1/2} , \qquad (2)$$

where S_w is the vertical velocity skewness. The first term on the r.h.s. of Eq. (1) is the turbulent advection term. It describes the contribution of the CBL-scale coherent structures to $\left\langle w^2\theta \right\rangle$. The second term is a correction to the first term due to vertical inhomogeneity of the mean temperature. The third term is a conventional diffusion term. As seen from Fig. 1, the proposed parameterisation is in good agreement with data from large-eddy simulation (LES) of the surface-heating-driven shear-free CBL capped by a strong temperature inversion (detailed description of the simulation is given in Zilitinkevich *et al.* 1997). Notice that "traditional" parameterisation based on the the quasi-normal approximation for the fourth-order moments is inconsistent with the data.

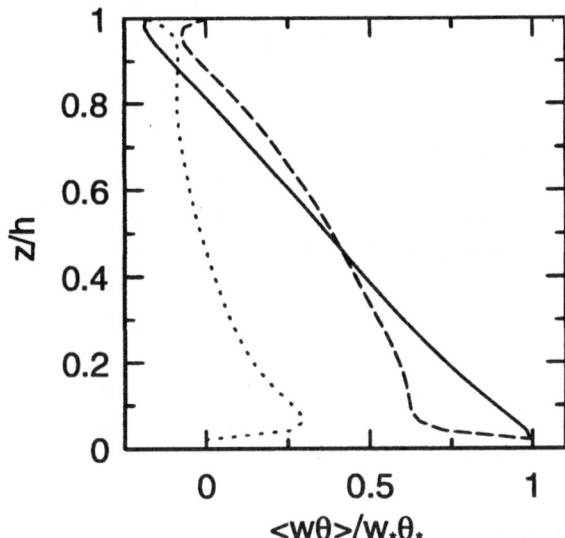

Figure 2.   Decomposition of the potential temperature flux, $\langle w\theta \rangle$, using the decomposition of the Green function, $G(z, z')$, into the contributions from below and from above the diagonal $z = z'$, viz., $G(z, z') = G(z, z')_b + G(z, z')_t$. Total flux is shown by the solid line, the bottom-up component, $\langle w\theta \rangle_b$, by the dashed line, and the top-down component, $\langle w\theta \rangle_t$, by the dotted line.

Employing the above parameterisation for the third-order flux, along with the standard parameterisation for the pressure-scalar covariance that accounts for the direct effect of buoyancy, the budget equation for the potential temperature flux provides a non-local turbulence closure for the flux in question. The solution to that equation in terms of the Green function is nothing but an integral closure. In particular cases it reduces to a number of closures proposed earlier. The counter-gradient correction closure (Deardorff 1972), the transport-asymmetry closure employing the second derivative of transported scalar (Wyngaard and Weil 1991), and an integral closure principally similar to that for passive scalars (Berkowicz and Prahm 1979) are among them. The Green-function solution to the flux budget equation provides mathematically rigorous procedure for decomposition of turbulence statistics into the bottom-up and top-down components. That decomposition resembles the bottom-up/top-down decomposition derived by Wyngaard (1983) from physical arguments. As illustrated in Fig. 2, the proposed Green-function decomposition exhibits unexpected, essentially non-linear profiles of the bottom-up and top-down components of the potential temperature flux in the CBL, in sharp contrast to universally adopted linear profiles.

The proposed advection + diffusion turbulence closure is likely to be applicable also to the passive scalar transport in near-neutral skewed turbulence. In that case, taking $\beta = 0$, *i.e.* considering the quantity θ as a passive scalar, our basic results remain in force. Furthermore, principally similar approach employing the advection + diffusion closure for third moments can be taken to consider the non-local momentum transport.

Physical ideas underlying our approach diverge from conventional turbulence-closure philosophy according to which a flux in question is expressed as a function of turbulence moments of the same or the lower order. Our parameterisation for the potential temperature flux (a second-order moment) involves not only the first-order moment (potential temperature gradient) and second-order moments (vertical velocity variance and temperature variance), but also a third-order moment, namely, the vertical velocity triple correlation, or, alternatively, the vertical velocity skewness.

Acknowledgements

This work was supported by the German Federal Ministry for Education, Research and Technology (Grant 03F08GUS), the German Co-ordinating Office of the World Ocean Circulation Experiment (Grant 03F0157A), the Russian Fund for Fundamental Investigations (Grant 95-05-14172), and the National Observatory of Athens, Greece (Contract No. JOR3-CT95-0008, JOULE Program). It is the contribution No. 1404 of the Alfred Wegener Institute for Polar and Marine Research, Bremerhaven, Germany.

References

Berkowicz, R. and Prahm, L. P. (1979) Generalisation of K-theory for turbulent diffusion. Part I: Spectral turbulent diffusivity concept, *J. Appl. Meteorol.* **18**, 266–272.

Canuto, V. M., Minotti, F., Ronchi, C. and Ypma, R. M. (1994) Second-order closure PBL model with new third-order moments: comparison with LES data, *J. Atmos. Sci.* **51**, 1605–1618.

Deardorff, J. W. (1972) Theoretical expression for the counter-gradient vertical heat flux, *J. Geophys. Res.* **77**, 5900–5904.

Wyngaard, J. C. (1983) Lectures on the planetary boundary layer, in *Mesoscale Meteorology – Theories, Observations and Models*, eds. D. K. Lilly and T. Gal-Chen, pp. 603–650, NATO ASI Series, D. Reidel, Dordrecht.

Wyngaard, J. C. and Weil, J. C. (1991) Transport asymmetry in skewed turbulence, *Phys. Fluids* A**3**, 155–162.

Zilitinkevich, S. S., Gryanik, V. M., Lykossov, V. N. and Mironov, D. V. (1997) *A Look at the Hierarchy of Non-Local Turbulence Closures for Convective Boundary Layers*, Berichte aus dem Fachbereich Physik, Report 81, Alfred-Wegener-Institut für Polar- und Meeresforschung, Bremerhaven, Germany, 31 pp.

ON THE ERTEL AND IMPERMEABILITY THEOREMS FOR SLIGHTLY VISCOUS CURRENTS IN STRATIFIED ROTATING SYSTEMS.

ETTORE SALUSTI

INFN - Phys. Dept.
University of Rome "La Sapienza"
Piazzale A.Moro 2, Roma, Italy

AND

ROBERTA SERRAVALL

Dept. of Mechanics and Aeronautics
University of Rome "La Sapienza"
Via Eudossiana 18, Roma, Italy

Abstract. In this note we make a theoretical analysis of how a mild fluid viscosity can affect the potential vorticity for stratified fluids in a rotating system. A generalization of the classical Ertel theorem is discussed and the law of conservation corresponding to novel invariants Π is obtained. These invariants do not have a classical form: indeed one example is mere classical potential vorticity multiplied by a function of time. Our results are also compared with those deriving from the recent Impermeability theorem by Haynes and McIntyre, a comparison that provides fresh insight into stratified fluid dynamics. These considerations can be of interest to determine the realistic time evolution of turbulent and of stratified water masses. It has to be stressed that similar relations hold for a large class of conserved quantities, such as tracers, entropy, etc. Of interest is also the relation with the Joukovsky theorem in classical mechanics.

U. Frisch (ed.), Advances in Turbulence VII, 395–398.

1. Introduction

The Ertel theorem concerning the time-evolution of potential vorticity (PV in the following) plays a central role in geophysical fluid dynamics (Ertel, 1942; Gill, 1982; Pedlosky, 1987; Haynes and McIntyre, 1987 and 1990; Danielsen, 1990; Müller 1995) and in magnetohydrodynamics (Hide, 1983 and 1996). Indeed fundamental vorticity and circulation theorems can be deduced from Ertel theorem, as nicely discussed by Müller (1995; M95 in the following). In addition, in some cases of interest PV is conserved. Finally in some classes of flow it is the PV that determines the system's evolution.

In such a context, it is not surprising that many realistic aspects of the dynamics of stratified fluids in a rotating system can be investigated in terms of EPV and its evolution. For magnetohydrodynamics a clear synthesis can be found in Hide (1996). For meteorological analyses a mature discussion can be found in Haynes et al. (1990). For physical oceanography a theoretical review is due to M95.

In this complex and stimulating context our study stemmed from a rather naive remark about an unnecessary stiffness in the original deduction of the Ertel theorem. This in turn allowed a generalization of this theorem to slightly viscous fluids, a generalization that can easily be applied to various examples of interest. Of particular importance is the generalization of similar ideas to non-ideal fluids, in the light of the Impermeability theorem (Haynes and McIntyre, 1987 and 1990). These considerations are of fundamental interest to determine times, zones, space scales of patches of strongly stratified and of more homogeneous fluids in order to identify allowed and forbidden zones for turbulent versus stratified flows.

2. The Ertel theorem and its generalizations.

We first sketch the classical Ertel theorem. Calling $\vec{u}_a$ the fluid absolute velocity and $\vec{\omega}_a$ the absolute vorticity, namely the sum of planetary and relative vorticities, we have:

$$\vec{\omega}_a \equiv \nabla \times \vec{u}_a \equiv 2\vec{\Omega} + \nabla \times \vec{u} \equiv 2\vec{\Omega} + \vec{\omega} \qquad (1)$$

where $\vec{u}$ is now the fluid relative velocity and $2\vec{\Omega}$ is the Coriolis term. For any regular scalar quantity $\lambda(x, y, z, t)$, from the momentum equation in classical fluid dynamics the Ertel theorem gives:

$$\frac{D}{Dt}\Pi_\lambda \equiv (\frac{\partial}{\partial t} + \vec{u} \cdot \nabla)\Pi_\lambda \equiv (\frac{\partial}{\partial t} + \vec{u} \cdot \nabla)(\frac{\vec{\omega}_a}{\rho} \cdot \nabla\lambda) =$$

$$= \frac{\vec{\omega}_a}{\rho} \cdot \nabla\frac{D\lambda}{Dt} + \frac{1}{\rho^3}\nabla\lambda \cdot \nabla\rho \times \nabla p + \frac{\nabla\lambda}{\rho} \cdot \nabla \times \frac{\vec{F}}{\rho} \equiv \quad (2)$$

$$\equiv \qquad \mathcal{E}_1 \qquad + \qquad \mathcal{E}_2 \qquad + \qquad \mathcal{E}_3$$

where $\Pi_\lambda \equiv \frac{\vec{\omega}_a}{\rho} \cdot \nabla\lambda$ is the λ-potential vorticity, ρ the fluid density, p the pressure, while $\vec{F}$ represents some forcing such as friction, external rotational forces or heating.

In the usual discussion (Pedlosky, 1987) Π_λ is conserved if and only if:

$$\mathcal{E}_1 = 0 \ , \mathcal{E}_2 = 0 \ , \mathcal{E}_3 = 0. \qquad (3)$$

which is easily recognized as an unnecessarily rigid request. Indeed a milder request is in general sufficient:

$$\frac{D}{Dt}\Pi_\lambda \equiv \mathcal{E}_1 + \mathcal{E}_2 + \mathcal{E}_3 = \alpha(t)\Pi_\lambda + \gamma(t) \qquad (4)$$

where $\alpha(t)$ and $\beta(t)$ are functions of time only, to give a new quantity

$$P_\lambda = \Pi_\lambda e^{-\int_{t_0}^t \alpha(t')dt'} + \beta(t) \qquad \frac{d}{dt}\beta(t) + \gamma e^{-\int_{t_0}^t \alpha(t')dt'} = 0 \quad (5)$$

that is a conserved quantity, namely $\frac{D}{Dt}P_\lambda = 0$. Many applications to realistic cases can be found in Salusti (1998).

3.   The Impermeability theorem

In the preceding section the classical PV is discussed for an ideal fluid. General forcings, such as heating, external rotational forces and friction can be discussed in this new formalism. The observation of Haynes and McIntyre (1987, 1990) that all these effects can be written as a flux divergence, is thus of particular interest:

$$\frac{\partial}{\partial t}(\rho\Pi_\lambda) + \nabla \cdot \vec{J}_\lambda = 0 \qquad (6)$$

where the λ-potential vorticity flux vector $\vec{J}_\lambda$ is:

$$\vec{J}_\lambda = \rho\Pi_\lambda \vec{u} - \frac{\lambda}{\rho^2}\nabla\rho \times \nabla p - \vec{F} \times \nabla\lambda - \vec{\omega}_a\frac{D\lambda}{Dt} \qquad (7)$$

Note that mass conservation is not necessary to obtain this relation.

It is also important to stress how (6) implies that $\rho\Pi_\lambda$ cannot cross λ-constant surfaces since

$$\frac{\vec{J}_\lambda}{\rho\Pi_\lambda} \cdot \nabla\lambda + \frac{\partial}{\partial t}\lambda = 0 \tag{8}$$

i.e. the "λ-velocity" normal to λ-surfaces exactly equals the surface normal velocity. Now, since $\rho\Pi_\lambda$ cannot cross λ-surfaces, it can only move along λ-surfaces and so can be created and destroyed only at λ-surface boundaries: consequently, if a fluid volume is bounded solely by λ-surfaces, then its global value cannot change with time and it can only be concentrated or diluted by intrusions of fluid.

It is worth noting that the frictional term in equations (7) produces an effective forcing in a direction orthogonal to both $\vec{F}$ and $\nabla\lambda$, and so $\rho\Pi_\lambda$ is deflected in this new direction. From equation (7) it is also clear that $\vec{J}_\lambda/\rho\Pi_\lambda$ is not the fluid velocity $\vec{u}$, and consequently a fluid parcel can cross any λ-surfaces, although leaving its own $\rho\Pi_\lambda$ before the surface. These considerations allow to infer, in the general time evolution of a stratified fluid, the evolution of regions of large and of small π_λ. So if in a patch of fluid particularly large values of π_λ, initially found finds this corresponds to large λ gradient for long times; the same hold for region of vanishing π_λ where it is natural to expect that a significant turbulence evolution can be found.

References

Danielsen E.F., 1990: In defense of Ertel's potential vorticity and its general applicability as a meteorological tracer. *Jour. Atm. Sci.*, **47**, 16, 2013-2020.

Gill A.E., 1982: Atmosphere-Ocean dynamics. *Academic Press*, pp.662.

Ertel H., 1942: Ein neuer hydrodynamischer Wirbelsatz. *Meteorol. Z.*, **59**, 277-281.

Haynes P.H., and McIntyre M.E., 1987: On the evolution of vorticity and potential vorticity in the presence of diabatic heating and frictional or other forces. *Jour.Atm. Sci.*, **44** (5), 828-840.

Haynes, P.H., and McIntyre M.E., 1990: On the conservation and Impermeability theorems for potential vorticity. *Jour. Atm. Sci.*, **47**, 16, 2021-2031.

Hide R., 1996: Potential magnetic field and potential vorticity in magneto hydrodynamics. *Geoph. J. Int.*, **125**, F1-F3.

Müller P. 1995: Ertel potential vorticity theorem in Physical Oceanography. *Reviews of Geophysics*, **33**, 1, 67-97.

Pedlosky J., 1987: Geophysical fluid dynamics. *Springer*, pp. 710.

Salusti E., 1998: On the Ertel and Impermeability theorem for sligthly viscous fluids, submitted to G.A.F.D.

AN IMPROVED 'FLYWHEEL' MODEL FOR CONVECTIVE TURBULENCE IN LIQUID METALS

R. VERZICCO[1] AND R. CAMUSSI[2]
[1] Università di Roma 'La Sapienza', Dipartimento di Meccanica e Aeronautica, via Eudossiana 18, 00184 Roma Italia.
[2] Università di Roma 'Tre', DIMI, via della Vasca Navale 79, 00146 Roma, Italia.

We present and discuss some results on the effect of the Prandtl (Pr) number in the heat transfer properties of a Rayleigh–Bénard flow. Generally the Nusselt number Nu (the non-dimensional heat transfer) depends on geometrical parameters (shape of the cell and boundary conditions) and upon the Rayleigh (Ra) and Prandtl numbers. If one is only concerned with the flow changes induced by the last two parameters, however, all the other factors have to be ruled out. This consideration induced us to perform all the simulations in a fixed cell geometry (a cylindrical cell of aspect ratio 1 as in [1] and [2]). The effect of Ra and Pr has been then analyzed using different series of numerical simulations in which these factors were changed separately. In particular, two series of simulations were performed at "low" and "high" Prandtl, $Pr = 0.022$ and $Pr = 0.7$ respectively, with Ra varied in such a way to obtain a sufficiently long power law range of the Nu vs Ra relation. In the third series, in contrast, Ra was fixed at $\simeq 6 \times 10^5$, while Pr covered the range $2.2 \times 10^{-3} \leq Pr \leq 15$. The analysis of the velocity and temperature fields, have shown that the fluid structures are different for low-Pr (≤ 0.3) and high-Pr flows. In the first case, the velocity and temperature fields are dominated by a large-scale flow filling the whole domain (see [3]). For high-Pr flow, on the contrary, the recirculating cell becomes much weaker and the temperature field is characterized by the appearance of plumes. These mechanisms result in different exponents β in the Nu vs Ra relation ($Nu \sim Ra^\beta$) that are shown in Fig. 1a and are in agreement with other results in literature. In the low Pr case, β ($= 0.25 \pm 0.004 \simeq 1/4$) is lower than that in high Pr ($\beta = 0.285 \pm 0.004 \simeq 2/7$). This difference might appear negligible, however, such a small deviation could imply errors in the predicted heat transfer

399

U. Frisch (ed.), Advances in Turbulence VII, 399–402.

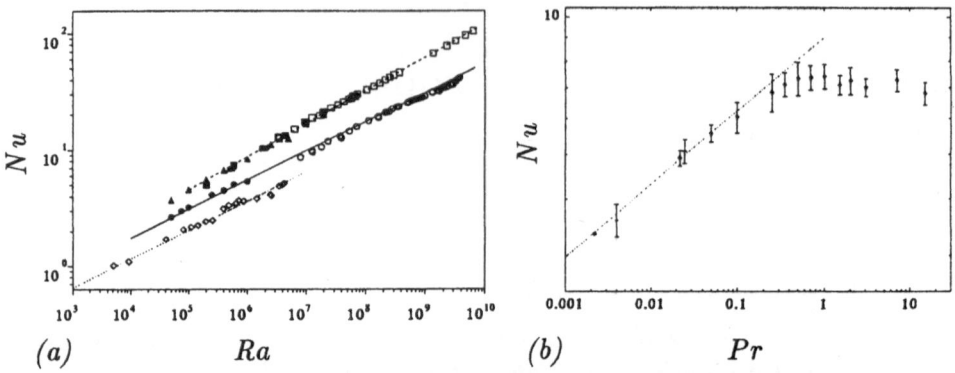

Figure 1. (*a*) Nu *vs* Ra relation for different Pr: • present numerical results at $Pr = 0.022$, ■ present numerical results at $Pr = 0.7$, filled ◇ present numerical results at $Pr = 0.35$, filled △ numerical results by [4] at $Pr = 0.7$, ○ experimental results by [1] at $Pr = 0.025$, □ experimental results by [1] and by [5] at $Pr = 4.0$, ◇ experimental results by [6] at $Pr = 0.005$ ---- fit $Nu \sim Ra^{0.285}$. ——— fit $Nu \sim Ra^{0.25}$. ········ fit $Nu \sim Ra^{0.25}$. (*b*) Nu *vs* Pr relation at $Ra = 6 \cdot 10^5$, ---- fit $Nu \sim Pr^{0.14}$: ◇ present numerical results, − experimental results from [6] and [7] for Sodium and Mercury, respectively (the value of Rossby has been corrected to account for the different aspect ratio of his cell.

above 100%. The value $\beta = 2/7$ for high Pr flows has been theoretically justified using the idea of the plumes (see [8]). In contrast, for low Pr, the plumes are not observed, therefore a different heat transfer model has to be considered. In this paper we propose a simple model to explain the smaller β starting from the 'flywheel' model first proposed by [9] and successively reconsidered by [10] for a laminar flow. In particular, the value $\beta = 1/4$ can be explained considering the presence of the recirculating cell and using dimensional arguments on the temperature equation. In fact, if we focus on the temperature equation inside the thermal boundary layer considering the structure of the mean flow (a large scale motion that induces a persisting horizontal current U sweeping the hot and cold plates) and the fact that for low Pr the thermal boundary layer is much thicker than the viscous layer, the temperature equation reads:

$$U\frac{\partial T}{\partial y} \approx \kappa \frac{\partial^2 T}{\partial x^2}, \tag{1}$$

where each term is intended averaged in time (the statistical steadiness of the flow is assumed). We assume, then, that the velocity U is proportional to the free–fall velocity $U \approx A\mathcal{U}$ and that the horizontal temperature difference Δ_h is proportional to the total temperature difference $\Delta_h \approx B\Delta$. Finally being the vertical temperature difference Δ between hot and cold plates mostly sustained within the thermal boundary layers δ_T we can write

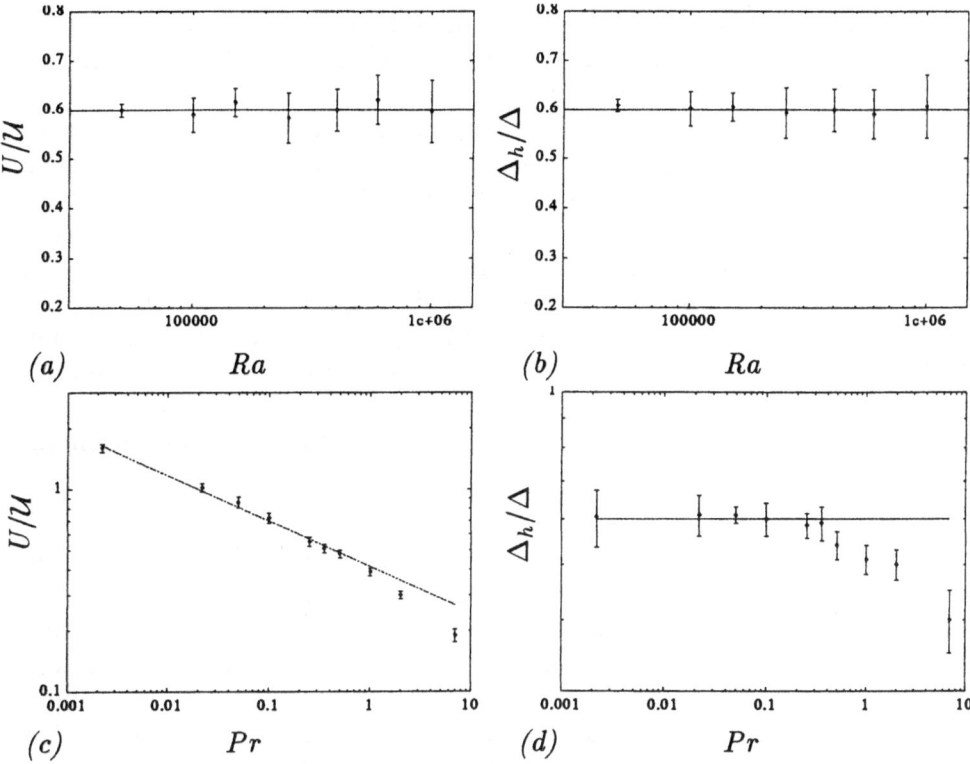

Figure 2. (*a*) Maximum vertical velocity *vs Ra* at $Pr = 0.022$ symbols, numerical results, line $U/\mathcal{U} \sim 0.6$ fit, (*b*) the same as (*a*) for the horizontal temperature difference, line $\Delta_h/\Delta \sim 0.6$. (*c*) Maximum vertical velocity *vs Pr* at $Ra = 6 \cdot 10^5$ symbols, numerical results, ---- $U/\mathcal{U} \sim Pr^{-0.22}$ fit, (*d*) the same as (*c*) for the horizontal temperature difference, ---- $\Delta_h/\Delta \sim 0.6$.

$Nu \approx d/(2\delta_T)$, with d the distance between tha plates, and the above equation becomes

$$A\sqrt{g\alpha\Delta dB}\frac{\Delta}{d} \approx \frac{\kappa\Delta}{2\delta_T^2} \quad \Rightarrow \quad Nu \approx \left(\frac{AB}{2}\right)^{1/2}(RaPr)^{1/4}. \quad (2)$$

This relation relies on two fundamental assumptions: (i) the viscous boundary layers are contained within the thermal ones (true only for low–Pr flows) (ii) the horizontal velocity U and temperature difference Δ_h induced by the cell are proportional respectively to the free-fall velocity and the vertical temperature difference through coefficients of order 1 and independent of Ra. On the other hand, these coefficients depend upon Pr in a way which gives the correct exponents for Pr in (2). The assumptions behind the model have been verified directly using the flow fields obtained from the numerical simulations and the results are shown in figure 2. In the first two panels it is shown that indeed for fixed Pr the factors A and B are independent of Ra. In addition, when the values of A and B provided by the fits of figures 2a-b are inserted into the second of (2) we obtain $Nu \simeq 0.17 Ra^{0.25}$ which is in very good agreement with the results of figure 1a. Figures 2c-d report respectively the ratio $U/\mathcal{U}$ and $\Delta_{\rm h}/\Delta$ as functions of Pr and from these plots the functional dependence of A and B on Pr can be achieved. As already anticipated by figures 1a-b we can see that there are two distinct flow regimes: in the first (for $Pr \leq 0.3$) the horizontal temperature difference is approximately constant with Pr while for $Pr > 0.3$ it decreases. However, in the high Pr regime our model is not applicable since, owing to the presence of thermal plumes, the large scale flow is not entirely responsible for the heat transfer. When the values for A and B of figures 2c-d are inserted in the second of (2) we obtain the formula

$$Nu \approx 0.27 Pr^{0.14} Ra^{0.25}. \tag{3}$$

This fit is in quantitative agreement with all the numerical and experimental results of figure 1a obtained in the low Pr regime.

The fact that the quantitative agreement of this formula is satisfactory also for the data outside the range of the present numerical simulations (like the high Ra experiments in mercury and the experiments in sodium) indicates that the essential features of the heat transfer mechanism in low Pr flows are correctly accounted by the present model.

References

1. Cioni, S., Ciliberto, S. and Sommeria, J. (1997) Strongly turbulent Rayleigh–Bénard convection in mercury: comparison with results at moderate Prandtl number, *J. Fluid Mech.*, **335**, pp. 150–181.
2. Verzicco, R., and Camussi, R. (1997) Transitional regimes of low-Prandtl thermal convection in a cylindrical cell, *Phys. of Fluids.*, 9, pp. 1287–1295.
3. Camussi, R., and Verzicco, R. (1998) Turbulent convection in mercury: scaling laws and spectra, To appear in *Phys. of Fluids*.
4. Kerr, R.M. (1996) Rayleigh number scaling in numerical convection *J. Fluid Mech.*, **310**, pp. 139–176.
5. Chillá, F., Ciliberto, S., Innocenti, C., & Pampaloni, E. (1993) Boundary layer and scaling properties in turbulent thermal convection *Il Nuovo Cimento*, 15(9), pp. 1229–1249
6. Horanyi, S., Krebs, L. & Müller, U. (1998) Turbulent Rayleigh–Bénard convection in low Prandtl–number fluids Submitted to International Journal of Heat and Mass Transfer.
7. Rossby, H.T. (1969) A study of Bénard convection with and without rotation *J. Fluid Mech.*, **36**, pp. 309–335.
8. Castaing, B., Gunaratne, G., Heslot, F., Kadanoff, L., Libchaber, A., Thomae, S., Wu, X.Z., Zaleski, S. and Zanetti, G. (1989) Scaling of hard thermal turbulence in Rayleigh–Bénard convection, *J. Fluid Mech.*, **204**, pp. 1–30.
9. Jones, C.A., Moore, D.R. and Weiss, N.O. (1973) Axisymmetric convection in a cylinder, *J. Fluid Mech.*, **73**, pp. 353–388.
10. Busse, F. H. and Clever, R. M. (1982) An asymptotic model of two dimensional convection in the limit of low Prandtl number, *J. Fluid Mech.*, **102**, pp. 75–83.

TURBULENT REGIMES OF A CONVECTION FROM A LOCAL SOURCE IN ROTATING FLUIDS

B.M. Boubnov
Institute of Atmospheric Physics RAS
109017, Moscow, Russia

Spatial inhomogeneity of heating of fluids in the gravity fields is the cause of all motions in nature: in the atmosphere and oceans on the Earth, in astrophysical and planetary objects. All natural objects rotate and convective motions in rotating fluids are of interest in many geophysical and astrophysical phenomena. Two simplest cases of the convective motions in rotating fluids are usually consider: the infinite plane layer with the temperature difference between the horizontal boundaries (vertical temperature gradient) and rotating annuli with the temperature differences between the vertical walls (horizontal temperature gradient) [1].

Convection motions from a local (but not small) source of density is the more usual case in geophysics, but the most complicated, because of the inhomogeneous density distribution in the horizontal, and vertical directions. In addition to the density distributions, the local source makes the geometry more complicated. Here we will consider different turbulent regimes of convective motions from the local source of density. Results of two different experiments on convection from a local source are consider: thermal convection from a heating rotated disk and a "dense" convection from a source of a salt water.

First let us consider convection from an isolated source of a salt water in a very large tank without rotation. For this case a horizontal scale H and a vertical scale L is larger than diameter of a source D. A picture of convective motions for this case is present in Fig. 1.

Below the disk there are three regions of turbulent flow. In region I ($D/z >> 1$, where z is a distance from a disk) the horizontal component of velocity is large (due to entrainment from the side). It can be defined as velocity scale from the local approximation $U \propto (BD)^{1/3}$, where B is a buoyancy flux per unit area $B = g^* V$, $g^* = (\Delta\rho/\rho)g$ and V is a velocity of a more dense fluid that come from source of buoyancy.

The diameter of the turbulent plume in region I decreases with height to a minimum value of approximately $0.4 D$. In region II, z is of $O(D)$ and the vertical component of velocity is much larger than the horizontal velocity and approximation $U \propto (Bz)^{1/3}$ [2]. At very large distances from the disk $z >> D$. This is region III, where the flow from the disk is similar to the a flow from a point source, with the velocity dependence $U \propto (B_0/z)^{1/3}$, where B_0 is a total buoyancy flux $B_0 = \pi D^2 B/4$.

Rotation of the fluid will change the flows in region I (and consequently in regions II and III), because the Coriolis force to force the flow to be horizontal. For small Rossby number $Ro = U/\Omega D$ according the Proudman-Taylor theorem suggests that horizontal entrainment flows could not exist, and convective motions below the

U. Frisch (ed.), Advances in Turbulence VII, 403–406.

disk are similar to the motions in a plane layer of fluid. A system of a vertical vortices arranged as 2D turbulent motions

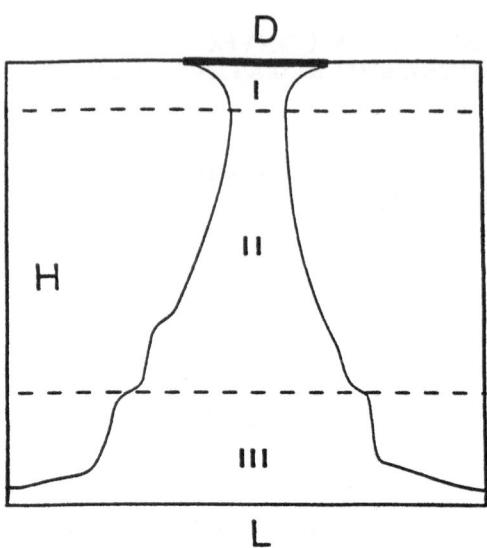

Figure1. A sketch for "dense" convection experiments from a local source. Regions of turbulent flow : I - horizontal, II - accelerated, III - decelerated.

For Ro≈ 1 a Coriolis force changes the direction of the entrainment flow and creates a cyclonic circulation below the disk. Depending on the buoyancy flux B (or the heat flux f) a single vortex with can originates below (in the case of heating) or above the disk (cooling or "dense" experiments with a salt water).

For a small buoyancy flux there is not enough energy to create an intensive single vortex. In this case there are three main regimes of flow in the space between disk and boundaries: a symmetric circulation (low rotation rate), baroclinic vortices with a size comparable to the gap between the disk and the side wall, and small turbulent vortices.

For thermal convection experiments a heating disk is mount at a bottom of a large tank. All system rotates with a constant angular velocity Ω around the vertical axis. For non-dimensional parameters it is convenient to use a rotational Peclet number $Pe_\Omega = C_1 (Ra_f/Ta)^2$, where $Ta = 4\Omega^2 D^4/\nu^2$ and $Ra_f = (\varepsilon D^4/k^2\nu)$. Here ε is a rate of kinetic energy dissipation, k and ν are thermodiffusivity and viscosity of fluid and $C_1 = 0.63$ [3]. A second non-dimensional parameter ε_T is a temperature difference between heating disk and ambient air $\varepsilon_T = \Delta T/T_0$, and T_0 is a temperature of ambient air.

There are four main regimes: convective vortices (CV), tornado-like vortices (TLV), anticyclonic jet (AJ), and turbulent convection (TC). Boundaries between the regimes weakly depend on ε_T. For fast rotation rate $(Pe_\Omega< 3)$ is a regime of convective vortices (CV). In this regime many vortices are below the disk and periodically propagate away from disk. For medium rotation rate $(3<Pe_\Omega< 20)$ a single intensive vortex periodically arrized below the disk and after some time moves to the side of tank. For small rotation rate $(20<Pe_\Omega< 150)$ a turbulent jet is similar to the jet in Fig.1, but it moves up at some angle to vertical axis and rotates around it in an anticyclonic direction. A turbulent jet is a main structure for very slow rotation rate and in a non-rotating fluid $(150<Pe_\Omega)$.

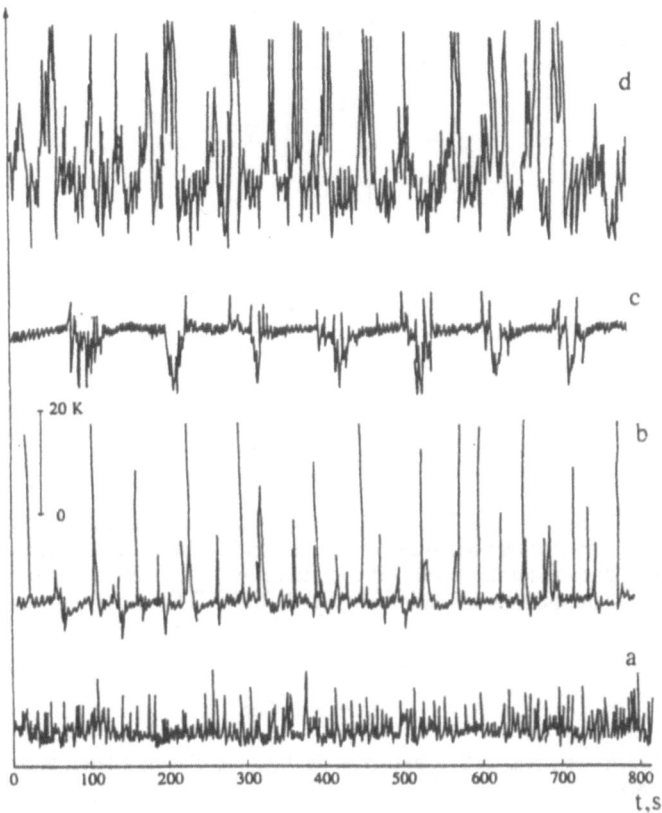

Figure2. A temperature records for $\varepsilon_T =0.932$ and a different rotation rate Ω: a-0, b-0.046 s^{-1}, c-0.146 s^{-1}, d-0.698 s^{-1}.

In Fig.2 temperature differences between two points at a height 13 cm above the disk are present for a different rotation rate. One point is above the center of a disk, the other is at distance 6.5 cm. A dependence of a period of a motion T from rotational Peclet number Pe_Ω for different ε_T is present in Fig.3 for three regimes. For regime of anticyclonic jet this period is a time for one revolution of a jet. For tornado-like regime this period is a time between an origin of intensive vortex.

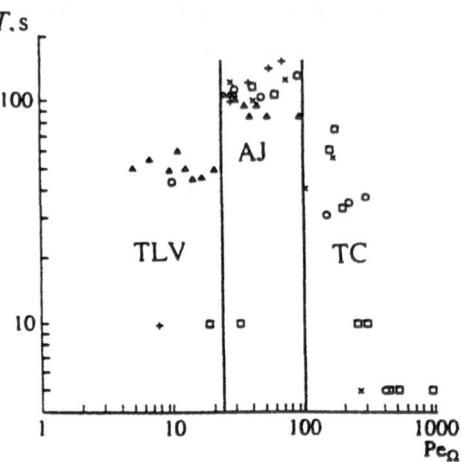

Figure 3. Dependence of a period of motion T from rotational Peclet number Pe_Ω for different ε_T: + - 0.263, x - 0.519, o - 0.932, Δ - 1.26, □ - 1.46.

For very small rotation rate and large Rossby number a rotation changes a stability of turbulent convective flow. Regimes of motion, as incline jet and tornado-like vortex, can be develop when an inertia force is the same order as Coriolis and buoyancy forces. This work was support by INCO-COPERNICUS (CT96- 0111) grant.

REFERENCES

1. Boubnov, B.M.& Golitsyn, G.S. *Convection in Rotating Fluids*. Kluwer Academic Publishers, Dordrecht, 1995.
2. Boubnov, B.M. and Fernando, H.J.S. Regimes of convection from isolated source of buoyancy in rotating fluids. *Izv. RAS, Atmos. Ocean Physics.*, (in press).
3. Boubnov, B.M. 1997 Convection from local source of heating in slow rotating fluid. *Izv. RAS, Atmos. Ocean Physics.*, 3 3(1997), No.6, 795-803.

MAGNETOHYDRODYNAMICS IN A TURBULENT SWIRLING FLOW

P. ODIER, J.-F. PINTON
Ecole Normale Supérieure, CNRS URA 1325, 69364 Lyon France

AND

S. FAUVE
Ecole Normale Supérieure, CNRS URA 1306, 75231 Paris France

We report an experimental study of the magnetic field fluctuations generated in a turbulent flow of liquid gallium, in the presence of an externally applied field. We consider the case of a weak 'seed' field $\vec{B}_0$, so that the Lorentz forces do not modify the flow [1]. The velocity gradients induce magnetic field fluctuations $\vec{b}$ at all scales, the description of which pertains to the dynamics of a 'passive vector' in turbulence, in analogy to the passive scalar case [2]. However this passive vector dynamics involves stretching of magnetic field lines by velocity gradients, analogous to stretching of vorticity lines, and is thus at an intermediate level of complexity between passive scalar advection and fully developed turbulence. In particular, stretching of magnetic field lines by velocity gradients may overcome Joule dissipation and generate a large scale magnetic field by amplification of weak initial disturbances– this is the dynamo effect.

We use the flow created in the gap between two coaxial rotating disks, the von Kármán swirling flow, as it is known to produce a very intense turbulence in a compact region of space [3]. In addition, this flow possesses many features, such as differential rotation or poloidal and toroidal mean flow components, which are known to favor dynamo action [4]. Two 11 kW ac-motors are used to drive the disks at a constant frequency Ω, adjustable in the range 5 to 50 Hz. The enclosing cylindrical vessel has a volume of 5.5 liters. It is filled with liquid gallium (density $\rho = 6.09 \times 10^3$ kg.m^{-3}), chosen for its high electrical conductivity ($\sigma = 3.68 \times 10^6$ ohm.m^{-1}). Its kinematic viscosity is $\nu = 3.1 \times 10^{-7}$ m^{-2}.s^{-1}. The integral kinematic and magnetic Reynolds numbers of the flow are defined as: $R_e = 2\pi R^2 \Omega / \nu \in [10^6, 10^7]$ and $R_m = 2\pi \mu_0 \sigma R^2 \Omega \in [1.3, 14]$. Two pairs of Helmholtz coils are set to

U. Frisch (ed.), Advances in Turbulence VII, 407–410.

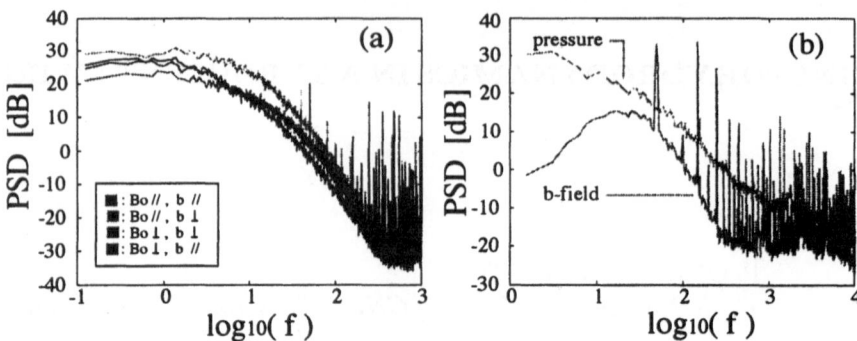

Figure 1. (a) magnetic spectra for different orientations $(\vec{B}_0, \vec{b})$ with respect to the rotation axis, of the applied and induced fields; $\Omega = 40$ Hz, $R_m = 10.8$; probe at $d = 1$ cm from the wall. (b) comparison with a spectrum of pressure fluctuations at the wall, in the same conditions. The low frequency cut-off of the magnetic spectrum is due to the AC filtering of the gaussmeter. These features are found for every value of R_m and d.

produce an external field $B_0 \sim 40$ gauss, either parallel or perpendicular to the rotation axis. Magnetic measurements are performed inside the vessel using directional and temperature compensated Hall probes.

Figure 1(a) shows typical power spectra of magnetic field fluctuations; they start with a flat frequency content followed by a steep cut-off region. The transition between the two regimes happens for a frequency of the order of Ω. The high frequency regions are quite similar and show an algebraic decay $\tilde{b}^2(f) \propto f^{-\alpha}$. Our measurements yield $\alpha = 3.8 \pm 0.2$ for all orientations and accessible values of R_m. This scaling is observed in a range of frequencies where the pressure fluctuations also follow a power law (figure 1(b)), i.e. in the inertial range of the velocity field [5]. These results can be understood as follows: in the presence of a uniform and constant applied field $\vec{B}_0$, magnetic field perturbations, $\vec{b}$, are governed by the equations [4]:

$$(\partial_t + \vec{u}.\vec{\nabla})\vec{b} = \left[(\vec{b} + \vec{B}_0).\vec{\nabla}\right]\vec{u} + \lambda\Delta\vec{b}, \quad \vec{\nabla}.\vec{b} = 0. \tag{1}$$

Since the magnetic diffusivity is orders of magnitude larger than the kinematic viscosity, $\vec{b}$ adiabatically follows $\vec{u}$. In this "quasistatic" approximation, one has for the leading order magnetic field perturbation $\vec{b}_0$:

$$\lambda\Delta\vec{b}_0 \approx -(\vec{B}_0.\vec{\nabla})\vec{u}, \tag{2}$$

provided that $b \ll B_0$, as is well verified in our experiment. Thus, $\vec{b}_0$ obeys a Poisson equation – this is analogous to the pressure field albeit second order derivatives of the velocity field are involved in that case. Keeping in mind that the flow is not modified by $\vec{B}_0$, dimensional analysis in the

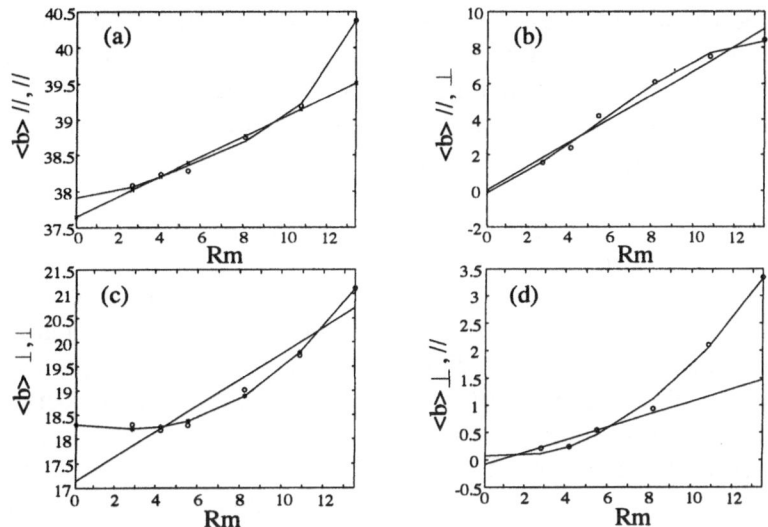

Figure 2.   Average value of magnetic for different orientations $(\vec{B}_0, \vec{b})$ with respect to the rotation axis, of the applied and induced field.

framework of Kolmogorov phenomenology then leads to [6]:

$$\tilde{b}_0^2(k) \propto k^{-2}\tilde{u}^2(k) \sim k^{-11/3} \ , \tag{3}$$

in the inertial range. This is consistent with our measurements, providing Taylor's hypothesis may be used. Note that, when EDQNM closures are applied to MHD equations, the relation $\tilde{b}^2(k) \propto k^{-2}\tilde{u}^2(k)$ between kinetic and magnetic energy subsists when the magnetic field is generated by the dynamo effect, although both spectra are steeper because of the effect of the Lorentz force [7].

We now consider the temporal average, $\bar{b}$ of the magnetic field $\vec{b}$. Equation (2) yields:

$$\bar{b}_0(\vec{r}) \approx g \star (\vec{B}_0.\vec{\nabla})\vec{u} \ , \tag{4}$$

where g is a Green function (equal to the free space one only for insulating boundary conditions at the vessel walls). $\bar{b}_0(\vec{r})$ is the leading order distorsion of the magnetic field lines; it is due to the non uniformity of the flow along $\vec{B}_0$ and varies linearly with R_m at small R_m [1,4]. This is evidenced in figure 2(b), where the azimutal component of the induced field is plotted against R_m, for an applied field parallel to the rotation axis z. Here an azimutal component b_θ is induced by the stretching term $\partial_z u_\theta$, i.e. by differential rotation (the disks rotate in opposite directions). Thus, one of the important source mechanisms for dynamos [4] occurs in our experimental flow configuration.

For higher values of R_m, we observe that the growth of the induced field is no longer linear – see figure 2; the change being quite clear for $R_m \geq 6$, particularly when the applied field is perpendicular to the rotation axis. We interpret this as follows: from (1), the next order correction $\bar{b}_1$ to $\bar{b}_0$ is governed by the equation

$$\lambda \Delta \bar{b}_1 \approx \overrightarrow{(\vec{u}.\vec{\nabla})\vec{b}_0} - \overrightarrow{(\vec{b}_0.\vec{\nabla})\vec{u}} \ , \tag{5}$$

showing that $\bar{b}_1$ is quadratic in R_m. Its sign depends on the direction of $\vec{B}_0$ and on the sign of the velocity gradients: our measurements show a saturation effect in the perpendicular direction when $\vec{B}_0$ is parallel to the rotation axis (figure 2(b)), but a quadratic growth in the direction of the applied field (figure 2(a,c)).

As regards to dynamo action, this non linear growth may lead to an increase of the net magnetic energy. The threshold for dynamo effect corresponds to a vertical asymptote for $< b^2 > / B_0^2$ (spatial averages here) as a function of R_m since a finite b would be obtained for $B_0 \to 0$. We cannot compute $< b^2 >$ without a complete 3D measurement of $\vec{b}$ throughout the flow. We can however try to get an estimate of the threshold R_m^c for the onset of dynamo action by interpolation of our results for $\bar{b}$ at higher magnetic Reynolds numbers. We compute $B_0 R_m / \bar{b}$ as a function of R_m: in the limit of a week seed field, $B_0 \to 0$, this quantity is the inverse of a "susceptibility" that characterizes the magnetic response of the turbulent fluid to the applied field $\vec{B}_0$. We observe that $B_0 R_m / \bar{b}$ decreases with R_m and yields $R_m^c \sim 25$ as an estimate for the onset of dynamo action in this flow. We note that numerical studies of flows with similar geometries [7, 8, 9] give comparable values for R_m^c. Work is underway to operate the same flow using liquid sodium as a working fluid; magnetic Reynolds numbers exceeding 50 will be reached.

References

1. Roberts P.H. (1967), *An introduction to MHD*, Longmans.
2. Monin A.S., Yaglom A.M. (1975), *Statistical fluid mechanics*, MIT Press.
3. Pinton J.-F., Labbé R. (1994), *J. Phys. II France*, **4**, pp. 1461–1468.
4. Moffatt H. K. (1978), *Magnetic field generation in conducting fluids*, Cambridge U. P.
5. Fauve S., Laroche C., Castaing B. (1993), *J. Phys. II France*, **3**, p. 271.
6. Moffat H. K. (1961), *J. Fluid Mech.*, **11**, pp. 625–635.
7. Léorat J., Pouquet A., Frisch U. (1981), *J. Fluid Mech.*, **104**,pp.419–443.
8. Dudley N.L., James R.W. (1989), *Proc. Roy. Soc. Lond.*, **A425**, p. 407.
9. Nore C., Brachet M.-E., Politano H., Pouquet A. (1997), *Phys. Plasmas*, **4**, pp. 1–3.

PROBABILITY DISTRIBUTION OF THE DENSITY FIELD IN ONE DIMENSIONAL GAS DYNAMICS

T. PASSOT[1] AND E. VÁZQUEZ-SEMADENI[2]
[1] CNRS, Observatoire de Nice, BP 4229, 06304 Nice Cedex 4, France.
[2] Instituto de Astronomía, UNAM, Apdo. Postal 70-264, México, D. F. 04510, México

A crucial ingredient in establishing a theory of compressible turbulence is the characterization of the density fluctuation statistics, an information particularly relevant in Cosmology and interstellar gas dynamics. A first step towards this characterization is the determination of the probability density function (PDF) of the density field. Previous work [1, 2] has found log-normal distributions in isothermal simulations of 2D and 3D compressible turbulence, whereas power-law regions have been reported at high densities in simulations of interstellar gas subject to complex heating and cooling processes [3]. To understand the reason of this difference we have performed one-dimensional simulations of a randomly accelerated barotropic gas, for different values of the polytropic index γ [4]. The equations for the non-dimensional density ρ and velocity u read

$$\frac{\partial \rho}{\partial t} + \frac{\partial(\rho u)}{\partial x} = 0 \tag{1}$$

and

$$\frac{\partial u}{\partial t} + u\frac{\partial u}{\partial x} = -\frac{1}{\gamma M^2 \rho}\frac{\partial \rho^\gamma}{\partial x} + \nu\frac{\partial^2 u}{\partial x^2} + a_r, \tag{2}$$

where M is the Mach number of the velocity unit. These runs start at rest, but are driven by a random acceleration a_r at wave numbers 1–19, with a correlation time $t_{\text{corr}} = 3 \times 10^{-3}$. The probability densities of $\log \rho$ are obtained by considering all grid points (typically 3072 or 6144 for the highest Mach numbers) and averaging over nearly 150 non-dimensional time units, sampling every $t = 0.1$. Thus, the histograms contain over 4.6 million points. It turns out that *a log-normal density PDF only appears when $\gamma = 1$*. In fact, as explained below, we suggest that the case $\gamma = 1$ may be

411

U. Frisch (ed.), Advances in Turbulence VII, 411–414.

singular in this respect. The histograms for $\log \rho$ are shown in Fig. 1 for three different simulations, all with $M = 3$, and $\nu = 0.003$, but with $\gamma = 0.3$ in (a), $\gamma = 1$ in (b) and $\gamma = 1.7$ in (c). Mass diffusion (a term of the form $\mu \Delta \rho$ in the RHS of eq. (1)) is used for the simulations at the highest Mach number due to the numerical difficulties created by strong shock collisions. Here $\mu = 5 \times 10^{-4}$ for the runs with $\gamma = 0.3$ and 1.7. The dotted line in Fig. 1b shows a least-squares fit to a log-normal curve. As can be seen the fit is excellent, suggesting that indeed the $\log \rho$ histogram is a true log-normal when $\gamma = 1$. Instead, in Fig. 1a, (resp. 1c) a clear power-law tail is seen to develop at large (resp. small) density fluctuations. The deviation from a log-normal towards power-law tails is more noticeable at large M. In the case $\gamma = 0$ (Burgers' equation) power-law tails are observed both for small and large densities, indicating the singular character of this limit. For the decaying Burgers' equation at small Reynolds number a power law is observed only for densities larger than the mean [5]. In order to interpret

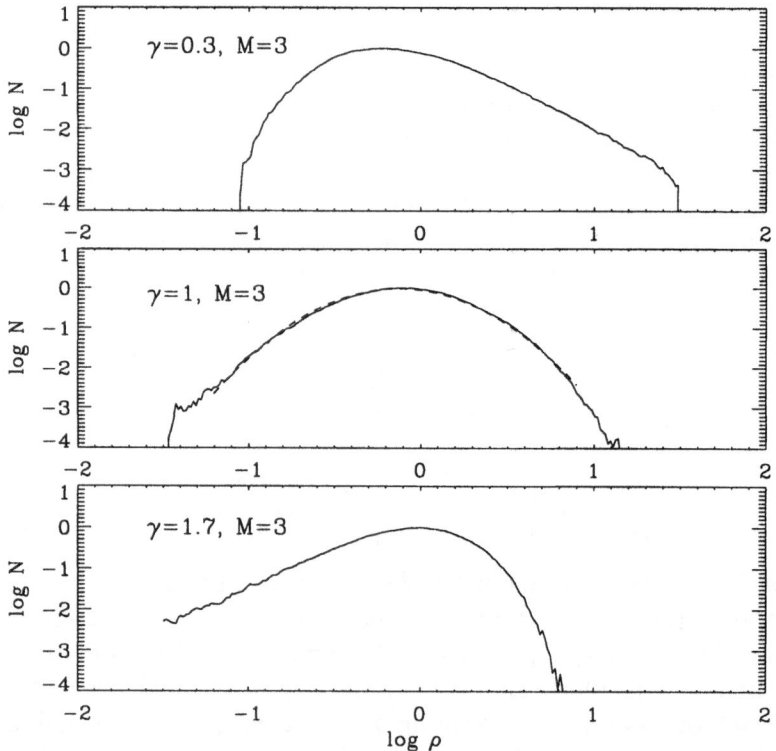

Figure 1. Logarithm of the histograms of $\log \rho$ for the runs with $M = 3$ at $\gamma = 0.3$ (top), $\gamma = 1$ (middle) and $\gamma = 1.7$ (bottom).

these results, it is useful to consider the equations written in terms of the density logarithm s (when $\gamma = 1$), or the variable $v = (1 - \gamma) \ln \rho$ (when $\gamma \neq 1$). In the former case, the system is invariant upon the change $s \to s+b$, where b is an arbitrary constant, reflecting the fact that the sound speed does not depend on the local density of the fluid. In the latter case, noting that the pressure term rewrites as $\frac{1}{(1-\gamma)M^2} \frac{\partial}{\partial x} e^{-v}$, we see that as v increases (resp. decreases), the pressure becomes exponentially small (resp. large), thus allowing (or severely hindering) the formation of density structures, peaks or voids according on the sign of $1 - \gamma$. It results that the PDF of $s = \ln \rho$ for $\gamma > 1$ will appear similar to that for $\gamma < 1$ after we change $s \to -s$.

The build up of these density fluctuations is a random multiplicative process which, at the level of the variable s, becomes additive. For $\gamma = 1$, the random variable s is thus the sum of individual random variables (the density fluctuations), each having the same probability density (individual jumps have the same average magnitude, related to the Mach number of the flow but independent of the local density). The sum of identical random processes is known to have a Gaussian distribution, due to the Central Limit Theorem, whatever the distribution of the individual processes. The PDF of s is thus expected to be normal :

$$P(s)ds = \frac{1}{\sqrt{2\pi\sigma_s^2}} \exp(-\frac{(s - s_0)^2}{2\sigma_s^2})ds, \tag{3}$$

where σ_s^2 denotes the variance of the distribution. Figure 2 (left panel) shows a plot of $\log(\sigma_s)$ vs. $\log(\tilde{M})$ obtained by combining data from several simulations with M ranging from 0.5 to 10. This plot shows that $\sigma_s^2 \approx \beta \tilde{M}^2$, with $\beta \approx 1$, with a very good accuracy, up to the highest Mach numbers reached in our simulations. On the other hand, we see in the right panel of the same figure, which displays $\log \sigma_\rho$ vs. $\log \tilde{M}$, that the density standard deviation also scales like $\tilde{M}$ for small values of $\tilde{M}$, while for $\tilde{M} \geq 0.5$ the points curve up, a reflection of the relation $\sigma_\rho^2 = e^{\sigma_s^2} - 1$ between the two variances when ρ obeys a log-normal distribution. The maximum of the distribution s_0 is simply related to σ_s due to the constraint of mass conservation. Writing $\langle \rho \rangle = \int_{-\infty}^{+\infty} e^s P(s)ds = 1$ we find $s_0 = -\frac{1}{2}\sigma_s^2$, a relation also well verified numerically [4].

When $\gamma \neq 1$, the local Mach number of the flow is no longer independent of the local density and there is no reason to expect a log-normal PDF for the density. We nevertheless propose a heuristic model, reproducing most of the features of the PDF's obtained in our simulations, which consists in taking the same functional form of the PDF as in the isothermal case, but replacing $\tilde{M}$ by $\hat{M}(s;\gamma)$ where $\hat{M}(s;\gamma) = u_{rms}/c(s) = Mu_{rms}e^{\frac{1-\gamma}{2}s}$ when $s_- \leq s \leq s_+$, and $\hat{M}(s;\gamma) = \hat{M}(s_+;\gamma)$ (resp. $\hat{M}(s_-;\gamma)$) for $s > s_+$

(resp. $s < s_-$). These cutoff mean that the probability of new fluctuations arising within previous peaks or voids decreases as the amplitudes of the latter become larger because the fraction of space they occupy decreases. The PDF thus reads

$$P(s; \gamma)ds = C(\gamma) \exp\left(-\frac{s^2}{2\hat{M}^2(s; \gamma)} - \alpha(\gamma)s\right)ds \qquad (4)$$

where $C(\gamma)$ is a normalizing constant. The parameter $\alpha(\gamma)$ is again determined by the constraint of mass conservation stating that the mean value of the density should be 1. For $\gamma < 1$, $\hat{M}(s; \gamma)$ grows exponentially with s for $s_- \leq s \leq s_+$ and as a consequence the PDF for $0 < s < \hat{M}(s; \gamma)$ is rapidly dominated by the power-law (in ρ) behavior $P(s; \gamma) \sim e^{-\alpha(\gamma)s}$, while the Gaussian-like decay again dominates for $s > \hat{M}(s; \gamma)$. For $s < 0$, the local turbulent Mach number decreases as s decreases and we expect a drop off of the PDF more rapid than when $\gamma = 1$. The behavior is exactly opposite when $\gamma > 1$.

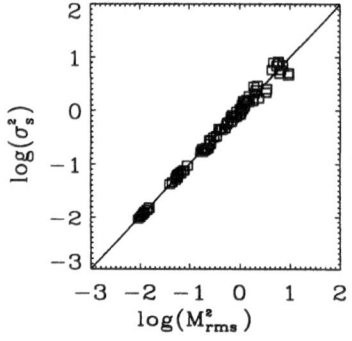 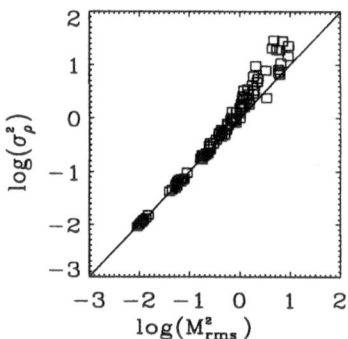

Figure 2. (Left) Variance of $s = \ln \rho$ vs. the mean square Mach number $\bar{M}^2$ for various simulations with $\gamma = 1$ (Right) Variance of ρ vs. $\bar{M}$.

References

1. Vázquez-Semadeni, E. (1994) Hierarchical Structure in Nearly Pressureless Flows as a Consequence of Self-Similar Statistics, Astrophys. J. **423**, pp. 681–692.
2. Padoan, P., Nordlund, A. and Jones, B.J.T. (1997), The Universality of the Stellar Initial Mass Function, Mon. Not. R. Ast. Soc. **288**, pp. 145–152.
3. Scalo, J., Vázquez-Semadeni, E., Chappel, D. and Passot, T. (1997) On the density probability function of galactic gas. I. Numerical simulations and the significance of the polytropic index, submitted to *Astrophys. J.*
4. Passot, T. and Vázquez-Semadeni, E. (1998) Density probability distribution in one-dimensional polytropic gas dynamics, submitted to Phys. Rev. E.
5. Gotoh, T. and Kraichnan, R.H. (1993) Statistics of decaying Burgers turbulence *Phys. Fluids A* **5**, pp. 445–457.

SELF-SIMILAR DECAY OF TWO-DIMENSIONAL TURBULENCE

J. R. CHASNOV[1] AND J. R. HERRING[2]
[1] *The Hong Kong University of Science and Technology, Clear Water Bay, Kowloon, Hong Kong.*
[2] *National Center for Atmospheric Research, Boulder, CO 80307 USA.*

Recent numerical simulations [2] of decaying two-dimensional homogeneous turbulence at high Reynolds numbers have exhibited an approximate self-similar evolution of the energy spectrum. We analyze here the theoretical implications of self-similarity.

We consider a two-dimensional velocity field $\mathbf{u} = (u_1, u_2, 0)$ with vorticity $\boldsymbol{\omega} = \nabla \times \mathbf{u} = (0, 0, \omega)$. The equations for the mean-square velocity ($2\times$ energy) and mean-square vorticity ($2\times$ enstrophy) are given by [1]

$$\frac{d}{dt}\langle \mathbf{u}^2 \rangle = -2\nu \langle \omega^2 \rangle, \qquad \frac{d}{dt}\langle \omega^2 \rangle = -2\nu \langle (\nabla \omega)^2 \rangle, \tag{1}$$

where ν is the kinematic viscosity of the fluid. Our main objective is to determine the long-time decay laws of the energy and enstrophy.

We begin with some relevant definitions. In two-dimensional turbulence, the following characteristic length scales λ and μ are of some importance:

$$\lambda = \langle \mathbf{u}^2 \rangle^{1/2}/\langle \omega^2 \rangle^{1/2}, \qquad \mu = \langle \omega^2 \rangle^{1/2}/\langle (\nabla\omega)^2 \rangle^{1/2}. \tag{2}$$

The Reynolds numbers R_λ and R_μ constructed from these length scales are defined as

$$R_\lambda = \frac{\langle \mathbf{u}^2 \rangle^{1/2}\lambda}{\nu}, \qquad R_\mu = \frac{\langle \mathbf{u}^2 \rangle^{1/2}\mu}{\nu}, \tag{3}$$

and their ratio ρ will play a pivotal role in our analysis:

$$\rho = R_\lambda/R_\mu = \lambda/\mu. \tag{4}$$

We will consider separately two distinct kinds of self-similar decay. First, we consider *complete self-similarity* for which the energy spectrum during

415

U. Frisch (ed.), *Advances in Turbulence VII*, 415–418.

the decay maintains its shape on log-log axes over all wave numbers. Second, we consider *partial self-similarity* for which the spectral shape is maintained only over scales directly unaffected by viscosity.

Using the length scale λ, we look for an energy spectrum of self-similar form

$$E(k, t) = \langle \mathbf{u}^2 \rangle \lambda \widehat{E}(\widehat{k}), \qquad \widehat{k} = k\lambda. \tag{5}$$

For complete self-similarity, the length scale ratio ρ is necessarily constant during the decay: physically, all length scales must grow at the same rate. It is also simple to determine that ρ is directly related to the time-independent self-similar spectrum via

$$\rho = \frac{\left[\int_0^\infty \widehat{E}(\widehat{k})d\widehat{k}\right]^{\frac{1}{2}} \left[\int_0^\infty \widehat{k}^4 \widehat{E}(\widehat{k})d\widehat{k}\right]^{\frac{1}{2}}}{\int_0^\infty \widehat{k}^2 \widehat{E}(\widehat{k})d\widehat{k}}. \tag{6}$$

We assume that for times $t \geq t_*$ following an initial transient, the spectrum undergoes complete self-similar decay with $\rho = \rho_*$ constant. Assuming constant ρ during the decay results in closure of (1), and an analytical solution for the decay laws is most easily determined by first obtaining an equation for λ:

$$\frac{d}{dt}\lambda^2 = 2\nu(\rho^2 - 1); \tag{7}$$

which may be integrated immediately from t_* to t:

$$\lambda^2 = 2\nu(\rho_*^2 - 1)(t - t_*) + u_*^2/\omega_*^2, \tag{8}$$

where u_* and ω_* are the root-mean-square values of the velocity and vorticity at $t = t_*$. The energy and enstrophy equations may then be subsequently integrated to obtain

$$\langle \mathbf{u}^2 \rangle = u_*^2 \left[1 + 2(\rho_*^2 - 1)R_{\lambda*}^{-1}\omega_*(t - t_*)\right]^{-1/(\rho_*^2 - 1)}, \tag{9}$$

$$\langle \omega^2 \rangle = \omega_*^2 \left[1 + 2(\rho_*^2 - 1)R_{\lambda*}^{-1}\omega_*(t - t_*)\right]^{-\rho_*^2/(\rho_*^2 - 1)}, \tag{10}$$

where $R_{\lambda*}$ is the value of R_λ at $t = t_*$. Fully-developed turbulence corresponds to large values of ρ_* signifying a wide separation of scales between λ and μ. For asymptotically large ρ_*, we see from (9) and (10) that the energy is conserved for finite times, (though for fixed ρ_* as $t \to \infty$, the energy decays to zero), and that the enstrophy decays as

$$\langle \omega^2 \rangle = \frac{u_*^2}{2\nu\rho_*^2}t^{-1}. \tag{11}$$

Previously [2], we have shown that complete self-similarity also occurs for decaying turbulence at constant R_λ. The Reynolds number R_λ can be shown to satisfy the equation

$$\frac{d}{dt}R_\lambda = (\rho^2 - 2)\langle\omega^2\rangle^{1/2}, \tag{12}$$

so that decay with constant R_λ corresponds to $\rho_*^2 = 2$. It can be further shown that the analytical results found in [2] can be recovered directly from (9) and (10).

The time-evolution equation for the energy spectrum is written as

$$\frac{\partial}{\partial t}E(k,t) + 2\nu k^2 E(k,t) = T(k,t), \tag{13}$$

where $T(k,t)$ is the nonlinear transfer spectrum. We now transform (13) into an equation for the self-similar spectrum $\widehat{E}(\widehat{k})$. Using (1), (5) and (7), and after some algebraic manipulations, we find

$$2(\widehat{k}^2 - 1)\widehat{E}(\widehat{k}) + \left(\rho^2 - 1\right)\left(\widehat{E}(\widehat{k}) + \widehat{k}\frac{d}{d\widehat{k}}\widehat{E}(\widehat{k})\right) = \langle\mathbf{u}^2\rangle^{-3/2}R_\lambda T(k,t). \tag{14}$$

For complete self-similar decay with $\rho = \rho_*$ constant, the transfer spectrum must thus evolve with self-similar form

$$T(k,t) = \langle\mathbf{u}^2\rangle^{3/2}R_\lambda^{-1}\widehat{T}(\widehat{k}), \qquad \widehat{k} = k\lambda, \tag{15}$$

which depends explicitly on the viscosity ν through the Reynolds number R_λ. This differs from standard two-point closure theories [4], for which the factor R_λ^{-1} is absent.

On the other hand, partial self-similarity assumes that the self-similar form (5) is valid only over wave numbers for which viscosity is negligible. The transfer scaling given by (15) is thus unsuitable because of its direct dependence on viscosity. Partial self-similar decay solutions may be obtained from (14) under the assumption $\rho \to \infty$, asymptotically. Equation (14) then reduces at long-times to

$$\left(\widehat{E}(\widehat{k}) + \widehat{k}\frac{d}{d\widehat{k}}\widehat{E}(\widehat{k})\right) = \langle\mathbf{u}^2\rangle^{-3/2}\frac{R_\lambda}{\rho^2}T(k,t). \tag{16}$$

In order for viscosity to cancel from the right-hand side of (16), ρ^2 must scale like

$$\rho^2 = cR_\lambda, \tag{17}$$

where c is a nondimensional proportionality constant. Together with (4), this implies $R_\lambda = cR_\mu^2$.

The relationship between ρ and R_λ given by (17) permits analytical closure of (1). We obtain for the enstrophy equation

$$\frac{d}{dt}\langle\omega^2\rangle = -2c\langle\omega^2\rangle^{3/2}, \tag{18}$$

which may be integrated from a time t_* after which partial self-similarity occurs:

$$\langle\omega^2\rangle = \omega_*^2\left[1 + c\omega_*(t - t_*)\right]^{-2}. \tag{19}$$

Further integration of the energy equation results in the solution

$$\langle\mathbf{u}^2\rangle = u_*^2\left[1 - \frac{2R_{\lambda_*}^{-1}\omega_*(t - t_*)}{1 + c\omega_*(t - t_*)}\right]. \tag{20}$$

The long-time asymptotic solutions of (19) and (20) are given by

$$\langle\mathbf{u}^2\rangle = u_*^2\left[1 - 2(cR_{\lambda_*})^{-1}\right], \quad \langle\omega^2\rangle = (ct)^{-2}. \tag{21}$$

The assumption of partial self-similarity thus results in the t^{-2} enstrophy decay law originally proposed by Batchelor [1], and found by standard two-point closures [4].

For partial self-similar decay the energy approaches a nonzero value asymptotically (apart from the special case $\rho_*^2 = 2$ discussed earlier, for which $\langle\mathbf{u}^2\rangle$ vanishes in (21)). This presents a significant physical difference between complete and partial self-similarity.

Our previous direct numerical simulations [2] seem to support complete self-similarity in decaying two-dimensional turbulence, where the asymptotic decay law of the enstrophy for high initial Reynolds numbers was found to be approximately $t^{-0.8}$. Numerical simulations currently in progress provide even stronger support in favor of complete self-similarity. Hence the partial self-similar decay obtained by standard two-point closure theories accounts for the disagreement found earlier between simulations and theory [3].

References

1. Batchelor, G. K. (1969) Computation of the energy spectrum in homogeneous two-dimensional turbulence, *Phys. Fluids Suppl.* II, **12**, 233–239.
2. Chasnov, J. R. (1997) On the decay of two-dimensional homogeneous turbulence, *Phys. Fluids* **9**, 171–180.
3. Herring, J. R. and McWilliams, J. C. (1985) Comparison of direct numerical simulation of two-dimensional turbulence with two-point closure: the effects of intermittency, *J. Fluid Mech.* **153**, 229–242.
4. Lesieur, M. (1990) *Turbulence in Fluids*, Kluwer Academic Publishers, Dordrecht.

VELOCITY FLUCTUATION PROPERTIES IN MERCURY CONVECTION

S. CIONI(*) AND J. SOMMERIA
Laboratoire de Physique, Ecole Normale Supérieure de Lyon (URA 1325 CNRS), 46, Allée d'Italie 69364 Lyon Cedex 07 France
(*) *email:* scioni@physique.ens-lyon.fr

Abstract. Direct measurements of both velocity and temperature fluctuations in turbulent convection have been carried out. The velocity field is obtained from the electric potential induced by a DC magnetic field. It was found that, for weak amplitude of the magnetic field (less than 500 gauss), the velocity field behaves as in ordinary turbulence. Using spectra and histograms, the statistical properties of both fields were analysed. The velocity fluctuation spectra were close to those of temperature. It was also found that, the probability density functions (PDF) of the velocity fluctuations were weakly non-Gaussian shaped whereas the PDF of the temperature showed clear exponential tails.

In spite of comparatively large number of studies of temperature measurements in thermal convection (fluid layer heated from bottom and cooled from above), reliable systematic results on the velocity field have not yet been obtained, due to the difficulty to measure velocity fluctuations. The knowledge of the velocity field is widely recognized as crucial to the full understanding of turbulent convection. This lack on the velocity field knowledge becomes deeper for the low Prandtl number convection: no velocity measurements exist for that case. Thus, the aim of this work is to provide measurements of velocity fluctuations in mercury ($Pr \simeq 0.025$) convection. In order to achieved this task we applied a weak external D.C. magnetic field as a mean to investigate velocity fluctuations by the induced potential difference between two electrodes (Tsinober *et al.* 1987).

The experimental configuration is shown schematically in Fig. 1. The convective cell, consisting of a vertical cylinder 213 mm high and aspect ratio $\Gamma = D/H = 1$ (see for more details Cioni *et al.* (1997)), is placed

U. Frisch (ed.), Advances in Turbulence VII, 419–422.

inside a cylindrical electromagnet. The inner radius of the magnet is 300 mm the total heigth about 800 mm. The strength of the magnetic field can be varied up to 0.4 Tesla with homogeneity about $B \pm 10$ Gauss over the whole volume. The intensity of the magnetic field was increased stepwise with the heat flux kept constant. In the present paper heat flux was equal 1 W/cm^2. In each step of a run, measurements were made after steady state was reached.

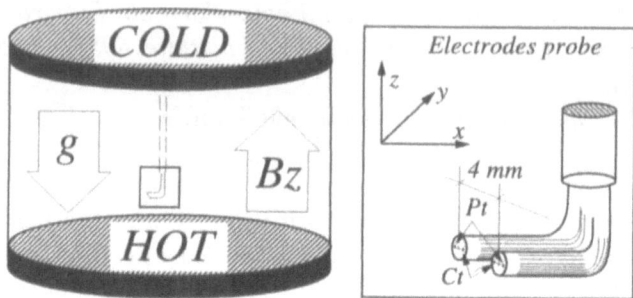

Figure 1.   Sketch of experimental set up configuration. The magnetic field is directed along the vertical axis (z) and the pick up electrodes (spaced 4 mm) were in the horizontal plane (xy). The junction between the Ct and Pt electrodes is ensured by the presence of mercury.

The external magnetic field was along the vertical axis (i.e. parallel to the gravity). Therefore we could in principle obtain data on the *horizontal* velocity fluctuations by measuring the potential difference between two electrodes placed inside the fluid (see Tsinober *et al.* (1987) and references within for a comprehensive description of the method and a review). Measurements were performed using a four-electrodes probe consisting of two Platinum-Mercury-Constantan thermocouples which allowed us to measure both temperature (2 point measurements) and velocity (between the Platinum electrodes). The Pt electrodes have been chosen in place of Ct ones because the thermoelectric power of Pt-Hg (0.6 $\mu V/°C$) is 10 times lower than that of Ct-Hg. A micrometric screw along the central axis of the cylinder allowed us to move it along the vertical direction with a positioning within 0.01 mm. In the present paper we report some preliminary results of direct measurements of both velocity and temperature fields. Figure 2 shows a typical time recording of both velocity and temperature fluctuating field. Note that the velocity signal is dominated by a periodic oscillation, which gives rise to the broad low frequency peak observed in the velocity spectra (Fig. 4a).

The fluctuation properties of those fields have been characterized by means of various statistical quantities such as the PDF as well the frequency power spectra.

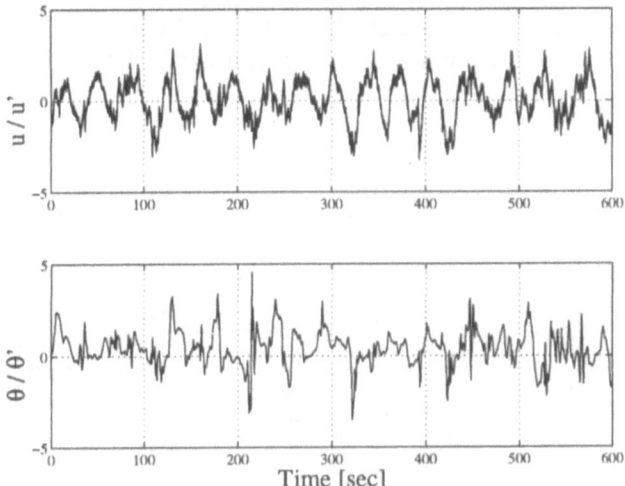

Figure 2. A typical time recording of both velocity (top) and temperature (bottom) fluctuations normalized by the root-mean-squared $h = H/2$ ($Ra = 3.4 \times 10^8$ and $B = 250$ *gauss*).

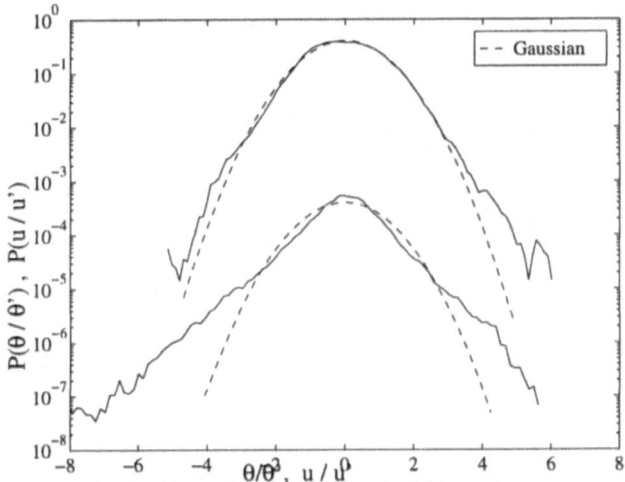

Figure 3. PDF of velocity (top) and temperature (bottom) fluctuations normalized by the rms, of the data shown in Fig.2. The lower PDF has been shifted 3 decades.

It has been found that the velocity fluctuations are close to those obtained in ordinary turbulence and their PDF are weakly non-Gaussian (see Fig. 3). This depart from the purely Gaussian shape is probably controlled by large scale fluctuations, which are influenced by the presence of the magnetic field as shown by Cioni (1998). By contrast, the heat flux is not modified by that field, indicating that the local turbulent fluctuations are not modified. On the other hand the temperature PDF still remain exponential shaped.

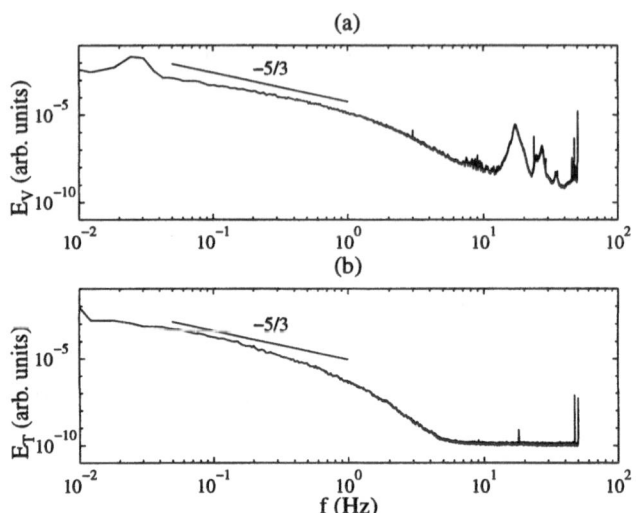

Figure 4. Frequency spectra of velocity (a) and temperature (b) fluctuations in the same conditions of the previous figures.

Figure 4 show both velocity (2a) and temperature (2b) fluctuation spectra. Velocity and temperature spectra have a similar inertial range. This confirms previous experimental results of Cioni *et al.* (1995) that is both temperature and velocity fields in low Prandtl number convection behave as in ordinary turbulence. Similar results has been obtained by Camussi and Verzicco (1998) with numerical simulations in mercury convection but at lower Rayleigh. However, E_T falls down faster at high frequency, probably due to the effect of thermal diffusivity, 50 times higher than viscosity.

References

Camussi, R. and Verzicco, R. (1998) Convective turbulence in mercury: scaling law and frequency spectra, *To appear in Physics of Fluids*.

Cioni, S. (1998) Turbulent natural convection at low Prandtl number, *To appear in Journal of Experimental Thermal and Fluid Science*.

Cioni, S., Ciliberto, S. and Sommeria, J. (1995) Temperature structure functions in turbulent convection at low Prandtl number, *Europhysics Letters* **32** (5), 413-418.

Cioni, S., Ciliberto, S. and Sommeria, J. (1997) Strongly turbulent Rayleigh-Bénard convection in mercury: comparison with results at moderate Prandtl number, *J. Fluid Mech.* **335**, 150-81;

Tsinober, A., Kit, E. and Teitel, M. (1987) On the relevance of the potential-differnce method for turbulence measurements, *Journal of Fluid Mech.* **175**, 447.

ACCELERATION OF HORIZONTAL MEAN CURRENTS IN DNS OF STRATIFIED TURBULENT SHEAR FLOWS

M. GALMICHE[1], O. THUAL[1] AND P. BONNETON[2]
[1] *Institut de Mécanique des Fluides de Toulouse*
Allée du Professeur Camille Soula
31400 Toulouse, France.
[2] *Département de Géologie et d'Océanographie*
URA CNRS 197
Université Bordeaux 1, 33405 Talence, France.

We present some results and comments on DNS of freely decaying horizontal parallel flows when initially affected by a turbulent velocity field in a strongly stratified fluid. Turbulent shear flows in stratified media have already been simulated by Jacobitz et al. [2] for instance. Experiments have been carried out in a thermally stratified wind tunnel by Piccirillo et al. [3]. However, these numerical and experimental studies generally focus on the influence of a horizontal mean flow on the stratified turbulence. In the present paper, we focus on the accelerating effect of stratified turbulence onto the horizontal mean flow. We solve the 3D Navier-Stokes equations under the Boussinesq approximation with a pseudo-spectral direct simulation code (O. Thual [4]). The flow evolves in a periodic cubic box with no external forcing, the initial condition being the sum of a horizontal parallel flow $\vec{U}(z, 0)$ in the x-direction and a homogeneous, isotropic turbulent velocity field $\vec{u}'(x, y, z, 0)$. The Reynolds number is 67 initially (based on the *rms* value u_0 of u', the integral lengthscale l and the kinematic viscosity). We choose $\vec{U}(z, 0)$ as a cosine with amplitude U_0 and a wavelength equal to the size of the box L_b. The value of U_0 is chosen such that the mean flow energy is initially one third of the total energy. In the stratified case, the Prandtl number is taken equal to unity and there is no potential energy initially. The initial Froude number $Fr = u_0/Nl$ (where N is the Brunt-Väisälä pulsation) is 0.13. Such a low value is typical of strongly stratified turbulence. Figure 1 shows the evolution of the mean flow energy normalized by the initial total energy. The time unit is the initial turnover timescale l/u_0. Three cases are considered : the turbulent-stratified case, the turbulent-non stratified case and the non turbulent case. On this plot,

U. Frisch (ed.), Advances in Turbulence VII, 423–424.
© 1998 *Kluwer Academic Publishers.*

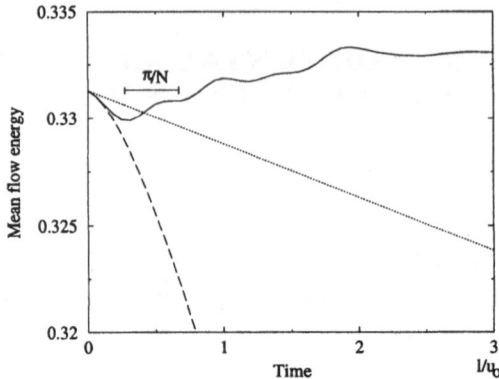

Figure 1. Evolution of the mean flow energy normalized by the initial total energy. ——— : stratified case ($Fr = 0.13$ initially). – – – : non stratified case. · · · · · : non turbulent, monodimensional case. The time unit is the initial turnover timescale l/u_0'. A strong discrepancy is observed between the stratified and non stratified cases.

we observe a strong discrepancy between the stratified and non stratified cases. In the non stratified case, the mean flow rapidly loses its coherence ans decays much faster than in the non turbulent case. In contrast, some king of oscillations are observed in the stratified case, associated with an acceleration of the mean flow. As a consequence, the mean current totally dominates the flow at large t (not shown on Fig. 1), its final value being higher than in the non turbulent, purely viscous case. These results lead us to two conclusions : (*i*) The classical concept of turbulent viscosity which is sometimes used in the case of non stratified turbulence, totally fails in the stratified case. (*ii*) Stratified turbulence is able to transfer energy to the mean flow on a time scale of order π/N. This is an evidence for the sensitivity of the mean flow to the wavy properties of the medium and leads us to study, on a more general point-of-view, possible horizontal shear production by the distorsion of an internal wave field. The authors will soon report on this subject. Further details on this work may be found in Ref. [1].

References

1. Galmiche, M., Thual, O., Bonneton, P. (1998) Production of horizontal mean flows in DNS of turbulent stratified shear flows. Brief Communication submitted to *Physics of Fluids.*
2. Jacobitz, F.G., Sarkar, S., Van Atta, W. (1997) Direct numerical simulations of the turbulence evolution in a uniformly sheared and stably stratified flow. *J. Fluid Mech.* **342**, 231-261.
3. Piccirillo, P., Van Atta, W. (1997) The evolution of a uniformly sheared thermally stratified turbulent flow. *J. Fluid Mech.* **334**, 61-86.
4. Thual, O. (1992) Zero-Prandtl number convection. *J. Fluid Mech.* **240**, 229-258.

HEAT TRANSFER IN RAYLEIGH-BÉNARD SYSTEMS

F. TOSCHI[1,2], R. TRIPICCIONE[3], R. BENZI[4]
[1] *Dipartimento di Fisica, Università di Pisa,*
Piazza Torricelli 2, I-56126, Pisa, Italy.
[2] *INFM, Unità di Tor Vergata.*
[3] *INFN, Sezione di Pisa.*
[4] *AIPA, Via Po 14, I-00100, Roma, Italy.*

Abstract. In this paper we discuss some theoretical aspects concerning the scaling laws of the Nusselt number versus the Rayleigh number in a Rayleigh-Bénard cell. We present a new set of numerical simulations, compare against the predictions of existing models and propose a new theory which relies on the hypothesis of Bolgiano scaling.

Many aspects of the dynamics of a Rayleigh-Bénard convective cell have been investigated for a long time. Two theories have been proposed for the dependence of the Nusselt number (Nu, the ratio between the turbulent heat flux and the purely conductive one) as a function of the Rayleigh number (Ra), by Castaing et al. [1] and by Shraiman and Siggia [2]. The former relies upon general scaling argument and dimensional analysis, while the latter starts from basic equations and a specific model to derive the scaling relations.

In this work, we consider again this problem, starting from the results of a number of numerical simulations that we have performed (see [3] for an extended review) and the results of a recent experiment [4]. The simulation code used a Lattice Boltzmann algorithm (LBE, see [9] for details) on a APE100 parallel computer [10]. We were able to investigate almost an order of magnitude in Rayleigh number (Ra) around $Ra \sim 10^6$. The spatial resolution was of 160^3 grid points, with aspect ratio 1 and Prandtl number $Pr = 1$.

We now briefly describe our main numerical findings, then we will turn to more theoretical considerations. Results relevant for what follows are:

i) The Nusselt number scales (versus the Rayleigh number) as $Nu \sim Ra^{2/7}$ (see Fig. 1) for Ra as moderate as $Ra \simeq 10^6 \div 10^7$. This result

425

U. Frisch (ed.), Advances in Turbulence VII, 425–428.

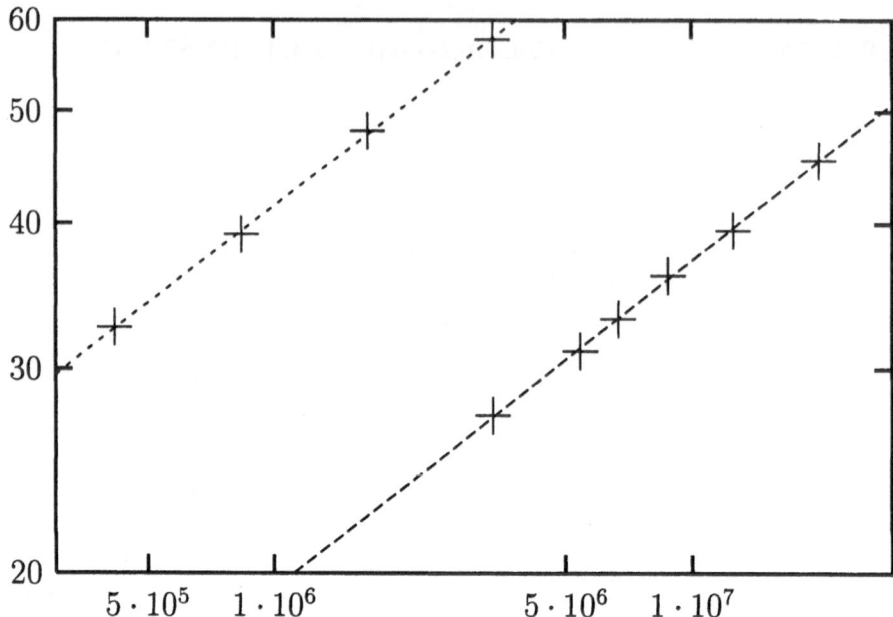

Figure 1.   Log-Log plot of Nu vs. Ra, for no-slip (lower data) and free-slip boundary conditions. The slope is $\sim 2/7$ in both cases.

is also predicted from theories [1, 2]. The normalization is in reasonable agreement with experimental data. Our values for Nu are slightly larger that the experimental ones, due to the fact that our vertical walls are free-slip, giving rise to a slightly enhanced heat transport.

ii) The same kind of scaling (with exponent 2/7) has been observed for a simulation where the boundary conditions for the velocity on the top/bottom walls were free-slip (see again Fig. 1). In this case the viscous boundary layer is completely absent and the velocity reaches its maximum value just on the walls.

iii) We have observed the presence of a Bolgiano scaling, in accordance with what found in [5]. We have measured the local Bolgiano length, $L_B(z)$, as a function of the distance from the isothermal walls:

$$L_B(z) \equiv \frac{\epsilon(z)^{5/4}}{N(z)^{3/4}(\alpha g)^{3/2}} \tag{1}$$

We have found that near the center of the cell its value is of the order of the cell size, while near the boundaries it is of the order of the thermal boundary layer thickness. $L_B(z)$ has its minimum value near the thermal boundary layer. We found that $L_B(\lambda) \simeq \lambda$ as Ra is varied (see Fig. 2). This point will be particularly important in the following.

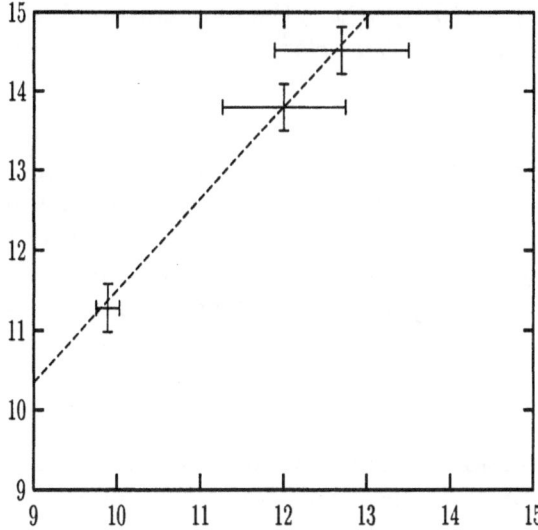

Figure 2. Boundary layer thickness, λ, vs. dissipative Bolgiano length, r_B, for different *Ra*. The straight line is a linear fit with slope ~ 1.15.

iv) Furthermore we have measured the behavior of the energy and temperature dissipation field as a function of the distance from the top/bottom walls. We have seen that the fraction of energy dissipated inside the thermal boundary layer is small (less than 20%) in the *Ra* interval that we have explored. We have also noticed that increasing the *Ra* number, a larger fraction of the energy tends to be dissipated in the boundary layers.

With the help of numeric findings, we propose a new explanation for the scaling of the Nusselt number versus the Rayleigh number. Our point of view is based on the relevance of Bolgiano scaling, and on the numerically observed result that the effects of buoyancy do depend on the distance from the isothermal walls. In particular we observed that the Bolgiano length is small near the top/bottom walls, while it becomes very large in the center of the cell. This implies that the dynamics of the fluid is dominated by buoyancy effects near the walls while near the center of the cell the fluids behave almost as ordinary homogeneous/isotropic turbulence.

To make a more quantitative model we define the Bolgiano dissipative scale, r_B, as the scale at which buoyancy forces are balanced by viscous effects. We will take as a working definition of the dissipative Bolgiano length, r_B, the minimum of the Bolgiano length, *i.e.* $r_B = L_B(\lambda)$. It can be shown, using appropriate version of the Von Karman equation (see Yakhot [6]) that

$$r_B \sim N^{-1/8} \tag{2}$$

where $N = (\chi/2) \cdot \int \sum_i (\partial_i T)^2 d^3x$ is the temperature dissipation which can be proved to scale as $N \sim Nu \cdot Ra^2$ (see [7] for a derivation). Our main ansatz is that:

$$\lambda \simeq r_B \qquad (3)$$

From this ansatz and from the known relation (see [3]) $Nu \sim 1/\lambda$ we are able to show, again, that $Nu \sim Ra^{2/7}$. Note that in this approach the viscous boundary layers does not play any important rôle, which is an explanation of why the Nu scaling is not changed in free-slip conditions.

Furthermore, assuming a linear velocity behaviour in the boundary layer, we have:

$$\epsilon_{\rm BL} = Ra^{1/7} \cdot \epsilon \qquad (4)$$

where $\epsilon_{\rm BL}$ is the fraction of energy dissipated in the boundary layers and ϵ is the energy dissipated in the whole convective cell.

This result predicts the rate at which energy is concentrated in the boundary layer as the Rayleigh number is increased and is consistent with numerical data (see [3]). For values of Ra of the order of 10^{11} the energy dissipation is almost all constrained in the boundary layers. Around this Ra value we estimate that our scaling argument will fail and a new kind of scaling for the Nu number should be observed [8].

Further questions, on which we are still working, remain opened, one particularly interesting being the dependence of the Nu vs. Ra scaling on the Prandtl number (Pr).

References

1. Castaing, B., Gunaratne, G., Heslot, F., Kadanoff, L., Libchaber, A., Thomae, S., Wu, X., Zalesky, S., Zanetti, G. (1989) *J. Fluid. Mech.*, **204**, 1.
2. Shraiman, B. I., Siggia, E. D. (1990) *Phys. Rev. A*, **42**, 3650.
3. Benzi, R., Toschi, F., Tripiccione, R. (1998) *Journ. of Stat. Phys.*, Kadanoff's issue, submitted.
4. Ciliberto, S., Cioni, S., Laroche, C. (1996) *Phys. Rev. E*, **54**, R5901.
5. Benzi, R. *et al.* (1994) *Europhys. Lett.*, **25**, 341;
   Benzi, R. *et al.* (1994) *Europhys. Lett.*, **28**, 231;
   Benzi, R. *et al.* (1997) *Atti convegno Lincei*, **131**, 41.
6. Yakhot, V. (1992) *Phys. Rev. Lett.*, **69**, 769.
7. Chillá, F., Ciliberto, S., Innocenti, C., Pampaloni, E. (1993) *Nuovo Cimento D*, **15D**, 1229.
8. Chavanne, X., Chillà, F., Castaing, B., Hébral, B., Chabaud, B., Chaussy, J. (1997) *Phys. Rev. Lett.*, **79**, 3648.
9. Benzi, R., Succi, S., Vergassola, M. (1992) *Phys. Rep. C*, **222**, 3.
10. Our numerical simulations were performed on the APE100 parallel supercomputer at I.N.F.N. of Pisa. See for details, Bartoloni, A. et al. (1993) *Int. Jour. Mod. Phys. C*, **4**, 993-1006.

INTERACTION OF TURBULENCE AND LARGE-SCALE VORTICES IN INCOMPRESSIBLE 2D FLUIDS

J.-P. LAVAL[1], B. DUBRULLE[2] AND S. NAZARENKO[3]

[1] *CEA/DAPNIA/Sap, l'Orme des Merisiers, 91191 Gif sur Yvette Cedex, France.*

[2] *CNRS, URA 2052 CEA/DAPNIA/Sap, l'Orme des Merisiers, 91191 Gif sur Yvette Cedex, France.*

[3] *Mathematics Institute University of Warwick COVENTRY CV4 7AL, UK*

In 2D fluid dynamic, the inverse energy cascade leads to the formation of intense coherent large-scale vortices. These large structures eventually start producing a significant feed back on small-scale turbulence [2] [7]. To study the interaction of turbulence with large-scale vortices, we consider a situation where such non-local interactions are dominant for the turbulence evolution, and local interactions between the small scales may be neglected. It is reasonable to think that the turbulence becomes non-local at a late stage of the evolution when most of the energy is condensed at the largest scales of the system. A formalism suitable for description of the non-local 2D turbulence was developed in [3]. In this paper, the large-scale are described by Euler equations with an additional term (the averaged Reynolds stress) due to the interaction between turbulence and large scales. The small scales of turbulence are just advected by the mean flow and their equation expresses the conservation of turbulent enstrophy in both physical and spectral space. One can view the small-scales dynamics in (x,k) space as the motion of a large number of wavepackets (particles). This model is exact and conserves both energy and enstrophy in the case of widely separated large and small scales.

The large scales are computed with a classical spectral code and the small ones are computed with a Particles In Cell method which consists in spreading the turbulent enstrophy on particles and computing the evolution of these particles in both physical and spectral space. By switching off the interaction term, particles behave like passive scalars, and we recover results about particles trajectories and probability distribution of space

429

U. Frisch (ed.), Advances in Turbulence VII, 429–432.

derivative of scalar field. With this two-fluid formulation, one can prove that the small-scale energy is preserved separately if the initial small-scales turbulence is locally isotropic [6]. Numerical simulations with this two-fluid model confirm the influence of local isotropy in the two-scale interaction [6]. The figure 1 shows the evolution of the two scales energy in the case of an isotropic initial distribution of particles in the (kx,ky) space (inset fig.1). This initial condition of particles correspond to a locally isotropic small scales turbulence and one can check that the energy transfer between the two scales is very small (less than 0.06% a t=50). By comparison, the energy transfer with anisotropic particles distribution becomes none negligible since the beginning of the simulation and it is 30 times more important at t=50. This conservation was proved analytically using the two-fluid model formulation in the simple case of a dipole with initial isotropic turbulence [3].

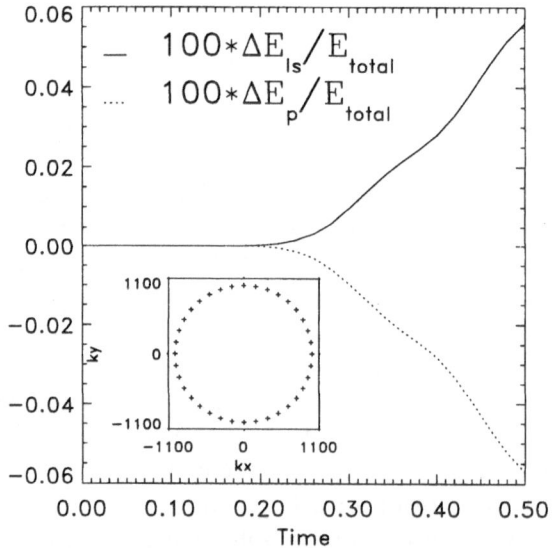

Figure 1.   Particles (ΔE_p) and Large Scales (ΔE_{ls}) energy variations in the case of isotropic initial distribution of particles in (kx,ky) space (inset). Large scales are computed on a 128×128 grid and small scales are modelised with approximately 600000 particles.

We use this formulation to build a two-fluid model for situations without scale separation when the distinction between the small and large scales is made with respect to the grid scale. A procedure is built to produce small-scale particles from the large-scale component when the latter is approaching to the grid scale and to transfer particles back to the large-scale component if their wavenumbers become sufficiently large. The two-fluid model can be tested by comparison with DNS at high resolution and low viscosity or with an hyperviscosity. We performed two simulations in the case of decaying turbulence where the energy is concentrated in the largest scales at initial time. The first simulation is a direct simulation of Navier Stokes equations on a 1024 × 1024 grid with an hyperviscosity to preserve a large inertial range, and the second one uses the two-fluid model on a 128 × 128 grid. For the same integration time, the second simulation is approximatly 20 times quicker than the direct simulation (approximatly 6 hours on a workstation). The figures 2 and 3 show the resulting vorticity at a time t=10. One can see that the large scales structures are very similar. Starting from an initial situation with no energy at small scale, we reach the situation where a continuous energy spectrum is established over the largest wavenumber of the spectral large scales simulation and a continuous spectrum of slope -3 is obtained over many decades (fig. 4). The value of this slope is in agreement with the statistical theory of Batchelor [1] for 2D turbulence but also with analysis of non-local energy transfer developed in [6]. One can see that the two energy spectra are very close in the non-dissipative range of the direct simulation (fig. 5). Several other comparisons of the two-fluid model performed in [4] for decaying or forced turbulence prove that the largest structures of the flow are well simulated.

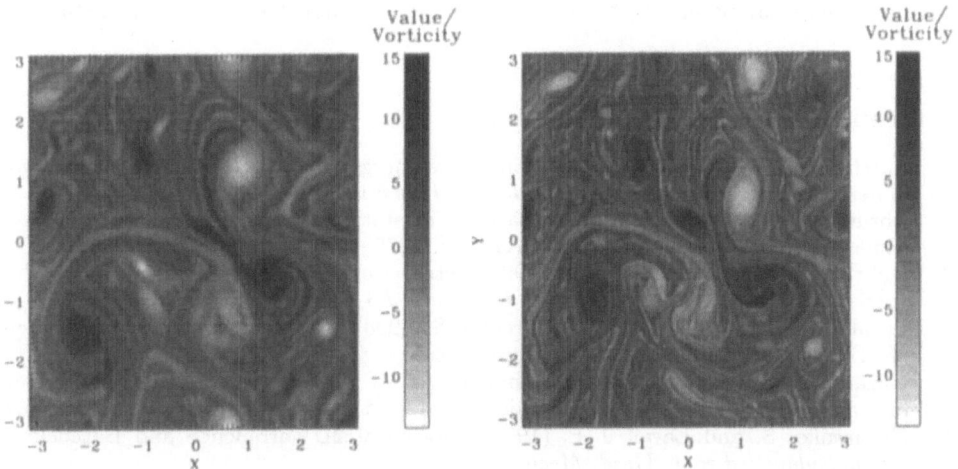

Figure 2. Vorticity field at t=10 obtained with our two-fluid model on a 128 × 128 grid.

Figure 3. Vorticity field at t=10 obtained via simulation of Navier Stokes equations with an hyperviscosity on a 1024 × 1024 grid.

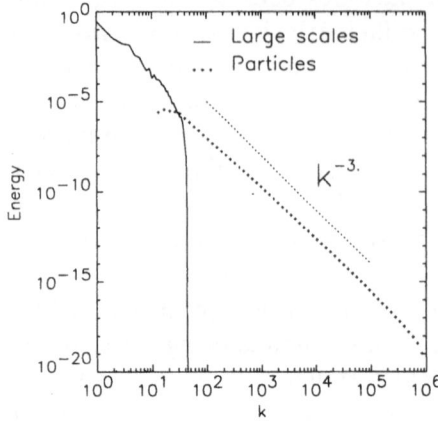

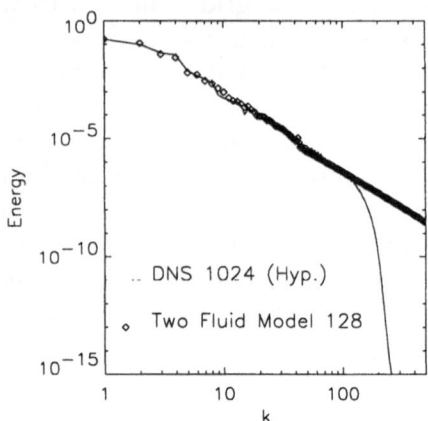

Figure 4. Energy spectra for the two scales of our model. The slope of the small scales energy spectrum is -3 over several decades.

Figure 5. Comparison of energy spectra for the direct simulation (fig. 3) and for our two-fluid model (fig. 2).

The interest of our model with respect to other classical models is that the small scales are computed rather than postulated. Therefore, small scales informations, like the energy or enstrophy distribution, are preserved. An other advantages is that it can be easily generalized to take into account effects of waves (see e.g. [5]). This model can be an interesting tool to simulate geophysical or astrophysical flows with realistic Reynolds numbers.

References

1. Batchelor, G. K. (1969) Computation of the energy spectrum in homogeneous two-dimensional turbulence, *Phys. Fluids Suppl. II* **A12**, pp, 233–239
2. Borue, V., (1994) Inverse energy cascade in stationary two-dimensional homogeneous turbulence, *Phys. Rev. Lett.* **A72** pp. 1475–1478.
3. Dubrulle, B. and Nazarenko, S (1997) Interaction of turbulence and large-scales vortices in incompressible 2D fluid, *Physica D* **A110**, pp. 123–138
4. Laval, J.-P. Dubrulle, B. and Nazarenko, S (1998) Two-fluid numerical simulation of 2D turbulence *submitted to J. Fluid. Mech.*
5. Nazarenko, S. Zabusky, N. and Schneidegger, T. (1995), Nonlinear sound-vortex interactions in inviscid isentropic fluid: a two fluid model, *submitted to Phys. Fluids.*
6. Nazarenko, S. .and Laval, J.-P. (1997), Non-local 2D turbulence and Batchelor's regime, *submitted to J. Fluid. Mech.*
7. Smith, L. M. and Yakhot, V., (1994) Finite-size effects in forced, two-dimensional turbulence *J. Fluid Mech.* **A274**, pp. 115–138.

ON GEOMETRICAL ALIGNMENT PROPERTIES OF TWO-DIMENSIONAL FORCED TURBULENCE

B. PROTAS[1], A. BABIANO[2]
[1] *Institute of Aeronautics and Applied Mechanics,*
Warsaw University of Technology,
Nowowiejska 24, 00-665 Warsaw, Poland
[2] *CNRS/LMD/ENS,*
24 rue Lhomond, 75231 Paris Cedex 5, France

1. Introduction and Basic Concepts

In the present work we investigate the physical mechanisms which underlie the formation and evolution of the enstrophy cascade in two-dimensional (2D) forced turbulence. Following the work [1], it is argued that analysis of vorticity gradient can capture important aspects of the enstrophy dynamics. An extensive discussion of the generation of strong vorticity jumps in 2D turbulence can be found in [2] (and references quoted therein). Intensity of the production of vorticity gradient depends on (i) its alignment with respect to the principal axes of strain and (ii) strain magnitude. It will be shown below that in the 2D forced turbulence vorticity gradient reveals a tendency towards alignment that results in its intensified amplification. This implies internal organization of the flow at small scales.

In the analysis we will use the equation for inviscid evolution of vorticity gradient $\nabla\omega$ (obtained by taking the gradient of the 2D vorticity equation):

$$\frac{d}{dt}\nabla\omega = \left(\frac{\partial}{\partial t} + V \cdot \nabla\right)\nabla\omega = -\nabla V \cdot \nabla\omega, \qquad (1)$$

where V represents the velocity field. In the absence of the typical "vortex stretching", the non-linear term $-\nabla V \cdot \nabla\omega$ is responsible for the formation of small scales and sustains the enstrophy cascade [1]. The local flow topology is characterized by the velocity gradient tensor ∇V. Its eigenvalues are given by $\lambda_\pm = \pm\frac{1}{2}\sqrt{s_{11}^2 + s_{12}^2 - \omega^2}$, where s_{11} and s_{12} denote the respective strain components and ω is vorticity. According to the relative importance

U. Frisch (ed.), Advances in Turbulence VII, 433–436.

of strain and vorticity they can be either purely real or purely imaginary. It
can be shown [1] that in the first regime (called *hyperbolic*) vorticity gradient is either exponentially amplified or damped, depending on its alignment
with respect to the eigenbasis of the strain tensor $D = \frac{1}{2}\left(\nabla V + \nabla V^T\right)$.
Thus from the point of view of the dynamics of the enstrophy cascade these
regions may be regarded as active. On the other hand, in the second case
(called *elliptic*) vorticity gradient merely rotates and oscillates, without any
amplification or damping. In our analysis we will focus on the hyperbolic
parts of the flow.

In the hyperbolic regime, vorticity gradient is amplified when it is
aligned possibly close to the compressing eigendirection d_2 of the strain
tensor D. Conversely, alignment with the stretching eigendirection d_1 results in damping. These effects are feasibly described by two parameters
[3]:

$$\kappa = \frac{d\ln|\nabla\omega|}{dt} = -m^T D m = \sqrt{s_{11}^2 + s_{12}^2}\,\cos\left(2\alpha\right), \qquad (2)$$

$$e = \frac{-m^T D m}{(D{:}D)^{\frac{1}{2}}} = \frac{\sqrt{2}}{2}\cos\left(2\alpha\right), \qquad (3)$$

where $m = \frac{\nabla\omega}{|\nabla\omega|}$ and $\alpha = \angle\left(\nabla\omega, d_2\right)$ is the angle between the vortic-
ity gradient and the compressing eigenvector of D. The former represents
combined effects of strain magnitude and geometrical alignment, whereas
the latter of geometrical alignment only. It will be shown below that the
alignment parameter e (or equivalently the angle α) has interesting statistical properties in 2D forced turbulence implying the presence of non-trivial
geometrical alignments.

2. Results

We study the alignment properties of vorticity gradient in a high resolution
(1728×1728) numerical simulation of 2D forced turbulence. Vorticity gradient is a small scale quantity, therefore in order to obtain reliable results we
use two dissipation models: (i) standard Newtonian (molecular) viscosity
and (ii) hyperviscosity with the order of the (hyper-)laplacian $p = 8$. We
also check that the alignments are dynamic, and not purely kinematic. To
this end we compare our results with the random fields which were obtained
by randomizing the phases in the coefficients of the Fourier expansions of
the vorticity fields. In the following, the three cases will be referred to as
Newtonian, *hyperviscous* and *random*, respectively.

Analysis of the field averaged data reveals that both hyperviscous and Newtonian cases are characterized by positive production of vorticity gradient. One can however see that the hyperviscous fields are much more efficient, both concerning the net production as well as the fraction of the flow domain where build-up of $\nabla\omega$ occurs. The random case is neutral in this respect. In Fig. 1 we show the PDF's of the angle α between the vorticity gradient and the compressing eigendirection $\mathbf{d_2}$. Strong tendency towards alignment resulting in amplification of vorticity gradient can be observed in both hyperviscous and Newtonian cases. In the former however this effect is much more pronounced.

Next we investigate how the alignment properties of vorticity gradient change with its magnitude. In Fig. 2 we show the average values of the alignment parameter e (cf. Eq. (3)) corresponding to different values of $|\nabla\omega|$. One can see that effective alignments are correlated with substantial magnitudes of $|\nabla\omega|$. This effect can also be observed in Figs. 3 and 4 where we present the PDF's of the alignment angle α conditioned to certain values of $|\nabla\omega|$. As the modulus $|\nabla\omega|$ increases, one can note the developing bias towards preferential alignment, although there are some differences between the hyperviscous and Newtonian cases.

3. Conclusions and Discussion

The positive correlation between the alignment parameter e and the magnitude of vorticity gradient is an objective manifestation of an internal structure in the investigated turbulent flows. The random fields turn out to be neutral as regards the characteristics that were considered which however does not preclude the presence of some other potentially possible alignments. This means that the alignments presented above are not merely a kinematic property of vector fields, but follow from the dynamics prescribed by the Navier-Stokes system. One may conclude that the hyperviscous case is characterized by a stronger tendency for preferential alignment of vorticity gradient than the Newtonian case. The source of this discrepancy is however not clear. Finally, it should be added that preliminary results concerning the influence of organized coherent motion on geometrical statistics of vorticity gradient were presented in [4].

4. Acknowledgments

We are grateful for enlightening discussions to N. K.-R. Kevlahan and J. Szumbarski. B.P. was supported by the French Embassy in Warsaw (Grant No. 76440) and the Foundation for Polish Science. Computations were performed at IDRIS.

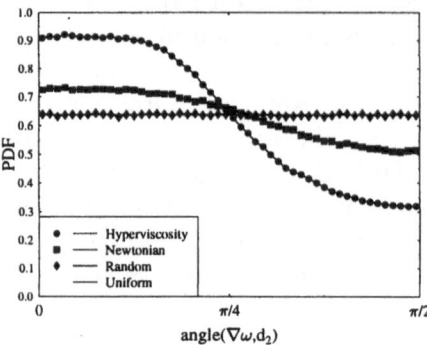

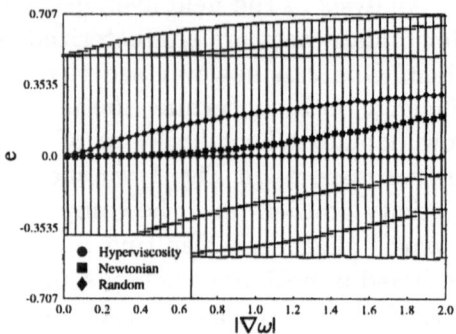

Figure 1. *PDF* of the angle $\angle\,(\nabla\omega, \mathbf{d_2})$ between the vorticity gradient $\nabla\omega$ and the compressing eigendirection $\mathbf{d_2}$ for the three cases discussed.

Figure 2. The alignment parameter e vs. the magnitude of $|\nabla\omega|$. Average values of e are indicated, with error bars representing the standard deviations.

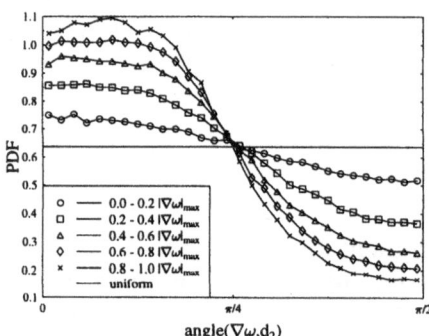

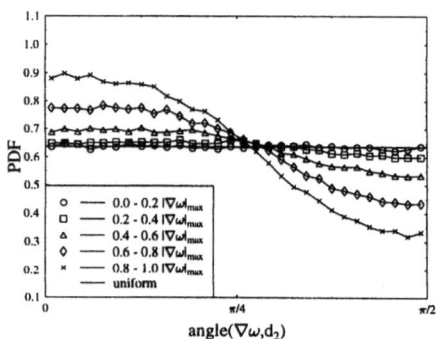

Figure 3. *PDF* of the alignment angle conditioned to various magnitudes of vorticity gradient for the hyperviscous case.

Figure 4. *PDF* of the alignment angle conditioned to various magnitudes of vorticity gradient for the Newtonian case.

References

1. Weiss, J. (1991) The dynamics of enstrophy transfer in two-dimensional turbulence, *Physica* D **48**, 273.
2. Yao, H.B., Zabusky, N.J., Dritschel, D.G. (1995) High gradient phenomena in two-dimensional vortex interactions, *Phys. Fluids* **7**, 539-548.
3. Ottino, J.M. (1995) *The kinematics of mixing: stretching, chaos and transport*, Cambridge University Press, Cambridge.
4. Protas, B., Wesfreid, J.-E. (1997) Stretching of vorticity gradients in two-dimensional wake flows, in *Proceedings of the 11th Symposium on Turbulent Shear Flows* **10-1**, Grenoble, September 1997.

GROWTH OF FINITE PERTURBATIONS

E. AURELL[1], G. BOFFETTA[2], A. CRISANTI[3], G. PALADIN
AND A. VULPIANI[3]
[1] *Department of Mathematics, Stockholm University, S-106 91
Stockholm, Sweden*
[2] *Istituto di Fisica Generale, Università di Torino, Via Pietro
Giuria 1, I-10125 Torino, Italy*
[3] *Dipartimento di Fisica, Università di Roma "La Sapienza"
P.le Aldo Moro 2, I-00185 Roma, Italy*

We investigate the growth of finite perturbations in fully developed turbulence. Consideration of these quantities is relevant for *a priori* discussions of the predictability problem of geophysical flows and weather forecasts. In an early pioneering investigation[4] E. Lorenz showed that under natural assumptions and assuming Kolmogorov scaling, the growth rate of finite perturbations scales with perturbation size as $\lambda(\delta) \sim \delta^{-\gamma}$, where the predictability scaling exponent γ is equal to 2.

The result of Lorenz can be questioned from two points of view. The standard characterization of the chaotic behavior of a dynamical system is given by the maximum Lyapunov exponent $\lambda_{\max}$, which, in a turbulent flow, inversely proportional to the shortest characteristic time of the system, the turnover time at the Kolmogorov scale[5]. If the growth of finite perturbations were ruled by the Lyapunov exponent then the the growth rate would be independent of the size of the perturbation ($\gamma = 0$). On the other hand the grow rate would increase with the Reynolds number of the flow as $\gamma(\mathrm{Re}) \sim \mathrm{Re}^{1/2}$.

A second possible objection stems from the effects of deviations from the Kolmogorov 1941 theory. In the multifractal approach one assumes that there exist several possible scaling behaviours of the field, $v_l \sim l^h$. The scaling exponents of velocity structure functions of order p, $< v_l^p > \sim l^{\zeta_p}$, are then given as a Legendre transform by $\zeta_p = \min_h [\, h\, p - D(h) + 3\,]$. In a similar manner the predictablity scaling exponent can be computed as $\gamma = \max_h [\frac{D(h)-2-h}{h}]$. It could therefore be surmised that the value of γ, as the values of most of the ζ_p's, is not determined *per se* in the multifractal approach, but would depend on the function $D(h)$.

U. Frisch (ed.), Advances in Turbulence VII, 437–438.

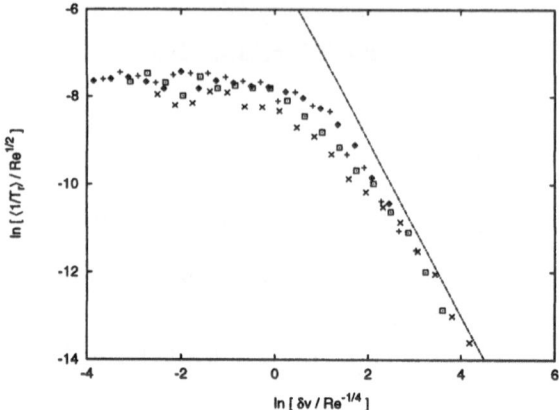

Figure 1. The growth rate $\langle 1/T_r(\delta v)\rangle$ versus error size δv in logarithmic scale at different Reynolds numbers Re $= \nu^{-1}$. Both quantities are scaled with powers of Reynolds number as predicted by Kolmogorov theory to match the dependence of Lyapunov exponent and Kolmogorov scale with Re. The number of shells (N) and viscosities (ν) in the different simulations are: $N = 24$ and $\nu = 10^{-8}$ (diamonds); $N = 27$ and $\nu = 10^{-9}$ (plus) $N = 32$ and $\nu = 10^{-10}$ (squares); $N = 35$ and $\nu = 10^{-11}$ (crosses). The straight line has slope -2. Small initial perturbations are followed as they grow, and the time to double an error (T_r) measured for a number of error thresholds r.

We show that γ is in fact equal to the Lorenz value of 2 also in the multifractal approach[1, 2]. This is a consequence of one of the admissibility conditions on the function $D(h)$, that it must reproduce the Kolmogorov law that third order structure functions scale linearly with increment, i.e. $\zeta_3 = 1$. We validate these results in numerical experiments on shell models of turbulence (see fig. 1). We also show that for small but finite perturbations the growth rate is well described by the Lypaunov exponent. The Ruelle and Lorenz scenarios hence hold for small and large perturbations respectively, compared with the velocity fluctuations on the Kolmogorov scale.

References

1. Aurell E., Boffetta G., Crisanti A., Paladin G. & Vulpiani A. (1996), *Phys. Rev. Lett.*, **77** pp. 1262.
2. Aurell E., Boffetta G., Crisanti A., Paladin G. & Vulpiani A.(1997), *J. of Physics*, **30** pp. 1-26.
3. Benzi R., Paladin G., Parisi G. & Vulpiani A. (1984), *J. Phys.* **17A**, pp. 3521; Parisi G. & Frisch U. (1985) in *Turbulence and Predictability of Geophysical Flows ans Climatic Dynamics* edited M. Ghil et al. (North-Holland, New York, 1985), pp. 84; Paladin G. & Vulpiani A. (1987), *Phys. Rep.* **156**, pp. 147.
4. Lorenz E.N. (1969), *Tellus* **21**, pp. 3.
5. D. Ruelle (1979), *Phys. Lett.* **72A**, pp. 81.

THE OBUKHOV-BOLGIANO INERTIAL RANGE
IN A SHELL MODEL OF TURBULENT CONVECTION

S.A. LOZHKIN AND P.G. FRICK
Institute of Continuous Media Mechanics,
Korolev 1, 614061, Perm, Russia

The very earliest studies of the peculiarities of developed turbulence in a medium with an inhomogeneous density, attributed to the essential influence of buoyancy forces on the inertial range dynamics, have been carried out by Obukhov [1] and Bolgiano [2]. Obukhov established the existence of a boundary scale l_B, above which the input of kinetic energy by buoyancy forces is essential in comparison with the Kolmogorov spectral flux. Bolgiano studied turbulence in a medium with a stable stratification, when the kinetic energy was transformed into the potential, and derived, by performing the dimensional analysis, the spectral laws for kinetic energy and energy of fluctuations of temperature, ($E_v(k) \sim k^{-11/5}$ and $E_t(k) \sim k^{-7/5}$).

However, the question of the validity of these laws in natural and laboratory convective flows still remains to be answered. There is the opinion that the buoyancy force does not influence the structure of developed convective turbulence far away from the boundary layers. In this case, the temperature behaves as a passive scalar and, therefore, the Kolmogorov state with the law $k^{-5/3}$ for the spectra of velocity and temperature fluctuations is established. The results of currently performed experiments on this problem are rather contradictory.

The shell models appeared as an attempt to describe the spectral transfer of energy in fully developed turbulence using a small number of variables U_n, each being a collective characteristic of the amplitudes of velocity fluctuations in the whole octave of wave numbers $2^n < |k| < 2^{n+1}$. The equations for U_n must reproduce the main features of the Navier-Stockes equations, namely, the same conservation laws and nonlinearity.

In this investigation, we present the shell model for fully developed turbulent convection of an incompressible fluid. The shell equations are based on the GOY shell model, to which the term describing the effect of buoyancy forces and the equations for temperature fluctuations are added.

439

U. Frisch (ed.), Advances in Turbulence VII, 439–440.

The latter have the same structure of nonlinear terms as the basic GOY equation.

Our main goal here is to study the formation of the Obukhov-Bolgiano inertial range, which is expected to appear in the large scale part of the spectrum (for scales $l > l_B$).

Numerical calculations have been performed to study the large Rayleigh numbers convective turbulence in both three- and two-dimensional cases. It has been found that at the earlier stages of its development the model of 3D turbulent convection shows an inertial range which corresponds to the Obukhov-Bolgiano state. At this time the input of kinetic energy provided by the buoyancy forces and the rate of spectral flow of kinetic energy follows the same law, and the buoyancy forces remain essential in an extended interval of scales. This state is, however, unstable and changes to the Kolmogorov-like state, in which the temperature behaves as a passive scalar in all scales excluding the largest one. The instability of the Obukhov state is caused by the loss of correlation between the velocity and temperature shell variables.

In 2D-turbulence, the direct transport of enstrophy arrests the transport of energy to the small scales thus causing the accumulation of kinetic energy in the stationary heated system. In order to obtain the stationary regime it is necessary to introduce a "large-scale viscosity" (or linear friction). Numerical solutions for 2D turbulent convection are considered at very large Grasshoff numbers, when a wide interval of excited wave numbers allows one to follow the spectrum formation on both sides of the Bolgiano scale. Qualitatively, the spectra obtained correspond to the expected spectral distribution for 2D turbulent convection, but the detailed analysis shows that the spectral laws differ from those, predicted for the Obukhov-Bolgiano range by the dimensional arguments. The analysis of flux testifies that the dynamics of convective range is mainly determined by linear friction.

Consideration is also given to the shell model with nonlocal spectral interaction which is of a vital importance in the event of the extreme value of the Prandtl number. The cascade processes in convective turbulence at a very low and large Prandtl number are examined. Different spectral inertial intervals are discussed.

A detailed presentation of this work may be found in Ref. [3], [4].

References

1. Obukhov A.M. (1959), *Dokl. Akad. Nauk SSSR*, **125:6**, pp. 1246-1248.
2. Bolgiano R. (1959), *J. Geophys. Res*, **64:12**, pp. 2226-2229.
3. Lozhkin S.A., Frick P.G. *Izv. RAN. Mekhanika Zhidkosti i Gaza*, in press.
4. Lozhkin S.A., Frick P.G. (1996) *Mat. Mod. Sist. i Proces.*, N 4. pp. 53-60.

HEAT TRANSPORT AND THERMAL FLUCTUATIONS IN TURBULENT THERMAL CONVECTION

S. CILIBERTO, S. CIONI, C. LAROCHE

Ecole Normale Supérieure de Lyon, Laboratoire de Physique,
C.N.R.S. URA1325, 46 Allée d'Italie, 69364 Lyon, France

Turbulent thermal convection in a fluid layer heated from below, that is Rayleigh Benard convection, presents several open problems which are not yet very well understood [1]. A very important one is the dependence of the Nusselt number Nu as a function of the Rayleigh number Ra [1][2][3]. Experimentally it is found that $Nu \propto Ra^{2/7}$ but the models constructed to explain this behaviour fail to correctly describe all the other experimental observations. One of the reasons for this descrepancy between theory and experiment is that it is not jet completely understood the role played by the mean circulation flow (MCF) and by thermal fluctuations on the heat transport. The MCF consists in a large scale convective roll which involves all the cell containing the convective fluid, thus it behaves like a large coherent structure of the turbulent flow. The MCF strongly perturbs the boundary layers. However it is not clear whether heat is transported mainly by the MCF or by the thermal plumes detaching from the boundary layers.

To investigate this problem, we performed several experiments in water (Prandtl number about 3) for $10^6 < Ra < 10^{11}$ (see ref.3 and 4 for experimental details). In these experiments we have perturbed either the MCF or the boundary layers. The direct perturbation of the MCF by means of small screens allows us to study its influence on the thermal boundary layer and on the heat transport properties[4]. We find that the MCF couples the upper and lower boundary layers. This coupling produces a coherent oscillation of the temperature field at a well defined frequency f_p. This frequency is clearly observed in the power spectrum plotted in fig.1a. The spatial coherence of the MCF flow is checked by computing the coherence function between two probes at two different locations inside the flow. The coherence function, plotted in fig.1b, is very close to 1 at f_p and its harmonics. In contrast when the MCF is suppressed the coherent oscillation disappears. The frequency f_p is not detected in any part of the cell, as can

441

U. Frisch (ed.), Advances in Turbulence VII, 441–444.
© 1998 *Kluwer Academic Publishers.*

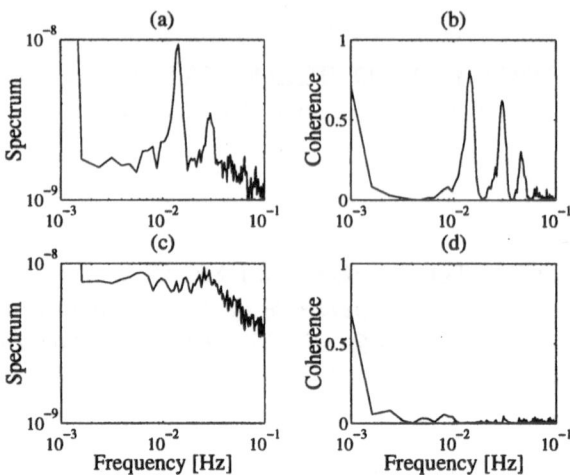

Figure 1. Spectra (a),(c) and coherence functions (b),(d) of the temperature signals recorded at $Ra = 6.4 \cdot 10^{10}$ (the depth d of the layer was 40 cm) by two probes P1 and P2. P1 was located at $d/9 \simeq 45mm$ from the bottom plate whereas P2 at 45 mm from the top. Only the low frequency part of the spectra is shown. The whole spectrum extends to roughly $10Hz$). Figures a) and b) correspond to measurements done when the mean flow was present whereas figures c) and d) to measurements done when the MCF was suppressed by the screens.

be observed in the power spectrum, plotted in fig.1c. Furthermore we show in fig.1d that the coherence function between the two probes is zero when the MCF is suppressed. These results show that the coupling between the two boundary layers, produced by the MCF, is an essential ingredient in order to generate the coherent oscillation.

In spite of this important perturbations produced by the MCF suppression the mean temperature profile is not influenced by the MCF. As a consequence the values of the Nusselt number Nu are not affected and no significant variations of the heat transport is produced by the screens. The heat flow measurements, done with the screens are compared with those done without screens in fig.2. A more sensible way of showing this is to measure at constant $Nu\ Ra$ (i.e. at constant heating power) the ratio Ra_{scr}/Ra, between the Rayleigh number Ra_{scr}, measured with screens, and Ra measured without screens. The ratio Ra_{scr}/Ra is plotted as a function of the applied heat flow $Nu\ Ra$, in fig.2c). We clearly see

that the ratio is close to one within ± 0.02, showing that the heat transport is not affected, within experimental errors, by the MCF suppression. This observation indicates that the relevant contribution to the heat transport is given by the coupling of the vertical velocity with the thermal fluctuations which cannot be neglected also inside the boundaries. Thus one may conclude that the most important characteristic length of the system

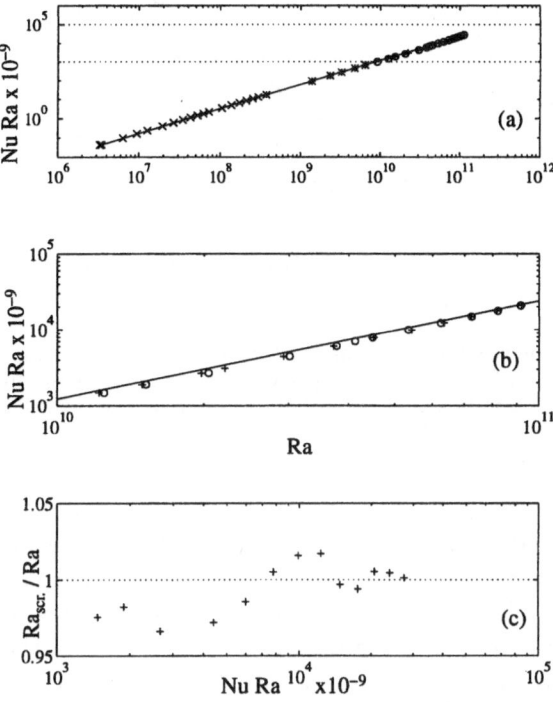

Figure 2. a) Dependence of the non-dimensional heat flow $Nu \cdot Ra$ as function of Ra in the range $(3.5 \times 10^6 < Ra < 1.1 \times 10^{11})$. (x) data from ref.3, (*) data from ref.5. Data of the present experiment (+) with MCF, (o) without MCF. The fit represented in the figure corresponds to the ensemble of the data. The slope of the best fit line is 1.28 ± 0.01. The two horizontal dashed lines indicate the region where the properties of the mean flow have been studied. An expanded view of this region is shown in b). c) Ratio Ra_{scr}/Ra as a function of $Nu \cdot Ra$. Here Ra_{scr} indicates the value of Ra measured without MCF.

is the size of the thermal plumes, typically of the order of the boundary layer thickness. If one can change this characteristic length then a variation of the heat flow should be observed.

Driven by these idea, we have perturbed the boundary layers by constructing a rough bottom plate with a mean roughness comparable to the thermal boundary layer thickness. The important point is that the height h of the roughness was distributed as a power law. that is $P(h) \propto h^{-\xi}$, where $P(h)$ is the probability distribution function of h. This allows us to perturb the characteristic length of the system without imposing a new one. We have studied the boundary layer profile and the dependence of Nu as a function of Ra, finding that when the mean roughness becomes comparable to the boundary layer thickness the boundary layer profile is strongly mod-

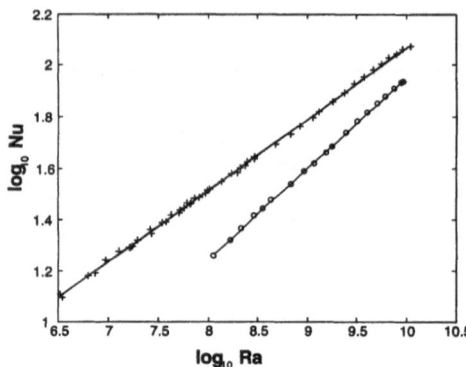

Figure 3. Dependence of the non-dimensional heat flow Nu as function of Ra in the range ($3.5 \times 10^6 < Ra < 10^{10}$): (o) with rough bottom plate, (+) with smooth bottom plate. The continuous line has slope 0.28 ± 0.01, whereas the dashed one is 0.35 ± 0.01. The roughness exponent ξ is 2 in this case.

ified and the dependence of Nu versus Ra changes. The dependence of Nu versus Ra measured in presence of a very rough bottom plate is compared in fig.1 to that measured in the same cell but with a smooth plate. We clearly see that in the smooth case the standard law $Nu \propto Ra^{2/7}$ is found whereas in the rough case we find $Nu \propto Ra^{0.35}$. Thus the fact of having modified the characteristic size of the plumes have completly changed the dependence of Nu versus Ra. The dependence of the heat trasport as a function of ξ is under investigation.

In conclusion the main result of our experiment is that a perturbation of the MCF does not change the heat transport across a Benard cell, whereas the pertubation of the characteristic size of the thermal fluctuations clearly changes the heat transport properties. This is very important in order to construct a reliable model of the heat transport which takes into account also the thermal fluctuations and their coupling with the MCF.

References

1. E. Siggia (1993) High Rayleigh number convection, *Ann. Rev. Fluid Mech* **28**, 137-168 .
2. B. Castaing, G. Gunaratne, F. Heslot, L. Kadanoff, A. Libchaber, S. Thomae, X. Wu, S. Zalesky., G. Zanetti (1989) Scaling of hard turbulence in Rayleigh-Bénard convection, *J. Fluid Mech.***204**, 1-29.
3. F. Chillá, S. Ciliberto, C. Innocenti, E. Pampaloni (1993) Boundary layer and scaling properties in turbulent thermal convection, *Nuovo Cimento D* **15**, 1229-1249 .
4. S. Ciliberto, S.Cioni, C. Laroche (1996), Large scale flow properties of turbulent thermal convection, *Phys.Rev.E* **54**, R5901-R5905.
5. S. Cioni, S. Ciliberto, J. Sommeria (1997) Strongly turbulent Rayleigh-Bénard convection in mercury: comparison with results at moderate Prandtl number, *J. Fluid. Mech.* **335**, 111-140

INVERSE ENERGY CASCADE IN 2D TURBULENCE: EXPERIMENTAL STUDY

J. PARET AND P. TABELING
Laboratoire de Physique Statistique, Ecole Normale Supérieure
24 rue Lhomond, 75231 Paris Cedex 5, France.

1. Introduction

Thirty years ago, Kraichnan [1] conjectured that the twin constraints of energy and enstrophy conservation for inviscid two-dimensional flows should impose the existence of two different inertial ranges for 2D turbulence forced at a given scale: an enstrophy transferring range extending from the input scale towards the viscous scale (the direct enstrophy cascade) and an energy transferring range extending from the input scale towards large scale, known as the inverse energy cascade. Kraichnan proposed that the latter range should display a Kolmogorov type scaling:

$$E(k) = C_k \varepsilon^{2/3} k^{-5/3}$$

as long as this scaling range does not extend to scales comparable to that of the fluid domain. When the cascade reaches the box size, he conjectured that the flow should experience a kind of Bose-Einstein condensation with a pile-up of kinetic energy in the lowest accessible mode. Since then, a number of numerical simulations have confirmed these conjectures [2, 3] but, aside from the early observations by Sommeria [4], there was no experimental evidence supporting them.

We provide here the results of an extensive experimental study of the two-dimensional inverse energy cascade which show that Kraichnan's conjectures were basically correct. Moreover, we investigate intermittency effects and the role of coherent structures in the inverse cascade regime and show that the inverse cascade is non-intermittent with quasi-gaussian statistics. Finally, the study of relative dispersion of pairs of passive particles shows that it is ruled by Richardson's law, $R^2 \sim t^3$.

The experimental set-up we use has been described in a number of papers (see Ref. [5] for a detailed description). It is made of a 15 cm × 15 cm

U. Frisch (ed.), Advances in Turbulence VII, 445–448.

PVC cell filled with two thin layers of electrolyte in a stable configuration (the heavier underlying the lighter). Permanents magnets are placed below the cell and the interaction of their vertical magnetic field with an electric current driven from one side of the cell to the other produces Laplace forces which drive the flow. The latter is visualized by latex particles placed at the free surface, recorded on a video tape and further processed by standard PIV techniques which allow us to obtain the complete velocity fields at any time. In the present experiments, we use total fluid depths of 5.5 mm (inverse cascade) and 7.5 mm (condensate), the injection scale l_i is 1.5 cm and the injection Reynolds number (based on the injection scale and the root-mean-square velocity) is around 100.

2. Inverse energy cascade

When the bottom friction (which is the main dissipation mechanism in this system) is not too low, we have found that, after a short transient, a stationary regime is achieved and that the flow is homogeneous and isotropic in this regime, with an energy spectrum (Fig. 1(a)) displaying a scaling range with scaling exponent very close to $-5/3$ [6]. Moreover, the Kolmogorov constant was determined to be $C_K \sim 6.5$, a value consistent with numerical estimates. Although the scaling range we observe is not very wide, it is well defined and it is then natural to wonder whether there exist any intermittency corrections as it is the case for 3D turbulence. We have performed this analysis using ensemble averaging over 5 experiments, each having a duration of about 40 turn-over times. Although the absolute scaling exponents for the structure functions are not well defined due to the modest size of our inertial range, relative exponents for longitudinal structure functions, determined using ESS, can be measured and they are equal, within experimental accuracy, to their Kolmogorov values, $\zeta_p = p/3$ (Fig. 2(a)). Moreover, normalized even-order moments $H_{2n} = S_{2n}/(S_2)^n$, either longitudinal or transverse, are almost undistinguishable from their gaussian values up to order 12. These results show that there is no intermittency in the inverse cascade range. We have also determined the distribution of vortex sizes (Fig. 2(b)). Vortices are characterized by thresholding the vorticity fields at ± 1.5 their rms value and keeping the region with vorticity levels above the threshold. We find that the distribution of vortex sizes has a peak for a size corresponding to the injection scale and then exponentially decreases at larger sizes. This observation shows that the role of coherent structures in the inverse cascade dynamics is probably not the naive one, involving the formation of larger and larger vortices through merging events. We have indeed observed that merging events are rare and that the dynamics are rather governed by the aggregation of same sign vortices into large recirculation regions.

3. Bose-Einstein condensation

By increasing the total fluid depth, we are able to decrease the bottom friction. When this friction is low enough, the condensation process is observed: the energy accumulates into the lowest accessible mode and the energy spectrum displays a sharp bump at this wave-number (Fig. 1(b)), with a scaling consistent with a k^{-3} energy spectrum although it is not developped enough to draw a definite conclusion. The flow displays a mean global rotation with weak superimposed fluctuations even if the forcing injects an approximately zero net circulation. Moreover, this mean rotation is organized around a central vortex with very high vorticity, an observation in agreement with the vortex intensification process observed numerically by Smith and Yakhot [3].

4. Richardson's law

Finally, we studied the relative dispersion of pairs of particles initially close to each other. This is done by numerically integrating the equations of motion for passive particles using the experimentally measured velocity fields. This technique allows to easily determine a large number of trajectories, which is necessary in order for the statistical analysis to make sense. We typically compute several tens of thousand trajectories and the mean squared separation between particles is found to follow Richardson's law $R^2 \sim t^3$ (Fig. 3).

5. Conclusion

This comprehensive experimental study of the inverse energy cascade shows that, depending on the external friction at large scales, two different stationary regimes can be achieved. The inverse cascade regime, with $-5/3$ scaling, appears to be non-intermittent and to be driven by an aggregation mechanism of same sign vortices. Moreover, for this regime, relative dispersion of pairs of particles is in agreement with Richardson's law. These results tend to confirm the picture of the inverse cascade advocated by Kraichnan. Moreover, the fact that the properties of the inverse cascade can be essentially derived from dimensional analysis leads us to wonder whether it could be accessible to a full theoretical treatment.

References

1. Kraichnan, R.H. (1967): Inertial ranges in two-dimensional turbulence, *Phys. Fluids* **10**, 1417–1423.
2. Maltrud, M.E. and Vallis, G.K. (1991): Energy spectra and coherent structures in forced two-dimensional and beta-plane turbulence, *J. Fluid Mech.* **228**, 321–342.
3. Smith, L.M. and Yakhot, V. (1994): Finite-size effects in forced two-dimensional turbulence, *J. Fluid Mech.* **274**, 115–138.
4. Sommeria, J. (1986): Experimental study of the two-dimensional inverse energy cascade in a square box, *J. Fluid Mech.* **170**, 139–168.
5. Cardoso, O., Marteau, D. and Tabeling, P. (1994): Quantitative experimental study of the free decay of quasi-two-dimensional turbulence, *Phys. Rev. E* **49**, 454–461.
6. Paret, J. and Tabeling, P. (1997): Experimental observation of the two-dimensional inverse energy cascade, *Phys. Rev. Lett.* **79**, 4162–4165.

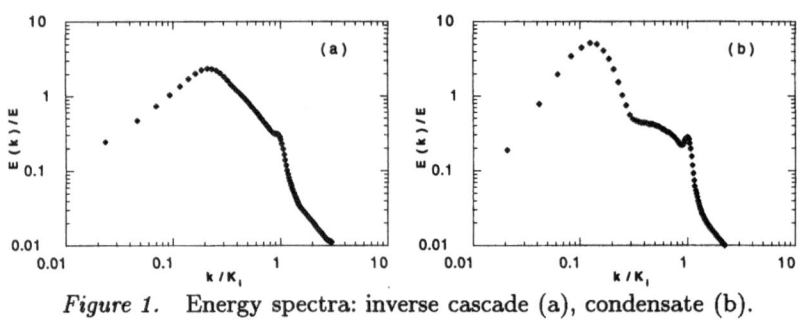

Figure 1.   Energy spectra: inverse cascade (a), condensate (b).

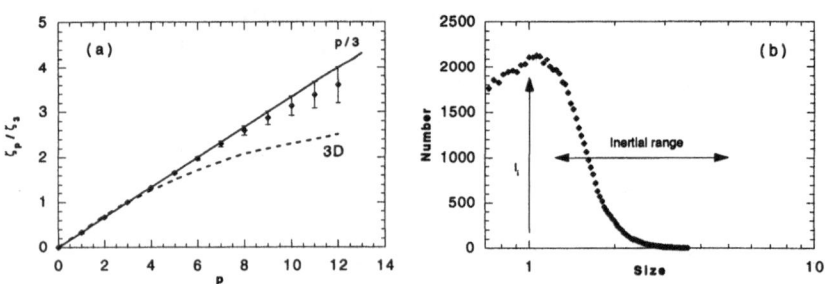

Figure 2.   (a): Evolution of $\tilde{\zeta}_p$ with p for $p \leq 12$. (b): Distribution of vortex sizes.

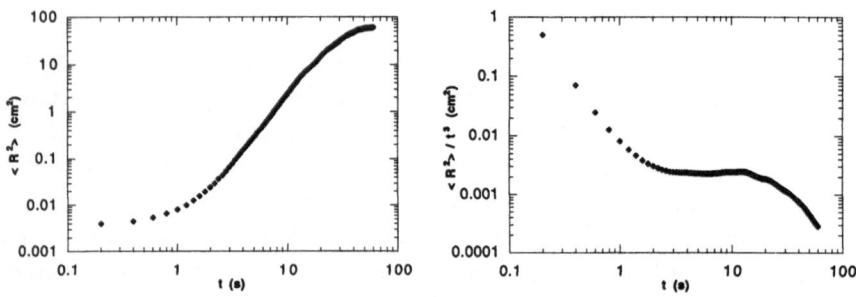

Figure 3.   Evolution of the mean squared separation $\langle R^2 \rangle$. (a): $\langle R^2 \rangle$ vs. t, (b): $\langle R^2 \rangle/t^3$ vs. t.

VERY LONG TIME SCALE SIGNATURE OF INTERMITTENC IN THE ATMOSPHERIC FLOW

P. D. DITLEVSEN
Niels Bohr Institute, Geophysical Department, Juliane Maries Vej 30, DK-2100 Copenhagen O, Denmark

The climate of the Earth is the long term mean of the state of the atmo-spheric – and oceanic flows. As a practice, more or less arbitrarily, 30 years averages defines climate. Is this long term mean at all influenced by the much shorter time – and spatial scale turbulence of the underlying dynam-ical system? The answer to this provocative question is that some indica-tions that it does can be found in paleoclimatic records. Detailed timeseries of climatic proxies have been obtained from cores drilled in the Greenland ice-sheet [1]. These records have a temporal resolution of about 1–5 years from present to 90 kyrs B.P. This enables us to separate, by spectral fil-tering, dynamics of long timescales changes (climate) and short timescales (atmosphere/ocean fluctuations). The analysis of the isotope temperature proxy shows that the state of the fast timescale flow was different in the glacial climate in comparison to the present climate [2]. Figure 1 shows the $\delta^{18}O$ temperature proxy signal. (a) is the full signal while (b) is the 100 years running mean and (c) is the residual. The envelop of the residual is proportional to the degree of glaciation. This indicates that the atmosphere was in in a more stormy and turbulent state in the glacial climate.

A higher temporal resolution parameter is the content of dust in the ice-core [3]. Dust taken up from the continents is passively advected in the atmosphere and deposited with precipitation on the ice. The dust record is an indirect proxy for the state of the climatic system and wind strengths in the atmosphere.

Standard models (stochastic climate models [4]) of the climatic changes are that they are noise triggered, where the noise is the fast atmospheric fluctuations forcing the climate from one stable state to another. This type of dynamics is described by a Langevin equation, $dy = -(dU/dy)dt + \sigma dL$.

This model can be tested against the dust record. Here the requirement that the drift-term can be neglected in comparison to the diffusion-term from one measurement point to the next is fulfilled. It turns out that this

U. Frisch (ed.), Advances in Turbulence VII, 449–452.
© *1998 Kluwer Academic Publishers.*

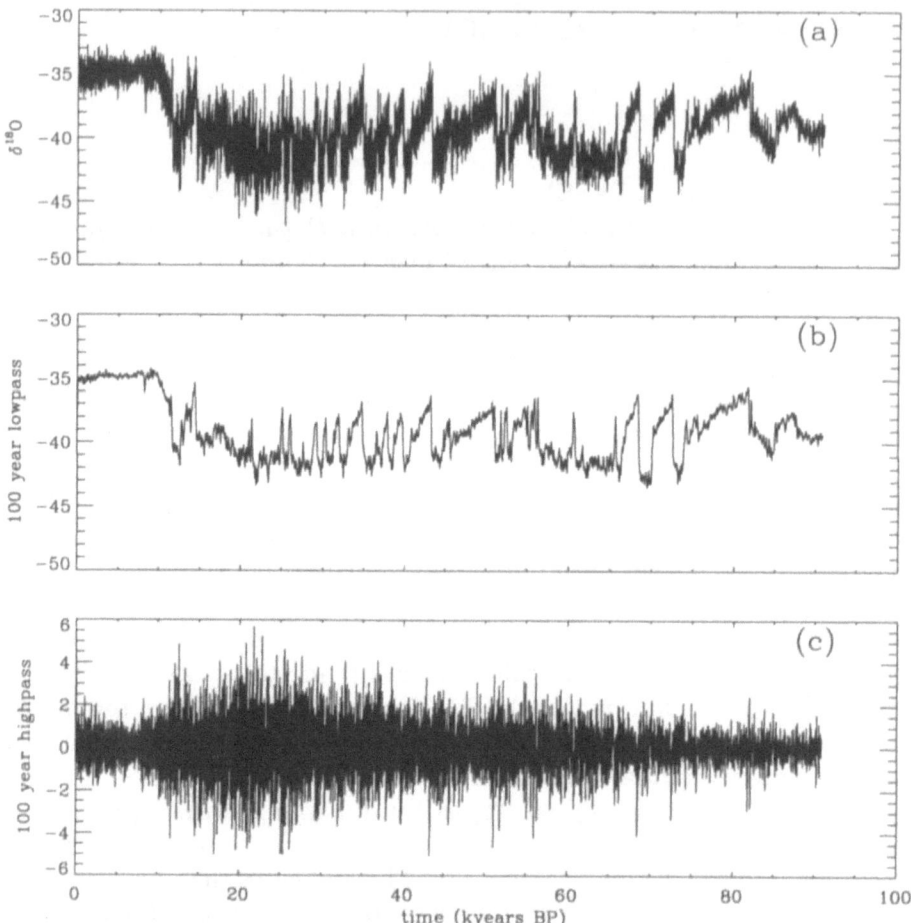

Figure 1. (a) The $\delta^{18}O$ in per mil deviation from standard mean ocean water. This is a paleo-temperature proxy. (b) is the 100 years running mean representing the slowly varying components of the climate. (c) is the residual ((a)–(b)) representing the fast components of the climate including the atmospheric temperature variations. The envelop indicates a more variable atmosphere in the glacial climate.

dynamics is consistent with the dust record. However, for this to be true, the noise is not gaussian white noise but a compound noise with an α-stable white noise component with $\alpha = 1.75$ [5]. The α-stable distributions are characterized by power-law tails, such that only moments of order less than α exists ($\langle |x|^{\beta} \rangle = \infty$ for $\beta \geq \alpha$). These distributions fulfill a generalized version of the central limit theorem, namely that sums of α-stable stochastic variables are again α-stable, with the same value of α.

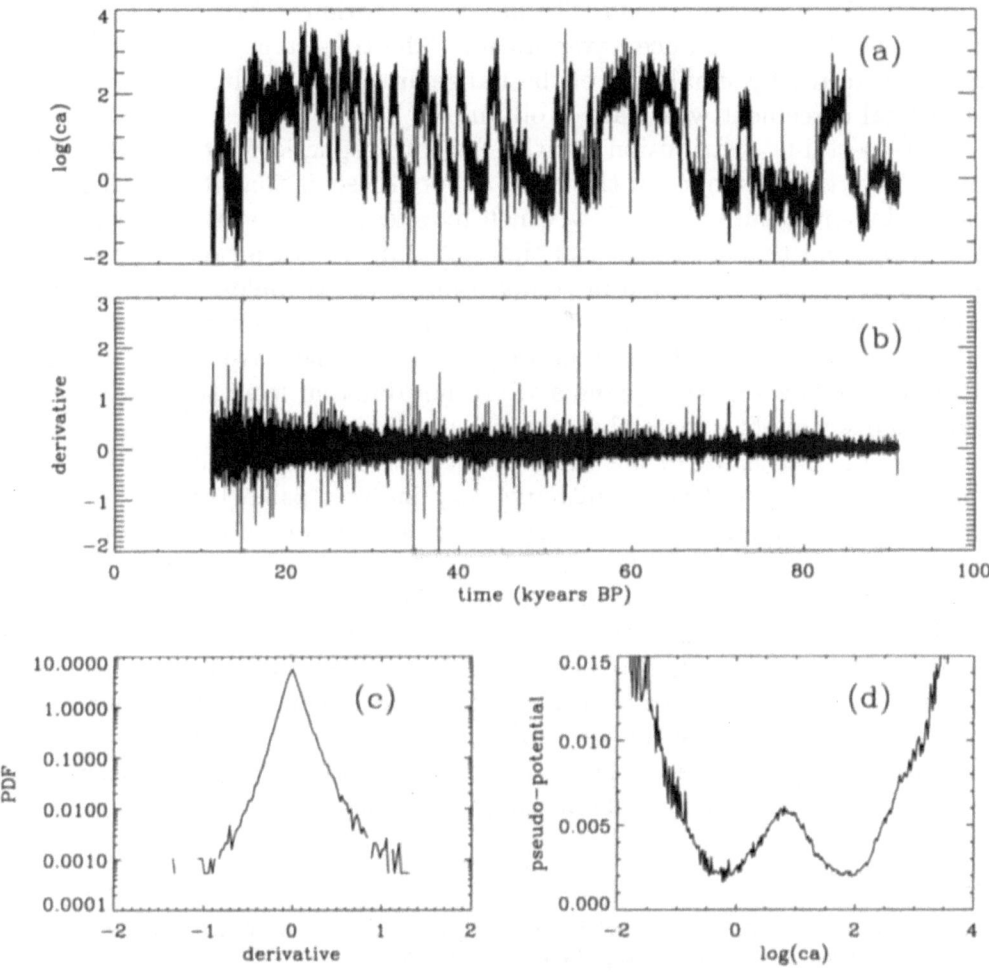

Figure 2.   (a) The logarithm of the calcium record from the GRIP ice-core. This record only covers the glacial period. The calcium originates from dust transported through the atmosphere and deposited with snow onto the ice-sheet. The resolution is roughly annual down to 90 kyr BP. (b) is the derivative of log(ca), represented by the annual increments. (c) is the probability density of (b), which is strongly intermittent. (d) is the climate pseudo-potential. The two climatic states represented by the minima are the full glacial state and the interstadials. They are clearly seen in figure 1 (b).

Figure 2 shows the log(ca) record and its derivative which approximates the noise component in the Langevin equation. Figure 2 (c) shows that this noise is very intermittent. The record has a bimodal distribution, with shifts between a cold glacial climate and warmer interstadials. Figure 2 (d) shows the climate pseudo-potential U. This is derived using the Fokker-Planck equation. The two stable states probably represents two states of the Atlantic ocean circulation. The ocean circulation is forced by the wind shear and the wind induced evaporation. The consistency of the analysis has been tested by simulation of the Langevin equation where an excellent statistical agreement with data is obtained.

The α-stable distribution for the fast atmospheric fluctuations originates from the dynamics of the flow. A hypothesis is that it is a result of the turbulent nature of the flow which manifests itself on these very long timescales. For this record the fast timescales are annual means. This is still an extremely long timescale in comparison to the turbulent flow. The direct measurements of the state of the atmosphere covers a period of roughly 100 years which is short on climatological time scales. We know that the atmospheric flow on other planets are rather different from the flow on the Earth, so we might expect that a different climate, like the glacial, could lead to a different state of the atmospheric flow. The only evidence we have for that is through these admittedly very indirect paleoclimatic records. It is therefore important to increase our understanding of the relationship between these records and the underlying climate dynamics.

If the interpretation presented in this analysis is correct it questions the validity of any of the present day climate models. All general circulation models, used for assessing climate change scenarios, are coarse resolution models for which all effects of turbulence have been smoothed out and parametrized by large scale eddy diffusion. No extreme event will ever be captured in these models. The findings here indicates that these extreme events actually can play a role in triggering climatic changes.

References

1. Dansgaard, W et. al. (1993) Evidence for general instability of past climate from a 250-kyr ice-core record. *Nature* **364**, pp. 218–220.
2. Ditlevsen P. D., Svensmark H. & Johnsen S. (1996) Contrasting atmospheric and climate dynamics of the last-glacial and Holocene periods. *Nature* **379**, pp. 810–812.
3. Fuhrer, K., Neftel, A., Anklin M. & Maggi, V. (1993) Continuous Measurements of hydrogen peroxide, formaldehyde, calcium and ammonium concentrations along the new GRIP ice core from Summit, central Greenland. *Atmos. Environ.* A, **27**, pp. 1873–1880.
4. Hasselman, K, Stochastic climate models (1976) *Tellus* **28**, pp. 473–485.
5. Ditlevsen P. D. (1998) Observation of α-stable noise induced millennial climate changes from an ice-core record. *to be published.*

INTERMITTENCY IN MHD FLOWS

S. GALTIER[1], T. GOMEZ[1,2], H. POLITANO[1] & A. POUQUET[1]
[1] *CNRS, UMR 6529, Observatoire de la Côte d'Azur, BP 4229,*
06304 Nice Cedex 4, France.
[2] *Laboratoire de Modélisation en Mécanique, Université de Paris*
VI, 4 Place Jussieu, 75252 Paris Cedex 5, France.

Magnetic fields are ubiquitous in the Universe, often coupled to turbulent motions that are compressible as well. Most models do not take compressibility into account except for those relying on a Burgers–like approach; restricting the variations to one space dimension, several extensions to MHD of the Burgers equation have been proposed with various degrees of complexity. When retaining only one component of the velocity u_x and of the magnetic field b_y [1], one in fact may recover two Burgers equations for $v^{\pm} = u_x \pm b_y$ with an extra dissipative term that couples them when the magnetic Prandtl number differs from unity.

When two components of the magnetic field are taken into account, the temporal evolution of the flow is slowed–down [2]. In the presence of external forcing (as would occur on the surface of the Sun because of foot–point motions of magnetic arches), a numerical integration of the equations using a pseudo–spectral code with periodic boundary conditions on a grid of $2,048$ points displays temporal intermittency as exemplified by the Joule dissipation $\mathcal{D}_M$. The temporal variation of the spatial maxima of the square magnetic current is shown in Figure 1 for a new run on $4,096$ points for over $1,000$ turn–over times based on the *r.m.s.* velocity. Histograms of the Joule dissipation display a power–law behavior [3] with $\mathcal{P}(\mathcal{D}_M) \sim \mathcal{D}_M^{-1.9}$, as observed in hard X–ray for the Sun; this sporadic dissipation may be responsible for the heating of the solar corona to more than 10^6 Kelvin. When now taking into account not only the maxima but *all* the localized peaks that form within the flow at lower intensity, and computing again the histogram of the spatio–temporal Joule dissipation, Figure 2 shows that, for weak events, a steeper spectrum emerges.

Spatial intermittency in astro–geophysical flows has been quantified in the solar wind [4], and detailed studies of the sun and the interstellar

U. Frisch (ed.), Advances in Turbulence VII, 453–456.

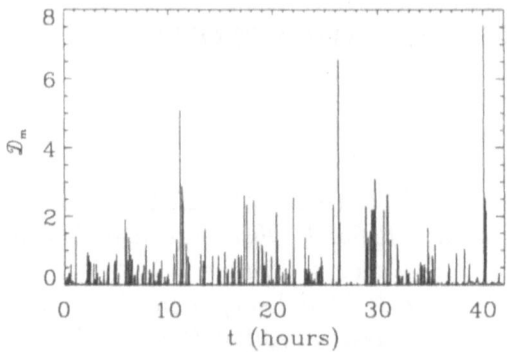

Figure 1.   Temporal variation of the spatial maxima of the Joule dissipation $\mathcal{D}_M$ for a 1D solar model.

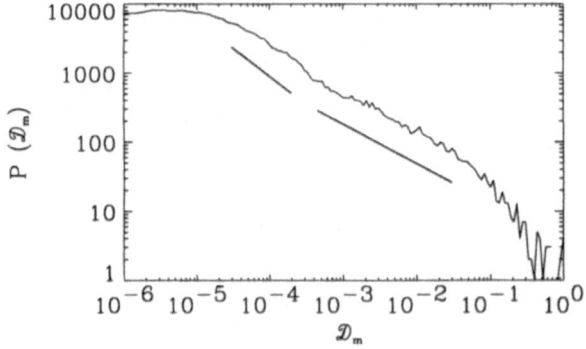

Figure 2.   Histogram of the spatio-temporal peaks of the Joule dissipation $\mathcal{D}_M$ in a 1D solar model. Straight lines with -1.81 and -1.57 slopes are also plotted.

medium are likely to give similar informations. A theoretical understanding of intermittency in such flows may be initialized with the derivation of the relationship equivalent to the "4/5" law of Kolmogorov [5] and its sequel [6], *but* for conducting fluids. The incompressible case in variable dimension is treated in [7] [8] without any assumption on the amount of correlation $\langle \mathbf{v} \cdot \mathbf{b} \rangle$ between the velocity $\mathbf{v}$ and the magnetic field $\mathbf{b}$. A von Kármán–Howarth equation for magnetohydrodynamic fluids is derived, the analysis of which shows that, as expected, the third–order cross correlations between the basic fields – velocity and magnetic fields, or the Elsässer variables $\mathbf{z}^{\pm} = \mathbf{v} \pm \mathbf{b}$ – play a fundamental role in the dynamical evolution of MHD flows. Furthermore, similarly to Yaglom for the case of the passive scalar [9], and defining the "Yaglom" fields $Y_p^{\pm} = \langle [\delta z_L^{\mp}(\mathbf{r})(\delta z_i^{\pm}(\mathbf{r}))^2]^{p/3} \rangle$, one obtains [8]

$$Y_3^{\pm} = \langle \delta z_L^{\mp}(\mathbf{r})(\delta z_i^{\pm}(\mathbf{r}))^2 \rangle = -\frac{4}{3}\epsilon^{\pm}r,$$

where summation over the repeated index i is understood, the subscript L denotes the longitudinal component of the fields, brackets indicate space–averaging and $\epsilon^{\pm}$ are the rates of the $z^{\pm}$ energy transfer to small scales. Using two–dimensional numerical simulations of incompressible MHD flows with periodic boundary conditions, a scaling study of the structure functions of the Elsässer variables indicates [10] that neither the She–Leveque model [11] (or SL) nor its extension to MHD [12] [13] works: MHD is more intermittent than neutral fluids, and dissipative structures are more complex, made up of filaments, sheets and ribbons.

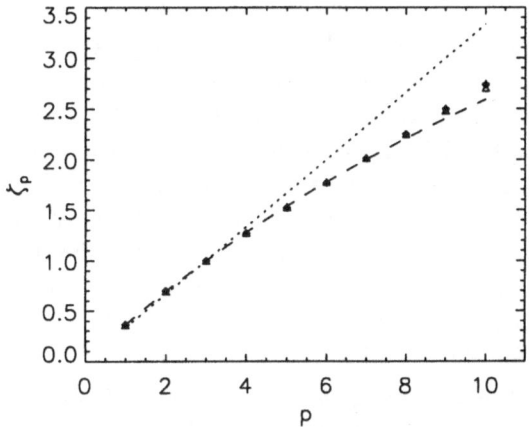

Figure 3. Anomalous exponents of structure functions for $Y_p^+ \sim r^{\zeta_p^+}$, computed with a 2D simulation of incompressible MHD turbulence; K41 law (dotted line) and SL model (dash). Different symbols refer to exponents obtained for various ranges of fit (see text).

On the other hand, when looking at the scaling of the fields $Y_p^{\pm} \sim r^{\zeta_p^{\pm}}$, one finds (using absolute values) that the fluid SL model [11] applies, as shown in Figure 3; the data stem from a 2D numerical simulation of incompressible MHD turbulence on a 512^2 grid, with 29 temporal snapshots accumulated, giving a total of $\sim 7 \times 10^6$ points. We also take advantage of the Extended Self–Similarity method [14] and, as suggested in [8], we make use of $Y_3^{\pm}$ instead of the distance r for computing $\zeta_p^{\pm}$; fits are done in the scale intervals $[0.04, 0.22]$, $[0.04, 0.14]$ and $[0.04, 0.09]$. This could be interpreted as the signature of advection. However, higher resolution runs must be performed before reaching a final conclusion as to the validity of the SL scaling for the Yaglom fields in MHD.

Note that the same SL–like scaling was found to apply for the passive scalar [15], when using the cross–correlator $Y_p^\theta = \langle[|\delta u_L(\mathbf{r})|(\delta\theta(\mathbf{r}))^2]^{p/3}\rangle \sim r^{\zeta_p^\theta}$ stemming from the Yaglom law [9].

Relationships similar to the 4/5 law of Kolmogorov can also be found in the 1D compressible case; extending this type of analysis to the simple configuration proposed in [2], one similarly finds

$$\langle(\delta u(r))^3\rangle + 3\langle\delta u(r)(\delta\mathbf{b}(r))^2\rangle = -12\epsilon^T r,$$

where ϵ^T is the mean rate of the total (kinetic plus magnetic) energy transfer. Furthermore, in this simple configuration, one straightforwardly obtains as well the equivalent scaling law for magnetic helicity $H^M = \langle\mathbf{a}\cdot\mathbf{b}\rangle$ where $\mathbf{b} = \nabla\times\mathbf{a}$, with $\mathbf{a}$ the magnetic potential. It reads

$$\langle\delta u(r)(\delta\mathbf{a}(r)\cdot\delta\mathbf{b}(r))\rangle - 2\langle(\mathbf{a}(x)\cdot\mathbf{b}(x))u(x+r)\rangle = -4\tilde{\epsilon}_m r,$$

where $\tilde{\epsilon}_m$ is the mean magnetic helicity transfer rate.

In conclusion, we have shown that simple MHD flows can reproduce statistically the observed histograms for solar flares. The intermittent behavior of such data led us to investigate intermittency properties of MHD flows in several cases. Generalizations of the "4/5" law to various cases for MHD have been obtained. Data analyzed in terms of the Yaglom fields seem to display a universal behavior compatible with the She–Leveque model [11], whereas the structure functions for the basic MHD fields differ from the neutral fluid scaling [10] in a way compatible with a phenomenology that incorporates Alfvén waves [16].

Numerical simulations were performed at IDRIS (Orsay) and on SIVAM (OCA); both centers are gratefully acknowledged, as well as CNRS-PNST.

References

1. Thomas J. 1968 *Phys. Fluids*, **11**, 1245.
2. Yanase S. 1997 *Phys. Plasmas*, **4**, 1010.
3. Galtier S. & Pouquet A. 1998 *Solar Phys.*, to appear.
4. Burlaga L. 1991 *J. Geophys. Res.*, **96**, 5847.
5. Kolmogorov A. 1941 *Dokl. Akad. Nauk SSSR*, **32**, 16.
6. Fulachier L. & Dumas R. 1976 *J. Fluid Mech.* **77**, 257.
7. Politano H. & Pouquet A. 1998 *Phys. Rev. E Rapid Comm.*, **57**, R21.
8. Politano H. & Pouquet A. 1998 *Geophys. Res. Lett.*, **25**, 273.
9. Yaglom A. 1949 *Dokl. Akad. Nauk SSSR* **69**, 743.
10. Politano H., Pouquet A. & Carbone V. 1998 submitted to *Europhys. Lett.*.
11. She Z.S. & Leveque E. 1994 *Phys. Rev. Lett.*, **72**, 336.
12. Grauer R., Krug J. & Marliani C. 1994 *Phys. Letters A*, **195**, 335.
13. Politano H. & Pouquet A. 1995 *Phys. Rev. E*, **52**, 636.
14. Benzi R. *et al.* 1993 *Phys. Rev. E*, **48**, R29.
15. Pinton J.F. *et al.* 1997 submitted to *Physica D*.
16. Iroshnikov P. 1963 *Sov. Astron.* **7**, 566; Kraichnan R. 1965 *Phys. Fluids* **8**, 1385.

INVERSE CASCADE AND ROSSBY WAVES IN THE KOLMO-GOROV FLOW ON THE BETA-PLANE.

B. VILLONE[1], B. LEGRAS[2] AND U. FRISCH[3]
[1] *Istituto di Cosmogeofisica, CNR, C. Fiume 4, 10133 Torino, Italy.*
[2] *CNRS/LMD/ENS, 24 rue Lhomond, 75231 Paris Cedex 5, France.*
[3] *CNRS, Observatoire de Nice, BP 4229, 06304 Nice Cedex 4, France.*

Large-scale geophysical flows are subject to the competing effects of quasi-two-dimensional turbulence and Rossby waves. It is well known from phenomenological arguments and numerical simulations that the inverse cascade which characterizes the large-scale dynamics of two-dimensional turbulence can be halted by Rossby wave dispersion. A particular example for which this effect is amenable to a detailed numerical and theoretical understanding is the supercritical large-scale dynamics of the Kolmogorov flow on the β-plane. This flow is governed by the one-dimensional "β-Cahn–Hilliard" equation [1], obtained by multiscale technique, with cubic non linearity

$$\partial_t v(x,t) = \partial_x \left(\left(\lambda_1 v^2 - \lambda_2 \right) \partial_x v \right) - \lambda_3 \partial_x^4 v - \beta \partial_x^{-1} v \tag{1}$$

In the absence of β, the solutions to this equation live essentially within a slow manifold of soliton-like solutions characterized by alternating kinks and antikinks. With periodic boundary conditions of period L, fixed points are obtained with N pairs of regularly spaced kinks and antikinks. These fixed points are unstable saddle points of a Lyapunov functional excepted for the gravest mode $N = 1$ which is a stable absolute minimum. The temporal evolution is a succession of annihilations of kinks and antikinks, leading eventually to the gravest mode (see, e.g., Ref. [2]). The uniqueness of the final solution is ensured by the existence of a Lyapunov functional. Although this dynamics borrows most of its character from the heteroclinic connection between unstable fixed points, the trajectory in phase-space does not generally proceed from the vicinity of one fixed point to

U. Frisch (ed.), Advances in Turbulence VII, 457–460.

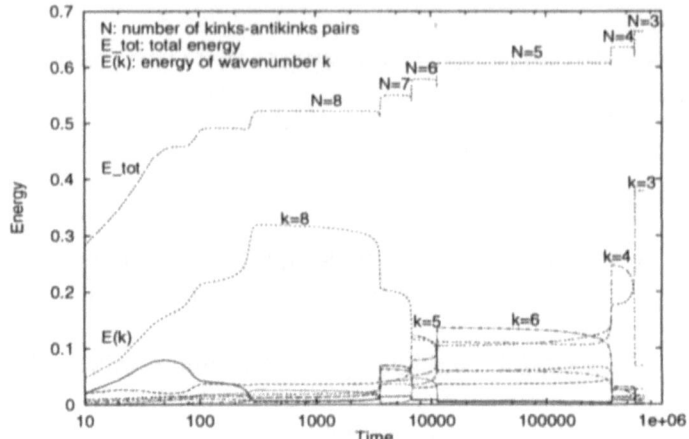

Figure 1.   Simulation of the Cahn–Hilliard equation with $\beta = 0$. An inverse cascade of kink-antikink pairs is observed until $N = 1$ is reached (not shown in the figure). Energy jumps during annihilations are constant, up to an exponentially small deviation.

another. Fast jumps from one branch of the slow manifold to the next may occur away from the fixed points. Indeed, it is observed that the dominant wavenumber does not necessarily decrease monotonically with time as the number of kinks and antikinks decrease (see Figure 1).

It is remarkable that, excepting the short annihilation episodes, the motion of kinks can be described by a simple ODE [3]

$$\mathcal{A}\dot{x}_j = \left(e^{-s(x_j - x_{j-1})} - e^{-s(x_{j+1} - x_j)}\right) \qquad (2)$$

where $s = \sqrt{\lambda_2/2\lambda_3}$, x_j is the location of the j-th kink and $\mathcal{A}$ is a Toeplitz matrix. The exponential coupling between adjacent kinks explains why the duration of the plateaus increases, on the average, exponentially as N decreases.

In the presence of β-effect, the dispersive action of the waves modifies the cascade. For large values of β, only resonant wave interactions survive. A new equation is obtained in the limit $\beta \to \infty$. This equation still has a Lyapunov functional but possesses multiple attracting steady-states, each with a single mode excited. A detailed presentation of this regime may be found in Ref. [1].

For smaller values of β, the solution retains the character of kink dynamics with superimposed propagating waves. The main result is that the inverse cascade is interrupted at a wavenumber which increases with β (see Figure 2). Notice the difference in the physical space between the asymptotic state for this case and the state having the same number of kink-antikink pairs for the case $\beta = 0$ in the Figure 3. The equation does not have a Lyapunov functional. In the small β limit, the solution can be studied by singular perturbation techniques. The fixed points of the Cahn–Hilliard equation are modified and stabilized by β-effect. The stabilization occurs at order β^2. Since the stabilizing term has also an algebraic Λ^4 dependence on the wavelength $\Lambda = L/N$, it can compensate the destabilizing coupling of the kinks which decreases exponentially with Λ. More precisely, we find, at leading order in Λ, that the period-N fixed point is marginally stable for

$$\beta_c^2 = 35,389,440 \, s^5 \lambda_3^2 \frac{e^{-s\Lambda}}{\Lambda^5} \frac{1}{4 + 9\tan^2 \frac{\pi}{2N}} . \tag{3}$$

It is not easy to verify this relation as the eigenvalues separate widely and the numerics become very stiff in the large Λ limit. Table 1 compares (3) with direct calculations of the solutions and their stability within an interval of length L=78.4 (10 unstable modes when $\beta = 0$). The calculations are done using 128 modes within the interval and a QR algorithm. The agreement weakens as N increases and the separation of kinks and antikinks get smaller but the order of magnitude is correctly predicted. On the other hand, the final states of numerical integrations starting from random initial conditions always fall on a wavenumber which is found stable with (3) but generally with a much larger value of β, an indication that the basin of attraction of the solutions does not span at once a large fraction of the slow manifold. In a recent work, Manfroi and Young [4] studied a closely related problem in which stabilization is provided by a friction coefficient μ. In this case, a Lyapunov functional is preserved which imposes also that marginal μ scales as Λ^{-3}. Therefore, the stabilizing effect of the β term is much more effective owing to the exponential dependence in (3). Notice that this scaling or that of Manfroi and Young differ from the phenomenological prediction of Rhines [5] that β_c scales as Λ^{-2}.

TABLE 1. Comparison of (3) with direct calculation of the stability.

N	direct	β_c
2	2.32×10^{-6}	1.45×10^{-6}
3	4.96×10^{-4}	2.20×10^{-4}
4	1.07×10^{-2}	3.2×10^{-3}
5	8.96×10^{-2}	1.8×10^{-2}

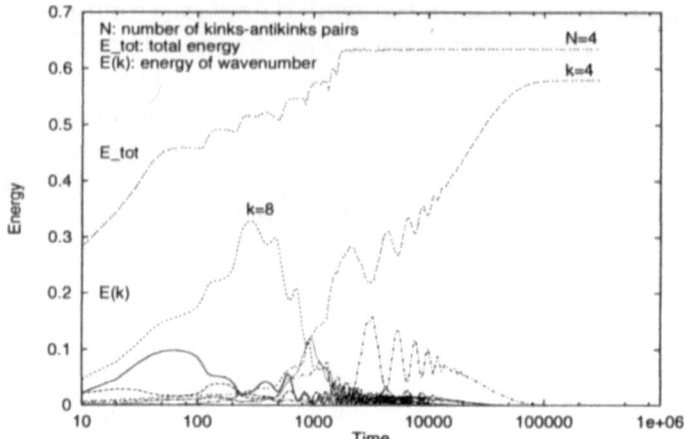

Figure 2.   Simulation of the β-Cahn–Hilliard equation with $\beta = 10^{-3}$. Propagating Rossby waves accelerate the cascade until it stops with four kink-antikink pairs. However the transients die out very slowly after the energy has reached its final plateau.

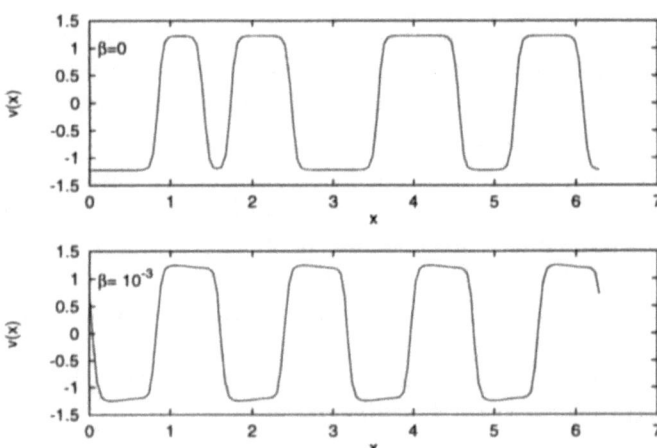

Figure 3.   Solutions with four kink-antikink pairs : for $\beta = 0$ in a slowly evolving state, for $\beta = 10^{-3}$ in the asymptotic state exhibiting distorted plateaus between kinks.

References

1. Frisch, U., Legras, B. and Villone, B. (1996) Large-scale Kolmogorov flow on the beta-plane and resonant wave interactions, *Physica* D**96**, 36–56.
2. She, Z.S. (1987) Metastability and vortex pairing in the Kolmogorov flow, *Phys. Lett.* A**124**, 161–164.
3. Kawasaki, K. & Ohta, T. (1982) Kink dynamics in one-dimensional nonlinear systems, *Physica* A**116**, 573–593.
4. Manfroi, A.J. & Young, W.R. (1997) Slow evolution of zonal jets on the beta-plane, *J. Atmos. Sci.*, in press.
5. Rhines, P. (1975) Waves and turbulence on a β-plane, *J. Fluid Mech.* **69**, 417–443.

THE TURBULENT REGIMES OF RAYLEIGH-BENARD CONVECTION

X. CHAVANNE, F. CHILLA[1], B. CASTAING,
B. CHABAUD, B. HEBRAL AND P. ROCHE

CNRS, Centre de Recherches sur les Très Basses Températures,
Laboratoire associé à l'Université Joseph Fourier,
BP 166, 38042 Grenoble-Cedex 9, France

[1] *Present address : Ecole Normale Supérieure de Lyon,*
46 Allée d'Italie, 69364 Lyon-Cedex 7, France

A Rayleigh-Bénard convection experiment is performed with Helium at low temperature, as the working one-phase fluid [1]. Helium is closed in a cylindrical cell 20 cm high and 10 cm in diameter. Three experimental parameters, independently adjustable, and accurately measured for each fixed condition, allow to study the Rayleigh Bénard problem from the conductive regime up to the turbulent convective one. These parameters are : i) the average fluid density in the cell, measured with a 0.7% precision (from 8 g/m^3 up to 135 kg/m^3 as well as close to the critical point : 69.6 kg/m^3, Tc = 5.1953 K, Pc = 2.275 bars) ; ii) the mean absolute temperature (from 2 K up to 8 K) controlled within a 0.1 mK stability and measured with a 2 mK absolute accuracy ; iii) the temperature gradient along the cell height, ΔT (from less than 1 mK up to 5 K), adjusted via the heating power applied on the bottom cell plate, and measured with a 20 mK resolution, 1% uncertainty thermocouple. Two thermometers of 200 μm size and 2 mm vertically apart of each other are located in the cell, at half height and half way between the centre and the wall.

We focused on the convection at high Rayleigh numbers (from Ra = 2.10^7 up to Ra = 2.10^{14} within Boussinesq conditions). Two kinds of studies were performed [2] :

- From calorimetric measurements and physical quantities relevant for convection, the dimensionless numbers of the convection are calculated : the Rayleigh number Ra, the Nusselt number Nu, and the Prandtl number Pr. Experimental points are plotted on Fig. 1 which represents the NuRa diagram.

U. Frisch (ed.), Advances in Turbulence VII, 461–464.
© 1998 *Kluwer Academic Publishers.*

- From the two local detectors, the cross spectrum of the two signals is calculated. The phase evidences a characteristic time, related to the time lag of thermal coherent structures between the thermometers. Hence a velocity of the flow in the cell, and the corresponding Reynolds number (Re), are deduced.

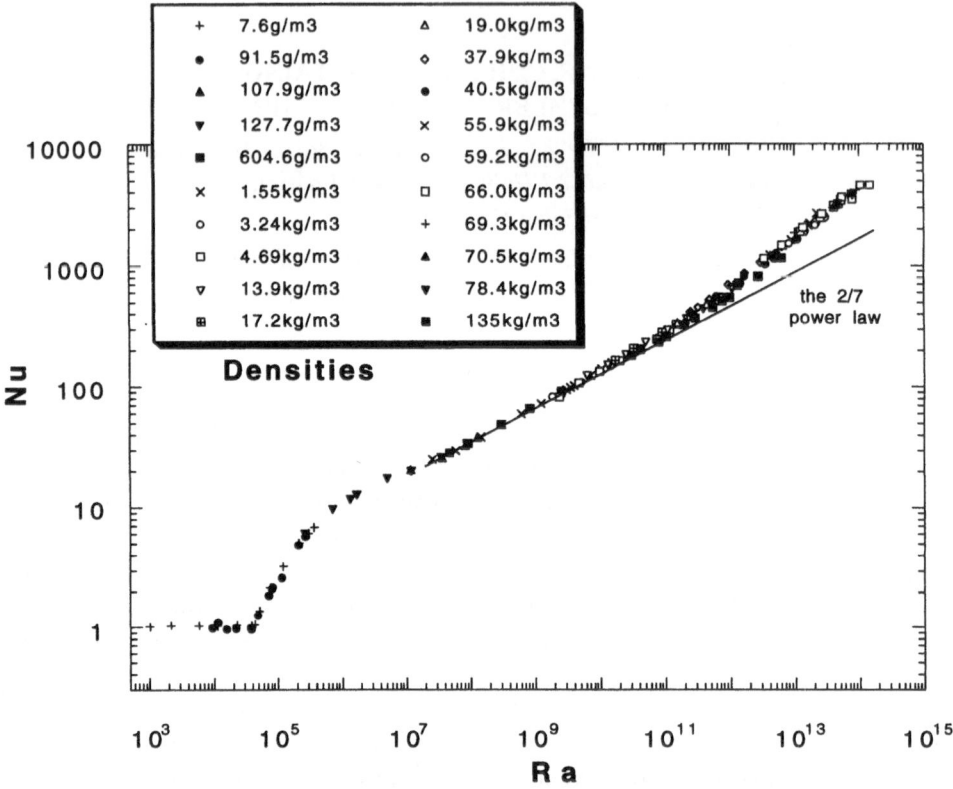

Figure 1. Experimental points in the NuRa diagram for the different densities, and taking into account the Boussinesq conditions.

On the NuRa diagram we have represented the theoretical 2/7 power law which fits well the experimental point from Ra ≈ 2.10³ up to about Ra ≈ 10¹¹. This behavior has already been noticed in previous works (for instance see Ref. [3]). But the departure from this 2/7 regime is observed for the first time. The new regime which sets up is characterized by a more efficient heat transfer : a power law may fit the new trend of the NuRa points with a 0.39 exponent.

With Re, deduced from local measurements, the flow behavior can be studied in relation with Ra in the central region, as well as near the wall. No Re change is noticed over the whole working range in the centre, whereas the estimated viscous dissipation due to the flow friction on the wall, exhibits a laminar turbulent transition for Ra about 10^{11}.

In short, both studies evidence a transition in the turbulent regime of convection, taking place near the wall at roughly Ra = 10^{11}. The two characteristics of the new regime - enhanced heat transfer and turbulent boundary layer -, suit well the ultimate regime of convection predicted by R. Kraichnan [4]. Having in mind the forced convection heat flow accross a turbulent boundary layer, the following relation should hold [5] :

$$Pr\ Re^* = A\ Nu\ (\ln Re^* + f(Pr))$$

where A is a constant, and Re = Re^* (2.5 ln Re^* + 5.2). Figure 2 shows that $f(Pr)$ dominates on ln Re^* as expected only for large values of Pr and that a power law well fits its variation.

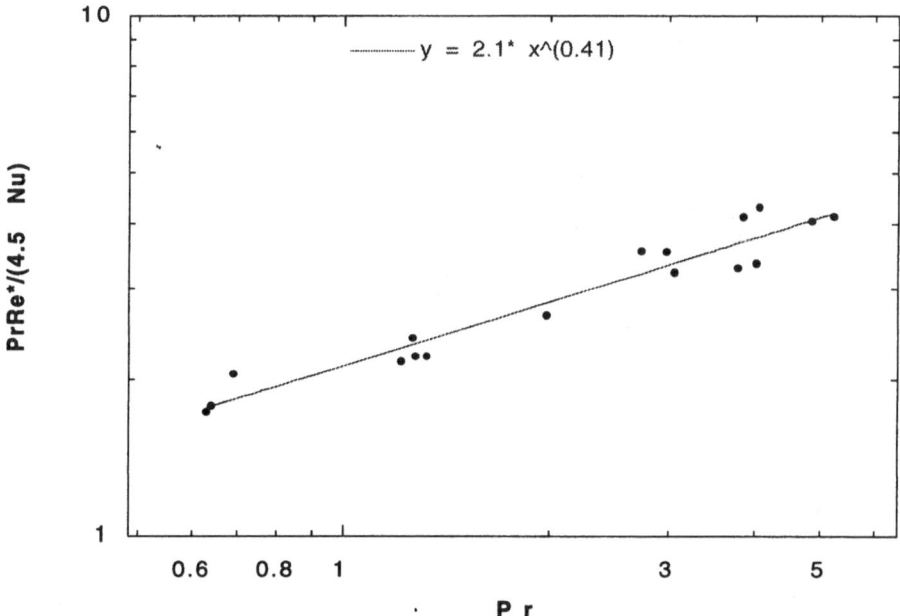

Figure 2. $\frac{Pr\ Re^*}{4.5\ Nu}$ versus Pr. A power law with 0.41 exponent well fits for Re > 30 000.

From calorimetric measurements, other studies are carried out. In Fig. 1, the experimental uncertainties on Ra and Nu are roughly twice smaller than the symboles size. The points scattering thus corresponds to a real effect of convection, that is the Nu dependence on Pr. Indeed in the new regime Pr may range from 0.7 to 30. A weak increase of Nu with Pr is noticed (about 15% as Pr rises by a factor of 10). Non Boussinesq effects at large ΔT have also been evidenced and studied on the NuRa curve, both with He as a dilute gas and near the critical point.

References

1. Chavanne X, Chillà F., Chabaud B., Castaing B., Chaussy J. and Hébral B. High Rayleigh number convection with gaseous helium at low temperature (1996) *J. Low Temp. Phys.* **104**, 109.
2. Chavanne X., Chillà F., Castaing B., Hébral B., Chabaud B. and Chaussy J. Observation of the ultimate regime in Rayleigh-Bénard convection (1997) *Phys. Rev. Letters* **79**, 3648.
3. Heslot F., Castaing B. and Libchaber A. (1987) *Phys. Rev. A* **36**, 5810.
4. Kraichnan R. H. (1962) *Phys. Fluids* **5**, 1374.
5. Landau L. and Lifchitz E. (1989) Mécanique des fluides, Ed. Mir.

NON-GAUSSIAN PROBABILITY DENSITY FUNCTIONS OF SMALL-SCALE FLUCTUATIONS IN A STABLY STRATIFIED MEDIUM

J.-R. ALISSE AND C. SIDI

Service d'Aéronomie du CNRS

BP3 Fort de Verrières F-91371 Verrières-le-Buisson

We present experimental results relative to the PDFs of temperature and velocity small-scale fluctuations in a stably stratified medium, namely the stable and clear free atmosphere.

There, vertical profiles of temperature and velocities small-scale fluctuations usually show a characteristic pattern of alternating agitated and calm layers, in accordance with the well-known "blini" structure of turbulent patches within the stably stratified atmosphere. Within agitated layers, kinetic energy (E_K) and available potential energy (E_P) spectra versus vertical wavenumber, m, often scale like $m^{-5/3}$ in the short wavelengths band $O(1 - 10 \text{ m})$. This scaling and apparent isotropy of the velocity fluctuations, which show nearly equal variances along different directions, suggest a turbulent close-to-inertial subrange, *à la Kolmogorov*. We recall that in the stably stratified atmosphere, E_P is proportional to temperature variance. Calm layers also evidence random fluctuating fields of atmospheric variables in the same wavelengths band. Besides apparent anisotropy of the velocity fluctuations variances, these fields often show energies spectra scaling like m^{-3}, down to the instrumental noise level. We study the PDFs of these small-scale fluctuating fields in both dynamical regimes. Our database consists of vertical profiles of atmospheric variables obtained by a balloon-borne instrumentation along four stratospheric balloons flights, from the clouds top up to $22 - 27$ km and two corresponding descents down to the tropopause. These flights occured during a field experiment described in Ref. [1] through very different meteorological conditions. Instrumentation devoted to turbulence measurements included high-resolution cold wire thermometers and a two-axis ionic anemometer, following the "differential sounding" technique (see Ref.[2]). Dimensions of sensors (a few cms) and effective vertical resolution of the measurements (from 0.1 to 0.2 m) are much larger than the dissipative scales.

U. Frisch (ed.), Advances in Turbulence VII, 465–468.

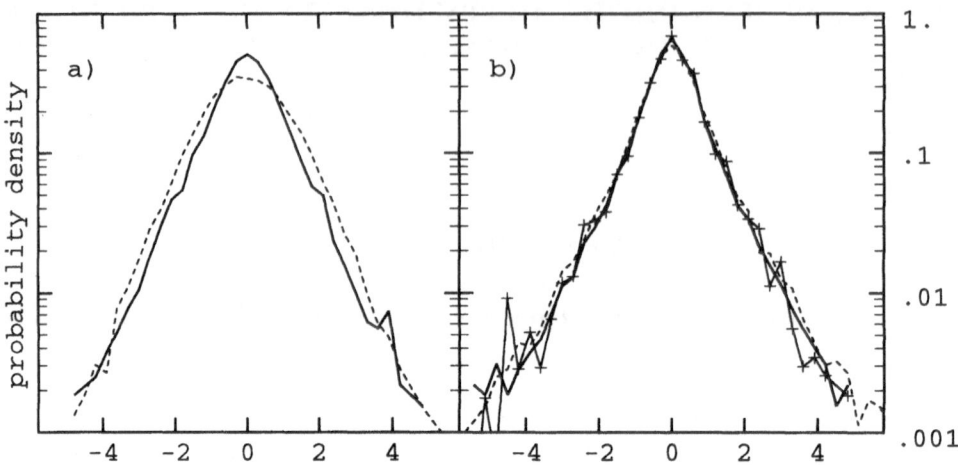

Figure 1. Synthetic PDF of normalized fluctuations of a) horizontal velocities, and b) temperature. Dark continuous lines correspond to data within turbulent layers, dashed ones to data within calm layers. Bin width is 0.3σ and the minimum amount of occurences is 5. Third curves (light lines and markers) is the reconstructed PDF using equation 1.

Hereafter, we consider band-filtered fluctuations in the spectral band [2 m-20 m]. The 2 m bound is chosen in order to reduce the noise contributions to the data statistics whilst the 20 m one corresponds to an approximate experimental upper limit of the $m^{-5/3}$ domains, when observed in the stratosphere. Thus, comparisons of the two dynamical regimes statistical properties are made in one and the same wavelengths band.

We showed in Ref. [3] that, to get reliable estimations of statistical parameters from vertical profiles of fluctuations in that spectral band, one has to consider homogeneous layers with a minimum depth of 200 m. We hence delineated 48 homogeneous layers, either turbulent (with E_K and E_P spectra scaling like $m^{-5/3}$) or calm (spectra scaling like m^{-3}). We thereupon compared the cumulative distribution functions (CDFs) of normalized fluctuations of temperature T, horizontal velocity U and vertical velocity W of each layer with all the others, using the Kolmogorov-Smirnov (KS) test. KS tests showed that nearly all the PDFs pertaining to one variable and one dynamical regime (but for W fluctuations within calm layers) do not significantly differ one from each others. These similarities allowed us to build up synthetic PDFs corresponding to collected normalized fluctuations. While horizontal velocities synthetic PDFs significantly differ (Fig. 1.a), temperature synthetic PDFs appear very close each other (Fig. 1.b), suggesting independence of T distributions upon the dynamical regime, and thus, upon the velocities PDFs (Ref. [3]). While turbulent W fluctuations PDFs also collapse towards a single PDF, nearly identical to the turbulent U fluctuations synthetic PDF, no conclusive results have been obtained as regards W fluctuations within calm layers. All these estimated PDFs are non-gaussian, showing extended close-to-exponential tails.

To further investigate the non-gaussian shape, we now turn to the analysis of the conditional means, as introduced by Ching in Refs. [4] and [5]. This author showed that the PDF of a homogeneous random field may be deduced fom the conditional means of the laplacian and square-gradient of the fluctuating field. Owing to the close-to-isotropy evidences within turbulent layers, these conditional means are proportional to those obtained when replacing the vector ∇ by the operator ∂_z. Using the same notations as in [4], we write:

$$P(x) = \frac{C}{\langle |\partial_z X|^2 | x \rangle} \exp \left(\int_0^x \frac{r(x')}{q(x')} dx' \right) \tag{1}$$

$$r(x) = \frac{\langle \partial_z^2 X | x \rangle}{\langle |\partial_z X|^2 \rangle} \tag{2}$$

$$q(x) = \frac{\langle |\partial_z X|^2 | x \rangle}{\langle |\partial_z X|^2 \rangle} \tag{3}$$

Here, X is the normalized fluctuation with value x. Notice that r, q and P basically depend upon ratios, thus unchanged as long as the above-mentionned proportionality holds. Then, using (1), we may "rebuild" the PDF of any homogeneous and isotropic field. As an example, we consider the PDFs of T fluctuations. The reconstructed PDF fits quite well with the original one (Fig 1.b). Some information may also be drawn from the experimental conditional means, $r(x)$ and $q(x)$, plotted in Fig. 2. Mi and Antonia (Ref. [6]) argued that the relation $r(x) = -x$ is a general property of all stationary process, while, in Ref.[5], Ching and Tsang showed that there may be some odd-power corrections to this relation. A weighted polynomial fit upon experimental $r(x)$ lead to a quasi-linear fit with slope -0.96. The present data set is too short to allow reliable estimations of any deviations from the linear prediction. Our results seem hence to fit quite well with the theoretical predictions of Mi and Antonia. As regards $q(x)$, it was argued in Ref. [7], that the non-constancy of $q(x)$ (constant $q(x)$ would lead to gaussian PDFs) is linked with intermittency, because the involved gradient terms are related to the dissipation. We must nevertheless be cautious in that way, since we are dealing, here, with band-filtered data in a spectral band well above the dissipative range, and hence have no access to the dissipative gradients. As a matter of fact, the band-filtered signals have an intermittent appearance, in that sense that local mean am-

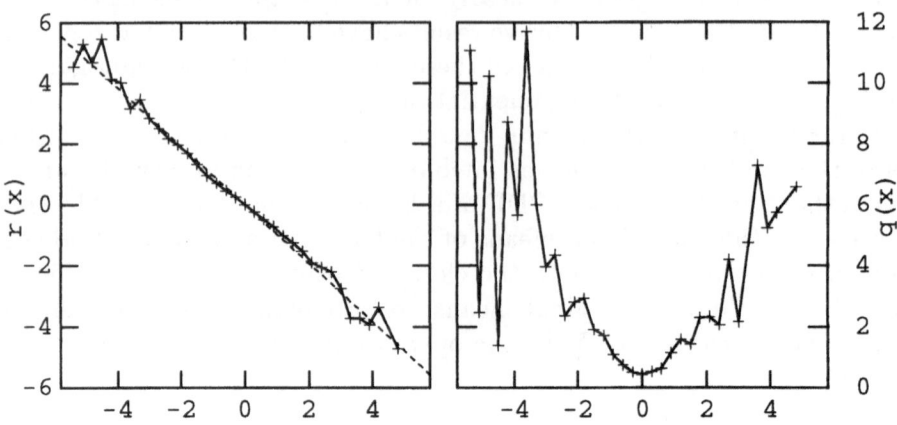

Figure 2. Conditional means $r(x)$ and $q(x)$ as a function of x for the temperature fluctuations of the turbulent layers.

plitudes may strongly differ from the overall mean. It is noteworthy that this intermittency of the signal leads to a non-constant $q(x)$, perhaps related to the dissipative processes. As a matter of fact, the shape of the PDF is mainly dependent upon the behaviour of the function $q(x)$. Further studies are needed to understand the physical meaning of $q(x)$, i.e. of the $|\nabla X|^2$ statistics, when considering band-filtered fluctuations in the inertial subrange.

References

1. Dalaudier, F., Sidi, C., Crochet, M. and Vernin, J. (1994) Direct evidence of "sheets" in the atmospheric temperature field, *J. Atmos. Sci.*, **Vol. 51**, pp. 237-248.
2. Barat, J. (1982) Some characteristics of clear-air turbulence in the middle atmosphere, *J. Atmos. Sci.*, **Vol. 39**, pp. 2553-2564.
3. Alisse, J.-R. and Sidi, C. (1998) Experimental probability density functions of small-scale fluctuations in a stably stratified medium, submitted to *J. Fluid Mech.*
4. Ching, E.S. (1996) General formula for stationary or statistically homogeneous probability density functions, *Phys. Rev. E*, **Vol. 53**, pp. 5899–5903.
5. Ching, E.S. and Tsang, Y.K. (1997) Passive scalar conditional statistics in a model of random advection, *Phys. Fluids*, **Vol. 9**, pp. 1353–1361.
6. Mi, J. and Antonia, R.A. (1995) General relation for stationary probability density functions, *Phys. Rev. E*, **Vol. 51**, pp. 4466-4468.
7. Yakhot, V. (1989) Probability distributions in high-Rayleigh-Bénard convection, *Phys. Rev. Lett.*, **Vol. 63**, pp. 1965–1967.

MULTIPLE SCALE DYNAMO

C.NARTEAU[1], J.L. LE MOUËL[1], M. SHNIRMAN[3], C.J. ALLÈGRE[1,2]
[1] *Institut de Physique du Globe de Paris, 4 Place Jussieu,*
75252 Paris Cedex 05, France.
[2] *Université Paris VII, 2 Place Jussieu, 75252 Paris Cedex 05,*
France.
[3] *International Institute of earthquake Prediction Theory and*
Mathematical Geophysics, Warshavs Koye sh. 79, Kor. 2, Moscow
113556, Russia.

The dynamo process in the Earth outer core is the most convincing theory to explain the origin and evolution of the geomagnetic field. This process relies on the one hand on an equilibrium between magnetic and Coriolis forces and, on the other hand on an equilibrium between induction and diffusion of the magnetic field. Let us recall some physical properties of the Earth's outer core : its viscosity ν is of the order of $10^{-6}m^2.s^{-1}$, the fluid velocity u is of the order of $10^{-4}m.s^{-1}$ and the angular velocity Ω is $7.3 rad.s^{-1}$. Among other dimensionless numbers, the system is characterized by a Reynold's number of the order of 10^8, an Ekman number ($E = \nu/2\Omega r^2$; r is the core radius) of the order of 10^{-15}, and a Prandtl magnetic number ($P_m = \nu/\eta$; η is the magnetic diffusivity) of the order of 10^{-5} [1]. Consequently, a turbulent flow is expected and it may be premature to describe it by a global numerical computation [2].

In general, the possibility for a small scale turbulence to generate of a large scale magnetic field has been estimated through a mean field theory. This mechanism is characterized by a scalar field α within the core which is given a priori in so-called nearly axisymmetrical dynamo models. We consider a different approach. The small scale turbulence is replaced by a fully developed turbulence. The fully developed turbulence produces a multiscale helical motion which generates, in presence of a large scale primary field, a secondary magnetic field of large scale.

More specifically, adopting a rather phenomenological point of view, we consider a hierarchical system made of embedded cyclones of increasing dimensions, untill the largest whose size is comparable with the size of the core. At each level of the hierarchical system, each cell can contain a left-

U. Frisch (ed.), Advances in Turbulence VII, 469–470.

handed or a right handed cyclone with a certain probability. We assume that energy sources (gravitational energy due to latent heat of crystallization or thermal convection) of the dynamo generate a random helicity flux in our hierarchical system. After introducing some simple rules of scale transfer, we describe shematically how electromagnetic and hydrodynamic forces act together to create or destroy cyclones; for example, we suppose that if two cyclones generate parallel electric currents (case of helicities of same sign), they attract each other and may coalesce to build a cyclone of upper dimension. Inversely, a cyclone of higher level may disappear and generate cyclones of lesser level. Basically we have a two-sided (direct and inverse) cascade of helicity. The magnetic fields act in turn on the helical motion (Lenz's law), which generates a feedback mechanism.

This effect, combined with the effect of a differential rotation, can make a dynamo, let say an $\alpha'\omega$ dynamo. An α'^2 dynamo can also work without the help of a differential rotation. We study different loops in which the poloïdal and toroïdal fields are built one from the other.

The amount of large scale symmetry breaking in the system is characterized by the constant polarity intervals durations and the intensity of the generated secondary magnetic field. The reversals frequency appears to depend on the efficiency of the cascading processes. The model displays the well-known features of the geomagnetic field : constant polarity intervals, excursions, reversals, secular variation. The duration/frequencies diagram can be made similar to the observed one.

References

1. Dormy, E. (1997) Modélisation Numérique de la Dynamo Terrestre, *I.P.G.P.*, Phd Thesis.
2. Glatzmaier, G.A., Roberts, P.H., 1995, A three-dimensional self-consistent computer simulation of a geomagnetic field reversal, *Nature*, **77**, pp. 203–209.
3. Krause F. and Rädler K.H. (1980) Mean-field magnetohydrodynamics and dynamo Theory, *Pergamon Press*, Great Britain.
4. Le Mouël J.L., C.J. Allègre, C. Narteau (1997) Multiple Scale Dynamo, *Proc. Natl. Acad. Sci. USA*, **94**, pp. 5510–5114.

VORTEX STRUCTURE OF A BREAKING WAVE IN STRATIFIED FLOW OVER AN OBSTACLE

O. EIFF[1], P. BONNETON[2]
[1] *Centre National de Recherches Météorologiques de Toulouse, Météo-France, 42 av G. Coriolis, 31057 Toulouse, France.*
[2] *Département de Géologie et Océanographie, Université Bordeaux 1, av des Facultés, 33405 Talence, France.*

1. Introduction

Under stably stratified atmospheric conditions, standing internal waves are generated when the flow passes over an obstacle such as a mountain. When the stratification is sufficiently strong, these waves can break, leading to a localised turbulent zone. Although the occurrence of mountain-wave breaking has been studied extensively [e.g., Smith (1985) and Castro & Snyder (1993)], only Laprise & Peltier (1989a, 1989b) have investigated the breaking mechanism, using two-dimensional linear stability analysis and numerical simulation. In slightly different wave breaking situations, however, the wave breaking was essentially found to be dominated by a three-dimensional instability mechanism with the formation of longitudinal vortices [Fritts & Isler (1994)]. In the case of breaking mountain waves, neither a three-dimensional instability analysis nor examination of the possible evolving vortex structures has been undertaken so far. It is the objective of this study to experimentally examine the underlying vortex structures in the mountain-wave breaking region.

2. Experimental Set-up

Two-dimensional Gaussian-shaped obstacles of height H were towed at uniform velocity (U) in linearly stratified saline solutions of Brunt-Väisälä frequency (N) of approximately 1 rad/sec. The Froude number $(F = U/NH = 0.6)$ was well below the critical value at which wave breaking occurs. The Reynolds number $(Re = UH/\nu)$ corresponding to these flow configurations was approximately 200.

U. Frisch (ed.), Advances in Turbulence VII, 471–474.

Cross-sectional planes of the flow where examined with particle tracking as well as fluorescent-dye imaging methods to reveal the topological features of the breaking-wave zone.

3. Results and Discussion

The particle-tracking results reveal that, even under controlled laboratory conditions, the topology of the breaking wave zone is unstable. In a vertical $x - z$ plane one can occasionally identify two counter-rotating foci as depicted in figure 1(a), but this topology does not necessarily subsist for long times (usually less than $Nt = 20$). Surprisingly, a topology with no critical points appears instead [figure 1(b)].

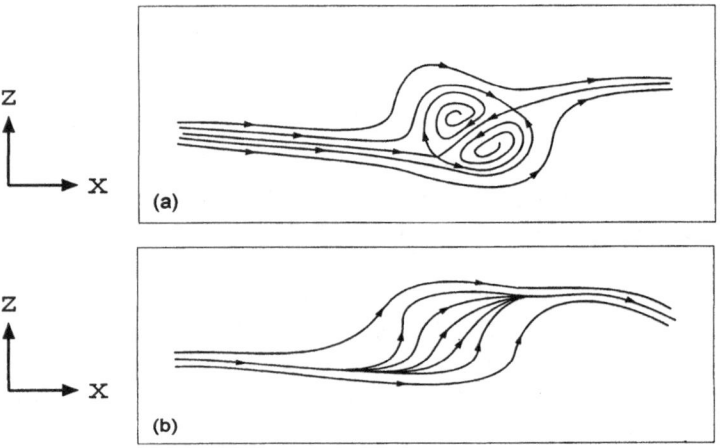

Figure 1. Sectional streamlines deduced from particle-tracking results in a vertical plane in the breaking wave zone. Two different topological patterns are observed: (a) two counter-rotating focal points with two saddle points; (b) no focal or saddle points.

Numerical investigations of wave breaking in critical levels by Fritts & Isler (1994) suggest that the flow is organised into counter-rotating streamwise vortices, i.e., with vorticity axes oriented in the streamwise (x) direction. The counter-rotating foci as shown in figure 1(a), on the other hand, reveal vortices aligned in the transverse (y) direction. Although the disappearance of these foci as shown in figure 1(b) could be explained by a relaminarisation of quasi two-dimensional transverse vortices, the possibility that it is due to the transverse advection of three-dimensional structures was considered.

Figure 2 shows a possible configuration of such three-dimensional structures embedded within the wave-breaking zone. The structures are toroidal and are inclined with respect to the vertical. Sectional streamlines in a vertical plane through the center of a torus would reveal the topology of figure 1(a) which was observed in the experiments. On the other hand, no foci but strong upward flow would be observed in a vertical plane positioned in the region between two adjacent structures [figure 1(b)]. Since in the experiment the plane of examination was fixed with respect to the mountain, the observed flow pattern depicted in figure 1(b) would therefore be the result of transverse advection of the structures.

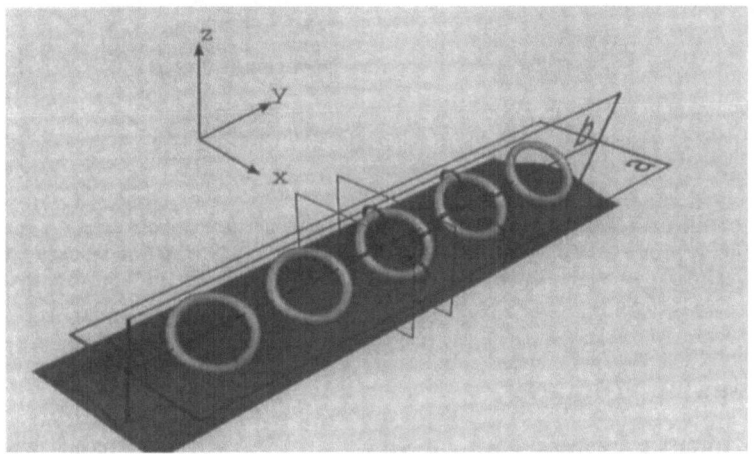

Figure 2.   Proposed vortex skeleton model of the dominant structures induced by wave-breaking over a mountain. The co-rotating toroidal structures are inclined and therefore exhibit strong vorticity components in all three directions.

In order to verify the model proposed in figure 2, the topology was studied carefully in additional planes of the flow, specifically, the horizontal $x - y$ plane labelled "a" in figure 2 as well as the transverse-vertical $y - z$ plane labelled "b". According to the model, the topology consists of counter-rotating foci in both these planes, as shown in figure 3(a).

Figure 3(b) shows an example of the flow pattern obtained in the horizontal $x - y$ plane at an elevation z that corresponds to the center of the structure in the vertical plane. The streamline pattern that can be deduced from these results reveals the same topology as the one depicted in figure 3(a). Similar experimental agreement with the model was also found in the $y - z$ plane and by the simultaneous examination of two orthogonal planes.

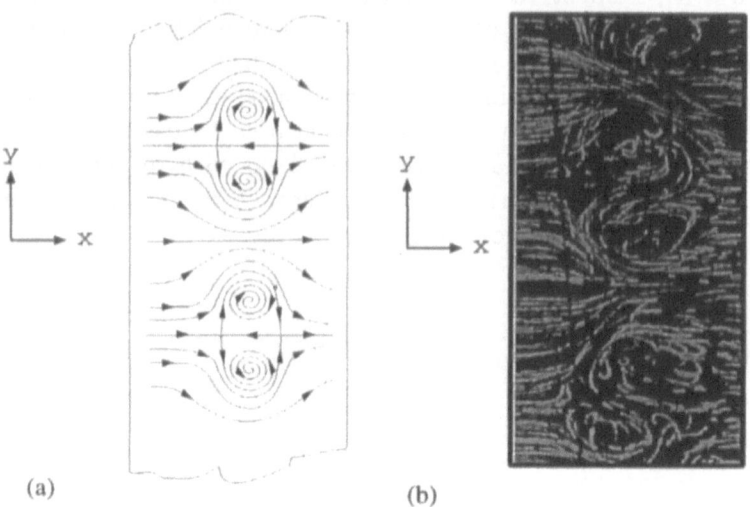

Figure 3. (a) Expected sectional streamline pattern in the horizontal $x - y$ plane which intersects the center of two toroidal structures. The same pattern is expected in the transverse-vertical $y - z$ plane. (b) Particle-tracking results in the equivalent horizontal $x - y$ plane. The topology is in agreement with (a), i.e., the presence of toroidal vortices.

4. Concluding Remarks

The experimental results confirm that toroidal vortices exist within the localised zone of breaking mountain waves. The characteristic scale of the vortices in both the vertical and horizontal directions is about $1.5H$. These vortices are expected to play a role in the mixing efficiency and turbulence characteristics of wave breaking.

5. References

Castro, I.P. & Snyder, W.H., 1993: Experiments on wave breaking in stratified flow over obstacles, *J. Fluid Mech.*, **225**, 195-211.
Fritts, D.C. & Isler, J.R., 1994: Gravity wave breaking in two and three dimensions. 2. Three-dimensional evolution and instability structure, *J. Geophys. Res.*, **99**, 8109-8123.
Laprise, R. & Peltier, W.R., 1989a: The linear stability of non-linear mountain waves: implications for the understanding of severe downslope windstorms, *J. Atmos. Sci*, **46**, 545-564.
Laprise, R. & Peltier, W.R., 1989b: The structure and energetics of transient eddies in a numerical simulation of breaking mountain waves, *J. Atmos. Sci*, **46**, 586-585.
Smith, R.B., 1985: On severe downslope winds. *Tellus*, **41A**, 401-415.

ANISOTROPY AND ENERGY DECAY IN MAGNETO-HYDRODYNAMIC TURBULENCE: THEORY AND SOLAR WIND OBSERVATIONS

S. OUGHTON[1], W. H. MATTHAEUS[2], S. GHOSH[3]

[1] *Department of Mathematics,*
University College London, London WC1E 6BT, UK.
[2] *Bartol Research Institute,*
University of Delaware, Newark, DE 19716, USA.
[3] *Space Applications Corporation*
901 Follin Lane, Suite 400 Vienna, VA 22180, USA.

Anisotropies in the spectra of magnetohydrodynamic (MHD) scale fluctuations are observed or inferred in many plasmas, *e.g.*, laboratory fusion machines, the solar wind and corona, and the interstellar medium (see [1, 2] for references). Anisotropy is expected since a mean magnetic field B_0 (unlike a mean flow), cannot be transformed away and thus provides a preferred direction. Understanding the evolution and development of this spectral anisotropy is a major theoretical challenge as nonlinear effects play a crucial role. Nonetheless, some progress has been made. Using numerical simulations and a *reduced* MHD (RMHD) approach, we have developed a relatively simple model of the process [1, 2], which we present below.

Simulations for freely-decaying 2D and 3D incompressible MHD turbulence [1, 2, 3, 4, 5] have shown that for $\delta B / B_0 \lesssim \frac{1}{2}$, spectral transfer in the parallel direction is essentially non-existent, whereas in the perpendicular direction the energy cascade continues out to the dissipation scale in the usual way (directions are relative to B_0, and δB is the rms value of the fluctuating magnetic field). This can be understood in terms of resonant triad interactions involving the Fourier modes. The (Fourier) eigenmodes are left or right propagating Alfvén waves—except that excitations with wavevector k perpendicular to B_0 are not eigenmodes in the usual sense, although the Alfvén waves can still couple with them. These latter modes are known as zero-frequency or non-propagating modes and can be interpreted as 2D turbulence (with the planes perpendicular to B_0). Calculation of the leading-order nonlinear corrections to the evolution reveals that two modes, k_1, k_2, will resonantly pump a third, k_3, only if one of the driving

U. Frisch (ed.), Advances in Turbulence VII, 475–478.

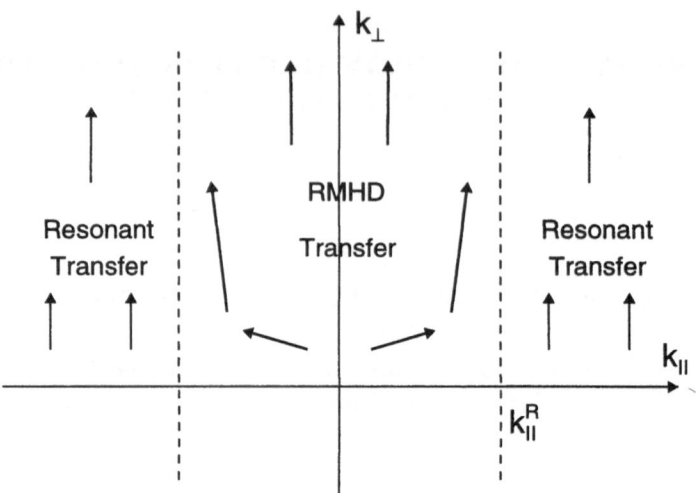

Figure 1.   Schematic diagram of the distinct dynamical regions in Fourier space associated with RMHD and resonant triad dynamics.

modes is a zero-frequency one [3]. Since, also, $\mathbf{k}_3 = \mathbf{k}_1 + \mathbf{k}_2$, it is possible to excite modes with higher $k_\perp$, but not higher $k_\parallel$. Thus, there is preferential transfer in the perpendicular directions. Clearly, the presence of a 2D component of the turbulence is crucial to anisotropy development via this mechanism. Despite recent suggestions [6], there appear to be no good *a priori* reasons to discard these components, and moreover caution should be exercised when drawing conclusions about full MHD from analyses based on RMHD [5, 7, 8, 9].

Resonant couplings are not the only way spectral anisotropy is produced, however. A strong $\mathbf{B}_0$ is associated with RMHD processes [10, 11, 5]. RMHD modes are those for which the Alfvén timescale exceeds the large-scale eddy-turnover time, *i.e.*, $1/|\mathbf{k}\cdot\mathbf{B}_0| \gtrsim 1/(k_c\delta B) \Rightarrow k_\parallel \lesssim k_c\delta B/B_0 = k_\parallel^R$, where k_c is the correlation scale for the turbulence. This inequality estimates the minimum parallel lengthscale dynamically excited by RMHD turbulence. For those RMHD modes which qualify, resonant interactions still occur. In addition, however, *all* RMHD modes engage in nonlinear interactions of the familiar Kolmogorov cascade kind, since B_0 (*e.g.*, wave effects) is relatively unimportant for these modes. Thus, excitations cascade out to the dissipation scale in the $\perp$ direction, but only to $k_\parallel^R$ parallel to $\mathbf{B}_0$. It follows that if the initial data contain excitations on both sides of $k_\parallel^R$, there will be two distinct regions in k-space, each governed by a different dynamics (Fig. 1). Modes with $|k_\parallel| \gtrsim k_\parallel^R$ are strongly influenced by Alfvén wave effects, with spectral transfer occuring mainly via the resonant process summarised above. In contrast, the RMHD modes behave more or less like standard MHD turbulence without a mean field, and attempt to maintain roughly isotropic transfer. However, they are stymied in their attempts at this in the parallel direction because of the dynamically imposed RMHD "wall" at $k_\parallel^R$. RMHD dynamics does not allow substantial spectral transfer past $k_\parallel^R$, since assumptions in the RMHD derivation [10] amount to the condition that $k_\parallel \ll k_\perp$ always. Thus, for the RMHD modes also, perpendicular spectral transfer is enhanced.

Using the above arguments one obtains an equation [1, 2] which predicts the degree of anisotropy as a function of $\delta B/B$ and initial condition parameters for $\frac{1}{10} \lesssim \frac{\delta B}{B} < 1$, namely $\cos^2 \theta = m(\delta B/B)^2 + c$, where B is the total field magnitude, m and c depend on the turbulence parameters, and θ is the angle between a mean wavenumber and $\mathbf{B}_0$. This linear scaling with $(\delta B/B)^2$ is supported by simulation results for a wide range of flows, including compressible, incompressible, decaying, and driven systems [1, 2].

Spectral anisotropy can also be produced in other ways. When B_0 is strong, for example, the viscosity is no longer isotropic, so that even *linear* dynamics can lead to pronounced anisotropy. Some consequences of this have been examined elsewhere [12, 13].

We anticipate that the above model for the development of spectral anisotropy in MHD turbulence will be useful in understanding the behaviour of many astrophysical and space physics systems. In particular, observations of the solar wind at length-scales of 10^3–10^7 km indicate that the MHD turbulence displays both spectral and variance anisotropy, [14, 15, 16]. Observations also indicate that, usually, $\delta B/B_0 \sim 1$, so that the solar wind plasma is often in the interesting scaling regime. Cosmic ray measurements and theory [17, 18] also suggest that the solar wind consists of two coupled components, namely 2D turbulence and Alfvén waves. The former has been observed [18] to account for as much as 80% of the fluctuation energy, consistent with dynamic evolution of the plasma towards a quasi-2D state as a consequence of anisotropic spectral transfer.

Anisotropy is likely to influence the solar wind plasma in many ways, including spatial transport of turbulence, cosmic ray scattering, and turbulent heating. An example is a simple transport theory [19], employing a local Taylor–Karman decay phenomenology appropriate for quasi-2D MHD, that accounts reasonably well for the radial distribution of turbulent energy from 1 to 40 AU in the low-latitude solar wind. Nevertheless, as noted above, observations seem also to require an additional ingredient, conveniently identified with nearly parallel propagating Alfvén waves. We suggest that a two component dynamical model of energy decay and spectral transfer [14], constrained by decay rates [20] and anisotropy scalings [1, 2] from numerical simulations, may serve as a useful improvement to existing phenomenologies for solar wind and coronal physics problems.

This work was supported by the NASA SPTP, the NSF, the Nuffield Foundation, and PPARC. Simulations were performed at the SDSC.

References

1. Oughton, S., Ghosh, S. & Matthaeus, W. H. (1997) Scaling of spectral anisotropy with magnetic field strength in decaying MHD turbulence. *Phys. Plasmas*, in press.
2. Matthaeus, W. H., Oughton, S., Ghosh, S. & Hossain, M. (1998) Scaling of anisotropy in hydromagnetic turbulence. *Phys. Rev. Lett.*, submitted.
3. Shebalin, J. V., Matthaeus, W. H. & Montgomery, D. (1983) Anisotropy in MHD turbulence due to a mean magnetic field. *J. Plasma Phys.* **29**, 525.
4. Oughton, S., Priest, E. R. & Matthaeus, W. H. (1994) The influence of a mean magnetic field on three-dimensional MHD turbulence. *J. Fluid Mech.* **280**, 95.
5. Kinney, R. & McWilliams, J. C. (1998) Turbulent cascades in anisotropic magnetohydrodynamics. *Phys. Rev. E*, in press.
6. Sridhar, S. & Goldreich, P. (1994) Toward a theory of interstellar turbulence: I. Weak Alfvénic turbulence. *Astrophys. J.* **432**, 612.
7. Montgomery, D. C. & Matthaeus, W. H. (1995) Anisotropic modal energy transfer in interstellar turbulence. *Astrophys. J.* **447**, 706.
8. Ng, C. S. & Bhattacharjee, A. (1996) Interaction of shear-Alfvén wave packets: Implications for weak magnetohydrodynamic turbulence in astrophysical plasmas. *Astrophys. J.* **465**, 845.
9. Goldreich, P. & Sridhar, S. (1997) Magnetohydrodynamic turbulence revisited. *Astrophys. J.* **485**, 680.
10. Montgomery, D. C. (1982) Major disruption, inverse cascades, and the Strauss equations. *Physica Scripta* **T2/1**, 83.
11. Kinney, R. & McWilliams, J. C. (1997) Magnetohydrodynamic equations under anisotropic conditions. *J. Plasma Phys.* **57**, 73.
12. Montgomery, D. C. (1992) Modifications of magnetohydrodynamics as applied to the solar wind. *J. Geophys. Res.* **97**, 4309.
13. Oughton, S. (1996) Ion parallel viscosity and anisotropy in MHD turbulence. *J. Plasma Phys.* **56**, 641.
14. Tu, C.-Y. & Marsch, E. (1995) MHD structures, waves and turbulence in the solar wind. *Space Sci. Rev.* **73**, 1.
15. Matthaeus, W. H., Bieber, J. W. & Zank, G. P. (1995) Unquiet on any front: Anisotropic turbulence in the solar wind. *Rev. Geophys. Supp.* **33**, 609.
16. Matthaeus, W. H., Ghosh, S., Oughton, S. & Roberts, D. A. (1996) Anisotropic three-dimensional MHD turbulence. *J. Geophys. Res.* **101**, 7619.
17. Bieber, J. W., Matthaeus, W. H., Smith, C. W., Wanner, W., Kallenrode, M. & Wibberenz, G. (1994) Proton and electron mean free paths: The Palmer consensus revisited. *Astrophys. J.* **420**, 294.
18. Bieber, J. W., Wanner, W. & Matthaeus, W. H. (1996) Dominant two-dimensional solar wind turbulence with implications for cosmic ray transport. *J. Geophys. Res.* **101**, 2511.
19. Zank, G. P., Matthaeus, W. H. & Smith, C. W. (1996) Evolution of turbulent magnetic fluctuation power with heliocentric distance. *J. Geophys. Res.* **101**, 17 093.
20. Hossain, M., Gray, P. C., Pontius Jr., D. H., Matthaeus, W. H. & Oughton, S. (1995) Phenomenology for the decay of energy-containing eddies in homogeneous MHD turbulence. *Phys. Fluids* **7**, 2886.

A REDUCED DESCRIPTION OF RAPIDLY ROTATING TURBULENT CONVECTION

K. JULIEN
Department of Applied Mathematics, University of Colorado, Boulder, CO 80309

E. KNOBLOCH
Department of Physics, University of California, Berkeley, CA 94720 and JILA, University of Colorado, Boulder, CO 80309

AND

J. WERNE
Colorado Research Associates, 3380 Mitchell Lane, Boulder, CO 80301

1. Introduction

The tendency towards two-dimensionality in rapidly rotating flows is described by the Taylor-Proudman theorem. In applications strict two-dimensionality is usually broken by the presence of boundary and/or thermal forcing, or by initial conditions. Both types of forcing are present in thermal convection in a rapidly rotating horizontal layer, described by the equations

$$\frac{D\mathbf{u}}{Dt} + \hat{\boldsymbol{\Omega}} \times \mathbf{u} = -\nabla\pi + E\nabla^2\mathbf{u} + \frac{Ra}{\sigma}E^2 T\hat{\mathbf{z}}, \qquad \nabla \cdot \mathbf{u} = 0. \tag{1}$$

$$\sigma\frac{DT}{Dt} = E\nabla^2 T. \tag{2}$$

Here $E \equiv \nu/2\Omega d^2$ is the Ekman number, assumed to be small, and Ra and σ are the Rayleigh and Prandtl numbers; distances have been expressed in units of the layer depth d and time in units of the Coriolis time $(2\Omega)^{-1}$. All other symbols have their usual meaning. In the following we employ a multiple scale expansion in both time and space, and focus on horizontal scales on which the horizontal and vertical velocities are comparable, but the horizontal velocity components are still in geostrophic balance.

U. Frisch (ed.), *Advances in Turbulence VII*, 479–482.

2. Reduced Interior Equations

We introduce "fast" horizontal variables $x' \equiv E^{-\frac{1}{3}}x$, $y' \equiv E^{-\frac{1}{3}}y$ and use the notation $D \equiv \partial_z$ to denote derivatives with respect to the "slow" variable z. In thermal convection these are the dominant scales selected by linear theory. We also introduce a slow time $t' \equiv E^{\frac{1}{3}}t$. Finally, we write $\mathbf{u} = \nabla \times \phi\hat{\mathbf{z}} + \nabla \times \nabla \times \psi\hat{\mathbf{z}}$ and scale the streamfunctions according to $\phi = E\phi'$, $\psi = E^{\frac{4}{3}}\psi'$. This rescaling implies that on the length scales of interest the vertical and horizontal velocities are of the same order, both $\mathcal{O}(E^{\frac{2}{3}})\Omega d$ in dimensional units. The local Rossby number is thus $\mathcal{O}(E^{\frac{1}{3}})$ and hence is small; i.e., even though the scales of interest are small, the flow is still rotation dominated. Moreover, with this scaling the horizontal velocity components are in geostrophic balance at leading order. Finally, we introduce the scaled Rayleigh number $Ra' \equiv E^{\frac{4}{3}}Ra$ and split the temperature T into its mean and fluctuating parts: $T(x,y,z,t) \equiv \overline{T}(z) + E^{\frac{4}{3}}\theta(x,y,z,t)$. Here, the overbar denotes a spatial average in the horizontal *and* a time-average. The time-averaging is an essential aspect of our decomposition, and allows us to close the problem (Julien and Knobloch 1998).

On dropping all primes we obtain the following reduced system,

$$\partial_t \nabla_\perp^2 \phi - J[\phi, \nabla_\perp^2 \phi] - D\nabla_\perp^2 \psi = \nabla_\perp^4 \phi + \mathcal{O}(E^{\frac{1}{3}}) , \tag{3}$$

$$\partial_t \nabla_\perp^2 \psi - J[\phi, \nabla_\perp^2 \psi] + D\phi = -\frac{Ra}{\sigma}\theta + \nabla_\perp^4 \psi + \mathcal{O}(E^{\frac{1}{3}}) , \tag{4}$$

i.e., a pair of equations for the vertical velocity $w \equiv -\nabla_\perp^2 \psi$ and the vertical vorticity $\zeta \equiv -\nabla_\perp^2 \phi$, coupled via vertical stretching and driven by thermal buoyancy:

$$\sigma(\partial_t \theta - J[\phi, \theta] - \nabla_\perp^2 \psi D\overline{T}) = \nabla_\perp^2 \theta + \mathcal{O}(E^{\frac{1}{3}}) . \tag{5}$$

Here $J[f,g] \equiv \partial_x f \partial_y g - \partial_y f \partial_x g$ is the horizontal Jacobian operator. The equations are closed using the first integral of the mean temperature equation:

$$D\overline{T} = -1 - \sigma\overline{(\nabla_\perp^2 \psi \theta - \langle \nabla_\perp^2 \psi \theta \rangle)} + \mathcal{O}(E^{\frac{1}{3}}) . \tag{6}$$

Here the angular brackets denote a vertical average. The resulting equations describe the dynamics in the bulk, outside of any Ekman boundary layers required by horizontal boundaries. In the present problem such boundaries are passive (Julien and Knobloch 1998).

3. Results

Equations (3–6) were solved numerically using a pseudo-spectral Petrov-Galerkin method in which field variables are represented with sines or

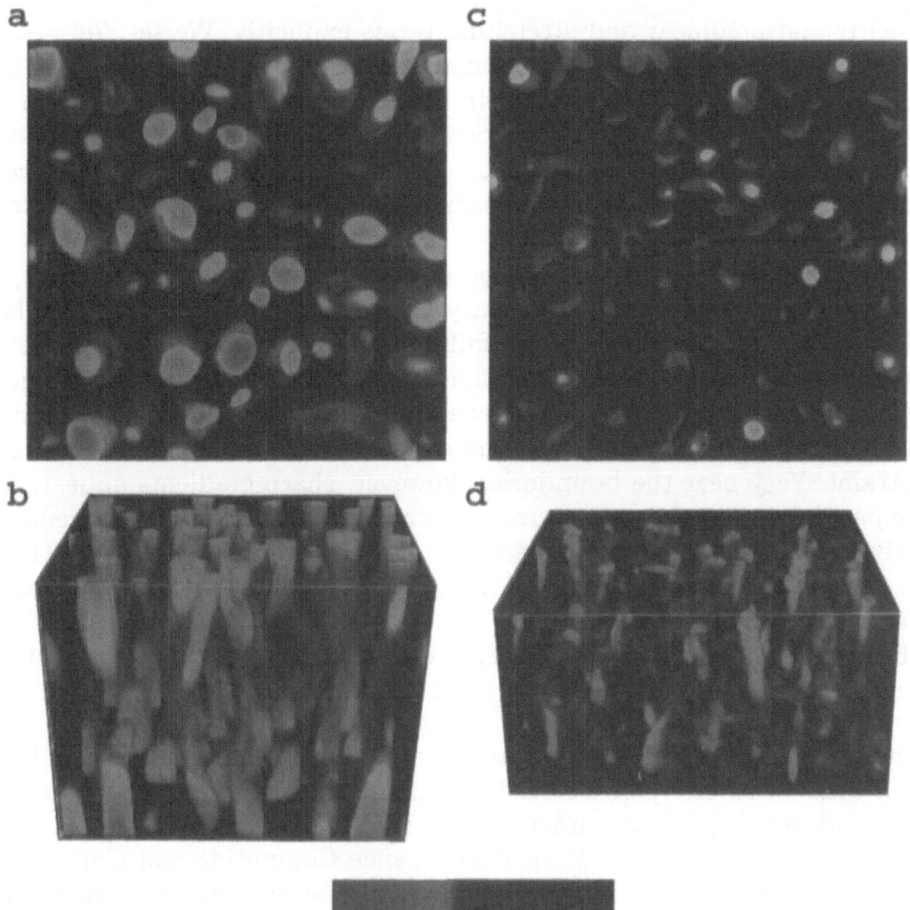

Figure 1. Figure 1. Results of direct numerical simulations of the reduced equations (a,b) and the full three-dimensional Boussinesq equations (c,d). Top views are depicted in (a) and (c), while perspective views are shown in (b) and (d). The visualization of vortical structures is achieved with the quantity λ_2, the intermediate eigenvalue of $R^2 + S^2$, where R and S are the rotation and strain matrices of the velocity field. The negative values shown in grey (see color bar) indicate vortex tubes (Jeong and Hussain, 1995). The reduced equations are simulated with $Ra = 20E^{-\frac{4}{3}}$ and an aspect ratio of $6E^{-\frac{1}{3}} \times 6E^{-\frac{1}{3}} \times 1$ (not to scale); they capture the coherent structures of the full Boussinesq equations (first reported in Julien et al 1996). The parameter values are $Ra = 1.0 \times 10^7$, $E = 9.4 \times 10^{-5}$, and an aspect ratio of $2 \times 2 \times 1$ (to scale); the effective Rossby number is 0.3. The aspect ratio selection for the reduced problem was based on the number of critical modes (as determined by linear theory) that fit into the full three-dimensional Boussinesq simulation.

cosines in the vertical and periodic Fourier modes in the horizontal. Time-stepping is via a mixed implicit/explicit 3rd-order Runge-Kutta scheme developed by Spalart et al (1991), with the diffusion and forcing terms treated implicitly and nonlinear and stretching terms explicitly. We set $Ra' = 20$ (2.2 times critical), $\sigma = 1$, and the domain width to 6 times the most unstable linear wavelength ($\lambda_\perp \approx 4.8154$) in both horizontal directions. Top and bottom boundary conditions are impenetrable/fixed temperature and side boundaries are periodic. The calculations were conducted with 64^3 spectral modes and were de-aliased in all spatial directions at each Runge-Kutta sub-timestep.

Fig. 1a–b shows a sample solution obtained using only spatial averaging for the mean terms: in a sufficiently large domain (as in Fig. 1a–b) the horizontal average of the rising and falling plumes becomes equivalent to a time-average. After a very short initial transient, vortical buoyant plumes emerge and mutually advect one another laterally. The plumes are columnar, spanning the layer depth, as one expects from the Taylor-Proudman constraint. Very near the boundaries, however, sharp gradients appear, as anticipated from the thermal boundary conditions. These features resemble closely those found in numerical solutions of the primitive equations (1,2) at large rotation rates (Fig. 1a–b). A detailed comparison of the two sets of solutions will appear elsewhere.

Equations (3,4) with $Ra = 0$ describe decaying rapidly rotating turbulence and merit study in their own right (Julien et al 1998).

Acknowledgment: This research was initiated while EK was a Visiting Fellow at JILA, University of Colorado, Boulder, and partially supported by the Department of Energy under grant DE-FG03-95ER-25251. JW is partially supported by an NSF High Performance Computing and Communication Grand Challenge grant in geophysical and astrophysical fluid turbulence (ECS9217394). The numerical calculations were conducted at the Pittsburgh Supercomputing Center under MetaCenter grant MCA935010P.

References

Jeong, J. and Hussain, F. (1995) On the identification of a vortex, *J. Fluid Mech.*, **285**, pp. 69–94.

Julien, K. and Knobloch, E. (1998) Strongly nonlinear convection cells in a rapidly rotating fluid layer: the tilted f-plane, *J. Fluid Mech.*, **360**, pp. 141–178.

Julien, K., Knobloch, E. and Werne, J. (1998) A new class of equations for rotationally constrained flows, *Theor. Comp. Fluid Dyn*, in press.

Julien, K., Legg, S., McWilliams, J. and Werne, J. (1996) Rapidly rotating turbulent Rayleigh-Bénard convection, *J. Fluid Mech.* **322**, pp. 243–273.

Spalart, P. R., Moser, R. D. and Rogers, M. M. (1991) Spectral methods for the Navier-Stokes equations with one infinite and two periodic directions, *J. Comput. Phys.*, **96**, pp. 297–324.

THE ONSET OF TURBULENCE IN A SHEAR FLOW ON THE BETA-PLANE. AN ASYMPTOTIC APPROACH BASED ON THE CRITICAL LAYER THEORY

V.P. REUTOV, S.V. SHAGALOV AND G.V. RYBUSHKINA
Institute of Applied Physics, RAS, 46 Uljanov Str.,
Nizhny Novgorod 603600, Russia

The paper is concerned with the onset of barotropic two-dimensional turbulence in a shear layer on the β-plane. An asymptotic approach involving a nonlinear theory of the critical layer (CL), which was elaborated earlier mainly for the single–mode instability problem, is extended to generation of the wave packets with a narrow spectrum.

Consider a weakly supercritical zonal flow with velocity profile $U = \tanh(y)$ and, following Hickernell (1984), write the planetary vorticity gradient β in the form $\beta = \beta_m + \epsilon\beta_1$, where $\beta_m = 4/\sqrt{27}$ is a marginal neutral value of β in the absence of Eckman dissipation and $\epsilon \ll 1$ characterizes the smallness of the supercriticality (standard normalizations are used). The Eckman dissipation parameter λ and the dimensionless viscosity (the inverse Reynolds number) are supposed to be small: $\lambda = \epsilon\lambda_1 = O(\epsilon)$, $\nu = \epsilon^{3/2}\nu_1 = O(\epsilon^{3/2})$. The flow velocity profile and scaling for the CL and instability range are drawn schematically in Fig. 1.

The streamfunction in the outer (with respect to the CL) flow region is sought in the form $\psi = \int U \, dy + \epsilon[A(\xi_1,\tau_1,\tau)\text{sech}(y)\exp(-cy+ik\xi)+\text{c.c.}] + \epsilon^{3/2}\psi^{(3/2)}(\xi,\xi_1,\tau_1,\tau,y) + ...$, where $\tau = \epsilon t$, $\tau_1 = \epsilon^{1/2}t$, $\xi = x - ct$, $\xi_1 = \epsilon^{1/2}\xi$; $k = \sqrt{2/3}$ and $c = -1/\sqrt{3}$ are the wave number and phase velocity of the marginal neutral wave; A is the slowly varying complex amplitude (wave envelope); and c.c. designates the complex conjugate. For the CL-region, the inner variable $Y = (y-y_c)/\epsilon^{1/2}$ is introduced and the absolute vorticity, that is a sum of the flow vorticity and the planetary vorticity, is sought as an expansion (without constant term) $\zeta_a = \epsilon^{3/2}\zeta_a^{(3/2)} + ...$ A new variable ζ, that meets boundary conditions $\zeta_Y \to 0$ at $Y \to \pm\infty$, is introduced by the relation $\zeta_a^{(3/2)} = (-1/6)U_c^{IV}Y^3 + \beta_1 Y - 2\text{Re}[6ck^2 Y A \exp(ik\xi) + 6ikA_{\xi_1}\exp(ik\xi)] + \zeta$ (the primes and the subscript c designate the deriva-

U. Frisch (ed.), Advances in Turbulence VII, 483–486.
© 1998 *Kluwer Academic Publishers.*

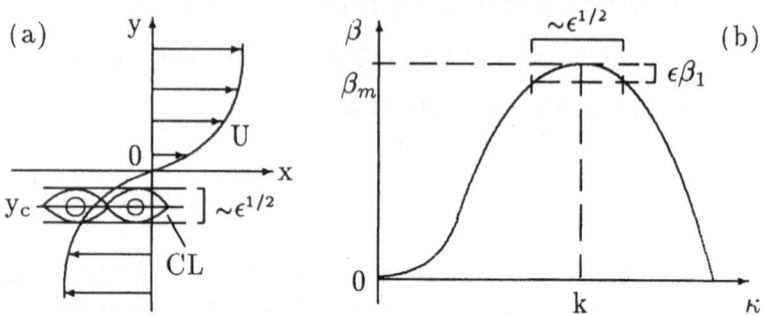

Figure 1. The flow velocity profile with the critical layer localization (a) and the neutral curve in the absence of Eckman friction (b) (κ is the wave number of disturbance).

tives of U with respect to y at $y = y_c$). This allowed the system of equations governing the evolution of the complex wave amplitude and the CL-vorticity to be written as follows

$$\frac{\partial A}{\partial \tau_1} + (c_g - c)\frac{\partial A}{\partial \xi_1} + i\alpha_1\epsilon^{1/2}\frac{\partial^2 A}{\partial \xi_1^2} = i\epsilon^{1/2}\alpha_2 \int_{-\infty}^{+\infty} < \zeta e^{-ik\xi} > dY - \epsilon^{1/2}\lambda_1 A,$$

$$\frac{\partial \zeta}{\partial \tau_1} + U_c'Y\frac{\partial \zeta}{\partial \xi} - 2k\mathrm{Im}\left(Ae^{ik\xi}\right)\frac{\partial \zeta}{\partial Y} = \qquad (1)$$

$$-2\mathrm{Im}\left[\beta_1 Ae^{ik\xi} - 6ck^3A^2e^{2ik\xi} + 6k(c - c_g)\frac{\partial^2 A}{\partial \xi_1^2}e^{ik\xi}\right] + \nu_1\frac{\partial^2 \zeta}{\partial Y^2},$$

where $c_g = 1/\sqrt{3}$ is a group velocity, $< \dots >$ designates local averaging over ξ, $\alpha_1 \approx 1.18$, and $\alpha_2 \approx 0.059$ Unlike similar equations for the purely temporal problem (Churilov, 1989), the effects associated with the finite bandwidth of the instability range and the wave packet propagation are taken into account in (1) through the spatial derivatives of the wave amplitude.

Equations (1) were integrated with the aid of a numerical scheme including the Fourier series expansion along the flow direction and the finite-difference approximation across the CL. The scheme was realized for periodic boundary conditions corresponding to the annular zonal flow. In the computations we specified the viscous CL-scale $l_\nu = (\nu/kU_c')^{1/3} = 0.173$, the offset of the beta-parameter $(\beta_m - \beta)/6k^2(c_g - c) = 0.1$, the number of wavelengths of the marginal wave at a spatial period $N = 12$ and the supercriticality $\Gamma = (\beta - \beta_m)\alpha_2\pi/U_c' - \lambda$ varying due to λ.

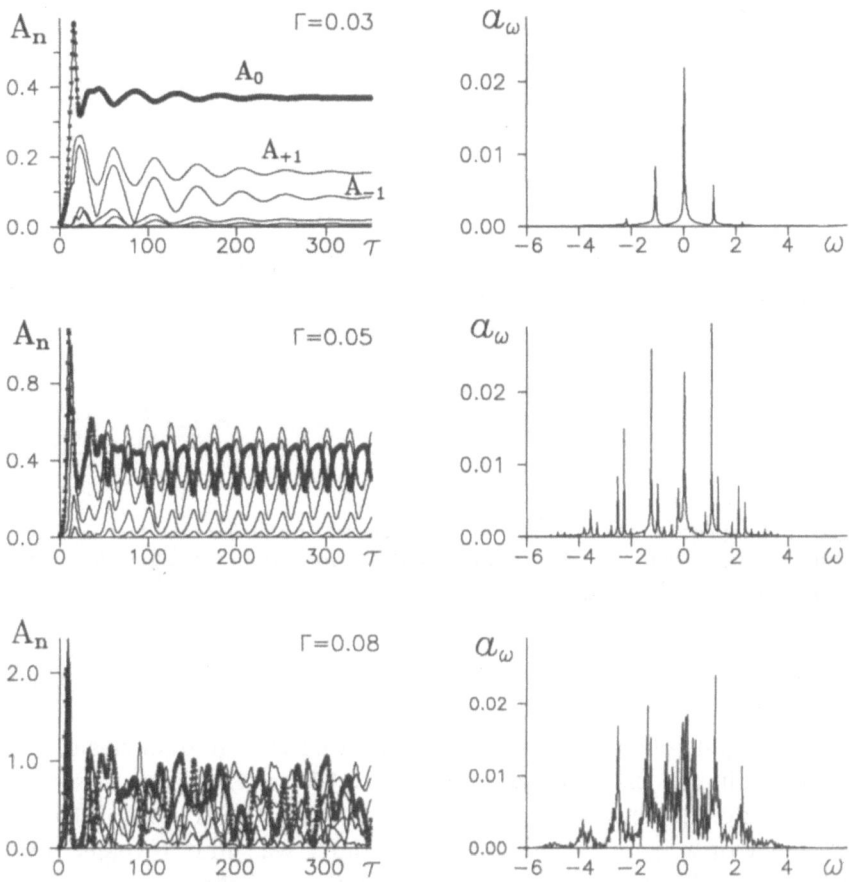

Figure 2. The time dependence of mode amplitudes A_n and the frequency spectra a_ω of the complex wave envelope (normalized plots for different supercriticalities Γ).

The flow regimes appearing successively as the supercriticality increases are given in Figs. 2,3. Figure 2 shows the amplitudes of the space harmonics A_n having wave numbers $\kappa_n = k + nk/N$ ($n = 0, \pm 1, \pm 2, \pm 3$) and the frequency spectra a_ω of the complex envelope A at fixed ξ. One can see that the stationary mode generation is established at $\Gamma = 0.03$. The discrete peaks of a_ω correspond to synchronized eigenfrequencies of the linear problem for $\kappa = \kappa_n$. When $\Gamma = 0.05$, the periodic oscillations of the space harmonic amplitudes producing satellites in the frequency spectrum are established and, finally, the chaotic behaviour occurs at $\Gamma = 0.08$. As follows from Fig. 3, the "cat's eyes" structures in the streamlines inside the CL

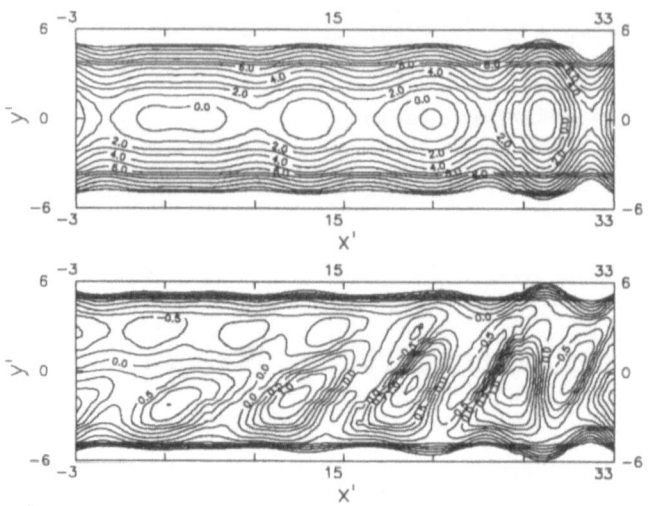

Figure 3. Snapshot of the streamlines (top picture) and of the absolute vorticity isolines (bottom picture) inside the CL at the one-half-period of the arrival flow in chaotic regime $(y' = (y - y_c)/l_\nu,\ x' = k\xi)$.

are chaotically modulated and absolute vorticity becomes quite confusive at $\Gamma = 0.08$.

The analysis of the phase portraits showed that frequencies of the satellites appearing in the spectrum at $\Gamma = 0.05$ are incommensurate. Therefore, we conclude that the system demonstrates the transition to chaos via breakdown of a quasiperiodic motion which was first revealed by Ruelle and Takens (1971). The physical mechanism of chaos is associated with arising of sophisticated behaviour of absolute vorticity in the wave packet field, when the local critical layers corresponding to different modes of the wave packet overlap.

Thus, we have revealed the transition to the chaotic behaviour due to narrow wave packet interaction with a nonlinear unsteady CL in a zonal shear flow.

This work was supported by the Russian Foundation for Basic Research (project code 98-05-64686).

References

1. Churilov, S.M. (1989) The nonlinear stabilization of a zonal shear flow instability, *Geophys. and Astrophys. Fluid Dynamics* **46**, 159–175.
2. Hickernell, F.J. (1984) The time–dependent critical layers in shear flows on the beta-plane, *J. Fluid Mech.* **142**, 431–449.
3. Ruelle, D. and Takens, F. (1971) On the nature of turbulence, *Commun. Math. Phys.* **20**, Springer–Verlag, Berlin, pp.167–192.

EFFECTS OF STABLE STRATIFICATION ON FLOW BEHIND A TWO-DIMENSIONAL FENCE SUBMERGED IN A TURBULENT BOUNDARY LAYER

N.STEGGEL AND I.P.CASTRO

EnFlo, School of Mechanical and Materials Engineering,
University of Surrey, Guildford GU2 5XH, UK.

1. Introduction

There is a considerable literature on the general problem of the flow around two and three-dimensional obstacles submerged in turbulent boundary layers. The most obvious context is that of the flow of the earth's atmospheric boundary layer around buildings and other structures with the consequent loads on the structures or changes in the nature of dispersion of pollutants. However, the overwhelming majority of the extant work has been concerned with the case of neutrally stable boundary layers despite the fact that atmospheric flows are usually characterised by at least some degree of non-neutral stability. A study has therefore been initiated in the Environmental Flow Research Centre with the objective of assessing the effects of stable stratification on the formation and development of wakes downstream of bluff obstacles submerged in turbulent boundary layers. Results from the first stage of this work, in which mean and turbulence velocities behind a two-dimensional obstacle have been studied, are reported below.

2. Experimental arrangements

The experiments were conducted in the stratified flow wind tunnel at En-Flo, in which simulated atmospheric boundary layers can be generated. A 100mm high, thin, flat plate was mounted spanwise on the rough surface, within both a neutrally stable boundary layer of depth about 1m and a moderately stratified layer whose Monin-Oboukhov length scale was around 170mm. The surface roughness was characterised by a roughness length, z_0, of 1.0mm and the turbulence properties of both flows were similar to corresponding atmospheric boundary layers. Defining a typical Froude Number,

U. Frisch (ed.), Advances in Turbulence VII, 487–490.

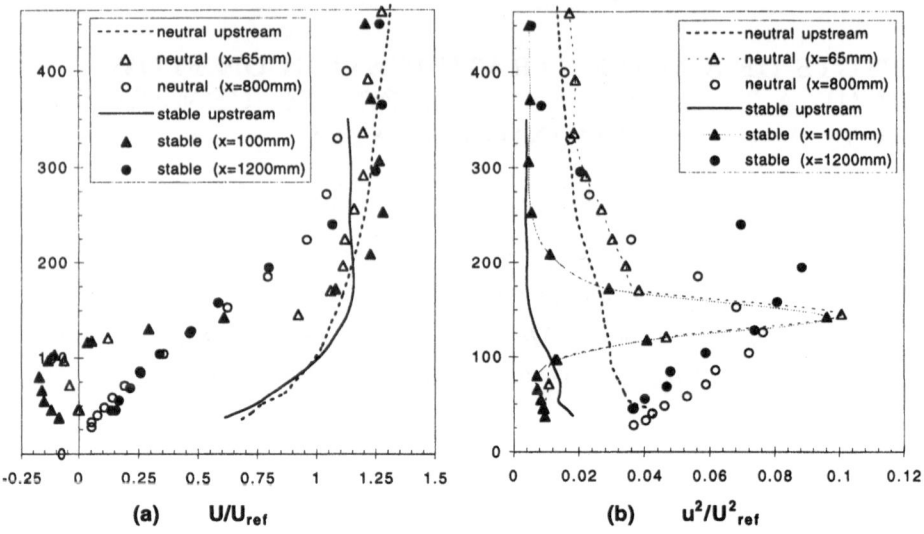

Figure 1. (a) z(mm) versus U/U_{ref}; (b) z(mm) versus $\overline{u^2}/U^2_{ref}$. $x/X_R \approx 0.083, 1.0$.

Fr, appropriate to the stratifed flow over the fence as U_{ref}/NH_F, where N is the Brunt-Väisälä frequency, we have $Fr \approx 1.5$. (U_{ref} and N are both taken, somewhat arbitrarily, as the values at the fence height, H_F). This is perhaps a little lower than might normally be expected for strongly stable (e.g. night-time) atmospheric flows over large buildings, so represents an extreme case. Mean velocity measurements and turbulence measurements were made using a two-component laser doppler anemometer which, in the case of the stratified flow (in which local temperatures exceeded 60°C), used a specially constructed, air-cooled fibre-optic probe.

3. Discussion of results

Fig. 1 shows vertical profiles of velocity and longitudinal turbulence at two downstream locations in both the stable and neutral boundary layers together with the undisturbed boundary layer profiles. X_R is defined as the reattachment length behind the fence and was found to be around 800mm in the neutral layer and 1200mm in the stable layer. (Comparison between the stresses in the two layers are thus given at similar values of x/X_R). This substantial difference in X_R can be explained on the basis of the lower external turbulence levels 'seen' by the seperated shear layer in the stable flow, both because of the generally lower stresses upstream and because the fence is a much larger fraction of the boundary layer depth in that case. Clear similarities can be observed in the profiles at the corresponding x/X_R

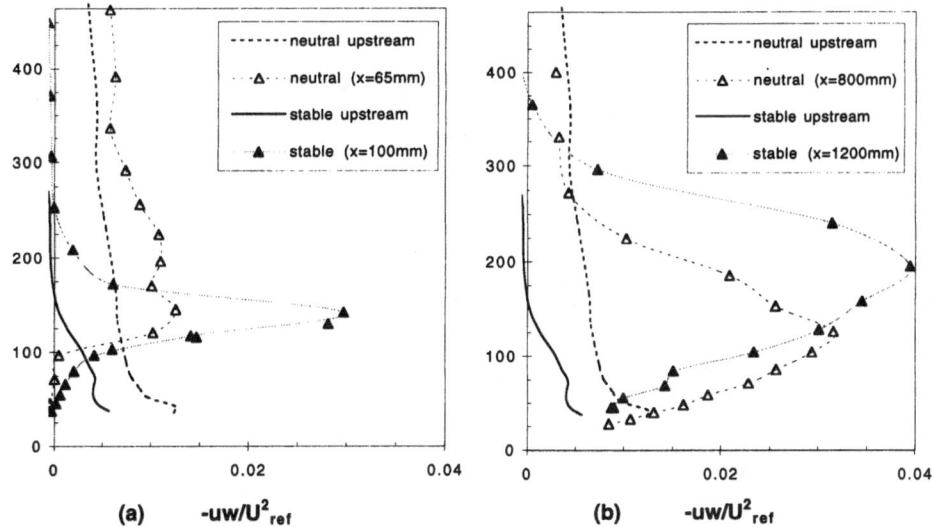

Figure 2. z(mm) versus $-\overline{uw}/U_{ref}^2$; (a) $x/X_R \approx 0.083$, (b) $x/X_R \approx 1.0$.

values although for longitudinal turbulence at $x/X_R \approx 1.0$ the disturbance under stable conditions appears to peak higher in the boundary layer.

Shear stress profiles are shown in Fig. 2. Just after separation, $x/X_R \approx 0.083$, the vertical (not shown), normal and shear stress levels are considerably higher in the stable layer than in the neutral case for reasons which are currently unclear; one might have anticipated the reverse, because of the stabilising effect of the temperature profile. Furthermore, all the stresses around reattachment peak at rather higher values in the stable case, as emphasised by Fig. 3, where the maximum stress levels are plotted versus downstream distance. Crudely, at reattachment one might expect that the stable flow case, in that it corresponds to weaker external turbulence, would lead to higher peak $\overline{w^2}$ and lower peak $\overline{u^2}$ (see, eg. [1]). But this appears not to be the case. However, the choice of reference velocity is not obvious and further analysis is necessary.

Despite these differences between the two flows, it is evident that the general stress levels in the separated shear layers are, in fact, much closer in the two flows than they are upstream. The clear implication is that the high intensity, large scale mixing associated with the separated region is not strongly affected by upstream stability, which seems largely to act only to change the overall bubble length, nor by the wall cooling within the separated flow. Temperature profiles (not shown) suggest that the separated flow is well-mixed in the stable case.

It may be more appropriate to make comparison between results for

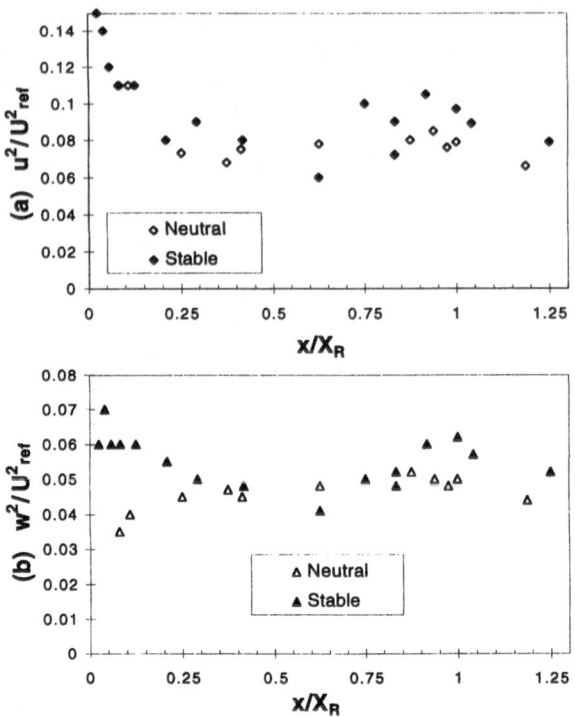

Figure 3. Maximum normal and shear stress (a) $\overline{u^2}/U_{ref}^2$; (b) $\overline{w^2}/U_{ref}^2$ versus x/X_R.

flows in which the ratio of the fence height to the boundary layer depth, H_F/δ, is similar before detailed conclusions on the effects of stable stratification can be deduced. It would not be sensible to reduce the fence height since the roughness elements used to simulate the rough wall are themselves 20mm high. The next stage of work will therefore involve the simulation of a neutrally stable boundary layer of similar depth to the stable layer described here. Subsequently, we anticipate studying wakes behind 3-D obstacles, particularly those which produce persistent axial vorticity, so that the effects of stability on, for example, far-fetch downwash can be assessed. The influence of stability on building-effected plume dispersion characteristics is also being studied - this is the eventual objective of the work.

References

1. Castro, I.P. & Haque, A. (1988) The structure of a shear layer bounding a separation region. Part 2. Effects of free-stream turbulence, *J. Fluid Mech.* **192**, pp. 577–595.

VIII

Transport of Passive Scalars

Transport de Scalaires Passifs

INTERMITTENCY OF PASSIVE ADVECTION

K. GAWEDZKI

CNRS, IHES, 91440 Bures-sur-Yvette, France

1. Introduction

An explanation of the origin of intermittency in the fully developed turbulence remains one of the main open problems of theoretical hydrodynamics. Quantitatively, intermittency is measured by deviations from Kolmogorov's scaling of the velocity correlators. Below, we shall describe a recent progress in the understanding of intermittency in a simple system describing the passive advection of a scalar quantity (temperature, density of a pollutant) by a random velocity field. The system, maintained in a stationary state by a large scale source, exhibits a down-scale energy cascade. If the molecular diffusion is small, an intermittent inertial range sets in [1]. We shall explain this phenomenon by identifying the origin of the anomalous scaling of scalar correlators in a simple model of passive advection due to Kraichnan [2].

2. Kraichnan model

The advection of a scalar quantity $\theta(t, \mathbf{r})$ by a velocity field $\mathbf{v}(t, \mathbf{r})$ is described by the linear differential equation

$$\partial_t \theta + \mathbf{v} \cdot \nabla \theta = f \tag{1}$$

where $f(t, \mathbf{r})$ represents the external source of the scalar. If the source vanishes then the scalar is carried along by the flow:

$$\theta(t, \mathbf{r}(t)) = \theta(t_0, \mathbf{r}_0) \tag{2}$$

where $\mathbf{r}(t)$ is the (Lagrangian) trajectory of the fluid particle located at time t_0 at $\mathbf{r}_0$:

$$\frac{d\mathbf{r}}{dt} = \mathbf{v}(t, \mathbf{r}), \qquad \mathbf{r}(t_0) = \mathbf{r}_0. \tag{3}$$

U. Frisch (ed.), Advances in Turbulence VII, 493–502.
© *1998 Kluwer Academic Publishers.*

The non-zero source f keeps creating the scalar along the Lagrangian trajectory and Eq. (2) is modified to

$$\theta(t, \mathbf{r}(t)) = \theta(t_0, \mathbf{r}_0) + \int_{t_0}^{t} ds \, f(s, \mathbf{r}(s)). \tag{4}$$

In the presence of diffusion of the scalar the above equations require small changes:

1. Eq. (1) picks up a term $\kappa \nabla^2 \theta$ on the right hand side with κ standing for the diffusion constant,

2. Brownian motions $\beta(t)$ should be superimposed on the Lagrangian trajectories by adding the term $\kappa \frac{d\beta}{dt}$ to the velocity in Eq. (3),

3. the right hand sides of Eqs. (2) and (4) should be averaged over β.

The Kraichnan model of the passive advection of the scalar [2] assumes velocities decorrelated at different times, and, for a fixed time, distributed as a Gaussian (i.e. non-intermittent) field with mean zero and a non-smooth typical behavior in space

$$|\mathbf{v}(t, \mathbf{r}) - \mathbf{v}(t, \mathbf{r}')| \ \sim \ |\mathbf{r} - \mathbf{r}'|^{\xi/2} \tag{5}$$

where ξ is a fixed parameter between 0 and 2. This may be achieved by imposing the velocity covariance

$$\left\langle \mathbf{v}^{\alpha}(t, \mathbf{r}) \, \mathbf{v}^{\beta}(t', \mathbf{r}') \right\rangle \ = \ \delta(t - t') \, \mathcal{D}^{\alpha\beta}(\mathbf{r} - \mathbf{r}') \tag{6}$$

with

$$\mathcal{D}^{\alpha\beta}(0) - \mathcal{D}^{\alpha\beta}(\mathbf{r}) \ = \ D_0 \, r^{\xi} \left[(d - 1 + \xi) \, \delta^{\alpha\beta} - \xi \frac{r^{\alpha} r^{\beta}}{r^2} \right] \equiv D^{\alpha\beta}(\mathbf{r}) \tag{7}$$

where d denotes the space dimension. The typical behavior (5) follows since $D^{\alpha\beta}(\mathbf{r}) \propto r^{\xi}$. The incompressibility $\nabla \cdot \mathbf{v} = 0$ of the velocity field is assured by the relation $\partial_{r^{\alpha}} \mathcal{D}^{\alpha\beta}(\mathbf{r}) = 0$.

We shall be interested in the effect caused by a steady injection of the scalar on distances longer than some large scale L. One may conveniently model such a source by a random Gaussian field $f(t, \mathbf{r})$ with mean zero and covariance

$$\langle f(t, \mathbf{r}) \, f(t', \mathbf{r}') \rangle \ = \ \delta(t - t') \, C(\tfrac{\mathbf{r} - \mathbf{r}'}{L}) \tag{8}$$

where $C(\frac{r}{L})$ is approximately constant for $r \ll L$ and decays fast for $r \gg L$. We shall assume f independent of the velocities $\mathbf{v}$.

3. Steady state

The evolution equations of the hydrodynamical type imply identities for the correlators of the evolving quantities known under the name of Hopf equations. These equation usually couple the correlators with different number of points. The case of the passive advection is no exception since its Hopf equations couple the scalar equal-time correlators

$$F_N(t; \underline{\mathbf{r}}) = \left\langle \prod_{n=1}^{N} \theta(t, \mathbf{r}_n) \right\rangle \tag{9}$$

to the mixed correlators involving both the scalar and the velocity field. However, in the Kraichnan model with temporarily decorrelated velocities, the mixed correlators may be expressed by the known 2-point function of the velocities and the correlators of the scalar alone. The resulting Hopf equations may be easily seen to take the form

$$\partial_t F_N = -M_N F_N + \mathcal{C} \otimes F_{N-2} \tag{10}$$

with the shorthand notation

$$(\mathcal{C} \otimes F_{N-2})(t; \underline{\mathbf{r}}) = \sum_{n<m} \mathcal{C}(\tfrac{\mathbf{r}_n - \mathbf{r}_m}{L}) \, F_{N-2}(t; \mathbf{r}_1, \underset{\hat{n}\ \hat{m}}{\ldots\ldots}, \mathbf{r}_N) \tag{11}$$

and with

$$M_N = -\tfrac{1}{2} \sum_{\substack{n,m \\ \alpha\beta}} \mathcal{D}^{\alpha\beta}(\mathbf{r}_n - \mathbf{r}_m) \, \partial_{r_n^\alpha} \partial_{r_m^\beta} \tag{12}$$

being a symmetric (due to the incompressibility), positive, singular elliptic differential operator of the 2nd order. In the presence of the diffusion, M_N should be modified by the subtraction of $\sum_n \kappa(\nabla_{\mathbf{r}_n})^2$. In the translationally invariant sector, i.e. for the homogeneous distributions of the scalar, the matrix $\mathcal{D}^{\alpha\beta}(\mathbf{r})$ in the definition of M_N may be replaced by $-D^{\alpha\beta}(\mathbf{r})$ since the contribution of $\mathcal{D}^{\alpha\beta}(0)$ drops out. Below, we shall use this scaling form of dimension $(length)^{\xi-2}$ of the operators M_N assuming also that the diffusion constant κ has been taken to zero.

The Hopf equations (10) may be solved by induction with respect to N. Denoting by $P_N(t; \underline{\mathbf{r}}, \underline{\mathbf{r}}')$ the Green functions $\mathrm{e}^{-tM_N}(\underline{\mathbf{r}}, \underline{\mathbf{r}}')$, we obtain

$$\begin{aligned}
F_N(t; \underline{\mathbf{r}}) = {} & \int P_N(t - t_0; \underline{\mathbf{r}}, \underline{\mathbf{r}}') \, F_N(t_0; \underline{\mathbf{r}}') \, d\underline{\mathbf{r}}' \\
& + \int_{t_0}^{t} ds \int P_N(t - s; \underline{\mathbf{r}}, \underline{\mathbf{r}}') \, (\mathcal{C} \otimes F_{N-2})(s, \underline{\mathbf{r}}') \, d\underline{\mathbf{r}}' . \tag{13}
\end{aligned}$$

For a concentrated initial distribution of θ with fast decaying N-point functions and for $t \to \infty$, the N-point functions $F_N(t; \underline{r})$ tend to the solution $F_N(\underline{r})$ of the stationary version of the Hopf equations

$$M_N F_N = \mathcal{C} \otimes F_{N-2} \tag{14}$$

inductively determined by the relations

$$F_N(\underline{r}) = \int G_N(\underline{r}, \underline{r}') \, (\mathcal{C} \otimes F_{N-2})(\underline{r}') \, d\underline{r}'. \tag{15}$$

where $G_N(\underline{r}, \underline{r}') = \int_0^\infty ds \, P_N(s; \underline{r}, \underline{r}')$ is the kernel of the inverse of the operator M_N. In particular, the limiting stationary distribution is independent of the initial one and has vanishing odd-point functions.

4. Zero mode dominance

We are interested in the scaling properties of the stationary N-point functions $F_N(\underline{r})$ for the injection scale L large w.r.t. point separations, i.e. in the inertial range. Since M_N have the scaling dimension of $(length)^{\xi-2}$ and $\mathcal{C}(\frac{\underline{r}}{L})$ is approximately constant for large L, we might expect that the solutions of the chain (14) of equations scale like

$$F_N(\lambda \underline{r}) = \lambda^{\frac{(2-\xi)N}{2}} F_N(\underline{r}). \tag{16}$$

This would be the normal Kolmogorov-Obukhov-Corrsin scaling.

In the 1994 paper [3], Kraichnan has argued in favor of the anomalous scaling of the scalar correlators. His paper steered a renewed interest in the problem which led to the discovery of a simple mechanism by which the correlators avoid the normal scaling. As was realized in [4] and [5], see also [6], for large L the normally scaling contributions to F_N are dominated by the ones of the *zero modes* of the operators M_N. The stationary Hopf equations (14) leave the freedom to add the solutions of their homogeneous version $M_N f_N = 0$. It is not obvious, however, whether the solution (15), which is defined unambiguously, contains the scaling zero modes which become leading in the inertial range. Such zero mode contributions were, indeed, found in the 2-point function. The latter may be calculated exactly [2] with the result

$$F_2(r) = A_0 L^{2-\xi} + A_1 r^{2-\xi} + \ldots \tag{17}$$

where the terms vanishing as inverse powers when $L \to \infty$ were omitted. The constant $\propto L^{2-\xi}$ is a zero mode of operator M_2 and it dominates for

$r \ll L$. The constant terms, as well as the ones independent of some of $\mathbf{r}_n$'s, however, do not contribute to the structure functions

$$S_N(r) = \left\langle (\theta(\mathbf{r}) - \theta(\mathbf{0}))^N \right\rangle \tag{18}$$

of the scalar. As a result, the 2-point structure function scales as $r^{2-\xi}$, i.e. normally.

What about the higher-point functions? In [4] and in [5] it was shown that the 4-point structure function is dominated for $r \ll L$ by the contribution of a zero mode of the operator M_4 for, respectively, small velocity exponent ξ and large space dimension d. In [7] and [8] these results have been extended to N-point functions with the result

$$S_N(r) \; \propto \; r^{\zeta_N} \quad \text{with} \quad \zeta_N = \frac{(2-\xi)N}{2} - \begin{cases} \frac{N(N-2)\xi}{2(d+2)} + \mathcal{O}(\xi^2) , \\[2mm] \frac{N(N-2)\xi}{2d} + \mathcal{O}(d^{-2}) . \end{cases} \tag{19}$$

The scaling dimensions of the relevant zero modes where found by restricting operators M_N to scaling functions. Upon such a restriction, M_N's become operators with discrete spectrum to which standard perturbative technics apply. A more difficult perturbative analysis around $\xi = 2$ [6][10][11] confirms the zero mode dominance of the inertial-range scaling also in this regime.

5. Slow modes

The Green functions $P_N(t; \mathbf{r}, \mathbf{r}') = \mathrm{e}^{-t\, M_N}(\mathbf{r}, \mathbf{r}')$ have a natural interpretation in the language of Lagrangian trajectories. They are the joint probability distribution functions (PDF's) of the time t positions $\mathbf{r}_n$ of the trajectories $\mathbf{r}_n(s)$ starting at time zero at points $\mathbf{r}'_n$. This follows from the $\mathcal{C} = 0$ version of Eq. (13) if we recall that, in the absence of the source, the scalar density is carried along the Lagrangian trajectories. This interpretation of $P_N(t; \mathbf{r}, \mathbf{r}')$ shows that, effectively, the (differences of the) Lagrangian trajectories undergo a simple diffusion process with the space dependent diffusion matrix $\frac{1}{2} D^{\alpha\beta}(\mathbf{r}_n - \mathbf{r}_m) \propto |\mathbf{r}_n - \mathbf{r}_m|^\xi$ so that they diffuse very slowly when they are close but faster and faster when, eventually, they separate.

Let us look closer at the effective diffusion of the Lagrangian trajectories. The scaling property of operators M_N implies that

$$P_N(\lambda^{2-\xi} t; \lambda\mathbf{r}, \lambda\mathbf{r}_0) = \lambda^{-Nd} P_N(t; \mathbf{r}, \mathbf{r}_0) . \tag{20}$$

Thus the *distances* scale as $(time)^{\frac{1}{2-\xi}}$ as compared to $(time)^{\frac{1}{2}}$ in the standard diffusion. We may probe the effective spread of Lagrangian trajectories with scaling functions f_N such that $f_N(\lambda \mathbf{r}) = \lambda^\sigma f_N(\mathbf{r})$, e.g. with the sum of squares of distances between $\mathbf{r}_n$'s. The average of f_N over the time zero positions of N Lagrangian trajectories which at time t pass by points $\mathbf{r}$ is[1]

$$\langle f_N \rangle_{t,\mathbf{r}} = \int P_N(t;\mathbf{r},\mathbf{r}') \, f_N(\mathbf{r}') \, d\mathbf{r}' = \lambda^{-\sigma} \int P_N(\lambda^{2-\xi}t; \lambda\mathbf{r},\mathbf{r}') \, f_N(\mathbf{r}') \, d\mathbf{r}'$$

$$= (\tfrac{t}{\tau})^{\frac{\sigma}{2-\xi}} \int P_N(\tau; (\tfrac{\tau}{t})^{\frac{1}{2-\xi}}\mathbf{r}, \mathbf{r}') \, f_N(\mathbf{r}') \, d\mathbf{r}' \tag{21}$$

where we have used the scaling properties of f_N and P_N and have set $\tau = \lambda^{2-\xi}t$. If we let t grow and keep τ constant, the last integral may be expected to tend to the limit $\int P_N(\tau; 0, \mathbf{r}') f_N(\mathbf{r}') \, d\mathbf{r}'$ so that, if the latter integral does not vanish,

$$\langle f_N \rangle_{t,\mathbf{r}} \quad \propto \quad (\tfrac{t}{\tau})^{\frac{\sigma}{2-\xi}} \tag{22}$$

This is the super-diffusive behavior which sets in for typical scaling probes f_N. Suppose, however, that f_N is a scaling zero mode of M_N. Since

$$\frac{d}{dt} \langle f_N \rangle_{t,\mathbf{r}} = - \int P_N(t;\mathbf{r},\mathbf{r}') \, (M_N f_N)(\mathbf{r}') \, d\mathbf{r}' = 0, \tag{23}$$

the average $\langle f_N \rangle_{t,\mathbf{r}}$ is constant in time instead of growing like $(\tfrac{t}{\tau})^{\frac{\sigma}{2-\xi}}$: the zero modes describe structures preserved in mean by the Lagrangian flow.

A constant or linear functions are the obvious examples of scaling zero modes but a closer analysis [12] of operators M_N shows that there is a whole discrete series $f_{N,i}$ of them with scaling dimensions $\sigma_i \geq 0$. Besides, each such scaling zero mode gives rise to a tower of *slow modes* $f_{N,ik}$, $k = 0, 1, \ldots$, starting at $f_{N,i0} = f_{N,i}$, with scaling dimensions $\sigma_i + (2 - \xi)k$. The averages $\langle f_{N,ik} \rangle_{t,\mathbf{r}}$ are polynomials of order k in t so that they grow slower than the typical behavior (22). All these scaling modes appear in the small λ asymptotic expansion [12]

$$P_N(\tau; \lambda\mathbf{r}, \mathbf{r}') = \sum_{i,k} \lambda^{\sigma_i+(2-\xi)k} \, f_{N,ik}(\mathbf{r}) \, g_{N,ik}(\tau;\mathbf{r}') \tag{24}$$

which, when inserted on the right hand side of Eq. (21), gives

$$\langle f_N \rangle_{t,\mathbf{r}} = \sum_{i,k} (\tfrac{t}{\tau})^{\frac{\sigma-\sigma_i}{2-\xi}-k} \, f_{N,ik}(\mathbf{r}) \int g_{N,ik}(\tau;\mathbf{r}') \, f_N(\mathbf{r}') \, d\mathbf{r}'. \tag{25}$$

[1]for the later convenience, we study the backward evolution in time

The functions $g_{N,ik}(\tau; \underline{\mathbf{r}})$ are finite and decay fast for large $\underline{\mathbf{r}}$. For generic f_N, the constant zero mode $f_{N,00}$ corresponding to $g_{N,00}(\tau; \underline{\mathbf{r}}') = P_N(\tau; \mathbf{0}, \underline{\mathbf{r}}')$ gives the dominant contribution leading to the behavior (22). This term (and many others) vanishes for $f_N = f_{N,jl}$ resulting in the slower growth.

Upon integration over t, the asymptotic expansion (24) induces the one for the kernels of the inverses of M_N's:

$$G_N(\lambda \underline{\mathbf{r}}, \underline{\mathbf{r}}') = \sum_i \lambda^{\sigma_i} f_{N,i}(\underline{\mathbf{r}}) \, h_{N,i}(\underline{\mathbf{r}}') \qquad (26)$$

(the contributions of the slow modes disappear under the time integration). This expansion is behind the zero-mode dominance of the short-distance behavior of the iterative solutions (15) for the stationary N-point functions of the scalar. The subtractions required in the passage from the correlation functions to the structure functions leave only the contributions of the zero modes $f_{N,i}$ fully dependent on all $\mathbf{r}_n$'s.

6. Fuzzy trajectories

The asymptotic expansion (24) describes what happens when the final points of the Lagrangian trajectories are taken together. It was established with mathematical rigor for the case of two Lagrangian trajectories and was checked in perturbative approaches for $N > 2$. At the first glance, it may look bizarre. Indeed, we may naively expect that if the final points $\mathbf{r}_n$ of the trajectories converge then their initial points should also do with the joint PDF $P_N(t; \lambda \underline{\mathbf{r}}, \underline{\mathbf{r}}')$ becoming proportional to $\prod \delta(\mathbf{r}'_n - \mathbf{r}'_{n+1})$ when $\lambda \to 0$. Instead, $P_N(t; \lambda \underline{\mathbf{r}}, \underline{\mathbf{r}}')$ tends to a regular limit $g_{N,00}(t; \underline{\mathbf{r}}')$. The mechanism by which $P_N(t; \underline{\mathbf{r}}, \underline{\mathbf{r}}')$ avoids the singularity at $\underline{\mathbf{r}} = 0$ is somewhat subtle [12]. Recall that the Lagrangian trajectories satisfy the differential equation (3). The uniqueness of solutions of such an equation requires the Lipschitz condition $|\mathbf{v}(t, \mathbf{r}) - \mathbf{v}(t, \mathbf{r}')| \sim |\mathbf{r} - \mathbf{r}'|$ for $\mathbf{r}'$ tending to $\mathbf{r}$. But our velocities are only Hölder continuous in $\mathbf{r}$ with exponent $\frac{\xi}{2} < 1$, see Eq. (5). The resulting non-uniqueness of the Lagrangian trajectories passing through a fixed point violates the Newton-Leibniz paradigm and allows for a continuum of trajectories with coincident final points. Although the trajectories keep collapsing continuously, the probabilistic quantities such as the joint PDF's $P_N(t; \underline{\mathbf{r}}, \underline{\mathbf{r}}')$ still make perfect sense. They reflect, however, the fuzzyness of the trajectories in their asymptotics (24).

A less rigorous-minded reader might object that the non-uniqueness of the Lagrangian trajectories is a mathematical nuisance since more realistic turbulent velocities, even when showing the behavior (5) with ξ close to $\frac{2}{3}$ in the inertial range, are smoothed on short distances by the viscous effects so that their Lagrangian trajectories are unique. A closer examination, shows, however, that such Lagrangian trajectories, although uniquely determined by sharp values at one time, exhibit a *sensitive dependence* on these values signaled by non-zero *Lyapunov exponents*. As a result, if the fixed-time positions of the trajectories have an ϵ-spread, the trajectories at different times are spread in a large region and will stay such if the viscosity is taken to zero first and the spread ϵ only next. The non-uniqueness of the trajectories caused by their continuous collapse is simply a useful mathematical abstraction describing a real physical phenomenon: a fast spread of the trajectories in each realization of the high Reynolds number velocity field. Such a spread makes the identification of the trajectories practically impossible. It should be remarked that in the incompressible case, the collapse of the trajectories must be accompanied symmetrically by their branching since $P_N(t; \underline{\mathbf{r}}, \underline{\mathbf{r}}') = P_N(t; \underline{\mathbf{r}}', \underline{\mathbf{r}})$. In compressible flows, this symmetry is broken. In particular, in the utmost compressible Burgers flows, the trajectories only collapse together (in a discrete process) sticking to the shocks.

7. Numerical results

The region of ξ neither close to 0 nor to 2 has up to now defied an exact analysis. The first numerical simulations of the system [13][14] were based on the direct solution of the scalar equation (1). They did not cover the region of small ξ and did not permit to discriminate between the small ($\propto \xi$) anomalous structure-function exponents implied by the perturbative zero-mode analysis and the different ($\propto 1$) predictions of a closure of the structure-function equations proposed in [3], see also [15].

Recently, a new numerical analysis [16] of the 4-point function of the scalar in two and three space dimensions has been performed for different values of ξ. It was based on a direct generation of samples of Lagrangian trajectories passing at time t by N fixed points. Such samples allow a direct simulation of the PDF's $P_N(t; \underline{\mathbf{r}}, \underline{\mathbf{r}}')$ as well as direct simulations of the stationary correlation functions F_N. Let us explain how the latter were done in [16]. Upon setting $t_0 = -\infty$ and $\theta(t_0) = 0$ in Eq. (4), one obtains for even N the relations

$$F_N(\underline{\mathbf{r}}) = \Big\langle \prod_{n=1}^{N} \int_{-\infty}^{t} ds \, f(s, \mathbf{r}_n(s)) \Big\rangle = \sum_{\Pi} \Big\langle \prod_{\{n,m\} \in \Pi} \mathcal{T}(\mathbf{r}_n, \mathbf{r}_m) \Big\rangle \quad (27)$$

where the sum, resulting from the average over the Gaussian source, is over the pairings Π of the indices $1, \ldots, N$ and where

$$\mathcal{T}(\mathbf{r}_n, \mathbf{r}_m) = \int_{-\infty}^{t} ds \, C\Big(\frac{\mathbf{r}_n(s) - \mathbf{r}_m(s)}{L}\Big) \quad (28)$$

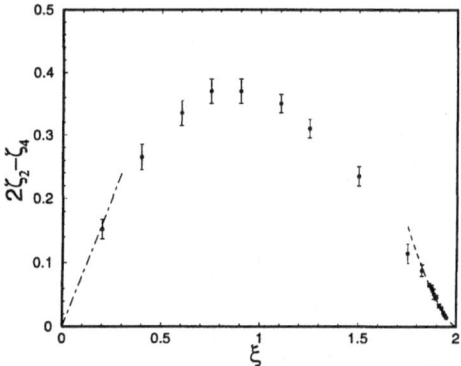

Figure 1.   Anomalous exponent of the 4-point structure function of the scalar as a function of ξ (U. Frisch, A. Mazzino and M. Vergassola, cond-mat/9802192)

is approximately equal to the time spent within distance L by the two trajectories with the end-points $\mathbf{r}_n$ and $\mathbf{r}_m$. The average of the product of such times may be simulated using the ensemble of trajectories passing through points $\mathbf{r}_n$. It is finite since the trajectories effectively separate as $(time)^{\frac{1}{2-\xi}}$. Due to the limited dependence of trajectories on the initial conditions discussed above, this remains true even if $\mathbf{r}_n$ and $\mathbf{r}_m$ tend to each other. The comparison of $F_N(\mathbf{r})$ calculated this way for different configurations of points $\mathbf{r}_n$ showed the dominance of the structure functions $S_N(r)$ by subleading terms of $F_N(\mathbf{r})$, as in the discussion at the end of Section 5. The resulting values of the anomalous 4-point-structure-function exponent $2\zeta_2 - \zeta_4$, presented in Fig. 1 borrowed from [16], confirm the predictions of the perturbative zero-mode analysis around $\xi = 0$ indicated in Fig. 1 by the broken-dotted line.

8. Conclusions

We have exhibited a mechanism behind intermittency of the scalar structure functions in the Kraichnan model of passive advection. The important point is that the anomalous scaling originated from a discrete series of solutions of the homogeneous Hopf equations without the forcing and diffusion terms. This, and the relation of such solutions to the effective fuzzyness of Lagrangian trajectories at high Reynolds numbers promise to be the stories that persist in other intermittent hydrodynamical systems. In particular, it has been indicated recently [17] how discrete solutions of the unforced Hopf equations may naturally give rise to a multiscaling picture of anomalous exponents in the Navier-Stokes case.

References

1. R. Antonia, E. Hopfinger, Y. Gagne and F. Anselmet: *Temperature Structure Functions in Turbulent Shear Flows*. Phys. Rev. **A 30** (1984), 2704-2707
2. R.H. Kraichnan: *Small-Scale Structure of a Scalar Field Convected by Turbulence*. Phys. Fluids **11** (1968), 945-963
3. R. H. Kraichnan: *Anomalous Scaling of a Randomly Advected Passive Scalar*. Phys. Rev. Lett. **72** (1994), 1016-1019
4. K. Gawędzki and A. Kupiainen: *Anomalous Scaling of the Passive Scalar*. Phys. Rev. Lett., **75** (1995), 3834-3837
5. M. Chertkov, G. Falkovich, I. Kolokolov and V. Lebedev: *Normal and Anomalous Scaling of the Fourth-Order Correlation Function of a Randomly Advected Scalar*. Phys. Rev. **E 52** (1995), 4924-4941
6. B. Shraiman and E. Siggia: *Anomalous Scaling of a Passive Scalar in Turbulent Flow*. C.R. Acad. Sci. **321** (1995), 279-284
7. D. Bernard, K. Gawędzki and A. Kupiainen: *Anomalous Scaling of the N-Point Functions of a Passive Scalar*. Phys. Rev. **E 54** (1996), 2564-2572
8. M. Chertkov and G. Falkovich: *Anomalous Scaling Exponents of a White-Advected Passive Scalar*. Phys. Rev Lett. **76** (1996), 2706-2709
9. B. Shraiman and E. Siggia: *Symmetry and Scaling of Turbulent Mixing*. Phys. Rev. Lett. **77** (1996), 2463-2466
10. O. Gat, V. S. L'vov and I. Procaccia: *Perturbative and Non-Perturbative Analysis of the 3'rd Order Zero Modes*. Phys. Rev. **E 56** (1997), 406-416
11. E. Balkovsky, G. Falkovich and V. Lebedev: *Three-Point Correlation Function of a Scalar Mixed by an Almost Smooth Random Velocity Field*. Phys. Rev. **E 55** (1997), R4881-R4884
12. D. Bernard, K. Gawędzki and A. Kupiainen: *Slow Modes in Passive Advection*. cond-mat/9706035
13. R. H. Kraichnan, V. Yakhot and S. Chen: *Scaling Relations for a Randomly Advected Passive Scalar Field*. Phys. Rev. Lett. **75** (1995), 240-243
14. A. L. Fairhall, B Galanti, V. S. L'vov and I. Procaccia: *Direct Numerical Simulations of the Kraichnan Model: Scaling Exponents and Fusion Rules*. Phys. Rev. Lett. **79** (1997), 4166-4169
15. R. H. Kraichnan: *Passive Scalar: Scaling Exponents and Realizability*. Phys. Rev. Lett. **78** (1979), 4922-4925
16. U. Frisch, A. Mazzino and M. Vergassola: *Intermittency in Passive Scalar Advection*. cond-mat/9802192
17. V. I. Belinicher, V. S. L'vov and I. Procaccia: *Computing the Scaling Exponents in Fluid Turbulence from First Principles: Demonstration of Multi-scaling*. chao-dyn/9708004

INSTANTON FOR THE KRAICHNAN MODEL

E. BALKOVSKY[1] AND V. LEBEDEV[1,2]
[1] *Department of Physics of Complex Systems,*
Weizmann Institute of Science, Rehovot 76100, Israel
[2] *Landau Institute for Theoretical Physics, RAS, Kosygina 2,*
117940, Russia.

Abstract. We consider high-order correlation functions of passive scalar in the Kraichnan model [1]. Using the instanton formalism we find the exponents ζ_n of the structure functions S_n for $n \gg 1$ at the condition $d\zeta_2 \gg 1$ (d is the dimensionality of space). At $n < n_c$ (where $n_c = d\zeta_2/[2(2-\zeta_2)]$) the exponents are $\zeta_n = (\zeta_2/4)(2n - n^2/n_c)$, whereas at $n > n_c$ they are n-independent: $\zeta_n = \zeta_2 n_c/4$. We also estimate n-dependent factors in S_n and critical behavior of S_n at n close to n_c.

We develop a technique based on the path-integral representation of the dynamical correlation functions of classical fields [2]. The main idea, formulated in [3], is related to the possibility to use the saddle-point approximation in the path integral at large order n of correlation functions. The saddle-point conditions are integro-differential equations describing an object which, by analogy with the quantum field theory, we call instanton. The formalism enables to find correlation functions of the passive scalar for arbitrary $n \gg 1$, provided $d\zeta_2 \gg 1$. Technical details are presented in [4].

The advection of the passive scalar θ is described by the equation

$$\partial_t \theta + \mathbf{v}\nabla\theta - \kappa\nabla^2\theta = \phi,$$

where κ is the diffusion coefficient. The passive scalar correlation functions are determined by the statistics of the advecting velocity $\mathbf{v}$ and the source ϕ, which in the frame of the Kraichnan model are assumed to be random, δ-correlated in time functions, described by Gaussian statistics homogeneous in space. The statistical properties of the fields are entirely characterized by their pair correlation functions. For the velocity we have

$$\langle v_\alpha(t_1, \mathbf{r}_1) v_\beta(t_2, \mathbf{r}_2) \rangle = \delta(t_1 - t_2)\{\mathcal{V}_0\delta_{\alpha\beta} - \mathcal{K}_{\alpha\beta}(\mathbf{r}_1 - \mathbf{r}_2)\}$$

U. Frisch (ed.), Advances in Turbulence VII, 503–506.

$$\mathcal{K}_{\alpha\beta}(\mathbf{r}) = \frac{D_*}{d(d-1)} r^{-\gamma} \left[(d+1-\gamma)r^2 \delta_{\alpha\beta} - (2-\gamma)r_\alpha r_\beta \right],$$

where d is the dimensionality of space and D_* is a constant characterizing the strength of velocity fluctuations. Correlation functions of ϕ are determined by a smooth function $\chi(r_{12}/L)$. One assumes that fluctuations of the velocity are strong enough to guarantee $D_* L^{2-\gamma} \gg \kappa$. Then there exists a convective interval of scales $r_d \ll r \ll L$, where r_d is the diffusive length $r_d^{2-\gamma} \sim \kappa/D_*$. We study the properties of correlation functions of θ only in the convective interval.

It is convenient to examine the anomalous scaling in terms of structure functions $S_n(r) = \langle |\theta(\mathbf{r}/2) - \theta(-\mathbf{r}/2)|^n \rangle$. One expects that they have a scaling behavior $S_n(r) \propto r^{\zeta_n}$. In [5, 6] S_n were analyzed perturbatively at the conditions $(2-\gamma) \ll 1$ or $d\gamma \gg 1$ where the statistics of θ is close to Gaussian. The analysis predicts anomalous scaling with

$$\zeta_n = \frac{n\gamma}{2} - \frac{2-\gamma}{2(d+2)} n(n-2). \tag{1}$$

The calculations leading to (1) are correct if the the anomalous contribution is much smaller than the normal one, which gives a condition $n \ll d\gamma/(2-\gamma)$. On the contrary, our approach allows to find the exponents ζ_n for any order $n \gg 1$ at the same condition $d\gamma \gg 1$ as in [6].

Statistics of classical fields in the presence of random forces can be examined with the help of the field technique formulated in [2]. The correlation functions are calculated as path integrals with the weight $\exp(i\mathcal{I})$, where $\mathcal{I}$ is the effective action related to dynamical equations for the fields. One could expect [3] that the limit $n \gg 1$ can be considered in the saddle-point approximation in the path integral. Unfortunately, in its naive form, the method is not suited for the problem, because of strong fluctuations on the background of the saddle-point solution [3, 4].

To get rid of these fluctuations, we transformed the action into a new form, so that the structure functions are given by the formula

$$S_n = \int \frac{dy\, d\vartheta}{2\pi} \int \mathcal{D}R \mathcal{D}m \, \exp\left(i\mathcal{I}_R - \mathcal{F}_\lambda - iy\vartheta + n\ln|\vartheta| \right), \tag{2}$$

$$i\mathcal{I}_R = i \int dt d\mathbf{r}_{12} m_{12} (\gamma^{-1}\partial_t R_{12}^\gamma + D_*) - \frac{D_*}{d} \int dt d\mathbf{r}_{1234} Q_{12,34} m_{12} m_{34},$$

$$\mathcal{F}_\lambda = \frac{y^2}{2} \int dt\, d\mathbf{r}_1\, d\mathbf{r}_2\, \chi(R_{12})\beta(\mathbf{r}_1)\beta(\mathbf{r}_2).$$

Here $R_{12}(t) \equiv R(t, \mathbf{r}_1, \mathbf{r}_2) = |\rho(t, \mathbf{r}_1) - \rho(t, \mathbf{r}_2)|$ is the absolute value of the distance between two Lagrangian trajectories, which are determined by the equation $\partial_t \rho = \mathbf{v}(t, \rho)$ and initial condition $\rho(0, \mathbf{r}) = \mathbf{r}$. The auxiliary field

$m_{12} \equiv m(t, \mathbf{r}_1, \mathbf{r}_2)$ is the conjugated field to R_{12}, and Q is some rather complicated function of R_{ij}. In the function $\beta(\mathbf{r}_1) = \delta_\Lambda(\mathbf{r}_1 - \mathbf{r}/2) - \delta_\Lambda(\mathbf{r}_2 + \mathbf{r}/2)$ we use δ-functions smeared over $\Lambda^{-1} \gg r_d$, since (2) is obtained in the diffusionless limit. Let us stress that due to the definition of R_{12} the triangle inequality $R_{12} + R_{23} > R_{13}$ should be satisfied. It is very hard to it take into account explicitly. We ignore the constrains at treating the instanton, which is correct at the condition $d\gamma \gg 1$.

The integral (2) is calculated in the saddle-point approximation regarding the number n large enough. It means that

$$S_n \sim \exp\left(i\mathcal{I}_R - \mathcal{F}_\lambda - iy\vartheta + n\ln\vartheta\right)|_{\text{inst}} ,$$

where solutions of the instantonic equations

$$i(\gamma^{-1}\partial_t R_{12}^\gamma + D_*) = 2\frac{D_*}{d}\int d\mathbf{r}_3\, d\mathbf{r}_4\, Q_{12,34}m_{34} , \tag{3}$$

$$iR_{12}^{\gamma-1}\partial_t m_{12} + \frac{y^2}{2}\chi'(R_{12})\beta(\mathbf{r}_1)\beta(\mathbf{r}_2) \tag{4}$$

$$+\frac{D_*}{d}\int d\mathbf{r}_3\, d\mathbf{r}_4\, \left\{2\frac{\partial Q_{12,34}}{\partial R_{12}}m_{12}m_{34} + 4\frac{\partial Q_{13,24}}{\partial R_{12}}m_{13}m_{24}\right\} = 0 ,$$

$$|y|^2 \int dt\, d\mathbf{r}_{12}\chi(R_{12})\beta(\mathbf{r}_1)\beta(\mathbf{r}_2) = n \tag{5}$$

should be substituted. Eqs. (3-5) are extrema conditions for (2).

It is possible to solve the instantonic equations and calculate the structure functions. First, at $n < n_c \equiv d\gamma/(2(2-\gamma))$ one gets

$$S_n \sim \left(\frac{n}{\gamma}\frac{\chi(0)C_1}{D_*}L^\gamma\right)^{n/2}\left(\frac{r}{L}\right)^{\zeta_n} . \tag{6}$$

Here $\zeta_n = n\gamma/2 - (2-\gamma)n^2/(2d)$. The quantity C_1 in (6) is a non-universal constant of order unity, depending on the details of the pumping.

In the case $n > n_c$ we obtained the following result

$$S_n \sim \left(\frac{n}{\gamma}\frac{\chi(0)C_2}{D_*}L^\gamma\right)^{n/2}\left(\frac{r}{L}\right)^{\zeta_c} . \tag{7}$$

The scaling exponents ζ_n appear to be n-independent and equal to the value $\zeta_c = d\gamma^2/(8(2-\gamma))$. The quantity C_2 in (7) is again a non-universal constant of order unity.

The vicinity of the critical value $n = n_c$ needs a special treatment. Though the anomalous exponents are not changed, the structure functions acquire critical dependence on $|n - n_c|$

$$S_n \sim \left(\frac{(n-n_c)^2}{\gamma n_c}\frac{\chi(0)C_\pm}{D_*}L^\gamma\right)^{n_c/2}\left(\frac{r}{L}\right)^{\zeta_n} , \tag{8}$$

Here the condition $n_c/\ln(L/r) \ll |n - n_c| \ll n_c$ is implied. The constants $C_\pm$ are non-universal constants, different for $n < n_c$ and $n > n_c$.

Let us summarize. First of all, we have established the n-dependence of the scaling exponents. Note the similar non-analytical behavior of ζ_n for Burgers' turbulence. For $n \ll n_c$ our result coincides with the answer obtained perturbatively at large d [6]. Surprisingly, the quadratic dependence of ζ_n on n is kept up to $n = n_c$. Such n-dependence of ζ_n is well-known from the log-normal distribution proposed by Kolmogorov. However, it is definitely inapplicable here. The reason is in n-dependence of the structure functions S_n which is also established in our work. The expressions (6,7) show that the behavior of r-independent prefactors in S_n is characteristic rather of Gaussian distribution. A natural explanation can be found in terms of zero mode ideology [5, 6]. We know that for $n > 2$ the main contribution to the structure function S_n in the convective interval is related to zero modes of the equation for the n-th order correlation function of the passive scalar. The exponents of the modes are determined by the equation (and could be very sensitive to the value of n) whereas the coefficients at the modes (determining their contribution to S_n) are determined by the pumping scale where the statistics of the passive scalar is nearly Gaussian.

Probably the most striking feature of our results is the critical behavior of S_n near $n = n_c$. The behavior, determined by (8) resembles mainly a critical behavior of, say, heat capacity near a second-order phase transition, the quantity $(n - n_c)/n_c$ playing the role of the reduced temperature. Then the inequality after (8) is the analog of the size restriction (the conventional critical behavior is observed if the critical radius is less than the size of the specimen). It would be very interesting to check the possibility of the existence of such critical behavior for real systems.

References

1. R. H. Kraichnan, Phys. Fluids **11**, 945 (1968).
2. P. C. Martin, E. Siggia, and H. Rose, Phys. Rev. A**8**, 423 (1973), C. de Dominicis, J. Physique (Paris) **37**, c01-247 (1976), H. Janssen, Z. Phys. B **23**, 377 (1976).
3. G. Falkovich, I. Kolokolov, V. Lebedev and A. Migdal, Phys. Rev. E**54**, 4896 (1996).
4. E. Balkovsky and V. Lebedev, in preparation.
5. K. Gawedzki and A. Kupiainen, Phys. Rev. Lett. **75**, 3608 (1995), D. Bernard, K. Gawedzki and A. Kupiainen, Phys. Rev. E **54**, 2564 (1996).
6. M. Chertkov, G. Falkovich, I. Kolokolov and V. Lebedev, Phys. Rev. E **52**, 4924 (1995), M. Chertkov and G. Falkovich, Phys. Rev. Lett. **76**, 2706 (1995).

ON THE COUPLING BETWEEN ANOMALOUS PASSIVE SCALAR AND VELOCITY INCREMENTS IN TURBULENCE

O. N. BORATAV[1] AND R. B. PELZ[2]
[1] *School of Chemical Engineering, Cornell University, Olin Hall, Ithaca, NY 14850, USA*
[2] *Mechanical and Aerospace Engineering Department, Rutgers University, Piscataway, NJ 08854-8058, USA*

In 1949, Yaglom[1] derived an exact relation for the mixed passive scalar/velocity structure function known as the four-thirds law:

$$S_3^{TU}(r) \equiv \left\langle (T(\boldsymbol{x}+\boldsymbol{r}) - T(\boldsymbol{x}))^2 \, (u(\boldsymbol{x}+\boldsymbol{r}) - u(\boldsymbol{x})) \right\rangle = -\frac{4}{3}\epsilon_T r \quad , \qquad (1)$$

where $S_3^{TU}(r)$ is the third order structure function of the mixed passive scalar (T) and velocity (u) increments, $\boldsymbol{r}$ is the separation between the two points, (where $r = |\boldsymbol{r}|$), $\boldsymbol{x}$ is the space location of a given point, and ϵ_T is the passive scalar dissipation rate. This expression can be generalized noting that the Euler equations are invariant under the scaling transformation: $\boldsymbol{x} \mapsto \lambda\boldsymbol{x}$, $\boldsymbol{u} \mapsto \lambda^h \boldsymbol{u}$, $t \mapsto \lambda^{1-h}t$, and using the four-fifths law[2], one obtains[3] $h = 1/3$. Then, since the passive scalar advection equation (for zero diffusivity) is invariant under the scaling transformation $T \mapsto \lambda^m T$, one obtains $m = h = 1/3$ using both the four-fifths and the four-thirds laws. This result can be generalized to the ($3ath$ order) passive scalar/velocity mixed structure function as:

$$S_{3a}^{TU}(r) \equiv \left\langle (T(\boldsymbol{x}+\boldsymbol{r}) - T(\boldsymbol{x}))^{2a} \, (u(\boldsymbol{x}+\boldsymbol{r}) - u(\boldsymbol{x}))^a \right\rangle = -\frac{4}{3}\epsilon_T r^a \, . \, . \qquad (2)$$

Now, given the scaling forms, $\langle (\Delta T^3)^a \rangle \sim r^\alpha$, and $\langle (\Delta u^3)^a \rangle \sim r^\beta$, one asks how γ given in $\langle (\Delta T^2 \Delta u)^a \rangle \sim r^\gamma$ can be related to α and β ? Using the Hölder inequality and simplification[4] lead to:

$$\gamma > \frac{2\alpha}{3} + \frac{\beta}{3} \qquad (3)$$

This result simply means[4] that γ can *not* be smaller than the smaller of α or β. According to this finding, given that α and/or β are anomalous, γ

U. Frisch (ed.), Advances in Turbulence VII, 507–510.

might *not* even be anomalous. Here, one has to define clearly what '*anomalous*' means. We use this term when a scaling exponent shows deviations from a prediction based on the generalization of an exact result (i.e. a result derived from the governing dynamical equations of the problem). For β and γ, what is meant by 'anomalous' is clear, since we have two exact results (4/5th and 4/3rd laws) on which we can base our generalization. On the other hand, for α, there is no exact result on which the generalization of the a-th order scaling exponent can be based. It is common in literature to take $\alpha = a$ (see the previously given notation), however, this prediction can only be approximate.

We perform direct numerical simulations by solving the three dimensional Navier-Stokes equation and the advection-diffusion equation for a passive scalar using a Fourier pseudospectral method with an effective resolution of 1024^3 or 1200^3 points with symmetries in a triply-periodic domain. The initial condition of the passive scalar is chosen such that

$$T(x_i, y_j, z_k) = \left(\sum_{even \; l,m,n=0}^{2N} + \sum_{odd \; l,m,n=1}^{2N-1} \right) \hat{T}_{l,m,n} \cos lx_i \cos my_j \cos nz_k,$$

where $x_i = i2\pi$, $y_j = j2\pi$, $z_k = k2\pi$ and $0 \le i, j, k < N$. A range of Prandtl numbers (Pr) is considered in simulations with $Pr = 0.5, 1, 3$ and 5. In this paper, we give results from $Pr = 1$ simulation only. The results of the passive scalar exponents in this paper are computed at time $t = 4.05$ (at this time, Taylor Reynolds number Re_λ is 141) and at $t = 5.04$ ($Re_\lambda = 99$). We also computed the velocity exponents at these particular times and also used our previous results[5] given at time $t = 6$.

We use the 4/5th and the 4/3rd law to determine the boundaries of the inertial/convective range. To compute the scaling exponents, we use the extended self similarity (ESS) method[6]. For the velocity increment exponents β, we plot the a-th order moment against $\langle |\Delta u^3| \rangle$ whereas for the mixed exponent γ we plot the moments against the quantity $\langle |\Delta T^2 \Delta u| \rangle$, from which we extract the slopes using least squares.

The uncertainty in the exponent calculation arises from the time dependence of the data, different criteria that can be used to determine the boundaries of the inertial range and the error in the ESS least squares fitting. We find that the fitting error is the least in magnitude among these three uncertainty sources.

We present the scaling exponents in Table 1. For all the orders investigated here, the passive scalar deviates more than the velocity, i.e. $\alpha < \beta$. The exponent of the mixed structure function γ is also anomalous and its relative magnitude is such that $\beta > \gamma > \alpha$. The magnitude of γ with respect to the other two is such that it is closer to the less anomalous one, i.e. closer to the velocity structure function exponent β, rather than the passive scalar structure function exponent α.

Now, let us consider the following question: Given that the velocity and

a	$\langle (\Delta T^3)^a \rangle$	$\langle (\Delta u^3)^a \rangle$	$\langle (\Delta v^3)^a \rangle$	$\langle (\Delta T^2 \Delta u)^a \rangle$	$\langle (\Delta T^2 \Delta v)^a \rangle$
$\frac{2}{3}$	0.63 ± 0.01	0.71 ± 0.01	0.73 ± 0.01	0.72 ± 0.01	0.72 ± 0.01
$\frac{4}{3}$	0.97 ± 0.02	1.29 ± 0.02	1.22 ± 0.02	1.25 ± 0.02	1.24 ± 0.02
$\frac{6}{3}$	1.08 ± 0.04	1.73 ± 0.04	1.53 ± 0.04	1.61 ± 0.04	1.58 ± 0.04

TABLE 1. Exponents ζ in the power law $\langle (..)^a \rangle \sim r^\zeta$, for different orders of a.

passive scalar increments exhibit anomalous scaling, can a mixed increment of velocity and passive scalar result in an even larger anomaly than the one that would be obtained when ΔT^{2p} and Δu^p act independent of one another? Let us denote the anomaly by the deviation μ in $\langle \Delta u^p \rangle \sim r^{p/3 - \mu_p^U}$, $\langle \Delta T^{2p} \rangle \sim r^{2p/3 - \mu_{2p}^T}$, $\langle \Delta u^p \Delta T^{2p} \rangle \sim r^{p - \mu_{3p}^{TU}}$. Here μ_p^U, μ_{2p}^T and μ_{3p}^{TU} denote the deviations (which can be positive or negative depending on the order p) from the Kolmogorov-Oboukhov-Yaglom scaling predictions.

If the anomalies in the passive scalar and velocity are statistically independent, one obtains:

$$\mu_p^U + \mu_{2p}^T = \mu_{3p}^{TU} \ . \tag{4}$$

To check whether this is the case, we consider $p = 2$ as an example. From Table 1 obtain, $\mu_2^U = -0.04$, $\mu_4^T = 0.36$ and $\mu_6^{TU} = 0.39$. Hence, $\mu_2^U + \mu_4^T = 0.32 < \mu_6^{TU}$. In other words, the anomaly of the mixed increments μ_6^{TU} is larger than the contribution of the anomaly of each of them acting independent of one another. We call this the *enhanced (coupled) anomaly*.

Finally, we note that, using refined similarity hypothesis[7] the right hand side of Eq.(2) can be expressed as the product of the r-averaged velocity and passive scalar moments[8]. This result implies that there is no room for *enhanced anomaly* under the refined similarity formalism.

Acknowledgements: We would like to thank Luca Biferale, Reginald Hill and Marc Nelkin for helpful comments.

References

1. Yaglom, A. M. (1949) Local Structure of the Temperature Field in a Turbulent Flow, *Dokl. Akad. Nauk SSSR*, **69(6)**, 743.
2. Kolmogorov, A. N. (1941) Dissipation of Energy Under Locally Isotropic Turbulence, *C. R. Acad. Sci.*, **32**, 16.
3. Frisch, U. (1995) *Turbulence: The Legacy of A. N. Kolmogorov*, Cambridge University Press.

4. Boratav, O. N. and Pelz, R. B. (1998) Coupling between anomalous velocity and passive scalar increments in turbulence, submitted to *Phys. Fluids*.
5. Boratav, O. N. and Pelz, R. B. (1997) Structures and Structure Functions in the Inertial Range of Turbulence, *Phys. Fluids*, , **9(5)**, 1400.
6. Benzi, R., Ciliberto, S., Baudet, C., Chavarria, G. R. and Tripiccione, R. (1993) Extended Self-Similarity in the Dissipation Range of Fully Developed Turbulence, *Europhys. Lett.*, **24(4)**, 275.
7. Kolmogorov, A. N. (1962) A refinement of previous hypotheses concerning the local structure of turbulence in a viscous incompressible fluid at high Reynolds number, *J. Fluid Mech.*, **13**, 82.
8. Nelkin, M. (1997) private communication.

TWO TYPES OF INTERMITTENCY OF PASSIVE SCALAR FLUCTUATIONS

T. ELPERIN, N. KLEEORIN AND I. ROGACHEVSKII

Department of Mechanical Engineering, Ben-Gurion University of the Negev, Beer-Sheva 84105, P. O. Box 653, Israel

I. Introduction. Turbulent transport of passive scalar (e.g., number density of particles) advected by an incompressible fluid flow in the case $U = v$ was studied in a large number of publications. Here U is the particle velocity and v is the velocity of fluid. Interesting features in turbulent transport of passive scalar appear when $U \neq v$. In this case new phenomena, e.g., turbulent thermal diffusion and turbulent barodiffusion [1], and self-excitation, i.e., exponential growth of fluctuations of the number density of inertial particles [2] occur. These effects are caused by inertia of particles which results in a divergent velocity field of particles. The self-excitation of the fluctuations results in the intermittency in spatial distribution of inertial particles [2].

There are two types of intermittency: the intermittency in the systems with and without external pumping. Intermittency in the systems without external pumping was predicted by Zeldovich, Molchanov, Ruzmaikin and Sokoloff in 1985 (see, e.g., [3], and references therein). In these systems under certain conditions there is a self-excitation of fluctuations of passive scalar or vector (magnetic) fields. The growth rate γ_s of the s-order correlation function of passive scalar (see [2]) and vector (see [3,4]) fields is given by $\gamma_s = s(s-1)\gamma_2/2$ [where γ_2 is the growth rate of the second-order correlation function of passive scalar or vector (magnetic) fields]. This implies that when $\gamma_2 > 0$ (i.e., fluctuations of passive fields are excited), higher moments grow faster than lower moments, i.e., $\gamma_s > \gamma_{s-1}$ and $\gamma_s > s\gamma_2/2$. This results in intermittency, i.e., the appearance of sharp peaks in which the main part of the field intensity is concentrated. Passive scalar fluctuations (e.g., fluctuations of the particles number density) can be excited when $U \neq v$ (e.g., for inertial particles) or when $U = v$ but div $v \neq 0$ (e.g., for a low-Mach-number compressible turbulent fluid flow) [2].

U. Frisch (ed.), Advances in Turbulence VII, 511–514.
© 1998 *Kluwer Academic Publishers.*

Intermittency in the systems with external pumping was predicted by Kraichnan in 1994 (see [5]). Here the fluctuations of passive scalar are sustained by an external source. In these systems a problem of anomalous scaling arises. For incompressible turbulent flow and when $\mathbf{U} = \mathbf{v}$ the anomalous scalings for scalar field can occur only beginning with the forth-order correlation function.

It is shown here that the anomalous scalings appear already in the second-order correlation function of the number density of inertial particles when the degree of compressibility $\sigma > 1/27$ (where σ is the ratio of the energies in compressible and incompressible component of the particles velocity). In this case there is no self-excitation of fluctuations of the number density of inertial particles, and these fluctuations are maintained by an external source. Which type of intermittency can occur in a system? It depends on Reynolds number in the case of inertial particles and on magnetic Reynolds number in the case of magnetic fluctuations. When Reynolds number is larger than a certain critical value, the first type of intermittency (without external pumping) occurs. On the other hand, when Reynolds number is smaller than this critical value, the second type of intermittency (with external pumping) appears.

II. Small-scale fluctuations. Number density $n_p(t, \mathbf{r})$ of small particles in a turbulent flow is determined by the equation: $\partial n_p / \partial t + \vec{\nabla} \cdot (n_p \mathbf{U}) = D \Delta n_p$, where $\mathbf{U}$ is a random velocity field of the particles which they acquire in a turbulent fluid velocity field, D is the coefficient of molecular diffusion. We consider the case of large Reynolds and Peclet numbers. To study the fluctuations of inertial particles concentration we derive equation for the second-order correlation function of particles concentration $\Phi = \langle \Theta(\mathbf{x}) \Theta(\mathbf{y}) \rangle$. We consider a homogeneous and isotropic turbulent velocity field. The particles velocity field is also homogeneous and isotropic, and it is compressible, i.e., $\vec{\nabla} \cdot \mathbf{U} \neq 0$ due to the particles inertia [2].

We seek a solution to the equation for Φ in the form: $\Phi(t, r) = \Psi(r) \exp[-\int_0^r \chi(x)\, dx] \exp(\gamma t)/r$, where unknown function $\Psi(r)$ is determined by equation $d^2\Psi/dr^2 - m(r)[\gamma + U_0(r)]\Psi = 0$, where $U_0(r) = [(2\chi/r + \chi^2 + \chi') - \kappa(r)]/m(r)$, and $1/m(r) = 2/\mathrm{Pe} + 2[1 - F - (rF_c)']/3$, and $\chi(r) = m(r)(10F_c' - F' + 2rF_c'')/3$, and $\kappa(r) = -2(8F_c'/r + 7F_c'' + rF_c''')/3$, and distance r is measured in units of l_0, time t is measured in units of $\tau_0 = l_0/u_0$, and $\cdot$Pe $= l_0 u_0/D \gg 1$ is the Peclet number. Here u_0 is the characteristic velocity in the maximum scale l_0 of turbulent motions. The function $F_c(r)$ describes the potential component whereas $F(r)$ corresponds to the vortical part of the turbulent velocity of particles [2].

We consider the case of large Schmidt numbers, Sc $= \nu_0/D \gg 1$, where ν_0 is the kinematic viscosity of the fluid. Solution of the equation for Ψ and the growth rate of fluctuations is obtained using an asymptotic analysis. In particular, the growth rate of fluctuations of particles concentration is

$$\gamma = \frac{2[c^2 + (q-a)^2]^2}{3q^4(3-p)^2 r_a^{2q}} \ln^2\left(\frac{\text{Re}}{\text{Re}^{(\text{cr})}}\right), \tag{1}$$

where $c = \sqrt{M(q,\sigma)}/2(q+3)(1+q\sigma)$, and σ is the parameter of compressibility, and $M(q,\sigma) = b_1 B^2 + b_2 B + b_3$, and $B(q,\sigma) = 2\sigma(q+3) - 1$, and $b_1(q) = 4(q-1)(9-2q^2) - q^2$, and $b_2(q) = 2(q+2)[2(q-1)(4q^2+9q+9)-3q]$, and $b_3(q) = 3(q+2)^2(2q-3)(2q+1)$, and $a = [q-\sigma(2q^2+3q-6)]/2(1+q\sigma)$, and $r_a = (\tau_p/\tau_0)^{1/(p-1)}$, $\text{Re} > \text{Re}^{(\text{cr})}$ and the critical Reynolds number $\text{Re}^{(\text{cr})}$ is given by $\text{Re}^{(\text{cr})} \simeq r_a^{p-3} \exp[(3-p)(\pi k + \arctan((q-a)/c))/c]$, where $k = 1, 2, 3, ...$, and $q = 2p - 1$, and p is the exponent in the spectrum of kinetic turbulent energy, τ_p is the characteristic time of coupling between the particle and surrounding fluid (Stokes time). Thus, the fluctuations of particle concentration can be excited without an external source. The physics of self-excitation (exponential growth) of fluctuations of particle concentration is as follows. The inertia causes particles inside the turbulent eddy to drift out to the boundary regions between eddies (the regions with maximum pressure of the fluid). Indeed, particles inertia results in $\vec{\nabla} \cdot \mathbf{U} \propto \tau_p \Delta P/\rho$, where P and ρ are the pressure and the density of a turbulent fluid. On the other hand, for large Peclet numbers $\vec{\nabla} \cdot \mathbf{U} \propto -dn_p/dt$. Therefore, $dn_p/dt \propto -\tau_p \Delta P/\rho$. Thus there is accumulation of inertial particles (i.e., $dn_p/dt > 0$) in regions with the maximum pressure of a turbulent fluid (i.e., where $\Delta P < 0$). Similarly, there is an outflow of inertial particles from the regions with the minimum pressure of fluid. This mechanism acts in a wide range of scales of a turbulent fluid flow. Turbulent diffusion results in relaxation of fluctuations of particle concentration in large scales. However, in small scales where turbulent diffusion is small, the relaxation of fluctuations of particle concentration is very weak. Therefore the fluctuations of particle concentration are localized in the small scales.

III. Anomalous scaling for fluctuations of particles concentration. Consider fluctuations of inertial particles concentration in the presence of an external source $I(r)$. We study the case when there is no self-excitation of the fluctuations of the number density of inertial particles, i.e., the case when $\text{Re} < \text{Re}^{(\text{cr})}$. Equation for the unknown function $\psi(t,r)$ is given by $\partial\psi/\partial t = (\partial^2\psi/\partial r^2)/m - U_0(r)\psi + \tilde{f}(r)$, where $\tilde{f}(r) = rI(r)\exp[\int_0^r \chi(x)\,dx]$. Hereafter we study a zero mode for this equation (i.e., the mode with $\gamma = 0$). The external source in these scales is chosen as follows: $I(r) = I_0(1 - r^s)$, where $s > 0$, and for $r > 1$, $I(r) = 0$. When $l_\nu \leq r < r_a$ the second moment $\Phi(r)$ is given by $\Phi = r^{1/2-a}(A_3 r^{|c|} +$

$A_4 r^{-|c|}) - I_- r^{3-q}$ [for $1/27 < \sigma < \min(\sigma_1, 1/7)$], and $\Phi = A_3 r^{-a} \cos(c \ln r + \varphi_1) - I_+ r^{3-q}$ (for $\sigma_1 < \sigma < 1/7$), where $I_\pm = I_0/2\beta_m[(3-q+a)^2 \pm c^2]$, and $\beta_m = (1+3\sigma)/3(1+\sigma)$, and $\sigma_1 = (B_1+1)/2(q+3)$, and B_1 is the smaller root of the equation $M = 0$, and the viscous scale is $l_\nu \sim \mathrm{Re}^{-1/(3-p)}$, where $\mathrm{Re} = l_0 u_0/\nu_0 \gg 1$ is the Reynolds number.

When $r_a < r \ll 1$ the second moment $\Phi(r)$ is given by $\Phi(r) = A_5 + A_6 r^{-q} - [I_0/2(3-q)]r^{3-q}$, and when $r \gg 1$ the function $\Phi(r) = A_7/r$. Matching functions $\Phi(r)$ and $\Phi'(r)$ at the boundaries of these regions yields the constant A_k. The term $\propto r^{3-q}$ in these equations corresponds to a normal scaling for the second moment of inertial particles concentration, whereas the term $\propto r^{-q}$ in the range $r_a < r \ll 1$ corresponds to the anomalous scaling. When $\sigma_1 < \sigma < 1/7$ the anomalous scaling in the range $l_\nu \le r < r_a$ is complex ($\propto r^{-a\pm i|c|}$).

IV. Conclusions. It is shown here that the anomalous scaling appears already in the second moment of the number density of inertial particles when the degree of compressibility of the particles velocity $\sigma > 1/27$. It is demonstrated that inertia of particles causes a self-excitation of fluctuations of particles concentration. The growth rates of the higher moments of particles concentration is higher than those of the lower moments, i.e., particles spatial distribution is intermittent. When the inertia effect is negligible (e.g., for small size of particles or gaseous admixture) but the fluid velocity field is divergent, the moments of the concentration field grow and there is intermittency without an external source of fluctuations of particles concentration. The self-excitation of fluctuations of particles concentration is observed in atmospheric turbulence, e.g., this effect causes formation of small-scale inhomogeneities in droplet clouds ("inch clouds") which were discovered recently [6]. Small-scale inhomogeneities in spatial distribution of inertial particles were observed also in laboratory [6].

References

1. T. Elperin, N. Kleeorin and I. Rogachevskii, Phys. Rev. Lett. **76**, 224 (1996); Phys. Rev. E **55**, 2713 (1997).

2. T. Elperin, N. Kleeorin and I. Rogachevskii, Phys. Rev. Lett. **77**, 5373 (1996).

3. Ya. B. Zeldovich, A. A. Ruzmaikin, and D. D. Sokoloff, *The Almighty Chance* (Word Scientific Publ., London, 1990).

4. I. Rogachevskii and N. Kleeorin, Phys. Rev. E **56**, 417 (1997).

5. R. Kraichnan, Phys. Rev. Lett. **72**, 1016 (1994).

6. B. Baker, J. Atmosph. Sci. **49**, 387 (1992); J. R. Fessler, J. D. Kulick, and J. K. Eaton, Phys. Fluids **6**, 3742 (1994).

TURBULENT TRANSPORT OF VECTOR (MAGNETIC) FIELD

N. KLEEORIN AND I. ROGACHEVSKII

Department of Mechanical Engineering, Ben-Gurion University of the Negev, Beer-Sheva 84105, P. O. Box 653, Israel

I. Introduction. The generation of magnetic fluctuations by homogeneous, isotropic and incompressible turbulence with zero mean magnetic field was mainly studied for magnetic Prandtl numbers $\Pr_m \geq 1$ (see, e.g., [1-3]). However, in astrophysics the magnetic Prandtl numbers is small. On the other hand, generation of magnetic fluctuations with zero mean field for $\Pr_m \ll 1$ were not observed in numerical simulations (see, e.g., [4]). In addition, the turbulent velocity field cannot be considered as a divergence-free in astrophysics (e.g., accretion disks, solar and stellar convection zones, galaxies).

We study here the generation of magnetic fluctuations with zero mean magnetic field for $\Pr_m \ll 1$. Turbulent fluid velocity field is assumed to be homogeneous and isotropic with a very short scale-dependent correlation time. We have found that magnetic fluctuations can be generated by turbulent motions of conducting fluid flow even with Kolmogorov energy spectrum $\Pr_m \ll 1$. Equation for the high-order correlation functions of the magnetic field is derived. It is shown that spatial distribution of the magnetic field is intermittent. In addition, we study the effect of compressibility (i.e., $\nabla \cdot \mathbf{u} \neq 0$) of the low-Mach-number turbulent fluid flow $\mathbf{u}$ on the generation of magnetic fluctuations. Problems of anomalous scalings for vector (magnetic) field is discussed as well.

II. Mathematical formulation of the problem. Consider dynamics of magnetic fluctuations $\mathbf{h}$ with zero mean field in a low-Mach-number compressible turbulent fluid flow. We derive equations for the high-order correlation functions of the magnetic field. To this purpose we use the method of path integrals (Feynman-Kac formula) [2,3,5]. The equation for the second-order correlation function $W(r,t) = \langle h_\tau(\mathbf{x},t) h_\tau(\mathbf{y},t) \rangle$ is given by $W(t,r) = \Psi(r)\sqrt{m}\exp(\gamma t)/r^2$, where the unknown function $\Psi(r)$ is determined by equation $\Psi'' - m(r)[\gamma + U(r)]\Psi = 0$, where h_τ is the pro-

U. Frisch (ed.), Advances in Turbulence VII, 515–518.

jection of magnetic field $\mathbf{h}$ on the direction $\mathbf{r} = \mathbf{x} - \mathbf{y}$, and $U(r) = (\chi^2 + 2\chi' + 4\kappa)/4m(r)$, and $\chi(r) = 4/r + (\ln m^{-1})'$, and $\kappa = 2m(f' + 2f'_c)/r$, and $f = F + rF'/3$, $\quad f_c = F_c + rF'_c/3$, and $1/m = 2/\mathrm{Rm} + 2[1 - F - (rF_c)']/3$, and $\mathrm{Rm} = u_0 l_0/\eta \gg 1$ is the magnetic Reynolds number, u_0 is the characteristic velocity in the maximum scale l_0 of turbulent motions, η is the magnetic diffusion, and $F' = dF/dr$. The function $F_c(r)$ describes the compressible (potential) component whereas $F(r)$ corresponds to the vortical part of the turbulence (see [5]). Here we consider a homogeneous, isotropic and reflectionally invariant (with zero mean helicity) compressible turbulent fluid velocity field. Equation the function $\Psi(r)$ is written in dimensionless variables: coordinates and time are measured in the units l_0 and τ_0, the velocity is measured in the units u_0, the magnetic field is measured in the units B_0. We also take into account the dependence of the momentum relaxation time τ of the turbulent velocity field on the scale of turbulent motions: $\tau(\mathbf{k}) = \tau_0(k/k_0)^{1-p}$, where p is the exponent in spectrum of kinetic turbulent energy, k is the wave number, $k_0 = l_0^{-1}$.

The technique of path integrals allows also to derive the equation for the high-order correlation function of the magnetic field [5]. The solution of this equation yields the growth rate of the s-order correlation function of the magnetic field: $\gamma_s = s(s - 1)\gamma_2/2$ [5]. This implies that if the second-order correlation function of the magnetic field grows ($\gamma_2 > 0$), than all high-order correlation functions grow. It is shown in Section III that under certain conditions $\gamma_2 > 0$. Note that the higher moments grow faster than the lower moments of magnetic field. Therefore spatial distribution of magnetic fluctuations is intermittent (i.e., $\gamma_s > s\gamma_2/2$). This is in agreement with a dynamo theorem [2].

III. Generation of Magnetic Fluctuations.

Consider the case of small magnetic Prandtl numbers $\Pr_m = \nu/\eta \ll 1$ which is typical for many astrophysical and geophysical applications (where ν is the kinematic viscosity). The solution of the equation for the function $\Psi(r)$ and the growth (or damping) rate $\gamma_2 \equiv \gamma$ of the magnetic fluctuations can be obtained using an asymptotic analysis [5]. The latter is given by $\gamma \simeq [4b^2 + (4 - q)^2] \ln(\mathrm{Rm}/\mathrm{Rm}^{\mathrm{cr}})/4(q - 1)$, where the critical magnetic Reynolds number $\mathrm{Rm}^{\mathrm{cr}}$ is given by $\mathrm{Rm}^{(\mathrm{cr})} \simeq \exp[(q - 1)(\pi k + \arctan[(4 - q)/2b] + S)/b]$, $S \simeq 1$, and $k = 1, 2, 3, \ldots$. Here $q = 2p - 1$, and $b^2 = (q^2 - 4)[3 + \sigma(4 - q)]/4(1 + q\sigma)$, and σ is the degree of compressibility. The critical magnetic Reynolds strongly depends on the parameter compressibility σ. Indeed, for incompressible fluid $\sigma = 0$ the critical magnetic Reynolds number $\mathrm{Rm}^{(\mathrm{cr})} = 412$. For compressible fluid flow, i.e., $\sigma = 0.1$ the value $\mathrm{Rm}^{(\mathrm{cr})} = 740$. For larger parameter of compressibility the critical magnetic Reynolds number increases sharply. For $\sigma \gg 1$ the value $\mathrm{Rm}^{(\mathrm{cr})} \to 10^6$. This implies that the compressibility impairs generation of magnetic fluctuations. Here we use $p = 5/3$ (Kolmogorov turbulence).

Note that magnetic fluctuations in a delta-correlated in time incompressible turbulent fluid flow were studied in [1] for $\text{Pr}_m \ll 1$. In the latter model the correlation time is assumed to be independent of the scale of turbulent motions and therefore $q = p$ and the necessary condition for the excitation of magnetic fluctuations is $p > 2$. The Kolmogorov turbulence with $p = 5/3$ and for $\text{Pr}_m \ll 1$ cannot generate magnetic fluctuations in the model with a scale-independent correlation time of the turbulent velocity field [1]. On the other hand, in the considered here model with a scale-dependent correlation time of the turbulent velocity field the magnetic fluctuations are excited for $3/2 < p < 3$.

IV. Anomalous scaling. Problems of anomalous scalings for vector (magnetic) and scalar (particles number density or temperature) fields passively advected by an turbulent fluid flow are a subject of active research in the last years. The anomalous scalings of the magnetic fluctuations were studied in [6] using the model of the delta-correlated in time incompressible turbulent velocity field. The correlation time in this model is independent of the scale of turbulent motions. Note that in this model the generation of magnetic fluctuations is possible only for $q = p > 2$. Taking into account the dependence of the correlation time on the scale of turbulent motions we have shown that the magnetic fluctuations can be generated for $3/2 < p < 3$.

Now we discuss the anomalous scaling for the model with the scale-dependent correlation time of turbulent motions. Consider the case when the magnetic Reynolds number $\text{Rm} < \text{Rm}^{\text{cr}}$ and the magnetic fluctuations are caused by an external source. The condition $\text{Rm} < \text{Rm}^{\text{cr}}$ implies that there is no self-excitation (i.e., exponential growth) of the magnetic fluctuations (see Section III). Solution for the correlation function $W(r)$ in the inertial range is given by $W(r) \sim r^{-(2p+1)/2} \cos(b \ln r + \varphi_0)$. This corresponds to the anomalous scaling of the magnetic fluctuations. The normal scaling for the second moment of the magnetic fluctuations is given by $W(r) \sim r^{4-2p} \propto r^{3-q}$, where $q = 2p - 1$.

The general solution of the equation with an external source for the correlation function W includes the solution describing the anomalous $W(r) \sim r^{-(2p+1)/2} \cos(b \ln r + \varphi_0)$ and normal $W(r) \propto r^{4-2p}$ scalings. The obtained anomalous scaling $W(r) \sim r^{-(2p+1)/2} \cos(b \ln r + \varphi_0)$ can be presented as the real part of the power-law function r^ζ with the complex exponent $\zeta = -1/2 - p - ib$. This anomalous scaling is significantly different from the normal scaling $W(r) \propto r^{4-2p}$, and corresponds to the deviation from the condition of the constant flux of magnetic fluctuations over the spectrum.

The obtained anomalous scaling is valid when the exponent of the energy spectrum of the turbulent velocity field is $3/2 < p < 3$. When $1 < p < 3/2$ the anomalous exponent in a low-Mach-number compressible turbulent flow is real, i.e., $\zeta = -1/2 - p + |b(\sigma, q)|$. In the case of incompressible turbulent flow ($\sigma = 0$) this result coincides with that obtained in [6].

V. Conclusions. In this study we have shown that the magnetic fluctuations can be generated for small magnetic Prandtl numbers. It is shown that the compressibility impairs generation of magnetic fluctuations (i.e., the threshold for generation of magnetic fluctuations by turbulent fluid flow with div $\mathbf{u} \neq 0$ is higher than that for the case of divergence-free fluid flow). The model of very short scale-dependent correlation time of the turbulent velocity field is considered. It is shown that the growth rates of the higher moments of the magnetic field is higher than those of the lower moments, i.e., the spatial distribution of the magnetic field is intermittent.

The presented analysis explains why in the numerical simulations [4] the magnetic fluctuations with zero mean magnetic field were not observed. The parameters in the numerical simulation [4] with zero mean field ($\mathbf{B} = 0$) are Rm ≤ 200, $\sigma = 0.01$. We have shown that even for incompressible turbulent fluid flow ($\sigma = 0$) the threshold of the excitation of the magnetic fluctuations Rm$^{\mathrm{cr}} = 412$ in the case of Pr$_m \leq 1$. Thus the magnetic fluctuations cannot be generated for the parameters used in [4]. Note that the value Rm$^{\mathrm{cr}} = 412$ also cannot be achieved in laboratory experiments. On the other hand, in astrophysical conditions Rm $\gg$ Rm^{cr} and Pr$_m \ll 1$, and therefore the magnetic fluctuations can be generated in the astrophysical conditions.

References

1. A. P. Kazantsev, Sov. Phys. JETP, **26**, 1031 (1968).
2. Ya. B. Zeldovich, A. A. Ruzmaikin, and D. D. Sokoloff, *The Almighty Chance* (Word Scientific Publ., London, 1990).
3. N. Kleeorin and I. Rogachevskii, Phys. Rev. E **50**, 493 (1994).
4. A. Brandenburg, R. L. Jennings, A. Nordlund, M. Rieutord, R. F. Stein and I. Tuominen, J. Fluid Mech. **306**, 325 (1996).
5. I. Rogachevskii and N. Kleeorin, Phys. Rev. E **56**, 417 (1997).
6. M. Vergassola, Phys. Rev. E **53**, 3021 (1996).

SCALING LAWS FOR A PASSIVE SCALAR IN A TURBULENT SWIRLING FLOW

F. PLAZA, J.-F. PINTON
École Normale Supérieure de Lyon, CNRS URA 1325,
46, allée d'Italie, 69364 Lyon cédex 7, France

AND

L. DANAILA, F. ANSELMET, P. LE GAL
I.R.P.H.E, 12, Av. Général Leclerc, 13003 Marseille, France

The understanding of the dynamics of a passive scalar in a homogeneous turbulent flow has made some remarkable theoretical progress recently [1,2]. However, experimental and numerical studies have shown the importance of large scale inhomogeneities in this problem [3,4] and the key role played by mixed velocity-scalar moments [5,6].

To study in detail these issues we have built an experiment where the large scale anisotropy in the scalar injection can be either enhanced or reduced. It consists in a swirling flow produced in the gap between two horizontal counter-rotating disks in air. Temperature fluctuations (playing the role of the passive scalar) are injected by heating of one disk while the other disk is cooled – figure (1-a). The lower disk is heated with a Thermocoax coil SEI-15/100 while the upper one is cooled by mean of dry ice placed inside it. Velocity and temperature are measured via a local dual hot wire / cold wire probe at various vertical positions in the flow. Both fields are then sampled simultaneously. The Reynolds number (based on the Taylor microscale) is about 300.

In the flow there exist regions of high temperature gradients close to the disks and in the middle a temperature mixing region where isotropy of the scalar is expected. This is experimentally verified by calculating the velocity / temperature increment joined probability $< \delta u \delta T > / \delta u_{rms} \delta T_{rms}$: it is clearly positive close to hot disk, negative close to cold disk. In the middle there exist a region where it is close to zero in the range from 5 to 150 Kolmogorov lengths η – figure (1-b). It is thus possible to have a region where velocity increments and temperature increments are decorrelated.

U. Frisch (ed.), Advances in Turbulence VII, 519–522.
© *1998 Kluwer Academic Publishers.*

We analyzed the statistical properties of velocity and temperature field in each typical region: hot, cold and "well mixed".

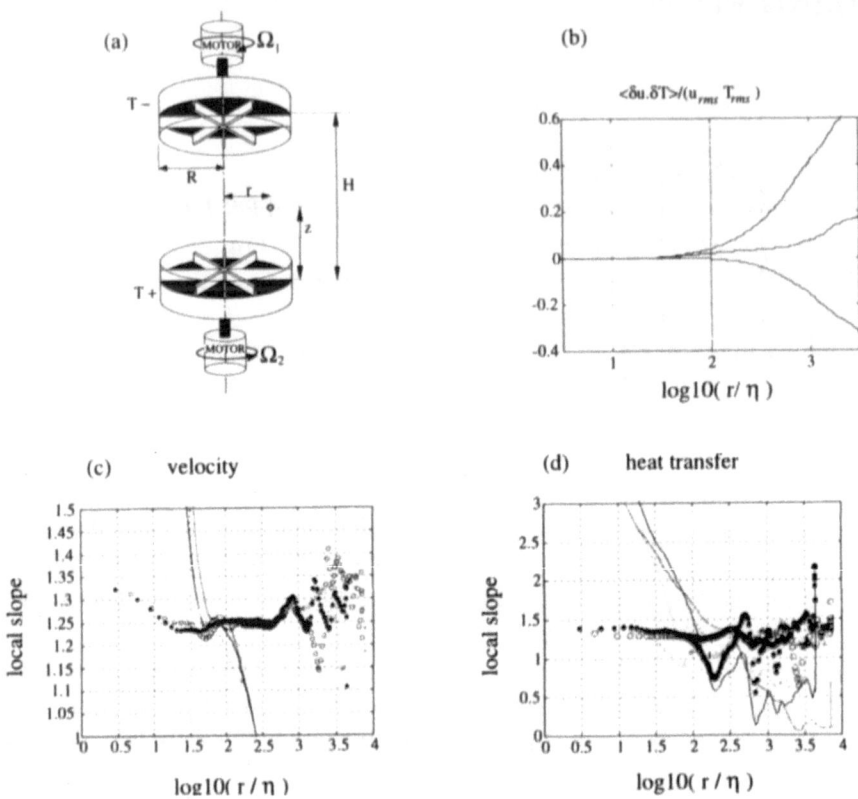

Figure 1.   (a) Sketch of the experimental set-up. (b) Temperature-velocity correlation for the three typical location of the probe in the flow. (c) ESS local slope for velocity (fourth moment): (+) close to lower hot disk; (*) close to upper cold disk; (o) in the mixed zone; plain lines correspond to the local slope of the moments. (d) ESS local slope for heat transport quantity (same location as (c)).

Statistics of velocity increments are in good agreement with now classical results in homogeneous turbulence, and are not influenced by the position of the probe (close to one disk or in the well mixed zone). Since the scaling range in real space is limited to less one decade at best at the Reynolds numbers under study, we concentrate on relative scaling as revealed by the Extended Self Similarity (ESS) property [7]. ESS for velocity is well verified and not influenced by the large scale temperature gradients. We measure relative exponents by computing the mean value of local slope on a $[20\eta - 600\eta]$ range, while a direct measurement on the moment itself is restricted to $[50\eta - 100\eta]$ – figure (3). These relative exponents have the values usually observed in homogeneous isotropic turbulence [8].

Concerning the increments of the temperature field, it is not possible to distinguish any scaling range wherever the probe is situated. ESS holds approximately in the mixed middle region. However, we do not observe a relative scaling in presence of large scale gradients. To test precisely this observation, we again compute the local slope of the $< |\delta T|^p >$ vs $< |\delta T|^2 >$ curve and do not detect any constant plateau, contrary to the case of velocity. We believe that the existence of scaling exponents for the temperature field is questionable. One possible reason could be that the cascade of the temperature field is influenced by both the intermittency in the velocity dissipation and in the scalar dissipation.

We have thus investigated the scaling properties of the mixed temperature-velocity moment reflecting the scalar transport $< |\delta u(r)(\delta T(r))^2| >$. As predicted by the Yaglom relation [10], there exists a limited range where $< \delta u(r)(\delta T(r))^2 >$ scales as the separation scale r; but here again such an 'inertial range' is too limited for the question of scaling in real space to be addressed properly. Nevertheless it is possible to detect that this relation is verified in a small scaling range of $[200\eta, 300\eta]$ in the middle zone, while it is absolutely not verified in presence of large temperature gradient. We again turn to relative scaling. We observe a scaling region when one moment is plotted as a function of another, although the scaling range is not as 'extended' as in the case of the velocity increments – figure (1-d). In addition, ESS is less affected by the large scale anisotropy than it is for temperature. It is thus possible to define scaling exponents for heat flux as:

$$< |\delta u(r)(\delta T(r))^2|^p > \sim < |\delta u(r)(\delta T(r))^2| >^{\xi_H(p)} \quad ,$$

which we can compare to the scaling exponents for kinetic transport:

$$< |\delta u(r)^3|^p > \sim < |\delta u(r)^3| >^{\xi_K(p)} \quad .$$

A remarkable fact is that $\xi_K(p) = \xi_H(p)$ in all situations, even not isotropic (up too our experimental precision) – see figure (2). This relation let us believe that there exists a degree of similarity between heat flux and kinetic energy cascades.

Bibliography

[1] Gawedzski K., Kupianainen A., *Phys. Rev. Lett.*, **75**, 3834, (1995).

[2] Pumir A., Shraiman B.I., Siggia E.D., *Phys. Rev. E*, **55**, 1263, (1997).

[3] Tong C., Warhaft Z., *Phys. Fluids*, **6**(6), 2165-2176, (1994).

[4] Pumir A., *Phys. Fluids*, **6**(6), 2118-2132, (1994), **6**(12), 3974-3984, (1994).

[5] Vaienti S., Ould-Rouis M., Anselmet F., Le Gal P., *Physica D*, **73**, 99, (1994).

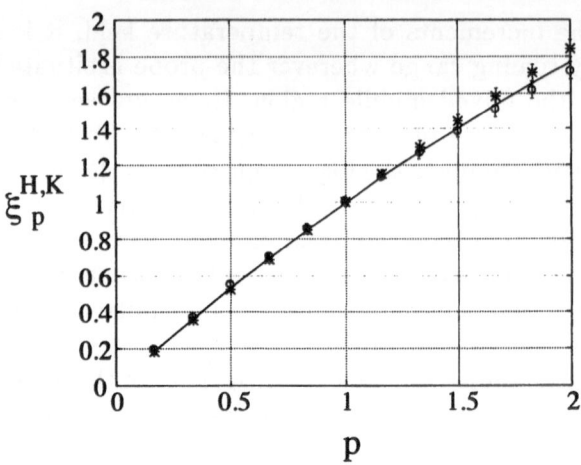

Figure 2.   ESS exponent calculated as the mean of the local slope

[6] Danaila L., Anselmet F., Le Gal P., Dusek C., Brun C., Pumir A., *Phys. Rev. Lett.*, to appear (1997).
[7] Benzi R., Ciliberto S., Tripiccione R., Baudet C., Succi S., *Phys. Rev. E.*, **48**, 29, (1993).
[8] Arnéodo A., et al. *Europhys. Lett.*, **34**(6), 411-416, (1996).
[9] Ruiz-Chavarria G., Baudet C., Ciliberto S., *Physica D*, **99**, 369-380,(1996).
[10] Monin A.S., Yaglom A.M.,*Statistical fluid mechanics*, MIT Press, (1975).

LOCALIZATION OF THE PASSIVE TRACER CLUSTERS IN 1D AND 2D COMPRESSIBLE TURBULENT MEDIA

A.I. SAICHEV AND I.S. ZHUKOVA

*Radiophysics Department, University of Nizhny Novgorod,
23 Gagarin Ave., Nizhny Novgorod 603600, Russia*

1. Introduction

The analysis of the passive tracer diffusion in a turbulent medium is of great interest for many problems of atmosphere and ocean (e.g., see [1]-[3]). The main attention has been drawn to the evolution of the tracer average density and to the analysis of the single particle's diffusion mainly for the case of incompressible medium. Meanwhile, the investigation of the tracer probability properties gives more adequate description of many significant peculiarities of the tracer density field, such as the appearance of the complicated layered structure of the tracer. The consideration of the medium compression is also important for the applications. For example, the floating tracer on the 3D incompressible medium's surface behaves similar to the tracer in 2D compressible one. The most distinctive peculiarity of the passive tracer in a such medium is the formation of the cluster structure, i.e., appearance of small regions of high density, surrounded by the broad regions of low density. The well known Large-Scale Structure of the mass distribution of the Universe is the good example of this structure [4]-[6]. Below we discuss some models of 1D and 2D compressible media illustrating arising and evolution of the cluster structure.

2. The Case of 1D Medium

Let us first consider the simplest case of 1D compressible medium whose density $\rho(x, t)$ depends only on the coordinate x. This model can describe the case of a narrow channel, where water moves only along the channel's walls. The motion of a floating tracer on the surface of the channel can be described as the motion of the passive tracer in a 1D compressible medium.

U. Frisch (ed.), Advances in Turbulence VII, 523–526.

In this case the floating tracer density field obeys to the 1D continuity equation

$$\frac{\partial \rho}{\partial t} + \frac{\partial}{\partial x}\left(v(x,t)\rho\right) = \mu \frac{\partial^2 \rho}{\partial x^2}, \rho(x,t=0) = \rho_0(x). \tag{1}$$

Here the molecular diffusion with the diffusion coefficient μ is included. At the quiet surface when the velocity identically equals zero: $v(x,t) \equiv 0$, solution of this equation in the case of initially placed together particles $(\rho_0(x) = M\delta(x))$

$$\rho(x,t) = \frac{M}{2\sqrt{\pi \mu t}} \exp\left(-\frac{x^2}{2\mu t}\right) \tag{2}$$

describes the monotonical spreading of the passive tracer due to the mutually independent Brownian motion of the particles. It turns out that in the presence of a chaotic surface motion, when the velocity $v(x,t)$ is a stationary homogeneous random field, influence of the chaotic compressions and dilations of the floating tracer gives rise to the new physical phenomenon - localization of the clusters due to the competition of the hydrodynamic chaotic motion and the molecular diffusion. As a result the density field acquires the form $\rho(x,t) = R(x - X(t),t)$, where $X(t)$ describes the purely hydrodynamic motion of the floating tracer and

$$R(x,t) = \frac{M}{\sqrt{2\pi}\sigma(t)} \exp\left(-\frac{x^2}{2\sigma^2(t)}\right) \tag{3}$$

is the "shape" of the cluster. Here $\sigma(t)$ is the cluster's effective width.

Further, we will concretize the velocity field $v(x,t)$ a Gaussian and delta-correlated in time with covariance function

$$\langle v(x,t)v(x+s,t+\tau)\rangle = a(s)\delta(\tau), a(s) = 2D - Bs^2 + \dots \tag{4}$$

It allows to find out the Fokker-Planck equation for the probability distribution function $g(\sigma,t)$ describing the fluctuations of the cluster's width

$$\frac{\partial g}{\partial t} + \mu \frac{\partial}{\partial \sigma}\left(\frac{g}{\sigma}\right) = B\frac{\partial^2}{\partial \sigma^2}\left(\sigma^2 g\right), g(\sigma;t=0) = \delta(\sigma). \tag{5}$$

It is easy to show that this equation has a stationary solution

$$g_\infty(\sigma) = \lim_{t\to\infty} g(\sigma;t) = \sqrt{\frac{2}{\pi}}\frac{l}{\sigma^2} \exp\left(-\frac{l^2}{2\sigma^2}\right), l = \sqrt{\frac{\mu}{B}}, \tag{6}$$

describing the statistical properties of the cluster's width in the case of large time when $Bt \gg 1$ (see also Fig. 1).

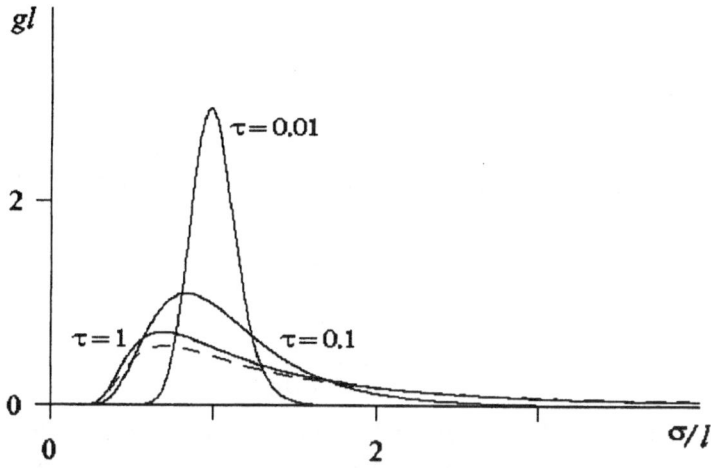

Figure 1. The fluctuations of the cluster's width for various time moments $\tau = Bt$. The dashed curve corresponds to the case of large time.

The existing of the stationary distribution (6) means that medium's chaotical compressions prevent from the molecular diffusion action and lead to the cluster localization. Indeed, the averaging of the cluster's "shape" (3) with the help of probability distribution function (6) one can obtain that the mean cluster's density profile

$$\langle R_\infty(x) \rangle = \lim_{t \to \infty} \langle R(x,t) \rangle = \frac{Ml}{\pi} \frac{1}{x^2 + l^2} \qquad (7)$$

is relatively well localized along the axis x.

3. The Case of 2D Medium

Now we consider that the 2D velocity field $v(\mathbf{x},t)$, which is Gaussian and delta-correlated in time with the covariance function

$$\langle v_i(\mathbf{x},t)v_j(\mathbf{x}+\mathbf{r},t+\tau) \rangle = \left\{ G(r)\delta_{ij} + \left[\delta_{ij} - \frac{r_i r_j}{r^2} \right] \frac{r}{2} \frac{d}{dr} G(r) \right\} \delta(\tau), \quad (8)$$

where $G(r)$ is the longitudinal components' covariance function of the 3D statistically isotropic field of the medium's velocity, decaing in Fourie series in the following manner: $G(r) = G - Dr^2/2$. Therefore the "shape" of the cluster has the form

$$R(\mathbf{x},\mathbf{y},t) = \frac{1}{2\pi\sqrt{I(\mathbf{y},t)}} \exp\left(-\frac{S_{22}(\mathbf{y},t)x_1^2 + S_{11}(\mathbf{y},t)x_2^2 - 2S_{12}(\mathbf{y},t)}{2I(\mathbf{y},t)} \right),$$
$$(9)$$

where $I(\mathbf{y},t) = S_{11}(\mathbf{y},t)S_{22}(\mathbf{y},t) - S_{12}^2(\mathbf{y},t)$, and $S_{\alpha\beta}$ satisfy to the equations

$$\frac{\partial S_{11}}{\partial t} = 2\mu + 2S_{11}u_{11} + 2S_{12}u_{12}, S_{11}(\mathbf{y},t=0) = 0,$$

$$\frac{\partial S_{22}}{\partial t} = 2\mu + 2S_{22}u_{22} + 2S_{12}u_{21}, S_{22}(\mathbf{y},t=0) = 0, \qquad (10)$$

$$\frac{\partial S_{12}}{\partial t} = S_{12}(u_{11}+u_{22}) + S_{11}u_{21} + S_{22}u_{12}, S_{12}(\mathbf{y},t=0) = 0,$$

where

$$u_{\alpha\beta}(\mathbf{x},t) = \frac{\partial v_\alpha}{\partial x_\beta}.$$

One can derive the Fokker-Planck equation for the probability distribution function $f(i,q,t) = \langle \delta(I(t) - i)\delta(Q(t) - q) \rangle$, where $Q = (S_{11}+S_{22})/2$, describing the diffusion and the "shape" (i.e. dispersion) of the cluster

$$\frac{\partial f}{\partial \tau} + 2\gamma q \frac{\partial f}{\partial i} + \gamma \frac{\partial f}{\partial q} - 2\frac{\partial}{\partial i}(if) + 2\frac{\partial}{\partial q}(qf) = 4\frac{\partial}{\partial i}\left(i\frac{\partial}{\partial i}(if)\right)$$

$$+4\frac{\partial}{\partial q}\left(q\frac{\partial}{\partial q}(qf)\right) + 4\frac{\partial}{\partial i}\left(i\frac{\partial}{\partial q}(qf)\right) - 3i\frac{\partial^2 f}{\partial q^2}, \qquad (11)$$

$$f(i,q,\tau=0) = \delta(i)\delta(q),$$

where $\tau = Dt/2$, $\gamma = 4\mu/D$. This equation enables us to find out the peculiarities of arising and evolution of the cluster's structure in a 2D compressible medium.

Acknowledgments

This work was supported in part by the Russian Foundation of Basic Research, Grants No. 97-02-16521, 95-IN-RU-723, 96-15-96722.

References

1. Csanady, G.T. (1973) *Turbulent Diffusion in the Environment*, Reidel, Boston.
2. Careta, A., Sagues, F., Ramirez-Piscina, L., and Sancho, J.M. (1993) Effective diffusion in a stochastic velocity field, *J. Stat. Phys.* **71**, 235-242.
3. Crisanty, A., and Vulpiani, A. (1993) On the effects of noise and drift on diffusion in fluids, *J. Stat. Phys.* **70**, 197-211.
4. Gurbatov, S.N., Malakhov, A.N., and Saichev, A.I. (1991) *Nonlinear Random Waves and Turbulence in Nondispersive Media: Waves, Rays and Particles*, Manchester University Press, Manchester.
5. Saichev, A.I., and Woyczynski, W.A. (1996) Density fields in Burgers and KdV-Burgers turbulence, *SIAM J. Appl. Math.* **56**, No. 4, 1008-1038.
6. Saichev, A.I., and Woyczynski, W.A. (1997) Advection of passive and reactive tracers in multi-dimensional Burgers' velocity field, *Physica D* **100**, 119-141.

SURFACE PATTERNS AND TURBULENT MIXING IN THE OCEAN SURFACE BOUNDARY LAYER

S. LEIBOVICH[1], T. HAEUSSER[2], AND G. YANG[1]
Sibley School of Mechanical & Aerospace Engineering[1]
Department of Physics[2]
Cornell University, Ithaca, NY, USA

1. Introduction

Wind generated stirring in the ocean and in lakes and ponds inevitably occurs in the presence of surface gravity waves. Waves can modify the turbulence and especially the mixing in significant ways. Energetic, persistent coherent motions, known as Langmuir circulation (Langmuir, 1938) arise by an instability mechanism due to the interaction of the waves and shear. Their imprint often is strikingly visible as surface windrows. These patterns may be thought of as the coherent structures in the mixed layer, but they are of much larger scale and persistence than those appearing in "ordinary" boundary layers.

In the surface layer of the ocean, the dominant motions typically are the orbital motions in surface gravity waves. Mean currents, especially those locally driven by wind, are usually an order of magnitude smaller than wave orbital speeds; and turbulent fluctuations are two orders of magnitude smaller than than r.m.s. wave orbital speeds. It is clearly imprudent to attempt to simulate the turbulence in this portion of the ocean without accounting for waves. In fact, when any motion is dominated by a *known* spatially and temporally varying motion that can be given a deterministic representation, as with waves near the ocean surface, it is clear that one should account for these motions at the outset. The effects of waves on the current structure and Langmuir circulation have been modelled by Craik and Leibovich (1976, henceforth CL). Langmuir circulation, and its theoretical treatment, is reviewed by Leibovich (1983). In the development of the theory, turbulence and heat exchanges were modelled by constant eddy diffusivities, and it is clear that use of the theory as a predictive tool requires a less *ad hoc* representation of the turbulent exchanges. In this

U. Frisch (ed.), Advances in Turbulence VII, 527–530.

paper, we summarize the results of Leibovich and Yang (1998, henceforth LY) for turbulent, wind/wave driven flow in a water layer of constant depth found by large eddy simulation (LES).

Pattern evolution occurs in large aspect ratio situations and is not amenable to computation from a "microscopic" description, as in LES. We present results from simulations of amplitude (Davey–Hocking–Stewartson) equations constructed from the full equations by perturbation methods. These show the formation of many defects, which move through the fluid in a chaotic manner, and which might be related to "Y-junctions" observed in the ocean.

2. LES Formulation

The LES computational model of Leibovich & Yang (1998) is based on the decomposition described by CL, in which the instantaneous velocity field is represented by $\mathbf{u}(\mathbf{x}, t) = \mathbf{u}_w(\mathbf{x}, t) + \mathbf{v}(\mathbf{x}, t)$, where $\mathbf{u}_w(\mathbf{x}, t)$ is an irrotational surface wave velocity field that is constructed to account for the displacement of the air–sea interface. With this decomposition introduced, an explicit time filter applied to the Navier–Stokes equations yields an equation set for resolved motion which requires two kinds of closures, one for wave-rotational interactions and one for "ordinary" subgrid scale motions. The "vortex force" of CL has been shown to be an asymptotically correct description of the wave-rotational interaction terms up to sixth order in surface wave slope, and is adopted for the LES. Both Smagorinsky and dynamic subgrid closures were used for the remaining terms, with the Smagorinsky model adopted for the principal runs. The result then is an "ordinary" LES with an additional vortex force term, $\mathbf{u_s} \times \boldsymbol{\omega}$ appearing as an apparent force due to the waves. Here $\mathbf{u_s}$ is the Stokes drift of the surface waves, and $\boldsymbol{\omega}$ is the resolved vorticity.

Water of finite depth d and constant density is assumed subject to a constant wind stress with friction velocity u_* applied at the mean free surface, $z = 0$, and the bottom $z = -d$ is a no-slip surface. A horizontally homogeneous wave field with a Stokes drift profile $\mathcal{U}_s(z)$ is assumed to exist parallel to the applied stress. The problem depends on a Reynolds number, $R = u_* d/\nu$, a Langmuir number, $\mathcal{U}_s(0)/u_*$, and a Froude number $u_*/\sqrt{gd}$. The last parameter appears through the Stokes drift with standard representations of the surface gravity wave spectrum. In the absence of the Stokes drift, the problem is a turbulent stress–driven Couette flow, and with Stokes drift, it yields turbulent Langmuir circulation.

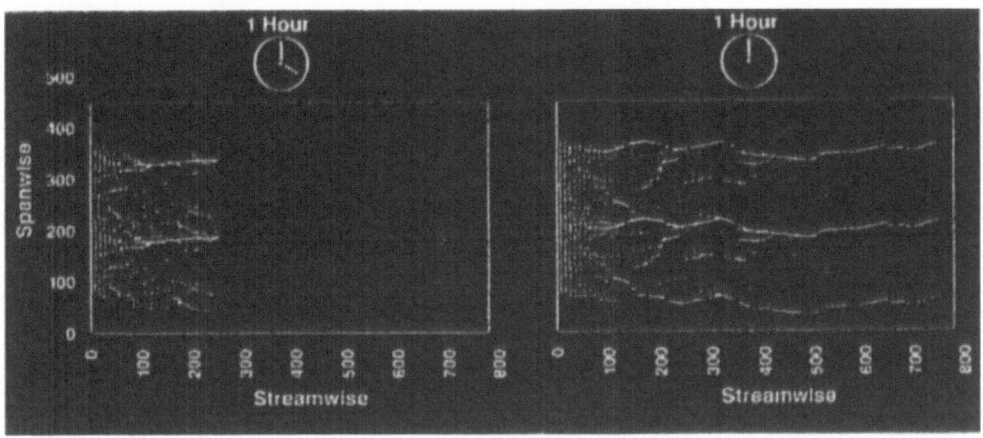

Figure 1.   Surface tracers released at a constant rate along a line at the left of figures, shown at 20 minutes and one hour after tracer release is initiated. Spanwise and streamwise distances are in meters.

3. Simulation Results

We describe features of a simulation in LY with the parameter choices $\mathcal{U}_s(0)/u_* = 11.6$, $u_*/\sqrt{gd} = 4.5 \times 10^{-4}$, and and $R = 590$. The first corresponds to a "fully developed sea", and the second to an example with depth of 50 meters and a wind speed of 9 m/s.

Persistent surface patterns (Fig. 1) form when waves (Langmuir case) are included, and are absent in the Couette case. These are manifestations of strong longitudinal coherent vortices extending throughout the entire layer and leading to strong mixing. A measure of mixing efficiency, the volume averaged vertical flux of resolved kinetic energy is three orders of magnitude larger in the Langmuir case. Statistical measures are quite different in the two cases. These are nearly Gaussian except near the surface and bottom in the Couette case, and quite non-Gaussian in the Langmuir case.

4. Amplitude Equations

In this section we assume an exponential Stokes drift profile $\mathcal{U}_s(z) = \mathcal{U}_s(0) \exp(z/\ell)$. We also include the Coriolis force, but neglect the horizontal component of the Earth's angular velocity. This introduces another dimensionless parameter, the Ekman number $E = \nu/(d^2 \Omega_v)$, where Ω_v is the magnitude of the vertical component of the Earth's angular velocity. We report results for the parameter choices $\mathcal{U}_s(0)/u_* = 10$, $\ell/d = 0.2$, and the two cases $E = \infty$ and $E = 1$.

When R exceeds the critical value R_c, the shear current is linearly unstable to rolls of wavenumber k_c that are at an angle α_c to the right of the

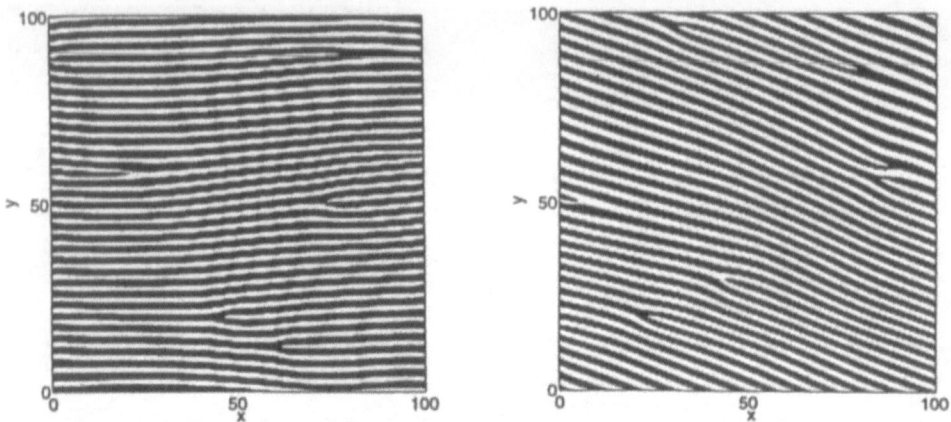

Figure 2. Snapshots of simulations of the amplitude equations for $E = \infty, R/R_c = 1.5$ (left) and $E = 1, R/R_c = 2$ (right). Distances are measured in units of d.

wind and travel at a phase speed v_c away from the wind (in the northern hemisphere). For $E = \infty$ the critical parameters $(R_c, k_c/d^{-1}, \alpha_c, v_c/u_*)$ are $(4.06, 2.03, 0°, 0)$, and for $E = 1$ they are $(4.70, 2.02, 19°, 1.66)$.

A perturbation analysis in $\epsilon = (R/R_c - 1)^{1/2}$ yields an equation for the amplitude A of the critical mode. Davey *et al.* (1974) recognized that continuity demands that this equation be coupled to a second equation, whenever a mean flow is induced at second order in ϵ, which is the case for our system:

$$(\partial_t + v_\mu \partial_\mu) A = c_1(R/R_c - 1) A + c_{\mu\nu}\partial_{\mu\nu} A - c_3|A|^2 A + A\, e_\mu \partial_\mu P,$$
$$\partial_{\mu\mu} P = f_\mu \partial_\mu |A|^2, \tag{1}$$

Greek indices run from 1 to 2, repeated indices are summed, and the coefficients are real or complex and must be computed. Figure 2 shows snapshots of simulations of these equations with coefficients corresponding to the two cases $E = \infty, R/R_c = 1.5$ and $E = 1, R/R_c = 2$. The snapshots show the sign of any physical field, e. g., the vertical velocity in the mid-plane.

References

Craik, A. D. D. & Leibovich, S. (1976) A rational model for Langmuir circulations. *J. Fluid Mech.* **73**, pp. 401–426.

Davey, A., Hocking, L. M., & Stewartson, K. (1974) On the nonlinear evolution of three-dimensional disturbances in plane Poiseuille flow. *J. Fluid Mech.* **63**, pp. 529–536.

Langmuir, I. (1938) Surface motion of water induced by wind. *Science* **87** pp. 119–123.

Leibovich, S. (1983) The form and dynamics of Langmuir circulations. *Ann. Rev. Fluid Mech.* **15** pp. 391–427.

Leibovich, S. & Yang, G. (1998) Turbulent Flow in natural water bodies driven by wind and surface waves. *J. Fluid Mech.* (submitted).

PHASE DIAGRAM FOR TURBULENT DISPERSION

A. C. FANNJIANG
Department of Mathematics, UC Davis
Davis, CA 95616-8633, U.S.A.

1. Introduction

The motion of passive scalar in turbulent flows is described by the equation

$$d\boldsymbol{x}(t) = \boldsymbol{v}(\boldsymbol{x}(t), t)dt + \sqrt{2\nu}d\boldsymbol{w}(t)$$

where $\nu \geq 0$ is the molecular diffusivity, $\boldsymbol{w}(t)$ the standard Brownian motion and $\boldsymbol{v}$ a time stationary, space homogeneous incompressible *Markovian* velocity field in d dimensions with the correlation given by the formula

$$\hat{R}_{ij}(\mathbf{k}, t) = \rho(|\mathbf{k}|^{2\beta}t)\frac{1}{|\mathbf{k}|^{2\alpha-1}}\frac{1}{|\mathbf{k}|^{d-1}}(\delta_{ij} - \frac{k_ik_j}{|\mathbf{k}|^2}), \quad \kappa_0 \leq |\mathbf{k}| \leq \kappa_1, \quad (1)$$

$\forall i, j = 1, ..., d$. We assume that $\kappa_0 = 0, \kappa_1 < \infty$ for $\alpha < 1$, and $\kappa_0 > 0, \kappa_1 = \infty$ for $\alpha > 1$ so that $\hat{R}_{ij}$ is integrable in $\mathbf{k}$ and the velocity is well defined. The time correlation function ρ is assumed to decay fast. The Kolmogorov-Obukhov spectrum corresponds to $\alpha = 4/3, \beta = 1/3$, in three dimensions. When ρ is a exponential function, the velocity field is Gaussian and is a Ornstein-Uhlenbeck process.

The scaling limit of turbulent dispersion concerns the asymptotic behavior of $\boldsymbol{x}(t)$ as a function of time for $t >> 1$. This can be conveniently formulated as convergence of the scaled motion $\boldsymbol{x}^{\varepsilon}(t) = \varepsilon\boldsymbol{x}(t/\varepsilon^{2q})$, with a small parameter $\varepsilon << 1$, for suitable $q > 0$. How does the scaling exponent q depend on the parameters α, β and the cut-offs κ_0, κ_1? The full answer to this amounts to the complete phase diagram for turbulent dispersion.

2. Results and Methods

For convenience, we shall state our results for $\nu = 0$ unless otherwise indicated. For $\alpha < 1$, the scaling behaviors can be conveniently divided into four scenerios.

U. Frisch (ed.), Advances in Turbulence VII, 531–534.
© 1998 *Kluwer Academic Publishers.*

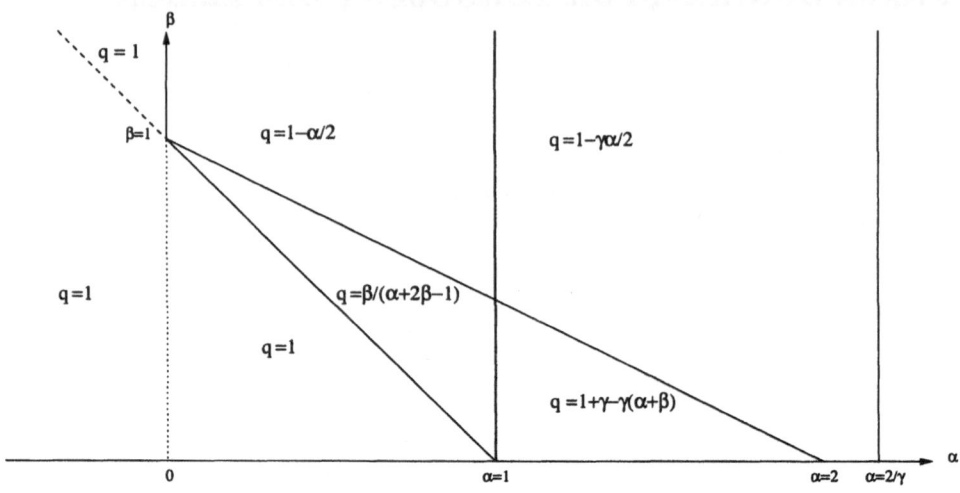

Phase diagram for γ=<1.

(a) The homogenization regime ($\alpha + \beta < 1$) where both space and time dependences of velocity are important.
(b) The homogenization regime with $\nu > 0$ ($\alpha + \beta > 1, \alpha < 0$).
(c) The frozen path regime ($\alpha + \beta > 1, \alpha + 2\beta < 2$) where the space dependence of velocity is negligible.
(d) The frozen field regime ($\alpha + 2\beta > 2, 0 < \alpha < 1$) where the time dependence is negligible.

For $\alpha > 1$ the scaling exponent q depends on the infrared cut-off $\kappa_0 = \varepsilon^\gamma, \gamma > 0$. Taking $\kappa_0(\varepsilon)$ into account and rescaling the equation of motion, the scaling behaviors fall into regimes similar to either (a) or (c) above. For details, we refer the reader to the figures and the paper [4].

The results are derived by using the following methods.

(i) The variational principle for the scale dependent eddy diffusivity, with $\nu = 0$, for Markovian flows with the time correlation function ρ having the form as in (1) (cf. [4] for details). This is used in the homogenization regime (a) as well as obtaining bounds for scaling exponents for other regimes.
(ii) An variational principle for the box diffusivity in general *steady* flows, with $\nu > 0$, established in [3]. This is used in the frozen field regime (d) and the homogenization regime (b).

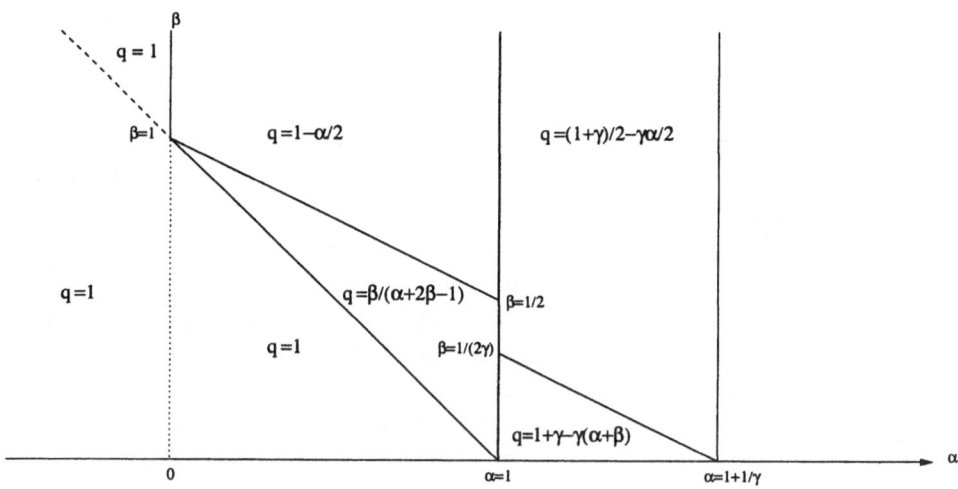

Phase diagram for γ>1.

(iii) A Stieltjes integral representation formula for the effective diffusivity, with $\nu > 0$, established in [2]. This is used in the homogenization regime (b).

(iv) An anomalous turbulent diffusion theorem proved in [5]. This is used in the frozen path regime (c).

(v) A classical turbulent diffusion theorem with the Taylor-Kubo formula, proved in [6]. This is used for $\alpha > 1$.

Scaling arguments are also used to clarify the implications of the mathematical theorems for the frozen path regime (c) and the frozen field regime (d).

One of the technical innovations of this study is a variational principle for the eddy diffusivity at the scale n for Markovian velocity fields when the molecular diffusion is absent:

$$D^n(\mathbf{e}) = \inf_{f \in C_p([0,n]^d)} \left\{ -\int_n \mathcal{A}ff - \int_n \mathcal{A}f'f' \right\} \tag{2}$$

with the function f' given by $\mathcal{A}f' + \mathbf{v} \cdot (\nabla f + \mathbf{e}) = 0$ where $C_p([0,n]^d)$ is the space of smooth periodic functions in $[0,n]^d$. Here the operator $\mathcal{A}$ is the generator of the Markovian velocity. For $\alpha < 1$, substituting the trivial trial function $f = 0$ into (2) we obtain easily the bound

$$D^n(\mathbf{e}) \leq c|\mathbf{e}|^2(1 + n^{2(\alpha+\beta-1)}), \quad c > 0, \quad \forall \mathbf{e} \in R^d,$$

as n tends to ∞, for the Markovian flows with the correlation (1). This translates into a lower bound $q \geq 2 - \alpha - \beta$ for $\alpha + \beta > 1$ and $q \geq 1$ for $\alpha + \beta \leq 1$.

3. Conclusion

It is noteworthy that the scaling exponent q is continuous across all phase boundaries except on $\alpha = 1$ where the transition from ultraviolet to infrared cut-off takes place.

Part of the phase diagram was first derived Avellaneda and Majda [1]. The main difference in assumption between this paper and [1] lies in the fact that *both* ultraviolet and infrared cut-offs, with the infrared cut-off exponent $\gamma = 1$, are made throughout the region $(0 < \beta < 1/2, 0 < \alpha < 2)$ considered in [1]. This difference explains why in [1] only three, instead of four, scaling laws were identified in the region $(0 < \beta < 1/2, 0 < \alpha < 2)$. With the infrared cut-off, formula $q = 1 + \gamma - \gamma(\alpha + \beta)$ holds for regime (c) as well as for $\alpha + 2\beta < 2, \alpha > 1$.

Positive molecular diffusivity $\nu > 0$ was used to obtain the exponents for two regimes: the frozen field regime $(\alpha + 2\beta > 2, \alpha > 0)$ and the homogenization regime $(\alpha + \beta > 1, \alpha < 0)$. We take the conventional wisdom that in three dimensions, the effect of molecular diffusion is insignificant and the same exponents should hold true. However, for steady flows in *two* dimensions or for steady flows with a *nonzero, but small mean* drift in any dimensions molecular diffusion can significantly affect the dispersion rate. Turbulent dispersion in flows with a nonzero mean drift has different phase diagrams and will be reported in a forthcoming paper.

References

1. Avellaneda, M. and Majda, A. J. (1992):*Renormalization theory for eddy diffusivity in turbulent transport*, Physical Review Letters, **68:20**, pp. 3028-31.
2. Avellaneda, M. and Vergassola, M. (1995): *"Stieltjes integral representation of effective diffusivities in time-dependent flows."*, Phys. Rev. E, **52:3B**, pp. 3249-3251.
3. Fannjiang, A. C. (1998): *Anomalous Diffusion in Random Flows*, IMA Proceedings Volume 99, Mathematics of Multiscale Materials, K. M. Golden, G. R. Grimmett, R. D. James, G. W. Milton and P. N. Sen, eds.
4. Fannjiang, A.C. (1998): *Phase diagram for turbulent dispersion in a Markovian flow*, submitted to Physica D.
5. Fannjiang, A. C. and Komorowski, T. (1998): *The Taylor-Kubo formula and anomalous turbulent diffusion for a non-mixing flow*, submitted to SIAM J. Appl. Math..
6. Kesten, H. and Papanicolaou, G. (1979): *A limit theorem in turbulent diffusion*, Comm. Math. Phys., **65**, pp. 97-128.
7. Komorowski, T. (1998): *Application Of The Parametrix Method To Diffusions In A Turbulent Gaussian Environment."*, Stoch. Proc. Appl. (in press).

MULTISCALING IN PASSIVE SCALAR ADVECTION AS STOCHASTIC SHAPE DYNAMICS

O. GAT[1,2] AND R. ZEITAK[2,3]
[1] *Dept. Physique Theorique, Univeriste de Geneve,*
32 Bld d'Yvoy 1211 Geneve 4, Switzerland
[2] *Dept. Chem. Physics, Weizmann Institute of Science,*
Rehovot 76100, Israel
[3] *Laboratoire de Physiqe Statistique, ENS, 24 rue Lhomond,*
75231 Paris Cedex 5, France

1. Introduction

The study of intermittency in fluid turbulence is hampered by the lack of understanding of its basic mechanism [1]. The Kraichnan model of passive scalar advection by a rapidly varying velocity [2] is an example of a simplified model which displays intermittency, which is believed to be related to intermittency in turbulence. A mechanism for the generation of multiscaling in the Kraichnan model has been identified, and multiscaling exponents have been calculated by several methods [3, 4, 5, 6, 7].

The evolution of scalar correlation functions in the Kraichnan model is determined by the Kraichnan operators. Anomalous scaling structures are commonly identified with scale invariant zero modes of the Kraichnan operators. his scaling picture, which is based on the field approach to the passive scalar problem, while mathematically understood, leaves the physical meaning of the zero modes somewhat obscure.

In this paper we present an alternative theoretical approach to the analysis of the Kraichnan model, which focuses on the Lagrangian paths of the advecting velocity field [8], replacing the stochastic partial differential equation for the scalar field by a small set of stochastic ordinary differential equations. We analyze the evolution of the shape defined by N points advected by the velocity field. We separate the overall scale, which is growing in time on the average, from the normalized shape which approaches a stationary asymptotic state. We show that the spectrum relaxation rates to this asymptotic state is none other than the spectrum of Nth order

U. Frisch (ed.), Advances in Turbulence VII, 535–538.
© 1998 *Kluwer Academic Publishers.*

anomalous exponents, and the zero-modes obtain a physical meaning as the adjoint family to the relaxation eigenstates.

The present approach has also been shown to be a useful calculational tool, both for analytic perturbative calculations [10], and in a numerical Monte-Carlo method [11, 12].

2. Lagrangian path formulation of passive scalar advection

The dynamics of a passive scalar θ advected by an incompressible velocity field $\mathbf{u}$ are described by

$$\partial_t \theta(\mathbf{r}, t) = \kappa \nabla^2 \theta(\mathbf{r}, t) - \mathbf{u} \cdot \nabla \theta(\mathbf{r}, t) + f(\mathbf{r}, t) \tag{1}$$

where κ is the molecular diffusivity, and f models the injection and extraction of scalar by external sources. The Kraichnan model specifies that $\mathbf{u}$ and f are Gaussian fields, δ-correlated in time. $\mathbf{u}$ is self-similar in space with characteristic exponent $0 \leq \xi \leq 2$, and f is large scale with a characteristic scale L, the integral scale.

The present analysis is based on the observation that, when $\kappa = 0$ the amount of scalar at some point is equal to the forcing accumulated along the Lagrangian path leading to this point. This observation may be formalized and generalized to nonzero κ by a path integral formalism [9], which implies

$$\theta(\mathbf{r}, t) = \int_{-\infty}^t dt' < f(\mathbf{r}'(t'), t') >_\eta \tag{2}$$

with the trajectory $\mathbf{r}'$ obeying

$$\partial_{t'} \mathbf{r}'(t') = \mathbf{u}(\mathbf{r}'(t'), t') + \sqrt{2\kappa}\eta(t'), \qquad \mathbf{r}'(t) = \mathbf{r} \tag{3}$$

and η is a centered Gaussian white random vector.

The representation (2) yields an expression for the scalar correlation functions in terms of the forcing correlation function Ξ:

$$\begin{aligned}
F_{2n}(\mathbf{r}_1, \ldots, \mathbf{r}_{2n}, t) &= \langle \theta(\mathbf{r}_1) \cdots \theta(\mathbf{r}_n) \rangle_{\mathbf{u}, f} = \langle \int_{-\infty}^t dt_1 \cdots dt_n \\
&\times \Big[\Xi(\mathbf{r}_1'(t_1) - \mathbf{r}_2'(t_1)) \cdots \Xi(\mathbf{r}_{2n-1}'(t_n) - \mathbf{r}_{2n}'(t_n)) + \text{permutations} \Big] \rangle_{\mathbf{u}, \eta},
\end{aligned} \tag{4}$$

while the odd moments vanish; each of the trajectories $\mathbf{r}_i'$ obeys an equation of the form (3).

The representation (4) expresses F_{2n} as the expectation value of the forcing correlation function accumulated over the Lagrangian trajectories. Since the forcing drops off sharply beyond L the integral measures the correlation of the times during which pairs of points were within distance L. In particular the two point moment F_2 is the average time which two trajectories which end in a given distance of each other stay within distance L. Excluding the case $d = 2$, $\xi = 0$, the trajectories will separate indefinitely with probability 1, and the integrals in (4) converge for each realization.

It is possible to turn eq. (4) into a recursive expression by ordering the times such that $t_1 > t_2 > \cdots > t_n$, and integrating out $t_2, \cdots, t_n$:

$$F_{2n}(\mathbf{r}_1, \ldots, \mathbf{r}_{2n}, t) = \langle \int_{-\infty}^{t} dt_1$$
$$\times \left[\Xi(\mathbf{r}_1'(t_1) - \mathbf{r}_2'(t_1)) F_{2n-2}(\mathbf{r}_3'(t_1), \cdots, \mathbf{r}_{2n}'(t_1)) + \text{permutations} \right] \rangle_{u,\eta}.$$
$$(5)$$

This form for F_{2n} is useful for the following analysis because it contains only a single time integration, with the integrand depending only on the simultaneous positions of $2n$ particles.

We intend to study the small κ limit of the problem. It can be argued that for $\xi < 2$, and correlation functions whose arguments are all distinct, this limit is achieved by simply setting $\kappa = 0$. We are going to concentrate on these cases, and put $\kappa = 0$ from now on.

3. Shape evolution operator and its relation to multiscaling

A component of the evolution of an initial configuration is a rescaling of all the coordinates which all increase on the average like t^{1/ζ_2}; this rescaling is analogous to Richardson diffusion for the present case [1]. The dynamic exponent $\zeta_2 = 2 - \xi$ is also the characteristic exponent of the second order structure function [2]. Anomalous scaling will result from the dynamics of the a normalized *shape*, from which overall expansion has been factored out

Consider an initial configuration described by the shape $\mathbf{Z}_0$ and overall scale s_0. We fix a scale $s > s_0$ and examine the shape when the configuration reaches the scale s for the first time (we know that this occurs in finite time with probability one). We define the shape evolution operator $\gamma(\mathbf{Z}_0, \mathbf{Z}, \frac{s}{s_0})$ as follows: If the initial distribution of shapes is $\rho_{s_0}(\mathbf{Z}_0)$, the shape distribution at scale s is

$$\rho_s(\mathbf{Z}) = \int d\mathbf{Z}_0 \rho_0(\mathbf{Z}_0) \gamma(\mathbf{Z}_0, \mathbf{Z}, \frac{s}{s_0}) \qquad (6)$$

For very large values of $\frac{s}{s_0}$ we expect the final shape distribution to approach an asymptotic distribution $\beta_0(\mathbf{Z})$. This distribution is invariant under the transformation (6) and is an eigenfunction of γ with eigenvalue 1. We further assume that there exists a spectrum of eigenfunctions β_n for γ with eigenvalues $(\frac{s}{s_0})^{-\lambda_n}$, and that the eigenfunctions $\beta_n(\mathbf{Z})$ form a complete set, so we can expand the operator γ

$$\gamma(\mathbf{Z}_0, \mathbf{Z}, \frac{s}{s_0}) = \sum_m \left(\frac{s}{s_0} \right)^{-\lambda_m} \beta_m(\mathbf{Z}) \mu_m(\mathbf{Z}_0). \qquad (7)$$

γ is a non-Hermitian operator, and the μ_m's are its left eigenvectors, so that β_m and μ_m are biorthogonal families of functions.

We now use the representation (5) for $F_n(s_0, \mathbf{Z}_0)$. We split the integration along the Lagrangian trajectory at the first point where the overall scale reaches the value s. It follows from the shape evolution equation (6) that we can write

$$F_n(s_0, \mathbf{Z}_0) = I(s_0, s, \mathbf{Z}_0) + \int d\mathbf{Z} \gamma \left(\mathbf{Z}_0, \mathbf{Z}, \frac{s}{s_0} \right) F_n(s, \mathbf{Z}) . \qquad (8)$$

I is the average of the contribution a specific forcing accumulated along the path from s_0 to s. The forcing is that of eq (5) so that for s small enough its L dependence is only through F_{n-2}. Averaging eq. (8) with $\beta_m(\mathbf{Z}_0)$ gives

$$\left[\bar{F}_n^{(m)}(s_0) - \left(\frac{s}{s_0} \right)^{\lambda_m} \bar{F}_n^{(m)}(s) \right] = \int \beta_m(\mathbf{Z}_0) I(s_0, s, \mathbf{Z}_0) d\mathbf{Z}_0 . , \qquad (9)$$

where $\bar{F}_n^{(m)}(s) = \int \beta_m(\mathbf{Z}) F_n(s\mathbf{Z}) d\mathbf{Z}$.

The term in the RHS of (9) is dependent on L only through F_{n-2}. Thus, this term cannot have anomalous $L^{(\frac{n}{2}(2-\xi)-\zeta_n)}$ scaling. Thus the only new (*i.e.*, not derived from lower order correlation functions) anomalous behavior that may be contained in $\bar{F}_n^{(m)}(s)$ must annihilate the LHS of (9). Hence, it must be proportional to s^{λ_m}. We conclude that the exponents λ_m, defined as the decay rates toward the invariant measure, are precisely the anomalous exponents of F_n.

Finally, it follows from (7) $\bar{F}_n^{(m)}(s)$ is the mth coefficient in an expansion of F_n in terms of μ_m that, so that the functions μ_m are the scaling structures associated with λ_m, identifying them with the zero modes.

References

1. U. Frisch, *Turbulence*, Cambridge University Press (1995).
2. R. H. Kraichnan, Phys. Fluids **11** 945 (1968)
3. R. H. Kraichnan, Phys. Rev. Lett., **72** 1016 (1994)
4. R. H. Kraichnan, V. Yakhot and S. Chen, Phys. Rev. Lett., **75**, 240 (1996).
5. K. Gawędzki and A. Kupiainen, Phys. Rev. Lett., **75**, 3834, (1995).
6. M.Chertkov, G. Falkovich, I. Kolokolov and V. Lebedev, Phys. Rev. E **52** 4924, (1995).
7. O. Gat, V.S. L'vov, and I. Procaccia, Phys. Rev. **E56**, 406 (1997).
8. A similar point of view is presented in D. Bernard, K. Gawedzki, A. Kupiainen, e-print cond-mat/9706035 (1997).
9. See for example, I. T. Drummond, J. Fluid Mech. **123**, 59 (1982).
10. O. Gat and R. Zeitak, Phys. Rev. E. in press (1998).
11. O. Gat, I. Procaccia, and R. Zeitak, in preparation (1998).
12. The idea of using paths for passive scalar Monte-Carlo simulation was developed independently by U. Frisch and M. Vergassola.

STUDY OF THE "LOCAL" LINK BETWEEN A PASSIVE SCALAR AND ITS DISSIPATION USING THE WAVELET DECOMPOSITION

A. BENAISSA[1], J. LEMAY[2] AND F. ANSELMET[3]

1 *Mechanical Engineering, Queen's University, Ontario, Canada*
2 *Département de Génie Mécanique, Université Laval, Québec, Canada*
3 *Institut de Recherche sur les Phénomènes Hors Equilibre, Marseille, France*

1. Introduction

It is now widely recognized that large coherent structures are key elements in the dynamics of turbulent flows, since it is known that they dominate the most important phenomena in these flows : turbulence production, turbulent diffusion, etc .... [1, 2]. The interest in these structures is growing because of their importance in understanding flow equilibrium. Statistical considerations such as conditional averages or conditional probability density functions (pdfs) have shown [3] that the inter-dependence between temperature and its dissipation is particularly strong in regions where turbulent flows are non isotropic or influenced by boundary conditions which make the temperature pdfs $P(\theta)$ asymmetric, whereas the correlation coefficient was found to result from frequencies between those associated with the integral scale and the Taylor micro-scale.

In the present paper, we investigate the statistical link between temperature and its dissipation with particular emphasis on the dominant coherent movements in turbulent boundary layers, namely, sweeps and ejections. For this purpose, a local cross-correlation analysis is carried out applying the wavelet transform in conjunction with the standard detection schemes WAG and VITA. The effect of coherent structures passage on local isotropy is investigated through conditional cross-spectra inferred from the wavelet analysis [4, 5] and transposed in the spectral domain to analyze the local link between θ and its dissipation ε_θ. We analyze data related to the three θ derivatives involved in the dissipative term of the transport equation for $1/2\ \overline{\theta^2}$. The main objective of the present investigation is to achieve a better understanding of the effects of the "local" link on the correlation between temperature and its dissipation.

2. Experimental arrangement and data analysis

Measurements were taken in the experimental working section of the S2 wind tunnel at IRPHE, [5, 6]. The flow under investigation is a flat plate turbulent boundary layer which develops along the lower floor of the working section. This flow was slightly heated from the beginning to obtain a constant uniform temperature difference of 10K with respect to the outer flow. The mean external longitudinal velocity (associated with

U. Frisch (ed.), Advances in Turbulence VII, 539–542.

x direction) is U_1 = 12.6 m/s and temperature acts as a passive contaninant. Measurements were performed at the downstream position (X=3.7 m), where the boundary layer thickness is 57 mm and the Reynolds number based on the momentum thickness is 5100. The topology of the organized motion has been investigated in terms of two specific events, coolings and heatings related to sweeps and ejections respectively [6].

In order to overcome the Fourier analysis limitations, the extension of the interspectrum notion to the wavelet analysis is developed to define an extensive wavelet interspectrum, as first introduced in [5] and recently used by M. Onorato et al. [7]. This complements Benaissa et al.'s [8] conditional spectrum, based on the wavelet decomposition and the detection of coherent motions.

Using the local spectral decomposition, the conditional spectrum is written as :

$$Ec(a) = \langle E(b,a) \rangle = \frac{1}{n} \sum_{m=1}^{n} E(b_{jm}, a) \qquad (1)$$

where j is the current index for detection instants t_j, and n is the total number of detections. This value represents the conditional spectrum; it can be used to evaluate the spectral contributions of coherent events at different scales.

The interspectrum is defined from the convolution of two functions f(t) and h(t) as in the case of the Fourier transform. Considering $T_f(a,b)$, the wavelet coefficients of f(t), and $T_h(a',b')$, those of h(t), we define the extensive wavelet interspectrum:

$$G_{fh}(b,b';a,a') = T_f(b,a).T_h^*(b',a') \qquad (2)$$
$$G_{fh}(b,b';a,a') = Co_{fh}(b,b';a,a') - jQo_{fh}(b,b';a,a')$$

with $Co_{fh}(b,b';a,a')$ and $Qo_{fh}(b,b';a,a')$ the extensive cospectrum and quadraspectrum respectively.

3. Results and discussion

This approach allows us to focus on the spectral contributions of sweeps and ejections to the linkage between temperature and its dissipation. The distributions of the wavelet global cospectra and those obtained with the Fourier transform at y^+=60 are in good agreement, as shown in figure 1.

The observations of conditional cospectrum and quadraspectrum lead to check "local" isotropy related to coherent events. Because the quadraspectrum is more sensitive to the linkage between temperature and the x- (or t-) temperature derivative, the conditional quadraspectrum is rather important compared to the mean (figure 2), and shifted toward the relatively high frequency range. This implies that sweeps have an important contribution to the linkage between θ and ε_θ through the strong negative front in the θ signature and the various small scales transported with these events.

The analysis of ejections gives a completely different situation in agreement with previous investigations [3]. One observes a weak deviation from zero of the cospectrum and quadraspectrum in a narrow range of frequencies around the maximum of the cospectrum and the minimum of the quadraspectrum. In the high frequency range the conditional quadraspectrum is slightly shifted towards lower frequencies.

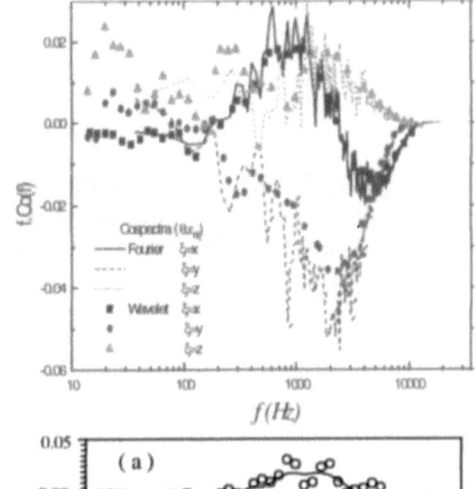

Figure 1. Fourier and mean wavelet cospectra between temperature and the contributions to the temperature dissipation in the three directions at $y^+=60$.

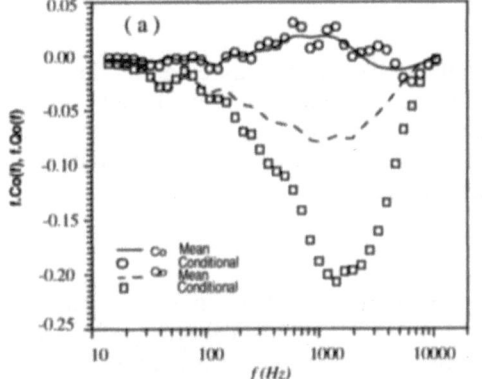
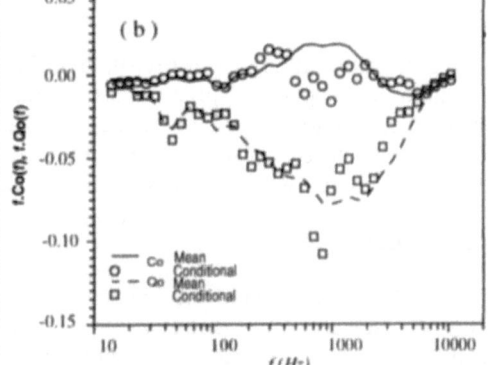

Figure 2. Mean and conditional cospectra and quadraspectra between θ and ε_θ at $y^+=60$ for cooling (a) and heating (b).

At a different position, in the intermittent region $y/\delta=0.88$ (or $y^+=1800$), with a temperature intermittency coefficient $\gamma=0.26$, where the three mean derivatives involved in the dissipation term are equal, only the isotropic temperature dissipation signal was processed. Mean and conditional co- and quadraspectra are presented in figure 3. The conditional quantities are related to high turbulent intensity events detected from local variance by VITA. The relative importance of the conditional cospectrum implies that these events play an important role in the link between temperature and its dissipation. Two peaks are observed in the conditional quadraspectrum, the first one corresponding to the large peak observed on the cospectrum and the other one to a higher frequency. The first frequency corresponds to the frequency occurrence of the studied events, while the second one is related to the size of the high variance events contained in the turbulent "spots". These observations lead to the conclusion that the local interaction between temperature and its dissipation in high turbulence events produces the link between these two quantities and reveals the imprint of two scale sizes in the dissipative field. At this position, the correlation coefficient between θ and $\varepsilon_{\theta x}$ is high and equal to 0.26, as compared to that at $y^+=60$ which is equal to 0.00.

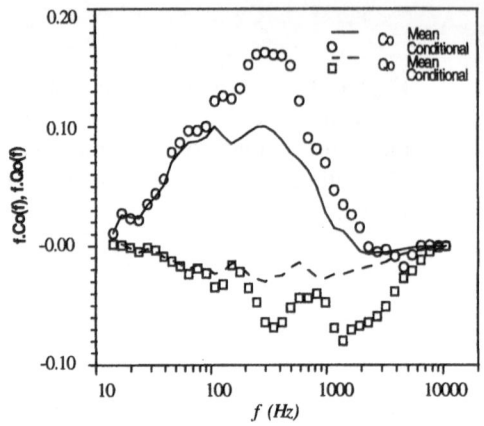

Figure 3. Mean and conditional cospectra and quadraspectra between θ and the isotropic dissipation $\varepsilon_{\theta x}$ at $y/\delta=0.88$.

The conditional approach reveals that the link between temperature and its dissipatio is related to "spots" of high levels of turbulence concentrated around high negativ temperature gradients and also, to a certain degree, to the frequency of occurrence of thes events. These decompositions indicate the scales of the flow which most contribute to th conditional link between temperature and its dissipation.

4. Conclusion

This paper has considered the impact of coherent structures in the context of local (in terms of space and time) isotropy. Indeed, local isotropy in a statistical sense is violated by the dominating events of the boundary layer, particularly sweeps. These events contribute to the linkage between temperature and its dissipation even in the central part of the boundary layer. Because of their intense energy, linked to the cold outer flow, they produce high temperature gradients. These events also carry a wide hierarchy of structures in the central part of the wall region.

For the external region of the boundary layer, $y/\delta=0.88$, the conditional approach reveals that the link between temperature and its dissipation is related to "spots" of high levels of turbulence concentrated around high negative temperature gradients and also, to a certain degree, to the frequency of occurrence of the events.

Acknowledgments

A.B. gratefully acknowledges the support of the NSERC of Canada.

References

[1]. Hussain, A.K.M.F. (1986) *J. Fluid Mech.*, Vol. 173.
[2]. Robinson, S. K. (1991) *Annual Rev. Fluid Mech.*, Vol. 23.
[3]. Anselmet, F., Djeridi, H. and Fulachier, L. (1994) *J. Fluid Mech.*, Vol. 280.
[4]. Farge, M. (1992) *Annual Rev. Fluid Mech.*, Vol. 24.
[5]. Benaissa, A. (1993) Ph. D. Thesis, Université d'Aix-Marseille II.
[6]. Antonia, R.A. and Fulachier, L. (1989) *J. Fluid Mech.*, Vol. 198.
[7]. Onorato, M., Salvetti, M. V., Buresti, G. and Petagna, P. (1997) *European Journal of Mechanics, B/Fluids*, Vol. 16, n° 4.
[8]. Benaissa, A., Liandrat, J. and Anselmet, F. (1995) *European Journal of Mechanics, B/Fluids*, Vol. 14, n° 6.

KINEMATIC SIMULATION FOR STRATIFIED FLOWS

F. NICOLLEAU AND J. C. VASSILICOS

*Department of Applied Mathematics and Theoretical Physics,
University of Cambridge, CB3 9EW, Cambridge, UK*

Kinematic Simulations (KS) are Lagrangian models of turbulent dispersion that are based on the integration of fluid element (or synonymously,
particle) trajectories in realisations of a prescribed turbulent-like Eulerian
velocity field. Small-scale turbulent-like flow structures exist in every re-
alisation of these fields and provide enough accuracy to predict correct
Lagrangian properties. Each trajectory is smooth and comparable in character to particle trajectories seen in nature and the laboratory.

KS has already been applied to homogeneous and isotropic turbulence
[1], and here we generalise KS to dispersion in stratified turbulent flows.
Our stratified version of KS is based on previous work by [3] which is itself
based on Rapid Distortion of KS fields. The KS results that we obtain for
one-particle dispersion are in agreement with experiments and Direct Numerical Simulation (DNS). In particular, Figure 1 shows the two classical
régimes $< x_1{}^2 > \sim t^2$ for $t \to 0$ and $< x_1{}^2 > \sim t$ for $t \to \infty$ obtained for
one-particle dispersion. We also clearly observe a confinement of particles
in the direction of stratification x_3 (see Figure 2).

The modelling of dispersion in stratified turbulent flow involves a deli-
cate balance between dispersion features in a quasi 2-D flow ($x_1 - x_2$ plane)
and dispersion features in the direction of stratification (x_3 direction). To
relate these dispersion features to the topological properties of the flow we
investigate two-particle dispersion in each direction and compare the dispersion laws to previous KS results obtained for 2-D isotropic turbulent-like
flows [2]. We also investigate the dependence of two-particle dispersion on
the power law of the energy spectrum $E(k) \sim k^{-p}$ of the velocity field used
in KS. In the plane orthogonal to the direction of stratification the law
proposed by [2] and based on the locality assumption is $< x_i{}^2 > \sim t^{4/(3-p)}$
and is observed in the plane orthogonal to the direction of stratification
confirming that dispersion in this plane is unaffected by stratification (see

U. Frisch (ed.), Advances in Turbulence VII, 543–546.
© 1998 *Kluwer Academic Publishers.*

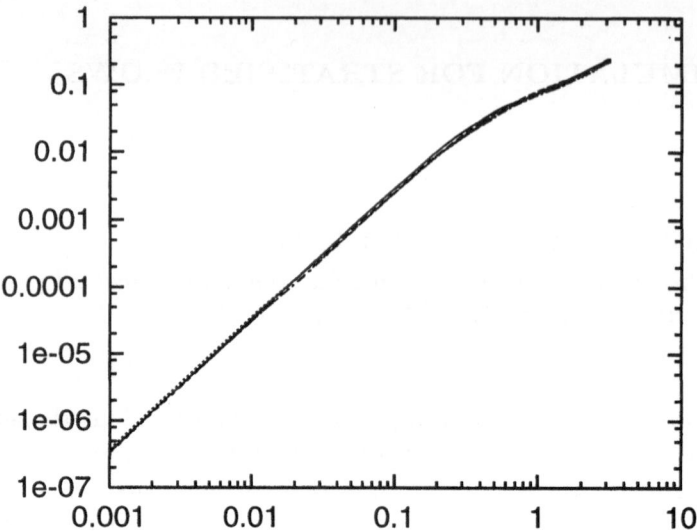

Figure 1.   One-particle dispersion in the plane orthogonal to the direction of stratification
($< x_1{}^2 >$) as a function of time t; Results are obtained for Brünt Väisälä frequencies
$N = 2\pi, 4\pi, 8\pi, 16\pi, 32\pi, 64\pi$ and 128π.

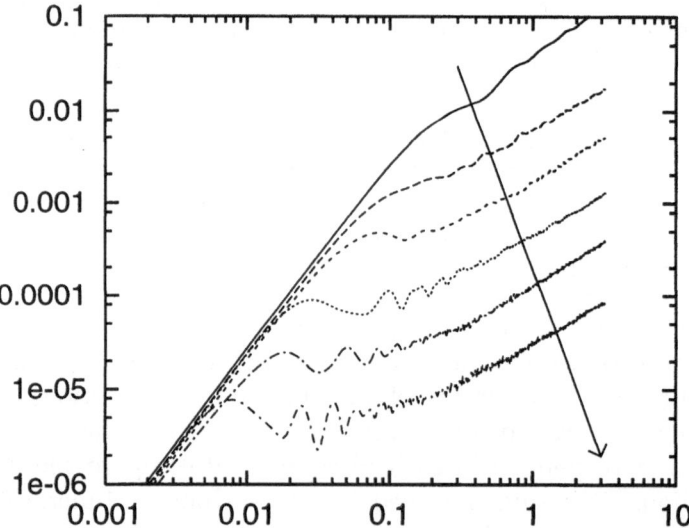

Figure 2.   One-particle dispersion in the direction of stratification, ($< \Delta x_3{}^2 >$) as a
functions of time t. From top to bottom, $N = 4\pi$, 8π, 16π, 32π, 64π and 128π.

Figure 3a-c). By contrast, in the direction of stratification there is no relation of this kind implying that the locality assumption does not hold in
the direction of stratification (see Figure 3b-d). Figure 3 shows 2-particle

dispersion for $p = 5/3$ and $p = 1.8$; straight lines correspond to the law predicted by [2] (3. and 3.33 respectively).

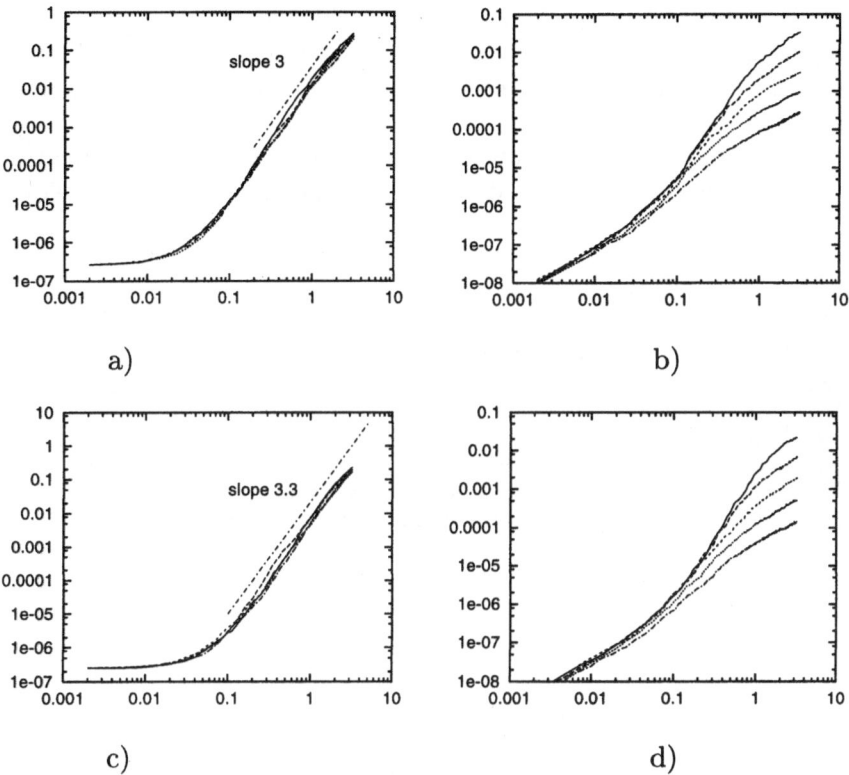

Figure 3. Two-particle dispersion left $< \Delta x_1 >$, right $< \Delta x_3 >$, as function of time t and for $N = 16\pi, 32\pi, 64\pi, 128\pi$ (upper to lower curves) for $p = 1.66$ a) and b), and for $p = 1.8$ c) and d).

From our first round of results we can already draw the following conclusions:

- The energy spectrum which is an input for any KS does not need to be calculated in detail using EDQNM or DNS. An analytically prescribed spectrum for stratified turbulence based on RDT is enough to obtain the correct dispersion properties.
- KS enables the study and prediction of dispersion in stratified turbulence for large values of N and large Reynolds numbers with only a small number of modes (between 500 and 2000 modes).
- Dispersion features in the plane orthogonal to the direction of stratification are unaffected by the stratification, in particular they are independent of the Brünt Väisällä frequency N (see Figures 1 and 3a,c).

- In the direction of stratification, the one-particle statistic $< x_3{}^2(t) >$ levels off at $t \approx \frac{\pi}{2N}$, and then the asymptotic law $< x_3{}^2(t) > \sim t$ is retrieved for $t > T_L$, but our model based on RDT and the Boussinesq approximation is not valid for times t larger than T_L.
- Two-particle dispersion is local in the plane orthogonal to the direction of stratification and obeys the scaling law proposed by [2] on the basis of locality.
- Two-particle dispersion is non-local in the direction of stratification; there is no power law in this case and the locality analysis developed in [2] does not apply.

References

1. Fung J.C.H., Hunt J.C.R., Malik N.A. & Perkins R.J. (1992). Kinematic Simulation of homogeneous turbulence by unsteady random Fourier modes. *J. Fluid Mech.* **236**, 281–317.
2. J. C. H. Fung and J. C. Vassilicos (1998), Two-particle dispersion in turbulentlike flows, *Phys. Rev. E* **52**(2), 1677-1690.
3. F. S. Godeferd, N. A. Malik, F. Nicolleau and C. Cambon (1997); Eulerian and Lagrangian statistics in homogeneous stratified flows to appear in ASR.

TURBULENT DIFFUSION IN THE PIPE FLOW BY THE STOCHASTIC LAGRANGIAN MODEL

Y. SAKAI[1], T. MUKOUYAMA[1], H. TSUNODA[2]
AND I. NAKAMURA[1]

[1] *Department of Mechano-Informatics & Systems, Nagoya
University, Furocho, Chikusaku, Nagoya 464-8603, Japan.*
[2] *Department of Mechanical System, Yamanashi
University, Takeda 4-3-11, Koufu 400-8511, Japan.*

1. Introduction

In this short paper, we present the results of the numerical simulations of
the scalar diffusion fields in a fully developed turbulent pipe flow by using
the Lagrangian stochastic process model. The model used here is a Generalized Langevin (GL) model expressed in the cylindrical coordinate system
[1]. The GL equation was first proposed by Haworth & Pope [2] in order
to model the joint PDF equation of the velocity, and then applied to simulate the velocity and scalar field with the scalar mixing model mainly for
self-similar turbulent flows [3][4]. With regard to the velocity field, recently
the relationship between stochastic Lagrangian models and second-moment
closures has been in detail examined [5], and the extension of the GL model
has been also made to include modelling of the near-wall turbulent flow [6].
However, to date, almost no applications of this technique to the scalar
diffusion field in a wall turbulent flow and inner flow such as pipe flow have
been made.

In the present study, the three types of axisymmetric scalar diffusion
fields from continuous sources in pipe flows have been simulated by the GL
model expressed in the cylindrical coordinates, then the simulation results
are compared with the experimental data and the theoretical results.

2. The GL Model and Simulation Method

The GL model for the increments of each velocity component of particle
dU_i^* in the cylindrical coordinates is given as follows [1],

$$dU_r^* = -\frac{1}{\rho}\frac{\partial\langle P\rangle}{\partial r}dt + \frac{1}{r^*}U_\theta^* dt + G_{rx}\left(U_x^* - \langle U_x\rangle\right)dt + \sqrt{C_0\varepsilon}dW_r ,$$

U. Frisch (ed.), Advances in Turbulence VII, 547–550.

$$dU_\theta^* = -\frac{U_r^* U_\theta^*}{r^*} U_\theta^* dt + G_{\theta\theta} U_\theta^* dt + \sqrt{C_0 \varepsilon} dW_\theta ,$$

$$dU_x^* = -\frac{1}{\rho}\frac{\partial \langle P \rangle}{\partial x} dt + \nu \frac{1}{r^*}\frac{d}{dr}\left(r^* \frac{d\langle U_x \rangle}{dr}\right) dt + G_{rx} U_r^* dt$$
$$+ G_{xx}\left(U_x^* - \langle U_x \rangle\right) dt + \sqrt{C_0 \varepsilon} dW_x , \qquad (1a,b,c)$$

where $*$ denotes the random variable attendant on the particle, ε : mean dissipation rate per unit mass, C_0 : Kolmogorov constant, dW_i : increment of an isotropic Wiener process in the i direction with the mean 0 and variance 1. The tensor G_{ij} is determined on the basis of the consistency condition [7] up to the second-order moments of the velocity field (for details on G_{ij}, see Ref.[1]). In this study, the dissipation and the first- and second-order moments of the Eulerian velocity field are specified as the known experimental data. The movement of the particle is modelled as follows,

$$dx_i^* = U_i^* dt + \sqrt{2\kappa} dW_i^\kappa , \qquad (2)$$

where κ is the molecular diffusivity and dW_i^κ is the increment of the isotropic Wiener process which is independent of dW_i in Eqs.(1a,b,c). The number density of particles tracked by Eq.(1a,b,c) and Eq.(2) is proportional to the scalar mean value.

3. Simulation Conditions

We made the simulations of the following three cases of axisymmetric scalar diffusion fields from continuous sources; CASE 1): the plume of smoke particles from a small disc source at the center of the pipe (Fig.1(a)), CASE 2): the plume temperature field from the axisymmetric narrow band source on

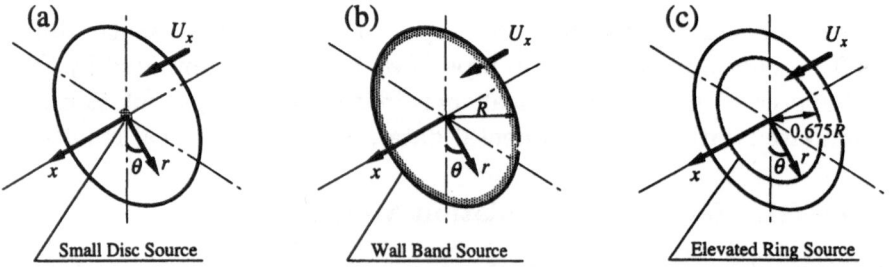

Figure 1. Coordinate systems and source configurations: (a) small disc source (uniform distribution of particles in the disc with radius $r = 0.01R$, (b) narrow band source on the wall (uniform distribution of temperature in the region of $0.995 \le r/R \le 1.00$), (c) elevated ring source (uniform distribution of temperature on the ring with radius $r = 0.675R$).

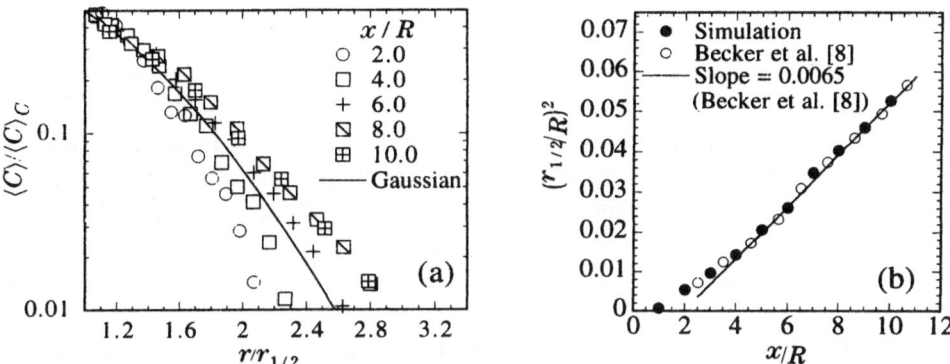

Figure 2. (a) Downsteam variation of radial profiles of the mean comcentration of the smoke particles in CASE 1, (b)Downstream change of the half-width $r_{1/2}$ in CASE 1.

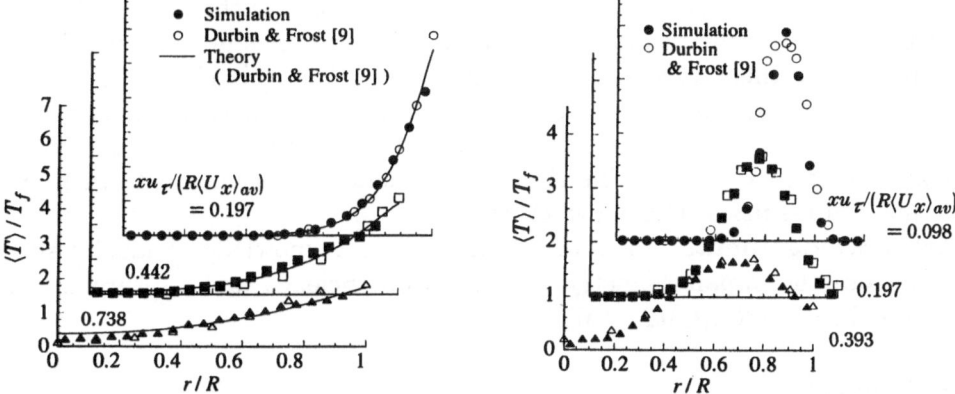

Figure 3. Downstream variation of radial profile of the mean temperature in CASE 2.

Figure 4. Downstream variation of the mean temperature profile in CASE 3.

the wall (Fig.1(b)), CASE 3): the plume temperature field from an elevated ring source above the wall (Fig.1(c)).

The simulation conditions are decided on the basis of the experiments by Becker et al. [8] in CASE 1 and Durbin & Frost [9] in CASE 2 and 3. They are summarized as follows;

CASE 1): Reynolds number $Re = 2\langle U_x \rangle_{av} R/\nu = 684,000$ (pipe radius $R = 10.05cm$, cross-sectional average of mean axial velocity $\langle U_x \rangle_{av} = 52.4m/s$), friction velocity $u_\tau = 11.2m/s$, molecular Schmidt number $Sc = \kappa/\nu = 38,000$, $C_0 = 3.8$, the perfect reflection on the wall.

CASE 2 and 3): $Re = 2\langle U_x \rangle_{av} R/\nu = 43,750 (R = 10.15cm, \langle U_x \rangle_{av} = 3.47m/s)$, $u_\tau = 0.18m/s$, molecular Prandtl number $Pr = \kappa/\nu = 0.71$, $C_0 = 2.5$, the perfect reflection on the wall.

4. Results

The downstream variation of the radial profile of the mean concentration $\langle C \rangle$ of the smoke particles in CASE 1 is shown in Fig.2(a), where $\langle C \rangle_C$ is the mean concentration on the plume centerline and $r_{1/2}$ is the half-width of the mean concentration profile. The figure gives only the data for large radii of $r/r_{1/2} > 1.1$. Becker et al. [8] showed that as going downstream and away from the source the foot of the concentration profile spreads in the radial direction. The variation of the present simulated profile consists with the results of Becker et al.[8]. Figure 2(b) shows the downstream change of $r_{1/2}$ with experimental data. It is found that the present simulation can predict very well the experimental downstream change of $r_{1/2}$. Figure 3 shows the downstream variation of the mean temperature radial profiles in CASE 2. In the figure, T_f is the flux temperature defined by $T_f = 2 \int_0^R \langle U_x \rangle (r) \langle T \rangle (r) \, r \, dr / (\langle U_x \rangle_{av} R^2)$, where $\langle U_x \rangle$ is the mean axial velocity and $\langle T \rangle$ is the mean temperature. The solid lines in the same figure represent the theoretical profiles by Durbin & Frost [9]. We find that the simulation results show good agreements with the experimental data and the theoretical lines. Figure 4 shows the downstream variation of the mean temperature in CASE 3. The simulation results almost coincide with experimental results of Durbin & Frost [9].

Acknowledgements : Authors would like to acknowledge financial support from the Japan Ministry of Education, Science and Culture under Grands-in-Aid (No.(C)(2)-09650183).

References

1. Sakai,Y. , Nakamura,I., Tsunoda,H. and Hanabusa,K. (1996) Diffusion in turbulent pipe flow using stochastic model, *JSME International J., Series B*, **39**-4, 667-675.
2. Haworth,D.C. and Pope,S.B. (1986) A generalized Langevin model for turbulent flows, *Phys. Fluids*, **29**-2, 384-405.
3. Haworth,D.C. and Pope,S.B. (1987) Monte Carlo solutions of a joint PDF equation for turbulent flows in general orthogonal coordinates, *J. Comp. Phys.*, **72**, 311-346.
4. Haworth,D.C. and Pope,S.B. (1987) A pdf modeling study of self-similar turbulent free shear flows, *Phys. Fluids*, **30**-4, 1026-1044.
5. Pope,S.B. (1994) On the relationship between stochastic Lagrangian models of turbulence and second-moment closure, *Phys. Fluids*, **6**-2, 973-985.
6. Dreeben,T.D. and Pope,S.B. (1997) Probability density function and Reynolds-stress modeling of near-wall turbulent flows, *Phys. Fluids*, **9**-1, 154-163.
7. Pope,S.B. (1987) Consistency conditions for random-walk models of turbulent dispersion, *Phys. Fluids*, **30**-8, 2374-2379.
8. Becker,H.A., Rosensweig,R.E. and Gwozdz,J.R. (1966) Turbulent dispersion in a pipe flow, *AIChE J.*, **12**-5, 964-972.
9. Durbin,P.A. and Frost,S. (1981) Dispersion from line sources in turbulent pipe flow, *Int. J. Heat Mass Transfer*, **24**-6, 969-982.

DIRECT NUMERICAL SIMULATION OF MIXING OF A PASSIVE SCALAR IN TURBULENT CHANNEL FLOW

G. BRETHOUWER, M.B.J.M. POURQUIÉ & F.T.M. NIEUWSTADT

Delft University of Technology, Laboratory for Aero- and Hydrodynamics, 2628 AL Rotterdamseweg 145 Delft, the Netherlands

1. Introduction

In various industrial applications and environmental processes like turbulent combustion and dispersion of air pollution, a dominant role is played by turbulent mixing. Therefore, it would be desirable to understand in detail the process of turbulent mixing. One approach is to develop closure models for the higher-order moments of the concentration. These can for instance be used in models for turbulent reacting flows based on the evolution of the scalar probability density function. An advantage of these so-called pdf models above other methods is that the source or reacting term is given in closed form. However, pdf-models still contain terms which need to be modelled [1]. These terms are difficult if not impossible to determine experimentally.

Numerical simulation such as DNS have proved to be an attractive tool to study the problem of turbulent mixing. Much insight has already been obtained in simulation studies limited mainly to mixing in homogeneous isotropic turbulence. In view of applications it seems interesting to extend the simulations to more realistic flows such as wall-bounded or free shear flows. The objectives of this work is to present results on turbulent mixing in a wall-bounded shear flow which are obtained by means of a DNS of a channel flow. In particular we will show that from these data the statistical quantities of pdf-models discussed above can be obtained.

First a short description of the numerical method is given and next some results are presented.

U. Frisch (ed.), Advances in Turbulence VII, 551–554.

2. Solution method and details of the computation

The incompressible Navier-Stokes equations together an advection-diffusion equation for a scalar are discretised on a three dimensional staggered grid with help of a finite volume method. The terms in the Navier-Stokes equations are discretised by central second-order schemes and the advection of the scalar a monotone scheme is used. The flow geometry is a channel. The boundary conditions for the flow field are periodic in the streamwise direction with a domain length of 5δ where δ is the channel height. The domain of the spatial developing scalar field has a length of 15δ which is obtained by putting three periodic flow domains after each other. The time integration has been done with an explicit second-order Adams-Bashforth scheme. The continuity in the form of a divergence-free velocity field is forced each timestep by means of a pressure correction method. The number of grid points in the normal, spanwise and streamwise direction is respectively $156 \times 120 \times 225$ for the flow field.

The scalar is released from a thin line source with a uniform distribution. The source is placed in the centre of the channel and its width is comparable to the Kolmogorov length scale. The scalar is passive so it has no effect on the flow field. The flux of the scalar through the wall is zero. The Schmidt number is set equal to one and the Reynolds number of the flow based on the mean velocity is 5600.

The mean velocity profiles and the velocity fluctuation profiles have been validated by comparison with the direct numerical simulations of channel flow from Kim et al. [2].

From the numerical data we obtained statistics of the concentration by averaging over time and the spanwise direction at each position in the streamwise direction x and in the wall-normal direction y.

3. Results

The profiles of the mean concentration as a function of y have a Gaussian shape. Only in the far field deviations from the Gaussian shape occur caused by the scalar reaching the wall. If the streamwise distance from the source is not too large the plume lies completely in the centre region of the channel. In this part of the channel the flow is almost homogeneous so that our results can be compared with experimental measurements of turbulent mixing behind line sources in homogeneous grid turbulence. Our results agree well with the available experimental data for the profile of the relative intensity of concentration fluctuations and for the second order moment, skewness and kurtosis of the concentration. As soon as the distance from the source is larger, the influence of the inhomogeneous near-wall region and the reflection of the scalar at the wall begins to influence the

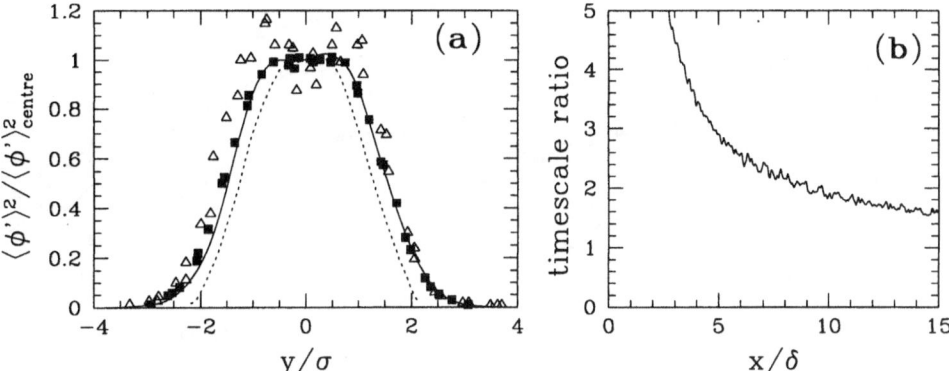

Figure 1. (a): The second moment of the concentration scaled with the second moment at the centre as a function of y/σ where σ is the standard deviation of the mean concentration in the y direction. Shown are our results at $x = 4\delta$ (solid line) and at $x = 12\delta$ (dashed line) and experimental data of mixing in the near field (■) and the far field (△). (b): The timescale ratio of the scalar dissipation to the energy dissipation in the centre of the channel.

statistics. This can be seen in Fig. 1.a which shows the second moment of the concentration scaled with the second moment in the centre of the channel. Here, two cross-stream profiles, one at $x = 4\delta$ and one at $x = 12\delta$ are shown. The distance y to the centre of the channel is scaled with the standard deviation of the mean concentration in the y-direction. Also plotted are experimental data of mixing behind a line source in homogeneous grid turbulence reported in [3]. Shown are experimental measurements in the near field and in the far field. The cross stream profile at $x = 4\delta$ is in good agreement with the near-field measurements. In the far field the experimental measurements show off centre peaks which are not present in our calculated profile. Furthermore, for large y/δ the concentration fluctuations are smaller in the channel than in homogeneous turbulence.

In most closure models the ratio of the time scale of kinetic energy dissipation to the time scale of the scalar dissipation is taken to be a constant given by $(\epsilon_\phi/\langle\phi'\phi'\rangle)/(\epsilon/k)$. Here, ϵ_ϕ is the dissipation of scalar fluctuations ϕ', ϵ the dissipation of turbulent kinetic energy and k the turbulent kinetic energy. The ratio of the two time scales is illustrated in Fig. 1.b as a function of x/δ at the centre of the channel. It is clear that this timescale ratio is not a constant as a function of the streamwise coordinate. Only at the end of the channel it appears that the time scale ratio approaches an asymptotic value of around 1.5. These observations agrees with the experimental measurements given in [4] where it is argued that close to the source the size of source affects the time scale ratio.

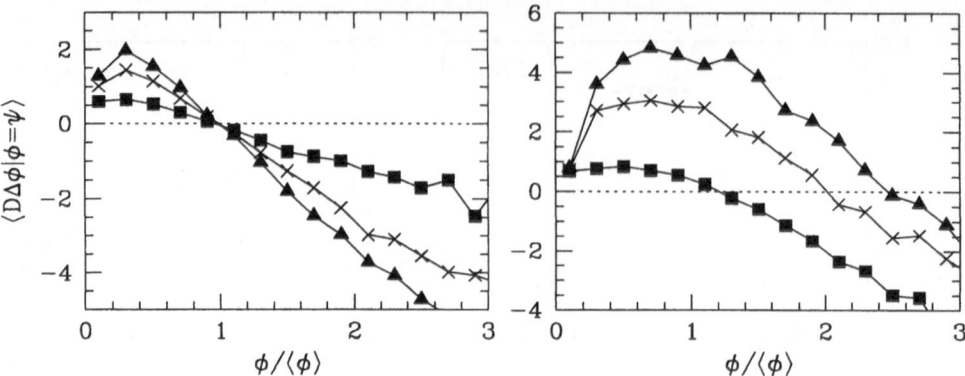

Figure 2.   The conditional expectation $< D\nabla^2\phi|\phi = \psi >$ at $x = 3.6\delta$ (▲), $x = 5\delta$ (×) and $x = 14\delta$ (■) as a function of the concentration scaled with the local mean concentration. Left in the centre of the channel and right off the centre at $1.3\sigma < y < 1.7\sigma$.

One of the terms in the transport equation of the scalar pdf for which a closure is needed, is the conditional expectation $< D\nabla^2\phi|\phi = \psi >$. This is the expectation of the molecular diffusivity $D\nabla^2\phi$ conditioned on the concentration ϕ being equal to ψ. Fig. 2 shows this conditional expectation as a function of the concentration ϕ, both are scaled with the local mean concentration. The figure at the left shows the conditional expectation in the centre of the channel at different streamwise positions. In the centre of the channel the molecular diffusivity term is positive if the concentration is lower than the local mean concentration which means that in this case the concentration increases due to molecular diffusion. If the concentration is higher than the mean concentration molecular diffusion decreases the concentration. This behaviour is consistent with the down the gradient behaviour of diffusion. The figure at the right shows the conditional expectation off the centre of the channel at different streamwise positions. Here one observes that close to the source the molecular diffusion can be positive for concentrations higher than the local mean concentration whereas some models for the conditional expectation assume that molecular diffusion drives the concentration always to the mean concentration [1].

References

1. Libby, P.A. & Williams, F.A. (1994) *Turbulent reacting flows*, Academic Press.
2. Kim, J., Moin, P. & Moser, R. (1987) Turbulence statistics in fully developed channel flow at low Reynolds number, *J. Fluid. Mech.* **177**, 133–166.
3. Sawford, B.L. & Sullivan, P.J. (1995) A simple representation of a developing contaminant concentration field, *J. Fluid Mech.* **289**, 141–157.
4. Li, J.D. & Bilger, R.W. (1996) The diffusion of conserved and reactive scalars behind line sources in homogeneous turbulence, *J. Fluid Mech.* **318**, 339–372.

PERSISTENCY OF MATERIAL ELEMENT DEFORMATION IN ISOTROPIC FLOWS AND GROWTH RATE OF LINES AND SURFACES

JÉRÔME DUPLAT AND EMMANUEL VILLERMAUX
LEGI/IMG B.P. 53 X 38041 Grenoble Cedex, France

Stretching is the key concept of mixing processes. Although it has been realized for a long time that material lines and surfaces grow, in the mean, when subjected to a series of uncorrelated, incompressible deformations, the question of the precise value of the net growth rate is still to be clarified.

We address this problem here by examining the value of the correlation time, or persistency, τ of a given stretching rate event γ.

We consider a material blob submersed in a random incompressible flow, undergoing a succession of independant stretching events. Each event is characterized by its three main axes of deformation **x,y,z**, the eigenvalues of the deformation tensor $(\gamma_x, \gamma_y, \gamma_z)$ (or (γ_x, γ_y) in two dimension) and the persistency τ, that is the correlation time during which $\gamma_x, \gamma_y, \gamma_z$ can be considered constant. The elongation is $\gamma\tau$ with $\gamma = \sqrt{\gamma_x^2 + \gamma_y^2 + \gamma_z^2}$.

We compute the effective growth rate of lines and surfaces averaged on an isotropic distribution of the axes of deformation, thus assuming a large enough number of events so that the entire distribution has been visited. Analytical results are derived for the 2D case, and for the special case of axisymmetric deformations in 3D (shown also to be a lower bound for the general 3D case). Numerical results are available for the general 3D case.

It is evidenced that the net growth rate of lines and surfaces is strongly affected by the persistency. Indeed, for both lines and surfaces the growth rate scales as $\frac{1}{15}\gamma^2\tau$ in the small $\gamma\tau$ limit (figure 1 and 2). This wanishingly small value is explained by the fact that, in the limit of small persistency, the material elements do not have time enough to align themselves with the direction of maximum stretch. This alignment condition is realized for $\gamma\tau \geq 10$, a value above which the growth rate reaches an asymptotic constant value.

U. Frisch (ed.), Advances in Turbulence VII, 555–556.
© 1998 *Kluwer Academic Publishers.*

Considering the values of the deformation rate tensor in turbulent flows as measured by Batchelor and Townsend (1956) and the data of line and surface growth rates extracted from Girimaji and Pope (1990) DNS calculations, we show that the elongation is most likely close to $\gamma\tau \approx 1$ in actual flows (see figure 2). This result has an important incidence for the kinetics of turbulent mixing when linking the number of stretching events with the net elongation undergone by a scalar blob (Villermaux et al. 1998).

References

Batchelor, G.K. and Townsend, A.A. 1956 Turbulent diffusion. in Batchelor, G.K. and Davis R.M. (eds), *Surveys in mechanics*, Cambridge University Press, 352–399

Girimaji S.S. and Pope S.B. Material-element deformation in isotropic turbulence *J. Fluid. Mech.*, **220**, 427–458

Villermaux, E., Innocenti, C. and Duplat, J. Histogramme des fluctuations scalaires dans le mlange turbulent transitoire, *C.R.Acad.Sci. Paris*, 1., **t. 326**, Serie IIb, 21–26

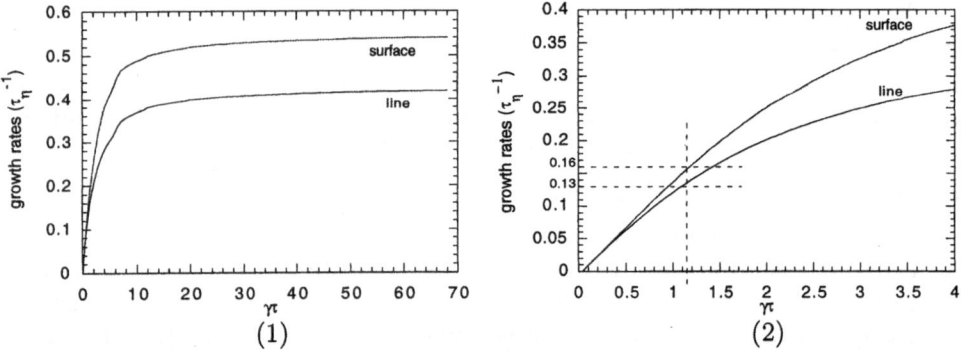

Figure 1. Effective growth rate of lines (lower curve) and surfaces (upper curve) in units of τ_η^{-1} as a function of the elongation $\gamma\tau$. This case corresponds to $(\gamma_x, \gamma_y, \gamma_z) = (0.43\tau_\eta^{-1}, 0.12\tau_\eta^{-1}, -0.55\tau_\eta^{-1})$ as measured by Batchelor and Townsend, where $\tau_\eta = \left(\frac{\nu}{\epsilon}\right)^{\frac{1}{2}}$ is the Kolmogorov timescale. (consistent with the Girimaji and Pope simulation).

Figure 2. Zoom of the previous figure at short persistencies. The values $0.13 \ \tau_\eta^{-1}$ and $0.16 \ \tau_\eta^{-1}$ are respectively the growth rates of lines and surfaces measured by Girimaji and Pope in a DNS simulation.

SMALL–SCALE STATISTICS OF THE PASSIVE SCALAR TURBULENT MIXING: A TEST OF THE INFLUENCE OF THE LARGE–SCALE PROPERTIES

L. DANAILA, F. ANSELMET, P. LE GAL
I.R.P.H.E., 12, Av. Général Leclerc, 13003 Marseille, France

C. BRUN, A. PUMIR
I.N.L.N., 1361 Route des Lucioles, 06560, Valbonne, France

AND

F.PLAZA, J.F. PINTON
E.N.S., 46 Allée d'Italie, 69364 Lyon, France

For very high Reynolds numbers, the Kolmogorov theory (K41) led to the conclusion that the small scale properties of turbulence are independent of the large injection scales, satisfy isotropy requirements and are statistically self-similar. Similar properties are expected to hold for a passive scalar mixed by a turbulent flow. The reported deviations from these universal statistics are attributed to the "internal intermittency" resulting from the dissipation rate fluctuations of both kinetic energy and scalar variance. As a consequence, the probability density functions (pdfs) associated with the inertial and dissipative scales are non Gaussian, presenting exponential tails. S. Vaienti and coworkers [1] have obtained an evolution equation for the pdfs of temperature increments $\Delta\theta(r) = \theta(x + r) - \theta(x) = X$. Writing this equation in a dimensionless form, the only parameter left in the equation is the Péclet number, $Pe_{\lambda_\theta} \equiv (\lambda_\theta\, u')/k_0$:

$$\frac{Pe_{\lambda_\theta}}{2}\left[\frac{2}{\tilde{r}} + \frac{\partial}{\partial\tilde{r}}\right](P(\tilde{r}, X)q_1(\tilde{r}, X)) + \frac{\partial^2}{\partial X^2}(P(\tilde{r}, X)q_2(\tilde{r}, X)) =$$

$$= \left[\frac{2}{\tilde{r}} + \frac{\partial}{\partial\tilde{r}}\right] \cdot \frac{\partial P(\tilde{r}, X)}{\partial\tilde{r}} \qquad (1)$$

where λ_θ is the Taylor scale for the temperature, and $\tilde{r} = r/\lambda_\theta$. This equation was derived using the assumptions of homogeneity and three-dimensional isotropy of the temperature and velocity fluctuations. The closure problem then consists in determining two conditional expectations:

U. Frisch (ed.), Advances in Turbulence VII, 557–560.

that of the longitudinal velocity increments conditioned by the temperature increments, $q_1(r, X) = \langle \Delta u / \Delta \theta \rangle$, and that of the square temperature gradient conditioned by the temperature increments, $q_2(r, X) = \langle (\nabla \theta)^2 / \Delta \theta \rangle$. These conditional expectations have been determined experimentally for a boundary layer over a heated wall [2]. Experimental observations and theoretical considerations based on the isotropy assumption then led to the conclusion that the velocity increments expectation has a "A" shape with two linear arms, and the square gradient conditional expectation q_2 has a parabolic behavior. The same features are obtained from DNS, [3], [4].

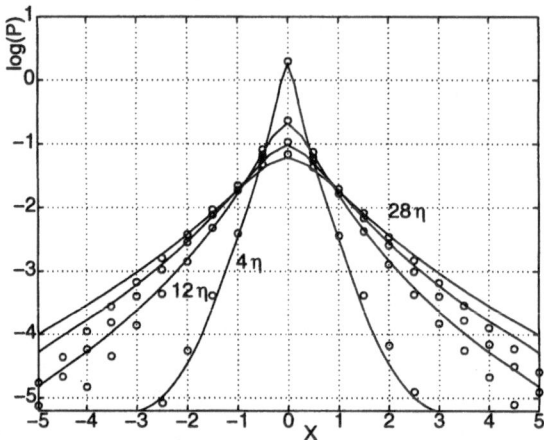

Figure 1. Our predictions (−) and the effective pdfs (o) for the scales: $4\eta, 12\eta, 20\eta$ and 28η. Natural logarithm representation.

The objective of the present work is to demonstrate that Eq.(1) successfully predicts the main features of the evolution of the pdf's as the scale varies. To this end, we first consider numerical ($R_\lambda = 82, Pr = 1$) and experimental ($R_\lambda = 200, Pr = 0.7$) data. We verify the balance of equation 1 using the numerical and experimental conditional expectations and pdfs and we determine in this way the equation validity domain. In a second time, starting from the largest scale of this validity domain and using the two conditional expectations, together with the initial pdf and its derivative, we integrate equation 1 to get the evolution of the pdfs that we compare with the pdfs actually obtained experimentally or numerically. To simulate isotropy, the equation verification and numerical resolution need the pdf and two conditional expectations symmetrization, especially for the experimental case [4]. We present in fig. 1 our numerical solution for DNS data, as compared to the real pdfs, using the natural logarithm scale. Indeed, the pdfs predicted by integration of equation Eq.(1) from $r = 32\eta$ ($6\lambda_\theta$) to $r = 4\eta$ are in excellent agreement with those obtained by DNS.

Therefore, equation 1 captures well the intermittent behavior of the turbulent mixing, and it is also a very sensitive test of local isotropy. Very good agreement is found between pdf's obtained when solving Eq.(1) and those measured for scales $r < 30\eta$ (DNS data), or for $r < 50\eta$ (experimental data). In fact, both flows considered here present a large scale mean temperature gradient, and we verified our theory using the increments, pdfs and conditional expectations computed in a direction normal to this large-scale non-homogeneous direction, in spite of the departure from local isotropy which is localized at large scales. In the direction parallel to the gradient, one does not expect such a good agreement, in view of the strong anisotropy persisting at small scales [5, 6, 7]. Thus, equation 1 can be used to predict the small scales statistics, starting from the pdf given at an initial scale, once a few conditional averages are known. For the large inertial scales, the passive scalar mixing presents large anisotropic structures, where the mixing is rather poor, so that equation 1, which does not take into account the mean scalar gradient term, is not verified. The physical significance of this large-scale discrepancy is that for these scales the scalar fluctuation dissipation is not balanced by turbulent advection only.

In the same framework, it may nevertheless be possible to control some non-isotropic effects. Such investigations are presently performed to predict the whole inertial range pdf evolution, as a function of the flow mixing characteristics. An alternative is to create experiments with more isotropic conditions, permitting us to test the validity of eq. 1 without anisotropic artifacts: it appears to be possible in the central region of a swirling flow between two counter-rotating disks. This experimental set-up presents some multiple advantages [8]: a large turbulent Reynolds number in a small space volume, a central region where the dynamic field is almost homogeneous and isotropic, and a double temperature source used as an original means to create some exotic properties of the mixing, when one of the disks is heated and the other one is cooled with respect to the ambient temperature. The rotation frequency of the disks is variable, between 20 and $40Hz$, and the turbulent Reynolds number is $R_\lambda \approx 400$. When the disks are spun, a stationary state is reached, with a maximum $3^{\circ}C$ temperature difference between the two disks. Simultaneous temperature-velocity measurements are performed using a classical hot wire $(5\mu m)$-cold wire $(0.6\mu m)$ probe. Thus, it is possible to create in the same flow the coexistence of two different mixing. This may create, in the central part, a zone where two isotropic criteria are respected: the velocity-temperature increments correlation coefficient is null, and the temperature increments skewness is small. In particular, the temperature derivative skewness is null (0.01) contrary to the value usually met in a single-source mean temperature gradient mixing (0.8). From these two points of view, the mixing is more isotropic. In the present case, the

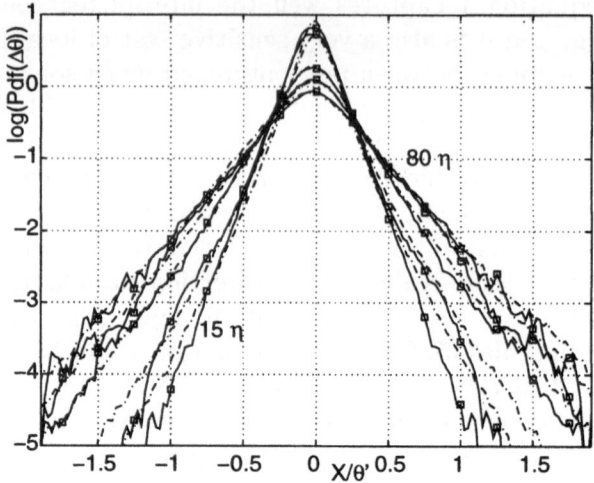

Figure 2. Numerical solution of 1 (− and small squares) compared to the real pdfs (−.), for the scales 80η, 60η, 40η, 20η and 15η. No symmetrization was necessary.

zero temperature increments skewness leads implicitly to symmetric pdfs for all the investigated scale range. These better isotropic conditions, and the larger $Pe_{\lambda_\theta} = 120$ number of the mixing, permits us to integrate numerically equation 1 with success, and to emphasize the intermittent behavior of the pdf, (see fig 2). This integration then needs no symmetrization for the pdfs and the two incoming closures, as was necessary in the boundary layer or in the jet [2], [4].

We acknowledge the support of DRET, under contract 95/2592 A, B,C, and of IDRIS for computer resources.

References

1. S. Vaienti, M. Ould-Rouis, F. Anselmet and P. Le Gal (1994) Statistics of temperature increments in fully developed turbulence: Part 1: Theory, *Physica D*, **73**, p. 99.
2. M. Ould-Rouis, F. Anselmet, P. Le Gal and S. Vaienti (1995) Statistics of temperature increments in fully developed turbulence: Part 2: Experiments, *Physica D*, **85**, p. 405.
3. A. Pumir, Small scale properties of scalar and velocity differences in three-dimensional turbulence (1994), *Phys. Fluids*, **6 (12)**, p. 3974.
4. L. Danaila, F. Anselmet, P.Le Gal, J.Dusek, C.Brun and A. Pumir (1997) Predictions of small-scale statistics, *Phys. Rev. Lett.*, **79**, p. 4577.
5. M. Holzer and E.D. Siggia (1994), Turbulent mixing of a passive scalar, *Phys. Fluids* **6**, p. 1820.
6. A. Pumir (1994) A numerical study of the mixing of the passive scalar in three dimensions in the presence of the a mean gradient, *Phys. Fluids* **6**, p. 2118.
7. C. Tong and Z. Warhaft (1994) On passive scalar derivative statistics in grid turbulence, *Phys. Fluids* **6**, p. 2165.
8. J.F. Pinton and R. Labbé (1994) Corrections to the Taylor hypothesis in swirling flows, *J. Phys. II France*, **4**, p. 1461.

CONDITIONAL VELOCITY STATISTICS IN TURBULENT JETS

J.F. LUCAS, L. PIETRI, M. AMIELH, F. ANSELMET
IRPHE, 12, Avenue Général Leclerc, 13003 Marseille, France

AND

A.A. BURLUKA, R. BORGHI
IRPHE, 38, rue Joliot Curie, 13451 Marseille Cedex 20, France

1. Introduction

The objective of the present work is to study the combined statistics of the velocity and scalar fields, in particular the conditional averages of velocity, which arise [1,2] in turbulent combustion models based on probability density functions (pdfs). When these approaches are applied on the single point statistics level, two kinds of processes need some closure assumptions, namely, the so-called micro-scale mixing, that is molecular diffusion enhanced by the small scale turbulence, and the turbulent convection. The micro-scale mixing was, and it still remains, a subject of numerous works the recent reviews of which can be found in [2,3]. The transfer of probability functions by the fluctuating turbulent velocity field in non-homogeneous flows, to which the present article is devoted, has gained less attention. Since the pioneering measurements of Bezuglov [4] and Scherbina [5], the experimental works on this subject remain scarce, which partly explains rather slow theoretical progress. The objective of this work is to present measurements associated with the transfer of iso-scalar lines by fluctuating velocity field in non-reacting jets and to provide comparison with the existing models.

2. Experimental setup

The facility [6,7] consists in a fully developed turbulent vertical pipe ($D_j = $ 26mm) flow of pure air which is slightly heated above the ambient temperature T_a (with $T_j - T_a = 20$K) discharging into ambient air in a slightly confined configuration (mean velocity $U_a = 0.9$m/s, square section 285x285 mm^2). In the basic nominal situation U_j is 12m/s and Re_j is 21000. Velocity

U. Frisch (ed.), Advances in Turbulence VII, 561–564.

measurements are performed with a two-component laser Doppler system (Argon 4W) fitted with fiber optics and two Burst Spectrum Analyzers operated with a Dantec acquisition software. The primary and secondary flows are simultaneously seeded using silicon oil particle diffusers (about $1\mu m$ in diameter). Simultaneous temperature measurements are performed using a $0.6\mu m$ cold wire. A specific home-made electronic device was designed for triggering a data wire acquisition each time a laser burst is detected.

3. Results and discussion

If one assumes that the thermochemistry of the flow is described by one scalar only, say $c(\mathbf{x}, t)$, then the transport equation for this scalar probability density function (pdf) $P(\hat{c}; \mathbf{x}, t)$ can be written as [1]:

$$\frac{\partial P(\hat{c}; \mathbf{x}, t)}{\partial t} + <u_j|\hat{c}> \cdot \frac{\partial P(\hat{c})}{\partial x_j} = -\frac{\partial^2 \chi_{\hat{c}} P(\hat{c})}{\partial \hat{c}^2} - \frac{\partial W(\hat{c}) P(\hat{c})}{\partial \hat{c}} \tag{1}$$

where the convective transfer of $P(\hat{c}; \mathbf{x}, t)$ is represented by $\langle u_j|\hat{c}\rangle(\mathbf{x}, t)$ averaged over the domains where $c(\mathbf{x}, t)$ equals $\hat{c}$. The first RHS term in Eq. (1) represents micro-scale mixing and the second stands for the chemical reactions, where $\chi_{\hat{c}}$ is the conditionally averaged scalar dissipation and $W(\hat{c})$ the chemical reaction rate.

The term with $\langle u_j|\hat{c}\rangle(\mathbf{x}, t)$ needs no closure if the joint pdf of velocity and scalar $P(\hat{\mathbf{u}}, \hat{c}; \mathbf{x}, t)$ is known. However, in the transport equation [8] for $P(\hat{\mathbf{u}}, \hat{c}; \mathbf{x}, t)$ there enter the non-closed terms containing the pressure gradients and turbulent kinetic energy dissipation averaged at simultaneous conditions $\mathbf{u}(\mathbf{x}, t) = \hat{\mathbf{u}}$ and $c(\mathbf{x}, t) = \hat{c}$ that substantially complicates the analysis. Besides, $P(\hat{\mathbf{u}}, \hat{c}; \mathbf{x}, t)$ possesses three more dimensions than $P(\hat{c}; \mathbf{x}, t)$, that renders the numerical simulations much more expensive.

To avoid the difficulties linked to the use of $P(\hat{\mathbf{u}}, \hat{c}; \mathbf{x}, t)$, it seems natural to use directly $\langle u_j|\hat{c}\rangle(\mathbf{x}, t)$ as it has the same number of dimensions as $P(\hat{c}; \mathbf{x}, t)$ itself. The first remarks on this way were made in [1,9], which consist in observing that

$$\langle u_j|\hat{c}\rangle(\mathbf{x}, t) = \langle u_j\rangle(\mathbf{x}, t) - \frac{\langle u'c'\rangle}{\langle c'^2\rangle}(\hat{c} - \langle c\rangle(\mathbf{x}, t)) \tag{2}$$

when $P(\hat{\mathbf{u}}, \hat{c}; \mathbf{x}, t)$ is jointly Gaussian. For the cases where the concentration field is not homogeneous, it was argued in [1] that this simple linear form (2) of $\langle u_j|\hat{c}\rangle(\mathbf{x}, t)$ needs some corrections as it implies very high velocities of transfer for scalar values far from the local mean. It was argued that such scalar values are convected by some large-scale structures, the velocities of which are finite, so $\langle u_j|\hat{c}\rangle(\mathbf{x}, t)$ should remain finite in a limit $\frac{|\hat{c}-\langle c\rangle|}{\sqrt{\langle c'^2\rangle}} \to \infty$.

Therefore, in [1], Eq. (2) was used with some rather complex modifications which are consistent with the normalization properties of $\langle u_j|\hat{c}\rangle(\mathbf{x}, t)$. In fact, some deviations from the linear form (2) were found experimentally by Scherbina [5] in heated round jets for $\frac{|\hat{c}-\langle c\rangle|}{\sqrt{\langle c'^2\rangle}} \geq 2.0$ though significant scatter of experimental points prevents from estimating quantitatively these deviations. We present on Fig. 1 (left) measurements of the conditional longitudinal velocity component for the radial position $r = 65mm$. It was found that near the jet axis Eq. (2) fits well the measurements when the scalar fluctuations are less than about two r.m.s. On the contrary, near the jet boundary, the curve, though having a central linear part, is not in agreement with (2) as its slope is notably greater than $R_{uc} = \frac{\langle u'c'\rangle}{\sqrt{\langle u'^2\rangle}\sqrt{\langle c'^2\rangle}}$. The possible explanation may be linked to the fact, that at this location a maximum is reached for the scalar flatness factor so that the intermittence of the jet boundary should play a role. However, this question needs further studies.

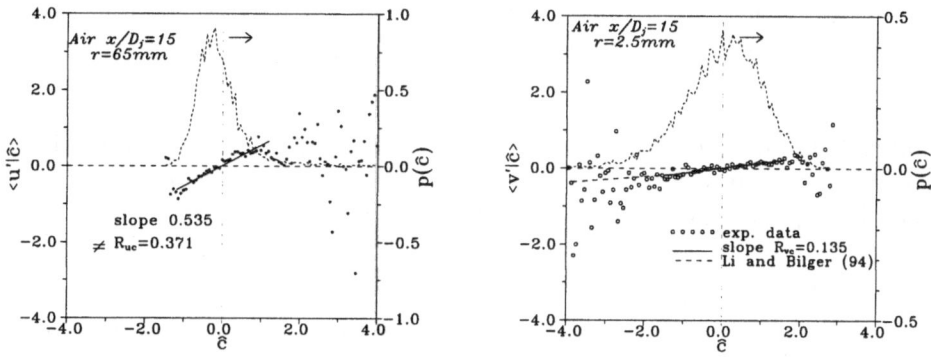

Figure 1. Conditionally averaged longitudinal velocity (left) and conditionally averaged radial velocity (right)

Recently, Li & Bilger [10] have performed $\langle u_j|\hat{c}\rangle(\mathbf{x}, t)$ measurements in a plane mixing layer in which all the velocity components, but the horizontal transverse $\langle V|\hat{c}\rangle$, are constant. They have found that Eq. (2) is "a reasonable approximation" to their measurements when $\frac{|\hat{c}-\langle c\rangle|}{\sqrt{\langle c'^2\rangle}} \lesssim .1.5$ and have proposed the following expression

$$\langle V|\hat{c}\rangle(y) = \langle V|\hat{c} = \langle c\rangle\rangle + \alpha\frac{v'}{\delta} \cdot (y - y_{\hat{c}}) \tag{3}$$

to account for the departures from the linear dependency. Here α is a constant, δ is the mixing layer width and $y_{\hat{c}}$ is the location where the mean scalar $\langle c\rangle$ is equal to $\hat{c}$. This expression yields the overall behavior of $\langle u_j|\hat{c}\rangle(\mathbf{x}, t)$ consistent with their experimental observations.

The comparison of Eq. (3) with the present measurements is presented on Fig. 1 (right). Even though Eq. (3) seems to hold near the axis of the jet, the general conclusion is that, similarly to Eq. (2), it is not able to predict correctly the evolution of the conditional velocity for the whole jet. An alternative to presume a particular form of $\langle u_j|\hat{c}\rangle(\mathbf{x}, t)$ has been developed in [11], using its independent transport equation. This equation for $\langle u_j|\hat{c}\rangle(\mathbf{x}, t)$ was implemented in calculations of a tubular chemical reactor in [12,13]. The dependence of $\langle u_j|\hat{c}\rangle(\mathbf{x}, t)$ on $\hat{c}$ was found to be essentially non-linear, with discontinuities near the limits $\hat{c} = 0$ and $\hat{c} = 1$ which are not yet explained.

4. Conclusions

Measurements of conditionally averaged velocities in a turbulent slightly heated jet were performed and compared with the existing theoretical and empirical predictions. It is suggested that deviations from these predictions may be partly caused by the intermittence of the jet boundaries. Future work will concentrate on developing models refined in this direction.

Acknowledgements
Financial support from EDF, GDF and INERIS is gratefully acknowledged. Considerable help was provided by L. Fulachier for the development of this research program and by S. Mucini for technical assistance.

References

1. Kuznetsov, V.R. and Sabel'nikov, V.A. (1992) *Turbulence and Combustion.* Hemisphere Publ., New York.
2. Borghi, R. (1988) *Progress in Energy and Combust. Sci.*, **Vol. 14**, pp. 245-292.
3. Dopazo, C. (1994) in *"Turbulent Reacting Flows"*, Libby, P.A. and Williams, F.A., Eds., Academic Press, London, pp. 375-474.
4. Bezuglov, V.A. (1974) in *Proc. of XXth Conf. of Moscow Institute of Physics and Technology*, Dolgoprudny, pp. 147-151 (in Russian).
5. Scherbina, Ju.A. (1982) *Izd-vo MFTI*, Dolgoprudny (in Russian).
6. Djeridane, T., Amielh, M., Anselmet, F. and Fulachier, L. (1996) *Phys. Fluids*, **Vol. 8**, pp. 1614-1630.
7. Pietri, L. (1997) Thèse de Doctorat, Université d'Aix-Marseille II.
8. Pope, S.B. (1981) *Phys. of Fluids*, **Vol. 24**, pp. 588-596.
9. Dopazo, C. (1975) *Phys. of Fluids*, **Vol. 18**, pp. 397-404.
10. Li, J.D. and Bilger, R.W. (1994) *Phys. of Fluids*, **Vol. 6**, pp. 605-610.
11. Burluka, A.A., Frost, V.A. and Meytlis, V. (1996) in *"Unsteady Combustion"*, Culick F. *et al.*, Eds., Kluwer Acad. Publ., London, pp. 543-552.
12. Kaminsky, V.A., Rabinovich, A.B. and Fedorov, A.Ya. (1996) in *"Computational Methods in Applied Sciences'96"*, John Wiley & Sons, 1996, pp. 372-378.
13. Fedorov, A.Ya., Frost, V.A. and Kaminsky, V.A. (1996) in *"Transport Phenomena in Combustion"*, Chan S.H., Ed., Taylor & Francis, San-Francisco, pp. 933-944.

SCALAR SUBGRID MODEL WITH FLOW STRUCTURE FOR LARGE-EDDY SIMULATIONS OF SCALAR VARIANCES

P. FLOHR, J.C. VASSILICOS
D.A.M.T.P., University of Cambridge,
Cambridge CB3 9EW, U.K.

Classical and current approaches to LES of scalar fields are based on the assumption of a turbulent diffusivity or a constant turbulent Prandtl number and allow only for the computation of the average concentration field. However, important properties such as mixing rates or possible peak concentrations are related to the concentration variance field, itself determined to a large extent by subgrid turbulent motions, and in this paper we seek a subgrid model that does not rely on the turbulent diffusivity assumption and that allows us therefore to predict the concentration variance field.

The approach that we take is based on the relation between the passive scalar field and the probability density for the positions of fluid elements. Such an approach allows us to directly model the subgrid turbulent velocity field and its advection of the scalar field (i.e. the fluid elements) below the grid. More importantly, our modelled subgrid velocity field can have well-defined spatio-temporal flow structure as well as obey certain fundamental turbulence statistical scalings such as Kolmogorov's $-5/3$ law. The novel subgrid model is based on kinematic simulations (KS) [2] which makes it possible to calculate the scalar variance with inertial range effects and flow structure explicitly resolved by the model flow, while the large-eddy simulation determines the value of the Lagrangian integral time-scale T_L. The kinematic subgrid field consists of a superposition of random Fourier modes that are coupled to the large-eddy field via the energy dissipation rate.

In the inertial time-range we find a well-defined Richardson t^3 scaling for the mean separation of particle pairs. We find that the Richardson constant $G_\Delta \approx 0.07$ which is small compared to the value obtained from stochastic models with the same T_L. This is due to the topology of Lagrangian trajectories in the KS model which is turbulent-like and accounts for small-scale turbulent-like flow structures. The probability density function of the

U. Frisch (ed.), Advances in Turbulence VII, 565–566.

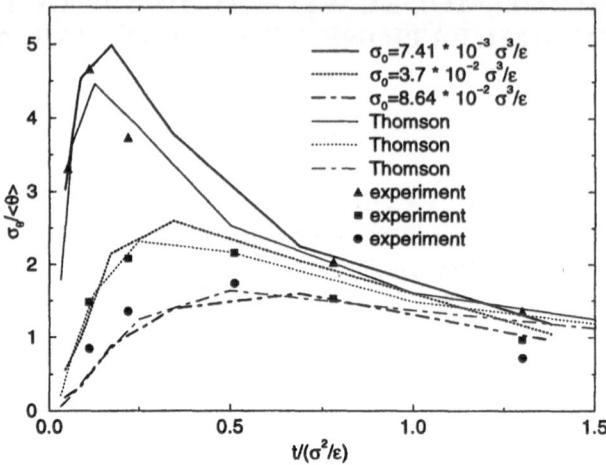

Figure 1. Concentration variance at the centre of the line source. We compare the numerical results from the KS subgrid model (thick lines) with a stochastic model [3] (thin lines) and experimental data [1] (symbols).

separation vector is highly non-gaussian in the inertial time-range. Its normalised shape compares well with findings from stochastic models [3] and for long times becomes gaussian. We also compute the concentration variance for the release from an instantaneous line source and compare our results with Thomson's stochastic model [3] and experiments [1], see figure 1. We stress that the results of Thomson's stochastic model highly depend on the choice of a model parameter which is linked to the lagrangian integral time-scale and which is not known experimentally with any certainty. In a KS model the lagrangian integral time-scale is given by construction and the results presented here are therefore free from tuning uncertain parameters.

References

1. Fackrell, J.E. and Robins, A.G. (1982) Concentration fluctuations and fluxes in plumes from point sources in a turbulent boundary layer, *J. Fluid. Mech.*, **117**, pp. 1–26.
2. Fung, J.C.H. and Vassilicos, J.C. (1998) Two-particle dispersion in turbulent-like flows, *Phys. Rev. E*, **57**(2), pp. 1677–1690.
3. Thomson, D.J. (1990) A stochastic model for the motion of particle pairs in isotropic high Reynolds numbers turbulence, and its application to the problem of concentration variance, *J. Fluid. Mech.*, **210**, pp. 113–153.

SMALL SCALE PASSIVE TEMPERATURE MEASUREMENTS IN FULLY DEVELOPED TURBULENCE

C. AURIAULT, Y. MALECOT AND Y. GAGNE

LEGI/IMG-CNRS, B.P.53X, 38041 Grenoble Cedex 9, France.

AND

B. CASTAING

CRTBT/CNRS/UJF, B.P.166X, 38042 Grenoble Cedex 9, France.

1. Introduction

It is well accepted that the passive scalar field in fully developped turbulence is more intermittent than the velocity one. In particular, the probability density function (pdf) of the scalar increments $\delta\theta(r)$ (hereafter the temperature, Pr$\approx$0.7) are more spread than the velocity pdf. However, as it is shown in figure 1, this temperature pdf becomes nearly gaussian when the temperature increments $\delta\theta(r)$ is conditionned to its local dissipation rate χ_r, whatever the scale r. Such an experimental feature suggests that the local variance transfer rate defined as: $\frac{\delta\theta^2(r)}{r/\delta u(r)}$ is also an appropriate quantity to study the small scale intermittency of the passive scalar (in agreement with [1]).

In this paper we experimentally study the behaviour of the variance transfer rate through the scales of different types of turbulent flows at several Reynolds numbers. In section 2, we show that it can be studied with an infinitely divisible cascade, equivalent to the E.S.S., even though the temperature itself is not fully developed. In section 3, we analyze the behaviour of the variance transfer rate with a cumulant expansion technique and give an estimation of the β_θ exponents which can characterize the passive scalar intermittency.

U. Frisch (ed.), Advances in Turbulence VII, 567–570.

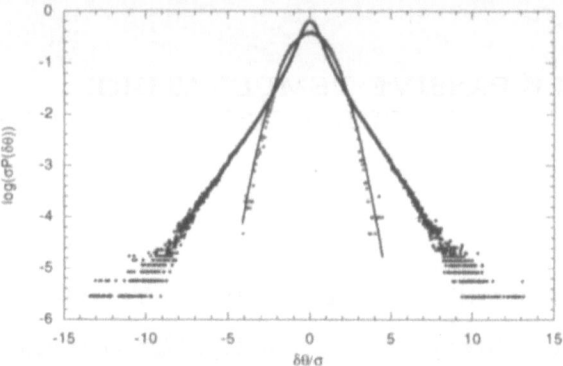

Figure 1. Pdf of $\delta\theta_r$ ($\bullet$) and pdf of $\delta\theta_{r/2<xr>}$ ($\circ$) for $l = 300\eta \approx 3.5\lambda$, obtained in the axis of the wind tunnel of ONERA in Modane ($R_\lambda = 1860$). Continuous line corresponds to the gaussian distribution with the same standard deviation σ.

2. Infinitely divisible cascade process

The variational model on the kinetic energy cascade, introduced by Castaing, can be adapted to the temperature variance cascade [2], which experimentally corresponds to $\tau(\mathbf{r}) = \delta u(\mathbf{r})\delta\theta^2(\mathbf{r})$. Such a model is based on two hypothesis. The first one consists to express any local pseudo transfer $\tau(\mathbf{r})$ with an independant multiplier random variable γ_r such as:

$$\tau(\mathbf{r}) = \gamma_r.\tau(l_0) \text{ where } l_0 \text{ is the integral scale}$$

which leads, in term of pdf, to:

$$P_r(\tau) = \int F_r(\ln\gamma)\frac{1}{\gamma}P_{l_0}\left(\frac{|\tau|}{\gamma}\right)d(\ln\gamma)$$

Where P_{l_0} (resp. F_r) is the pdf of $\tau(l_0)$ (resp. $\ln(\gamma)$).
Note that this analysis only considers the symetrical part of the $\tau(\mathbf{r})$ increments. The second hypothesis, which is the strongest one, assumes that the F distributions are infinitely divisible. At finite Reynolds number, this assumption is equivalent to the Extended Self Similarity. On figure 2, we observe that the E.S.S. is less verified by the passive temperature $\delta\theta(\mathbf{r})$ itself than by its pseudo transfer rate $\tau(\mathbf{r})$. This feature has been already noticed by Ruiz et al [3]. Moreover, at such Reynolds number, (namely $R_\lambda = 760$) the temperature spectra roughly exhibits a power law scaling range with an exponent value less than -5/3. As Sreenivasan has already remarked [4], the inertial slopes of temperature spectra reach the -5/3 value only for $R_\lambda \geq 1000$.

Therefore, figure 2 suggests that the E.S.S. works rather well for $\tau(r)$ even though the temperature field itself is not yet fully developed.

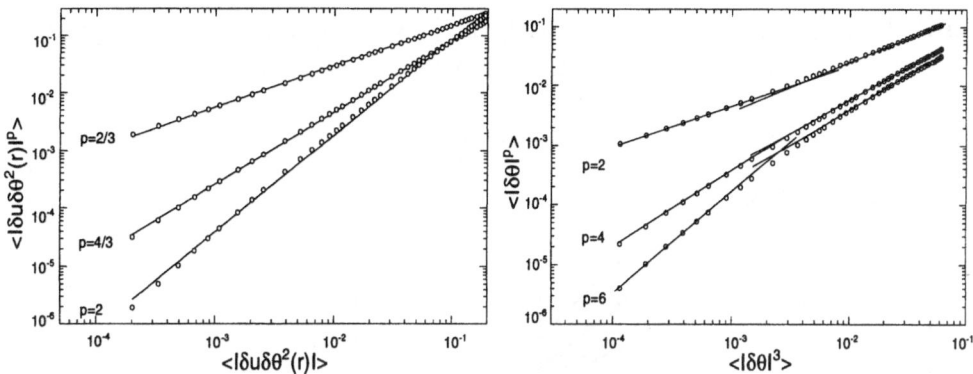

Figure 2. Relative dependance of $<| \delta u(r)\delta\theta^2(r)) |^P>$ on $<| \delta u(r)\delta\theta^2(r) |>$ on the left and of $<| \delta\theta(r) |^P>$ on $<| \delta\theta(r) |^3>$ on the right for the turbulent air jet ($R_\lambda = 760$). Lines correspond to power fit.
A clear scaling law appears in the case of the variance transfer rate whereas we observe two different slopes in the case of the temperature structure function.

3. Passive scalar intermittency

From an experimental point of view, we can measure the structure functions of $\tau(r) = \delta u(r)\delta\theta^2(r)$, which are linked to the cumulant generating function of the $F(r)$ distribution. By definition:

$$\ln < (\gamma_r)^P > \; = \; \ln\left(\frac{<|\tau(r)|^P>}{<|\tau(l_0)|^P>}\right)$$
$$= \; C_{1\tau(r)}\cdot p + C_{2\tau(r)}\cdot\frac{p^2}{2} + ... + C_{i\tau(r)}\cdot\frac{p^i}{i!} + ...$$

The variational model which takes into account both the integral scale and the viscous scale, induces a dependance on the ratio l_0/η and predicts:

$$C_i(r) \; = \; k_i\left(\left(\frac{r}{l_0}\right)^{-\beta_\theta} - 1\right) \quad (1)$$

$$\frac{1}{\beta_\theta} \; = \; \frac{1}{\beta_{\theta 0}}ln\left(\frac{R_\lambda}{R_\star}\right) \quad (2)$$

Measurements have been performed in several types of flows at different Reynolds numbers (see [5] for experimental details). The main results are:

(i) Beyond the third cumulant $C_{3\tau(r)}$, which is very small, all the other cumulants are neglictible (whatever the scale r).

(ii) The three non zero cumulants are nearly proportionnal according to the infinite divisibility hypothesis.

(iii) The cumulants follow a scaling power law, in agreement with the eq. (1), with the same β_θ exponent at a given Reynolds number (cf figure 3).

(iiii) This β_θ exponent behaves with the Reynolds number like the prediction of eq. (2), at least $300 \leq R_\lambda \leq 3000$ (cf figure 3).

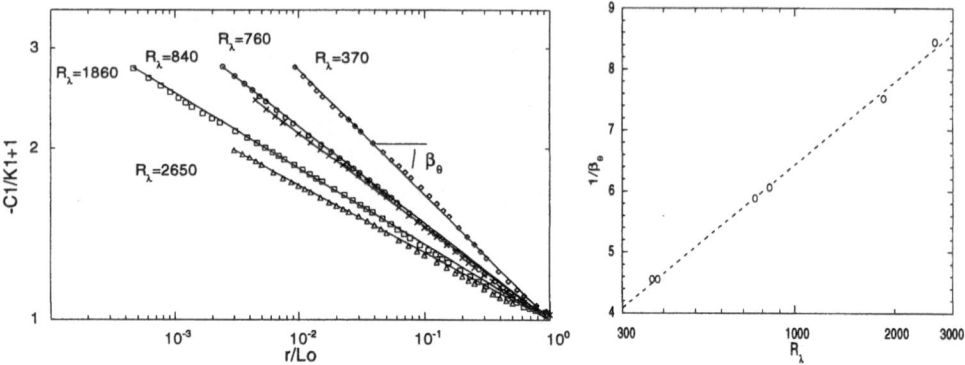

Figure 3.   Experimental scaling laws of the first cumulant C_1 of $\delta u(r)\delta\theta^2(r))$ obtained in the Modane wind tunnel. ($R_\lambda = 1860$ and $R_\lambda = 2650$) and in an air jet ($R_\lambda = 370$, $R_\lambda = 760$ and $R_\lambda = 840$) on the left. On the right, $\frac{1}{\beta_\theta}$ versus R_λ. Lines correspond to the variationnal prediction.

With our Reynolds numbers, the best fit leads to $\beta_{\theta 0} \approx 0.5$ and $R_\star \approx 40$. Of course it is necessary to continue this study with lower R_λ values and other types of flows.

Aknowledgements

This work is supported by the DSP contract 97/1045.

References

1. Leveque, E., Ciliberto, C., Baudet, C., & Ruiz-Chavarria, G. Cascade structures and scaling laws for the mixing of a passive scalar by fully developped turbulence, preprint.
2. Castaing, B. (1994) Scalar intermittency in the variational theory of turbulence *in Physica* D **73**, pp. 31–37
3. Ruiz-Chavarria, G., Baudet, C. & Ciliberto, S. (1996) Scaling laws and dissipation scale of passive scalar in fully developed turbulence *in Physica* D **99**, pp. 369–380..
4. A. K. Sreenivasan (1996) The passive scalar spectrum and the Obukhov-Corrsin constant *in Phys. Fluids* **8**, pp. 189–196.
5. Kahalerras, H., Malécot, Y., Gagne, Y., Castaing, B., (1998) Intermittency and reynolds number *in Phys. Fluids*.

SCALAR FLUCTUATIONS PDF'S IN TRANSIENT TURBULENT MIXING

E. VILLERMAUX, C. INNOCENTI AND J. DUPLAT

LEGI–CNRS, B.P. 53 X 38041 Grenoble Cedex, France

The nature of the interplay between a scalar field and the underlying turbulent velocity field is advantageously studied through a transitory situation, which evidences the pertinent timescales of the mixing process.

We follow the transient mixing of a scalar blob introduced at the origin of time in the far field of a high Reynolds number ($Re = u'L/\nu \approx 10^4$) turbulent jet, in the generic situation where the injection scale d is smaller than the local stirring (integral) scale L. The injection scale ranges from $d/L = 0.05$ to 0.16 and the Schmidt number $Sc = \nu/D$ of the scalar is varied over a broad range going from $Sc = 0.7$ (temperature in air), $Sc = 7$ (temperature in water) to $Sc = 2000$ (disodium fluorescein in water).

The scalar concentration fluctuations PDF, $P\left(\frac{C}{C_0}\right)$, exhibits rapidly an exponential shape whose argument is linearly increasing with time t as (figure 1)

$$P\left(\frac{C}{C_0}\right) \sim (\gamma t_s)^{-1/2} \exp\left(-\frac{t}{t_s}\frac{C}{C_0}\right),$$

for $t > 6d/u$. We show how the exponential PDF shape results from the distribution $\mathcal{P}(n) \sim \sigma^{-1} \exp\left(-\frac{(n-\bar{n})^2}{\sigma^2}\right)$ of the number of cumulated stretchings n experienced by the scalar sheets at a given instant of time when the average "sollicitation" number $\bar{n}$ approches the number $n_s = \frac{1}{2}\ln(BSc)$ needed to reach the dissipation scale.

The mixing time t_s is found to be

$$t_s = A\frac{d}{u}\ln(BSc)$$

with $A = 0.12$ and $B = 5$, independently of the Reynolds number; it represents the time needed to bring the sheets of scalar peeled-off from

U. Frisch (ed.), Advances in Turbulence VII, 571–572.
© 1998 *Kluwer Academic Publishers.*

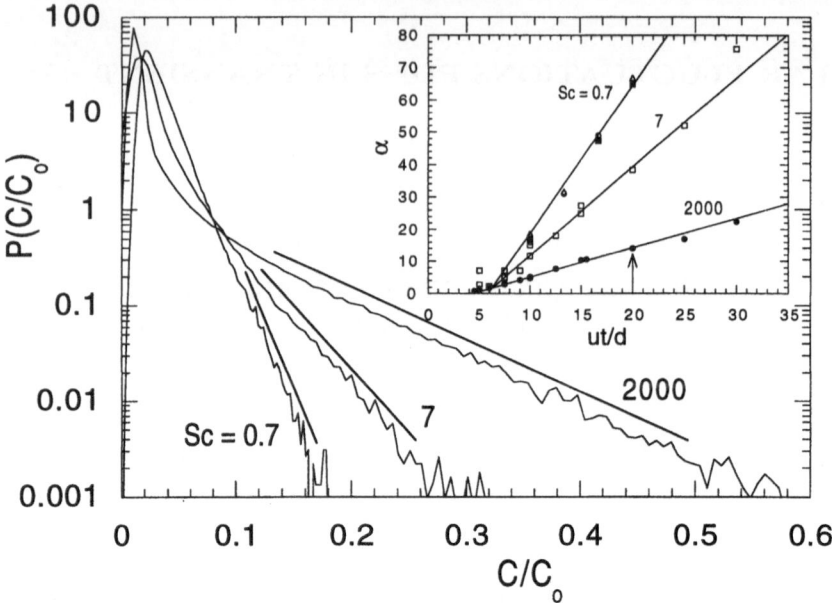

Figure 1. Fluctuations PDF's, normalized by the initial concentration (temperature), recorded for the three Schmidt numbers 20 diameters d downstream from the injection point. The PDFs exhibit an exponential decay (eq. (1)) of the form $P\left(\frac{C}{C_0}\right) \sim \exp\left(-\alpha\frac{C}{C_0}\right)$. Insert : The argument of the exponentials for different distances $\frac{x}{d} = \frac{ut}{d}$ and three Schmidt numbers is such that $\alpha = \frac{ut/d}{0.12\ln(5Sc)}$. • : $Sc = 2000, Re = 6000$ and $12000, d/L = 0.05, 0.1, 0.6$.

□ : $Sc = 7, Re = 6000, d/L = 0.05, 0.1, 0.6$. ■ : $Sc = 0.7, \quad d/L = 0.08$, $Re = 23000$; ○,△ : $Sc = 0.7, \quad d/L = 0.08, Re = 45000$.

the blob just after its formation, whose transverse size scales as the Taylor microscale and is about $\sqrt{5\frac{\nu}{\gamma}}$, down to the dissipation scale $\sqrt{D/\gamma}$ under the action of a constant rate of stretch related to the injection scale of the whole blob $\gamma \approx \frac{4u}{d}$.

The transient turbulent mixing of scalars thus "bypasses" the hierarchy of scales of the Kolmogorov cascade, the reduction process of the transverse scalar sheet thickness, initially pertaining to the dissipative range of scales, involving a strong, but constant stretching rate.

We underline the consequence of the dominance of the injection timescale d/u on the rate of the scalar fluctuations decay for the spectrum exponent, close to -1 for all scales smaller than d.

STATISTICAL PROPERTIES OF THE ENERGY FLOW OF A PASSIVE SCALAR IN FULLY DEVELOPED TURBULENCE

E. LEVEQUE, S. CILIBERTO AND C. BAUDET
CNRS, Laboratoire de Physique, E.N.S. de Lyon,
46 Allée d'Italie, 69364 LYON, France.

AND

G. RUIZ-CHAVARRIA
Departamento de Física, Facultad de Ciencias, U.N.A.M.,
04510 México D.F., Mexico.

A commonly-used approach in the statistical study of a passive scalar $\theta(\vec{r}, t)$ transported by a turbulent flow $\vec{u}(\vec{r}, t)$, is the analysis of the structure functions $\langle |\delta\theta_\ell|^p \rangle = \langle |\theta(\vec{r} + \vec{\ell}, t) - \theta(\vec{r}, t)|^p \rangle$, as a function of the scale separation ℓ. In the fully developed turbulent regime, dissipative dynamics are concentrated at such very small scales (compared to the integral scale of the flow) that there exists an intermediate range of scales, called inertial range, where only non-linear interactions prevail. Structure functions are then expected to follow power-law dependence on ℓ in that range: $\langle |\delta\theta_\ell|^p \rangle \sim \ell^{\xi_p}$. Experimental observations have ensured the existence of such scaling laws [1] and motivate today many efforts to predict the scaling exponents ξ_p.

An important difficulty in this study is the inexistence of an exact scaling relation for the scalar, analogous to *Kolmogorov's 4/5 law* for the velocity [2]. On the other hand, a scaling equation can be established for the scalar energy flux [3]:

$$\langle \delta u_\ell \delta\theta_\ell^2 \rangle = -4/3 \, N\ell \tag{1}$$

where $\delta u_\ell = [\vec{u}(\vec{r} + \vec{\ell}, t) - \vec{u}(\vec{r}, t)].\vec{\ell}/\ell$ is the longitudinal velocity increment, $N = \chi \langle |\nabla\theta|^2 \rangle$ is the (constant) mean scalar dissipation rate and χ is the diffusivity of the fluid. An other major difference between the velocity and scalar concerns the property of Extended Self Similarity [4]. Scaling laws, as mentioned previously, are strictly defined in the fully developed turbulent regime. In practice, this asymptotic regime is approached but never achieved in the sense that, one can never exactly exhibit a finite inertial-range of scales with pure power-law dependence. Deviations are observed

U. Frisch (ed.), Advances in Turbulence VII, 573–576.
© 1998 *Kluwer Academic Publishers.*

and are related to small but non-negligible dissipative effects. When considering structure functions against others, these deviations mainly cancel out such that one recovers power-law scalings over an extended range. In the case of $\langle |\delta\theta_\ell|^p \rangle$, it has been recently demonstrated experimentally that the range of self-similarity is not as extended as for the velocity structure functions [1]. These experimental and theoretical facts motivate us to pursue a different approach by focusing on the scaling properties of the structure functions of the scalar energy flux: $\langle |\delta u_\ell \delta\theta_\ell^2|^{p/3} \rangle$. Experimental results are presented and provide a new insight to the scalar statistical properties. The passive scalar is here temperature fluctuation.

The experimental data have been measured in the wake of a cylinder of diameter $D = 10$cm standing in a wind tunnel of size 50cm × 50cm × 300cm [1]. The fluid is air ($Pr = \nu/\chi \simeq 0.7$) and the Reynolds number of the flow is $Re = U_{rms}D/\nu \simeq 36000$. Various Reynolds number flows have been studied and have led to consistent similar results. The temperature fluctuations are triggered by the electrical heating of an array of parallel fine wires. This array, known as mandoline, permits an effective mixing of heat without mean temperature gradient. The mandoline stands vertically 40cm downstream the cylinder. In order to measure simultaneously the velocity and temperature fluctuations, two probes have been placed apart at a distance of 1mm. The temperature probe works as a cold wire, whereas the velocity probe works as a hot wire. A special attention has been paid to avoid interference between the two probes. These measurements have been made 3m downstream the cylinder. Various distances have been studied and have led to similar results. Local time measurements have been transformed into spatial measurements by use of Taylor's *frozen flow* hypothesis, justified for a turbulence rate $\mathcal{T} \simeq 10\%$.

As expected, the structure functions $\langle |\delta u_\ell \delta\theta_\ell^2|^{p/3} \rangle$, computed here as time average, do not exhibit a clear inertial-range with power-law dependence on ℓ. We rather focus on the relative scalings by studying $\langle |\delta u_\ell \delta\theta_\ell^2|^{p/3} \rangle$ as a function of $\langle |\delta u_\ell \delta\theta_\ell^2| \rangle$. In a log-log coordinate system, a clear scaling range is observed in *Figure 1*. In the dissipative range, the trivial scaling exponent $p/3$ is recovered. The property of E.S.S. is well satisfied for the structure functions of the energy scalar flux. This observation suggests that E.S.S. rather applies to fluxes: $\delta u_\ell^3/\ell$ and $\delta u_\ell \delta\theta_\ell^2/\ell$. Thus, it appears to be closely related to the process of energy transfer. It enables to access to an accurate estimation of the *asymptotic* scaling exponents Ψ_p, by taking into account that $\Psi_3 = 1$ from (1). We have actually checked that $\langle \delta u_\ell \delta\theta_\ell^2 \rangle \sim \langle |\delta u_\ell \delta\theta_\ell^2| \rangle$. We have evaluated the relative scaling exponents Ψ_p/Ψ_3 by least squares fitting, with an uncertainty of the order of one percent. The Ψ_p's are reported in TABLE 1.

In the third line of TABLE 1, the relative scaling exponents for the

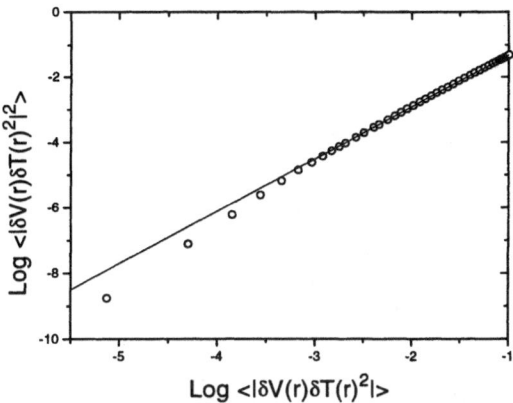

Figure 1.   Relative scalings for the structure functions of the scalar energy flux.

TABLE 1.   Scaling exponents of scalar energy flux and velocity.

order p	1	2	3	4	5	6	7	8
Ψ_p	0.40	0.73	1.00	1.23	1.43	1.60	1.80	1.95
ζ_p	0.36	0.70	1.00	1.27	1.53	1.77	2.01	2.23

velocity structure functions are very closed, as expected, to those usually measured in homogeneous and isotropic fully developed turbulence. In this experiment $\Psi_p \neq \zeta_p$; the kinetical and scalar energy transfers do not follow similar scalings.

We are now encline to focus on the coarse-grained thermal energy dissipation rate, approximated by the 1D surrogate $N_\ell = 1/\ell \int_r^{r+\ell} N(r')dr'$ where $N(r) = 3\chi(\partial u/\partial r)^2$, and test whether N_ℓ satisfies the closure condition proposed in the model of hierarchy [5]:

$$\frac{N_\ell^{(p+1)}}{N_\ell^{(\infty)}} \sim \left(\frac{N_\ell^{(p)}}{N_\ell^{(\infty)}} \right)^\beta , \tag{2}$$

where β is a constant independent of ℓ and $N_\ell^{(p)} = \langle N_\ell^{p+1} \rangle / \langle N_\ell^p \rangle$.

In order to test (2), we display $Y_{n+1} = \log_{10}(N_{\ell_2}^{(n+1)}/N_{\ell_1}^{(n+1)})$ as a function of Y_n, for different scale ratios ℓ_2/ℓ_1 and various n [6]. In *Figure 2*, the data points are aligned for fixed scale ratios ℓ_2/ℓ_1, which indicates that the

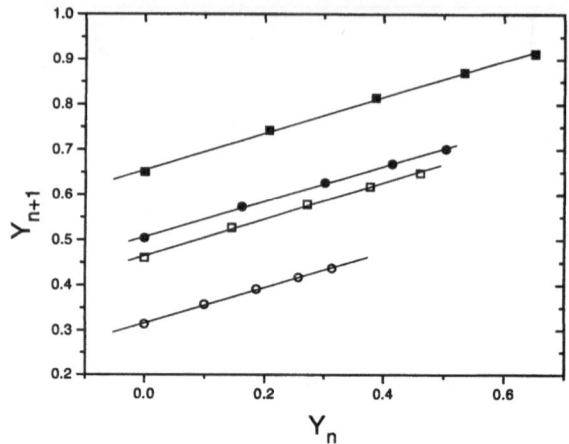

Figure 2. Y_{n+1}, as defined in text, versus Y_n for various arbitrary scale ratios ℓ_2/ℓ_1. The values of n are 0, 0.25, 0.50, 0.75 and 1.00.

closure relation (2) is well satisfied, with β given by the slope. The different lines, corresponding to different ratios ℓ_2/ℓ_1, are parallel which means that β is constant and independent of scale ratios. A least squares fitting leads to the value $\beta = 0.4 \pm 0.02$.

These results provide a new insight on the scalar statistics by focusing on the cascade-related quantities $\delta u_\ell \delta \theta_\ell^2$ and N_ℓ. Robust properties, shared with the velocity turbulent field, are then observed: the exact scaling relation (1), the property of Extended Self-Similarity and the statistical closure relation (2). Finally, beyond these results on the scalar, this study augurs the existence of a unified framework to describe the scaling properties of a turbulent system.

References

1. Ruiz-Chavarria, G., Baudet, C. and Ciliberto, S. (1996) Scaling laws and dissipation of a passive scalar in fully developed turbulence, *Physica D* **99**, 369.
2. Kolmogorov, A. N. (1941) Dissipation of energy in locally isotropic turbulence, *Dokl. Akad. Nauk SSSR* **32**, 16-18.
3. Monin, A. S. and Yaglom, A. M. (1975) *Statistical Fluid Mechanics*, M.I.T. Press, Cambridge.
4. Benzi, R., Ciliberto, S., Tripiccione, R., Baudet, C., Massaioli, F. and Succi, S. (1993) Extended self-similarity in turbulent flows, *Phys. Rev. E* **48**, R29-R32.
5. She, Z.-S. and Lévêque, E. (1994) Universal scaling laws in fully developed turbulence, *Phys. Rev. Lett.* **72**, 336-339.
6. Ruiz-Chavarria, G., Baudet, C. and Ciliberto, S. (1995) Hierarchy of the energy dissipation moments in fully developed turbulence, *Phys. Rev. Lett.* **74**, 1986-1989.

ON THE THREE-POINT CORRELATION FUNCTION OF A
PASSIVE SCALAR MIXED BY A TURBULENT FLOW

A. PUMIR
*Institut Non Linéaire de Nice, CNRS, 1361, route des Lucioles,
F-06560, Valbonne, France*

Mixing of a passive scalar, Θ, by a turbulent flow is a problem of obvious practical importance, and provides an interesting way to test our understanding of turbulence itself [1]. Theoretical considerations, inspired by the Kolmogorov K41 theory [2], suggest that the second order structure function behaves as $r^{2/3}$, as indeed found experimentally [3]. Higher order structure functions show significant departures from the predictions of the K41 theory. A particularly strong effect has been observed in a weakly heated turbulent boundary layers. The temperature signal shows a ramp like structure [4] resulting in a strong assymetry in the scalar derivative, and in a skewness of order 1 of $(\partial_x \Theta)$, x being the coordinate along the downstream direction. The third order structure function, $< (\Theta(x) - \Theta(0))^3 >$ behaves as x^{λ_3}, $\lambda_3 \approx 1$, whereas considerations based on the K41 phenomenology would rather suggest a value of order 5/3. A similar anisotropy of the scalar derivatives has been observed experimentally [5] and numerically [6, 7] when a large scale gradient is imposed. The numerical results suggest that the effect does not crucially depend on the statistical properties of the velocity field. This suggests that much can be understood with the help of simplified models.

Mixing by a random, Gaussian, white in time velocity field, scaling in space with a power law : $< v_a(\vec{r}, t) v_b(\vec{r}', t') >= C_{ab}^\epsilon(\vec{r} - \vec{r}' \delta(t - t')$, with :

$$D_{ab}^\epsilon(\vec{r}) \equiv (C_{ab}^\epsilon(0) - C_{ab}^\epsilon(\vec{r})) = D_0((d + 1 - \epsilon)\delta_{ab} - (2 - \epsilon)\frac{r^a r^b}{|\vec{r}|^2})|\vec{r}|^{2-\epsilon} \quad (1)$$

(d is the space dimension) has been introduced by Kraichnan [8, 9]. In this problem, the steady state N-point correlation function obeys a closed (Hopf) equation of the form [10] :

$$L(d, \epsilon) < \theta(\vec{r}_1)...\theta(\vec{r}_N) >= RHS \quad (2)$$

U. Frisch (ed.), Advances in Turbulence VII, 577–580.

where

$$L(d, \epsilon) \equiv \sum_{i \neq j}^{N} D_{ab}^{\epsilon}(\vec{r}_i - \vec{r}_j)\partial_{r_i}^a \partial_{r_j}^b \tag{3}$$

The right hand side term in Eq.(2) depends on the scalar forcing. In order to recover the behavior of the 2-point correlation function when two points are getting very close, one must choose $\epsilon = 2/3$.

In order to take into account the finite correlation time of the velocity field, Shraiman and Siggia [11] have proposed to replace in the Hopf equation Eq.(2) the operator $L(d, \epsilon)$ by

$$\mathcal{L}(d, \alpha) \equiv L(d, \epsilon = 0) + \alpha \mathcal{L}_D \tag{4}$$

This model may be justified by decomposing the action of the velocity field on a set of coordinates $(\vec{r}_1, ..., \vec{r}_N)$ as a part due to the effect of the velocity field (i) at scales of order $\sim R(\sim |\vec{r}_i - \vec{r}_j|)$, modelled by the operator $L(d, \epsilon = 0)$, (ii) at scales much larger than R (the forcing term) and (iii) at scales much smaller than R, leading to the Richardson diffusion $\mathcal{L}_D$. We consider here : $\mathcal{L}_D = \frac{3d}{2} R_g^2 \left(\frac{r_{12}^2 r_{23}^2 r_{31}^2}{R_g^6} \right)^{2/3} \nabla^2$ which respects the Richardson scaling. The parameter α is a free parameter, *a-priori* small.

These models are of the form $L \cdot \psi = RHS$. The RHS of these equations scales with an exponent, whereas the operator L is homogeneous in space. As a consequence (dimensional) power counting predicts that the solution ψ has a "naive" scaling law. An important remark is that anomalous scaling may arise from the existence of zero modes of the operator L, with a smaller scaling exponent than the dimensional one. This leads us to focus on the properties of the zero modes of the operators, $L(d, \epsilon)$ and $\mathcal{L}(d, \alpha)$ [11, 12, 13].

In the limit where ϵ or α is small, both operators reduce to $L(d, \epsilon = 0)$, which can be exactly diagonalized by group theoretical methods for the 3-point correlation function, for $d = 2$ and $d = 3$. The problem is invariant under any rotation of space ($SO(d)$ group), and under any rearrangement of the three points : $(\vec{r}_2 - \vec{r}_1, \vec{r}_3 - \vec{r}_1) = A(\vec{r}_2 - \vec{r}_1, \vec{r}_3 - \vec{r}_1)$, where A is a 2×2 "isospin" matrix with determinant unity ($SL(2)$ group). In our problem, the 3-point correlation function is *odd* in space. We choose to work in the representation $l = 1$ of the $SO(d)$ group. The smallest possible scaling exponent of the correlation function turns out to be 1, with an infinite degeneracy.

A perturbation theory has been developped to construct the zero modes of the operator at small values of α (or ϵ). One shows the exponent in the Kraichnan model behaves as $\lambda = 1 + \delta(\epsilon)\epsilon$ with $\delta(\epsilon) \to 0$ when $\epsilon \to 0$, [14] (this conclusions applies also to the model introduced in [11] [15]). In addition, the perturbation theory provides a way to compute the eigenfunction, provided the 3 points are not close to colinear [16]. Introducing the parametrization of the triangle $\rho_1 = (\vec{r}_1 - \vec{r}_2)/\sqrt{2}$ and $\vec{\rho}_2 = (\vec{r}_1 + \vec{r}_2 - 2\vec{r}_3)/\sqrt{6}$, and the Euler type coordinates $\rho_i^a \equiv \sum_{i'} \mathbf{R}_{ii'}(\chi)\xi_{i'}\hat{\eta}_{i'}^a$, where $(\hat{\eta}_1, \hat{\eta}_2)$ are two orthogonal, unit vectors that span the plane of the triangle $(\vec{r}_1, \vec{r}_2, \vec{r}_3)$ and χ is a (pseudospace) rotation angle, one obtains :

$$< \theta(\vec{r}_1)\theta(\vec{r}_2)\theta(\vec{r}_3) > \equiv \hat{G} \cdot \int_{-\pi}^{+\pi} \frac{\cos\phi\xi_1\hat{\eta}_1 + \sin\phi\xi_2\hat{\eta}_2}{[\cos(\phi)^2\xi_1^2 + \sin(\phi)^2\xi_2^2]^{(1-\lambda)/2}} f(\phi - \chi)d\phi$$

(5)

where the function $f(\chi)$ is determined by the perturbation expansion ($\hat{G}$ is the imposed scalar gradient).

In order to determine the zero modes of $L(d, \epsilon)$ or $\mathcal{L}(d, \alpha)$ numerically [17], we write the correlation function $< \theta(\vec{r}_1)\theta(\vec{r}_2)\theta(\vec{r}_3) > \equiv R^\lambda \hat{G} \cdot \vec{F}(\vec{\rho}_i/R)$, with $R^2 \equiv \sum_i \vec{\rho}_i^2$. The function $\vec{F}$ is solution of an elliptic boundary value problem, which can be reduced to a banded matrix using finite differences. The scaling exponent λ plays the role of a "nonlinear" eigenvalue.

The values of λ determined numerically generally agree with the predictions of the perturbation analysis : the exponenet is found to be very close to 1 at small values of α. In this sense, the model of [11] is consistent with the experimental values of the exponent. Note that the values found for the Kraichnan model with $\epsilon = 2/3$ is $\lambda \approx 1.38$, much higher than the experimental value.

The 3-point correlation function may be determined in a wind tunnel experiment, using two probes separated by a distance y in the direction of the gradient [18]. Taylor's hypothesis then permits to measure the scalar fiedl at the points $A = (x_1 + x_2/2, y)$, $B = (x_2, 0)$ and $C = (0, 0)$. Assuming scaling, the correlation function depends on x_1/y and x_2/y, and is even in these two variables.

An example of the correlation function is shown in Fig.1a, with $\alpha = 0.05$ ($\lambda \approx 1.02$) in $d = 3$. The integral representation corresponding to the same parameters is shown in Fig.1b. The agreement is very good except for configurations where the 3 points are near colinearity, and improves as the value of α diminishes.

Direct comparison with experiments are under way, and should permit us to test thoroughly the predictions of our simplified models, and we hope, to shed new lights on the intricate issues of turbulence.

It is a pleasure to acknowledge many discussions on these issues with L. Mydlarski, B. Shraiman, E. Siggia and Z. Warhaft.

References

1.   Frisch, U. (1995) *Turbulence : The legacy of AN Kolmogorov* , Cambridge University Press, Cambridge.

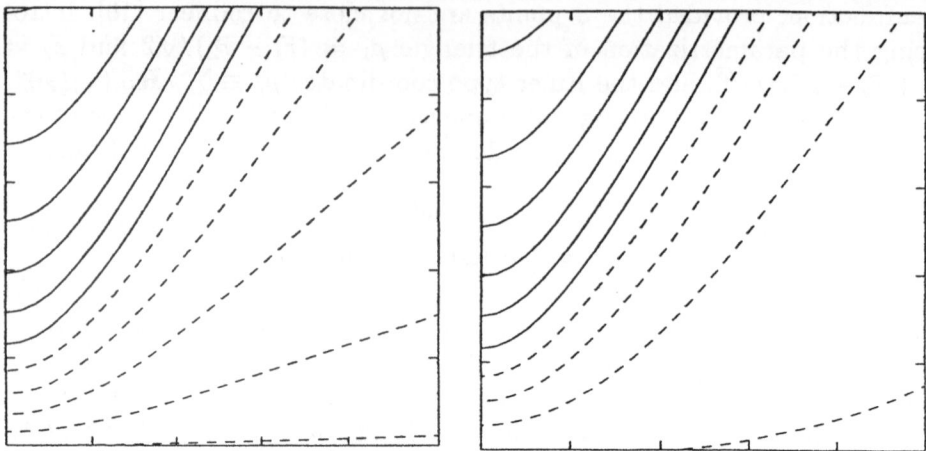

Figure 1. Contour plot of the computed 3-point correlation function $< \theta(A)\theta(B)\theta(C) >$ for 3 coplanar points $A = (x1 + x_2/2, y)$, $B = (x_2, 0)$ and $C = (0,0)$, as a function of x_1/y (horizontal) and x_2/y (vertical) in the range $0 \leq x_1/y$, $x_2/y \leq 5$ (integer values of are indicated by ticks along the perimeter). Fig.1a shows solution of the model with a value $\alpha = 0.25$. Fig.1b is the fit of the 3-point correlation function by the integral representation Eq.(5) with $\lambda = 1.02$. (integer values of are indicated by ticks along the perimeter).

2. Kolmogorov, A.N. (1941) Dokl. Akad. Nauk SSSR **30**, 9-13.
3. Mydlarski, L. & Warhaft, Z. (1998) *J. Fluid Mech.*, to appear.
4. Mestayer, P. (1983) *J. Fluid Mech.*, **125**, 475-503.
5. Tong, C. & Warhaft, Z. (1994) *Phys. Fluids*, **6**, 2165-2176.
6. Holzer, M. & Siggia, E.D. (1994) *Phys. Fluids*, **6**, 1820-1837.
7. Pumir, A. (1994) *Phys. Fluids*, **6**, 2118-2132.
8. Kraichnan, R.H. (1968) *Phys. Fluids*, **11**, 945-953.
9. Kraichnan, R.H. (1974) *J. Fluid Mech.* **64**, 737-762.
10. Shraiman, B.I. & Siggia, E.D. (1994) *Phys. Rev. E*, **49**, 2912-2927.
11. Shraiman, B.I. & Siggia, E.D. (1995) *C. R. Acad. Sci.* **321**, 279-285.
12. Gawędzki, K. & Kupiainen, A. (1995) *Phys. Rev. Lett.*, **75**, 3834-3837.
13. Chertkov, M., Falkovich, G., Kolokolov, I. & Lebedev, V. (1995), *Phys. Rev. E* **52**, 4924-4941.
14. Pumir, A., Shraiman, B.I. & Siggia, E.D. (1997) *Phys. Rev. E* , **55**, R1263-1266.
15. Shraiman, B.I. & Siggia, E.D. (1998) *Phys. Rev. E* **57** (to appear).
16. Shraiman, B.I. & Siggia, E.D. (1996) *Phys. Rev. Lett.* **77**, 2463-2464.
17. Pumir, A. (1997) *Europh. Lett* **37**, 529-534; see also Pumir, A. (1998) *Phys. Rev. E* **77**, to appear.
18. Mydlarski, L. & Warhaft, Z. (1998) in preparation, and Mydlarski, L., Warhaft, Z., Shraiman, B., Siggia, E. & Pumir, A. (1998) in preparation.

KINEMATIC SIMULATIONS

J. C. VASSILICOS
Department of Applied Mathematics and Theoretical Physics,
University of Cambridge, CB3 9EW, Cambridge, UK

N. A. MALIK
Department of Mathematics, The Hong Kong University of
Science and Technology, Clear Water Bay, Hong Kong

Department of Applied Mathematics and Theoretical Physics,
University of Cambridge, CB3 9EW, Cambridge, UK

J. DÁVILA
Group of Fluid Mechanics, University of Seville, Av. Reina
Mercedes s/n, 41012, Seville, Spain

AND

Y. SAKAI
School of Engineering, Nagoya University, Furo-cho, Chikusa-
ku, Nagoya 464-01, Japan

There are two aspects to Kinematic Simulations.

(i) Kinematic Simulations (KS) are Lagrangian models of turbulent dispersion that incorporate small-scale turbulent-like flow structure. Unlike stochastic models of dispersion, KS is not based on Langevin-type equations and Wiener processes. In KS fluid element trajectories (fluid element and particle are synonymous here) are obtained by integrating

$$dx(t) = u(x(t), t)dt$$

in many different realisations of a turbulent-like Eulerian velocity field $u(x(t), t)$.
The prescribed Eulerian properties of this field are

- incompressibility,

U. Frisch (ed.), Advances in Turbulence VII, 581–584.
© 1998 *Kluwer Academic Publishers.*

- its energy spectrum,

- the existence of eddying, straining and streaming flow structures,

- and the time-dependence of these structures.

The Lagrangian integral scale is an output of the model, finite Reynolds number effects are easily incorporated with a fast high-wavenumber drop-off in the energy spectrum, and the same KS model can be used unaltered for both one-particle and two-particle dispersion.

A comparison of Lagrangian statistics of two-particle dispersion in isotropic turbulence obtained with KS on the one hand and DNS on the other at $Re_\lambda \approx 91$ shows good agreement for all these statistics. In particular, the flatness factor $\mu_4(t)$ of two-particle relative Lagrangian velocities which is grossly underestimated by two-particle stochastic models ([2]) is well predicted within a factor of 1.3 to 1.5 by KS.
It should also be noted that the number of modes used in KS is $O(10^2)$ whereas the equivalent DNS requires $O(10^6)$. KS can therefore be used to calculate $\mu_4(t)$ for a variety of Reynolds numbers larger than those accessible with DNS and it is found that $\mu_4(t)$ is very significantly larger than the gaussian value 3 and is a decreasing function of time in the inertial range of times. Stochastic models adapted for finite Reynolds numbers predict that $\mu_4(t) \approx 3$ for times larger than the Kolmogorov time-scale.

KS particle-pairs move along-side each other for quite some time until they meet a straining region and suddenly separate, whereas the separation of particle-pairs in stochastic models does not happen in bursts but continuously because of the Wiener processe's separating action. Hence, the flatness $\mu_4(t)$ is much smaller in stochastic models than in KS. For the same reason, Richardson's constant is much larger in stochastic models (with reasonable input parameter specification) than in KS.

(ii) KS also provides a conveniently maleable tool with which to numerically engineer small-scale turbulence flow structure at will and thereby explore how various properties of this flow structure relate to or determine turbulence statistics.
    Standard KS flow fields consist of a superposition of random incompressible Fourier modes with a prescribed energy spectrum. Realisations of such fields readily exhibit eddying, straining and streaming

regions but the time dependence of these structures is also important, and different models of flow unsteadiness lead to different Eulerian and Lagrangian frequency spectra.

Given a $-5/3$ energy wavenumber spectrum, either frequency spectrum can exhibit a ω^{-2} or a $\omega^{-5/3}$ power law in either the inertial or the inertial-advective ranges simply by modifying the unsteadiness properties of the flow. It is interesting however that only these possibilities exist for all the different unsteadiness models investigated.

This KS flow field can be further modified by adding small-scale vortex tubes that are advected by the surrounding random Fourier modes. The vortex tubes have a k^{-1} energy spectrum which cannot be seen in the measured wavenumber spectra because it is drowned under the very energetic superposition of the $k^{-5/3}$ random Fourier modes. Nevertheless the -1 signature of these vortex tubes appears clearly in that part of the inertial-advective range of the Eulerian frequency spectrum that is below the inertial range. The pressure wavenumber spectrum scales as k^{-2} over the entire range of scales and not as $k^{-7/3}$, which would have been the Kolmogorov prediction, even though the energy wavenumber spectrum of the flow obeys Kolmogorov scaling by construction. The vortex tubes at the very smallest scales of the flow therefore determine the pressure spectrum over the entire self-similar wavenumber range. The pressure wavenumber spectrum scales as $k^{-7/3}$ when the small number of vortex tubes is removed.

Fung and Vassilicos 1998 ([1]) observed that 2-D KS turbulent-like velocity fields made of random incompressible Fourier modes with a k^{-p} energy wavenumber spectrum have a fractal eddy structure if $1 < p < 3$ which consists of cats' eyes within cats' eyes.

Cats eyes correspond to saddle points or straining regions in 2-D and we find that the number per unit area of saddle points increases as the flow's microscale η decreases according to a η^{-D_s} law where $0 < D_s < 2$. The same finding holds in 3-D KS turbulent-like flows but with $0 < D_s < 3$, and we also find that

$$p + D_s = 3$$

in 2-D whereas

$$3p + 2D_s = 9$$

in 3-D. The power p of the energy spectrum's wavenumber scaling determines, therefore, the number of straining regions in the turbulent-like flow.

References

1. J. C. H. Fung and J. C. Vassilicos (1998), Two-particle dispersion in turbulentlike flows, *Phys. Rev. E* **52**(2), 1677-1690.
2. B. M. O. Heppe 1998 Generalized Langevin equation for relative turbulent dispersion. *J. Fluid Mech.* **357**, 167-198.

HIGH SCHMIDT NUMBER PASSIVE SCALAR IN A
TURBULENT NEAR-WAKE

H. REHAB, R. A. ANTONIA AND L. DJENIDI
Department of Mechanical Engineering, University of Newcastle,
N.S.W., 2308, Australia

Quantitative measurements of the instantaneous concentration field of a high Schmidt number passive scalar (fluorescein : $Sc = \nu/D \approx 2000$, ν and D are respectively the fluid viscosity and the scalar diffusivity) in the near-wake of a cylinder are made using the Laser Induced Fluorescence (LIF) technique. These measurements are relevant to the study of turbulent mixing, in particular for elucidating the role of molecular diffusion. The experiment is carried out in a closed circuit constant head water tunnel. The cylinder (aspect ratio of 40) is inserted in the working section with both ends supported on opposite walls of the test section. The tracer is injected through a slit (less than 0.5 mm thick) machined along the full cylinder length, at the forward stagnation point of the cylinder. The Reynolds number based on the free stream velocity, U_0, and the cylinder diameter, d, varies from $Re \approx 100$ (laminar) to $Re \approx 500$ (turbulent). The spatial resolution of the images, essentially fixed by the laser sheet thickness ($\approx 250\mu$m) is of the order of the Kolmogorov scale for $Re \approx 500$. At $Re \approx 100$, most of the dye resides at the centres of the Kármán vortices. The concentration at the centres remains approximately equal to C_0, the concentration of the dye at the injection location. Similarly, the rms concentration C' remains almost constant further downstream because both the dye molecular diffusion and the dilution effect by entrainment of the ambient clear fluid are quite weak (Figure 1a). At $Re \approx 500$ and beyond one diameter downstream of the cylinder, the maximum mean and rms concentration values are reached on the wake axis. This is mainly due to the strong dispersion of the scalar created by the turbulent velocity field (Figure 1b). The mean gradient of the concentration fluctuations normal to the axis (y direction) is typically twice as large as that in the x direction.

The evolution on the wake axis of $C'/\overline{C}$ ($\overline{C}$ is the local mean concentration) for $Re \approx 100$ and $Re \approx 500$ are shown in Figure 2. For $Re \approx 100$,

U. Frisch (ed.), Advances in Turbulence VII, 585–588.
© *1998 Kluwer Academic Publishers.*

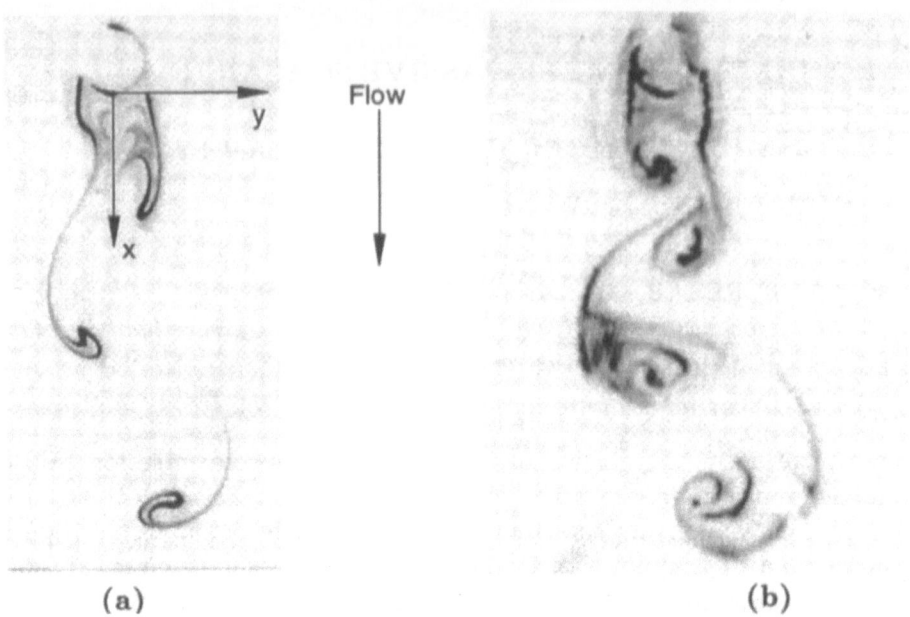

Flow

(a)   (b)

Figure 1. Instantaneous pictures in the near wake for (a) $Re \approx 100$ and (b) $Re \approx 500$. The flow extends $8d$ downstream. The fluorescein dye is injected at the forward stagnation point of the cylinder. The dark regions represent peak concentration values.

$C'/\overline{C}$ increases monotonically, the initial increase being especially rapid. For $Re \approx 500$, $C'/\overline{C}$ increases to a maximum near $x/d \approx 5$, before decreasing at larger x/d. If we neglect molecular diffusion ($Sc \gg 1$), and assume that the instantaneous concentration on the axis has two possible values : 0 or C_0, then it is easy to show that $C'/\overline{C} = [(C/C_0)^{-1} - 1]^{1/2}$. Downstream of the initial recirculation region, the measured values of $C'/\overline{C}$ (Figure 2) are in good agreement with this relation for $Re \approx 100$. For $Re \approx 500$, the measured values are below those given by this relation. The decay of $C'/\overline{C}$ in the turbulent case is mostly due to the large scale entrainment and stretching by the Kármán vortices and, to a lesser degree, to the turbulent diffusion. The homogenisation of the scalar in the early stages of mixing is dominated by the turbulence associated with the velocity field. This decay is accompanied by smoothing of the mean gradients of the concentration fluctuations in both x and y directions. For a heated cylinder and at $Re \approx 2000$, the temperature variance values are weaker (Mi, 1997), indicating an effect of the Reynolds number.

The second-order moment of the concentration increment $(\delta C_L)' = \langle \{ [C(x + r, t) - C(x, t)]^2 \} \rangle^{1/2}$ (where r is the longitudinal separation between the two points and C is the instantaneous concentration) has been evaluated along the wake centreline for both $Re \approx 100$ and 500 (Figure

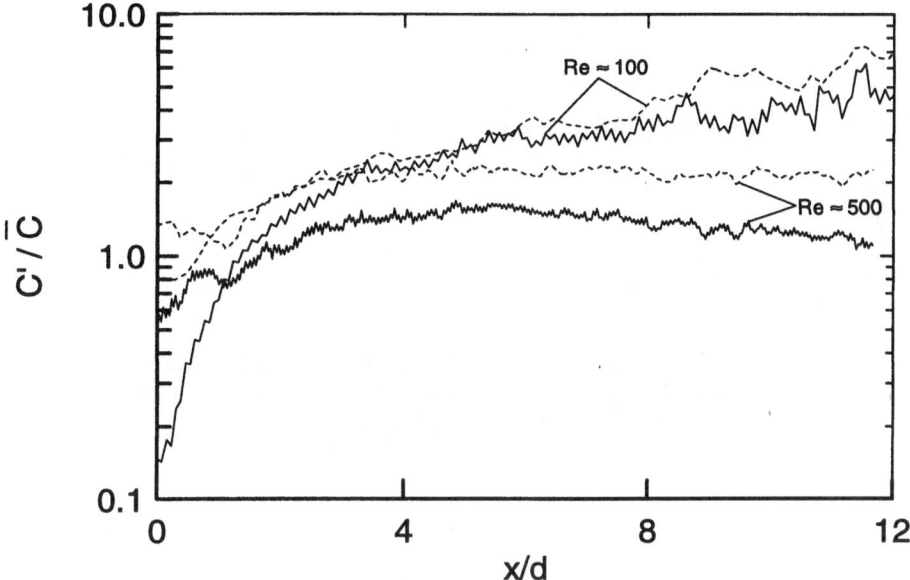

Figure 2.   Centreline distributions of the normalized scalar rms for the laminar ($Re \approx 100$) and turbulent ($Re \approx 500$) wakes. - - -, $[(C/C_0)^{-1} - 1]^{1/2}$: Centreline scalar rms calculated by assuming zero molecular diffusion.

3). In each case, $(\delta C_L)'/C'$ increases rapidly in the initial recirculation region then reaches a plateau beyond $x/d \approx 2$. The strong dependence of $(\delta C_L)'/C'$ on r in the initial recirculation zone reflects the non-homogeneous mixing there. Downstream of this zone, $(\delta C_L)'/C'$ is weakly dependent on r. As for $C'/\overline{C}$ (Figure 2), the values of $(\delta C_L)'/C'$ are weaker for $Re \approx 500$, reflecting the much greater homogeneity of the scalar field in the turbulent flow. The evolutions of $(\delta C_T)'/C'$ [$(\delta C_T)' = \langle\{[C(y+r,t) - C(y,t)]^2\}\rangle^{1/2}$] at $x/d = 10$ are shown in Figure 4. For $Re \approx 100$, $(\delta C_T)'/C'$ has two peaks corresponding to high concentration values at the centres of Kármán vortices while for $Re \approx 500$, the concentration increments are weaker and distributed from the centre towards the wake edges. This is also a measure of the improved mixing as the Reynolds number increases.

Further, the centreline probability density function (pdf) of the normalized concentration fluctuations ($C - \overline{C}/C'$ is strongly non-Gaussian, with a peak at negative values of $(C - \overline{C})/C'$ in good agreement with temperature results in a heated cylinder wake (Sreenivasan, 1981). The effect of the Schmidt number (or the Prandtl number for the temperature, $Pr = \nu/k \approx 0.7$, k is the scalar conductivity) is expected to be felt further downstream where the molecular diffusion becomes more effective.

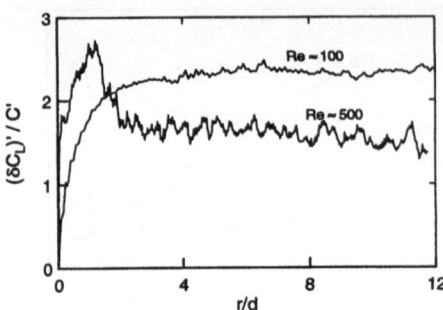

Figure 3. The centreline evolution of $(\delta C_L)'$, for $Re \approx 100$ and 500. The normalization is by the value of C' at $x = 0$.

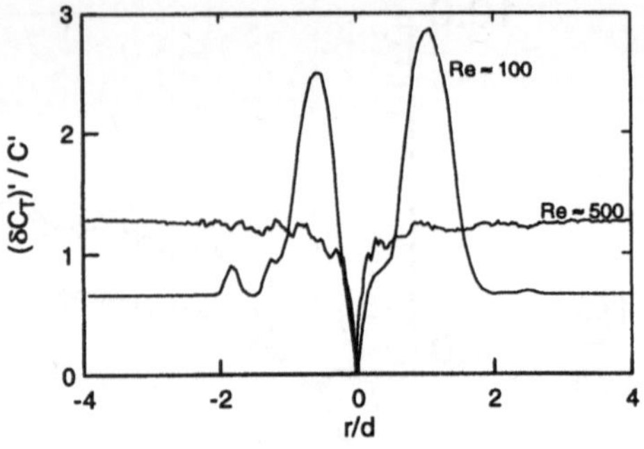

Figure 4. The y evolution of $(\delta C_T)'$ at $x/d = 10$ for $Re \approx 100$ and 500. The normalization is by the value of C' at $y = 0$ and $x/d = 10$.

References

MI, J. 1997. Private communication.
SREENIVASAN, K. R. 1981. Evolution of the centerline probability density function of temperature in a plane turbulent wake, *Phys. Fluids*, **24**, 1232-1234.

ANOMALOUS SCALING FOR PASSIVELY ADVECTED MAGNETIC FIELDS

M. VERGASSOLA[1]

[1] *CNRS, Observatoire de la Côte d'Azur, B.P. 4229, 06304 Nice Cedex 4, France.*

The Kraichnan model [1] for passive scalar advection is by now considered a paradigm for intermittency problems. A first systematic understanding of the scalar field scaling properties has been attained. The crucial property of the model is that equal-time correlation functions obey closed equations of motion. This stems from the small correlation time of the velocity field v advecting the scalar. It has been recognized that a general mechanism for intermittency and anomalous scaling is the presence of nontrivial zero modes of the closed equations satisfied by the correlation functions [2, 3, 4]. Such mechanism can be illustrated in its simplest form by considering the following linear differential equation

$$\frac{d^2 y(r)}{dr^2} + \frac{a}{r}\frac{dy(r)}{dr} + b\frac{y(r)}{r^2} = f\left(\frac{r}{L}\right), \qquad (1)$$

where r varies on the positive axis, $y(r)$ is the unknown function, a and b are two constants and $f(r/L)$ varies over distances of order L and is almost constant for $r \ll L$. In the physical situations, L is the integral scale and the mechanisms maintaining the turbulence are supposed to act at this scale. As for intermittency in Navier-Stokes turbulence, we are interested in the scaling behavior in the inertial range, i.e. for $r \ll L$. Note that the l.h.s. of (1) is invariant under scale transformations, reflecting the scale-invariance of v. The general solution of (1) is given by a non-homogeneous and a linear combination of two homogeneous solutions. The combination is fixed by the conditions that $y(r)$, being a correlation function, should be regular and vanish at infinity. The scaling of the non-homogeneous solution is easily predicted by dimensional arguments: the l.h.s. and the r.h.s. in (1) are balanced, giving the equivalent of the Kolmogorov predictions for Navier-Stokes turbulence. Here, it is clear that the validity of such arguments can

U. Frisch (ed.), Advances in Turbulence VII, 589–590.

be spoiled by the presence of the homogeneous solutions (zero modes). As the r.h.s. vanishes, no balance is indeed possible and dimensional arguments are uncapable of predicting their scaling behavior. The presence of zero modes provides therefore a simple mechanism for anomalous scaling and intermittency. In the Kraichnan model of scalar advection, a differential equation of the type (1) is found for the second-order correlations of the scalar field θ. The constant b in (1) however vanishes and no relevant zero mode is found. More precisely, there is a constant zero mode, but it disappears in the second-order structure function $S_2(r) = 2\left(\langle\theta^2\rangle - \langle\theta(r)\theta(0)\rangle\right)$. The vanishing of b is physically associated with the absence of stretching of θ and the conservation of θ^2 in the absence of forcing and dissipation. Nontrivial scaling appears from the fourth-order correlations, but (1) becomes then a partial differential equation, not solvable analytically.

For passively advected magnetic fields $\boldsymbol{B}$, nontrivial zero modes and anomalous scaling appear already at the level of second-order correlations [5]. The equation for the correlation function can be reduced to the form (1). The coefficient b now does not vanish on account of the dynamo stretching mechanism of magnetic field lines which, in analogy with vortex lines in Navier-Stokes turbulence, can lead to the amplification of $\boldsymbol{B}$. Transforming the equation to a Schrödinger-like form and determining the ground-state energy permits to show that no unbounded amplification of the magnetic energy (dynamo) takes place in 3-D if the velocity scaling exponent $\xi < 1$. The presence of a large-scale forcing will then make the system relax to a non-trivial stationary state. The corresponding anomalous inertial-range exponent γ of second-order correlations is calculated analytically and non-perturbatively: $\gamma = \frac{1+\xi}{2} - \frac{3}{2}\sqrt{1 - \frac{\xi(\xi+2)}{3}}$. The relationship between anomalies and conservation laws has been further stressed in Ref. [6] showing that, for the same model, the scaling behavior of magnetic helicity, which is conserved in the absence of forcing and dissipation, is normal.

References

1. Kraichnan, R.H. (1994) Anomalous Scaling of a Randomly Advected Passive Scalar, *Phys. Rev. Lett.* **52**, pp. 1016–1019.
2. Chertkov, M., Falkovich, G., Kolokolov, I. & Lebedev, V. (1995) Normal and anomalous scaling of the fourth-order correlation function of a randomly advected passive scalar, *Phys. Rev. E* **52**, pp. 4924–4941.
3. Gawędzki, K. & Kupiainen, A. (1995) Anomalous Scaling of the Passive Scalar, *Phys. Rev. Lett.* **75**, pp. 3834–3837.
4. Shraiman, B. & Siggia, E. (1995) Anomalous Scaling of a Passive Scalar in Turbulent Flow, *C. R. Acad. Sci.* **321**, pp. 279-284.
5. Vergassola, M. (1996) Anomalous Scaling for Passively Advected Magnetic Fields, *Phys. Rev. E* **53**, pp. R3021-3024.
6. Borue, V. & Yakhot, V. (1996) Normal and Anomalous Scaling in a Problem of a Passively Advected Magnetic Field, *Phys. Rev. E* **53**, pp. R5576-5579.

STRUCTURES AND INTERMITTENCY
IN A ONE–DIMENSIONAL PASSIVE SCALAR MODEL

A. MAZZINO[1] AND M. VERGASSOLA[1]
[1] CNRS, Observatoire de Nice, BP 4229, 06304 Nice Cedex 4, France.

Anomalous scaling behavior in the inertial range of scales and the presence of well localized coherent structures seem to be generic features of fully developed turbulent systems. The compatibility of structures with scale invariance and their dynamical relevance for intermittency has often been discussed in the literature. This issue has been explored [1] in the Kraichnan model [2] for passive scalar advection, where some understanding of intermittency has been recently achieved. The perturbation theories [3, 4] proposed for the calculation of intermittency scaling exponents assume indeed scale invariance. Furthermore, despite the short correlation in time of the velocity advecting the scalar, structures are likely to be present in the scalar field. It is then of interest to test the validity of perturbation theories by comparing their predictions with the results of numerical simulations. For numerical purposes it is convenient to consider the following one-dimensional model for the gradient field $\omega(x, t)$:

$$\partial_t \omega + \partial_x (v \omega) = \kappa \partial_x^2 \omega + f, \tag{1}$$

where both the (compressible) velocity field v and the large-scale injection f are Gaussian and δ-correlated in time, as in the incompressible Kraichnan model. The velocity has zero mean and correlation function $\langle v(x, t) v(0, t') \rangle = |x|^\xi \delta(t - t')$ and the injection is concentrated at the integral scale L, assumed large. The field ω is stretched by the gradients of the velocity field and is strongly amplified in the regions where the gradients are large and negative. Two key properties of the Kraichnan model are retained : (i) equal-time correlation functions obey closed equations of motion ; (ii) the limiting case $\xi = 0$ is a Gaussian limit of the model. This is crucial for developing the perturbation theory, that closely follows the one [3] for the Kraichnan model. Specifically, we obtain

U. Frisch (ed.), Advances in Turbulence VII, 591–592.

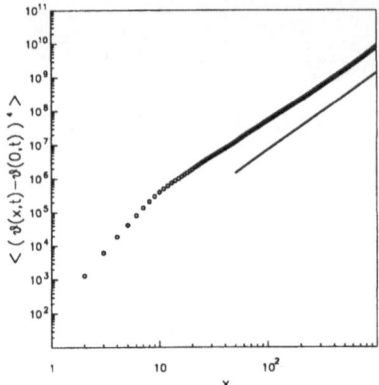 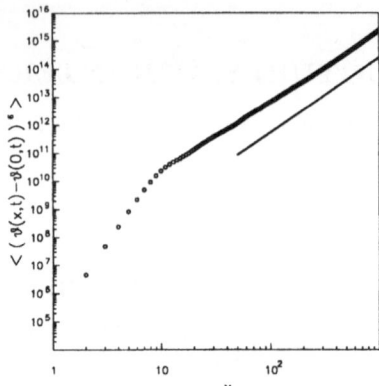

Figure 1.   The measured fourth (left) and sixth (right) order structure functions for
$\xi = 0.5$. Solid lines have the slopes predicted by the perturbative calculation: $\zeta_4 \simeq 2.29$
(left) and $\zeta_6 \simeq 2.67$ (right).

a scale–invariant zero mode at first order in ξ of the homogeneous equa-
tions for the correlations $C_p = \langle \omega(x_1) \ldots \omega(x_p) \rangle$ and scaling exponents for
the structure functions $\langle |\theta(x,t) - \theta(0,t)|^p \rangle \sim |x|^{\zeta_p}$ are reconstructed using
$\theta(x,t) = \int^x \omega(y,t)\, dy$. The prediction for the fourth and the sixth order are
$\zeta_4 = 4 - 6\xi + O(\xi^2) \simeq 4/(1+1.5\,\xi)$ and $\zeta_6 = 6 - 15\xi + O(\xi^2) \simeq 6/(1+2.5\,\xi)$,
where the classical procedure of Padé approximants has been used. The
validity of the resummation has been checked for ζ_4 by pushing the expansion at $O(\xi^2)$ and checking the stability of the approximant. For $\xi = 0.5$
the perturbative calculations predict $\zeta_4 \simeq 2.29$ and $\zeta_6 \simeq 2.67$. These values
are in in agreement (see Fig. 1) with the numerical integration of (1) us-
ing a pseudo-spectral code with resolution $N = 2^{14}$ and periodic boundary
conditions.

Strongly localized peaks are observed for ω. They have a finite life-time,
but can last for very long and their life-times are related to the arc-sine law
for Brownian motion. These events are responsible for the long tails of the
p.d.f. and the intermittency observed in the model.

References

1. Vergassola, M. & Mazzino, A. (1997) Structures and intermittency in a passive
   scalar model, *Phys. Rev. Lett.* **79**, pp. 1849–1852.
2. Kraichnan, R.H. (1994) Anomalous Scaling of a Randomly Advected Passive Scalar,
   Phys. Rev. Lett. **52**, pp. 1016–1019.
3. Gawędzki, K. & Kupiainen, A. (1995) Anomalous Scaling of the Passive Scalar,
   Phys. Rev. Lett. **75**, pp. 3834–3837.
4. Chertkov, M., Falkovich, G., Kolokolov, I. & Lebedev, V. (1995) Normal and anomalous scaling of the fourth-order correlation function of a randomly advected passive
   scalar, *Phys. Rev. E* **52**, pp. 4924–4941.

INVERSE CASCADE AND INTERMITTENCY OF PASSIVE SCALAR IN 1D SMOOTH FLOW; INVERSE CASCADE IN MULTIDIMENSIONAL COMPRESSIBLE FLOWS.

M. CHERTKOV
Physics Department, Princeton University, Princeton,
NJ 08544, USA

I. KOLOKOLOV
Budker Institute of Nuclear Physics, Novosibirsk 630090, Russia

AND

M. VERGASSOLA
CNRS, Observatoire de Nice, B.P. 4229, 06304 Nice Cedex 4,
France.

We considere random advection of Lagrangian tracer scalar field $\theta(t, y)$ by a compressible, spatially smooth and short-correlated in time velocity field [1, 2], see also [3]. Scalar fluctuations are maintained by a source $\chi(x)$ concentrated at the integral scale L.

In one-dimensional case the statistical properties of both scalar differences on the spatial distance x and the dissipation field can be analytically determined, exploiting the dynamical formulation of the model. The Gaussianity known to be present at small scales for incompressible velocity fields emerges here at large scales ($x \gg L$). These scales are shown to be excited by an inverse cascade of θ^2 and the probability distribution function $\mathcal{P}(\delta\theta_x)$ (PDF) of the corresponding scalar differences $\delta\theta_x$ to approach the Gaussian form, as larger and larger scales are considered:

$$\mathcal{P}(\delta\theta_x) = \frac{1}{2\sqrt{\pi\chi[0]\ln[x/L]}} \times$$

$$\times \left\{ \begin{array}{ll} \exp\left[-\delta\theta_x^2/(4\chi[0]\ln[x/L])\right], & |\delta\theta_x| \ll \ln[x/L], \\ \exp\left[-|\delta\theta_x|/(2\sqrt{\chi[0]})\right], & |\delta\theta_x| \gg \ln[x/L]. \end{array} \right. \tag{1}$$

U. Frisch (ed.), Advances in Turbulence VII, 593–594.

Small scales ($x \ll L$) statistics is shown to be strongly non-Gaussian. Collapse of scaling exponents for scalar structure functions takes place: moments of order $p \geq 1$ scale all linearly, independently of the order p. Smooth scaling x^p is found for $-1 < p < 1$. Tails of scalar differences PDF are exponential while, at the center, a cusped shape tends to develop when smaller and smaller ratios x/L are considered:

$$\mathcal{P}_*(\delta\theta_x) \rightarrow \begin{cases} \frac{1}{\pi}\frac{x}{L}\frac{1}{\delta\theta_x^2}, & 1 \gg |\delta\theta_x| \gg x/L, \\ \sim \frac{x}{L}\exp[-|\delta\theta_x|/2], & |\delta\theta_x| \gg 1, \end{cases} \qquad (2)$$

The same tendency is present for scalar gradients PDF $\mathcal{P}^\omega(\omega)$ with respect to the inverse of the Péclet number Pe (the pumping-to-diffusion scale ratio). The tails of the latter PDF are however much more extended, decaying as a stretched exponential of exponent 2/3, smaller than unity:

$$\mathcal{P}^\omega(\omega) \sim \frac{\kappa}{\text{Pe}}\exp\left[-(|\omega|\sqrt{\kappa/\epsilon_0})^{2/3}\right], \text{ for } \omega \gg 1/\sqrt{\kappa}. \qquad (3)$$

Here κ is the diffusion coefficient and $\epsilon_0 = <\kappa\left(\partial_y\theta(t;y)\right)^2> = <\kappa\omega^2>$ is the mean dissipation. This slower decay is physically associated with the strong fluctuations of the dynamical dissipative scale.

In multidimensional case it is shown that, depending on the dimensionality d of space and the degree of compressibility of the smooth advecting velocity field, the cascade of the scalar is direct or inverse. If $d > 4$, the cascade is always direct. For small enough degree of compressibility, the cascade is direct again. Otherwise it is inverse, i.e. very large scales are excited. The dynamical hint for the direction of the cascade is the sign of the Lyapunov exponent for particles separation. Positive Lyapunov exponents are associated to direct cascade and Gaussianity at small scales. Negative Lyapunov exponents lead to inverse cascade, Gaussianity at large scales and strong intermittency at small scales.

The detailed exposition of the results presented here can be found in [4, 5].

References

1. G.K. Batchelor (1959), Small scale variation of convected quantities like temperature in turbulent fluid, *J. Fluid Mech.* **5**, 113 .
2. R. Kraichnan (1968), Small-scale structure of a scalar convected by turbulence, *Phys. Fluids* **11** , pp. 945-953.
3. M. Chertkov, G.Falkovich, I. Kolokolov, V.Lebedev (1995), Statistics of Passive Scalar Advected by a Large-Scale 2D Velocity Field: Analytic Solution, *Physical Review* E, **51**, pp.5609-5627.
4. M. Chertkov, I. Kolokolov, M. Vergassola (1997), Inverse cascade and intermittency of passive scalar in 1d smooth flow, *Physical Review* E, **56**, pp.5483-5499.
5. M. Chertkov, I. Kolokolov, M. Vergassola (1998), Inverse versus direct cascades in turbulent advection, *Physical Review Letters*, **80**, p.512

TOPOLOGY OF DYNAMICALLY PASSIVE SCALAR FIELDS IN ISOTROPIC HOMOGENEOUS TURBULENCE

L. VALIÑO[1], B. CRESPO[1], J. SORIA [2], J. MARTÍN[3] AND C. DOPAZO[1,3]

[1]*LITEC/CSIC, María de Luna 3, Zaragoza 50015, Spain*
[2]*Monash University, Wellington Road, Clayton, Victoria 3168, Australia*
[3]*Universidad de Zaragoza, María de Luna 8, Zaragoza 50015, Spain*

Introduction

Transport of scalars by turbulent fluid motion is of fundamental importance and is encountered in applications such as pollutant formation, mass and heat transfer and chemical reactions. The physics of many of these applications is extremely complex and not well understood due to the intriguing complex topology of the fluid motions and scalar fields. In this paper, a new formulation is proposed to study the detailed topology of diffusing scalars and their relationship to the turbulent flow topology. Direct numerical simulations (DNS) are used to study those topologies in homogeneous isotropic turbulence, using an object-oriented code (Crespo 1994).

Yoda *et. al* (Yoda 1994) used a modified concentration gradient field to investigate the topology of the scalar field by calculating the invariants of $\frac{\partial^2 c}{\partial x_i \partial x_j}$, where c is the scalar concentration. This formulation has two serious drawbacks: it cannot be cast in terms of a dynamical system evolution, and no clear physical interpretation emerges from it.

1. New Formulation

The evolution of the vector, $\mathbf{y}(t)$, between two neighboring points, $\mathbf{y}(t)$, located over moving isoscalar surfaces, is given by:

$$\frac{dy_i}{dt} = \frac{\partial V_i}{\partial x_j} y_j,$$ (1)

where $\mathbf{V}$ is the isoscalar surface velocity. In the case of a non-diffusing, non-reacting scalar, $\mathbf{V}$ coincides with the flow velocity $\mathbf{u}$, as c is just convected and obviously, a constant scalar point is a material one. In the case of a diffusing/reacting scalar, c verifies

$$\frac{\partial c}{\partial t} + u_j \frac{\partial c}{\partial x_j} = D \frac{\partial^2 c}{\partial x_j \partial x_j} + S(c).$$ (2)

U. Frisch (ed.), Advances in Turbulence VII, 595–598.

The velocity of an isoscalar surface verifies by definition

$$\frac{\partial c}{\partial t} + V_j \frac{\partial c}{\partial x_j} = 0. \tag{3}$$

It is apparent that $\mathbf{V} \neq \mathbf{u}$ due to the existence of diffusion and/or chemical reaction. $\mathbf{V}$ may be decomposed as

$$\mathbf{V} = \mathbf{u} + \mathbf{V_c}, \tag{4}$$

where $\mathbf{V_c}$ is the velocity of the isoscalar surface relative to the fluid. This decomposition was introduced by Candel and Poinsot, in the context of flame propagation (Chong 1994).

There is a degree of freedom which avoids, in principle, a direct application of this method to the scalar topology. The motions inside the surface are irrelevant regarding the transport of the surface as a whole. However, it is proposed in this paper that only the component of the velocity normal to the surface (parallel to the scalar-gradient) is of interest, which agrees with a Fickian diffusion hypothesis. Hence, the natural choice for $\mathbf{V_c}$ is to be normal to the isosurface; the mass flux due to diffusion is parallel to ∇c and the reaction contribution is due to the existence of ∇c. Then, $\mathbf{V}$ can be considered as the velocity of a (non-fluid) particle moving with constant scalar value. The dynamical system defined by Eq. 1 has then a clear physical interpretation.

From the scalar gradient direction chosen for $\mathbf{V_c}$, Eq. 3 and Eq. 4, one readily obtains:

$$V_c = \frac{-D \nabla^2 c - S(c)}{|\nabla c|}. \tag{5}$$

128^3 DNS have been conducted to study the topology induced by $\mathbf{V}$. The non-reacting scalar has been implemented. The velocity field is forced ($Re_\lambda \simeq 47$). The initial scalar field is a "blob" in the middle of the cube with a value $c = 1$. The rest, $c = 0$. The mean is 0.5. Sc number 1 has been considered. Results are shown at the time non-dimensionalized by the "turn-over time" 2.8.

The invariants P, Q, R of each tensor $\frac{\partial V_{ci}}{\partial x_j}$, $\frac{\partial V_i}{\partial x_j}$, $\frac{\partial u_i}{\partial x_j}$ and $\frac{\partial^2 c}{\partial x_i \partial x_j}$, which define the topologies of $\mathbf{V_c}$, $\mathbf{V}$, $\mathbf{u}$ and ∇c (used in Yoda's et al. paper) respectively, have been calculated. The PDF of these quantities is shown for $\mathbf{V_c}$ in Figure 1. Observe that values for P different from zero appear, which means that $\nabla \cdot \mathbf{V_c}$ can be different from zero. That means that the physical volume conatained among any two iso-surfaces is not preserved, although $\mathbf{u}$ is divergence-free ($P = 0$ for all points).

Using the classification scheme given by Chong et. al (Chong 1994), the following regions appear: 1 stable node - stable node - stable node, 2 unstable node - unstable node - unstable node, 11 stable node - saddle - saddle, 12 unstable node - saddle - saddle, 18 stable focus - stretching, 19 unstable focus - stretching, 20 stable focus - contracting, 21 unstable focus - contracting.

As the scalar topology shows "compressible"-like behavior, new topologies are accessible (1, 2, 19 and 20), as shown in Figure 2. As $\mathbf{V_c}$ includes scalar-gradients in its definition, its topological regions shoud have a smaller characteristic length than the topology associated to $\mathbf{u}$. A dominant effect of $\mathbf{V_c}$ on the contribution to $\mathbf{V}$ is then expected, which is in agreement with the figure. As a consequence, statistical independence among these two topologies should exist, as confirmed by Figure 3.

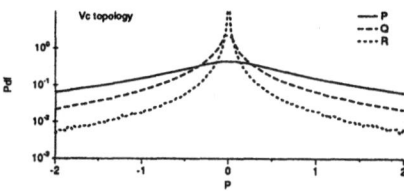

Figure 1.   PDF of P, Q and R for the inert scalar field excluding convection ($\mathbf{V_c}$). All quantities shown are normalized by one-fourth of the enstrophy.

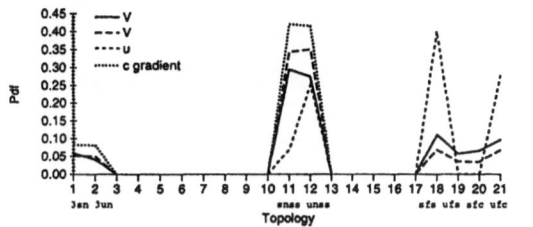

Figure 2.   PDF of the appearance of each topological region associated to u, $\mathbf{V_c}$, n V, and ∇c

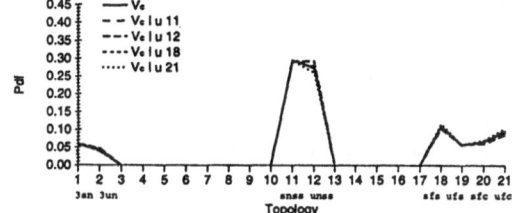

Figure 3.   PDF of the appearance of each topological region associated to $\mathbf{V_c}$, conditional on the u topology.

Finally, isocontours of the PDF of R, Q conditional on different values of P are shown for $\mathbf{V_c}$. In Figure 4, the case $P = 0$ is shown, comparing with the u topology. The line $D = 0$ at constant P, where D is the discriminant of the characteristic equation, is also shown. No asymmetries arise for the scalar topology, which is in accordance with the geometrical nature of $\mathbf{V_c}$ and the statistical homogeneity of the turbulent field. When $P \neq 0$, the simmetrical situation is broken: for $P > 0$ ("compression"), stable regions are more probable, while for $P > 0$, ("expansion"), the situation is just the opposite. This is clearly confirmed by Figure 5.

Conclusions

- A new sound physical description of the topology of the scalar has been deduced via a velocity associated with isoscalar surfaces
- This scalar topology includes the effect of diffusion and reaction through scalar derivatives, which implies topological regions of smaller characteristic lengths. As a consequence, there is statistical independence between scalar (excluding convection) and fluid velocity topologies. The scalar topology is then apparently governed by the geometrical isosurface configuration at a given time

The first author gratefully aknowledges the Spanish Ministry of Education for supporting his sabbatical stay in Monash University (PR95-388), where part of this work was done. This research has been supported in part by the the project P-78/96 of the regional government of Aragon

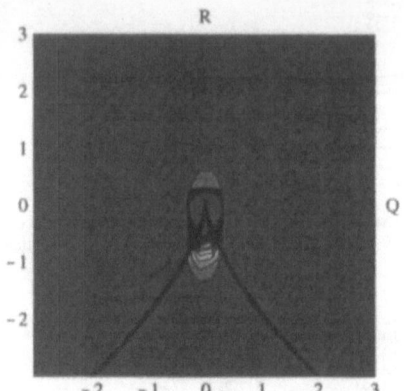
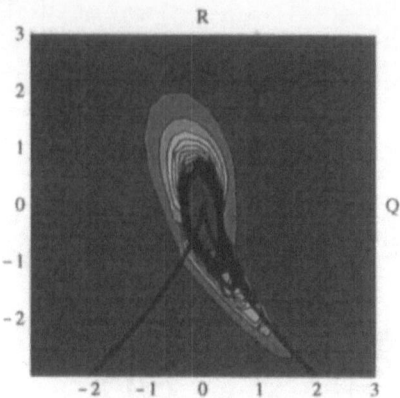

Figure 4. Isocontours of the PDF of R, Q conditional on $P = 0$ for the topology associated to $\mathbf{V_c}$ (left) and for the topology associated to $\mathbf{u}$ (right). The line $D = 0$ is shown.

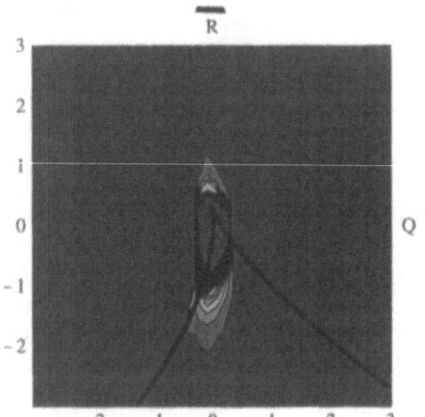
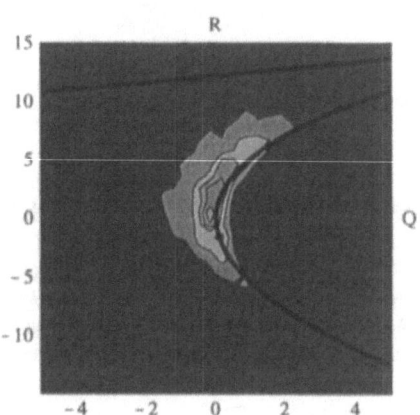

Figure 5. Isocontours of the PDF of R, Q conditional on $P = -1$("small expansion") (left) and $P = 7$ ("compression") (right) for the topology associated to $\mathbf{V_c}$ (u is incompressible, $P = 0$). The line $D = 0$ is shown.

(Spain), and by projects PB96-0407 and PB96-0719 of Spanish DGES.

References

B. Crespo, "DNS of inert and reacting scalars", Graduate project (in Spanish), 1994.

M. Yoda, L. Hesselink, and M. G. Mungal, "Instantaneous three-dimensional concentration measurements in the self-similar region of a round high-schmidt-number jet", J. Fluid Mech. **279**, 313 (1994).

M. S. Chong, A. E. Perry, and B. J. Cantwell, "A general classification of three-dimensional flow fields", Physics of Fluids A **5**, 765 (1990).

CHARACTERISTIC TIME DISTRIBUTIONS IN SCALAR MIXING

CÉSAR DOPAZO, JESÚS MARTÍN, LUIS VALIÑO(*)
Universidad de Zaragoza/(*)LITEC(CSIC), María de Luna 8, Zaragoza 50015, Spain

1. Introduction and Formulation

Turbulent mixing is characterised by multiple length and time scales over which scalar fields significantly fluctuate. Apart from being convected, the scalar field evolves under the random straning and rotation created by the turbulence. Scalar heterogeneities are smeared out by molecular diffusion enhanced by stretching and folding of isoscalar surfaces[1,2]. Turbulent mixing has its counterpart in the turbulent kinetic energy decay process, the latter being also influenced by the pressure isotropizing action.

For an inert scalar the mean value, $< C >$, is a constant and the scalar fluctuations, c , obey the equation[1]

$$\frac{\partial c}{\partial t} + u_j \frac{\partial c}{\partial x_j} = D\nabla^2 c. \tag{1}$$

where c is statistically homogeneous, D is the Fickian diffusion coefficient of c in the mixture and **u** is a zero-mean statistically homogeneous random velocity field governed by

$$\frac{\partial u_i}{\partial t} + u_j \frac{\partial u_i}{\partial x_j} = -\frac{1}{\rho}\frac{\partial p}{\partial x_i} + \nu\nabla^2 u_i \quad ; \quad \frac{\partial u_j}{\partial x_j} = 0 \tag{2}$$

where p, ρ and ν are the fluid pressure, density and kinematic vicosity, respectively.

The rational formulation of stochastic molecular mixing models crucially depends on the ability to parameterize the right hand side of (1) in terms of c, D and knowable information pertaining both to the scalar and to the turbulence fields. $\nabla^2 c$ is the first invariant of the second rank tensor $c_{,ij}$[3], which, to some extent might play for the scalar field a role similar to that of $u_{i,j}$ for the velocity field near critical points. It is, thus, logical searching for some correlation of the molecular diffusion term in (1) and the invariants of $u_{i,j}$ or, alternatively, those of its symmetric part, S_{ij}, the strain rate tensor, and its skew-symmetric part, W_{ij}, the rotation rate tensor.

The invariants of $u_{i,j}$ are defined as[4] $Q = -(1/2)u_{i,j}u_{j,i}$, $R = -(1/3)u_{i,j}u_{j,k}u_{k,i}$. The equivalent ones for S_{ij} and W_{ij} are $Q_S = -(1/2)S_{ij}S_{ji}$, $R_S = -(1/3)S_{ij}S_{jk}S_{ki}$, $Q_W = -\frac{1}{2}W_{ij}W_{ji}$. R_W is null.

U. Frisch (ed.), Advances in Turbulence VII, 599–602.
© 1998 *Kluwer Academic Publishers.*

Equation (1) can be alternatively rephrased as

$$\frac{\partial c^2}{\partial t} + u_j \frac{\partial c^2}{\partial x_j} = D\nabla^2 c^2 - 2\epsilon_c. \tag{3}$$

where $\epsilon_c = Dc_{,i}c_{,i}$ is the local/instantaneous scalar fluctuation dissipation rate. A diffusive characteristic time for the smearing out of the scalar heterogeneities, c^2, can be defined as

$$\tau_C \equiv \frac{c^2}{2\epsilon_c} \tag{4}$$

Ensemble averaging (3) yields

$$\frac{d<c^2>}{dt} = -2<\epsilon_c>. \tag{5}$$

A characteristic time of decay of $<c^2>$ is similarly defined by

$$T_C = \frac{<c^2>}{2<\epsilon_c>} \tag{6}$$

In the next section the distribution of τ_C will be presented from DNS data and the relative values of T_C, $<\tau_C>$ and $<1/\tau_C>^{-1}$ will be established.

Apart from the pressure term, the transport equation for c^2 is analogous to that for $u_i u_i$, twice the turbulent kinetic energy, namely,

$$\frac{\partial u_i u_i}{\partial t} + u_j \frac{\partial u_i u_i}{\partial x_j} = -\frac{2}{\rho}\frac{\partial p u_i}{\partial x_i} + \nu\nabla^2 u_i u_i - 2\epsilon. \tag{7}$$

where $\epsilon = \nu u_{i,j} u_{i,j}$ is the local/instantaneous turbulent kinetic energy dissipation rate. The local/instantaneous characteristic viscous decay time for $u_i u_i$ is

$$\tau = \frac{u_i u_i}{2\epsilon} \tag{8}$$

Averaging (7) leads to $d<u_i u_i>/dt = -2<\epsilon>$. The characteristic viscous decay time for $<u_i u_i>$ is thus

$$T = \frac{<u_i u_i>}{2<\epsilon>} \tag{9}$$

The transport equation for ϵ_c is readily derived, obtaining

$$\frac{\partial \epsilon_c}{\partial t} + u_j \frac{\partial \epsilon_c}{\partial x_j} = -2Dc_{,i}\,c_{,j}\,u_{j,i} + D\nabla^2 \epsilon_c - 2D^2 c_{,ij}\,c_{,ij}. \tag{10}$$

The first term on the right hand side of Eq. (10) is the production of ϵ_c by straining of scalar gradients; $u_{j,i}$ may be replaced by S_{ij} and then this term is related to that part of the strain aligned with the local and instantaneous scalar gradient vector. The last two terms of (10) are the diffusive transport and dissipation of ϵ_c, respectively.

Two local/instantaneous characteristic times can be defined from (10). One for production of ϵ_c through straining of $c_{,i}$, and other for diffusive dissipation of ϵ_c. Their definitions are

$$\tau_{SC} \equiv -\frac{\epsilon_c}{(2Dc_{,i}\,c_{,j}\,u_{j,i})} \quad ; \quad \tau_{DC} \equiv \frac{\epsilon_c}{2D^2 c_{,ij}\,c_{,ij}} \tag{11}$$

τ_C, τ_{SC} and τ_{DC} will be related to the invariants of $u_{i,j}$, S_{ij} and R_{ij} in the next section.

2. Results and Conclusions

Data fields from 128^3 DNS runs with an inert scalar have been used to obtain some of the variables defined in section 1. The velocity field is forced with $Re_\lambda \simeq 47$. The initial scalar distribution is a double Dirac delta, taking the region in the center of the cube a scalar value $C = 1$ and the rest $C = 0$. The scalar mean is 0.5. The value of the Schmidt number, Sc, is 1.0. The results presented here correspond to a time where the pdf of the scalar is almost uniformly distributed.

Figure 1 displays $< D\nabla^2 c|c >$, the conditional diffusion of c, for several ranges of Q_S values. It is pertinent to recall that $(-Q_S)^{-1/2}$ is a characteristic turbulent straining time. The larger $-Q_S$ the more intense the stretching process. The molecular mixing is thus enhanced by increasing $-Q_S$. A functional dependence of the conditional diffusion on c and on $-Q_S$ can easily be proposed.

Figures 2 and 3 present the probability density functions of τ and τ_C, respectively. The customarily used characteristic decay times, T and T_C, are plotted as well as the averaged times $< \tau >$, $< 1/\tau >^{-1}$, $< \tau_C >$, $< 1/\tau_C >^{-1}$. As a reference, the Kolmogorov time microscale and the integral time scale are also indicated.

Figure 4 is a scatter plot of τ_C versus τ. The various velocity and scalar characteristic times are plotted as a reference. The commonly used Spalding's hypothesis, $T_C = T/2$, is clearly incorrect at this stage of the scalar evolution. Not a clear physical picture emerges from this figure.

The characteristic average times $< \tau_\alpha| - Q_S >$ and $< \tau_\alpha|N >$ for $\alpha = C, SC, DC$, normalized with their respective standard deviations, are given in figures 5 and 6. While $-Q_S$ seems to have a mild influence upon τ_{SC} and τ_{DC}, its effect on τ_C is crucial, the more intense the straining action the smaller the diffusive decay time. For negative values of $N = log_{10}(Q_W/-Q_S)$, strain dominates over rotation while for positive values the opposite is true. Figure 6 indicates that mixing is more intense (small τ_C) in strain dominated regions.

The mixing intensity depends on Q_S and Q_W. In order to complete the picture introduced in this paper, the time evolution must be followed. A final objective of this research will be to parametrize the scalar mixing as a function of c, the Schmidt number, Q_S and Q_W.

The authors are grateful to the EU funding under the BRITE/EURAM Programme, Project BE95-1927.

References

1. Dopazo, C. 1994. "Recent Developments in PDF methods", In *Turbulent reacting flows*, Eds. Libby, P.A. and Williams, F.A., Ch 7, 375, Academic Press.
2. Ottino, J. M. 1989. "The Kinematics of Mixing: Stretching, Chaos and Transport", Cambridge Univ. Press, N.Y.

3. Vervisch, L., Bidaux, E., Bray, K. and Kollmann, W. 1995. "Surface density function in premixed turbulent combustion", *Phys. Fluids*, **7**, 2496-2503.
4. Chong, M.S., Perry, A.E., and Cantwell, B.J. 1990. "A general classification of three-dimensional flow fields", *Phys. Fluids A*, 2, **5**, 765-777.

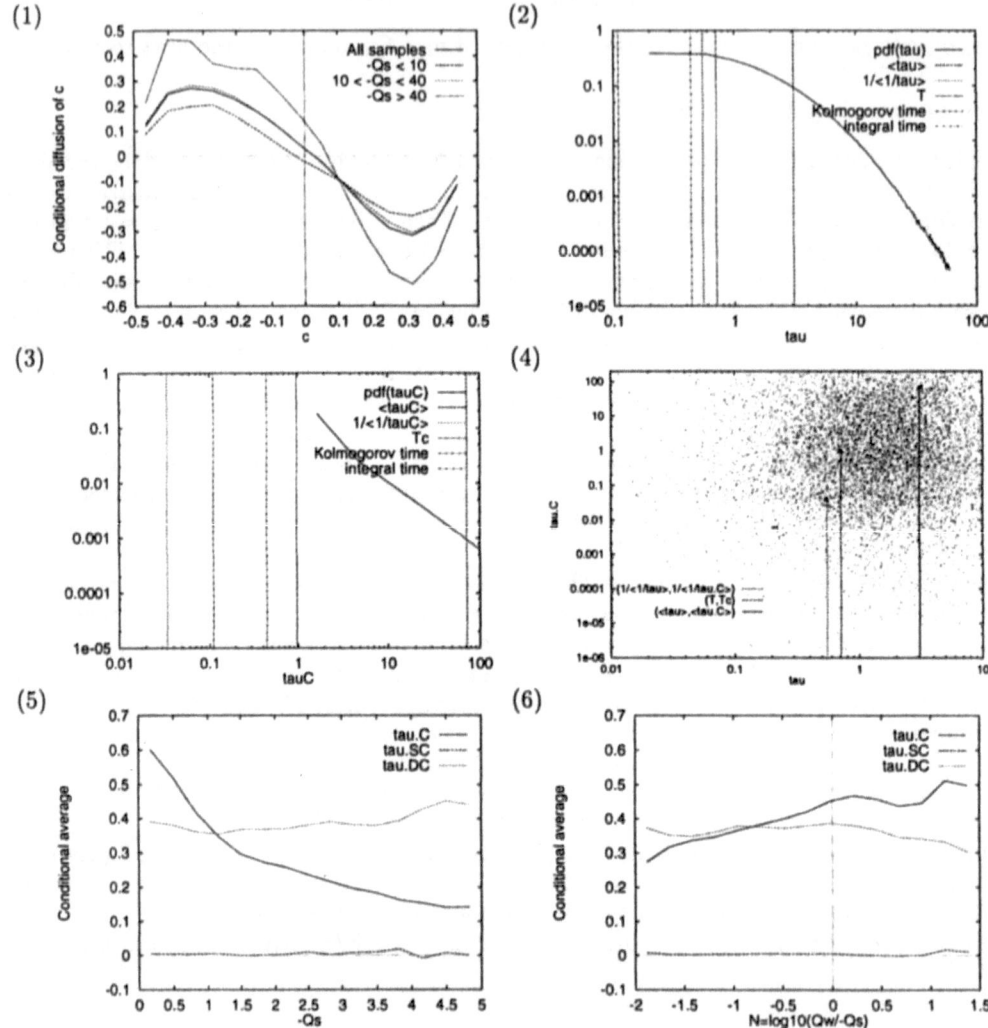

Special Invited Lecture

*Conférence Invitée
Spéciale*

NEW REMARKS ABOUT OLD IDEAS OF KOLMOGOROV

A.M. YAGLOM
Department of Aeronautics and Astronautics,
Massachusetts Institute of Technology,
Cambridge, Mass. 02139, USA,
and
Institute of Atmospheric Physics,
Russian Academy of Sciences,
Moscow 109017, Russia

The famous Kolmogorov theory of universal small-scale structure of developed turbulence was presented in 1941 in two short notes. An important development of this theory was outlined in 1962 in one more short note which turned out to be his last paper on turbulence. Between the two mentioned notes of 1941 and the note of 1962 Kolmogorov published also seven less important notes devoted to this subject (all of them are referenced in my paper about A.N. Kolmogorov (Yaglom 1994)); only three of these notes were recommended by Kolmogorov for inclusion in his Selected Works published with commentaries by him and some of his students and colleagues in three volumes in Russian in 1985-87 by Nauka Press (English translation Kolmogorov (1991); all the papers on turbulence are in vol. 1 "Mathematics and Mechanic"). In spite of their small number, brevity and late years of appearance, his publications on turbulence continue until now to provoke numerous lengthly comments and discussions (see, e.g., the book edited by J.C.R. Hunt, O.M. Phillips and D. Williams (1991) and the book by Frisch (1995)). However, all the new publications did not produce a completely clear and fully satisfactory reformulation of the old ideas by Kolmogorov; actually, some of the new developments of Kolmogorov's theory led to contradictory conclusions. Therefore, it seems reasonable to address here some questions relating to the present state of Kolmogorov's theory of developed turbulence.

In the paper Kolmogorov (1941a) he based his arguments on a refined form of the purely physical model introduced by Richardson of an energy cascade process with transfer along the spectrum of scales of turbulent fluctuations and also on the use of the general method of dimensional analysis.

U. Frisch (ed.), Advances in Turbulence VII, 605–610.

Kolmogorov's main refinement of the qualitative Richardson model consisted of the assumptions that transition to smaller scales is accompanied by isotropization of the statistical structure of fluctuations and that, at small scales, this structure depends only on the rate of the spectral energy transfer ε (which is equal to the mean rate of energy dissipation) and on the kinematic viscosity ν. In the paper Kolmogorov (1941b) the dynamic equations were also used; here the famous relation connecting the second- and third-order velocity structure functions (and leading to the so-called four-fifths law for the third-order structure function) was presented.

Kolmogorov's derivation of this equation was based on the assumption that the large-scale structure of the flow is isotropic; this assumptions seemed to be quite acceptable since, according to the Richardson–Kolmogorov cascade model, the large-scale structure determines only the rate of the energy transfer and otherwise does not affect the small-scale structure. The mathematically rigorous statistical description of the small-scale structure of Kolmogorov's locally isotropic turbulence required the construction of a mathematical theory of locally isotropic random fields; such a theory (which is in fact a natural generalization of Kolmogorov's results of 1940 relating to random processes with stationary increments) was developed by the present author (see, Sec. 13 of Monin and Yaglom, 1975). It was then natural to try and derive Kolmogorov's four-fifths law without the isotropy assumption. Such an attempt was carried out by Monin (1959; see also Sec. 22.1 of Monin and Yaglom, 1975). A more straightforward explanation of the same arguments was given by Frisch (1995; Sec. 6.2.1). Recently Lindborg (1996) found that Monin's derivation, as presented in his 1959 paper and in the book with Yaglom, contains a defect connected with the dubious neglect of terms containing pressure fluctuations. Later Hill (1997) proposed another method to prove that these terms nevertheless vanish in any locally isotropic turbulence, but his result contradicts the conclusion of Lindborg. Therefore, the derivation of dynamic equations for locally isotropic (but non-isotropic) turbulence apparently requires an additional careful analysis. Such analysis is also important to prove that the small-scale structure of any turbulent flow with high enough Reynolds number Re is independent of any features of the large-scale structure, except the rate of the energy transfer (provided isotropy of small-scale fluctuations is also proved in some way). One may hope that such analysis will also provide a simple quantitative interpretation of cascade models of energy transfer in terms of generally non-local triad interactions of spectral components of turbulence.

Kolmogorov's theory of 1941 leads to very simple equations for many statistical characteristics of developed turbulence relating to inertial-range scales (first of all, for velocity and scalar structure functions of various orders). However, for the dissipation-range characteristics of small-scale turbulent fluctuations, the theory gives equations containing some universal functions of unknown shape. To determine these functions many special non-rigorous hypotheses were proposed, beginning from the simplest speculative models of an energy-transfer function (many such models were considered in the old book by Monin and Yaglom) and finishing with quite complicated analytical theories of turbulence. During recent years there were several attempts to determine the shape of low-order velocity structure functions beyond the small-scale limit of the inertial range, either by DNS (direct numerical simulation) or by computations based on specific models for small-scale vortices making the main contribution to velocity differences (see, e.g., the survey by Pullin and Saffman (1998) devoted to the later approach). This later approach is related to studies of the influence of organized structures of various types on the small-scale turbulent characteristics; it seems quite promising in combination with independent study of vortical structures by DNS and experimental methods.

Let us now turn to Kolmogorov's ideas about the inevitable deviations from the laws predicted by him in 1941. This is due to small-scale intermittency, leading to strong fluctuations of the energy dissipation rate ε. Devoted to this topic, Kolmogorov's (1962) paper interpreted and represented in more general form the results of his former student Obukhov (1962) who first discovered strong fluctuations in atmospheric measurements of the longitudinal velocity structure function $D_{11}(\mathbf{r}) = \langle [u_1(\mathbf{x}+\mathbf{r}) - u_1(\mathbf{x})]^2 \rangle$ (where angular brackets denote statistical averaging) at a fixed value of the separation $r = |\mathbf{r}|$ (belonging to the inertial range of distances) and who then correctly explained this experimental result by fluctuations of the dissipation rate ε. To estimate qualitatively, if somewhat roughly, the fluctuations of structure functions, Obukhov assumed that the mean square velocity difference depends not on the constant mean dissipation rate $\langle \varepsilon \rangle$ but on the quantity ε_r which is the spatial average of the local dissipation $\varepsilon(\mathbf{x})$ over the volume of the sphere having $\mathbf{x}+\mathbf{r}$ and $\mathbf{x}$ as diametrically opposite points. ε_r is clearly a random variable having probability density function (pdf) depending on r. This pdf was assumed by Obukhov to be lognormal and averaging over it was included in the definition of the true mean value of squared velocity difference. Kolmogorov (1962) gave more precise mathematical meaning to Obukhov's assumptions presenting them in a form of three similarity hypotheses which generalize hypotheses formulated in his earlier paper (Kolmogorov 1941a).

The 1962 papers by Kolmogorov and Obukhov led to the appearance of an enormous literature (including many hundreds of titles) devoted to the study and theoretical modeling of small-scale intermittency of turbulence, particularly in the last few years (see, e.g., the above mentioned books by

Hunt, Phillips and Williams (1991) and by Frisch (1995), the more recent
book edited by Boratav *et al.* (1997) or any physics-oriented fluid-mechanics
journal, e.g., *J. Fluid Mech.*, *Phys. Fluids*, *Phys. Rev. Lett.* or *Phys. Rev.
E*). Note, however, that all this literature only briefly concerns the two most
general similarity hypotheses suggested by Kolmogorov at the very end of
his 1962 paper as the second alternative to the two hypotheses formulated
in Kolmogorov (1941a). (In fact, much more attention was given to the
third hypothesis of 1962 about the lognormal probability distribution of
dissipation rate, which was found to be only a crude first approximation.)
Meanwhile, just the final two hypotheses of the paper of 1962 (stating
that fluid-mechanical fields at high Re are random fields with probability
distributions for ratios of differences of field values at two pairs of points
that are invariant under all spatial similarity transformations) seem to be
very interesting and describe quite a new type of similarity which can have
a number of applications (possibly, not only in fluid mechanics). The low
popularity of the final suggestions by Kolmogorov is due to the fact that
while the mathematical theory of locally isotropic random fields, which first
appeared in Kolmogorov's paper (1941a), proved to be constructed rather
easily, the development of the theory of fields with the new type of similarity
(introduced in 1962) requires the overcoming of serious difficulties and does
not exist until now.

The 1962 work of Kolmogorov and Obukhov is worth of a few more comments. The validity of Kolmogorov's 1941 hypotheses, which postulated the
isotropy of the small-scale statistical structure of any developed turbulence
with high value of Re and the independence of this structure from any flow
parameters except ε and ν, is of course not proved rigorously but seems
to be physically natural and consistent with physical intuition. In contrast,
Obukhov's assumptions of 1962 and their reformulation by Kolmogorov
(1962) – in the form of hypotheses postulating that at fixed value of ε_r
the conditional mean values of all characteristics of turbulence relating to
spatial scale r depend only on ε_r and ν while their absolute (unconditional)
mean values must be obtained by averaging of conditional values over the
pdf for ε_r – are far less evident and physically convincing. Of course, their
use by Obukhov and Kolmogorov for crude quantitative estimation of the
influence on the turbulence characteristics of the fluctuations of ε describing the intermittency of turbulence was a brilliant piece of work deservedly
giving rise to a lot of subsequent continuations, refinements and developments. Nevertheless, at the end of his 1962 paper Kolmogorov set up a
problem of elimination of the rather special physical quantity ε_r from the
general theory considered by him and Obukhov and, in just this connection,
he proposed his final hypotheses mentioned above. Kolmogorov's problem
was only partially solved by the known multifractal formalism by Parisi

and Frisch (1985) where the quantity ε_r is not mentioned explicitly but can be understood implicitly. Final remarks of Kolmogorov's 1962 paper intended to sketch a most radical solution of the stated problem by indicating the possibility to base the theory on very general similarity hypotheses relating to small-scale structure of turbulence which include only two-point differences of fluid dynamic quantities but no physical parameters. The last sentence of his paper shows that Kolmogorov understood that his final hypotheses are insufficient for development of a complete theory but hoped that the suggested approach can be made more precise by including into it some additional general assumptions. Unfortunately, this very interesting idea of Kolmogorov did not get further development up to now.

Let us now mention two attempts to overcome the limits of both of Kolmogorov's theories related to small-scale turbulence. The first is the so-called extended self-similarity (ESS) hypothesis of Benzi et al. (1993, 1995) which postulates self-similarity (i.e., a power-law dependence) when a structure function of a certain order is expressed in terms of a structure function of another order, and this over a range of scales extending far beyond the small-scale limit of the inertial range (within this range ESS follows just from Kolmogorov's theory of 1962). ESS agrees well with numerous experimental data but its satisfactory theoretical explanation is not yet found (see the discussion ot this topic by Meneveau, 1996). The second related but wider approach is the incomplete similarity (IS) assumption by Barenblatt et al. (see Barenblatt and Goldenfeld (1995), Barenblatt and Chorin (1996, 1997), and the general discussion in the book by Barenblatt (1996)). The IS also means a power-law dependence (with exponents depending on Re) of the statistical characteristics on the independent variable, which extends over a wide range and is inexplicable by traditional dimensional analysis (in the opposite instance the similarity is called complete). In most (but not all) cases the satisfactory theoretical justification of IS is missing and therefore the main role is played by its confirmation through experimental data. However, Barenblatt et al. stress the wide prevalence of IS in nature and its relation to some features of solutions for a number of specific nonlinear equations; they also indicate some cases where a theoretical explanation was found and formulated their belief in the universality of IS. It seems doubtful that a universal proof of the validity of IS will apply to a great number of quite different physical problems; apparently the search for a proof must be carried out for each problem separately. In the case of developed turbulence such a proof, if it exists, for both ESS and IS must be based on some similarity features (maybe representing some general properties of solutions of the Navier–Stokes equations) of organized structures determining the shapes of the considered statistical characteristics. Such features must definitely exist if IS or/and ESS are valid.

References

Barenblatt, G.I.: 1996, Scaling, Self-similarity, and Intermediate Asymptotics, Cambridge University Press.

Barenblatt, G.I., and Chorin, A.J.: 1996, Small viscosity asymptotics for the inertial range of local structure and for the wall region of wall bounded turbulence, Proc. Nat. Acad. Sci., 93, 6749-6752.

Barenblatt, G.I., and Chorin, A.J.: 1997, Scaling laws and vanishing viscosity limits for wall-bounded shear flows and for local structure of developed turbulence, Comm. Pure Appl. Math., 50, 381-398.

Barenblatt, G.I., and Goldenfeld, N.: 1995, Does fully developed turbulence exist? Reynolds number dependence versus asymptotic covariance, Phys. Fluids, 7, 3078-3082.

Benzi, R., Ciliberto, S., Baudet, C., Chavarria, G.R., and Tripiccione, R.: 1993, Extended self-similarity in the dissipation range of fully developed turbulence, Europhys. Lett., 24, 275- 279.

Benzi, R., Ciliberto, S., Baudet, C., and Chavarria, G.R.: 1995, On the scaling of three-dimensional homogeneous and isotropic turbulence, Physica D, 80, 385-398.

Boratav, O.N., Eden, A., and Erzan, A., eds., (1997) Turbulence Modelling and Vortex Dynamics, Springer Lecture Notes in Physics, vol. 491.

Frisch, U.:1995, Turbulence. The Legacy of A.N. Kolmogorov, Cambridge University Press.

Hill, R. J.: 1997, Applicability of Kolmogorov's and Monin's equations of turbulence, J. Fluid Mech., 353, 67-81.

Hunt, J.C.R., Phillips, O.M., and Williams, D. (eds.): 1991, Turbulence and Stochastic Processes. Kolmogorov's Ideas 50 Years on, The Royal Society of London.

Kolmogorov, A.N.: 1940, Curves in Hilbert space which are invariant with respect to a one-parameter group of motions, Dokl. Akad. Nauk SSSR, 26, 6-9.

Kolmogorov, A.N.: 1941a, Local structure of turbulence in an incompressible fluid at very high Reynolds numbers, Dokl. Akad. Nauk SSSR,30, 229-303.

Kolmogorov, A.N.: 1941b, Energy dissipation in locally isotropic turbulence, Dokl. Akad. Nauk SSSR, 32, 19-21.

Kolmogorov, A.N.: 1962, A refinement of previous hypotheses concerning the local structure of turbulence in a viscous incompressible fluid at high Reynolds numbers, J. Fluid Mech., 13, 82-85.

Kolmogorov, A.N.: 1991, Collected Works. Vol. 1. Mathematics and Mechanics (ed. by V.M. Tikhomirov), Kluwer.

Lindborg, E.: 1996, A note on Kolmogorov's third-order structure-function law, the local isotropy hypothesis and the pressure-velocity correlation, J. Fluid Mech., 326, 343-356.

Meneveau, C. : 1996, Transition between viscous and inertial- range scaling of turbulence structure functions, Phys. Rev. E, 54, 3657-3663.

Monin, A.S.: 1959, Theory of locally isotropic turbulence, Dokl. Akad. Nauk SSSR, 125, 515-518.

Monin, A.S., and Yaglom, A.M. : 1975, Statistical Fluid Mechanics, vol. 2, the MIT Press.

Obukhov, A.M.: 1962, Some specific features of atmospheric turbulence, J. Fluid Mech., 13, 77-81.

Parisi, G., and Frisch, U. (1985) On the singularity structure of fully developed turbulence, in Turbulence and Predictability in Geophysical Fluid Dynamics (M. Ghil, R. Benzi and G. Parisi, eds.), pp.84-87, North-Holland.

Pullin, D.J., and Saffman, P.G. : 1998, Vortex dynamics in turbulence, Ann. Rev. Fluid Mech., 30, 31-51.

Yaglom, A.M.: 1994, A.N.Kolmogorov as a fluid mechanician and founder of a school in turbulence research, Ann. Rev. Fluid Mech., 26, 1-22.

AUTHOR INDEX
INDEX DES AUTEURS

Mechanics

FLUID MECHANICS AND ITS APPLICATIONS
Series Editor: R. Moreau

Aims and Scope of the Series

The purpose of this series is to focus on subjects in which fluid mechanics plays a fundamental role. As well as the more traditional applications of aeronautics, hydraulics, heat and mass transfer etc., books will be published dealing with topics which are currently in a state of rapid development, such as turbulence, suspensions and multiphase fluids, super and hypersonic flows and numerical modelling techniques. It is a widely held view that it is the interdisciplinary subjects that will receive intense scientific attention, bringing them to the forefront of technological advancement. Fluids have the ability to transport matter and its properties as well as transmit force, therefore fluid mechanics is a subject that is particularly open to cross fertilisation with other sciences and disciplines of engineering. The subject of fluid mechanics will be highly relevant in domains such as chemical, metallurgical, biological and ecological engineering. This series is particularly open to such new multidisciplinary domains.

Kluwer Academic Publishers – Dordrecht / Boston / London

Mechanics

FLUID MECHANICS AND ITS APPLICATIONS
Series Editor: R. Moreau

21. J.P. Bonnet and M.N. Glauser (eds.): *Eddy Structure Identification in Free Turbulent Shear Flows*. 1993   ISBN 0-7923-2449-8
22. R.S. Srivastava: *Interaction of Shock Waves*. 1994   ISBN 0-7923-2920-1
23. J.R. Blake, J.M. Boulton-Stone and N.H. Thomas (eds.): *Bubble Dynamics and Interface Phenomena*. 1994   ISBN 0-7923-3008-0
24. R. Benzi (ed.): *Advances in Turbulence V*. 1995   ISBN 0-7923-3032-3
25. B.I. Rabinovich, V.G. Lebedev and A.I. Mytarev: *Vortex Processes and Solid Body Dynamics*. The Dynamic Problems of Spacecrafts and Magnetic Levitation Systems. 1994
                                                                 ISBN 0-7923-3092-7
26. P.R. Voke, L. Kleiser and J.-P. Chollet (eds.): *Direct and Large-Eddy Simulation I*. Selected papers from the First ERCOFTAC Workshop on Direct and Large-Eddy Simulation. 1994
                                                                 ISBN 0-7923-3106-0
27. J.A. Sparenberg: *Hydrodynamic Propulsion and its Optimization*. Analytic Theory. 1995
                                                                 ISBN 0-7923-3201-6
28. J.F. Dijksman and G.D.C. Kuiken (eds.): *IUTAM Symposium on Numerical Simulation of Non-Isothermal Flow of Viscoelastic Liquids*. Proceedings of an IUTAM Symposium held in Kerkrade, The Netherlands. 1995   ISBN 0-7923-3262-8
29. B.M. Boubnov and G.S. Golitsyn: *Convection in Rotating Fluids*. 1995   ISBN 0-7923-3371-3
30. S.I. Green (ed.): *Fluid Vortices*. 1995   ISBN 0-7923-3376-4
31. S. Morioka and L. van Wijngaarden (eds.): *IUTAM Symposium on Waves in Liquid/Gas and Liquid/Vapour Two-Phase Systems*. 1995   ISBN 0-7923-3424-8
32. A. Gyr and H.-W. Bewersdorff: *Drag Reduction of Turbulent Flows by Additives*. 1995
                                                                 ISBN 0-7923-3485-X
33. Y.P. Golovachov: *Numerical Simulation of Viscous Shock Layer Flows*. 1995
                                                                 ISBN 0-7923-3626-7
34. J. Grue, B. Gjevik and J.E. Weber (eds.): *Waves and Nonlinear Processes in Hydrodynamics*. 1996   ISBN 0-7923-4031-0
35. P.W. Duck and P. Hall (eds.): *IUTAM Symposium on Nonlinear Instability and Transition in Three-Dimensional Boundary Layers*. 1996   ISBN 0-7923-4079-5
36. S. Gavrilakis, L. Machiels and P.A. Monkewitz (eds.): *Advances in Turbulence VI*. Proceedings of the 6th European Turbulence Conference. 1996   ISBN 0-7923-4132-5
37. K. Gersten (ed.): *IUTAM Symposium on Asymptotic Methods for Turbulent Shear Flows at High Reynolds Numbers*. Proceedings of the IUTAM Symposium held in Bochum, Germany. 1996   ISBN 0-7923-4138-4
38. J. Verhás: *Thermodynamics and Rheology*. 1997   ISBN 0-7923-4251-8
39. M. Champion and B. Deshaies (eds.): *IUTAM Symposium on Combustion in Supersonic Flows*. Proceedings of the IUTAM Symposium held in Poitiers, France. 1997
                                                                 ISBN 0-7923-4313-1
40. M. Lesieur: *Turbulence in Fluids*. Third Revised and Enlarged Edition. 1997
                                             ISBN 0-7923-4415-4; Pb: 0-7923-4416-2

Kluwer Academic Publishers – Dordrecht / Boston / London

Mechanics

FLUID **MECHANICS AND ITS APPLICATIONS**

Series Editor: R. Moreau

Kluwer Academic Publishers – Dordrecht / Boston / London

ICASE/LaRC Interdisciplinary Series in Science and Engineering

1. J. Buckmaster, T.L. Jackson and A. Kumar (eds.): *Combustion in High-Speed Flows.* 1994   ISBN 0-7923-2086-X
2. M.Y. Hussaini, T.B. Gatski and T.L. Jackson (eds.): *Transition, Turbulence and Combustion.* Volume I: Transition. 1994
                                                ISBN 0-7923-3084-6; set 0-7923-3086-2
3. M.Y. Hussaini, T.B. Gatski and T.L. Jackson (eds.): *Transition, Turbulence and Combustion.* Volume II: Turbulence and Combustion. 1994
                                                ISBN 0-7923-3085-4; set 0-7923-3086-2
4. D.E. Keyes, A. Sameh and V. Venkatakrishnan (eds): *Parallel Numerical Algorithms.* 1997   ISBN 0-7923-4282-8
5. T.G. Campbell, R.A. Nicolaides and M.D. Salas (eds.): *Computational Electromagnetics and Its Applications.* 1997   ISBN 0-7923-4733-1
6. V. Venkatakrishnan, M.D. Salas and S.R. Chakravarthy (eds.): *Barriers and Challenges in Computational Fluid Dynamics.* 1998   ISBN 0-7923-4855-9
7. U. Frisch (ed.): *Advances in Turbulence VII.* Proceedings of the Seventh European Turbulence Conference, held in Saint-Jean Cap Ferrat, France, 30 June – 3 July 1998. 1998   ISBN 0-7923-5115-0

KLUWER ACADEMIC PUBLISHERS – DORDRECHT / BOSTON / LONDON